Elena Cordero, Luigi Rodino
Time-Frequency Analysis of Operators

De Gruyter Studies in Mathematics

Volume 75

Elena Cordero, Luigi Rodino

Time-Frequency Analysis of Operators

DE GRUYTER

Mathematics Subject Classification 2010
42B10, 81S30, 42B37, 35S05

Authors
Prof. Dr. Elena Cordero
Dipartimento di Matematica
Università degli Studi di Torino
Via Carlo Alberto 10
10123 Torino
Italy
elena.cordero@unito.it

Prof. Dr. Luigi Rodino
Dipartimento di Matematica
Università degli Studi di Torino
Via Carlo Alberto 10
10123 Torino
Italy
luigi.rodino@unito.it

ISBN 978-3-11-053035-3
e-ISBN (PDF) 978-3-11-053245-6
e-ISBN (EPUB) 978-3-11-053060-5
ISSN 0179-0986

Library of Congress Control Number: 2020942033

Bibliographic information published by the Deutsche Nationalbibliothek
The Deutsche Nationalbibliothek lists this publication in the Deutsche Nationalbibliografie; detailed bibliographic data are available on the Internet at http://dnb.dnb.de.

Typesetting: VTeX UAB, Lithuania
Printing and binding: CPI books GmbH, Leck

www.degruyter.com

To Elisa and Francesco – E.C.
To Mia – L.R.

Preface

The main topics of this book are the applications of the time–frequency analysis methods to the study of the partial differential equations, with the emphasis on the theory of the pseudodifferential and Fourier integral operators reconsidered in terms of Gabor transform and Gabor frames.

The material comes largely from a series of papers of the authors in collaboration with Fabio Nicola. Although our friend Fabio does not appear as a coauthor of the present volume, much depends here on his results.

In the last ten years several other people contributed to the field, and we believe a book reporting, at least in part, on their studies is useful, despite the fact that research is going on and a large part of the work has still to be done. Our ambition is also to attract people working in partial differential equations from the theoretical and numerical point of view, convincing them that the time–frequency approach can be useful. Besides, the material is organized starting from scratch so that also nonexperts and young students can enter the subject.

We wish to thank the many friends and colleagues who have contributed in this monograph in various ways. In particular, two groups of scientists working on time–frequency analysis, one at the University and Politecnico of Torino and the other in Vienna (NuHAG), had an essential role; their research results, reported in the bibliography, give essential background to the book.

We wish also to express our gratitude to Stevan Pilipović and Nenad Teofanov (Novi Sad University), to Hans G. Feichtinger and Maurice de Gosson (University of Vienna), to Joachim Toft and Patrik Wahlberg (Linnaeus University of Växjö), to Federico Bastianoni and S. Ivan Trapasso (Politecnico of Torino), and to Jörg Seiler (University of Torino) for suggesting several improvements of the text.

Many classical and recent volumes have influenced the writing, the choice of topics and the proofs. Those that have had the most profound influence on this book are as follows:

K. Gröchenig, *Foundations of time–frequency analysis* [160],

C. Heil, *A basis theory primer* [183].

We greatly appreciate these texts and encourage the reader to consult them. Additional papers are listed in the references.

We wish to thank Elsevier, holding the copyright, for granting permission to include 13 figures in Chapter 6. Finally, we thank De Gruyter for all the support and patient waiting of the completion of the monograph.

Torino, April 2020 Elena Cordero, Luigi Rodino

https://doi.org/10.1515/9783110532456-201

Introduction

Two different names, "micro-local analysis" and "time–frequency analysis", are both used in Fourier analysis to identify the simultaneous study of a function in the space (or time) variable and in the dual variable, looking for possible localization of the frequencies. The two terms are not exactly synonymous though. In fact, aim of the micro-local analysis is the study of the partial differential equations, the main concern being the investigation of local singularities of the solutions and related issues of the general theory. In this perspective, we may refer to the monumental work of Lars Hörmander in the years 1970–2000, providing in particular the setting of the pseudodifferential and Fourier integral operator theory. Because of the problems under consideration, attention is limited to high frequencies, and final information is usually expressed in terms of asymptotic expansions rather than numerical values. The functional framework is given by the classical Lebesgue–Sobolev spaces.

The history of time–frequency analysis is more recent, despite the fact that its roots are in Fourier analysis used widely in the applied sciences. The main present motivation is the signal theory, with the aim of analyzing the local frequencies of the function (called signal in this context) by means of the short-time Fourier transform defined in the spirit of Dennis Gabor 1946. It is then evident that numerical aspects are dominant, with attention to low or, say, bounded frequencies. In fact, the short-time Fourier transform is often replaced by its discrete version, the Gabor frames. Pseudodifferential operators still appear, but with the new role of filters of the signal, with limited attention to differential equations. The new functional setting is given by modulation spaces and Wiener amalgam spaces.

The present book is devoted to exploring the interface between micro-local and time–frequency analysis, with intentions as before. Namely, we want to apply the time–frequency methods to study the partial differential equations. The usual micro-local methods are not abandoned, but the standard Fourier transform is systematically replaced by time–frequency representations as the short-time Fourier transform, the Wigner distribution, and Gabor frames. Correspondingly, Sobolev spaces are substituted by modulation and Wiener amalgam spaces.

This new point of view presents advantages and disadvantages. Precisely, the applications to partial differential equations in this book, reflecting the present state of the literature, are essentially limited to constant coefficient equations, with wave equation as a basic example, and Schrödinger equations. In fact, we cannot reproduce the classical results of Hörmander concerning propagation of singularities for variable-coefficient hyperbolic equations simply because short-time Fourier transform and Gabor frames ignore the exact location of the singularities generated by high frequencies. An advantage of our approach is the possibility of numerical applications by using Gabor frames. Other advantages are given by the use of modulation spaces which eliminate in some sense the poisoned part of the Lebesgue spaces L^p. Estimates in their framework are easy for some problems where L^p estimates fail.

https://doi.org/10.1515/9783110532456-202

Let us describe in short the contents of the six chapters of the book. For details we address to the respective introductions, providing exhaustive summary, historical notes, and motivations.

The first three chapters are addressed to the nonexperts, presenting the basics of the time–frequency analysis. They have a preliminary character with respect to the sequel and can be read independently as a short textbook, showing the state-of-the-art in the field. In particular, Chapter 1 treats the short-time Fourier transform and the Wigner distribution, as well as other related time–frequency representations. The symplectic and metaplectic groups are reviewed, having applications in the following chapters of the book. Chapter 2 is devoted to modulation and Wiener amalgam spaces. Chapter 3 addresses Gabor frames and kernel theorems in modulation spaces. The subsequent three chapters contain the aforesaid applications to partial differential equations. Precisely, Chapter 4 is devoted to pseudodifferential operators, playing the role of filters for signal theory and resolvent operators for equations. The reader may be surprised that the standard classes of the micro-local analysis, and related symbolic calculus, do not appear here. In fact, the approach of the time–frequency analysis prevails, and we mainly address the boundedness properties in modulation and Wiener amalgam spaces. Chapter 4 contains also a rather detailed presentation of the Born–Jordan quantization and related pseudodifferential calculus; the material there represents a novelty for textbooks. Chapter 5 concerns applications to operators with constant coefficients. In particular, Strichartz estimates are revisited, together with the use of Wiener amalgam spaces allowing an improvement of the classical results obtained in the framework of Lebesgue spaces. The second part of Chapter 5 considers hyperbolic and parabolic-type equations with constant coefficients. The use of Gabor matrix representations of the corresponding propagators provides optimally sparse decompositions. Chapter 6 is devoted to the study of general Schrödinger equations. To this end, the calculus of the Fourier integral operators in $\mathbb{R}^d$ is reconsidered in terms of the time–frequency analysis. Beside solving the problem of the global representation of the classical theory, this again allows for numerical applications. In Appendix A we collect some basic results from functional analysis, that we use throughout the volume. Concerning the bibliography, we have tried to be complete as far as possible, but some relevant omissions are certainly possible, and we apologize for them.

Contents

0 Preliminaries

In this part we present the notation used in this book and collect some basic results which will be useful in the sequel.

0.1 Notation and background

We use the standard set-theoretic notation, with one peculiarity: set-theoretic inclusion $A \subset B$ does not exclude equality. Thus proper inclusion has to be stated explicitly: $A \subset B$ *and* $A \neq B$.

We set $\mathbb{N} = \{0, 1, 2, \dots\}$, $\mathbb{N}_+ = \{1, 2, 3, \dots\}$, whereas $\mathbb{Z}$, $\mathbb{R}$, $\mathbb{C}$ stand for the set of all integer, real, and complex numbers, respectively. We define $\mathbb{R}_+ = \{t \in \mathbb{R} : t > 0\}$, $\mathbb{R}_- = \{t \in \mathbb{R} : t < 0\}$.

For $R > 0$, $x_0 \in \mathbb{R}^d$, we denote by $B_R(x_0)$ the ball centered at x_0 with radius R:

$$B_R(x_0) = \{y \in \mathbb{R}^d : |y - x_0| < R\}.$$

If X is a nonempty set and $f, g : X \to [0, +\infty)$, we set

$$f(t) \lesssim g(t), \quad t \in X,$$

if there exists $C > 0$ such that $f(t) \le Cg(t)$, for all $t \in X$. Moreover, if f and g depend on a further variable $z \in Z$, the statement that, *for all* $z \in Z$,

$$f(t, z) \lesssim g(t, z), \quad t \in X,$$

means that for every $z \in Z$ there exists a real number $C_z > 0$ such that $f(t, z) \le C_z g(t, z)$ for every $t \in X$. It may happen that $\sup_{z \in Z} C_z = \infty$. We set

$$f(t) \asymp g(t), \quad t \in X,$$

if $f(t) \lesssim g(t)$ and $g(t) \lesssim f(t)$, $t \in X$, and similarly as above if the functions depend on a further parameter.

The dimension of the space is denoted by d. We will employ the multiindex notation. Given $\alpha, \beta \in \mathbb{N}^d$ and $t \in \mathbb{R}^d$, we set

$$\begin{aligned}
&|\alpha| = \alpha_1 + \dots + \alpha_d, \quad \alpha! = \alpha_1! \cdots \alpha_d!,\\
&\alpha \le \beta \iff \alpha_j \le \beta_j, \quad \text{for } j = 1, \dots, d,\\
&\alpha < \beta \iff \alpha \neq \beta \quad \text{and} \quad \alpha \le \beta,\\
&\binom{\alpha}{\beta} = \prod_{j=1}^{d} \binom{\alpha_j}{\beta_j} = \frac{\alpha!}{\beta!(\alpha - \beta)!}, \quad \beta \le \alpha,\\
&t^\alpha = t_1^{\alpha_1} \cdots t_d^{\alpha_d}.
\end{aligned}$$

https://doi.org/10.1515/9783110532456-001

Observe that

$$\sum_{\beta\le\alpha}\binom{\alpha}{\beta}=2^{|\alpha|}, \tag{0.1}$$

and

$$|\alpha|!\le d^{|\alpha|}\alpha!. \tag{0.2}$$

Functions are always understood to be complex-valued, if not stated otherwise. Partial derivatives are denoted by

$$\partial_j=\partial_{t_j}=\frac{\partial}{\partial t_j},\quad j=1,\dots,d.$$

More generally, we set

$$\partial^\alpha=\partial_1^{\alpha_1}\cdots\partial_d^{\alpha_d}=\partial_t^\alpha=\partial_{t_1}^{\alpha_1}\cdots\partial_{t_d}^{\alpha_d}.$$

If $t,\xi\in\mathbb{R}^d$, we set

$$t\xi=t\cdot\xi=\sum_{j=1}^d t_j\xi_j,\quad |t|=\left(\sum_{j=1}^d t_j^2\right)^{1/2},$$

$$t^2=|t|^2=\sum_{j=1}^d t_j^2,\quad \langle t\rangle=(1+t^2)^{1/2}.$$

Observe that $\langle t\rangle$ is a smooth function satisfying $\partial_{t_j}\langle t\rangle = t_j/\langle t\rangle$. In general, one can verify that

$$|\partial_t^\alpha\langle t\rangle|\le 2^{|\alpha|+1}|\alpha|!\langle t\rangle^{1-|\alpha|}, \tag{0.3}$$

for all $t\in\mathbb{R}^d$ and all $\alpha\in\mathbb{N}^d$.

The following inequality, also called *Peetre's inequality*, will be used throughout the book:

$$\langle x+y\rangle^s\le 2^s\langle x\rangle^s\langle y\rangle^{|s|},\quad x,y\in\mathbb{R}^d,\ s\in\mathbb{R}. \tag{0.4}$$

The power of multiindex notation is well explained by the following formulas. For an open subset X of $\mathbb{R}^d$ and $m\in\mathbb{N}_+$, consider the spaces $C^m(X)$ of functions having continuous partial derivatives of order $\le m$, and $C^\infty(X)=\bigcap_{m\in\mathbb{N}_+}C^m(X)$. Given $f,g\in C^m(X)$, we have *Leibniz' formula:*

$$\partial^\alpha(fg)=\sum_{\beta\le\alpha}\binom{\alpha}{\beta}\partial^\beta f\partial^{\alpha-\beta}g,\quad |\alpha|\le m. \tag{0.5}$$

If we assume furthermore that X is convex, we have the *Taylor's formula:*

$$f(y) = \sum_{|\alpha|<m} \frac{1}{\alpha!} \partial^\alpha f(x)(y-x)^\alpha + \sum_{|\alpha|=m} \frac{m}{\alpha!}(y-x)^\alpha \int_0^1 (1-t)^{m-1} \partial^\alpha f((1-t)x+ty)\,dt, \quad \text{for } x,y \in X.$$

0.2 Function spaces and Fourier transform

Given two normed spaces A and B, we write $A \hookrightarrow B$ for the continuous embedding of A into B. We denote by $\mathcal{C}(\mathbb{R}^d)$ the space of continuous functions on $\mathbb{R}^d$, by $\mathcal{C}_c(\mathbb{R}^d)$ the space of continuous functions on $\mathbb{R}^d$ with compact support, by $\mathcal{C}_b(\mathbb{R}^d)$ the space of bounded continuous functions on $\mathbb{R}^d$, whereas $\mathcal{C}_0(\mathbb{R}^d)$ stands for the space of continuous functions vanishing at infinity:

$$\mathcal{C}_0(\mathbb{R}^d) = \left\{ f \in \mathcal{C}(\mathbb{R}^d) : \lim_{|t|\to+\infty} |f(t)| = 0 \right\}. \tag{0.6}$$

We let $\mathcal{C}^\infty(\mathbb{R}^d)$ be the space of smooth functions. For $m \in \mathbb{N}_+$, we define

$$\mathcal{C}^m(\mathbb{R}^d) = \{ f \in \mathcal{C}(\mathbb{R}^d) : \partial^\alpha f \in \mathcal{C}(\mathbb{R}^d),\ \forall \alpha \in \mathbb{N}_+^d, |\alpha| \le m \}$$

and similarly $\mathcal{C}_c^m(\mathbb{R}^d)$, $\mathcal{C}_b^m(\mathbb{R}^d)$, $\mathcal{C}_0^m(\mathbb{R}^d)$. We denote by $\mathcal{D}(\mathbb{R}^d)$ the space of smooth functions with compact support, which is also called *the space of test functions.*

The space $L^p(\mathbb{R}^d)$, $1 \le p \le \infty$, is the Banach space of measurable functions $f : \mathbb{R}^d \to \mathbb{C}$ satisfying

$$\|f\|_{L^p(\mathbb{R}^d)} = \left(\int_{\mathbb{R}^d} |f(t)|^p\,dt \right)^{1/p} < \infty,$$

if $p < \infty$ (where two functions define the same element if they coincide outside a set of Lebesgue measure zero). For $p = \infty$, $L^\infty(\mathbb{R}^d)$ is the space of essentially bounded measurable functions (defined up to sets of Lebesgue measure zero) with the norm

$$\|f\|_\infty = \operatorname{ess\,sup}_{t\in\mathbb{R}^d} |f(t)|.$$

We write $L^p_{\text{loc}}(\mathbb{R}^d)$ for the space of measurable functions $f : \mathbb{R}^d \to \mathbb{C}$ such that $\|f\chi_K\|_p < \infty$, for any compact set K of $\mathbb{R}^d$ and where χ_K is the characteristic function of the set K.

The convolution of two functions f, g is defined as

$$(f * g)(t) = \int_{\mathbb{R}^d} f(x-t)g(t)\,dt = \int_{\mathbb{R}^d} f(t)g(x-t)\,dt$$

whenever the integrals make sense. If $f \in L^p(\mathbb{R}^d), g \in L^q(\mathbb{R}^d)$, and

$$\frac{1}{p} + \frac{1}{q} = \frac{1}{r} + 1, \quad 1 \leq p, q, r \leq \infty,$$

then $f * g \in L^r(\mathbb{R}^d)$, and the following Young inequality is valid:

$$\|f * g\|_r \leq \|f\|_p \|g\|_q.$$

For $p = 2$, $L^2(\mathbb{R}^d)$ is a Hilbert space with inner product

$$\langle f, g\rangle = \int_{\mathbb{R}^d} f(t)\overline{g(t)}\, dt,$$

and $\overline{g(t)}$ is the complex conjugate of $g(t)$, $t \in \mathbb{R}^d$. For $1 \leq p \leq \infty$, we denote by p' the conjugate exponent of p, that is,

$$\frac{1}{p} + \frac{1}{p'} = 1.$$

We write $\mathcal{U}(L^2(\mathbb{R}^d))$ for the group of unitary operators acting on $L^2(\mathbb{R}^d)$.

We shall denote sequences of vectors in Banach/Hilbert spaces by $\{x_n\}$, whereas sequences of real or complex scalars will be denoted by (c_n).

For $1 \leq p < \infty$, the spaces $\ell^p(\mathbb{Z}^d)$ comprise sequences $c = (c_n)$ such that

$$\|c\|_{\ell^p} = \Big(\sum_{n \in \mathbb{Z}^d} |c_n|^p\Big)^{1/p} < \infty.$$

The space $\ell^\infty(\mathbb{Z}^d)$ is formed of sequences such that $\|c\|_{\ell^\infty} = \sup_{n\in\mathbb{Z}^d} |c_n| < \infty$. The space $c_0(\mathbb{Z}^d) \subset \ell^\infty(\mathbb{Z}^d)$ contains all sequences such that $\lim_{|n|\to\infty} |c_n| = 0$. The space $c_{00}(\mathbb{Z}^d) \subset c_0(\mathbb{Z}^d) \subset \ell^\infty(\mathbb{Z}^d)$ comprises sequences having only a finite number of nonvanishing coefficients.

The Schwartz space $\mathcal{S}(\mathbb{R}^d)$ of the set of rapidly decaying functions $f \in C^\infty(\mathbb{R}^d)$ such that

$$\sup_{t\in\mathbb{R}^d} |t^\beta \partial^\alpha f(t)| < \infty, \quad \text{for all } \alpha, \beta \in \mathbb{N}^d.$$

It is a Fréchet space with the obvious seminorms. Its topological dual $\mathcal{S}'(\mathbb{R}^d)$ is the space of the temperate distributions. We write $\langle \cdot, \cdot \rangle$ for the duality between temperate distributions and Schwartz functions; such duality is meant conjugate linear in the second component, thus extending the inner product on $L^2(\mathbb{R}^d)$. The Fourier transform is normalized to be

$$\hat{f}(\xi) = \mathcal{F}f(\xi) = \int_{\mathbb{R}^d} f(t)e^{-2\pi i t\cdot\xi}\, dt.$$

It defines an isomorphism of $\mathcal{S}(\mathbb{R}^d)$, which extends to an isomorphism of $\mathcal{S}'(\mathbb{R}^d)$, and an isometry of $L^2(\mathbb{R}^d)$. We recall Plancherel's formula

$$\|f\|_2 = \|\hat{f}\|_2$$

and Parseval's equality

$$\langle f, g\rangle = \langle \hat{f}, \hat{g}\rangle.$$

The inverse Fourier transform is

$$\check{f}(t) = \mathcal{F}^{-1}f(t) = \int_{\mathbb{R}^d} e^{2\pi i t\cdot\xi} f(\xi)\, d\xi. \tag{0.7}$$

The following identities are valid for functions in $\mathcal{S}(\mathbb{R}^d)$ and distributions in $\mathcal{S}'(\mathbb{R}^d)$:

$$\mathcal{F}((-2\pi i t)^\alpha f)(\xi) = \partial^\alpha \mathcal{F}f(\xi), \tag{0.8}$$

$$\mathcal{F}(\partial^\alpha f)(\xi) = (2\pi i\xi)^\alpha \mathcal{F}f(\xi). \tag{0.9}$$

We have also

$$\widehat{f * g} = \hat{f}\,\hat{g}, \tag{0.10}$$

$$\widehat{fg} = \hat{f} * \hat{g}, \tag{0.11}$$

for f and g in appropriate subspaces of $\mathcal{S}'(\mathbb{R}^d)$, e. g., $f \in \mathcal{S}(\mathbb{R}^d)$, $g \in \mathcal{S}'(\mathbb{R}^d)$, or vice-versa, for which fg and $f * g$ are well defined. Given a linear subspace $A \subset \mathcal{S}(\mathbb{R}^d)$, we shall denote by $\mathcal{F}A$ the linear subspace of all $f \in \mathcal{S}'(\mathbb{R}^d)$ such that $f = \hat{g}$, for $g \in A$.

We recall the Schwartz kernel theorem, one of the most fundamental results of distribution theory.

Theorem 0.2.1 (Schwartz kernel theorem). *Let A be a continuous linear operator which maps $\mathcal{S}(\mathbb{R}^d)$ into $\mathcal{S}'(\mathbb{R}^d)$ with the weak*-topology. Then there exists a tempered distribution $k \in \mathcal{S}'(\mathbb{R}^{2d})$ such that*

$$\langle Af, g\rangle = \langle k, g \otimes \overline{f}\rangle, \quad \forall f, g \in \mathcal{S}(\mathbb{R}^d).$$

The potential Sobolev spaces $H^s(\mathbb{R}^d)$, $s \in \mathbb{R}$, are defined by means of the Fourier transform as

$$H^s(\mathbb{R}^d) := \{f \in \mathcal{S}'(\mathbb{R}^d) : \langle\xi\rangle^s \hat{f}(\xi) \in L^2(\mathbb{R}^d)\},$$

with $\|f\|_{H^s(\mathbb{R}^d)} = \|\langle\cdot\rangle^s \hat{f}\|_{L^2(\mathbb{R}^d)}$. When $s = k \in \mathbb{N}$, an equivalent norm is given by

$$\|f\|_{H^k(\mathbb{R}^d)} \asymp \sum_{|\alpha|\le k} \|\partial^\alpha f\|_{L^2(\mathbb{R}^d)}, \quad f \in H^k(\mathbb{R}^d).$$

Moreover, we recall the Sobolev embedding

$$H^s(\mathbb{R}^d) \hookrightarrow \mathcal{C}_0(\mathbb{R}^d), \quad \text{for } s > d/2. \tag{0.12}$$

Finally, we recall that the space $H^s(\mathbb{R}^d)$, for $s > d/2$, is an algebra with respect to pointwise multiplication, and the following *Schauder estimates* hold:

$$\|fg\|_{H^s(\mathbb{R}^d)} \leq C_s \|f\|_{H^s(\mathbb{R}^d)} \|g\|_{H^s(\mathbb{R}^d)}, \quad f, g \in H^s(\mathbb{R}^d). \tag{0.13}$$

0.2.1 Fundamental operators

We introduce the fundamental operators in time–frequency analysis: the translation operator T_x of a function or distribution f on $\mathbb{R}^d$

$$T_x f(t) = f(t-x), \quad t, x \in \mathbb{R}^d$$

and the modulation operator

$$M_\xi f(t) = e^{2\pi i \xi \cdot t} f(t), \quad \xi, t \in \mathbb{R}^d.$$

Translation and modulation operators satisfy the so-called *commutation relations*

$$T_x M_\xi = e^{-2\pi i x \cdot \xi} M_\xi T_x. \tag{0.14}$$

In fact, for any function/distribution f on $\mathbb{R}^d$, we can write

$$T_x M_\xi f(t) = e^{2\pi i \xi \cdot (t-x)} f(t-x) = e^{-2\pi i \xi \cdot x} e^{2\pi i \xi \cdot t} f(t-x) = e^{-2\pi i \xi \cdot x} M_\xi T_x f(t).$$

We denote a point in the time–frequency space $\mathbb{R}^{2d}$ as $z = (x, \xi) \in \mathbb{R}^{2d}$, and the corresponding phase-space shift (time–frequency shift) acting on a function or distribution as

$$\pi(z) f(t) = M_\xi T_x f(t) = e^{2\pi i \xi \cdot t} f(t-x). \tag{0.15}$$

Using the commutation relations it is straightforward to compute the adjoint π^* of the time–frequency shift π as

$$\pi^*(z) = e^{-2\pi i z_1 \cdot z_2} \pi(-z), \quad z = (z_1, z_2) \in \mathbb{R}^{2d}. \tag{0.16}$$

Similarly, the composition of two time–frequency shifts is given by

$$\pi(z)\pi(w) = e^{-2\pi i z_1 \cdot w_2} \pi(z+w), \quad z = (z_1, z_2), w = (w_1, w_2). \tag{0.17}$$

The *reflection operator* is given by

$$\mathcal{I} f(t) = f(-t).$$

We define the *involution operator* as

$$f^*(t) = \overline{f(-t)}, \quad \forall t \in \mathbb{R}^d. \tag{0.18}$$

For $\lambda > 0$, the *dilation operator* is given by

$$D_\lambda f(t) = \lambda^{d/2} f(\lambda t), \quad t \in \mathbb{R}^d. \tag{0.19}$$

Observe that

$$\|D_\lambda f\|_2 = \|f\|_2, \quad \forall f \in L^2(\mathbb{R}^d).$$

The previous operators enjoy the following properties:

– The operators T_x and M_ξ are isometries on $L^p(\mathbb{R}^d)$, $1 \le p \le \infty$. In particular,

$$\|M_\xi T_x f\|_p = \|T_x M_\xi f\|_p = \|f\|_p, \quad f \in L^p(\mathbb{R}^d).$$

– For $f \in L^1(\mathbb{R}^d)$, we have

$$\widehat{T_h f}(\xi) = M_{-h}\hat{f}(\xi), \quad \forall \xi, h \in \mathbb{R}^d, \tag{0.20}$$

$$\widehat{M_\eta f}(\xi) = T_\eta \hat{f}(\xi), \quad \forall \xi, \eta \in \mathbb{R}^d, \tag{0.21}$$

$$\widehat{f^*}(\xi) = \overline{\hat{f}(\xi)}, \quad \forall \xi \in \mathbb{R}^d, \tag{0.22}$$

$$\widehat{D_\lambda f}(\xi) = \lambda^{-d/2}\hat{f}(\xi/\lambda), \quad \forall \xi \in \mathbb{R}^d,\ \lambda > 0. \tag{0.23}$$

0.2.2 Gaussian functions

In the following we shall use profusely rescaled Gaussian functions and their behavior under convolution and Fourier transform. From now on, we set $\varphi(t) = e^{-\pi t^2}, t \in \mathbb{R}^d$ (recall $t^2 = t \cdot t$). For $a > 0$, let us denote

$$\varphi_{\sqrt{a}}(t) = \varphi(\sqrt{a}t) = e^{-\pi a t^2}, \quad t \in \mathbb{R}^d. \tag{0.24}$$

Then,

$$\widehat{\varphi_{\sqrt{a}}}(\xi) = a^{-d/2}\varphi_{1/\sqrt{a}}(\xi). \tag{0.25}$$

In particular, for $a = 1$, $\widehat{(e^{-\pi t^2})}(\xi) = e^{-\pi \xi^2}$, that is, the Gaussian $e^{-\pi t^2}$ is an eigenfunction of the Fourier transform. We recall the *semigroup property* of Gaussians: for $a, b > 0$,

$$\varphi_{1/\sqrt{a}} * \varphi_{1/\sqrt{b}}(t) = \left(\frac{ab}{a+b}\right)^{d/2} \varphi_{1/\sqrt{a+b}}(t), \quad t \in \mathbb{R}^{2d}. \tag{0.26}$$

We recall from [145, page 257] the following well-known formula for Gaussian integrals.

Lemma 0.2.2. *Let A be a $d \times d$ complex matrix such that $A = A^*$ and $\operatorname{Re} A$ is positive definite. Then for every $z \in \mathbb{C}^n$,*

$$\int e^{-\pi x\cdot Ax - 2\pi i z\cdot x}\,dx = (\det A)^{-1/2} e^{-\pi z\cdot A^{-1}z}, \tag{0.27}$$

where the branch of the square root is determined requiring that $(\det A)^{-1/2} > 0$ when A is real and positive definite.

In particular, we have the following identity.

Lemma 0.2.3. *For $\operatorname{Re} c \geq 0$, $c \neq 0$. Let $\phi^{(c)}(t) = e^{-\pi c t^2}$, then*

$$\widehat{\phi^{(c)}}(\xi) = c^{-d/2}\phi^{(1/c)}(\xi), \quad \xi \in \mathbb{R}^d, \tag{0.28}$$

where the square root is chosen to have a positive real part.

Lemma 0.2.4. *For $c \in \mathbb{C}$, $c \neq 0$, consider the complex Gaussian function $\phi^{(c)}(t) = e^{-\pi c t^2}$, $t \in \mathbb{R}^d$. For every $c_1, c_2 \in \mathbb{C}$, with $\operatorname{Re} c_1 \geq 0$, $\operatorname{Re} c_2 > 0$, we have*

$$\phi^{(c_1)} * \phi^{(c_2)} = (c_1 + c_2)^{-d/2}\phi^{(\frac{c_1 c_2}{c_1+c_2})}. \tag{0.29}$$

Proof. Using equality (0.27),

$$\begin{aligned}
(\phi^{(c_1)} * \phi^{(c_2)})(t) &= \int e^{-\pi c_1 (t-y)^2 - \pi c_2 y^2}\,dt \\
&= e^{-\pi c_1 t^2}\int e^{-\pi(c_1+c_2)y^2 - 2\pi c_1 t\cdot y}\,dy \\
&= (c_1+c_2)^{-d/2} e^{-\pi c_1 t^2} e^{-\frac{\pi c_1^2 t^2}{c_1+c_2}}, \\
&= (c_1+c_2)^{-d/2}\phi^{(\frac{c_1 c_2}{c_1+c_2})}(t),
\end{aligned}$$

as desired. □

1 Basics of time–frequency representations and related properties

The first chapter of the book is devoted to the presentation of the basic tools of the time–frequency analysis, with emphasis on the short-time Fourier transform. To introduce the nonexpert readers to the subject, we shall start with a brief discussion on the so-called problem of the instantaneous frequencies. Namely, when considering the Fourier transform of a function or distribution f,

$$\hat{f}(\xi) = \mathcal{F}f(\xi) = \int f(t)e^{-2\pi i t\cdot\xi}\,dt,$$

and evaluating $|\hat{f}(\xi)|$, one obtains a precise description of the relevance of each frequency ξ in the Fourier decomposition. However, $|\hat{f}(\xi)|$ does not give information on which instants t contributed to such frequency. Roughly speaking, we may have different oscillations at different points t, and their presence is revealed by $|\hat{f}(\xi)|$, but we cannot detect any more their origin in the t-space.

Searching then for instantaneous frequencies, one is led to cut off $f(t)$ at the point $t = x$ of interest, and then take the Fourier transform of this restriction. The first problem of technical nature to this end is the definition of suitable cut-off functions $\varphi(t)$. A natural choice of $\varphi(t)$ is the characteristic function $\chi_Q(t)$ of a neighborhood Q of x, so that we have simply to perform the Fourier transform of the segment $\chi_Q(t)f(t)$ of $f(t)$ near x. However, one realizes that the jump of $\chi_Q(t)$ at the boundary of Q causes the presence of unexpected frequencies in the Fourier transform $\widehat{\chi_Q f}(\xi)$. The solution of the problem is to consider a more regular cut-off. As a curiosity, observe the use of a trapezoidal cut-off in the book of E. C. Titchmarsh [285], providing a Lipschitz $\varphi(t)$. Nowadays we may, of course, refer to $\varphi(t)$ in $\mathcal{D}(\mathbb{R}^d)$, the space of smooth functions with compact support on $\mathbb{R}^d$, and the additional frequencies contributed by such cut-off functions are negligible. So the search for instantaneous frequencies seems successful by taking $\varphi \in \mathcal{D}(\mathbb{R}^d)$ with very small support, to isolate the behavior of $f(t)$ at x.

At this moment we have to face a second problem: the Heisenberg uncertainty principle. There are several mathematical formulations of it. In the present context, let us limit ourselves to saying that by shrinking the support of $\varphi(t)$ we loose precision in determining frequencies at x. Actually, one cannot beat the Heisenberg principle, but it is possible to compromise, by contenting oneself with a description of the asymptotic behavior of the frequency spectrum in suitable unbounded domains of $\mathbb{R}^d_\xi$. A milestone in this connection is given by the definition of wave front set (WF) by L. Hörmander [191]. Namely, we consider a distribution f in $\mathbb{R}^d$ (or in a open subset of $\mathbb{R}^d$), the more interesting case being now $d > 1$, with t representing a space variable rather than time. We say that $(x, \xi_0) \notin \mathrm{WF}f$, $\xi_0 \neq 0$, if there exists $\varphi \in \mathcal{D}(\mathbb{R}^d)$, $\varphi(t) = 1$

https://doi.org/10.1515/9783110532456-002

for t in a neighborhood of x (it will be actually sufficient to assume $\varphi(x) \neq 0$), and there exists a conic neighborhood Γ_{ξ_0} of ξ_0 in $\mathbb{R}^d_\xi$ such that

$$|\widehat{\varphi f}(\xi)| \leq C_N \langle \xi \rangle^{-N} \quad \text{for } \xi \in \Gamma_{\xi_0},$$

for every $N > 0$, and a suitable constant $C_N > 0$. This definition had an enormous impact on the general theory of the partial differential equations, in particular, concerning propagation of singularities of solutions. In fact, the growth of the spectrum is responsible for a singularity of f at x, and we can use WFf to describe which high frequency directions are exactly causing the singularity. However, WFf does not give quantitative information on the spectrum of f and may turn out to be of limited interest in some applications. For example, in psychoacoustics, high frequencies are not audible!

In time–frequency analysis, in particular, in the applications to signal processing, one prefers to bargain with Heisenberg principle in a different way, by considering the Gabor transform [152]

$$Vf(x,\xi) = \int_{\mathbb{R}^d} f(t) e^{-\pi|t-x|^2} e^{-2\pi i t\cdot\xi}\, dt,$$

and accepting $Vf(x,\xi)$ as a substitute for the instantaneous frequency spectrum at x. One can raise immediately the objection that the Gaussian $\varphi(t) = e^{-\pi|t-x|^2}$ cannot be accepted as cut-off function since its support is not compact. On the other hand, the Gaussian decreases so rapidly at infinity that its numerical approximation vanishes outside a neighborhood of x. So the contributions to the spectrum from other points are not excluded in principle, but their value is extremely reduced. In favor of Gabor transform, with respect to the cut-off functions in $\mathcal{D}(\mathbb{R}^d)$, we may say that the regularity properties of the Gaussian are optimal as it is an entire function, and symbolic computations are easier. Compared to the wave front set of Hörmander, the localization proceeding is not so precise, but relevant frequencies are satisfactorily identified, with possibility of practical applications.

The short-time Fourier transform (STFT), which is a basic tool of the time–frequency analysis, is defined by replacing the Gaussian with a general *window* $g(t)$, namely

$$V_g f(x,\xi) = \int_{\mathbb{R}^d} f(t)\overline{g(t-x)} e^{-2\pi i t\cdot\xi}\, dt, \quad (x,\xi) \in \mathbb{R}^{2d}.$$

We shall choose $g \in \mathcal{S}(\mathbb{R}^d)$, or $g \in L^2(\mathbb{R}^d)$, or even g belonging to more general function spaces. Our applications to partial differential equations will be essentially based on the STFT. We shall also produce, in the last chapter of the book, the counterpart of the WF of Hörmander, namely the so-called Gabor wave front set WF$_G$, defined in terms of the STFT.

Besides the STFT, in the present chapter we shall also consider other time–frequency representations, in particular, the Wigner distribution [314]

$$Wf(x,\xi) = \int_{\mathbb{R}^d} f(x+t/2)\overline{f(x-t/2)}e^{-2\pi i t\cdot\xi}\,dt.$$

Because of the manifold applications, $Wf(x,\xi)$ is maybe the most popular time–frequency representation. It can be related easily to the STFT: $Wf(x,\xi)$ is roughly the STFT of $f(t)$ with window given by f itself. This has the advantage of giving a self-contained definition of the spectrum, without the help of the window g as mediator. More generally, one considers the Cohen representation [53] of f

$$Q_\sigma f = Wf * \sigma,$$

for some function or distribution σ. By convolving with suitable σ, one may obtain more flexible representations. Wigner distributions Wf will play an important role as well in the sequel of the book.

Let us describe more precisely the contents of the next sections. In Section 1.1 we report on basic facts concerning symplectic matrices, metaplectic operators, and symplectic Fourier transform. These topics can be seen as natural complements of the definition of Fourier transform, and metaplectic operators will play an important role in the applications to partial differential equations. Section 1.2 is the core of the chapter. We introduce here the definition of the STFT $V_g f$ and prove its properties. Attention is mainly fixed on the $L^2(\mathbb{R}^d)$ setting, with proof of the orthogonality relations and the analysis of the conjugation by metaplectic operators. The inversion formula is also first proved in $L^2(\mathbb{R}^d)$. By taking windows $g \in \mathcal{S}(\mathbb{R}^d)$, the properties of $V_g f$ are then extended to the case of a distribution $f \in \mathcal{S}'(\mathbb{R}^d)$. Finally, a formula is proved connecting $V_{g_1} f$ and $V_{g_2} f$ for different windows g_1, g_2; this will be frequently used in the sequel of the book. Section 1.3 addresses the analysis of the ambiguity function

$$A(f,g)(x,\xi) = e^{\pi i x\cdot\xi} V_g f(x,\xi).$$

The classical properties of the Wigner distribution are then reviewed, in particular, Moyal's identity

$$\|Wf\|_2 = \|f\|_2^2, \quad f \in L^2(\mathbb{R}^d),$$

the connection with metaplectic operators and the relations with STFT and ambiguity function. Cohen representations are also discussed in this section, extending the properties of the Wigner distribution.

Finally, attention is given to the τ-Wigner distribution, $\tau \in [0,1]$, defined by

$$W_\tau f(x,\xi) = \int_{\mathbb{R}^d} f(x+\tau t)\overline{f(x-(1-\tau)t)}e^{-2\pi i t\cdot\xi}\,dt.$$

The τ-Wigner distributions can be seen as Cohen representations, for a suitable choice of the kernel σ. The final part of the section is devoted to the explicit computations of the STFT in the case when f or g is a Gaussian function.

In conclusion, we observe that it turned out to be difficult for us to determine the priority of results, concerning the basics of time–frequency representations. For this reason, in this chapter we decided to omit precise references. We refer the reader to the textbooks [53, 105, 160, 183, 305] for a more complete bibliography.

1.1 Hints on symplectic and metaplectic groups

1.1.1 The symplectic group

Let us first introduce the *standard symplectic matrix* J, defined by

$$J = \begin{pmatrix} 0_d & I_d \\ -I_d & 0_d \end{pmatrix}, \tag{1.1}$$

where 0_d is the $d \times d$ zero matrix and I_d is the $d \times d$ identity matrix. Observe that

$$J^2 = -I_{2d}, \quad J^T = J^{-1} = -J$$

(the superscript ^{T} denotes the transposition).

Definition 1.1.1. A bilinear form on $\mathbb{R}^{2d}$ is called a symplectic form if it is antisymmetric and nondegenerate. The antisymmetric bilinear form σ on $\mathbb{R}^{2d}$ defined by

$$\sigma(z_1, z_2) = \xi_1 \cdot x_2 - \xi_2 \cdot x_1, \quad z_1 = (x_1, \xi_1), \quad z_2 = (x_2, \xi_2) \in \mathbb{R}^{2d}, \tag{1.2}$$

is symplectic; it is called the *standard symplectic form* on $\mathbb{R}^{2d}$.

To be precise, let us recall that antisymmetry means $\sigma(z_1, z_2) = -\sigma(z_2, z_1)$, for every $z_1, z_2 \in \mathbb{R}^{2d}$, and nondegeneracy means

$$\sigma(z_0, z) = 0 \quad \forall z \in \mathbb{R}^{2d} \Longleftrightarrow z_0 = 0.$$

The definition in (1.2) can be rewritten using the standard symplectic matrix

$$\sigma(z_1, z_2) = Jz_1 \cdot z_2 = z_2^T J z_1.$$

Definition 1.1.2. The symplectic group $Sp(d, \mathbb{R})$ is the subgroup of $2d \times 2d$ invertible matrices $\mathrm{GL}(2d, \mathbb{R})$, defined by

$$Sp(d, \mathbb{R}) = \{\mathcal{A} \in \mathrm{GL}(2d, \mathbb{R}) : \mathcal{A}^T J \mathcal{A} = J\}. \tag{1.3}$$

Observe that if $\mathcal{A}$ satisfies the identity in (1.3), then also the transpose $\mathcal{A}^T$ and the inverse $\mathcal{A}^{-1}$ fulfill the same identity, and so are symplectic matrices as well. Writing $\mathcal{A} \in Sp(d, \mathbb{R})$ in the following $d \times d$ block decomposition:

$$\mathcal{A} = \begin{pmatrix} A & B \\ C & D \end{pmatrix}, \tag{1.4}$$

from (1.3) we have that the four blocks must satisfy the following conditions:

$$D^T A - B^T C = I_d, \tag{1.5}$$
$$A^T C - C^T A = 0_d, \tag{1.6}$$
$$D^T B - B^T D = 0_d. \tag{1.7}$$

Moreover, since also

$$\mathcal{A}^{-1} = \begin{pmatrix} D^T & -B^T \\ -C^T & A^T \end{pmatrix},$$

is a symplectic matrix, relations (1.6) and (1.7) for $\mathcal{A}^{-1}$ give

$$CA^{-1} - A^{-T} C^T = 0_d, \tag{1.8}$$
$$-AB^T + BA^T = 0_d. \tag{1.9}$$

Algebraic arguments give then the following two propositions.

Proposition 1.1.3. *The subsets*

$$D = \left\{ \begin{pmatrix} L^{-1} & 0_d \\ 0_d & L^T \end{pmatrix} : L \in \mathrm{GL}(d, \mathbb{R}) \right\}, \quad N = \left\{ \begin{pmatrix} I_d & 0_d \\ C & I_d \end{pmatrix} : C \in M(d, \mathbb{R}),\ C = C^T \right\}$$

of $\mathrm{GL}(2d, \mathbb{R})$ *are subgroups of* $Sp(d, \mathbb{R})$.

Recall that a group G is generated by some elements $g_1, \dots, g_n \in G$ if every element of the group can be expressed as a combination (under the group operation) of finitely many elements g_j, $j = 1, \dots, n$, and their inverses.

Proposition 1.1.4. *With the notation of the previous proposition, the symplectic group* $Sp(d, \mathbb{R})$ *is generated by* $D \cup N \cup \{J\}$.

Definition 1.1.5. The symplectic algebra $\mathfrak{sp}(d, \mathbb{R})$ is the set of all $2d \times 2d$ real matrices $\mathcal{A}$ such that $e^{t\mathcal{A}} \in Sp(d, \mathbb{R})$ for all $t \in \mathbb{R}$.

1.1.2 The metaplectic group

There are many constructions of the metaplectic group $Mp(d, \mathbb{R})$, doubly covering the symplectic group $Sp(d, \mathbb{R})$, and the metaplectic representation of $Mp(d, \mathbb{R})$ into

$\mathcal{U}(L^2(\mathbb{R}^d))$, the group of unitary operators acting on $L^2(\mathbb{R}^d)$. In [145] G. B. Folland introduced the metaplectic representation μ of $Mp(d, \mathbb{R})$ as an intertwining operator. In short, in terms of the time–frequency shift operator $\pi(z)$ in (0.15), $z = (x, \xi) \in \mathbb{R}^{2d}$, for $\mathcal{A} \in Sp(d, \mathbb{R})$, one defines $\mu(\mathcal{A}) \in \mathcal{U}(L^2(\mathbb{R}^d))$ by

$$\pi(\mathcal{A}z) = c_{\mathcal{A}}\mu(\mathcal{A})\pi(z)\mu(\mathcal{A})^{-1} \tag{1.10}$$

where $c_{\mathcal{A}}$, $|c_{\mathcal{A}}| = 1$, is a phase factor. To be precise, recall first that the Heisenberg group $\mathbb{H}^d$ is the group obtained by defining on $\mathbb{R}^{2d+1}$ the product law

$$(z,t) \cdot (z',t') = \left(z + z', t + t' + \frac{1}{2}\sigma(z,z')\right), \quad z, z' \in \mathbb{R}^{2d}, \ t, t' \in \mathbb{R},$$

where σ is the standard symplectic form defined in (1.2).

The Schrödinger representation of the group $\mathbb{H}^d$ on $L^2(\mathbb{R}^d)$ is then defined by

$$\rho(q,p,t)f(x) = e^{2\pi i t}e^{-\pi i p\cdot q}e^{2\pi i p\cdot x}f(x-q), \quad x, q, p \in \mathbb{R}^d, \ t \in \mathbb{R}.$$

Observe that, for $t = 0$, we have

$$\rho(q,p,0)f(x) = e^{-\pi i p\cdot q}e^{2\pi i p\cdot x}f(x-q) = e^{-\pi i p\cdot q}M_pT_qf(x) = e^{-\pi i p\cdot q}\pi(z)f(x), \tag{1.11}$$

with $z = (q,p) \in \mathbb{R}^{2d}$. The symplectic group acts on $\mathbb{H}^d$ via automorphisms that leave the center $\{(0,t) : t \in \mathbb{R}\} \subset \mathbb{H}^d$ of $\mathbb{H}^d$ pointwise fixed

$$A \cdot (z,t) = (Az, t).$$

Therefore for any fixed $\mathcal{A} \in Sp(d, \mathbb{R})$, there is a representation

$$\rho_{\mathcal{A}} : \mathbb{H}^d \to \mathcal{U}(L^2(\mathbb{R}^d)), \quad (z,t) \mapsto \rho(\mathcal{A} \cdot (z,t))$$

whose restriction to the center is a multiple of the identity. By the Stone–von Neumann theorem (cf., e. g., Theorem 9.3.1 in [160]), $\rho_{\mathcal{A}}$ is equivalent to ρ. So there exists an intertwining unitary operator $\mu(\mathcal{A}) \in \mathcal{U}(L^2(\mathbb{R}^d))$ such that

$$\rho_{\mathcal{A}}(z,t) = \mu(\mathcal{A}) \circ \rho(z,t) \circ \mu(\mathcal{A})^{-1} \quad (z,t) \in \mathbb{H}^d. \tag{1.12}$$

By Schur's lemma (cf., e. g., Lemma 9.3.2 in [160]), μ is determined up to a phase factor e^{is}, $s \in \mathbb{R}$. Actually, the phase ambiguity is only a sign, so that μ lifts to a representation of the (double cover of the) symplectic group. This gives a precise meaning to (1.10).

Another construction of the metaplectic group $Mp(d, \mathbb{R})$ was presented by M. de Gosson in [105]. Since the metaplectic representation turns out to be faithful (one-to-one), the metaplectic group elements can be thought as unitary operators themselves.

Hence $Mp(n,\mathbb{R})$ is defined as a subgroup of the unitary group $\mathcal{U}(L^2(\mathbb{R}^d))$ and the corresponding representation is just the inclusion. A difficult point is to prove the existence of a projection

$$\tilde{\pi} : Mp(d,\mathbb{R}) \to Sp(d,\mathbb{R})$$

which makes $Mp(d,\mathbb{R})$ the double covering of $Sp(d,\mathbb{R})$.

We recall here the main points of the construction, and we refer to [105] for details.

It can be proved that the symplectic group $Sp(d,\mathbb{R})$ is generated by the so-called free symplectic matrices

$$S = \begin{pmatrix} A & B \\ C & D \end{pmatrix} \in Sp(d,\mathbb{R}), \quad \det B \neq 0.$$

To each such matrix we associate the generating function

$$W(x,\xi) = \frac{1}{2}DB^{-1}x\cdot x - B^{-1}x\cdot\xi + \frac{1}{2}B^{-1}A\xi\cdot\xi.$$

Conversely, to every polynomial of the type

$$W(x,\xi) = \frac{1}{2}Px\cdot x - Lx\cdot\xi + \frac{1}{2}Q\xi\cdot\xi$$

with

$$P = P^T, Q = Q^T$$

and

$$\det L \neq 0,$$

we can associate a free symplectic matrix, namely

$$S_W = \begin{pmatrix} L^{-1}Q & L^{-1} \\ PL^{-1}Q - L^T & PL^{-1} \end{pmatrix}.$$

Now, given S_W as above and $m \in \mathbb{Z}$ such that

$$m\pi \equiv \arg\det L \mod 2\pi,$$

we define the operator $\widehat{S}_{W,m}$ by setting, for $\psi \in \mathcal{S}(\mathbb{R}^d)$,

$$\widehat{S}_{W,m}\psi(x) = \frac{1}{i^{d/2}}\Delta(W)\int_{\mathbb{R}^d} e^{2\pi i W(x,\xi)}\psi(\xi)\,d\xi \tag{1.13}$$

(to be clear, $i^{d/2} = e^{i\pi d/4}$) where

$$\Delta(W) = i^m\sqrt{|\det L|}.$$

The operator $\widehat{S}_{W,m}$ is called a *quadratic Fourier transform* associated to the free symplectic matrix S_W. The class modulo 4 of the integer m is called *Maslov index* of $\widehat{S}_{W,m}$. Observe that if m is one choice of Maslov index, then $m+2$ is another equally good choice. Hence to each function W, we associate two operators, namely $\widehat{S}_{W,m}$ and $\widehat{S}_{W,m+2} = -\widehat{S}_{W,m}$.

The quadratic Fourier transform corresponding to the choices $S_W = J$ and $m = 0$ is denoted by $\widehat{J}$. The generating function of J being simply $W(x,\xi) = -x\cdot\xi$, it follows that

$$\widehat{J}\psi(x) = \frac{1}{i^{d/2}}\int_{\mathbb{R}^n} e^{-2\pi i x\cdot\xi}\psi(\xi)\,d\xi = \frac{1}{i^{d/2}}\mathcal{F}\psi(x) \tag{1.14}$$

for $\psi \in \mathcal{S}(\mathbb{R}^n)$, where $\mathcal{F}$ is the usual Fourier transform.

The quadratic Fourier transforms $\widehat{S}_{W,m}$ form a subset of the group $\mathcal{U}(L^2(\mathbb{R}^d))$ of unitary operators acting on $L^2(\mathbb{R}^d)$, which is closed under the operation of inversion, and they generate a subgroup of $\mathcal{U}(L^2(\mathbb{R}^d))$ which is, by definition, the metaplectic group $Mp(d,\mathbb{R})$.

Every $\widehat{S} \in Mp(d,\mathbb{R})$ is thus, by definition, a product

$$\widehat{S}_{W_1,m_1}\cdots\widehat{S}_{W_k,m_k}$$

of metaplectic operators associated to free symplectic matrices.

In fact, it can be shown that every $\widehat{S} \in Mp(d,\mathbb{R})$ can be written as a product of only two quadratic Fourier transforms $\widehat{S} = \widehat{S}_{W,m}\widehat{S}_{W',m'}$.

Furthermore (cf. Theorem 114 in [105]), the map

$$\widehat{S}_{W,m} \longmapsto S_W$$

extends to a surjective group homomorphism

$$\tilde{\pi} : Mp(d,\mathbb{R}) \longrightarrow Sp(d,\mathbb{R}),$$

with kernel

$$\ker\tilde{\pi} = \{-I, +I\}.$$

Hence $\tilde{\pi}$ is a twofold covering of the symplectic group.

We also observe that each metaplectic operator is, by construction, a unitary operator in $L^2(\mathbb{R}^d)$, but also an automorphism of $\mathcal{S}(\mathbb{R}^d)$ and of $\mathcal{S}'(\mathbb{R}^d)$.

Alternatively, $Mp(d,\mathbb{R})$ is generated by the following three type of operators (see Proposition 1.1.3 and [193]): $\widehat{J}$ in (1.14) and the operators

$$\mu\left(\begin{pmatrix} L^{-1} & 0_d \\ 0_d & L^T \end{pmatrix}\right) f(x) = \sqrt{|\det L|}\, f(Lx), \tag{1.15}$$

$$\mu\left(\begin{pmatrix} I_d & 0_d \\ C & I_d \end{pmatrix}\right)f(x) = e^{\pi i Cx\cdot x}f(x). \tag{1.16}$$

Another definition of metaplectic operators is also presented in [284]. This involves a time–frequency representation, the so-called Wigner distribution Wf of a function $f \in L^2(\mathbb{R}^d)$ defined in the following Section 1.3.2.

1.1.3 The symplectic Fourier transform

We recall here the notion and main properties of the symplectic Fourier transform for function/distributions defined on the phase space $\mathbb{R}^{2d}$.

Definition 1.1.6. The symplectic Fourier transform $\mathcal{F}_\sigma$ is defined, for $a \in \mathcal{S}(\mathbb{R}^{2d})$, by

$$\mathcal{F}_\sigma a(z) = \int_{\mathbb{R}^{2d}} e^{-2\pi i \sigma(z,z')} a(z')\,dz'. \tag{1.17}$$

We will often use the shorthand notation $a_\sigma = \mathcal{F}_\sigma a$.

We list here the main properties of the operator $\mathcal{F}_\sigma$.

Proposition 1.1.7. *We have*

(i) *The Fourier transform and symplectic Fourier transform are related by*

$$\mathcal{F}_\sigma a(z) = \mathcal{F}a(Jz) = \mathcal{F}(a \circ J)(z), \tag{1.18}$$

for every $a \in \mathcal{S}(\mathbb{R}^{2d})$.

(ii) *The symplectic Fourier transform $\mathcal{F}_\sigma$ is a topological isomorphism from $\mathcal{S}(\mathbb{R}^{2d})$ onto itself, which extends by duality to a topological isomorphism on $\mathcal{S}'(\mathbb{R}^{2d})$.*

(iii) *$\mathcal{F}_\sigma$ is involutive, that is, $\mathcal{F}_\sigma^2 = I$.*

(iv) *For every $a \in L^2(\mathbb{R}^{2d})$, $\mathcal{F}_\sigma$ is a unitary operator on $L^2(\mathbb{R}^{2d})$*

$$\|\mathcal{F}_\sigma a\|_{L^2(\mathbb{R}^{2d})} = \|a\|_{L^2(\mathbb{R}^{2d})}. \tag{1.19}$$

(v) *We have*

$$\mathcal{F}_\sigma(a * b) = \mathcal{F}_\sigma a \cdot \mathcal{F}_\sigma b, \quad a, b \in \mathcal{S}(\mathbb{R}^{2d}). \tag{1.20}$$

Proof. (i) Using the definition of the standard symplectic form $\sigma(z,z') = Jz \cdot z'$, we can write, for every $a \in \mathcal{S}(\mathbb{R}^{2d})$,

$$\mathcal{F}_\sigma a(z) = \int_{\mathbb{R}^{2d}} e^{-2\pi i Jz\cdot z'} a(z')\,dz' = \mathcal{F}a(Jz).$$

The second equality in (1.18) follows by observing that $Jz \cdot z' = z \cdot J^{-1}z'$ and performing the change of variables $J^{-1}z' = w$ ($dz' = dw$) in the above integral.

(ii) It immediately follows by the previous claim since $\mathcal{F} : \mathcal{S}(\mathbb{R}^{2d}) \to \mathcal{S}(\mathbb{R}^{2d})$ is a topological isomorphism which extends by duality on $\mathcal{S}'(\mathbb{R}^{2d})$.
(iii) Using (i), $\mathcal{F}_\sigma^2 a(z) = \mathcal{F}(\mathcal{F}a)(-z) = a(z)$, for every $z \in \mathbb{R}^{2d}$ and $a \in \mathcal{S}(\mathbb{R}^{2d})$.
(iv) It follows by (i) and the fact that $\mathcal{F}$ is a unitary operator on $L^2(\mathbb{R}^{2d})$.
(v) It is a consequence of (i). □

From the previous property (iii), we have that the symplectic Fourier transform satisfies $\mathcal{F}_\sigma^{-1} = \mathcal{F}_\sigma$, hence it coincides with its own inverse.

Let us study the behavior of $\mathcal{F}_\sigma$ under the action of the symplectic group.

Proposition 1.1.8. *For $a \in \mathcal{S}'(\mathbb{R}^{2d})$, $\mathcal{A} \in Sp(d, \mathbb{R})$, we have*

$$\mathcal{F}_\sigma a(\mathcal{A}z) = \mathcal{F}_\sigma(a \circ \mathcal{A})(z), \quad z \in \mathbb{R}^{2d}. \tag{1.21}$$

Proof. It is enough to consider $a \in \mathcal{S}(\mathbb{R}^{2d})$ and then extend the result by duality. Using the fact that symplectic matrices leave invariant the standard symplectic form, we can write

$$\sigma(\mathcal{A}z, z') = \sigma(\mathcal{A}\mathcal{A}^{-1}z, \mathcal{A}^{-1}z') = \sigma(z, \mathcal{A}^{-1}z'),$$

then making the change of variables $\mathcal{A}^{-1}z' = w$ in the integral expression of $\mathcal{F}_\sigma a(\mathcal{A}z)$ (observe $d\mathcal{A}z' = dw$), we immediately obtain (1.21). □

1.2 Time–frequency representations

Many results presented in this section are borrowed from the textbook [160].

1.2.1 The short-time Fourier transform

In what follows we often will call a function $f \in L^2(\mathbb{R}^d)$ a signal, according to the language of signal analysis. In order to obtain information about the *local frequency spectrum* of a signal f, we restrict f to an interval centered at the instant x which is the object of interest and then take the Fourier transform of this restriction. Such a localization in time is made by multiplying f with a smooth cut-off function g, called the *window function*. We shall work mainly with $g \in \mathcal{S}(\mathbb{R}^d)$. This is the idea under the construction of the short-time Fourier transform.

Definition 1.2.1. Fix a window function $g \in L^2(\mathbb{R}^d) \setminus \{0\}$. The short-time Fourier transform (STFT) of a function $f \in L^2(\mathbb{R}^d)$ with respect to g is defined by

$$V_g f(x, \xi) = \langle f, M_\xi T_x g \rangle = \int_{\mathbb{R}^d} f(t)\overline{g(t-x)}e^{-2\pi i t\cdot\xi}\, dt, \quad x, \xi \in \mathbb{R}^d. \tag{1.22}$$

Remark 1.2.2. Assume that the window g is a compactly supported function centered at the origin; then $V_g f(x,\cdot)$ involves the Fourier transform of a segment of f centered around x. As x varies, the window g slides along the x-axis to different positions, and $V_g f(x,\cdot)$ is a substitute for the instantaneous frequency spectrum at x (impossible to be determined, by the uncertainty principle!)

We can rewrite expression (1.22) in terms of the time–frequency shift $\pi(z)$ (cf. (0.15)), $z=(x,\xi)\in\mathbb{R}^{2d}$, as follows:

$$V_g f(z)=\langle f,\pi(z)g\rangle,\quad z\in\mathbb{R}^{2d}. \tag{1.23}$$

Consider $f\in L^p(\mathbb{R}^d)$, $g\in L^{p'}(\mathbb{R}^d)$, then by Hölder's inequality, the integral in (1.22) is absolutely convergent and the STFT is well defined, see the subsequent Proposition 1.2.10. Easy computations yield the following equivalent formulae for the STFT:

$$\begin{aligned}
V_g f(x,\xi) &= \widehat{(f\cdot T_x\bar{g})}(\xi) && (1.24)\\
&= \langle \hat{f}, T_\xi M_{-x}\hat{g}\rangle && (1.25)\\
&= e^{-2\pi i x\cdot\xi}\widehat{(\hat{f}\cdot T_\xi\bar{\hat{g}})}(-x) && (1.26)\\
&= e^{-2\pi i x\cdot\xi}V_{\hat{g}}\hat{f}(\xi,-x) && (1.27)\\
&= e^{-2\pi i x\cdot\xi}(f*M_\xi g^*)(x) && (1.28)\\
&= (\hat{f}*M_{-x}\hat{g}^*)(\xi) && (1.29)\\
&= e^{-\pi i x\cdot\xi}\int_{\mathbb{R}^d} f\left(t+\frac{x}{2}\right)\bar{g}\left(t-\frac{x}{2}\right)e^{-2\pi i t\cdot\xi}\,dt. && (1.30)
\end{aligned}$$

In particular, formula (1.27),

$$V_g f(x,\xi)=e^{-2\pi i x\cdot\xi}V_{\hat{g}}\hat{f}(\xi,-x), \tag{1.31}$$

is called *the fundamental identity of time–frequency analysis*. In fact, it combines both f and $\hat{f}$ into a joint time–frequency representation. Since

$$\begin{pmatrix}\xi\\-x\end{pmatrix}=\begin{pmatrix}0_d & I_d\\ -I_d & 0_d\end{pmatrix}\begin{pmatrix}x\\ \xi\end{pmatrix}=J\begin{pmatrix}x\\ \xi\end{pmatrix},$$

formula (1.31) can be rewritten using the standard symplectic matrix as

$$V_g f(z)=e^{-2\pi i x\cdot\xi}V_{\hat{g}}\hat{f}(Jz),\quad z=(x,\xi)\in\mathbb{R}^{2d}.$$

In this identity the switch to Fourier transforms corresponds to a rotation of the time–frequency plane by an angle of $\pi/2$.

Lemma 1.2.3 (Switching f and g). *For $f,g\in L^2(\mathbb{R}^d)$, we have*

$$V_f g(x,\xi)=e^{-2\pi i x\cdot\xi}\overline{V_g f(-x,-\xi)}. \tag{1.32}$$

Proof. It is a straightforward computation. In details

$$V_f g(x,\xi) = \langle f, M_\xi T_x g\rangle = \overline{\langle M_\xi T_x g, f\rangle} = \overline{\langle g, T_{-x} M_{-\xi} f\rangle} = e^{-2\pi i x\cdot\xi}\overline{\langle g, M_{-\xi} T_{-x} f\rangle}. \qquad \square$$

Another easy computation gives the relation below.

Lemma 1.2.4 (Fourier transform of the STFT). *For $f, g \in L^2(\mathbb{R}^d)$, we have*

$$\widehat{V_g f}(y,\eta) = e^{2\pi i y\cdot\eta} \hat{f}(-\eta)\bar{\hat{g}}(y). \tag{1.33}$$

Remark 1.2.5. The STFT is linear in f and conjugate-linear in g. Keeping the window g fixed, we have that

$$V_g : f \mapsto V_g f$$

maps functions f on $\mathbb{R}^d$ to functions $V_g f$ on $\mathbb{R}^{2d}$.

We shall now exhibit the basic properties of the STFT. To reach this goal, we first establish some features of the translation operator T_h.

Theorem 1.2.6. *If $1 \le p < \infty$, $h \in \mathbb{R}^d$, then*

$$\lim_{h\to 0} \|T_h f - f\|_p = 0, \quad \forall f \in L^p(\mathbb{R}^d). \tag{1.34}$$

Proof. (*First step*) We prove (1.34) when $f \in C_c(\mathbb{R}^d)$. We can limit ourselves to the case $h \in \mathbb{R}^d$, $|h| \le 1$. The set $F = \{\operatorname{supp} f + h, |h| \le 1\}$ is bounded since $\operatorname{supp} f$ is compact, hence there exists a compact set $K \subset \mathbb{R}^d$ such that $F \subset K$. Of course, $\operatorname{supp} T_h f = \operatorname{supp} f + h$, hence

$$\begin{aligned}\|T_h f - f\|_p^p &= \int_K |T_h f(t) - f(t)|^p \, dt \\ &\le \Big(\sup_{t\in K} |f(t-h) - f(t)|\Big)^p \operatorname{meas}(K).\end{aligned}$$

The function f is continuous on the compact set K, hence uniformly continuous on K by the Heine–Cantor's theorem, that is,

$$\lim_{h\to 0} \sup_{t\in K} |f(t-h) - f(t)| = 0,$$

and it follows $\lim_{h\to 0} \|T_h f - f\|_p^p = 0$.
(*Second step*) Consider $f \in L^p(\mathbb{R}^d)$. We shall prove

$$\forall \epsilon > 0\ \exists \delta > 0, \quad \forall h \in \mathbb{R}^d, |h| < \delta \Longrightarrow \|T_h f - f\|_p < \epsilon.$$

Given $\epsilon > 0$, by the density of $C_c(\mathbb{R}^d)$ in $L^p(\mathbb{R}^d)$, there exists $g \in C_c(\mathbb{R}^d)$ such that $\|f-g\|_p < \epsilon/3$. By the first step, there exists a $\delta > 0$ such that $\forall h$, $|h| < \delta$, $\|T_h g - g\|_p < \epsilon/3$. Then

$$\begin{aligned}\|T_h f - f\|_p &= \|T_h f - T_h g + T_h g - g + g - f\|_p \\ &\leq \|T_h f - T_h g\|_p + \|T_h g - g\|_p + \|g - f\|_p = 2\|f - g\|_p + \|T_h g - g\|_p \\ &< 2\frac{\epsilon}{3} + \frac{\epsilon}{3} = \epsilon,\end{aligned}$$

for every $h \in \mathbb{R}^d$ with $|h| < \delta$. □

Remark 1.2.7. The function

$$\omega_p(h) = \|T_h f - f\|_p, \quad h \in \mathbb{R}^d,$$

is often called the L^p**-modulus of continuity** of f.

If $f \in L^\infty(\mathbb{R}^d)$, then f satisfies (1.34) with $p = \infty$ if and only if f is uniformly continuous, up to a null set. In particular,

$$\lim_{h \to 0} \|T_h f - f\|_\infty = 0, \quad \forall f \in C_0(\mathbb{R}^d). \tag{1.35}$$

Corollary 1.2.8. *Consider $h \in \mathbb{R}^d$. Then*

$$\lim_{h \to 0} \|M_h f - f\|_2 = 0, \quad \forall f \in L^2(\mathbb{R}^d).$$

Proof. By Plancherel identity,

$$\|M_h f - f\|_2 = \|T_h \hat{f} - \hat{f}\|_2,$$

and the claim is a consequence of Theorem 1.2.6 for $p = 2$. □

Remark 1.2.9. If we consider the linear mapping

$$\tau : \mathbb{R}^d \to \mathcal{L}(L^p(\mathbb{R}^d)), \quad \tau : h \mapsto T_h \tag{1.36}$$

for $1 \leq p < \infty$, we have

$$\lim_{h \to s} \|T_h f - T_s f\|_p = 0, \quad \forall f \in L^p(\mathbb{R}^d),$$

that is, $\tau(h) \to \tau(s)$ in the strong operator topology when $h \to s$. This follows by (1.34), with the change of variables $h \to h - s$ (hence $h \to 0 \iff h \to s$), and by

$$\|T_{h-s} f - f\|_p = \left\|T_s(T_{h-s} f - f)\right\|_p = \|T_h f - T_s f\|_p.$$

We may therefore say that $\{T_h\}_{h \in \mathbb{R}^d}$ is a strongly continuous one-parameter family of operators on $L^p(\mathbb{R}^d)$, $1 \leq p < \infty$. Analogously, $\{T_h\}_{h \in \mathbb{R}^d}$ is a strongly continuous one-parameter family of operators on $C_0(\mathbb{R}^d)$.

The information given by Corollary 1.2.8 extends easily to $L^p(\mathbb{R}^d)$, namely the same arguments as in Theorem 1.2.6 yield

$$\lim_{h\to 0} \|M_h f - f\|_p = 0, \quad \forall f \in L^p(\mathbb{R}^d),\ 1 \le p < \infty \tag{1.37}$$

and

$$\lim_{h\to 0} \|M_h f - f\|_\infty = 0, \quad \forall f \in \mathcal{C}_0(\mathbb{R}^d). \tag{1.38}$$

Hence $\{M_h\}_{h\in\mathbb{R}^d}$ is a strongly continuous one-parameter family of operators on $L^p(\mathbb{R}^d)$, $1 \le p < \infty$, and on $\mathcal{C}_0(\mathbb{R}^d)$.

Proposition 1.2.10. *Consider* $1 < p < \infty$. *If* $f \in L^p(\mathbb{R}^d)$, $g \in L^{p'}(\mathbb{R}^d)$, *with* $1/p + 1/p' = 1$, *then* $V_g f$ *is uniformly continuous on* $\mathbb{R}^{2d}$. *The same claim is valid for* $f \in L^1(\mathbb{R}^d)$, $g \in L^\infty(\mathbb{R}^d)$ *and uniformly continuous on* $\mathbb{R}^d$, *or vice versa. Moreover, we have*

$$\|V_g f\|_\infty \le \|f\|_p \|g\|_{p'}. \tag{1.39}$$

Proof. Consider $f \in L^p(\mathbb{R}^d)$, $g \in L^{p'}(\mathbb{R}^d)$, $1 \le p \le \infty$. By Hölder's inequality,

$$|V_g f(x,\xi)| \le \|f\|_p \|M_\xi T_x g\|_{p'} = \|f\|_p \|g\|_{p'}$$

so that $V_g f$ is defined everywhere on $\mathbb{R}^{2d}$ and we obtain (1.39). Let us check that $V_g f$ is uniformly continuous for $1 < p < \infty$. We assume $f \in L^p(\mathbb{R}^d) \setminus \{0\}$, $g \in L^{p'}(\mathbb{R}^d) \setminus \{0\}$, otherwise $V_g f(x,\xi) = 0$ and we are done. For $h = (h_1, h_2) \in \mathbb{R}^{2d}$,

$$\begin{aligned}
&|V_g f(x - h_1, \xi - h_2) - V_g f(x,\xi)| \\
&\quad = |V_g f(x - h_1, \xi - h_2) - V_g f(x - h_1, \xi) + V_g f(x - h_1, \xi) - V_g f(x,\xi)| \\
&\quad = |\langle f, M_{\xi - h_2} T_{x-h_1} g\rangle - \langle f, M_\xi T_{x-h_1} g\rangle + \langle f, M_\xi T_{x-h_1} g\rangle - \langle f, M_\xi T_x g\rangle| \\
&\quad = |\langle M_{h_2} f - f, M_\xi T_{x-h_1} g\rangle + \langle M_{-\xi} f, (T_{x-h_1} g - T_x g)\rangle| \\
&\quad \le \|M_{h_2} f - f\|_p \|M_\xi T_{x-h_1} g\|_{p'} + \|M_{-\xi} f\|_p \big\|T_x(T_{-h_1} g - g)\big\|_{p'} \\
&\quad = \|M_{h_2} f - f\|_p \|g\|_{p'} + \|f\|_p \|T_{-h_1} g - g\|_{p'}.
\end{aligned}$$

Now, take $\epsilon > 0$. By Theorem 1.2.6, there exists $\delta_1 > 0$ such that, if $|h_1| < \delta$, $\|T_{-h_1} g - g\|_{p'} < \epsilon/(2\|f\|_p)$, and analogously, by (1.37), there exists $\delta_2 > 0$ such that if $|h_2| < \delta_2$, then $\|M_{h_2} f - f\|_p < \epsilon/(2\|g\|_{p'})$. Taking $\delta = \min\{\delta_1, \delta_2\}$, if $|h| < \delta$, we have the claim. The proof in the case $p = 1$ and $p' = \infty$, with g uniformly continuous (and vice versa) is analogous. □

As we shall see in the next sections, the domain of the STFT can be extended to other function spaces. Indeed, the formula

$$V_g f(x,\xi) = \langle f, M_\xi T_x g\rangle \tag{1.40}$$

makes $V_g f$ well defined whenever f and g belong to any pair of dual spaces, extending the duality of $L^2(\mathbb{R}^d)$, namely linear in the first component and (conjugate)-linear in the second. We need, of course, the continuity of the time–frequency shifts in such spaces. For example, we can consider $f \in \mathcal{S}'(\mathbb{R}^d)$ and a window function $g \in \mathcal{S}(\mathbb{R}^d)$. We will see below that the time–frequency shifts are continuous on $\mathcal{S}(\mathbb{R}^d)$ and on $\mathcal{S}'(\mathbb{R}^d)$.

Moreover, the STFT can be defined for both $f, g \in \mathcal{S}'(\mathbb{R}^d)$. This is a consequence of interpreting the STFT as a sesquilinear form $(f, g) \mapsto V_g f$. Define $f \otimes g$ to be the tensor product

$$f \otimes g(x, y) = f(x)g(y), \quad x, y \in \mathbb{R}^d.$$

Let $F(x, y)$ be a function on $\mathbb{R}^{2d}$. Consider the linear operator $\mathfrak{T}_a$ (change of coordinates), defined by

$$\mathfrak{T}_a F(x, y) = F(y, y - x) \quad x, y \in \mathbb{R}^d. \tag{1.41}$$

Let $\mathcal{F}_2$ be the partial Fourier transform with respect to the second variable y:

$$\mathcal{F}_2 F(x, \xi) = \int_{\mathbb{R}^d} e^{-2\pi i y \cdot \xi} F(x, y)\, dy. \tag{1.42}$$

Using these operators in the definition (1.22), we obtain another expression for the STFT; for $f, g \in L^2(\mathbb{R}^d)$,

$$V_g f = \mathcal{F}_2 \mathfrak{T}_a (f \otimes \bar{g}). \tag{1.43}$$

The claim is obtained by writing down explicitly formula (1.24). Indeed,

$$\begin{aligned} V_g f(x, \xi) &= \int_{\mathbb{R}^d} e^{-2\pi i y \cdot \xi} f(y)\bar{g}(y - x)\, dy \\ &= \int_{\mathbb{R}^d} e^{-2\pi i y \cdot \xi} \mathfrak{T}_a (f \otimes \bar{g})(x, y)\, dy \\ &= \mathcal{F}_2 \mathfrak{T}_a (f \otimes \bar{g})(x, \xi). \end{aligned}$$

Thanks to (1.43), we can extend the domain of STFT to the space of tempered distributions $\mathcal{S}'(\mathbb{R}^d)$. Let us first recall a few definitions. The complex conjugate $\overline{f} \in \mathcal{S}'(\mathbb{R}^d)$ of a distribution $f \in \mathcal{S}'(\mathbb{R}^d)$ is defined by

$$\langle \overline{f}, \varphi \rangle = \overline{\langle f, \overline{\varphi} \rangle}, \quad \varphi \in \mathcal{S}(\mathbb{R}^d).$$

The tensor product of functions $f, g \in \mathcal{S}'(\mathbb{R}^d)$ is a distribution in $\mathcal{S}'(\mathbb{R}^{2d})$ defined by the rule

$$\langle f \otimes g, \Phi \rangle = \langle f_x, \overline{\langle g, \Phi(x, \cdot) \rangle} \rangle, \quad \Phi \in \mathcal{S}(\mathbb{R}^{2d}),$$

where f_x indicates that f acts only on variable x. In particular, $f \otimes g$ is the unique tempered distribution such that

$$\langle f \otimes g, \varphi_1 \otimes \varphi_2 \rangle = \langle f, \varphi_1 \rangle \langle g, \varphi_2 \rangle,$$

for every $\varphi_1, \varphi_2 \in \mathcal{S}(\mathbb{R}^d)$. Now, it is easy to check that both operators $\mathfrak{T}_a$ and $\mathcal{F}_2$ are topological isomorphisms on the Schwartz class $\mathcal{S}(\mathbb{R}^{2d})$ and on its dual space $\mathcal{S}'(\mathbb{R}^{2d})$. Therefore if $f, g \in \mathcal{S}'(\mathbb{R}^d)$, the tensor product $f \otimes g$ is a well-defined distribution in $\mathcal{S}'(\mathbb{R}^{2d})$, and consequently

$$V_g f = \mathcal{F}_2 \mathfrak{T}_a (f \otimes \bar{g}) \in \mathcal{S}'(\mathbb{R}^{2d}).$$

We note that when f and g are tempered distributions, the STFT $V_g f$ is no longer a function but a tempered distribution in $\mathcal{S}'(\mathbb{R}^{2d})$.

1.2.2 Orthogonality relations

The orthogonality relations for the STFT correspond to Parseval's formula for the Fourier transform, which is indeed the key ingredient to obtain such a result.

Theorem 1.2.11 (Orthogonality relations for the STFT). *Consider $f_i, g_i \in L^2(\mathbb{R}^d)$, $i = 1, 2$. Then $V_{g_i} f_i \in L^2(\mathbb{R}^{2d})$ and*

$$\langle V_{g_1} f_1, V_{g_2} f_2 \rangle_{L^2(\mathbb{R}^{2d})} = \langle f_1, f_2 \rangle_{L^2(\mathbb{R}^d)} \overline{\langle g_1, g_2 \rangle_{L^2(\mathbb{R}^d)}}. \tag{1.44}$$

Proof. Using that the operators $\mathcal{F}_2$ and $\mathfrak{T}_a$ are unitary on $L^2(\mathbb{R}^d)$, we can write

$$\begin{aligned}\langle V_{g_1} f_1, V_{g_2} f_2 \rangle &= \langle \mathcal{F}_2 \mathfrak{T}_a (f_1 \otimes \bar{g}_1), \mathcal{F}_2 \mathfrak{T}_a (f_2 \otimes \bar{g}_2) \rangle \\ &= \langle f_1 \otimes \bar{g}_1, f_2 \otimes \bar{g}_2 \rangle \\ &= \langle f_1, f_2 \rangle \langle \bar{g}_1, \bar{g}_2 \rangle,\end{aligned}$$

yielding the claim. □

Corollary 1.2.12. *For $f, g \in L^2(\mathbb{R}^d)$, we have*

$$\|V_g f\|_2 = \|f\|_2 \|g\|_2. \tag{1.45}$$

In particular, if $\|g\|_2 = 1$, the STFT becomes an isometry from $L^2(\mathbb{R}^d)$ into $L^2(\mathbb{R}^{2d})$

$$\|V_g f\|_2 = \|f\|_2, \quad \forall f \in L^2(\mathbb{R}^d). \tag{1.46}$$

Observe that V_g is not onto. Indeed, recall from Proposition 1.2.10 that $V_g f$ is a continuous function, whereas in $L^2(\mathbb{R}^{2d})$ there are functions which are not continuous.

1.2.3 Short-time Fourier transform and symplectic matrices

We use the orthogonality relations in (1.44) to prove the following properties of the STFT.

Proposition 1.2.13. *For $f, f_i, g, g_i \in L^2(\mathbb{R}^d)$, $i = 1, 2$, we have*

(i) *If $w \in \mathbb{R}^{2d}$,*

$$V_{\pi(w)g}(\pi(w)f)(z) = e^{-2\pi i\sigma(z,w)} V_g f(z), \quad z \in \mathbb{R}^{2d} \tag{1.47}$$

where σ is the standard symplectic form and $\pi(w)$ is the time–frequency shift of $w \in \mathbb{R}^{2d}$.

(ii) *Fourier transform of a product of STFTs:*

$$\mathcal{F}(V_{g_1} f_1 \overline{V_{g_2} f_2})(z) = (V_{f_2} f_1 \overline{V_{g_2} g_1})(-Jz), \quad z \in \mathbb{R}^{2d}, \tag{1.48}$$

where J is the standard symplectic matrix.

Proof. (i) It is a straightforward computation. Indeed,

$$\begin{aligned} V_{\pi(w)g}(\pi(w)f)(z) &= \langle \pi(w)f, \pi(z)\pi(w)g\rangle = \langle f, \pi^*(w)\pi(z)\pi(w)g\rangle \\ &= e^{2\pi i w_1\cdot w_2}\langle f, \pi(-w)\pi(z)\pi(w)g\rangle \\ &= e^{2\pi i w_1\cdot w_2} e^{2\pi i z_1\cdot w_2}\langle f, \pi(-w)\pi(z+w)g\rangle \\ &= e^{-2\pi i\sigma(z,w)}\langle f, \pi(z)g\rangle, \end{aligned}$$

which gives (1.47).

(ii) Observe that $\sigma(w, -Jz) = -z \cdot w$, so that using (1.47) and the orthogonality relations (1.44) gives

$$\begin{aligned} \mathcal{F}(V_{g_1} f_1 \overline{V_{g_2} f_2})(z) &= \int_{\mathbb{R}^{2d}} V_{g_1} f_1(w) \overline{V_{g_2} f_2(w)} e^{2\pi i\sigma(w,-Jz)}\, dw \\ &= \langle V_{g_1} f_1, V_{\pi(-Jz)g_2} \pi(-Jz) f_2\rangle \\ &= \langle f_1, \pi(-Jz)f_2\rangle \overline{\langle g_1, V_{\pi(-Jz)g_2}\rangle}, \end{aligned}$$

as desired. □

Proposition 1.2.14 (Symplectic covariance formula for the STFT). *Consider $f, g \in L^2(\mathbb{R}^d)$, the metaplectic operator $\widehat{\mathcal{A}}$ with projection $\mathcal{A} \in Sp(d, \mathbb{R})$. For $z = (x, \xi) \in \mathbb{R}^{2d}$, we set $\mathcal{A}^{-1}z = z' = (x', \xi')$. Then*

$$V_{\widehat{\mathcal{A}}g}(\widehat{\mathcal{A}}f)(z) = e^{\pi i(x'\cdot\xi' - x\cdot\xi)} V_g f(\mathcal{A}^{-1}z). \tag{1.49}$$

Proof. Using the intertwining property of the metaplectic representation with respect to the Schrödinger one (1.12) and the relation between time–frequency shifts and Schrödinger representation (1.11), we calculate

$$\begin{aligned}
V_{\widehat{\mathcal{A}}g}(\widehat{\mathcal{A}}f)(z) &= \langle \widehat{\mathcal{A}}f, \pi(z)\widehat{\mathcal{A}}g\rangle = \langle f, \widehat{\mathcal{A}}^{-1}\pi(z)\widehat{\mathcal{A}}g\rangle \\
&= e^{-\pi i x\cdot\xi}\langle f, \widehat{\mathcal{A}^{-1}}\rho(z,0)\widehat{\mathcal{A}}g\rangle = e^{-\pi i x\cdot\xi}\langle f, \rho(\mathcal{A}^{-1}z,0)g\rangle \\
&= e^{\pi i(x'\cdot\xi'-x\cdot\xi)}\langle f, \pi(\mathcal{A}^{-1}z)g\rangle = e^{\pi i(x'\cdot\xi'-x\cdot\xi)}V_gf(\mathcal{A}^{-1}z).
\end{aligned}$$

This concludes the proof. □

Let us see the action of the STFT on a time–frequency shift.

Proposition 1.2.15. *Whenever V_gf is defined, we have*

$$V_g(M_vT_uf)(x,\xi) = e^{-2\pi i u\cdot(\xi-v)}V_gf(x-u,\xi-v), \quad u,v,x,\xi \in \mathbb{R}^d. \tag{1.50}$$

In particular,

$$|V_g(\pi(z)f)(w)| = |T_zV_gf(w)|, \quad z,w \in \mathbb{R}^{2d}. \tag{1.51}$$

Proof. Using that M_v and T_u are unitary operators, we can write

$$\begin{aligned}
V_g(M_vT_uf)(x,\xi) &= \langle M_vT_uf, M_\xi T_xg\rangle = \langle f, T_{-u}M_{-v}M_\xi T_xg\rangle \\
&= \langle f, T_{-u}M_{\xi-v}T_xg\rangle = e^{-2\pi i u\cdot(\xi-v)}\langle f, M_{\xi-v}T_{x-u}g\rangle,
\end{aligned}$$

proving the claim. □

The formula in (1.50) is also known as the *covariance property* of the STFT.

1.2.4 Reproducing formula for the short-time Fourier transform

In what follows we are going to show that every function in $L^2(\mathbb{R}^d)$ can be represented by means of the STFT. Namely, once we have computed the STFT of a signal f, we can go back to the original signal f. Let us first give the definition of vector-valued integrals, here understood in a weak sense. If $g : \mathbb{R}^d \to B$, where B is a Banach space, then

$$f = \int_{\mathbb{R}^d} g(x)\,dx$$

means that

$$\ell(h) = \langle f, h\rangle = \int_{\mathbb{R}^d} \langle g(x), h\rangle\,dx, \quad \forall h \in B^*.$$

If the mapping $\ell : B^* \to \mathbb{C}$ is a bounded (conjugate)-linear functional on B^*, then ℓ defines a unique element $f \in B^{**}$, the bidual of B. We shall work with reflexive Banach spaces, so that $B^{**} = B$. Relevant choice will be $B = L^2(\mathbb{R}^d)$.

The vector-valued integrals object of our study are given by *superpositions* of time–frequency shifts

$$f = \int_{\mathbb{R}^{2d}} F(x,\xi) M_\xi T_x g \, dx \, d\xi. \tag{1.52}$$

Assuming $g \in L^2(\mathbb{R}^d)$, we have also $M_\xi T_x g \in L^2(\mathbb{R}^d)$, for every $(x,\xi) \in \mathbb{R}^{2d}$. If we choose $F \in L^2(\mathbb{R}^{2d})$, the (conjugate)-linear functional

$$\ell(h) = \int_{\mathbb{R}^{2d}} F(x,\xi) \overline{\langle h, M_\xi T_x g\rangle} \, dx \, d\xi \tag{1.53}$$

is bounded on $L^2(\mathbb{R}^d)$. Indeed, by Cauchy–Schwarz inequality and the orthogonality relations in (1.46), we estimate

$$|\ell(h)| \leq \|F\|_2 \|V_g h\|_2 = \|F\|_2 \|h\|_2 \|g\|_2, \quad \forall h \in L^2(\mathbb{R}^d). \tag{1.54}$$

Hence ℓ defines a unique function

$$f = \int_{\mathbb{R}^{2d}} F(x,\xi) M_\xi T_x g \, dx \, d\xi \in L^2(\mathbb{R}^d),$$

with norm

$$\|f\|_2 = \sup_{h \in L^2, \|h\|_2 = 1} |\ell(h)| \leq \|F\|_2 \|g\|_2,$$

satisfying $\ell(h) = \langle f, h\rangle$.

Theorem 1.2.16 (Inversion formula for the STFT). *Assume $g, \gamma \in L^2(\mathbb{R}^d)$ with $\langle g, \gamma\rangle \neq 0$. Then for all $f \in L^2(\mathbb{R}^d)$,*

$$f = \frac{1}{\langle \gamma, g\rangle} \int_{\mathbb{R}^{2d}} V_g f(x,\xi) M_\xi T_x \gamma \, dx \, d\xi. \tag{1.55}$$

Proof. By Theorem 1.2.11, the STFT $V_g f$ is in $L^2(\mathbb{R}^{2d})$. This implies that the vector-valued integral

$$\tilde{f} = \frac{1}{\langle \gamma, g\rangle} \int_{\mathbb{R}^{2d}} V_g f(x,\xi) M_\xi T_x \gamma \, dx \, d\xi$$

is a well-defined function $\tilde{f} \in L^2(\mathbb{R}^d)$. It remains to prove that $\tilde{f} = f$. Using the orthogonality relations (1.44), for every $h \in L^2(\mathbb{R}^d)$,

$$\begin{aligned}\langle \tilde{f}, h\rangle &= \frac{1}{\langle \gamma, g\rangle} \int_{\mathbb{R}^{2d}} V_g f(x,\xi)\langle M_\xi T_x \gamma, h\rangle \, dx\, d\xi \\ &= \frac{1}{\langle \gamma, g\rangle} \int_{\mathbb{R}^{2d}} V_g f(x,\xi)\overline{V_\gamma h(x,\xi)} \, dx\, d\xi \\ &= \langle f, h\rangle\end{aligned}$$

hence $f = \tilde{f}$, as desired. □

The adjoint of the short-time Fourier transform. Fix $g \in L^2(\mathbb{R}^d)$ and consider the linear operator A_g defined by

$$A_g F = \int_{\mathbb{R}^{2d}} F(x,\xi) M_\xi T_x g \, dx\, d\xi.$$

By (1.54), the operator A_g is a bounded operator from $L^2(\mathbb{R}^{2d})$ onto $L^2(\mathbb{R}^d)$. Moreover, $A_g = V_g^*$, where $V_g : L^2(\mathbb{R}^d) \to L^2(\mathbb{R}^{2d})$. Indeed,

$$\begin{aligned}\langle A_g F, h\rangle &= \int_{\mathbb{R}^{2d}} F(x,\xi)\langle M_\xi T_x g, h\rangle \, dx\, d\xi \\ &= \langle F, V_g h\rangle = \langle V_g^* F, h\rangle\end{aligned}$$

for every $h \in L^2(\mathbb{R}^d)$, $F \in L^2(\mathbb{R}^{2d})$. The inversion formula (1.55) can be then rephrased as

$$f = \frac{1}{\langle \gamma, g\rangle} V_\gamma^* V_g f, \quad f \in L^2(\mathbb{R}^d). \tag{1.56}$$

1.2.5 Short-time Fourier transform of tempered distributions

We come now to the study of the STFT on the Schwartz class and on the space of tempered distributions.

In what follows we will make heavily use of the Schwartz seminorms, for which there are equivalent definitions. Namely, for $m, n \in \mathbb{N}$, the Schwartz seminorms can be defined, for $f \in \mathcal{S}(\mathbb{R}^d)$, by

$$\|f\|_{m,n} = \sum_{|\alpha|\le m, |\beta|\le n} \sup_{t\in\mathbb{R}^d} |t^\alpha \partial^\beta f(t)| < \infty. \tag{1.57}$$

The linear space $\mathcal{S}(\mathbb{R}^d)$, endowed with the family of seminorms $\{\|\cdot\|_{m,n}\}$, $m, n \in \mathbb{N}$, is a Fréchet space. Other equivalent families of seminorms are provided by

$$p_{\alpha,\beta}(f) = \sup_{t\in\mathbb{R}^d}|t^\alpha\partial^\beta f(t)| = \|t^\alpha\partial^\beta f\|_\infty, \quad \alpha,\beta \in \mathbb{N}^d,$$

or

$$\|f\|_k = \sum_{|\alpha|\le k}\|(1+|t|^k)\partial^\alpha f\|_\infty, \quad k \in \mathbb{N}. \tag{1.58}$$

The following result shows how to exchange differential and power operators with time–frequency shifts.

Lemma 1.2.17. *For $g \in \mathcal{S}(\mathbb{R}^d)$, for every $\alpha, \beta \in \mathbb{N}^d$, we have*

$$t^\alpha\partial^\beta(M_\xi T_x g) = \sum_{\delta\le\alpha,\gamma\le\beta}\binom{\alpha}{\delta}\binom{\beta}{\gamma}(2\pi i\xi)^\gamma x^\delta M_\xi T_x(t^{\alpha-\delta}\partial^{\beta-\gamma}g). \tag{1.59}$$

Proof. Using

$$t^\alpha = (t-x+x)^\alpha = \sum_{\delta\le\alpha}\binom{\alpha}{\delta}x^\delta(t-x)^{\alpha-\delta}$$

and Leibniz' formula (0.5)

$$\partial_t^\beta(e^{2\pi i\xi\cdot t}g(t-x)) = \sum_{\gamma\le\beta}\binom{\beta}{\gamma}(2\pi i\xi)^\gamma e^{2\pi i\xi\cdot t}(\partial^{\beta-\gamma}g)(t-x),$$

we immediately obtain (1.59). □

Corollary 1.2.18. *The operator-valued map*

$$(x,\xi) \mapsto M_\xi T_x$$

is strongly continuous on $\mathcal{S}(\mathbb{R}^d)$ and $\mathcal{S}'(\mathbb{R}^d)$ (with the weak-topology).*

Proof. For $\varphi \in \mathcal{S}(\mathbb{R}^d)$, we have to show that

$$\lim_{|x|,|\xi|\to 0}\|t^\alpha\partial^\beta(M_\xi T_x\varphi - \varphi)\|_\infty = 0, \quad \forall\alpha,\beta \in \mathbb{N}. \tag{1.60}$$

Using (1.59),

$$\begin{aligned}\|t^\alpha\partial^\beta(M_\xi T_x\varphi - \varphi)\|_\infty &\le \|M_\xi T_x(t^\alpha\partial^\beta\varphi) - t^\alpha\partial^\beta\varphi\|_\infty \\ &\quad + \sum_{\substack{\delta\le\alpha,\gamma\le\beta,\\(\delta,\gamma)\ne(0,0)}}\binom{\alpha}{\delta}\binom{\beta}{\gamma}|(2\pi i\xi)^\gamma x^\delta|\|M_\xi T_x(t^{\alpha-\delta}\partial^{\beta-\gamma}g)\|_\infty.\end{aligned}$$

Observe that $t^\alpha \partial^\beta \varphi \in \mathcal{S}(\mathbb{R}^d) \subset \mathcal{C}_0(\mathbb{R}^d)$, hence

$$\lim_{|x|,|\xi|\to 0} \|M_\xi T_x(t^\alpha \partial^\beta \varphi) - t^\alpha \partial^\beta \varphi\|_\infty = 0, \quad \forall \alpha, \beta \in \mathbb{N},$$

in view of (1.35) and (1.38). Obviously,

$$\lim_{|x|,|\xi|\to 0} |(2\pi i \xi)^\gamma x^\delta| \|t^{\alpha-\delta} \partial^{\beta-\gamma} g\|_\infty = 0, \quad \text{whenever } (\delta, \gamma) \neq (0, 0).$$

This proves (1.60).

For $f \in \mathcal{S}'(\mathbb{R}^d)$, for every $\varphi \in \mathcal{S}(\mathbb{R}^d)$,

$$\begin{aligned} \lim_{|x|,|\xi|\to 0} \langle M_\xi T_x f, \varphi \rangle &= \lim_{|x|,|\xi|\to 0} \langle f, T_{-x} M_{-\xi} \varphi \rangle = \lim_{|x|,|\xi|\to 0} e^{2\pi i x \cdot \xi} \langle f, M_{-\xi} T_{-x} \varphi \rangle \\ &= \langle f, \varphi \rangle, \end{aligned}$$

by the first part of the proof. □

For $f \in \mathcal{S}'(\mathbb{R}^d)$, $g \in \mathcal{S}(\mathbb{R}^d) \setminus \{0\}$, let us define pointwise

$$V_g f(x, \xi) = \langle f, M_\xi T_x g \rangle, \quad (x, \xi) \in \mathbb{R}^{2d}.$$

An easy consequence of Corollary 1.2.18 is the continuity of the STFT.

Corollary 1.2.19. *For $g \in \mathcal{S}(\mathbb{R}^d) \setminus \{0\}$, $f \in \mathcal{S}'(\mathbb{R}^d)$, the STFT $V_g f$ is a continuous function on $\mathbb{R}^{2d}$.*

Proof. Set $h = (h_1, h_2) \in \mathbb{R}^{2d}$. We have, for every $(x, \xi) \in \mathbb{R}^{2d}$,

$$\begin{aligned} \lim_{|h_1|,|h_2|\to 0} V_g f(x + h_1, \xi + h_2) &= \lim_{|h_1|,|h_2|\to 0} \langle f, M_{\xi+h_2} T_{x+h_1} g \rangle \\ &= \lim_{|h_1|,|h_2|\to 0} \langle M_{-\xi} f, M_{h_2} T_{x+h_1} g \rangle \\ &= \lim_{|h_1|,|h_2|\to 0} e^{-2\pi i h_2 \cdot x} \langle T_{-x} M_{-\xi} f, M_{h_2} T_{h_1} g \rangle \\ &= \lim_{|h_1|,|h_2|\to 0} \langle T_{-x} M_{-\xi} f, M_{h_2} T_{h_1} g \rangle \\ &= \langle T_{-x} M_{-\xi} f, g \rangle = V_g f(x, \xi), \end{aligned}$$

where the result of the last but one equality follows by Corollary 1.2.18. □

We now investigate the behavior at infinity of the STFT.

We recall the notation

$$\langle (x, \xi) \rangle^s = (1 + |x|^2 + |\xi|^2)^{s/2}, \quad \text{for } s \in \mathbb{R}.$$

Proposition 1.2.20. *For $f \in \mathcal{S}'(\mathbb{R}^d)$, $g \in \mathcal{S}(\mathbb{R}^d) \setminus \{0\}$, there exist a constant $C > 0$ and an integer $N \in \mathbb{N}$ such that*

$$|V_g f(x, \xi)| \leq C \langle (x, \xi) \rangle^N, \quad (x, \xi) \in \mathbb{R}^{2d}. \tag{1.61}$$

Proof. For $f \in \mathcal{S}'(\mathbb{R}^d)$, $g \in \mathcal{S}(\mathbb{R}^d) \setminus \{0\}$, $M_\xi T_x g \in \mathcal{S}(\mathbb{R}^d) \setminus \{0\}$, hence there exist $C > 0$, $m, n \in \mathbb{N}$, such that

$$\begin{aligned}
|\langle f, M_\xi T_x g\rangle| &\le C \sum_{|\alpha|\le m, |\beta|\le n} \|t^\alpha \partial^\beta (M_\xi T_x g)\|_\infty \\
&\le C \sum_{|\alpha|\le m, |\beta|\le n} \sum_{\delta\le\alpha, \gamma\le\beta} \binom{\alpha}{\delta}\binom{\beta}{\gamma} |(2\pi i\xi)^\gamma x^\delta| \|M_\xi T_x (t^{\alpha-\delta}\partial^{\beta-\gamma} g)\|_\infty \\
&\le C \sum_{|\alpha|\le m, |\beta|\le n} \sum_{\delta\le\alpha, \gamma\le\beta} C_{\alpha,\beta,\gamma,\delta} |x|^{|\delta|} |\xi|^{|\gamma|} \\
&\le K \langle (x,\xi)\rangle^N
\end{aligned}$$

for $N = m + n$ (in the second inequality we applied (1.59)). □

Let us now show a pair of examples where it is easy to compute the STFT of a tempered distribution.

Example 1.2.21.

(i) In dimension $d = 1$, consider the delta distribution $\delta \in \mathcal{S}'(\mathbb{R})$. For any $g \in \mathcal{S}(\mathbb{R}) \setminus \{0\}$,

$$V_g\delta(x,\xi) = \langle \delta, M_\xi T_x g\rangle = \overline{(M_\xi T_x g)(0)} = \overline{g(-x)}.$$

(ii) Consider now $\delta' \in \mathcal{S}'(\mathbb{R})$. Then for every $(x,\xi) \in \mathbb{R}^{2d}$

$$\begin{aligned}
V_g\delta'(x,\xi) &= \langle \delta', M_\xi T_x g\rangle = -\langle \delta, (M_\xi T_x g)'\rangle \\
&= -(\langle \delta, (2\pi i\xi) M_\xi T_x g\rangle + \langle \delta, M_\xi T_x g'\rangle) \\
&= (2\pi i\xi)\overline{g(-x)} - \overline{g'(-x)}.
\end{aligned}$$

In what follows we study a vector-valued integral, obtained by multiplying the time–frequency shifts with a function F of rapid decay on the phase space.

Theorem 1.2.22. *Fix $g \in \mathcal{S}(\mathbb{R}^d) \setminus \{0\}$ and consider a measurable function F on $\mathbb{R}^{2d}$ with rapid decay*

$$\forall n \ge 0\ \exists C_n > 0, \quad |F(x,\xi)| \le C_n \langle (x,\xi)\rangle^{-n}, \ \forall (x,\xi) \in \mathbb{R}^{2d}. \tag{1.62}$$

Then the integral

$$f(t) = \int_{\mathbb{R}^{2d}} F(x,\xi) M_\xi T_x g(t)\, dx\, d\xi \tag{1.63}$$

defines a function $f \in \mathcal{S}(\mathbb{R}^d)$.

Proof. For every fixed $t \in \mathbb{R}^d$,

$$\int_{\mathbb{R}^{2d}} |F(x,\xi)||M_\xi T_x g(t)|\,dx\,d\xi \le C_n \|g\|_\infty \int_{\mathbb{R}^{2d}} \frac{1}{\langle (x,\xi)\rangle^n}\,dx\,d\xi.$$

Taking $n > 2d$, we see that the integral above converges to a number $f(t)$. Moreover, for $1 \le j \le d$,

$$\partial_{t_j} f(t) = \int_{\mathbb{R}^{2d}} F(x,\xi)\partial_{t_j}(M_\xi T_x g)(t)\,dx\,d\xi.$$

Differentiation is legitimate since the obtained integral is uniformly convergent. In fact, writing

$$A(x,\xi,t) = \partial_{t_j}(M_\xi T_x g)(t) = (2\pi i \xi_j) M_\xi T_x g(t) + M_\xi T_x \partial_{t_j} g(t),$$

and taking $n > 2d + 1$ in (1.62), we have

$$\int_{\mathbb{R}^{2d}} |A(x,\xi,t)F(x,\xi)|\,dx\,d\xi \le C \int_{\mathbb{R}^{2d}} (1+|\xi|)|F(x,\xi)|\,dx\,d\xi < \infty.$$

Generalizing the argument, we derive that $f \in \mathcal{C}^\infty(\mathbb{R}^d)$ and, using (1.59), obtain

$$\begin{aligned} t^\alpha \partial^\beta f(t) &= \int_{\mathbb{R}^{2d}} F(x,\xi) t^\alpha \partial^\beta (M_\xi T_x g)(t)\,dx\,d\xi \\ &= \sum_{\delta\le\alpha}\sum_{\gamma\le\beta} \binom{\alpha}{\delta}\binom{\beta}{\gamma} \int_{\mathbb{R}^{2d}} F(x,\xi)(2\pi i\xi)^\gamma x^\delta M_\xi T_x (t^{\alpha-\delta}\partial^{\beta-\gamma} g)\,dx\,d\xi. \end{aligned}$$

Hence

$$\|t^\alpha \partial^\beta f\|_\infty \le \sum_{\delta\le\alpha}\sum_{\gamma\le\beta} \binom{\alpha}{\delta}\binom{\beta}{\gamma} \int_{\mathbb{R}^{2d}} |F(x,\xi)||(2\pi i\xi)^\gamma x^\delta| \|t^{\alpha-\delta}\partial^{\beta-\gamma} g\|_\infty\,dx\,d\xi.$$

If we set

$$c_{\alpha,\beta} := \max\{\|t^{\alpha-\delta}\partial^{\beta-\gamma} g\|_\infty, \delta \le \alpha, \gamma \le \beta\},$$

we infer

$$\|t^\alpha \partial^\beta f\|_\infty \le c_{\alpha,\beta} \int_{\mathbb{R}^{2d}} |F(x,\xi)| p(x,\xi)\,dx\,d\xi, \tag{1.64}$$

where

$$p(x,\xi) = \sum_{\delta\le\alpha}\sum_{\gamma\le\beta} \binom{\alpha}{\delta}\binom{\beta}{\gamma} |(2\pi\xi)^\gamma x^\delta| = \prod_{j=1}^d (1+|x_j|)^{\alpha_j}(1+2\pi|\xi_j|)^{\beta_j}.$$

Taking $n > 2d + |\alpha| + |\beta|$ in (1.62), we conclude the convergence of the integral. This gives $f \in \mathcal{S}(\mathbb{R}^d)$. □

Thanks to the previous results, we can now provide a characterization of the Schwartz class $\mathcal{S}(\mathbb{R}^d)$ in terms of the STFT as follows.

Theorem 1.2.23. *Fix $g \in \mathcal{S}(\mathbb{R}^d) \setminus \{0\}$. For $f \in \mathcal{S}'(\mathbb{R}^d)$, the following conditions are equivalent:*
(i) $f \in \mathcal{S}(\mathbb{R}^d)$;
(ii) $V_g f \in \mathcal{S}(\mathbb{R}^{2d})$;
(iii) *For every $n \geq 0$, there exists $C_n > 0$ such that*

$$|V_g f(x,\xi)| \leq C_n \langle (x,\xi) \rangle^{-n}, \quad (x,\xi) \in \mathbb{R}^{2d}. \tag{1.65}$$

Proof. (i) $\Longrightarrow$ (ii). We use the representation of the STFT in (1.43) $V_g f = \mathcal{F}_2 \mathfrak{T}_a (f \otimes \bar{g})$. If $f, g \in \mathcal{S}(\mathbb{R}^d)$, then $f \otimes \bar{g} \in \mathcal{S}(\mathbb{R}^{2d})$ and $\mathfrak{T}_a$ and $\mathcal{F}_2$ are topological isomorphisms on $\mathcal{S}$. This gives $V_g f \in \mathcal{S}(\mathbb{R}^{2d})$.

(ii) $\Longrightarrow$ (iii). This is trivial since members of the Schwartz class are rapidly decreasing functions.

(iii) $\Longrightarrow$ (i). Set

$$\tilde{f}(t) = \frac{1}{\|g\|_2^2} \int_{\mathbb{R}^{2d}} V_g f(x,\xi) M_\xi T_x g(t)\, dx\, d\xi.$$

Theorem 1.2.22 gives $\tilde{f} \in \mathcal{S}(\mathbb{R}^d) \subset L^2(\mathbb{R}^d)$. The inversion formula for the STFT in $L^2(\mathbb{R}^d)$ (1.56) reads

$$\frac{1}{\|g\|_2^2} V_g^* V_g f = f,$$

so that $f = \tilde{f}$ in $L^2(\mathbb{R}^d)$ and hence $f \in \mathcal{S}(\mathbb{R}^d)$. □

Corollary 1.2.24. *The STFT can be interpreted as a continuous mapping $\mathcal{S}(\mathbb{R}^d) \times \mathcal{S}(\mathbb{R}^d) \to \mathcal{S}(\mathbb{R}^{2d})$, namely*

$$(f, g) \mapsto V_g f,$$

in the respective topology.

Proof. The proof follows from the argument (i) $\Longrightarrow$ (ii) of Theorem 1.2.23. □

The proof of Theorem 1.2.23 shows also that the inversion formula

$$\frac{1}{\|g\|_2^2} V_g^* V_g = f$$

is valid in $\mathcal{S}(\mathbb{R}^d)$.

Corollary 1.2.25. *Fix $g \in \mathcal{S}(\mathbb{R}^d) \setminus \{0\}$. Then the collection*

$$\|V_g f\|_{L^\infty_{v_s}} = \sup_{z \in \mathbb{R}^{2d}} \langle z \rangle^s |V_g f(z)|, \quad s \geq 0, \tag{1.66}$$

is an equivalent family of seminorms for $\mathcal{S}(\mathbb{R}^d)$.

Proof. Define

$$\widetilde{\mathcal{S}}(\mathbb{R}^d) = \{f \in \mathcal{S}'(\mathbb{R}^d) : \|V_g f\|_{L^\infty_{v_s}} < \infty\}.$$

Theorem 1.2.23 gives

$$\widetilde{\mathcal{S}}(\mathbb{R}^d) = \mathcal{S}(\mathbb{R}^d)$$

as an equality between sets. Let us show that they have the same topology. Consider $\alpha, \beta \in \mathbb{N}^d$. Using the inversion formula

$$f = \frac{1}{\|g\|^2} \int_{\mathbb{R}^{2d}} V_g f(z)\pi(z) g \, dz$$

and proceeding exactly as in the proof of Theorem 1.2.22 (cf. estimate (1.64)), we can write, for a sufficiently large n,

$$\begin{aligned}
\|t^\alpha \partial^\beta f\|_\infty &\leq C_{\alpha,\beta,g} \int_{\mathbb{R}^{2d}} |V_g f| p(x,\xi)\, dx\, d\xi \\
&\leq C' \int_{\mathbb{R}^{2d}} |V_g f(z)| \langle z\rangle^n \, dz \\
&\leq C' \|V_g f\|_{L^\infty_{v_{n+2d+\epsilon}}} \int_{\mathbb{R}^{2d}} \frac{1}{\langle z\rangle^{2d+\epsilon}} \, dz \\
&\leq C'' \|V_g f\|_{L^\infty_{v_{n+2d+\epsilon}}},
\end{aligned}$$

for every $\epsilon > 0$. Hence for every $\alpha, \beta \in \mathbb{N}^d$, there exists an $s > 0$ such that

$$\|t^\alpha \partial^\beta f\|_\infty \lesssim \|V_g f\|_{L^\infty_{v_s}}.$$

This implies that the identity operator $I : \widetilde{\mathcal{S}}(\mathbb{R}^d) \to \mathcal{S}(\mathbb{R}^d)$ is continuous. By the open mapping theorem (valid also for Fréchet spaces), the identity I is a topological isomorphism. The claim is proved. □

Theorem 1.2.26. *Assume $g, \gamma \in \mathcal{S}(\mathbb{R}^d) \setminus \{0\}$.*
(i) *If there exist $C, N > 0$ such that*

$$|F(x,\xi)| \leq C\langle(x,\xi)\rangle^N, \tag{1.67}$$

then

$$f = \int_{\mathbb{R}^{2d}} F(x,\xi) M_\xi T_x g \, dx\, d\xi \in \mathcal{S}'(\mathbb{R}^d),$$

that is, for every $\varphi \in \mathcal{S}(\mathbb{R}^d)$,

$$\langle f, \varphi\rangle = \int_{\mathbb{R}^{2d}} F(x,\xi)\langle M_\xi T_x g, \varphi\rangle \, dx\, d\xi.$$

(ii) *In particular, if $F = V_g f$ for some $f \in \mathcal{S}'(\mathbb{R}^d)$, then*

$$f = \frac{1}{\langle \gamma, g \rangle} \int_{\mathbb{R}^{2d}} V_g f(x,\xi) M_\xi T_x \gamma \, dx \, d\xi \tag{1.68}$$

for $\langle \gamma, g \rangle \neq 0$. Extending the notation (1.56), *we can write*

$$f = \frac{1}{\langle \gamma, g \rangle} V_\gamma^* V_g f. \tag{1.69}$$

(Observe that this implies the injectivity of the STFT in $\mathcal{S}'(\mathbb{R}^d)$.)

Proof. (i) The integral

$$\int_{\mathbb{R}^{2d}} F(x,\xi) \langle M_\xi T_x g, \varphi \rangle \, dx \, d\xi$$

converges absolutely because $|F(x,\xi)| = O((\langle (x,\xi) \rangle)^N)$ for $|x|, |\xi| \to +\infty$ and $\langle M_\xi T_x g, \varphi \rangle = \overline{V_\varphi g(x,\xi)} \in \mathcal{S}(\mathbb{R}^{2d})$. Indeed,

$$\begin{aligned}
|\langle f, \varphi \rangle| &\leq C \int_{\mathbb{R}^{2d}} |V_g \varphi(x,\xi)| \langle (x,\xi) \rangle^N \, dx \, d\xi \\
&\leq C \| V_g \varphi \langle \cdot \rangle^{N+2d+1} \|_\infty \int_{\mathbb{R}^{2d}} \frac{1}{\langle (x,\xi) \rangle^{2d+1}} \, dx \, d\xi \\
&= C' \| V_g \varphi \langle \cdot \rangle^{N+2d+1} \|_\infty .
\end{aligned}$$

Finally, using $V_g \varphi(x,\xi) = \mathcal{F}_2 \mathfrak{T}_a (\varphi \otimes \bar{g})(x,\xi)$ and the continuity of the operators $\mathcal{F}_2$ and $\mathfrak{T}_a$ on $\mathcal{S}(\mathbb{R}^{2d})$, we obtain that there exists an $m \in \mathbb{N}$ such that

$$\begin{aligned}
\| V_g \varphi(x,\xi) \langle (x,\xi) \rangle^{N+2d+1} \|_\infty &\leq C \sum_{|\alpha|,|\beta| \leq m} \| \partial_x^\alpha \partial_\xi^\beta (\varphi \otimes \bar{g}) \langle (x,\xi) \rangle^m \|_\infty \\
&\leq C \sum_{|\alpha| \leq m, |\beta| \leq m} \| \partial^\alpha \varphi(x) \langle x \rangle^m \|_\infty \| \partial^\beta \bar{g}(\xi) \langle \xi \rangle^m \|_\infty \\
&= C' \sum_{|\alpha| \leq m} \| \partial^\alpha \varphi(x) \langle x \rangle^m \|_\infty ,
\end{aligned}$$

hence $f \in \mathcal{S}'(\mathbb{R}^d)$.

(ii) By the previous part of the proof and Proposition 1.2.20, the integral (1.68) defines an element $\tilde{f} \in \mathcal{S}'(\mathbb{R}^d)$. Now combining the inversion formula (1.55) with (1.63), we obtain the identity in $\mathcal{S}(\mathbb{R}^d)$

$$\varphi = \frac{1}{\langle g, \gamma \rangle} \int_{\mathbb{R}^{2d}} V_\gamma \varphi(x,\xi) M_\xi T_x g \, dx \, d\xi .$$

Inserting the previous equality in (1.68), we infer

$$\begin{aligned}\langle f,\varphi\rangle &= \frac{1}{\langle g,\gamma\rangle}\int_{\mathbb{R}^{2d}} \overline{V_\gamma\varphi(x,\xi)}\langle f, M_\xi T_x g\rangle\,dx\,d\xi \\ &= \frac{1}{\langle \gamma,g\rangle}\int_{\mathbb{R}^{2d}} V_g f(x,\xi)\overline{V_\gamma\varphi(x,\xi)}\,dx\,d\xi \\ &= \langle \tilde{f},\varphi\rangle, \quad \forall \varphi\in\mathcal{S}(\mathbb{R}^d),\end{aligned}$$

and so $f = \tilde{f}$. For $g \in \mathcal{S}(\mathbb{R}^d)$, the notation of the adjoint of the STFT V_g: $V_g^* : L^2(\mathbb{R}^{2d}) \to L^2(\mathbb{R}^d)$, defined by

$$V_g^* F = \int_{\mathbb{R}^{2d}} F(x,\xi) M_\xi T_x g\,dx\,d\xi, \quad F\in L^2(\mathbb{R}^{2d}),$$

can be extended in this setting; it maps functions F on $\mathbb{R}^{2d}$ satisfying (1.67) to distributions $f = V_g^* F \in \mathcal{S}'(\mathbb{R}^d)$. With this notation we can therefore write the reproducing formula (1.69). □

Remark 1.2.27.

(i) The STFT is not surjective on $\mathcal{S}'(\mathbb{R}^{2d})$. Indeed, we recall that $V_g f$ has polynomial growth and it is continuous, whereas there are tempered distributions in $\mathcal{S}'(\mathbb{R}^{2d})$ which are neither continuous nor of polynomial growth. Take, for instance, in dimension $d = 2$ the tempered distribution

$$F(x,\xi) = e^x\cos(e^x)\delta_\xi = \partial_x\partial_\xi G(x,\xi), \quad (x,\xi)\in\mathbb{R}^2$$

where $G(x,\xi) = \sin(e^x)H(\xi)$ and $H(\xi)$ is the Heaviside function, defined by $H(\xi) = 0$ for $\xi \le 0$, $H(\xi) = 1$, for $\xi > 0$.

(ii) Observe that if F satisfies (1.67) and $g \in \mathcal{S}(\mathbb{R}^d)$, with the notation of Theorem 1.2.26, for every $\psi \in \mathcal{S}(\mathbb{R}^d)$,

$$\langle V_g^* F, \psi\rangle = \langle F, V_g\psi\rangle,$$

where the brackets are meant as duality between $\mathcal{S}'$ and $\mathcal{S}$.

1.2.6 Decay and further properties

The properties of the STFT on the Schwartz class are the key ingredient to show the following result.

Proposition 1.2.28. *For $1 < p < \infty$, $f \in L^p(\mathbb{R}^d)$, $g \in L^{p'}(\mathbb{R}^d)$, the STFT $V_g f$ is in $C_0(\mathbb{R}^{2d})$. The same is true whenever $f \in L^1(\mathbb{R}^d)$ and $g \in C_0(\mathbb{R}^d)$ (and vice versa).*

Proof. Assume first $f \in L^p(\mathbb{R}^d)$, $g \in L^{p'}(\mathbb{R}^d)$, with $1 < p < \infty$. Since the Schwartz class is dense in both spaces, there exist sequences $\{f_n\}$, $\{g_n\} \subset \mathcal{S}(\mathbb{R}^d)$ such that $f_n \to f$ in $L^p(\mathbb{R}^d)$ and $g_n \to g$ in $L^{p'}(\mathbb{R}^d)$. Observe that $V_{g_n} f_n \in \mathcal{S}(\mathbb{R}^{2d}) \subset \mathcal{C}_0(\mathbb{R}^{2d})$ by Theorem 1.2.23, hence

$$\begin{aligned} \|V_{g_n} f_n - V_g f\|_\infty &= \|V_{g_n} f_n - V_g f_n + V_g f_n - V_g f\|_\infty \\ &= \left\|V_{(g_n - g)} f_n + V_g (f_n - f)\right\|_\infty \\ &\le \|f_n\|_p \|g_n - g\|_{p'} + \|g\|_p \|f_n - f\|_{p'}, \end{aligned}$$

where the last inequality is due to Proposition 1.2.10. Since the sequence $\{\|f_n\|_p\}$ is bounded, we infer

$$\lim_{n\to\infty} \|V_{g_n} f_n - V_g f\|_\infty = 0.$$

This implies $V_g f \in \mathcal{C}_0(\mathbb{R}^{2d})$, since $\mathcal{C}_0(\mathbb{R}^{2d})$ is a Banach space with respect to the supnorm.

The same argument applies to the latter case. □

We end up with the short-time Fourier transform by showing a very important inequality, which will be exploit heavily in the following chapters.

Lemma 1.2.29. *Consider $g, h, \gamma \in \mathcal{S}(\mathbb{R}^d) \setminus \{0\}$ such that $\langle h, \gamma\rangle \neq 0$ and $f \in \mathcal{S}'(\mathbb{R}^d)$. Then*

$$\left|V_g f(x,\xi)\right| \le \frac{1}{|\langle h,\gamma\rangle|}(|V_h f| * |V_g \gamma|)(x,\xi), \quad \forall (x,\xi) \in \mathbb{R}^{2d}. \tag{1.70}$$

Proof. Using the inversion formula (1.68) with the windows h, γ,

$$f = \frac{1}{\langle h,\gamma\rangle} V_\gamma^* V_h f,$$

we obtain

$$V_g f = \frac{1}{\langle h,\gamma\rangle} V_g V_\gamma^* V_h f. \tag{1.71}$$

Now, the covariance of the STFT (1.50) gives

$$\begin{aligned} V_g V_\gamma^* V_h f(x,\xi) &= \langle V_\gamma^* V_h f, \pi(x,\xi) g\rangle \\ &= \langle V_h f, V_\gamma(\pi(x,\xi) g)\rangle \\ &= \int_{\mathbb{R}^{2d}} V_h f(x,\xi) e^{2\pi i x\cdot(\eta-\xi)} \overline{V_\gamma g(y-x,\eta-\xi)}\, dy\, d\eta \\ &= \int_{\mathbb{R}^{2d}} V_h f(x,\xi) e^{2\pi i y\cdot(\eta-\xi)} \overline{V_g \gamma(x-y,\xi-\eta)}\, dy\, d\eta \end{aligned}$$

where in the last equality we used (1.32). Taking the modulus in (1.71) gives

$$\begin{aligned}|V_g V_\gamma^* V_h f(x,\xi)| &\le \frac{1}{|\langle h,\gamma\rangle|}\int_{\mathbb{R}^{2d}} |V_h f||V_g\gamma(x-y,\xi-\eta)|\,dy\,d\eta \\ &= \frac{1}{|\langle h,\gamma\rangle|}(|V_h f| * |V_g\gamma|)(x,\xi), \quad \forall (x,\xi)\in\mathbb{R}^{2d},\end{aligned}$$

as desired. □

1.3 Ambiguity function, Wigner distribution, and related topics

1.3.1 The ambiguity function

We are now going to introduce another time–frequency representation: the ambiguity function, which is used in radar identifications and other applications. For background and motivation, we refer to [320].

Definition 1.3.1. Given $f \in L^2(\mathbb{R}^d)$, the *ambiguity function* Af is defined by

$$Af(x,\xi) = A(f,f)(x,\xi) = \int_{\mathbb{R}^d} f\left(t+\frac{x}{2}\right)\overline{f\left(t-\frac{x}{2}\right)} e^{-2\pi i t\cdot\xi}\,dt. \tag{1.72}$$

Its *polarized version* is called *a cross-ambiguity function*

$$A(f,g)(x,\xi) = \int_{\mathbb{R}^d} f\left(t+\frac{x}{2}\right)\overline{g\left(t-\frac{x}{2}\right)} e^{-2\pi i t\cdot\xi}\,dt, \quad f,g\in L^2(\mathbb{R}^d). \tag{1.73}$$

Remark 1.3.2.

(i) By the change of variables $t \mapsto t - \frac{x}{2}$ in (1.72) and (1.73), we obtain

$$Af(x,\xi) = e^{\pi i x\cdot\xi} V_f f(x,\xi), \quad A(f,g)(x,\xi) = e^{\pi i x\cdot\xi} V_g f(x,\xi) \tag{1.74}$$

so that "up to a chirp" the cross-ambiguity function is just the STFT.

(ii) We observe that the (cross)-ambiguity function is defined only up to phase factors, i. e., $c \in \mathbb{C}$, with $|c| = 1$. In fact, a straightforward computation gives

$$A(cf,cg)(x,\xi) = A(f,g)(x,\xi),$$

for every $c \in \mathbb{C}$, with $|c| = 1$.

Relations (1.74) imply that the cross-ambiguity function enjoys most of the properties of the STFT. In what follows we limit ourselves to listing the main issues, the proofs are easy consequences of the previous remark.

Proposition 1.3.3. *The cross-ambiguity function has the following properties:*

(i) (Moyal's identity) *For $f_i, g_i \in L^2(\mathbb{R}^d)$, $i = 1, 2$,*

$$\langle A(f_1, g_1), A(f_2, g_2)\rangle = \langle f_1, f_2\rangle\overline{\langle g_1, g_2\rangle}. \tag{1.75}$$

In particular,

$$\|Af\|_2 = \|f\|_2^2, \quad f \in L^2(\mathbb{R}^d).$$

(ii) *For $1 < p < \infty$, $f \in L^p(\mathbb{R}^d)$, $g \in L^{p'}(\mathbb{R}^d)$, the cross-ambiguity function is in $\mathcal{C}_0(\mathbb{R}^{2d})$, with*

$$\|A(f, g)\|_\infty \leq \|f\|_p\|g\|_{p'}.$$

The same holds true if $f \in L^1(\mathbb{R}^d)$ and $g \in \mathcal{C}_0(\mathbb{R}^d)$ (and vice versa).

(iii) *For $f, g \in L^2(\mathbb{R}^d)$,*

$$\overline{A(f, g)(x, \xi)} = A(\mathcal{I}f, \mathcal{I}g)(x, \xi), \quad (x, \xi) \in \mathbb{R}^{2d}, \tag{1.76}$$

(we use the notation $\mathcal{I}h(t) = h(-t)$).

(iv) *For $f, g \in L^2(\mathbb{R}^d)$, the involution is given by*

$$(A(f, g))^*(x, \xi) = \overline{A(f, g)(-x, -\xi)} = A(f, g)(x, \xi), \quad (x, \xi) \in \mathbb{R}^{2d}. \tag{1.77}$$

(v) *It is a continuous mapping from $\mathcal{S}(\mathbb{R}^d) \times \mathcal{S}(\mathbb{R}^d)$ into $\mathcal{S}(\mathbb{R}^{2d})$.*

1.3.2 The Wigner distribution

The Wigner distribution (or function) is a time–frequency representation closely related to the ambiguity function. We will see in the sequel that these representations are symplectic Fourier transforms of each other.

It was introduced by E. Wigner in 1932 [314] in the field of quantum mechanics and then employed in signal analysis by J. Ville [300]. Since then the Wigner distribution has been applied in several fields of mathematics, engineering, and physics, and it is for sure the most popular time–frequency representation.

Definition 1.3.4. For $f \in L^2(\mathbb{R}^d)$, the Wigner distribution Wf is defined by

$$Wf(x, \xi) = W(f, f)(x, \xi) = \int_{\mathbb{R}^d} f\left(x + \frac{t}{2}\right)\overline{f\left(x - \frac{t}{2}\right)}e^{-2\pi i t\cdot\xi}\,dt. \tag{1.78}$$

Similarly, the cross-Wigner distribution is

$$W(f, g)(x, \xi) = \int_{\mathbb{R}^d} f\left(x + \frac{t}{2}\right)\overline{g\left(x - \frac{t}{2}\right)}e^{-2\pi i t\cdot\xi}\,dt, \quad f, g \in L^2(\mathbb{R}^d). \tag{1.79}$$

The relation between the cross-Wigner distribution and the STFT is provided in what follows.

Lemma 1.3.5. *For $f,g \in L^2(\mathbb{R}^d)$, we have*

$$W(f,g)(x,\xi) = 2^d e^{4\pi i x\cdot\xi} V_{\mathcal{I}g} f(2x, 2\xi), \quad (x,\xi) \in \mathbb{R}^{2d}. \tag{1.80}$$

Proof. We make the change of variables $u = x + \frac{t}{2}$ in (1.79) so that $dt = 2^d du$ and

$$\begin{aligned} W(f,g)(x,\xi) &= 2^d \int_{\mathbb{R}^d} f(u)\overline{g(2x-u)} e^{-2\pi i 2(u-x)\cdot\xi}\, du. \\ &= 2^d e^{4\pi i x\cdot\xi} \int_{\mathbb{R}^d} f(u)\overline{(\mathcal{I}g)(u-2x)} e^{-2\pi i u\cdot(2\xi)}\, du \\ &= 2^d e^{4\pi i x\cdot\xi} V_{\mathcal{I}g} f(2x, 2\xi), \end{aligned}$$

as desired. □

Observe that, using the connection between Schrödinger representation and time–frequency shifts in (1.11), formula (1.80) can be rephrased as

$$W(f,g)(x,\xi) = 2^d \langle f, \rho(2x, 2\xi, 0)\mathcal{I}g\rangle. \tag{1.81}$$

Using this result, many properties of the STFT can be carried over to the cross-Wigner distribution as well. In what follows we list the most relevant ones.

Proposition 1.3.6. *The (cross-)Wigner distributions has the following properties:*

(i) (Moyal's identity) *For $f_i, g_i \in L^2(\mathbb{R}^d)$, $i = 1,2$,*

$$\langle W(f_1,g_1), W(f_2,g_2)\rangle = \langle f_1,f_2\rangle\overline{\langle g_1,g_2\rangle}. \tag{1.82}$$

In particular,

$$\|Wf\|_2 = \|f\|_2^2, \quad f \in L^2(\mathbb{R}^d).$$

(ii) *For $1 < p < \infty$, $f \in L^p(\mathbb{R}^d)$, $g \in L^{p'}(\mathbb{R}^d)$, the Wigner function is in $C_0(\mathbb{R}^{2d})$, with*

$$\|W(f,g)\|_\infty \le 2^d \|f\|_p \|g\|_{p'}.$$

The same holds true if $f \in L^1(\mathbb{R}^d)$ and $g \in C_0(\mathbb{R}^d)$ (and vice versa).

(iii) *For f,g such that $W(f,g)$ is well defined,*

$$W(f,g) = \overline{W(g,f)}.$$

In particular,

Wf is always a real-valued function.

(iv) *Consider f,g such that $W(f,g)$ is well defined. Recall that, for $z=(x,\xi)$, $\pi(z)=M_\xi T_x$ is the time–frequency shift operator. Then for $z_0,z_1\in\mathbb{R}^{2d}$, we have*

$$W(\pi(z_0)f,\pi(z_1)g)(z)=e^{-2\pi i[\sigma(z,z_0-z_1)+\sigma(z_0,z_1)]}W(f,g)\Big(z-\frac{z_0+z_1}{2}\Big).$$

In particular, for $z_0=z_1$,

$$W(\pi(z_0)f,\pi(z_0)g)(z)=W(f,g)(z-z_0)=T_{z_0}W(f,g)(z). \tag{1.83}$$

(v) *For $f,g\in L^2(\mathbb{R}^d)$,*

$$W(\hat f,\hat g)(x,\xi)=W(f,g)(-\xi,x),\quad (x,\xi)\in\mathbb{R}^{2d}. \tag{1.84}$$

(vi) *It is a continuous mapping from $\mathcal{S}(\mathbb{R}^d)\times\mathcal{S}(\mathbb{R}^d)$ into $\mathcal{S}(\mathbb{R}^{2d})$.*

Proof. The proofs are straightforward. Let us only detail claim (v). Observe that

$$\widehat{(\mathcal{I}g)}(\xi)=\int_{\mathbb{R}^d}e^{-2\pi i t\cdot\xi}g(-t)\,dt=\int_{\mathbb{R}^d}e^{-2\pi i(-t)\cdot\xi}g(t)\,dt=\hat g(-\xi)=\mathcal{I}\hat g(\xi).$$

Hence, by Parseval's formula (for the inverse Fourier transform) and the commutation relations (0.14),

$$\begin{aligned}W(\hat f,\hat g)(x,\xi)&=2^de^{4\pi i x\cdot\xi}\langle\hat f,M_{2\xi}T_{2x}\mathcal{I}\hat g\rangle\\&=2^de^{4\pi i x\cdot\xi}\langle f,T_{-2\xi}M_{2x}\mathcal{I}g\rangle\\&=2^de^{4\pi i x\cdot(-\xi)}\langle f,M_{2x}T_{-2\xi}\mathcal{I}g\rangle\\&=W(f,g)(-\xi,x).\end{aligned}$$

□

Relation (1.84) means that the Fourier transform produces a rotation by $\pi/2$ of the Wigner distribution in the time–frequency plane. Setting $z=(x,\xi)$, relation (1.84) can be rewritten in terms of the standard symplectic matrix as follows:

$$W(\hat J f,\hat J g)(z)=W(f,g)(J^{-1}z),\quad f,g\in L^2(\mathbb{R}^d),$$

with $J,\hat J$ as in (1.1), (1.14). This can be seen as a particular case of the symplectic covariance property for the Wigner distribution.

Proposition 1.3.7 (Covariance property for the Wigner distribution). *If $\widehat{\mathcal{A}}$ is a metaplectic operator with projection the symplectic matrix $\mathcal{A}\in Sp(d,\mathbb{R})$, then*

$$W(\widehat{\mathcal{A}}f,\widehat{\mathcal{A}}g)(z)=W(f,g)(\mathcal{A}^{-1}z),\quad f,g\in L^2(\mathbb{R}^d). \tag{1.85}$$

Proof. Observe that the reflection operator $\mathcal{I}g(t) = g(-t)$ is a metaplectic operator with projection the symplectic matrix $-I_{2d} \in Sp(d, \mathbb{R})$, hence we can write $\widehat{(-I_{2d})}g = \mathcal{I}g$.

Let us show first the following relation: for every $\mathcal{A} \in Sp(d, \mathbb{R})$,

$$\mathcal{I}(\widehat{\mathcal{A}}g) = \widehat{\mathcal{A}}(\mathcal{I}g), \quad \forall g \in L^2(\mathbb{R}^d). \tag{1.86}$$

This follows from the previous observation and from the fact that $-I_{2d}$ commutes with any matrix $\mathcal{A} \in Sp(d, \mathbb{R})$, so that

$$\begin{aligned}\mathcal{I}(\widehat{\mathcal{A}}g) &= \widehat{(-I_{2d})}(\widehat{\mathcal{A}}g) = (\widehat{-I_{2d}\mathcal{A}})g = (\widehat{\mathcal{A}(-I_{2d})})g \\ &= \widehat{\mathcal{A}}\widehat{(-I_{2d})}g = \widehat{\mathcal{A}}(\mathcal{I}g).\end{aligned}$$

Using (1.81) and (1.86), we now can compute

$$\begin{aligned}W(\widehat{\mathcal{A}}f, \widehat{\mathcal{A}}g)(x,\xi) &= 2^d \langle \widehat{\mathcal{A}}f, \rho(2x, 2\xi, 0)\mathcal{I}(\widehat{\mathcal{A}}g)\rangle \\ &= 2^d \langle f, \widehat{\mathcal{A}}^{-1}\rho(2x, 2\xi, 0)\widehat{\mathcal{A}}(\mathcal{I}g)\rangle \\ &= 2^d \langle f, \widehat{\mathcal{A}^{-1}}\rho(2x, 2\xi, 0)\widehat{\mathcal{A}}(\mathcal{I}g)\rangle \\ &= 2^d \langle f, \rho(\mathcal{A}^{-1}(2x, 2\xi), 0)\mathcal{I}g\rangle \\ &= W(f,g)(\mathcal{A}^{-1}(x,\xi)).\end{aligned}$$

This concludes the proof. □

The Wigner distribution possesses a factorization similar to the STFT in Lemma 1.43.

Definition 1.3.8. We define $\mathfrak{T}_s$ to be the symmetric coordinate change

$$\mathfrak{T}_s F(x,y) = F\left(x + \frac{y}{2}, x - \frac{y}{2}\right), \quad x, y \in \mathbb{R}^d. \tag{1.87}$$

Observe that the inverse change $\mathfrak{T}_s^{-1}$ is given by

$$\mathfrak{T}_s^{-1} F(x,y) = F\left(\frac{x+y}{2}, x - y\right). \tag{1.88}$$

Lemma 1.3.9. *For $f, g \in L^2(\mathbb{R}^d)$,*

$$W(f,g) = \mathcal{F}_2 \mathfrak{T}_s (f \otimes \bar{g}), \tag{1.89}$$

where $\mathcal{F}_2$ is the partial Fourier transform with respect to the second variable.

Proof. It follows by the definition of the cross-Wigner distribution in (1.79). □

Remark 1.3.10. Moyal's formula (1.82) can be proved also by using Lemma 1.3.9 and following the pattern of Theorem 1.2.11 (second proof).

As for the STFT, using Lemma 1.3.9, we can enlarge the domain of the Wigner distribution to $\mathcal{S}'(\mathbb{R}^d)$. Indeed, the operators $\mathcal{F}_2$ and $\mathfrak{T}_s$ are isometries on $\mathcal{S}(\mathbb{R}^{2d})$ and extend to isometries on $\mathcal{S}'(\mathbb{R}^{2d})$. Hence $\mathcal{F}_2\mathfrak{T}_s(f \otimes \bar{g})$ is well defined as a tempered distribution on $\mathbb{R}^{2d}$ whenever $f, g \in \mathcal{S}'(\mathbb{R}^d)$, and we may define $W(f,g) = \mathcal{F}_2\mathfrak{T}_s(f \otimes \bar{g})$.

Recall that we consider a distribution (conjugate)-linear instead of linear, so that the duality ${}_{\mathcal{S}'}\langle\cdot,\cdot\rangle_{\mathcal{S}}$ extends the inner product in $L^2(\mathbb{R}^d)$. In particular, the Dirac's delta δ is defined by

$$\langle \delta, f\rangle = \overline{f(0)}.$$

Let us compute the Wigner distribution of the Dirac's delta δ defined above. Let $\varphi \in \mathcal{S}(\mathbb{R}^{2d})$, then

$$\begin{aligned}
\langle W\delta, \varphi\rangle &= \langle \mathcal{F}_2\mathfrak{T}_s(\delta \otimes \bar{\delta}), \varphi\rangle = \langle \delta \otimes \bar{\delta}, \mathfrak{T}_s^{-1}\mathcal{F}_2^{-1}\varphi\rangle \\
&= \overline{\mathfrak{T}_s^{-1}\mathcal{F}_2^{-1}\varphi(0,0)} = \overline{\int_{\mathbb{R}^d} \varphi\left(\frac{x+\xi}{2}, t\right) e^{2\pi i t\cdot(x-\xi)}\, dt}\Big|_{x=\xi=0}. \\
&= \int_{\mathbb{R}^d} \overline{\varphi(0,t)}\, dt = \langle 1, \varphi(0,\cdot)\rangle \\
&= \langle \delta \otimes 1, \varphi\rangle.
\end{aligned}$$

Therefore $W\delta = \delta \otimes 1 \in \mathcal{S}'(\mathbb{R}^{2d})$.

We now show the link between the Wigner distribution and the ambiguity function.

Lemma 1.3.11. *For any $f, g \in \mathcal{S}'(\mathbb{R}^{2d})$,*

$$A(f,g) = \mathcal{F}_\sigma W(f,g). \tag{1.90}$$

Proof. We prove the result for $f, g \in \mathcal{S}(\mathbb{R}^d)$. The general case follows by duality arguments. Define by $\mathcal{F}_1 F(x,\xi) = \int_{\mathbb{R}^d} F(t,\xi) e^{-2\pi i t\cdot x}\, dt$ the partial Fourier transform in the first variable, so that the Fourier transform can be factorized as

$$\mathcal{F} = \mathcal{F}_2\mathcal{F}_1, \quad \mathcal{F}^{-1} = \mathcal{F}_1^{-1}\mathcal{F}_2^{-1}.$$

Moreover, from (1.18),

$$\mathcal{F}_\sigma W(f,g)(z) = \mathcal{F}W(f,g)(Jz) = \mathcal{F}^{-1}W(f,g)(-Jz).$$

For $z = (x,\xi)$, we have $-Jz = (-\xi, x)$ and using definition (1.89), we can write

$$\begin{aligned}
\mathcal{F}_\sigma W(f,g)(x,\xi) &= \mathcal{F}_1^{-1}\mathcal{F}_2^{-1}\mathcal{F}_2\mathfrak{T}_s(f \otimes \bar{g})(-\xi, x) \\
&= \int_{\mathbb{R}^d} \mathfrak{T}_s(f \otimes \bar{g})(t, x) e^{2\pi i t\cdot x}\, dt
\end{aligned}$$

$$= \int_{\mathbb{R}^d} f\left(t + \frac{x}{2}\right)\bar{g}\left(t - \frac{x}{2}\right)e^{2\pi i t\cdot x}\,dt$$
$$= A(f, g)(x, \xi),$$

as desired. □

Lemma 1.3.12. *Consider the Gaussian function* $\varphi_{1/\sqrt{a}}(x) = e^{-\pi x^2/a}$, $x \in \mathbb{R}^d$. *Then*

$$W\varphi_{1/\sqrt{a}}(x,\xi) = (2a)^{d/2}\varphi_{\sqrt{\frac{2}{a}}}(x)\varphi_{\sqrt{2a}}(\xi). \tag{1.91}$$

In particular, for $\varphi(x) = e^{-\pi x^2}$ *we have* $W\varphi(x,\xi) = 2^{d/2}e^{-2\pi(x^2+\xi^2)}$.

Proof. It is an easy computation, cf. (0.25),

$$\begin{aligned} W\varphi_{1/\sqrt{a}}(x,\xi) &= \int_{\mathbb{R}^d} e^{-\frac{\pi}{a}[(x+t/2)^2+(x-t/2)^2]}e^{-2\pi i t\cdot\xi}\,dt \\ &= \int_{\mathbb{R}^d} e^{-\frac{\pi}{a}(2x^2+t^2/2)}e^{-2\pi i t\cdot\xi}\,dt \\ &= e^{-2\pi\frac{x^2}{a}}\int_{\mathbb{R}^d} e^{-2\pi i t\cdot\xi}e^{-\frac{\pi}{2a}t^2}\,dt \\ &= \varphi_{\frac{a}{2}}(x)\widehat{\varphi_{2a}}(\xi) = (2a)^{\frac{d}{2}}\varphi_{\sqrt{\frac{2}{a}}}(x)\varphi_{\sqrt{2a}}(\xi), \end{aligned}$$

as required. □

1.3.3 Marginal densities

The fundamental problem of time–frequency analysis and quantum mechanics is to discover a good mathematical device able to simultaneously represent a given signal in terms of its intensity in time and frequency. Indeed, introducing Wf in quantum mechanics, Wigner wanted to find a joint probability density function of the position and momentum variables. The quantity $Wf(x,\xi)\Delta x\Delta\xi$ is interpreted as the probability that the particle in the quantum state $f \in L^2(\mathbb{R}^d)$, $\|f\|_2 = 1$, is located in $[x, x+\Delta x]$ with momentum in $[\xi, \xi+\Delta\xi]$. The marginal densities implied by a joint probability density are also implied by the Wigner distribution.

Proposition 1.3.13. *Consider* $f, \hat{f} \in L^1(\mathbb{R}^d) \cap L^2(\mathbb{R}^d)$, *then*

$$\int_{\mathbb{R}^d} Wf(x,\xi)\,d\xi = |f(x)|^2 \quad x \in \mathbb{R}^d, \tag{1.92}$$

$$\int_{\mathbb{R}^d} Wf(x,\xi)\,dx = |\hat{f}(\xi)|^2 \quad \xi \in \mathbb{R}^d. \tag{1.93}$$

In particular,

$$\int_{\mathbb{R}^d}\int_{\mathbb{R}^d} Wf(x,\xi)\,dx\,d\xi = \|f\|_2^2. \tag{1.94}$$

Proof. First, observe that by the identity (1.80) and the representation of the STFT in (1.29), we can write

$$|Wf(x,\xi)| = 2^d|V_{\mathcal{I}f}f(2x,2\xi)| = 2^d|\hat{f} * M_{-2x}\bar{\hat{f}}|(2\xi) \le 2^d|\hat{f}| * |\hat{f}|(2\xi).$$

Since $\hat{f} \in L^1(\mathbb{R}^d)$, by Young's inequality, we have $|\hat{f}| * |\hat{f}| \in L^1(\mathbb{R}^d)$, which is invariant under dilations. Hence, for every fixed $x \in \mathbb{R}^d$, $Wf(x,\cdot)$ is in $L^1(\mathbb{R}^d)$, and we can apply the inverse Fourier formula to obtain

$$\begin{aligned}\int_{\mathbb{R}^d} Wf(x,\xi)\,d\xi &= \mathcal{F}_2^{-1}[\mathcal{F}_2\mathfrak{T}_s(f\otimes\bar{f})](x,0)\\ &= \mathfrak{T}_s(f\otimes\bar{f})(x,0) = |f(x)|^2.\end{aligned}$$

By Proposition 1.3.6 (v), $W\hat{f}(\xi,-x) = Wf(x,\xi)$, so we can apply the previous computations to obtain

$$\int_{\mathbb{R}^d} Wf(x,\xi)\,dx = \int_{\mathbb{R}^d} W\hat{f}(\xi,-x)\,dx = \int_{\mathbb{R}^d} W\hat{f}(\xi,x)\,dx = |\hat{f}(\xi)|^2$$

for every $\xi \in \mathbb{R}^{2d}$, and so (1.93) is attained. The last formula is trivial. □

Observe that for the marginal property (1.92) we only require that the signal f is in $L^2(\mathbb{R}^d)$ with $\hat{f} \in L^1(\mathbb{R}^d)$, whereas formula (1.93) needs $f \in L^1(\mathbb{R}^d) \cap L^2(\mathbb{R}^d)$. Both formulas (1.92) and (1.93) (with Plancherel's equality) give (1.94), but be careful that in both cases the integrals in (1.94) are only understood as iterated integrals. In general, the assumption $f, \hat{f} \in L^1(\mathbb{R}^d)$ does not imply $Wf \in L^1(\mathbb{R}^{2d})$.

From the point of view of quantum mechanics, $\int_{\mathbb{R}^d} Wf(x,\xi)\,d\xi$ gives the probability density $|f(x)|^2$ of the position variable and $\int_{\mathbb{R}^d} Wf(x,\xi)\,dx$ yields the probability density $|\hat{f}(\xi)|^2$ of the momentum variable.

So far we have seen the nice properties of the Wigner distribution that would make it a nice candidate for being a joint probability density. Observe that, if the Wigner distribution were a joint probability density, then it would be nonnegative. Unfortunately, this is not the case for the Wigner distribution, in general. For instance, take $f \in L^2(\mathbb{R}^d)$ odd, then

$$Wf(0,0) = \int_{\mathbb{R}^d} f\left(\frac{t}{2}\right)\overline{f\left(-\frac{t}{2}\right)}\,dt = -\int_{\mathbb{R}^d} \left|f\left(\frac{t}{2}\right)\right|^2 dt = -2^d\|f\|_2^2 < 0.$$

Moreover, the following theorem by Hudson [198] shows that the Wigner distribution is positive only for f being a generalized Gaussian.

Theorem 1.3.14. *Assume $f \in L^2(\mathbb{R}^d)$. Then $Wf(x,\xi) > 0$ for all $(x,\xi) \in \mathbb{R}^{2d}$ if and only if f is a generalized Gaussian of the form*

$$f(x) = e^{-\pi x\cdot Ax+2\pi b\cdot x+c}, \tag{1.95}$$

where $A \in \mathrm{GL}(d,\mathbb{C})$ (the space of invertible $d\times d$ complex matrices) is a symmetric matrix with positive definite real part ($A = A^T$, $\Re A > 0$), $b \in \mathbb{C}^d$, $c \in \mathbb{C}$.

We refer to [198] or [160] for the proof of Hudson's theorem, cf. Lemma 1.3.12 for the Wigner of the rescaled Gaussian function φ.

For recent applications of the Wigner distribution to PDEs, we refer the reader to [27, 39]. For links between the Wigner distribution and Paley–Wiener theorems, we point to [29].

1.3.4 The Cohen class

Since Wf fails to be positive in general, a remedy could be the averaging of Wf by convolving it with suitable smooth functions σ, obtaining a more useful time–frequency representation. This is the idea under the definition of the Cohen class, introduced by L. Cohen in [53]: a time–frequency representation belongs to the Cohen class if it is of the form

$$\mathcal{Q}_\sigma f = Wf * \sigma, \quad \forall f \in \mathcal{S}(\mathbb{R}^d), \tag{1.96}$$

for some function or distribution $\sigma \in \mathcal{S}'(\mathbb{R}^{2d})$, called *kernel* of the Cohen distribution. Observe that if $f \in \mathcal{S}(\mathbb{R}^d)$ then $Wf \in \mathcal{S}(\mathbb{R}^{2d})$ by Proposition 1.3.6 (*vi*), and the convolution $Wf * \sigma$ is a well-defined distribution in $\mathcal{S}'(\mathbb{R}^{2d})$.

The Wigner distribution trivially belongs to the Cohen class since we may take as kernel the Dirac distribution $\sigma = \delta \in \mathcal{S}'(\mathbb{R}^{2d})$.

Proposition 1.3.15. *For $f \in L^2(\mathbb{R}^d)$ and $\sigma \in L^1(\mathbb{R}^d)$, the element of the Cohen class $\mathcal{Q}_\sigma f$ is well defined in $L^2(\mathbb{R}^{2d})$.*

Proof. The condition $f \in L^2(\mathbb{R}^d)$ implies that $Wf \in L^2(\mathbb{R}^{2d})$ by Moyal's formula, and consequently $\mathcal{Q}_\sigma f = Wf * \sigma \in L^2(\mathbb{R}^d)$ by Young's inequality. □

It is hard task to determine those kernels σ for which $Wf * \sigma$ is always positive. A simple case is provided by Gaussian kernels. The distribution $\mathcal{Q}_\sigma$ in (1.96) with kernel $\sigma(z) = e^{-2\pi|z|^2}$ is called *Husimi distribution* since it was introduced by K. Husimi in his paper [199].

Theorem 1.3.16. *Define*

$$\sigma_{a,b}(x,\xi) = e^{-2\pi(\frac{x^2}{a}+\frac{\xi^2}{b})} = \varphi_{\sqrt{\frac{2}{a}}}(x)\varphi_{\sqrt{\frac{2}{b}}}(\xi), \quad a,b > 0,\ (x,\xi) \in \mathbb{R}^{2d}. \tag{1.97}$$

Then
(i) *If* $ab = 1$, *then* $Wf * \sigma_{a,b} \geq 0$ *for all* $f \in L^2(\mathbb{R}^d)$.
(ii) *If* $ab > 1$, *then* $Wf * \sigma_{a,b} > 0$ *for all* $f \in L^2(\mathbb{R}^d) \setminus \{0\}$.
(iii) *If* $0 < ab < 1$, *then* $Wf * \sigma_{a,b}$ *may be negative.*

In the case $a = b = 1$, $\sigma_{1,1}$ is the Husimi's kernel. The proof of the above theorem needs the following lemma, which is an easy consequence of the semigroup property of Gaussian functions (0.26).

Lemma 1.3.17. *Choose real numbers* a, b, c, d *such that* $0 < c < a$, $0 < d < b$ *and* $cd = 1$. *Then it holds*

$$\sigma_{a,b} = \left(\frac{4ab}{(a-c)(b-d)}\right)^{\frac{d}{2}} \sigma_{c,d} * \sigma_{a-c,b-d}. \tag{1.98}$$

Proof. Using the semigroup property (0.26),

$$\begin{aligned}
&\left(\frac{4ab}{(a-c)(b-d)}\right)^{\frac{d}{2}} \sigma_{c,d} * \sigma_{a-c,b-d}(x,\xi) \\
&= \left(\frac{4ab}{(a-c)(b-d)}\right)^{\frac{d}{2}} (\varphi_{\sqrt{\frac{2}{c}}} * \varphi_{\sqrt{\frac{2}{a-c}}})(x)(\varphi_{\sqrt{\frac{2}{d}}} * \varphi_{\sqrt{\frac{2}{b-d}}})(\xi) \\
&= \left(\frac{4ab}{(a-c)(b-d)}\right)^{\frac{d}{2}} \left(\frac{c(a-c)}{2a}\right)^{\frac{d}{2}} \left(\frac{d(b-d)}{2b}\right)^{\frac{d}{2}} \varphi_{\sqrt{\frac{2}{a}}}(x)\varphi_{\sqrt{\frac{2}{b}}}(\xi) \\
&= \sigma_{a,b}(x,\xi),
\end{aligned}$$

that is, (1.98) holds. □

We can now prove the theorem above.

Proof of Theorem 1.3.16. First, we observe the following action of the reflection operator on the Wigner distribution:

$$\mathcal{I}(Wf) = W(\mathcal{I}f), \tag{1.99}$$

which follows from this straightforward computation:

$$\begin{aligned}
\mathcal{I}(Wf)(x,\xi) = Wf(-x,-\xi) &= \int_{\mathbb{R}^d} f\left(-x+\frac{t}{2}\right)\overline{f\left(-x-\frac{t}{2}\right)} e^{2\pi i \xi \cdot t}\, dt \\
&= \int_{\mathbb{R}^d} f\left(-x-\frac{t}{2}\right)\overline{f\left(-x+\frac{t}{2}\right)} e^{-2\pi i \xi \cdot t}\, dt = W(\mathcal{I}f)(x,\xi).
\end{aligned}$$

(i) If $ab = 1$, that is, $b = 1/a$, we can write the kernel $\sigma_{a,b}$ as Wigner distribution, using the Lemma 1.3.12,

$$\sigma_{a,b} = \varphi_{\sqrt{\frac{2}{a}}}(x)\varphi_{\sqrt{2a}}(\xi) = (2a)^{-\frac{d}{2}} W\varphi_{\frac{1}{\sqrt{a}}}(x,\xi).$$

By means of formula (1.99), the covariance of Wf, cf. Proposition 1.2.15, and Moyal's formula, cf. Proposition 1.3.6, we can write

$$\begin{aligned}(Wf * \sigma_{a,\frac{1}{a}})(x,\xi) &= (\sigma_{a,\frac{1}{a}} * Wf)(x,\xi) = \int_{\mathbb{R}^{2d}} \sigma_{a,\frac{1}{a}}(t,v) Wf(x-t,\xi-v)\,dt\,dv \\ &= \int_{\mathbb{R}^{2d}} Wf(x-t,\xi-v)(2a)^{-\frac{d}{2}} W\varphi_{\frac{1}{\sqrt{a}}}(t,v)\,dt\,dv \\ &= (2a)^{-\frac{d}{2}} \int_{\mathbb{R}^{2d}} T_{(x,\xi)}(\mathcal{I}Wf)(t,v) W\varphi_{\frac{1}{\sqrt{a}}}(t,v)\,dt\,dv \\ &= (2a)^{-\frac{d}{2}} \int_{\mathbb{R}^{2d}} W(M_\xi T_x \mathcal{I}f)(t,v)\overline{W\varphi_{\frac{1}{\sqrt{a}}}(t,v)}\,dt\,dv \\ &= (2a)^{-\frac{d}{2}} \langle W(M_\xi T_x \mathcal{I}f), W\varphi_{\frac{1}{\sqrt{a}}}\rangle \\ &= (2a)^{-\frac{d}{2}} |\langle M_\xi T_x(\mathcal{I}f), \varphi_{\frac{1}{\sqrt{a}}}\rangle|^2 \geq 0,\end{aligned}$$

and so we obtain the claim.

(ii) Assume $ab > 1$. Then we can choose $c, d \in \mathbb{R}$ such that $0 < c < a$ and $0 < b < d$ with $cd = 1$. By Lemma 1.3.17, we have

$$Wf * \sigma_{a,b} = \left(\frac{4ab}{(a-c)(b-d)}\right)^{\frac{d}{2}} (Wf * \sigma_{c,d}) * \sigma_{a-c,b-d}.$$

The right-hand side is strictly positive since it is the convolution of the function $Wf * \sigma_{c,d}$, nonnegative by the previous part of the proof (observe that $Wf * \sigma_{c,d} \geq 0$ is not the zero function, since f is not null) and a strictly positive one, $\sigma_{a-c,b-d} > 0$.

(iii) For the case $0 < ab < 1$, we exhibit a counterexample. If we choose $f(x) = xe^{-\pi x^2}$ on $\mathbb{R}$, we can verify that $Wf * \sigma_{a,b}(0,0) < 0$. Indeed, by direct calculation, we have that $Wf(x,\xi) = \sqrt{2}e^{-2\pi x^2}e^{-2\pi\xi^2}(x^2+\xi^2-1/4\pi)$ and hence

$$(Wf * \sigma_{a,b})(0,0) = \frac{\sqrt{2}}{8\pi}\sqrt{\frac{ab}{(a+1)(b+1)}}\left(\frac{ab-1}{(a+1)(b+1)}\right) < 0. \qquad \square$$

Coming back to general Cohen kernels, let us first observe that we can use the cross-Wigner distribution to provide an equivalent definition of the Cohen class.

Definition 1.3.18. A sesquilinear form $\mathcal{Q}_\sigma : \mathcal{S}(\mathbb{R}^d) \times \mathcal{S}(\mathbb{R}^d) \to \mathcal{S}(\mathbb{R}^{2d})$ belongs to the Cohen class if there exists a kernel $\sigma \in \mathcal{S}'(\mathbb{R}^{2d})$ such that

$$\mathcal{Q}_\sigma(f,g) = W(f,g) * \sigma, \quad \forall f,g \in \mathcal{S}(\mathbb{R}^d). \tag{1.100}$$

For $f = g$ from (1.100), we obtain (1.96). On the other hand, starting from (1.96), we may define $\mathcal{Q}_\sigma(f,g)$ uniquely. In fact, the polarization identity for the cross-Wigner distribution

$$W(f,g) = \frac{1}{4}(W(f+g) - W(f-g) + iW(f+ig) - iW(f-ig))$$

gives

$$\begin{aligned} W(f,g) * \sigma &= \frac{1}{4}(W(f+g) - W(f-g) + iW(f+ig) - iW(f-ig)) * \sigma \\ &= \frac{1}{4}(\mathcal{Q}_\sigma(f+g) - \mathcal{Q}_\sigma(f-g) + i\mathcal{Q}_\sigma(f+ig) - i\mathcal{Q}_\sigma(f+g)) \\ &= \mathcal{Q}_\sigma(f,g), \end{aligned}$$

that is, (1.100) is true.

We list the main properties for members of the Cohen class. We say that $\mathcal{Q}_\sigma$ is real-valued if $\mathcal{Q}_\sigma f$ is real-valued for every $f \in \mathcal{S}(\mathbb{R}^d)$.

Proposition 1.3.19. *The members of the Cohen class satisfy the following properties:*

(i) $\mathcal{Q}_\sigma$ *is covariant, namely, for every* $z \in \mathbb{R}^{2d}$,

$$\mathcal{Q}_\sigma(\pi(z)f) = T_z\mathcal{Q}_\sigma f, \quad \forall f \in \mathcal{S}(\mathbb{R}^d). \tag{1.101}$$

(ii) $\mathcal{Q}_\sigma$ *is real-valued if and only if* $\overline{\sigma} = \sigma$.

(iii) *(Marginal properties). The conditions*

$$\int_{\mathbb{R}^d} \mathcal{Q}_\sigma f(x,\xi)d\xi = |f(x)|^2, \quad \int_{\mathbb{R}^d} \mathcal{Q}_\sigma f(x,\xi)dx = |\hat{f}(\xi)|^2,$$

hold if and only if

$$\int_{\mathbb{R}^d} \sigma(t,\nu)\,d\nu = \delta(t), \quad \int_{\mathbb{R}^d} \sigma(t,\nu)\,dt = \delta(\nu), \tag{1.102}$$

respectively. Moreover,

$$\int_{\mathbb{R}^{2d}} \mathcal{Q}_\sigma f(x,\xi)\,dx\,d\xi = \|f\|_2^2$$

holds if and only if

$$\int_{\mathbb{R}^{2d}} \sigma(t,\nu)\,dx\,d\nu = 1. \tag{1.103}$$

(iv) *(Moyal's formula) Assume* $\hat{\sigma} \in L^\infty(\mathbb{R}^{2d})$. *Then for every* $f,g \in L^2(\mathbb{R}^d)$,

$$\langle \mathcal{Q}_\sigma f, \mathcal{Q}_\sigma g\rangle_{L^2(\mathbb{R}^{2d})} = |\langle f,g\rangle|^2, \quad \forall f,g \in L^2(\mathbb{R}^d)$$

if and only if

$$|\hat{\sigma}(x,\xi)| = 1 \quad a.e.\ (x,\xi) \in \mathbb{R}^{2d}. \tag{1.104}$$

Remark 1.3.20. The integrals in (1.102) and (1.103) have no meaning in general for $\sigma \in \mathcal{S}'(\mathbb{R}^{2d})$. Concerning $\int_{\mathbb{R}^{2d}} \sigma(t,\nu)\,dt\,d\nu$, it makes sense, of course, if $\sigma \in L^1(\mathbb{R}^{2d})$.

Otherwise, for $\sigma \in \mathcal{S}'(\mathbb{R}^{2d})$, we may read (1.103) as $\langle \sigma, 1\rangle = 1$, provided $\langle \sigma, 1\rangle$ is well defined for σ in suitable subclasses of $\mathcal{S}'(\mathbb{R}^{2d})$, for example, for $\sigma \in \mathcal{E}'(\mathbb{R}^{2d})$, the space of the distributions with compact support.

Similarly, we may argue on the meaning of (1.102). Note that for $\sigma \in \mathcal{S}'(\mathbb{R}^{2d})$ equivalent formulations to (1.102) are obtained by imposing on the Fourier transform $\hat{\sigma} \in \mathcal{S}'(\mathbb{R}^{2d})$ the requirement

$$\hat{\sigma}(x,0) = 1, \quad \hat{\sigma}(0,\xi) = 1 \tag{1.105}$$

provided again $\hat{\sigma}(x,0)$ and $\hat{\sigma}(0,\xi)$ make sense.

The computations in the subsequent proof of (iii) will be valid in all the cases, the integrals to be understood in the weak sense whenever σ is a distribution.

Proof of Proposition 1.3.19. (i) Using the covariant property of Wigner distribution (1.50) and the property, namely $T_z(F * G) = (T_zF) * G = F * (T_zG)$, we have

$$\begin{aligned} \mathcal{Q}_\sigma(\pi(z)f)(u) &= (W(\pi(z)f) * \sigma)(u) = (T_z Wf) * \sigma(u) \\ &= T_z(Wf * \sigma)(u) = T_z \mathcal{Q}_\sigma f(u). \end{aligned}$$

(ii) Using the fact that Wf is a real function,

$$\begin{aligned} \overline{\mathcal{Q}_\sigma f} &= \overline{Wf * \sigma} \\ &= \overline{Wf} * \overline{\sigma} \\ &= Wf * \overline{\sigma}. \end{aligned}$$

Hence $\mathcal{Q}_\sigma$ is real-valued if and only

$$Wf * \overline{\sigma} = Wf * \sigma, \quad \forall f \in \mathcal{S}(\mathbb{R}^d).$$

Taking the symplectic Fourier transform of both sides, this is equivalent to saying

$$Af \cdot \mathcal{F}_\sigma(\sigma) = Af \cdot \mathcal{F}_\sigma(\bar{\sigma}) \tag{1.106}$$

for every $f \in \mathcal{S}(\mathbb{R}^d)$, cf. Lemma 1.3.11 and Proposition 1.18 (v). Consider the Gaussian $f(t) = \varphi(t) = e^{-\pi t^2}$. The Wigner distribution is $W\varphi(z) = 2^{d/2}e^{-2\pi z^2}$ by Lemma 1.3.12 and the ambiguity function is given by $A\varphi(z) = 2^{-d/2}e^{-\pi z^2/2}$. Writing (1.106) for $f = \varphi$ and dividing by $A\varphi$, we obtain $\mathcal{F}_\sigma(\sigma) = \mathcal{F}_\sigma(\bar{\sigma})$, which gives $\sigma = \bar{\sigma}$, as desired.

(iii) We argue according to Remark 1.3.20:

$$\begin{aligned} \int_{\mathbb{R}^d} \mathcal{Q}_\sigma f(x,\xi)\,d\xi &= \int_{\mathbb{R}^d} (Wf * \sigma)(x,\xi)\,d\xi \\ &= \int_{\mathbb{R}^d}\Big(\int_{\mathbb{R}^{2d}} Wf(x-t,\xi-v)\sigma(t,v)\,dt\,dv\Big)\,d\xi \end{aligned}$$

$$= \int_{\mathbb{R}^{2d}} \sigma(t,v)\Big(\int_{\mathbb{R}^d} Wf(x-t,\xi-v)\,d\xi\Big)\,dt\,dv$$
$$= \int_{\mathbb{R}^{2d}} \sigma(t,v)\Big(\int_{\mathbb{R}^d} Wf(x-t,\xi)\,d\xi\Big)\,dt\,dv$$
$$= \int_{\mathbb{R}^{2d}} \sigma(t,v)|f(x-t)|^2\,dt\,dv$$
$$= \int_{\mathbb{R}^d} |f(x-t)|^2\Big(\int_{\mathbb{R}^d} \sigma(t,v)\,dv\Big)\,dt,$$

where we have made the change of coordinate $\xi \mapsto \xi - v$ and used the marginal property of the Wigner distribution (1.92). We deduce that

$$\int_{\mathbb{R}^d} \mathcal{Q}_\sigma f(x,\xi)\,d\xi = |f(x)|^2 \iff \int_{\mathbb{R}^d} \sigma(t,v)\,dv = \delta(t).$$

The second marginal property is proved in the same way as the previous one, employing the second marginal property of the Wigner distribution. By combining the preceding arguments,

$$\int_{\mathbb{R}^{2d}} \mathcal{Q}_\sigma f(x,\xi)\,dx\,d\xi = \int_{\mathbb{R}^{2d}} (Wf * \sigma)(x,\xi)\,dx\,d\xi$$
$$= \Big(\int_{\mathbb{R}^{2d}} Wf(t,v)\,dt\,dv\Big)\Big(\int_{\mathbb{R}^{2d}} \sigma(t',v')\,dt'\,dv'\Big)$$
$$= \|f\|_2^2 \int_{\mathbb{R}^{2d}} \sigma(t',v')\,dt'\,dv',$$

and the equality $\int_{\mathbb{R}^{2d}} \mathcal{Q}_\sigma f(x,\xi)\,dx\,d\xi = \|f\|_2^2$ holds for every $f \in \mathcal{S}(\mathbb{R}^d)$ if and only if

$$\int_{\mathbb{R}^{2d}} \sigma(t',v')\,dt'\,dv' = 1.$$

(iv) If $\hat{\sigma} \in L^\infty(\mathbb{R}^{2d})$ and $f \in L^2(\mathbb{R}^d)$, the convolution $Wf * \sigma = \mathcal{F}^{-1}(\widehat{Wf} \cdot \hat{\sigma})$ is a well-defined function in $L^2(\mathbb{R}^{2d})$, since $Wf \in L^2(\mathbb{R}^{2d})$ and

$$L^2(\mathbb{R}^{2d}) \cdot L^\infty(\mathbb{R}^{2d}) \hookrightarrow L^2(\mathbb{R}^{2d}).$$

Using Parseval's formula, we deduce that

$$\langle \mathcal{Q}_\sigma f, \mathcal{Q}_\sigma g\rangle_{L^2(\mathbb{R}^{2d})} = \langle Wf * \sigma, Wg * \sigma\rangle_{L^2(\mathbb{R}^{2d})}$$
$$= \langle \mathcal{F}^{-1}(\widehat{Wf} \cdot \hat{\sigma}), \mathcal{F}^{-1}(\widehat{Wg} \cdot \hat{\sigma})\rangle_{L^2(\mathbb{R}^{2d})}$$
$$= \langle \widehat{Wf} \cdot \hat{\sigma}, \widehat{Wg} \cdot \hat{\sigma}\rangle_{L^2(\mathbb{R}^{2d})}$$
$$= \langle \widehat{Wf}, \widehat{Wg}|\hat{\sigma}|^2\rangle_{L^2(\mathbb{R}^{2d})}.$$

If $|\hat{\sigma}(x,\xi)| = 1$ a. e., then we obtain the claim by Parseval's formula and Moyal's identity (1.82):

$$\langle Q_\sigma f, Q_\sigma g\rangle_{L^2(\mathbb{R}^{2d})} = \langle \widehat{Wf}, \widehat{Wg}|\hat{\sigma}|^2\rangle_{L^2(\mathbb{R}^{2d})} = \langle Wf, Wg\rangle_{L^2(\mathbb{R}^{2d})} = |\langle f,g\rangle|^2.$$

On the other hand, if we impose $\langle Q_\sigma f, Q_\sigma g\rangle_{L^2(\mathbb{R}^{2d})} = |\langle f,g\rangle|^2$, we obtain

$$\begin{aligned}\langle \widehat{Wf}, \widehat{Wg}|\hat{\sigma}|^2\rangle_{L^2(\mathbb{R}^{2d})} &= \langle Q_\sigma f, Q_\sigma g\rangle_{L^2(\mathbb{R}^{2d})} = |\langle f,g\rangle|^2\\ &= \langle Wf, Wg\rangle_{L^2(\mathbb{R}^{2d})} = \langle \widehat{Wf}, \widehat{Wg}\rangle_{L^2(\mathbb{R}^{2d})},\end{aligned}$$

and thus $\langle \widehat{Wf}, \widehat{Wg}|\hat{\sigma}|^2\rangle_{L^2(\mathbb{R}^{2d})} = \langle \widehat{Wf}, \widehat{Wg}\rangle_{L^2(\mathbb{R}^{2d})}$, for every $f,g \in L^2(\mathbb{R}^d)$. This easily gives the condition $|\hat{\sigma}(x,\xi)| = 1$ a. e. (take $f(t) = g(t) = \varphi(t) = e^{-t^2}$, $t \in \mathbb{R}^d$), which ends the proof. □

Hence not every element of Cohen class satisfies the marginal properties. For example, the member of Cohen class $Wf * \sigma_{a,b}$, where $\sigma_{a,b} = \exp(-2\pi(x^2/a + \xi^2/b))$, does not satisfy conditions (1.102) and thus it does not verify the marginal properties, whereas (1.103) is obviously satisfied after normalization.

The next proposition gives a sufficient condition for being a member of the Cohen class.

Theorem 1.3.21. *Suppose that a quadratic time-frequency representation Q, i. e., a sesquilinear form $Q : \mathcal{S}(\mathbb{R}^d) \times \mathcal{S}(\mathbb{R}^d) \to \mathcal{S}(\mathbb{R}^{2d})$, is covariant* (1.101) *and weakly continuous, that is,*

$$|Q(f,g)(0,0)| \le \|f\|_2^2\|g\|_2^2 \tag{1.107}$$

for all $f,g \in L^2(\mathbb{R}^d)$, or $f,g \in \mathcal{S}(\mathbb{R}^d)$. Then there exists a tempered distribution $\sigma \in \mathcal{S}'(\mathbb{R}^{2d})$ such that

$$Qf = Q(f,f) = Wf * \sigma \tag{1.108}$$

for all $f \in \mathcal{S}(\mathbb{R}^d)$.

Proof. The assumption (1.107) means that the map

$$(f,g) \longmapsto Q(f,g)(0,0)$$

is a bounded sesquilinear form. Hence, there exists a bounded operator A on $L^2(\mathbb{R}^d)$ such that

$$Q(f,g)(0,0) = \langle Af,g\rangle.$$

(cf. Theorem A.0.10 in Appendix A). Using the covariance property, we obtain that Qf can be expressed as

$$Qf(z) = T_{(-z)}Qf(0,0) = Q(\pi(-z)f)(0,0) = \langle A(\pi(-z)f), \pi(-z)f\rangle.$$

It follows, in particular, that A maps $\mathcal{S}(\mathbb{R}^d)$ into $\mathcal{S}'(\mathbb{R}^d)$, we can then apply Schwartz kernel theorem and thus there exists a distribution $k \in \mathcal{S}'(\mathbb{R}^{2d})$ such that

$$\langle Af,g\rangle = \langle k, g\otimes \bar{f}\rangle, \quad \forall f,g \in \mathcal{S}(\mathbb{R}^d).$$

Recalling that the unitary operators $\mathcal{F}_2$ and $\mathfrak{T}_s$ are continuous bijections on $\mathcal{S}(\mathbb{R}^{2d})$ and $\mathcal{S}'(\mathbb{R}^{2d})$, and the Wigner distribution is real-valued, we deduce by Lemma 1.3.9 that

$$\begin{aligned} Qf(0,0) &= \langle Af,f\rangle = \langle k, f\otimes \bar{f}\rangle = \langle \mathcal{F}_2\mathfrak{T}_s k, \mathcal{F}_2\mathfrak{T}_s(f\otimes \bar{f})\rangle \\ &= \langle \mathcal{F}_2\mathfrak{T}_s k, Wf\rangle = \langle \mathcal{F}_2\mathfrak{T}_s k, \overline{Wf}\rangle. \end{aligned}$$

Similarly,

$$\begin{aligned} Qf(z) &= Q(\pi(-z)f)(0,0) = \langle A(\pi(-z)f), \pi(-z)f\rangle = \langle \mathcal{F}_2\mathfrak{T}_s k, \overline{W(\pi(-z)f)}\rangle \\ &= \langle \mathcal{F}_2\mathfrak{T}_s k, T_{(-z)}\overline{Wf}\rangle = Wf * \mathcal{I}(\mathcal{F}_2\mathfrak{T}_s k). \end{aligned}$$

The desired kernel is $\mathcal{I}(\mathcal{F}_2\mathfrak{T}_s k) \in \mathcal{S}'(\mathbb{R}^{2d})$, and this concludes the proof. □

1.3.5 τ-Wigner distributions

We discuss here a generalization of the Wigner distribution, depending on a parameter $\tau \in [0,1]$. For more general perturbations of the Wigner distribution, we refer to [8].

Definition 1.3.22. For $\tau \in [0,1]$, $f,g \in L^2(\mathbb{R}^d)$, we define the (cross-)$\tau$-Wigner distribution by

$$W_\tau(f,g)(x,\xi) = \int_{\mathbb{R}^d} e^{-2\pi i t\cdot\xi} f(x+\tau t)\overline{g(x-(1-\tau)t)}\,dt. \tag{1.109}$$

For $f=g$, $W_\tau f := W_\tau(f,f)$ is called the τ-Wigner distribution of f.

So Wf is a particular case of τ-Wigner distribution corresponding to the value $\tau = 1/2$.

Proposition 1.3.23. *For $f,g \in L^2(\mathbb{R}^d)$, the cross-$\tau$-Wigner distribution has the following properties:*
(i) $W_{1-\tau}(f,g) = \overline{W_\tau(g,f)}$ *for every* $\tau \in [0,1]$.
(ii) $W_\tau f(x,\xi) = W_{1-\tau}\hat{f}(\xi,-x)$ *for every* $\tau \in [0,1]$. *Equivalently,*

$$W_\tau \hat{f}(z) = W_{1-\tau} f(Jz).$$

Proof. (i) It is an immediate consequence of definition of τ-Wigner distribution. Indeed,

$$\begin{aligned} W_{1-\tau}(f,g)(x,\xi) &= \int_{\mathbb{R}^d} e^{-2\pi i t\cdot\xi} f(x+(1-\tau)t)\overline{g(x-\tau t)}\,dt \\ &= \overline{\int_{\mathbb{R}^d} e^{2\pi i t\cdot\xi} g(x-\tau t)\overline{f(x+(1-\tau)t)}\,dt} \end{aligned}$$

$$= \overline{\int_{\mathbb{R}^d} e^{-2\pi i t\cdot\xi} g(x+\tau t)\overline{f(x-(1-\tau)t)}\,dt}$$
$$= \overline{W_\tau(g,f)}(x,\xi).$$

(ii) We detail the computations for $f \in \mathcal{S}(\mathbb{R}^d)$, so that all the integrals below are absolutely convergent. The general case follows by standard density arguments. We have

$$\begin{aligned}
W_{1-\tau}\hat{f}(\xi,-x) &= \int_{\mathbb{R}^d} e^{2\pi i t\cdot x}\hat{f}(\xi+(1-\tau)t)\overline{\hat{f}(\xi-\tau t)}\,dt\\
&= \int_{\mathbb{R}^3} e^{-2\pi i\xi\cdot(y-z)} e^{2\pi i t\cdot(x-y+\tau(y-z))} f(y)\overline{f(z)}\,dt\,dy\,dz\\
&= \int_{\mathbb{R}^3} e^{-2\pi i\xi\cdot\tilde{z}} e^{2\pi i t\cdot(x+\tau\tilde{z})} e^{-2\pi i t\cdot y} f(y)\overline{f(y-\tilde{z})}\,dt\,dy\,d\tilde{z}\\
&= \int_{\mathbb{R}^{2d}} e^{-2\pi i\xi\cdot\tilde{z}} e^{2\pi i t\cdot(x+\tau\tilde{z})}\left[\int_{\mathbb{R}^d} e^{-2\pi i t\cdot y} f(y)\overline{T_{\tilde{z}}f(y)}\,dy\right]dt\,d\tilde{z}\\
&= \int_{\mathbb{R}^d} e^{-2\pi i\xi\cdot\tilde{z}}\left[\int_{\mathbb{R}^d} e^{2\pi i t\cdot(x+\tau\tilde{z})}\widehat{f\cdot\overline{T_{\tilde{z}}f}}(t)\,dt\right]d\tilde{z}\\
&= \int_{\mathbb{R}^d} e^{-2\pi i\xi\cdot\tilde{z}}\mathcal{F}^{-1}(\widehat{f\cdot\overline{T_{\tilde{z}}f}})(x+\tau\tilde{z})\,d\tilde{z}\\
&= \int_{\mathbb{R}^d} e^{-2\pi i\xi\cdot\tilde{z}} f(x+\tau\tilde{z})\overline{f(x-(1-\tau)\tilde{z})}\,d\tilde{z} = W_\tau f(x,\xi),
\end{aligned}$$

as desired. The last formula immediately follows by observing that

$$W_{1-\tau}(\mathcal{F}^2 f)(z) = W_{1-\tau}(\mathcal{I}f)(z) = W_{1-\tau}f(-z). \qquad \square$$

From Proposition 1.3.23, we infer

$$W_\tau f + W_{1-\tau}f = 2\Re(W_\tau f), \quad W_\tau f - W_{1-\tau}f = 2\Im(W_\tau f).$$

The τ-Wigner distribution $W_\tau f$ possesses similar properties to those of the classical Wigner distribution as stated in the next proposition.

Proposition 1.3.24. *The τ-Wigner distribution satisfies the marginal properties for every $\tau \in [0,1]$, namely, for every $f \in \mathcal{S}(\mathbb{R}^d)$,*

$$\int_{\mathbb{R}^d} W_\tau f(x,\xi)\,dx = |\hat{f}(\xi)|^2, \qquad \int_{\mathbb{R}^d} W_\tau f(x,\xi)\,d\xi = |f(x)|^2.$$

Proof. Since $f \in \mathcal{S}(\mathbb{R}^d)$, all the integrals below are absolutely convergent, and we can make the changes of coordinate $y \mapsto x + \tau t$ and subsequently $s \mapsto y - t$ as follows:

$$\begin{aligned}\int_{\mathbb{R}^d} W_\tau f(x,\xi)\,dx &= \int_{\mathbb{R}^{2d}} e^{-2\pi i t\cdot\xi} f(x+\tau t)\overline{f(x-(1-\tau)t)}\,dt\,dx \\ &= \int_{\mathbb{R}^{2d}} e^{-2\pi i t\cdot\xi} f(y)\overline{f(y-t)}\,dy\,dt \\ &= \int_{\mathbb{R}^d} f(y)\Big[\int_{\mathbb{R}^d} e^{-2\pi i t\cdot\xi}\overline{f(y-t)}\,dt\Big]\,dy \\ &= \int_{\mathbb{R}^d} e^{-2\pi i y\cdot\xi} f(y)\,dy\,\overline{\int_{\mathbb{R}^d} e^{-2\pi i s\cdot\xi} f(s)\,ds} = |\hat{f}(\xi)|^2.\end{aligned}$$

Using Proposition 1.3.23 (ii), we obtain the second equality as well:

$$\begin{aligned}\int_{\mathbb{R}^d} W_\tau f(x,\xi)\,d\xi &= \int_{\mathbb{R}^d} W_{1-\tau}\hat{f}(\xi,-x)\,d\xi \\ &= |\hat{\hat{f}}(-x)|^2 = |\mathcal{F}^{-1}(\hat{f})(x)|^2 = |f(x)|^2.\end{aligned}$$

□

Moreover, $W_\tau(f,g)$ has a factorization similar to that of Wigner distribution.

Proposition 1.3.25. *For $f,g \in L^2(\mathbb{R}^d)$,*

$$W_\tau(f,g)(x,\xi) = \mathcal{F}_{t\to\xi}[T^{[x]}_{(\frac{1}{2}-\tau)t}\mathfrak{T}_s(f\otimes\overline{g})](x,\xi), \tag{1.110}$$

where $\mathfrak{T}_s$ is the symmetric coordinate change (1.87) and $T^{[x]}$ denotes the translation operator acting on the x-variable.

Proof. By definition of $W_\tau(f,g)$, we have

$$\begin{aligned}W_\tau(f,g)(x,\xi) &= \int_{\mathbb{R}^d} e^{-2\pi i t\cdot\xi} f(x+\tau t)\overline{g(x-(1-\tau)t)}\,dt \\ &= \int_{\mathbb{R}^d} e^{-2\pi i t\cdot\xi} T^{[x]}_{(\frac{1}{2}-\tau)t}\mathfrak{T}_s(f\otimes\overline{g})(x,t)\,dt \\ &= \mathcal{F}_{t\to\xi}[T^{[x]}_{(\frac{1}{2}-\tau)t}\mathfrak{T}_s(f\otimes\overline{g})](x,\xi),\end{aligned}$$

as desired. □

Observe that, since the linear operators appearing in (1.110) are isomorphisms on $\mathcal{S}(\mathbb{R}^{2d})$ and $L^2(\mathbb{R}^{2d})$, we immediately obtain

Corollary 1.3.26. *For $\tau \in [0,1]$, we have*
(i) $f,g \in \mathcal{S}(\mathbb{R}^d) \Longrightarrow W_\tau(f,g) \in \mathcal{S}(\mathbb{R}^{2d})$;
(ii) $f,g \in L^2(\mathbb{R}^d) \Longrightarrow W_\tau(f,g) \in L^2(\mathbb{R}^{2d})$.

Proposition 1.3.27. *The τ-Wigner distribution belongs to the Cohen class for every $\tau \in [0,1]$. In particular,*

$$W_\tau(f,g) = W(f,g) * \sigma_\tau, \quad \forall f,g \in \mathcal{S}(\mathbb{R}^d) \tag{1.111}$$

where the kernel $\sigma_\tau \in \mathcal{S}'(\mathbb{R}^{2d})$ is given by

$$\sigma_\tau(x,\xi) = \begin{cases} \frac{2^d}{|2\tau-1|^d} e^{2\pi i \frac{2}{2\tau-1} x\cdot\xi}, & \tau \neq \frac{1}{2}, \\ \delta, & \tau = \frac{1}{2}. \end{cases} \tag{1.112}$$

Proof. If $\tau = \frac{1}{2}$, $W_\tau(f,g)$ coincides with the classical cross-Wigner distribution and so the claim becomes trivial. For $\tau \neq \frac{1}{2}$, applying the inverse Fourier transform to (1.111), the claim turns into

$$\mathcal{F}^{-1}(W_\tau(f,g))(v,t) = (\mathcal{F}^{-1}W(f,g)) \cdot (\mathcal{F}^{-1}\sigma_\tau)(v,t), \tag{1.113}$$

for all $f,g \in \mathcal{S}(\mathbb{R}^d)$. Let us to compute $\mathcal{F}^{-1}\sigma_\tau(v,t)$. For short, we use integrals in the weak sense, application to test functions in $\mathcal{S}(\mathbb{R}^d)$ being understood. Since $\tau \neq \frac{1}{2}$, by (1.112),

$$\begin{aligned} \mathcal{F}^{-1}\sigma_\tau(v,t) &= \frac{2^d}{|2\tau-1|^d} \int_{\mathbb{R}^{2d}} e^{2\pi i v\cdot x} e^{2\pi i t\cdot\xi} e^{2\pi i \frac{2}{2\tau-1} x\cdot\xi}\, dx\, d\xi \\ &= \frac{2^d}{|2\tau-1|^d} \int_{\mathbb{R}^{2d}} e^{2\pi i (\frac{2}{2\tau-1}x+t)\cdot(\xi+\frac{2\tau-1}{2}v)} e^{-\pi i(2\tau-1)t\cdot v}\, dx\, d\xi. \end{aligned}$$

By making the change of variables $\frac{2}{2\tau-1}x + t = y$ and $\xi + \frac{2\tau-1}{2}v = \rho$, we get

$$\mathcal{F}^{-1}\sigma_\tau(v,t) = e^{-\pi i(2\tau-1)t\cdot v} \int_{\mathbb{R}^{2d}} e^{2\pi i y\cdot\rho}\, dy\, d\rho,$$

consequently

$$\mathcal{F}^{-1}\sigma_\tau(v,t) = e^{-\pi i(2\tau-1)t\cdot v}, \tag{1.114}$$

since $\int_{\mathbb{R}^{2d}} e^{2\pi i y\cdot\rho}\, dy\, d\rho = 1$.

Let us now compute the left-hand side of (1.113). To this end we use the factorization (1.110), the factorization of Wigner distribution (1.89), and the decomposition of the inverse Fourier transform $\mathcal{F}^{-1} = \mathcal{F}_1^{-1}\mathcal{F}_2^{-1}$ as follows:

$$\begin{aligned} \mathcal{F}^{-1}(W_\tau(f,g))(v,t) &= \mathcal{F}^{-1}_{x\to v}\mathcal{F}^{-1}_{\xi\to t}\mathcal{F}_{t\to\xi}[T^{[x]}_{(\frac{1}{2}-\tau)t}\mathfrak{T}_s(f\otimes g)](v,t) \\ &= \mathcal{F}^{-1}_{x\to v}[T^{[x]}_{(\frac{1}{2}-\tau)t}\mathfrak{T}_s(f\otimes g)](v,t) \end{aligned}$$

$$
\begin{aligned}
&= M^{[v]}_{(\frac{1}{2}-\tau)t}\mathcal{F}^{-1}_{x\to v}(\mathfrak{T}_s(f\otimes g))(v,t)\\
&= e^{-\pi i(2\tau-1)t\cdot v}\mathcal{F}^{-1}_{x\to v}\mathcal{F}^{-1}_{\xi\to t}(\mathcal{F}_{t\to\xi}(\mathfrak{T}_s(f\otimes g)))(v,t)\\
&= e^{-\pi i(2\tau-1)t\cdot v}\mathcal{F}^{-1}(W(f,g))(v,t).
\end{aligned}
$$

In view of (1.114), this completes the proof. □

Thanks to this proposition, we can give an alternative proof of the marginal properties of W_τ. Since the τ-Wigner distribution is a member of Cohen class, applying Proposition 1.3.19 and Remark 1.3.20, we can easily see that the kernel σ_τ in (1.112) satisfies conditions (1.105). Indeed, for $\tau \neq \frac{1}{2}$,

$$
\begin{aligned}
\mathcal{F}\sigma_\tau(x,0) &= \frac{2^d}{|2\tau-1|^d}\int_{\mathbb{R}^{2d}} e^{-2\pi i x\cdot t}e^{2\pi i\frac{2}{2\tau-1}t\cdot\xi}\,dt\,d\xi\\
&= \frac{2^d}{|2\tau-1|^d}\int_{\mathbb{R}^{2d}} e^{2\pi i t\cdot(\frac{2}{2\tau-1}\xi-x)}\,dt\,d\xi\\
&= \int_{\mathbb{R}^{2d}} e^{2\pi i t\cdot\omega}\,dt\,d\omega = 1,
\end{aligned}
$$

where again integrals are in the weak sense. Similarly, one can check that $\mathcal{F}\sigma(0,\xi) = 1$.

Corollary 1.3.28 (Moyal's formula for τ-Wigner distributions). *For $\tau \in [0,1]$, the (cross-)τ-Wigner distribution satisfies Moyal's formula*

$$\langle W_\tau(f_1,g_1), W_\tau(f_2,g_2)\rangle = \langle f_1,f_2\rangle\overline{\langle g_1,g_2\rangle}, \quad f_1,f_2,g_1,g_2 \in L^2(\mathbb{R}^d). \tag{1.115}$$

Proof. It is enough to prove (1.115) for $f_1 = f_2$, $g_1 = g_2$ and then apply the polarization identity. The Cohen kernel σ_τ of $W_\tau f$ is given by (1.112). Hence, using (1.114), we compute

$$\mathcal{F}(\sigma_\tau)(x,\xi) = \mathcal{F}^{-1}(\sigma_\tau)(-x,-\xi) = \mathcal{F}^{-1}(\sigma_\tau)(x,\xi) = e^{-\pi i(2\tau-1)x\cdot\xi}$$

so that

$$|\widehat{\sigma_\tau}(x,\xi)| = 1, \quad \forall (x,\xi) \in \mathbb{R}^{2d},$$

and condition (1.104) is satisfied. This completes the proof. □

We now want to find a relation between the τ-Wigner distribution and the STFT which generalizes Lemma 1.3.5 for the Wigner distribution. To this aim, we need to introduce a new operator depending on parameter $\tau \in [0,1]$, which plays the same role of the reflection operator in (1.80).

Definition 1.3.29. For $\tau \in (0,1)$, we define the operator A_τ as follows:

$$A_\tau : f(t) \longmapsto \mathcal{I}f\left(\frac{1-\tau}{\tau}t\right). \tag{1.116}$$

For $\tau \in (0,1)$ and $f \in L^p(\mathbb{R}^d)$, $1 \le p \le \infty$, we obtain

$$\|A_\tau f\|_{L^p(\mathbb{R}^d)} = \frac{|\tau|^{\frac{d}{p}}}{|\tau - 1|^{\frac{d}{p}}} \|f\|_{L^p(\mathbb{R}^d)},$$

with the convention $d/\infty = 0$.

Proposition 1.3.30. *For $\tau \in (0,1)$, we have*

$$W_\tau(f,g)(x,\xi) = \frac{1}{\tau^d} e^{2\pi i \frac{1}{\tau} x\cdot\xi} V_{A_\tau g} f\left(\frac{1}{1-\tau}x, \frac{1}{\tau}\xi\right), \quad f,g \in L^2(\mathbb{R}^d). \tag{1.117}$$

Proof. Using definition of τ-Wigner distribution and performing the change of variables $y = x + \tau t$, we can write

$$\begin{aligned} W_\tau(f,g)(x,\xi) &= \int_{\mathbb{R}^d} e^{-2\pi i t\cdot\xi} f(x+\tau t)\overline{g(x-(1-\tau)t)}\,dt \\ &= \frac{1}{\tau^d} e^{2\pi i \frac{1}{\tau}\xi\cdot x} \int_{\mathbb{R}^d} e^{-2\pi i q\cdot\frac{\xi}{\tau}} f(y)\overline{\mathcal{I}g\left(\frac{1-\tau}{\tau}y - \frac{x}{\tau}\right)}\,dy \\ &= \frac{1}{\tau^d} e^{2\pi i \frac{1}{\tau}\xi\cdot x} \int_{\mathbb{R}^d} e^{-2\pi i y\cdot\frac{\xi}{\tau}} f(y)\overline{A_\tau g\left(y - \frac{1}{\tau-1}x\right)}\,dy \\ &= \frac{1}{\tau^d} e^{2\pi i \frac{1}{\tau}\xi\cdot x} V_{A_\tau g} f\left(\frac{1}{1-\tau}x, \frac{1}{\tau}\xi\right), \end{aligned}$$

as required. □

It remains to consider the cases $\tau = 0$ and $\tau = 1$. For $\tau = 0$, $W_0(f,g)$ is the well-known (cross-)Rihaczek distribution. In detail,

$$W_0(f,g)(x,\xi) = R(f,g)(x,\xi) = e^{-2\pi i x\cdot\xi} f(x)\overline{\hat{g}(\xi)}, \quad f,g \in L^2(\mathbb{R}^d). \tag{1.118}$$

For $\tau = 1$, $W_1(f,g)$ is the conjugate-(cross-)Rihaczek distribution $R^*(f,g)$ given by

$$W_1(f,g)(x,\xi) = R^*(f,g)(x,\xi) = \overline{R(g,f)} = e^{2\pi i x\cdot\xi}\overline{g(x)}\hat{f}(\xi), \quad f,g \in L^2(\mathbb{R}^d). \tag{1.119}$$

Moreover, we have

$$\begin{aligned} R(f,g) &= W(1,\delta) * W(f,g) = e^{-4\pi i x\cdot\xi} * W(f,g)(x,\xi), \\ R^*(f,g) &= W(\delta,1) * W(f,g) = e^{4\pi i x\cdot\xi} * W(f,g)(x,\xi), \end{aligned}$$

hence recapturing the Cohen kernels σ_0, σ_1 in (1.112).

Using (1.33), we obtain the following relationship between the Fourier transform of the STFT and the cross-Rihaczek distribution:

$$\widehat{V_g f}(x,\xi) = W_0(f,g)(-\xi,x) = W_0(f,g)(-J(x,\xi)), \quad (x,\xi) \in \mathbb{R}^{2d}. \tag{1.120}$$

In terms of the symplectic Fourier transform (cf. (1.18)),

$$\mathcal{F}_\sigma V_g f(z) = W_0(f,g)(z), \quad z \in \mathbb{R}^{2d}, \quad f,g \in L^2(\mathbb{R}^d), \tag{1.121}$$

or, in terms of the conjugate-(cross-)-Rihaczek distribution,

$$\mathcal{F}_\sigma V_g f(z) = \overline{W_1(g,f)(z)}, \quad z \in \mathbb{R}^{2d}, \quad f,g \in L^2(\mathbb{R}^d). \tag{1.122}$$

It is then natural to determine the symplectic Fourier transform of a τ-Wigner distribution. In what follows we write $f_\lambda(t) = f(\lambda t)$, $\lambda \in \mathbb{R}$, for short.

Lemma 1.3.31. *For $\tau \in (0,1)$, $f,g \in L^2(\mathbb{R}^d)$, we have*

$$\mathcal{F}_\sigma W_\tau(f,g)(y,\eta) = \tau^d(1-\tau)^d W_{1-\tau}(f_\tau, (\mathcal{I}g)_{1-\tau})(y, \tau(1-\tau)\eta), \quad (y,\eta) \in \mathbb{R}^{2d}. \tag{1.123}$$

Proof. Using the inversion formula of the Fourier transform in $\mathcal{S}'(\mathbb{R}^d)$, the relation $\mathcal{F}_\sigma W_\tau(f,g)(y,\eta) = \mathcal{F}W_\tau(f,g)(\eta,-y) = \mathcal{F}_1\mathcal{F}_2 W_\tau(f,g)(\eta,-y)$, and the factorization (1.110), we can write

$$\begin{aligned}
\mathcal{F}_\sigma W_\tau(f,g)(y,\eta) &= \mathcal{F}W_\tau(f,g)(\eta,-y) = \mathcal{F}_1\mathcal{F}_2^2[T^{[t]}_{(\frac{1}{2}-\tau)(-y)}\mathfrak{T}_s(f\otimes g)](\eta,-y)\\
&= \int_{\mathbb{R}^d} e^{-2\pi i\eta\cdot t} f(t+\tau y)\overline{g(t-(1-\tau)y)}\,dt\\
&= \int_{\mathbb{R}^d} e^{-2\pi i\eta\cdot t} f\Big(\tau\Big(y+\frac{1}{\tau}t\Big)\Big)\overline{g\Big(-(1-\tau)\Big(y-\frac{1}{1-\tau}t\Big)\Big)}\,dt\\
&= \tau^d(1-\tau)^d \int_{\mathbb{R}^d} e^{-2\pi i\eta\cdot\tau(1-\tau)t'} f(\tau(y+(1-\tau)t'))\overline{g(-(1-\tau)(y-\tau t'))}\,dt'
\end{aligned}$$

where we made the change of variables $t = \tau(1-\tau)t'$. □

Corollary 1.3.32. *For $f,g \in L^2(\mathbb{R}^d)$, we have the following formulas:*
(i)

$$\mathcal{F}_\sigma W(f,g)(y,\eta) = 2^{-2d} W(f_{\frac{1}{2}}, (\mathcal{I}g)_{\frac{1}{2}})\Big(y, \frac{\eta}{4}\Big), \quad (y,\eta) \in \mathbb{R}^{2d}; \tag{1.124}$$

(ii)

$$\mathcal{F}_\sigma W(f,g)(y,\eta) = 2^{-d} W(f, \mathcal{I}g)\Big(\frac{y}{2}, \frac{\eta}{2}\Big), \quad (y,\eta) \in \mathbb{R}^{2d}. \tag{1.125}$$

Proof. (i) It follows by choosing $\tau = 1/2$ in formula (1.123).

(ii) Using the inversion formula of the Fourier transform,

$$\begin{aligned}\mathcal{F}_\sigma W(f,g)(y,\eta) &= \int_{\mathbb{R}^d} e^{-2\pi i\eta\cdot t} f\left(t+\frac{y}{2}\right)\overline{g\left(t-\frac{y}{2}\right)}\,dt \\ &= 2^{-d}\int_{\mathbb{R}^d} e^{-2\pi i\eta\cdot\frac{t'}{2}} f\left(\frac{y}{2}+\frac{t'}{2}\right)\overline{g\left(-\left(\frac{y}{2}-\frac{t'}{2}\right)\right)}\,dt' \\ &= 2^{-d}\int_{\mathbb{R}^d} e^{-2\pi i\frac{\eta}{2}\cdot t'} f\left(\frac{y}{2}+\frac{t'}{2}\right)\overline{(\mathcal{I}g)\left(\frac{y}{2}-\frac{t'}{2}\right)}\,dt' \\ &= W(f,\mathcal{I}g)\left(\frac{y}{2},\frac{\eta}{2}\right),\end{aligned}$$

as desired. □

For more general perturbations of the Wigner distribution and a characterization of the Cohen class, we refer to [90].

1.3.6 Time–frequency representations of Gaussians

In what follows we shall compute STFT and Wigner distributions related to Gaussian functions. Such formulas will be very useful in the next chapters to prove the sharpness of continuity properties. To shorten notation, in what follows we write $x\cdot\xi = x\xi$ and $x^2 = x\cdot x$.

Proposition 1.3.33. *Given $a,b,c>0$, consider the generalized Gaussian function*

$$f(x,\xi) = f_{a,b,c}(x,\xi) = e^{-\pi a x^2} e^{-\pi b\xi^2} e^{2\pi i c x\xi}, \quad (x,\xi)\in\mathbb{R}^{2d}. \tag{1.126}$$

For $\Phi(x,\xi) = e^{-\pi(x^2+\xi^2)}$, $z=(z_1,z_2)$, $\zeta=(\zeta_1,\zeta_2)\in\mathbb{R}^{2d}$, we obtain

$$\begin{aligned}V_\Phi f(z,\zeta) &= \frac{1}{[(a+1)(b+1)+c^2]^{\frac{d}{2}}} e^{-\pi\frac{[a(b+1)+c^2]z_1^2+[(a+1)b+c^2]z_2^2+(b+1)\zeta_1^2+(a+1)\zeta_2^2-2c(z_1\zeta_2+z_2\zeta_1)}{(a+1)(b+1)+c^2}} \\ &\quad\times e^{-\frac{2\pi i}{a+1}[z_1\zeta_1+(cz_1-(a+1)\zeta_2)\frac{c\zeta_1+(a+1)z_2}{(a+1)(b+1)+c^2}]}.\end{aligned} \tag{1.127}$$

Proof. We write

$$\begin{aligned}V_\Phi f(z,\zeta) &= \int_{\mathbb{R}^{2d}} e^{-\pi a x^2-\pi b\xi^2+2\pi i c x\xi} e^{-2\pi i(\zeta_1 x+\zeta_2\xi)} e^{-\pi[(x-z_1)^2+(\xi-z_2)^2]}\,dx\,d\xi \\ &= e^{-\pi(z_1^2+z_2^2)}\int_{\mathbb{R}^d}\left(\int_{\mathbb{R}^d} e^{-\pi[(a+1)x^2-2xz_1]} e^{-2\pi i x(\zeta_1-c\xi)}\,dx\right) \\ &\quad\times e^{-\pi[(b+1)\xi^2-2\xi z_2]} e^{-2\pi i\zeta_2\xi}\,d\xi\end{aligned}$$

$$= e^{-\pi(1-\frac{1}{a+1})z_1^2-\pi z_2^2} \int\limits_{\mathbb{R}^d} \Bigg(\int\limits_{\mathbb{R}^d} e^{-\pi[\sqrt{a+1}x-\frac{z_1}{\sqrt{a+1}}]^2} e^{-2\pi i x(\zeta_1-c\xi)}\, dx \Bigg) \times e^{-\pi[(b+1)\xi^2-2\xi z_2]} e^{-2\pi i\zeta_2\xi}\, d\xi.$$

With the change of variables $\sqrt{a+1}x - \frac{z_1}{\sqrt{a+1}} = t$, $dx = \frac{dt}{(a+1)^{d/2}}$, we obtain

$$V_\Phi f(z,\zeta) = \frac{1}{(a+1)^{d/2}} e^{-\pi\frac{a}{a+1}z_1^2-\pi z_2^2} \int\limits_{\mathbb{R}^d} \Bigg(\int\limits_{\mathbb{R}^d} e^{-\pi t^2} e^{-2\pi i(\frac{t}{\sqrt{a+1}}+\frac{z_1}{a+1})(\zeta_1-c\xi)}\, dt \Bigg) \times e^{-\pi[(b+1)\xi^2-2\xi z_2]} e^{-2\pi i\zeta_2\xi}\, d\xi$$

$$= \frac{1}{(a+1)^{d/2}} e^{-\pi\frac{a}{a+1}z_1^2-\pi z_2^2-2\pi i\frac{z_1\zeta_1}{a+1}} \int\limits_{\mathbb{R}^d} e^{-\pi\frac{(\zeta_1-c\xi)^2}{a+1}} e^{2\pi i\frac{cz_1}{a+1}\xi} \times e^{-\pi[(b+1)\xi^2-2\xi z_2]} e^{-2\pi i\zeta_2\xi}\, d\xi$$

$$= \frac{1}{(a+1)^{d/2}} e^{-\pi\frac{a}{a+1}z_1^2-\pi z_2^2-2\pi i\frac{z_1\zeta_1}{a+1}-\pi\frac{\zeta_1^2}{a+1}} \times \int\limits_{\mathbb{R}^d} e^{-\pi[(b+1)\xi^2+\frac{c^2}{a+1}\xi^2-2\frac{c\zeta_1}{a+1}\xi-2\xi z_2]} e^{2\pi i(\frac{cz_1}{a+1}-\zeta_2)\xi}\, d\xi. \tag{1.128}$$

The last integral can be computed as follows:

$$I := \int\limits_{\mathbb{R}^d} e^{-\pi[\frac{(a+1)(b+1)+c^2}{a+1}\xi^2-2\frac{c\zeta_1+z_2(a+1)}{a+1}\xi]} e^{2\pi i(\frac{cz_1}{a+1}-\zeta_2)\xi}\, d\xi$$

$$= \int\limits_{\mathbb{R}^d} e^{-\frac{\pi}{a+1}\{[(a+1)(b+1)+c^2]\xi^2-2[c\zeta_1+z_2(a+1)]\xi\}} e^{2\pi i(\frac{cz_1}{a+1}-\zeta_2)\xi}\, d\xi$$

$$= \int\limits_{\mathbb{R}^d} e^{-\frac{\pi}{a+1}[\sqrt{(a+1)(b+1)+c^2}\xi-\frac{c\zeta_1+z_2(a+1)}{\sqrt{(a+1)(b+1)+c^2}}]^2} e^{\frac{\pi}{a+1}\frac{[c\zeta_1+z_2(a+1)]^2}{(a+1)(b+1)+c^2}} e^{2\pi i(\frac{cz_1}{a+1}-\zeta_2)\xi}\, d\xi$$

$$= e^{\frac{\pi}{a+1}\frac{[c\zeta_1+z_2(a+1)]^2}{(a+1)(b+1)+c^2}} \int\limits_{\mathbb{R}^d} e^{-\frac{\pi}{a+1}[\sqrt{(a+1)(b+1)+c^2}\xi-\frac{c\zeta_1+z_2(a+1)}{\sqrt{(a+1)(b+1)+c^2}}]^2} e^{2\pi i(\frac{cz_1}{a+1}-\zeta_2)\xi}\, d\xi.$$

Making the change of variables $t = \dfrac{\sqrt{(a+1)(b+1)+c^2}\xi-\frac{c\zeta_1+z_2(a+1)}{\sqrt{(a+1)(b+1)+c^2}}}{\sqrt{a+1}}$, so that

$$d\xi = \left[\frac{a+1}{(a+1)(b+1)+c^2}\right]^{d/2} dt,$$

we can write

$$I = \frac{(a+1)^{\frac{d}{2}}}{[(a+1)(b+1)+c^2]^{\frac{d}{2}}} e^{\frac{\pi}{a+1}\frac{[c\zeta_1+z_2(a+1)]^2}{(a+1)(b+1)+c^2}} \times \int\limits_{\mathbb{R}^d} e^{-\pi t^2} e^{2\pi i(\frac{cz_1}{a+1}-\zeta_2)[\frac{\sqrt{a+1}}{\sqrt{(a+1)(b+1)+c^2}}t+\frac{c\zeta_1+(a+1)z_2}{(a+1)(b+1)+c^2}]}\, dt$$

$$= \frac{(a+1)^{\frac{d}{2}}}{[(a+1)(b+1)+c^2]^{\frac{d}{2}}} e^{\frac{\pi}{a+1}\frac{[c\zeta_1+z_2(a+1)]^2}{(a+1)(b+1)+c^2}} e^{-2\pi i(\frac{cz_1}{a+1}-\zeta_2)\frac{c\zeta_1+(a+1)z_2}{(a+1)(b+1)+c^2}} \times e^{-\pi[\frac{a+1}{(a+1)(b+1)+c^2}(\frac{cz_1}{a+1}-\zeta_2)^2]}.$$

The result then follows by substituting the value of the integral I in (1.128). □

Lemma 1.3.34. *Consider $\varphi(t) = e^{-\pi t^2}$ and its rescaled version $\varphi_{\sqrt{\lambda}}(t) = e^{-\pi\lambda t^2}$. Then*

$$W(\varphi, \varphi_{\sqrt{\lambda}})(x,\xi) = \frac{2^d}{(\lambda+1)^{\frac{d}{2}}} e^{-\frac{4\pi\lambda}{\lambda+1}x^2} e^{-\frac{4\pi}{\lambda+1}\xi^2} e^{-4\pi i\frac{\lambda-1}{\lambda+1}x\xi}. \tag{1.129}$$

Proof. The proof is a straightforward computation. Indeed,

$$\begin{aligned} W(\varphi, \varphi_{\sqrt{\lambda}})(x,\xi) &= \int_{\mathbb{R}^d} e^{-\pi(x+\frac{t}{2})^2-\pi\lambda(x-\frac{t}{2})^2} e^{-2\pi it\xi}\,dt \\ &= e^{-\pi(1+\lambda)x^2} \int_{\mathbb{R}^d} e^{-\frac{\pi}{4}[(1+\lambda)t^2+4(1-\lambda)xt]} e^{-2\pi it\xi}\,dt \\ &= e^{-\pi(1+\lambda)x^2} \int_{\mathbb{R}^d} e^{-\frac{\pi}{4}(\sqrt{1+\lambda}t+2\frac{(1-\lambda)}{\sqrt{1+\lambda}}x)^2} e^{\pi\frac{(1-\lambda)^2}{1+\lambda}x^2} e^{-2\pi it\xi}\,dt \\ &= e^{-\pi[(1+\lambda)-\frac{(1-\lambda)^2}{1+\lambda}]x^2} \int_{\mathbb{R}^d} e^{-\frac{\pi}{4}(\sqrt{1+\lambda}t+2\frac{(1-\lambda)}{\sqrt{1+\lambda}}x)^2} e^{-2\pi it\xi}\,dt. \end{aligned}$$

Making the change of variables

$$2u = \sqrt{1+\lambda}t + 2\frac{(1-\lambda)}{1+\lambda}x, \quad t = \frac{2u}{\sqrt{1+\lambda}} - 2\frac{(1-\lambda)}{\sqrt{1+\lambda}}x, \quad dt = \frac{2^d}{(1+\lambda)^{d/2}}du,$$

$$\begin{aligned} W(\varphi, \varphi_{\sqrt{\lambda}})(x,\xi) &= e^{-\pi\frac{4\lambda}{1+\lambda}x^2} \int_{\mathbb{R}^d} e^{-\pi u^2} e^{-2\pi i(\frac{2u}{\sqrt{1+\lambda}}-\frac{2(1-\lambda)}{1+\lambda}x)\xi} \frac{2^d}{(1+\lambda)^{d/2}}du \\ &= \frac{2^d}{(1+\lambda)^{d/2}} e^{-\pi\frac{4\lambda}{1+\lambda}x^2} e^{4\pi i\frac{1-\lambda}{1+\lambda}x\xi} \int_{\mathbb{R}^d} e^{-\pi u^2} e^{-2\pi iu(\frac{2\xi}{\sqrt{1+\lambda}})}du \\ &= \frac{2^d}{(1+\lambda)^{d/2}} e^{-\pi\frac{4\lambda}{1+\lambda}x^2} e^{-\pi\frac{4}{1+\lambda}\xi^2} e^{4\pi i\frac{1-\lambda}{1+\lambda}x\xi} \end{aligned}$$

as desired. □

We now get

Corollary 1.3.35. *Consider φ and $\varphi_{\sqrt{\lambda}}$ as in the assumptions of Lemma 1.3.34. Then*

$$\mathcal{F}W(\varphi, \varphi_{\sqrt{\lambda}})(\zeta_1,\zeta_2) = \frac{1}{(\lambda+1)^{\frac{d}{2}}} e^{-\frac{\pi}{\lambda+1}\zeta_1^2} e^{-\frac{\pi\lambda}{\lambda+1}\zeta_2^2} e^{-\pi i\frac{\lambda-1}{\lambda+1}\zeta_1\zeta_2}. \tag{1.130}$$

Proof. Formula (1.130) is easily obtained from (1.129) using well-known Gaussian integral formulas; it can also be painlessly obtained from (1.129) and (1.125), we leave the details to the interested reader. □

Observe that $\mathcal{F}_\sigma F(\zeta) = \mathcal{F}F(J\zeta)$. Hence the symplectic Fourier transform of the Wigner distribution (1.129) is given by

$$\mathcal{F}_\sigma W(\varphi, \varphi_{\sqrt{\lambda}})(\zeta_1, \zeta_2) = \frac{1}{(\lambda+1)^{\frac{d}{2}}} e^{-\frac{\pi\lambda}{\lambda+1}\zeta_1^2} e^{-\frac{\pi}{\lambda+1}\zeta_2^2} e^{\pi i \frac{\lambda-1}{\lambda+1}\zeta_1\zeta_2}. \tag{1.131}$$

We now study the STFT of a rescaled Gaussian by a matrix $A \in \mathrm{GL}(d, \mathbb{R})$. Given a function f on $\mathbb{R}^d$ and $A \in \mathrm{GL}(d, \mathbb{R})$, we set

$$f_A(t) = f(At).$$

We also consider the unitary operator $\mathcal{U}_A$ on $L^2(\mathbb{R}^d)$ defined by

$$\mathcal{U}_A f(t) = |\det A|^{1/2} f(At) = |\det A|^{1/2} f_A(t). \tag{1.132}$$

Lemma 1.3.36. *Let $A \in \mathrm{GL}(d, \mathbb{R})$, $\varphi(t) = e^{-\pi t^2}$, then*

$$V_\varphi \varphi_A(x, \xi) = c_A e^{-\pi(I-(A^TA+I)^{-1})x^2} M_{-((A^TA+I)^{-1})x} e^{-\pi(A^TA+I)^{-1}\xi^2},$$

where $c_A = (\det(A^TA + I))^{-1/2}$.

Proof. By definition of the STFT,

$$\begin{aligned} V_\varphi \varphi_A(x, \xi) &= \int_{\mathbb{R}^d} e^{-\pi Ay\cdot Ay} e^{-2\pi i \xi\cdot y} e^{-\pi(y-x)^2}\, dy \\ &= e^{-\pi x^2} \int_{\mathbb{R}^d} e^{-\pi(A^TA+I)y\cdot y + 2\pi x\cdot y} e^{-2\pi i\xi\cdot y}\, dy. \end{aligned}$$

Now, we rewrite the generalized Gaussian above using the translation and dilation operators, that is,

$$e^{-\pi(A^TA+I)y^2+2\pi xy} = c_A^{1/2} e^{\pi(A^TA+I)^{-1}x^2} (T_{(A^TA+I)^{-1}x} \mathcal{U}_{(A^TA+I)^{1/2}})\varphi(y),$$

and use the properties $\mathcal{F}\mathcal{U}_B = \mathcal{U}_{(B^T)^{-1}}\mathcal{F}$, for every $B \in \mathrm{GL}(d, \mathbb{R})$, and $\mathcal{F}T_x = M_{-x}\mathcal{F}$. Thereby,

$$\begin{aligned} V_\varphi \varphi_A(x, \xi) &= c_A^{1/2} e^{-\pi(I-(A^TA+I)^{-1})x^2} \mathcal{F}(T_{(A^TA+I)^{-1}x} \mathcal{U}_{(A^TA+I)^{1/2}})\varphi(\xi) \\ &= c_A e^{-\pi(I-(A^TA+I)^{-1})x^2} M_{-(A^TA+I)^{-1}x} e^{-\pi(A^TA+I)^{-1}\xi^2}, \end{aligned}$$

as desired. □

1.3.7 Short-time Fourier transform of time–frequency representations

To investigate the local properties of the STFT and the Wigner distribution, we will need to compute their STFTs. We remark that the STFT of a function on $\mathbb{R}^{2d}$ is a function on $\mathbb{R}^{4d}$, so one has to be careful when dealing with the STFT $V_gf(x,\xi), (x,\xi) \in \mathbb{R}^{2d}$, of $f \in \mathcal{S}'(\mathbb{R}^d)$, and the STFT $V_\Phi F(z,\zeta), (z,\zeta) \in \mathbb{R}^{4d}$, of $F \in \mathcal{S}'(\mathbb{R}^{2d})$. We write $z = (z_1, z_2) \in \mathbb{R}^{2d}$ and $\zeta = (\zeta_1, \zeta_2) \in \mathbb{R}^{2d}$, when necessary.

Lemma 1.3.37. *Fix a nonzero $\varphi \in \mathcal{S}(\mathbb{R}^d)$ and let $f, g \in \mathcal{S}(\mathbb{R}^d)$.*
(i) *Set $\Phi = V_\varphi\varphi \in \mathcal{S}(\mathbb{R}^{2d})$. Then the STFT of V_gf with respect to the window Φ is given by*

$$V_\Phi(V_gf)(z,\zeta) = e^{-2\pi i z_2\cdot\zeta_2}\overline{V_\varphi g(-z_1-\zeta_2,\zeta_1)}V_\varphi f(-\zeta_2, z_2+\zeta_1), \tag{1.133}$$

where $z = (z_1,z_2), \zeta = (\zeta_1,\zeta_2) \in \mathbb{R}^{2d}$.
(ii) *Let $\Phi = W(\varphi,\varphi) \in \mathcal{S}(\mathbb{R}^{2d})$. Then the STFT of $W(g,f)$ with respect to the window Φ is given by*

$$V_\Phi(W(g,f))(z,\zeta) = e^{-2\pi i z_2\cdot\zeta_2}\overline{V_\varphi g\left(z_1-\frac{\zeta_2}{2}, z_2+\frac{\zeta_1}{2}\right)}V_\varphi f\left(z_1+\frac{\zeta_2}{2}, z_2-\frac{\zeta_1}{2}\right), \tag{1.134}$$

where $z = (z_1,z_2), \zeta = (\zeta_1,\zeta_2) \in \mathbb{R}^{2d}$.

Proof. Before we calculate $V_\Phi(V_gf)(z,\zeta) = \langle V_gf, M_\zeta T_z\Phi\rangle = (V_gf\overline{T_z\Phi})^\wedge(\zeta)$, we rewrite the time–frequency shift $M_\zeta T_z\Phi$. We use Proposition 1.2.15 to evaluate $T_z\Phi$ and find that

$$T_z\Phi(x,\xi) = V_\varphi\varphi(x-z_1,\xi-z_2) = e^{2\pi i(\xi-z_2)\cdot z_1}V_\varphi(M_{z_2}T_{z_1}\varphi)(x,\xi).$$

Now we substitute this expression into the formula for $V_\Phi(V_gf)$, and after rearranging some terms, we apply Proposition 1.2.13 (*ii*) to get

$$\begin{aligned}
V_\Phi(V_gf)(z,\zeta) &= \langle V_gf, M_\zeta T_z\Phi\rangle \\
&= \iint_{\mathbb{R}^{2d}} V_gf(x,\xi)\overline{V_\varphi(M_{z_2}T_{z_1}\varphi)(x,\xi)}e^{-2\pi i[(x\cdot\zeta_1+\xi\cdot\zeta_2)+(\xi-z_2)\cdot z_1]}\,dx\,d\xi \\
&= e^{2\pi i z_1\cdot z_2}\iint_{\mathbb{R}^{2d}} V_gf(x,\xi)\overline{V_\varphi(M_{z_2}T_{z_1}\varphi)(x,\xi)}e^{-2\pi i[\zeta_1\cdot x+(z_1+\zeta_2)\cdot\xi]}\,dx\,d\xi \\
&= e^{2\pi i z_1\cdot z_2}(V_gf\cdot\overline{V_\varphi(M_{z_2}T_{z_1}\varphi)})^\wedge(\zeta_1, z_1+\zeta_2) \\
&= e^{2\pi i z_1\cdot z_2}(V_{(M_{z_2}T_{z_1}\varphi)}f\cdot\overline{V_\varphi g})(-z_1-\zeta_2,\zeta_1) \\
&= e^{-2\pi i z_2\cdot\zeta_2}(V_\varphi f)(-\zeta_2, z_2+\zeta_1)\overline{V_\varphi g(-z_1-\zeta_2,\zeta_1)}.
\end{aligned}$$

This proves (i).

We now work out the case (ii). Using Lemma 1.3.5 and formulas (1.83) and (1.47), we compute

$$\begin{aligned}
V_\Phi(W(g,f))(z,\zeta) &= \langle W(g,f), M_\zeta T_z \Phi\rangle = \langle W(g,f), M_\zeta T_z W(\varphi,\varphi)\rangle \\
&= \int_{\mathbb{R}^{2d}} W(g,f)(x,\xi)\overline{W(\varphi,\varphi)(x-z_1,\xi-z_2)}e^{-2\pi i(x\cdot\zeta_1+\xi\cdot\zeta_2)}\,dx\,d\xi \\
&= 2^{2d}\int_{\mathbb{R}^{2d}} V_{\mathcal{I}f}g(2x,2\xi)\overline{V_{\mathcal{I}T_{z_1}M_{z_2}\varphi}(T_{z_1}M_{z_2}\varphi)(2x,2\xi)} \\
&\quad\times e^{-2\pi i(x\cdot\zeta_1+\xi\cdot\zeta_2)}\,dx\,d\xi \\
&= \int_{\mathbb{R}^{2d}} V_{\mathcal{I}f}g(x,\xi)e^{2\pi i(-x\cdot\frac{\zeta_1}{2}-\xi\cdot\frac{\zeta_2}{2})}\overline{V_{\mathcal{I}T_{z_1}M_{z_2}\varphi}(T_{z_1}M_{z_2}\varphi)(x,\xi)}\,dx\,d\xi \\
&= \int_{\mathbb{R}^{2d}} V_{T_{\frac{\zeta_2}{2}}M_{-\frac{\zeta_1}{2}}\mathcal{I}f}(T_{\frac{\zeta_2}{2}}M_{-\frac{\zeta_1}{2}}g)(x,\xi) \\
&\quad\times \overline{V_{\mathcal{I}T_{z_1}M_{z_2}\varphi}(T_{z_1}M_{z_2}\varphi)(x,\xi)}\,dx\,d\xi \\
&= \langle V_{T_{\frac{\zeta_2}{2}}M_{-\frac{\zeta_1}{2}}\mathcal{I}f}(T_{\frac{\zeta_2}{2}}M_{-\frac{\zeta_1}{2}}g), V_{\mathcal{I}T_{z_1}M_{z_2}\varphi}(T_{z_1}M_{z_2}\varphi)\rangle.
\end{aligned}$$

We can apply the orthogonality relations to obtain

$$V_\Phi(W(g,f))(z,\zeta) = \langle T_{\frac{\zeta_2}{2}}M_{-\frac{\zeta_1}{2}}g, T_{z_1}M_{z_2}\varphi\rangle\overline{\langle T_{\frac{\zeta_2}{2}}M_{-\frac{\zeta_1}{2}}\mathcal{I}f, \mathcal{I}T_{z_1}M_{z_2}\varphi\rangle}.$$

Applying the commutation relations (0.14) to the first factor on the right-hand side, we have

$$\langle T_{\frac{\zeta_2}{2}}M_{-\frac{\zeta_1}{2}}g, T_{z_1}M_{z_2}\varphi\rangle = e^{2\pi i z_2\cdot(z_1-\frac{\zeta_2}{2})}V_\varphi g\left(z_1-\frac{\zeta_2}{2}, z_2+\frac{\zeta_1}{2}\right).$$

Whereas, observing that the operator $\mathcal{I}$ is unitary, the second factor becomes

$$\begin{aligned}
\langle T_{\frac{\zeta_2}{2}}M_{-\frac{\zeta_1}{2}}\mathcal{I}f, \mathcal{I}T_{z_1}M_{z_2}\varphi\rangle &= \langle f, M_{-\frac{\zeta_1}{2}}T_{z_1+\frac{\zeta_2}{2}}M_{z_2}\varphi\rangle \\
&= e^{2\pi i z_2\cdot(z_1+\frac{\zeta_2}{2})}V_\varphi f\left(z_1+\frac{\zeta_2}{2}, z_2-\frac{\zeta_1}{2}\right),
\end{aligned}$$

thereby proving our claim. □

We now generalize formula (1.134) to τ-Wigner distributions defined in (1.109). An easy computation shows the following commutation relation between the operators A_τ in (1.116) and the time–frequency shifts $\pi(y,\eta)$, $(y,\eta)\in\mathbb{R}^{2d}$:

$$\pi(y,\eta)A_\tau = A_\tau\pi\left(-\frac{1-\tau}{\tau}y, -\frac{\tau}{1-\tau}\eta\right), \quad \tau\in(0,1). \tag{1.135}$$

Lemma 1.3.38. *Consider $\tau \in (0,1)$. Let $\varphi_1, \varphi_2 \in \mathcal{S}(\mathbb{R}^d)$ $f, g \in \mathcal{S}(\mathbb{R}^d)$ and set $\Phi_\tau = W_\tau(\varphi_1, \varphi_2)$. Then*

$$V_{\Phi_\tau} W_\tau(g,f)(z,\zeta) = e^{-2\pi i z_2\cdot\zeta_2} V_{\varphi_1} g(z_1 - \tau\zeta_2, z_2 + (1-\tau)\zeta_1) \times \overline{V_{\varphi_2} f(z_1 + (1-\tau)\zeta_2, z_2 - \tau\zeta_1)} \tag{1.136}$$

where $z = (z_1, z_2), \zeta = (\zeta_1, \zeta_2) \in \mathbb{R}^{2d}$.

Proof. The argument follows the pattern of the $\tau = 1/2$-case, already shown in Lemma 1.3.37 (ii). Using the covariant property (1.101) and the representation of the τ-Wigner distribution as an STFT in (1.117),

$$\begin{aligned}
&V_{\Phi_\tau} W_\tau(g,f)(z,\zeta)\\
&= \langle W_\tau(g,f), M_\zeta T_z W_\tau(\varphi_1,\varphi_2)\rangle\\
&= \langle W_\tau(g,f), M_\zeta W_\tau(\pi(z)\varphi_1, \pi(z)\varphi_2)\rangle\\
&= \frac{1}{\tau^{2d}} \int_{\mathbb{R}^{2d}} V_{A_\tau f} g\Big(\frac{1}{1-\tau}x, \frac{1}{\tau}\xi\Big) e^{-2\pi i (x,\xi)\cdot(\zeta_1,\zeta_2)} \overline{V_{A_\tau \pi(z)\varphi_2} \pi(z)\varphi_1\Big(\frac{1}{1-\tau}x, \frac{1}{\tau}\xi\Big)}\, dx\, d\xi\\
&= \frac{(1-\tau)^d}{\tau^d} \int_{\mathbb{R}^{2d}} V_{A_\tau f} g(x,\xi) e^{-2\pi i (\zeta_1,\zeta_2)\cdot((1-\tau)x, \tau\xi)} \overline{V_{A_\tau \pi(z)\varphi_2} \pi(z)\varphi_1(x,\xi)}\, dx\, d\xi.
\end{aligned}$$

To shorten notation, we write

$$c_\tau = \frac{(1-\tau)^d}{\tau^d}.$$

Using formula (1.47), the orthogonality relations (1.44), and the commutation relations between π and A_τ in (1.135), we can write

$$\begin{aligned}
&V_{\Phi_\tau} W_\tau(g,f)(z,\zeta)\\
&= c_\tau \int_{\mathbb{R}^{2d}} V_{\pi(\tau\zeta_2, -(1-\tau)\zeta_1) A_\tau f} \pi(\tau\zeta_2, -(1-\tau)\zeta_1) g \overline{V_{A_\tau \pi(z)\varphi_2} \pi(z)\varphi_1(x,\xi)}\, dx\, d\xi\\
&= c_\tau \langle \pi(\tau\zeta_2, -(1-\tau)\zeta_1) g, \pi(z_1,z_2)\varphi_1\rangle \overline{\langle \pi(\tau\zeta_2, -(1-\tau)\zeta_1) A_\tau f, A_\tau \pi(z_1,z_2)\varphi_2\rangle}\\
&= c_\tau e^{-2\pi i \tau z_2\cdot\zeta_2} \langle g, \pi(z_1 - \tau\zeta_2, z_2 + (1-\tau)\zeta_1)\varphi_1\rangle\\
&\quad \times \overline{\langle A_\tau f, \pi(-\tau\zeta_2, (1-\tau)\zeta_1) A_\tau \pi(z_1,z_2)\varphi_2\rangle}\\
&= c_\tau e^{-2\pi i \tau z_2\cdot\zeta_2} \langle g, \pi(z_1 - \tau\zeta_2, z_2 + (1-\tau)\zeta_1)\varphi_1\rangle\\
&\quad \times \overline{\langle A_\tau f, A_\tau \pi((1-\tau)\zeta_2, -\tau\zeta_1)\pi(z_1,z_2)\varphi_2\rangle}\\
&= e^{-2\pi i \tau z_2\cdot\zeta_2} e^{-2\pi i (1-\tau) z_2\cdot\zeta_2} \langle g, \pi(z_1 - \tau\zeta_2, z_2 + (1-\tau)\zeta_1)\varphi_1\rangle
\end{aligned}$$

$$\times \overline{\langle f, \pi(z_1 + (1-\tau)\zeta_2, z_2 - \tau\zeta_1)\varphi_2\rangle}$$
$$= e^{-2\pi i z_2\cdot\zeta_2} V_{\varphi_1} g(z_1 - \tau\zeta_2, z_2 + (1-\tau)\zeta_1)\overline{V_{\varphi_2} f(z_1 + (1-\tau)\zeta_2, z_2 - \tau\zeta_1)}.$$

The claim is proved. □

We now consider the case $\tau = 0$ and compute the STFT of the (cross-)Rihaczek distribution defined in (1.118).

Lemma 1.3.39 (STFT of the Rihaczek distribution). *Let $\varphi_1, \varphi_2 \in \mathcal{S}(\mathbb{R}^d)$, $f, g \in \mathcal{S}(\mathbb{R}^d)$ and set $\Phi_0 = W_0(\varphi_1, \varphi_2)$. Then*

$$V_{\Phi_0} W_0(g,f)(z,\zeta) = e^{-2\pi i z_2\cdot\zeta_2} V_{\varphi_1} g(z_1, z_2 + \zeta_1)\overline{V_{\varphi_2} f(z_1 + \zeta_2, z_2)} \tag{1.137}$$

where $z = (z_1, z_2)$, $\zeta = (\zeta_1, \zeta_2) \in \mathbb{R}^{2d}$.

Proof. We use the definition in (1.118) and formula (1.47) in the following computations:

$$\begin{aligned}
V_{\Phi_0} W_0(g,f)(z,\zeta) &= \langle W_0(g,f), M_\zeta T_z W_0(\varphi_1,\varphi_2)\rangle \\
&= \int_{\mathbb{R}^{2d}} e^{-2\pi i x\cdot\xi} g(x)\overline{\hat{f}(\xi)} e^{-2\pi i (x\cdot\zeta_1 + \xi\cdot\zeta_2)} e^{2\pi i (x - z_1)\cdot(\xi - z_2)} \overline{\varphi_1(x - z_1)\widehat{\varphi_2}(\xi - z_2)}\, dx\, d\xi \\
&= e^{2\pi i z_1\cdot z_2} \int_{\mathbb{R}^d} e^{-2\pi i x\cdot(z_2 + \zeta_1)} g(x)\overline{\varphi_1(x - z_1)}\, dx \int_{\mathbb{R}^d} \overline{\hat{f}(\xi)} e^{-2\pi i \xi\cdot(z_1 + \zeta_2)} \widehat{\varphi_2}(\xi - z_2)\, d\xi \\
&= e^{2\pi i z_1\cdot z_2} V_{\varphi_1} g(z_1, z_2 + \zeta_1)\overline{V_{\widehat{\varphi_2}} \hat{f}(z_2, -(z_1 + \zeta_2))} \\
&= e^{2\pi i z_1\cdot z_2} V_{\varphi_1} g(z_1, z_2 + \zeta_1)\overline{V_{\varphi_2} f(z_1 + \zeta_2, z_2)} e^{-2\pi i z_2\cdot(z_1 + \zeta_2)} \\
&= e^{-2\pi i z_2\cdot\zeta_2} V_{\varphi_1} g(z_1, z_2 + \zeta_1)\overline{V_{\varphi_2} f(z_1 + \zeta_2, z_2)},
\end{aligned}$$

as desired. □

Corollary 1.3.40 (STFT of the conjugate-Rihaczek distribution). *Let $\varphi_1, \varphi_2 \in \mathcal{S}(\mathbb{R}^d)$, $f, g \in \mathcal{S}(\mathbb{R}^d)$ and set $\Phi_1 = W_1(\varphi_1, \varphi_2)$. Then*

$$V_{\Phi_1} W_1(g,f)(z,\zeta) = e^{-2\pi i z_2\cdot\zeta_2} V_{\varphi_1} g(z_1 - \zeta_2, z_2)\overline{V_{\varphi_2} f(z_1, z_2 - \zeta_1)}, \tag{1.138}$$

where $z = (z_1, z_2)$, $\zeta = (\zeta_1, \zeta_2) \in \mathbb{R}^{2d}$.

Proof. Using the connection between the Rihaczek and the conjugate-Rihaczek distribution in (1.119) and the result of Lemma 1.3.39, we can write

$$\begin{aligned}
V_{\Phi_1} W_1(g,f)(z,\zeta) &= \langle W_1(g,f), M_\zeta T_z W_1(\varphi_1,\varphi_2)\rangle \\
&= \langle \overline{W_0(f,g)}, M_\zeta T_z \overline{W_0(\varphi_2,\varphi_1)}\rangle \\
&= \overline{\langle W_0(f,g), M_{-\zeta} T_z W_0(\varphi_2,\varphi_1)\rangle}
\end{aligned}$$

$$
\begin{aligned}
&= \overline{V_{W_0(\varphi_2,\varphi_1)} W_0(f,g)(z,-\zeta)} \\
&= \overline{e^{2\pi i z_2\cdot\zeta_2} V_{\varphi_2} f(z_1, z_2-\zeta_1)\overline{V_{\varphi_1} g(z_1-\zeta_2, z_2)}} \\
&= e^{-2\pi i z_2\cdot\zeta_2} V_{\varphi_1} g(z_1-\zeta_2, z_2)\overline{V_{\varphi_2} f(z_1, z_2-\zeta_1)}.
\end{aligned}
$$

The proof is completed. □

Remark 1.3.41.

(i) Heuristically, formulas (1.137) and (1.138) can be inferred by putting $\tau = 0$ and $\tau = 1$, respectively, into expression (1.136).

(ii) The STFT of a multilinear version of the Rihaczek distribution was computed in [13, Lemma 3.3], cf. formula (3.3). However, there is a flaw in the phase factor of that formula. Indeed, the exponential $e^{2\pi i u_0\cdot(u_1+\cdots+u_m)}$ should be replaced by $e^{2\pi i \sum_{i=1}^m u_i\cdot v_i}$, as the linear case $m = 1$ in (1.137) shows.

2 Function spaces

The modern functional analysis requires a precise setting of the involved function spaces as a preliminary step in the study of mathematical problems. The Lebesgue spaces L^p, $1 \leq p \leq \infty$, play a relevant role in this connection. Once the short-time Fourier transform (STFT) $V_g f(x, \xi)$ is introduced as a basic tool of the time–frequency analysis, it is then natural to define Lebesgue spaces involving simultaneously x and ξ variables. Namely, for a function or distribution f, one defines for a couple of indices p, q, with $1 \leq p, q \leq \infty$, and a weight function $m(x, \xi)$ in $\mathbb{R}^{2d}$, the mixed Lebesgue norm

$$\|f\|_{M_m^{p,q}(\mathbb{R}^d)} = \left(\int_{\mathbb{R}^d} \left(\int_{\mathbb{R}^d} |V_g f(x, \xi)|^p m(x, \xi)^p \, dx \right)^{\frac{q}{p}} d\xi \right)^{\frac{1}{q}}.$$

According to H. G. Feichtinger, in the early 1980s, the corresponding $M_m^{p,q}(\mathbb{R}^d)$ were called modulation spaces. If $p = q$, the notation is shortened to $M_m^p(\mathbb{R}^d)$. If $m \equiv 1$, the spaces are simply denoted by $M^{p,q}(\mathbb{R}^d)$, $M^p(\mathbb{R}^d)$.

It is important to acknowledge that H. G. Feichtinger was the scientist introducing the word *modulation* into the mathematical standard terminology. Unlike engineers who modulate a given frequency by changing the amplitude, he was thinking of imposing a particular frequency on a given function and looking at *modulation invariant Segal algebras*, see, e. g., [118] or, in connection with the factorization problem of Segal algebras, the paper [129], where this property played an important role.

The STFT of f with window g can be rewritten as a convolution of f with a modulated version of g, see formula (1.28) in the previous chapter. So modulation spaces got their name because they *capture the behavior/effect of modulation* of the window function g in the filtering process (convolution).

They were used for signal theory in the past years, and still they are subject of intensive study in this context. In the present book we adopt them for the analysis of the partial differential operators, with a detailed investigation of the properties necessary in this new perspective.

Let us shortly compare $M_m^{p,q}(\mathbb{R}^d)$ with other function spaces in the literature. The Sobolev spaces $W_s^p(\mathbb{R}^d)$, based on Lebesgue L^p spaces, have nowadays a dominant role in the theory of the partial differential equations. In the case $p = q = 2$, the modulation space $M_m^2(\mathbb{R}^d)$ with a suitable weight m may reproduce the standard Sobolev spaces $H^s(\mathbb{R}^d) = W_s^2(\mathbb{R}^d)$, see Proposition 2.3.6 in the sequel. The situation is drastically different in the case of potential Sobolev spaces $W_s^p(\mathbb{R}^d)$ with $p \neq 2$, for which we cannot establish exact identities, but only inclusions, see Proposition 2.3.21 (for more general inclusion properties, we refer to the paper by Sugimoto and Tomita [274]).

In fact, as it concerns local regularity, we may say that $f \in M^{p,q}(\mathbb{R}^d)$ is of class $\mathcal{F}L^q(\mathbb{R}^d)$. More precisely, in Proposition 2.3.26 we shall see that if $f \in \mathcal{S}'(\mathbb{R}^d)$ is compactly supported, then we have $f \in M^{p,q}(\mathbb{R}^d)$ if and only if its Fourier transform $\hat{f}$ belongs to $L^q(\mathbb{R}^d)$. From this point of view, ancestors of modulation spaces are the spaces

https://doi.org/10.1515/9783110532456-003

$B_{q,k}(\mathbb{R}^d)$, used by Hörmander (see [196, Section 10.1]) and others in the study of the fundamental solutions of the equations with constant coefficients. Namely, $f \in B_{q,k}(\mathbb{R}^d)$ means that $\hat{f}k \in L^q(\mathbb{R}^d)$ (definition of weight function $m(x,\xi) = k(\xi)$), so modulation spaces and $B_{q,k}$ spaces locally coincide, apart from notation and definition of weight functions. The properties of functions or distributions in $\mathcal{F}L^q(\mathbb{R}^d)$ represent one of the main problems of harmonic analysis, but once the mysterious $\mathcal{F}L^q$ regularity is assumed as true definition, several computations simplify. In fact, the key idea for the modulation spaces is to join the local $\mathcal{F}L^q$ regularity with an L^p behavior at infinity; see again Proposition 2.3.26, where it is also proved that if $\hat{f}$ is compactly supported then $f \in M^{p,q}(\mathbb{R}^d)$ if and only if $f \in L^p(\mathbb{R}^d)$. Such an interplay between local and asymptotic properties provides a nice theory for $M^{p,q}$ spaces, a basic example being the well-known Feichtinger algebra $M^1(\mathbb{R}^d)$, the smallest Banach algebra invariant with respect to pointwise and convolution products, see Propositions 2.4.19 and 2.4.23.

Let us also emphasize the action of the Fourier transform on $M^{p,q}_m(\mathbb{R}^d)$, namely for $p = q$ we have (cf. Proposition 2.3.28)

$$\mathcal{F}M^p(\mathbb{R}^d) = M^p(\mathbb{R}^d).$$

In this chapter we shall also consider the so-called Wiener amalgam spaces. Though not directly related to Fourier transform, they enter the picture in a natural way. Namely, one defines the Wiener amalgam space $W(B,C)$, with B and C being Banach spaces satisfying certain assumptions, as the set of all distributions $f \in \mathcal{S}'(\mathbb{R}^d)$ with local regularity expressed by B and such that for a window $g \in \mathcal{D}(\mathbb{R}^d)$ the function $F_g(x) = \|fT_xg\|_B$ belongs to the global space C, see Definition 2.4.7 for details. Wiener amalgam spaces (by a different name then) in full generality have been introduced in [122]. The role of the function g as a window is strictly related to that of g in the definition of the STFT V_gf, and the basic connection is given by

$$\mathcal{F}M^{p,q} = W(\mathcal{F}L^p, L^q).$$

So that, for $p = q$, we obtain

$$M^p = W(\mathcal{F}L^p, L^p),$$

and in particular

$$M^1 = W(\mathcal{F}L^1, L^1),$$

see the examples following Definition 2.4.7. By using properties of the Wiener amalgams, we can infer results for inclusion, convolution and multiplication relations for $M^{p,q}_m$.

The next chapters of the book will show that the use of modulation spaces offers relevant advantages in the study of partial differential operators, with respect to the standard frame of Sobolev spaces. Natural applications of $M_m^{p,q}$ will concern, in particular, equations globally defined on $\mathbb{R}^d$ and requiring a precise Fourier analysis of the solutions. On the other hand, drawbacks may obviously originate from the more involved definition of $M_m^{p,q}$. In particular, a complication in the $M_m^{p,q}$ setting is represented by the behavior of the norm under dilations. The issue is essential for the applications to linear and nonlinear partial differential equations, and we shall devote to it a large part of the chapter, see Theorem 2.6.6.

Let us list more precisely the contents of the next sections. In Section 2.1 we give the definition of the weight functions under our consideration and study their properties. In harmonic analysis the term *weight function* is usually reserved to a submultiplicative function

$$v(z_1 + z_2) \leq v(z_1)v(z_2), \quad \forall z_1, z_2 \in \mathbb{R}^{2d},$$

basic example being $v(z) = (1 + |z|)^s$, $s \geq 0$. In the definition of modulation spaces, we shall refer, more generally, to v-moderate weights $m(z)$, $z \in \mathbb{R}^{2d}$, satisfying for some $C > 0$ the inequality

$$m(z_1 + z_2) \leq Cv(z_1)m(z_2), \quad \forall z_1, z_2 \in \mathbb{R}^{2d}.$$

Note that in the theory of partial differential equations this is also named *temperance property*, see Hörmander [196, Section 10.1] and Nicola-Rodino [236, Section 1.1].

Submultiplicative weights (also called *Beurling weights*) are quite important. However, the pointwise inverse of such a weight is not submultiplicative anymore. But any positive or negative power of a submultiplicative function is, of course, a moderate weight. It is not wrong to think of moderate functions as mostly being of that type.

In Section 2.2 we consider mixed-norm Lebesgue spaces. In fact, the definition of $M_m^{p,q}$ involves a repeated integration of this type, with different exponents in the x and ξ variables if $p \neq q$. So, as a preparation for the study of $\|f\|_{M_m^{p,q}}$, we collect here the properties of the mixed-norms used in the sequel.

Section 2.3 is the core of the chapter: the definition of $M_m^{p,q}$ is given here and basic properties are proved, in particular, it is shown that the definition does not depend on the choice of the window g in the expression of the STFT $V_g f$. Banach properties, relations with Sobolev spaces, density of the Schwartz class $\mathcal{S}$, duality, and interpolation are discussed. An equivalent definition of $M_m^{p,q}(\mathbb{R}^d)$ is also given in terms of a smooth partition of unity in the dual space $\mathbb{R}^d_\xi$. This provides alternative norms for $M_m^{p,q}(\mathbb{R}^d)$, which turn out to be useful in the proof of the local $\mathcal{F}L^q$ regularity and the global L^p behavior. Finally, the action of the Fourier transform on modulation spaces is considered. As observed before, this introduces in a natural way to the Wiener amalgam spaces, studied in the subsequent section.

The first part of Section 2.4 is a short survey on the general theory of the Wiener amalgams. Relevant examples are $W(L^p, L^q_m)$, $1 \leq p, q \leq \infty$, that for $p = q$ reduce to the standard Lebesgue spaces, $W(L^p, L^p_m) = L^p_m$. Coming back to modulation spaces, besides the fundamental identities $\mathcal{F}M^{p,q} = W(\mathcal{F}L^p, L^q)$, $M^p = W(\mathcal{F}L^p, L^p)$, it is observed that if $f \in M^{p,q}_m(\mathbb{R}^d)$, then the STFT $V_g f$ is in the Wiener amalgam $W(\mathcal{F}L^1, L^{p,q}_m)(\mathbb{R}^{2d})$.

The rather long analysis of norms of dilations begins in Section 2.5, concerning Wiener amalgams, and it is concluded in Section 2.6, on modulation spaces. The problem is to study the dependence on $\lambda > 0$ of $\|f_\lambda\|_{M^{p,q}}$, where $f_\lambda(t) = f(\lambda t)$. More generally, one may consider $\|f_A\|_{M^{p,q}}$, $A \in \mathrm{GL}(d, \mathbb{R})$, with $f_A(t) = f(At)$. The main result of the section gives sharp estimates of the type

$$\|f_\lambda\|_{M^{p,q}} \lesssim \lambda^r \|f\|_{M^{p,q}},$$

where r depends on d, p, q and takes different values for $\lambda \to 0^+$, $\lambda \to \infty$.

Section 2.7 proves the sharpness of the results of Section 2.3 for convolution, inclusion, and multiplication in modulation spaces. Here the definition of $M^{p,q}_m$ is also extended to the case $0 < p, q < 1$ in the quasi-Banach setting.

2.1 Weight functions

A weight function m on $\mathbb{R}^{2d}$ is a positive, locally bounded, and measurable function on $\mathbb{R}^{2d}$, that is, $m(z) > 0$, for every $z \in \mathbb{R}^{2d}$, and $m\chi_K \in L^\infty(\mathbb{R}^{2d})$, for every compact set $K \subset \mathbb{R}^{2d}$, χ_K being the characteristic function of the set K.

Weights on the time–frequency space $\mathbb{R}^{2d}$ may quantify the growth and smoothness of functions on $\mathbb{R}^d$, as we shall see in the sequel.

We say that two weights m_1, m_2 on $\mathbb{R}^{2d}$ are **equivalent** if there exist positive constants C_1, C_2 such that

$$C_1 m_1(z) \le m_2(z) \le C_2 m_1(z), \quad \forall z \in \mathbb{R}^{2d}. \tag{2.1}$$

Among many possible choices of weights, we shall focus on the following types.

Definition 2.1.1. A weight function v on $\mathbb{R}^{2d}$ is named **submultiplicative** if

$$v(z_1 + z_2) \le v(z_1)v(z_2), \quad \forall z_1, z_2 \in \mathbb{R}^{2d}. \tag{2.2}$$

For a fixed submultiplicative weight v, a weight function m on $\mathbb{R}^{2d}$ is called v**-moderate** if there exists a positive constant C such that

$$m(z_1 + z_2) \le Cv(z_1)m(z_2), \quad \forall z_1, z_2 \in \mathbb{R}^{2d}. \tag{2.3}$$

The set of all v-moderate weights on $\mathbb{R}^{2d}$ is denoted by $\mathcal{M}_v(\mathbb{R}^{2d})$.

Important examples of submultiplicative weights are the functions $v(z) = (1+|z|)^s$, for $s \ge 0$. In fact, using the triangle inequality $|z_1 + z_2| \le |z_1| + |z_2|$, $z_1, z_2 \in \mathbb{R}^{2d}$, we infer

$$1 + |z_1 + z_2| \le 1 + |z_1| + |z_2| \le (1 + |z_1|)(1 + |z_2|)$$

so that

$$(1+|z_1+z_2|)^s \le (1+|z_1|)^s(1+|z_2|)^s, \quad \forall z_1, z_2 \in \mathbb{R}^{2d}.$$

Other examples are the exponential weights $v(z) = e^{a|z|}$, for any $a \ge 0$, as it follows immediately from $|z_1+z_2| \le |z_1| + |z_2|$, and the subexponential weights $v(z) = e^{a|z|^b}$, $a \ge 0$, $0 \le b < 1$. The last case can be proved in this way. First, we write $|z_1+z_2|^b \le (|z_1|+|z_2|)^b$ and then we notice that

$$(|z_1|+|z_2|)^b \le |z_1|^b + |z_2|^b.$$

The claim follows by observing that the function $f(t) = (t+1)^b - t^b - 1$ for $t \ge 0$ is a decreasing function whenever $b < 1$.

Further examples of submultiplicative weights are the logarithmic functions

$$v(z) = \log(e+|z|)^t, \quad t \ge 0,\ z \in \mathbb{R}^{2d}.$$

If v_1 and v_2 are submultiplicative weights, then it is straightforward to check that the product $v_1 v_2$ enjoys the same property. Therefore the functions

$$v_{a,b,s,t}(z) = (1+|z|)^s e^{a|z|^b} \log(e+|z|)^t, \quad a,s,t \ge 0,\ 0 \le b \le 1$$

are submultiplicative weights. It can be checked easily that the weights

$$m_{a,b,s,t}(z) = (1+|z|)^s e^{a|z|^b} \log(e+|z|)^t, \quad a,s,t \in \mathbb{R},\ 0 \le b \le 1$$

are $v_{a,b,s,t}$-moderate.

In the sequel we shall mainly work with the weight function $\langle z\rangle^s = (1+z^2)^{s/2}$, which is equivalent to $(1+|z|)^s$, but $\langle z\rangle^s$ fails to be submultiplicative for $s \ge 0$. In fact, Peetre's inequality (0.4) (which is sharp!) gives

$$\langle z_1+z_2\rangle^s \le 2^s \langle z_1\rangle^s \langle z_2\rangle^s, \quad s \ge 0.$$

Lemma 2.1.2. *If a weight function v on $\mathbb{R}^{2d}$ satisfies*

$$v(z_1+z_2) \le Cv(z_1)v(z_2), \quad \forall z_1, z_2 \in \mathbb{R}^{2d} \tag{2.4}$$

for a suitable constant $C > 0$, then $\tilde{v}(z) = Cv(z)$ is a submultiplicative weight.

Proof. We have, by definition of $\tilde{v}$ and using (2.4),

$$\tilde{v}(z_1+z_2) = Cv(z_1+z_2) \le C^2 v(z_1)v(z_2) = \tilde{v}(z_1)\tilde{v}(z_2),$$

as desired. □

Hence the function $v(z) = 2^s\langle z\rangle^s$, $s \geq 0$, is a submultiplicative weight. Since the constant 2^s will be irrelevant when considering weighted function spaces, we shall mainly use the weights

$$v_s(z) = \langle z\rangle^s \tag{2.5}$$

and, by abuse of notation, we shall denote by $\mathcal{M}_{v_s}(\mathbb{R}^{2d})$ the class of v_s-moderate weights, which indeed coincides with the class $\mathcal{M}_{2^s v_s}(\mathbb{R}^{2d})$.

For $s \leq 0$, the weights v_s are $v_{|s|}$-moderate by Peetre's inequality

$$\langle z_1 + z_2\rangle^s \leq 2^s\langle z_1\rangle^{|s|}\langle z_2\rangle^s,$$

i. e., $v_s \in \mathcal{M}_{v_{|s|}}(\mathbb{R}^{2d})$.

Equivalent weights originate the same modulation spaces. This suggests considering only functions v which are continuous, since any submultiplicative weight is equivalent to a continuous function.

Lemma 2.1.3. *Consider a submultiplicative weight v and a function ψ such that $\psi \in C_c(\mathbb{R}^{2d}) \setminus \{0\}$, $\psi \geq 0$, then*

$$\tilde{v}(z) = v * \psi(z)$$

is a weight function equivalent to v. Moreover, $\tilde{v}$ is v-moderate.

Proof. Observe that $\tilde{v} > 0$ is well defined and continuous by Young's inequality and the smoothing property of the convolution. Moreover, $\tilde{v}$ is equivalent to v. Let K be the support of ψ, then

$$\tilde{v}(z) = \int_K v(z-y)\psi(y)\,dy \leq v(z)\int_K v(-y)\psi(y)\,dy \leq v(z)\|v\|_{L^\infty(-K)}\|\psi\|_{L^1(K)} = Cv(z),$$

with $C = \|v\|_{L^\infty(-K)}\|\psi\|_{L^1(K)}$. Vice versa, consider a compact set $\tilde{K} \subset K$ where $\psi(z) > 0$ for every $z \in \tilde{K}$, then

$$\begin{aligned}\int_{\tilde{K}} v(z)\,dy &= \int_{\tilde{K}} v(z-y+y)\frac{\psi(y)}{\psi(y)}\,dy \leq \int_{\tilde{K}} v(z-y)\psi(y)\frac{v(y)}{\psi(y)}\,dy\\ &\leq \|v/\psi\|_{L^\infty(\tilde{K})}\int_{\tilde{K}} v(z-y)\psi(y)\,dy \leq \|v/\psi\|_{L^\infty(\tilde{K})}\int_K v(z-y)\psi(y)\,dy\\ &= \|v/\psi\|_{L^\infty(\tilde{K})}\tilde{v}(z)\end{aligned}$$

hence $v(z) \leq C\tilde{v}(z)$, with $C = (1/\mathrm{meas}(\tilde{K}))\|v/\psi\|_{L^\infty(\tilde{K})}$.

Finally, observe that $\tilde{v}$ is v-moderate:

$$\tilde{v}(z_1+z_2) = \int_K v(z_1+z_2-y)\psi(y)\,dy \leq v(z_1)\tilde{v}(z_2), \quad z_1, z_2 \in \mathbb{R}^{2d}.$$ □

Observe that if v is continuous and satisfies (2.2), then $v(0) = v(0+0) \leq v(0)^2$ and, dividing by $v(0) > 0$, it follows that $v(0) \geq 1$.

Another desirable property is the following symmetry: for $z = (x, \xi) \in \mathbb{R}^{2d}$,

$$m(x, \xi) = m(-x, \xi) = m(x, -\xi) = m(-x, -\xi)$$

enjoyed by the weights $m_{a,b,s,t}$.

Other weights can be constructed by restricting weight functions on $\mathbb{R}^{2d}$ to $\mathbb{R}^m$, $m < 2d$. For instance, we can consider only weights in the time (resp. frequency) variables

$$m(x, \xi) = \langle x \rangle^s, \quad m(x, \xi) = \langle \xi \rangle^s, \; x, \xi \in \mathbb{R}^d, \; s \in \mathbb{R}.$$

Further examples we shall use are

$$m(x, \xi) = \langle x \rangle^s \langle \xi \rangle^t, \quad s, t \in \mathbb{R}.$$

We end this section by observing that submultiplicative weights grow at most exponentially.

Lemma 2.1.4. *Let v be a submultiplicative weight on $\mathbb{R}^{2d}$. Then v grows at most exponentially. That is, there exist constants $a \geq 0$, $C > 0$ such that*

$$v(z) \leq Ce^{a|z|}, \quad z \in \mathbb{R}^{2d}.$$

Proof. Since any submultiplicative function is equivalent to a continuous one by Lemma 2.1.3, we assume without loss of generality that v is continuous.

Define a by $e^a = \sup_{|z| \leq 1} v(z)$. Because v is continuous, it satisfies $v(0) \geq 1$ so that $a \geq 0$. Take any $z \in \mathbb{R}^{2d}$ and select $n \in \mathbb{N}_+$ such that $n - 1 < |z| \leq n$. Then $|z|/n \leq 1$ and using the submultiplicativity property

$$v(z) = v\left(n\frac{z}{n}\right) \leq \left(v\left(\frac{z}{n}\right)\right)^n \leq e^{an} < e^{a(|z|+1)} = Ce^{a|z|},$$

with $C = e^a$, as desired. □

In the sequel we shall only work with submultiplicative weights v which are even, that is, $v(-z) = v(z)$, for $z \in \mathbb{R}^{2d}$. Under such an assumption, v-moderate weights satisfy the following inequalities.

Lemma 2.1.5. *If $m \in \mathcal{M}_v(\mathbb{R}^{2d})$ then there exists a constant $C > 0$ such that*

$$\frac{1}{C}\frac{m(z)}{v(t)} \leq m(z - t) \leq Cv(t)m(z), \quad t, z \in \mathbb{R}^{2d}. \tag{2.6}$$

In particular,

$$\frac{1}{C}\frac{1}{v(z)} \leq m(z) \leq Cv(z), \quad z \in \mathbb{R}^{2d},$$

and for every $z \in l + [0,1]^{2d}$, $l \in \mathbb{Z}^{2d}$,

$$\frac{1}{C'} m(l) \leq m(z) \leq C' m(l). \tag{2.7}$$

Proof. Using (2.3), we write $m(z-t+t) \leq Cv(t)m(z-t)$ and obtain the left-hand side inequality in (2.6). Similarly, $m(z-t) \leq Cv(-t)m(z) = Cv(t)m(z)$, since v is even, and the right-hand side inequality in (2.6) is shown. The latter inequalities follow from (2.6), exchanging the roles of the variables t and z and setting $t = 0$. Finally, writing $z = l+z'$, with $z' \in [0,1]^{2d}$, we have $m(z'+l) \leq Cm(l)v(z')$ and $m(l) \leq Cm(l+z')v(-z')$. So we can choose $C' = C \max_{z' \in [0,1]^{2d}} v(z')$ and obtain (2.7). □

2.2 Mixed norm spaces

The Lebesgue mixed norm spaces were studied in detail by Benedek and Panzone in their work [10]. Such spaces can be considered as an extension of the classical L^p-spaces, enjoying similar properties, which we summarize below. For the proofs we refer mainly to the original paper [10].

We limit ourselves to presenting their definition in the Euclidean space $\mathbb{R}^d$, $d \in \mathbb{N}_+$, endowed with the Lebesgue measure $dx_1 \cdots dx_d$. This can be extended to more general measure spaces, in particular, we recall the discrete case with the counting measure. Given $p_i \in (0,\infty]$, $i = 1,\dots,d$, and a weight function m on $\mathbb{R}^d$, we say that a measurable function $f : \mathbb{R}^d \to \mathbb{C}$ belongs to $L_m^{p_1,\dots,p_d}(\mathbb{R}^d)$ if the following norm is finite:

$$\|f\|_{L_m^{p_1,p_2,\dots,p_d}} = \left(\int_{\mathbb{R}} \cdots \left(\int_{\mathbb{R}} |f(x_1,\dots,x_d)|^{p_1} m(x_1,\dots,x_d)^{p_1}\, dx_1 \right)^{\frac{p_2}{p_1}} \cdots dx_d \right)^{\frac{1}{p_d}} \tag{2.8}$$

with standard modification when $p_i = \infty$ for some i. That is, we assume that the number obtained by taking successively the p_1-norm in x_1, the p_2-norm in x_2, …, the p_d-norm in x_d, in that order for the function fm, is finite.

Observe that if $p_i = p$ for every $i = 1,\dots,d$, $m(x) \equiv 1$, then $L_m^{p_1,p_2,\dots,p_d}(\mathbb{R}^d) = L^p(\mathbb{R}^d)$, the standard Lebesgue space. If $p_i = p$ for every $i = 1,\dots,d$, then $L_m^{p_1,p_2,\dots,p_d}(\mathbb{R}^d) = L_m^p(\mathbb{R}^d)$ is the weighted L^p-space of measurable functions f such that

$$\|f\|_{L_m^p} = \|fm\|_{L^p} < \infty.$$

If we replace $\mathbb{R}$ with a countable set X_i, $i = 1,\dots,d$, endowed with the counting measure, then the weighted mixed norm spaces are denoted by $\ell_m^{p_1,\dots,p_d}(X)$, $X = \prod_{i=1}^d X_i$, $m = (m_\lambda) : X \to \mathbb{C}$, and are spaces of sequences $a = (a_\lambda)$, $\lambda = (\lambda_1,\dots,\lambda_d) \in X$ such that

$$\|a\|_{\ell_m^{p_1,\dots,p_d}} = \left(\sum_{\lambda_1 \in X_1} \cdots \left(\sum_{\lambda_d \in X_d} |(am)(\lambda_1,\dots,\lambda_d)|^{p_1} \right)^{\frac{p_2}{p_1}} \cdots \right)^{\frac{1}{p_d}} < \infty.$$

We next recall a particular instance of such discrete mixed norm spaces that we shall often use in the sequel.

Definition 2.2.1. For $1 \le p, q \le \infty$, $m \in \mathcal{M}_v(\mathbb{R}^{2d})$, the space $\ell^{p,q}_m(\mathbb{Z}^{2d})$ consists of all sequences $a = (a_{k,n})_{k,n\in\mathbb{Z}^d}$ such that the norm

$$\|a\|_{\ell^{p,q}_m} = \Bigg(\sum_{n\in\mathbb{Z}^d}\Bigg(\sum_{k\in\mathbb{Z}^d}|a_{k,n}|^p m(k,n)^p\Bigg)^{\frac{q}{p}}\Bigg)^{\frac{1}{q}}$$

is finite.

Remark 2.2.2. Writing $\lambda = (k,n) \in \mathbb{Z}^{2d}$, by (2.7) one can easily show the norm-equivalence

$$\Bigg\|\sum_{\lambda\in\mathbb{Z}^{2d}} a_\lambda T_\lambda \chi_{[0,1]^{2d}}\Bigg\|_{L^{p,q}_m} \asymp \|a\|_{\ell^{p,q}_m}. \tag{2.9}$$

Next, let us list the main properties of the mixed norm spaces. We exhibit the Euclidean case, the discrete one is analogous. To simplify notations, we use the multiindex convention

$$P = (p_1, \dots, p_d), \quad P' = (p_1', \dots, p_d'), \quad \frac{1}{p_i} + \frac{1}{p_i'} = 1,\ i = 1, \dots, d.$$

We write $1 \le P \le \infty$ if $1 \le p_i \le \infty$ for every $i = 1, \dots, d$ (and similarly if the inequalities are strict). Hence the mixed norm space $L^{p_1,\dots,p_d}_m(\mathbb{R}^d)$ is shortened to $L^P_m(\mathbb{R}^d)$.

Theorem 2.2.3. *Consider a weight function m on $\mathbb{R}^d$. Then*

(i) *For $1 \le P \le \infty$, $L^P_m(\mathbb{R}^d)$ is a Banach space, and every sequence converging in $L^P_m(\mathbb{R}^d)$ contains a subsequence convergent almost everywhere to the limit function.*

(ii) *If f is a measurable function over $\mathbb{R}^d$, then*

$$\|f\|_{L^P_m} = \sup_{g\in L^{P'}_{\frac{1}{m}},\ \|g\|=1}\Bigg|\int_{\mathbb{R}^d} f(x)g(x)\,dx\Bigg|, \tag{2.10}$$

where $dx = dx_1 \cdots dx_d$.

(iii) *For $1 < P < \infty$, $L^P_m(\mathbb{R}^d)$ is reflexive and for $1 \le P < \infty$, $(L^P_m(\mathbb{R}^d))' = L^{P'}_{\frac{1}{m}}(\mathbb{R}^d)$.*

(iv) (Hölder's inequality) *If $1 \le P \le \infty$, $f \in L^P_m(\mathbb{R}^d)$, $g \in L^{P'}_{1/m}(\mathbb{R}^d)$, then $fg \in L^1(\mathbb{R}^d)$, with*

$$\Bigg|\int_{\mathbb{R}^n} f(x)g(x)\,dx\Bigg| \le \|f\|_{L^P_m}\|g\|_{L^{P'}_{\frac{1}{m}}}. \tag{2.11}$$

(v) (Young's inequality) *Let m be a v-moderate weight. Consider $1 \le P, Q, R \le \infty$ such that*

$$\frac{1}{P} + \frac{1}{Q} = 1 + \frac{1}{R},$$

that is, for $P=(p_1,\dots,p_d)$, $Q=(q_1,\dots,q_d)$, $R=(r_1,\dots,r_d)$,

$$\frac{1}{p_i}+\frac{1}{q_i}=1+\frac{1}{r_i},\qquad i=1,\dots d.$$

If $f\in L^P_v(\mathbb{R}^d)$, $g\in L^Q_m(\mathbb{R}^d)$ *then* $f*g\in L^R_m(\mathbb{R}^d)$, *with*

$$\|f*g\|_{L^R_m}\le C\|f\|_{L^P_v}\|g\|_{L^Q_m}. \tag{2.12}$$

Proof. The proofs when $m\equiv 1$ are carried out in [10] by using similar arguments as for standard L^p-spaces. The weight m does not change the proofs of (i)–(iv), one just has to replace the function f with fm and follow the standard pattern.

Let us spend a few words for Young's inequality. It is enough to prove it for non-negative functions f and g. Here m is a v-moderate weight, so that in view of (2.3),

$$(f*g)(x)m(x)=\int_{\mathbb{R}^d} m(x-y+y)f(x-y)g(y)\,dy \tag{2.13}$$

$$\le C\int_{\mathbb{R}^d} v(x-y)m(y)f(x-y)g(y)\,dy \tag{2.14}$$

$$=C[(fv)*(mg)](x). \tag{2.15}$$

Hence using the unweighted Young's inequality,

$$\|f*g\|_{L^R_m}\le C\|(fv)*(mg)\|_{L^R}\le C\|fv\|_{L^P}\|mg\|_{L^Q}=C\|f\|_{L^P_v}\|g\|_{L^Q_m},$$

as desired. □

If we choose $P=Q=R=1$, $m=v$ submultiplicative, then Young's inequality gives

$$L^1_v(\mathbb{R}^d)*L^1_v(\mathbb{R}^d)\hookrightarrow L^1_v(\mathbb{R}^d),$$

that is, $L^1_v(\mathbb{R}^d)$ *is a Banach algebra under the convolution product.*

The following result will be useful in the sequel.

Lemma 2.2.4. *Consider* $1\le P\le\infty$ *and* $m\in\mathcal{M}_v$. *For* $z\in\mathbb{R}^d$, *the translation operator* T_z *is bounded on* $L^P_m(\mathbb{R}^d)$ *with*

$$\|T_zf\|_{L^P_m}\le Cv(z)\|f\|_{L^P_m},\quad \forall f\in L^P_m(\mathbb{R}^d). \tag{2.16}$$

Proof. Consider $f\in L^P_m(\mathbb{R}^d)$, then using $m(x)\le Cv(z)m(x-z)$,

$$\|T_zf\|_{L^P_m}=\|f(\cdot-z)m(\cdot)\|_{L^P}\le Cv(z)\|f(\cdot-z)m(\cdot-z)\|_{L^P}$$
$$=Cv(z)\|fm\|_{L^P}=Cv(z)\|f\|_{L^P_m}.$$

This concludes the proof. □

2.3 Modulation spaces

Modulation spaces were introduced and studied by H. G. Feichtinger in the early 1980s. The original setting was that of locally compact Abelian (LCA) groups [117]. The main point was to define smoothness over LCA groups, where one cannot use dilation or differentiation. The corresponding spaces in Euclidean spaces were introduced somewhat later [126], to model the family of the Besov spaces from H. Triebel [297] and J. Peetre [240]. The conditions formulated in the original paper [117, 126] already include spaces of ultra-distributions introduced by N. Teofanov in [282], as noted in [127].

The subject is so vast that an entire book should be devoted to its study. Here we limit ourselves to considering the case of weights with at most polynomial growth, so that we remain in the field of tempered distributions.

Notice that a large part of the results we shall present stay valid also for weights with at most exponential growth, replacing the Schwartz class and the space of tempered distributions with suitable Gelfand–Shilov classes and their duals.

From now on, we assume that there exist $s_0, C > 0$ such that the weight m satisfies

$$m(z) \leq C\langle z\rangle^{s_0} = Cv_{s_0}(z), \quad z \in \mathbb{R}^{2d}. \tag{2.17}$$

The STFT $V_g f$ plays the central role in the definition of modulation spaces: the measure of its decay on the phase space provides information on smoothness and decay of the function f.

Definition 2.3.1. Fix $g \in \mathcal{S}(\mathbb{R}^d) \setminus \{0\}$, a weight function $m \in \mathcal{M}_v$. Consider the indices

$$P = (p_1, \dots, p_d), \quad Q = (q_1, \dots, q_d), \quad p_i, q_i \in [1, \infty], \; i = 1, \dots, d. \tag{2.18}$$

Then

$$M_m^{P,Q}(\mathbb{R}^d) = \{f \in \mathcal{S}'(\mathbb{R}^d) : \|f\|_{M_m^{P,Q}(\mathbb{R}^d)} < \infty\}, \tag{2.19}$$

where

$$\|f\|_{M_m^{P,Q}(\mathbb{R}^d)} = \|V_g f\|_{L_m^{p_1,\dots,p_d,q_1,\dots,q_d}}.$$

If $p = p_1 = \dots = p_d, q = q_1 = \dots = q_d$, we write $M_m^{p,q}$ in place of $M_m^{P,Q}$. In this case we have

$$\|f\|_{M_m^{p,q}(\mathbb{R}^d)} = \left(\int_{\mathbb{R}^d}\left(\int_{\mathbb{R}^d} |V_g f(x,\xi)|^p m(x,\xi)^p \, dx\right)^{\frac{q}{p}} d\xi\right)^{\frac{1}{q}} \tag{2.20}$$

(with obvious modifications when $p = \infty$ or $q = \infty$). These are the original modulation spaces introduced by Feichtinger in [126]. Roughly speaking, a weight in ξ regulates the smoothness of $f \in M_m^{p,q}(\mathbb{R}^d)$, whereas a weight in x regulates the decay of $f \in M_m^{p,q}(\mathbb{R}^d)$. If $p = q$, the notation $M_m^{p,p}$ is shortened to M_m^p. In this case, $\|f\|_{M_m^p(\mathbb{R}^d)} = \|(V_g f)m\|_{L^p(\mathbb{R}^{2d})}$.

If $m \equiv 1$, the space $M_m^{P,Q}$ is simply denoted by $M^{P,Q}$.

The extension of the original modulation spaces in Definition 2.3.1 was widely studied and applied to the investigation of boundedness and sampling properties for pseudodifferential operators in [228, 229, 241].

Proposition 2.3.2. *The definition of $M_m^{P,Q}(\mathbb{R}^d)$ is independent of the window $g \in \mathcal{S}(\mathbb{R}^d)$. Different windows yield equivalent norms on $M_m^{P,Q}(\mathbb{R}^d)$.*

Proof. Set $P = (p_1, \dots, p_d)$, $Q = (q_1, \dots, q_d)$. Let g, g_0 be two different window functions in $\mathcal{S}(\mathbb{R}^d)$. Take first g_0 as fixed window function to measure the modulation space norm of $f \in M_m^{P,Q}(\mathbb{R}^d)$. Applying Lemma 1.2.29 with $h = \gamma = g_0$, we obtain the pointwise estimate

$$|V_g f(x,\xi)| \le \frac{1}{\|g_0\|_2^2}(|V_{g_0} f| * |V_g g_0|)(x,\xi), \quad (x,\xi) \in \mathbb{R}^{2d}.$$

Young's inequality (2.12) with $R = (p_1, \dots, p_d, q_1, \dots, q_d)$, $P = 1$ and hence $Q = (p_1, \dots, p_d, q_1, \dots, q_d)$, gives

$$\|V_g f\|_{L_m^{p_1,\dots,p_d,q_1,\dots,q_d}} \lesssim \|V_{g_0} f\|_{L_m^{p_1,\dots,p_d,q_1,\dots,q_d}} \|V_g g_0\|_{L_v^1}$$

since, by Theorem 1.2.23, $V_g g_0 \in \mathcal{S}(\mathbb{R}^{2d}) \hookrightarrow L_v^1(\mathbb{R}^{2d})$. Switching the roles of g and g_0 yields the reverse inequality

$$\|V_{g_0} f\|_{L_m^{p_1,\dots,p_d,q_1,\dots,q_d}} \lesssim \|V_g f\|_{L_m^{p_1,\dots,p_d,q_1,\dots,q_d}} \|V_{g_0} g\|_{L_v^1}$$

so that $\|V_{g_0} f\|_{L_m^{p_1,\dots,p_d,q_1,\dots,q_d}}$ and $\|V_g f\|_{L_m^{p_1,\dots,p_d,q_1,\dots,q_d}}$ are equivalent norms. The proof is completed. □

Let us present a further extension of Definition 2.3.1, introduced by S. Bishop in [19], in her study of Schatten p-class properties for integral operators. This is a natural generalization of the previous definition.

Assume that c is a permutation of the set $\{1, \dots, 2d\}$. We identify c with the linear bijection

$$\tilde{c} : \mathbb{R}^{2d} \to \mathbb{R}^{2d}, \quad \tilde{c}(x_1, \dots, x_{2d}) = (x_{c(1)}, \dots, x_{c(2d)}). \tag{2.21}$$

Definition 2.3.3 (Mixed modulation spaces). Consider $g \in \mathcal{S}(\mathbb{R}^d) \setminus \{0\}$, indices P, Q as in (2.18), and let c be a permutation corresponding to the map $\tilde{c}$ as above. Then $M(c)_m^{P,Q}(\mathbb{R}^d)$ is the **mixed modulation space** of tempered distributions $f \in \mathcal{S}'(\mathbb{R}^d)$ for which

$$\|f\|_{M(c)_m^{P,Q}} = \|V_g f \circ \tilde{c}\|_{L_m^{P,Q}} < \infty. \tag{2.22}$$

If $p = p_1 = p_2 = \cdots = p_d, q = q_1 = q_2 = \cdots = q_d$, we simply write $M(c)^{p,q}_m$; if $p = p_1 = p_2 = \cdots = p_d = q_1 = q_2 = \cdots = q_d$, we shorten to $M(c)^p_m$.

Remark 2.3.4.

(i) If c is the identity permutation and $P = (p_1, \dots, p_d), Q = (q_1, \dots, q_d)$, we come back to the modulation spaces in Definition 2.3.1.

(ii) If the indices satisfy

$$p = p_1 = p_2 = \cdots = p_d = q_1 = q_2 = \cdots = q_d,$$

and $m \equiv 1$, given any permutation c, a simple change of variables yields

$$M(c)^p(\mathbb{R}^d) = M^p(\mathbb{R}^d).$$

(iii) In general, the mixed modulation spaces differ from the classical ones of Definition 2.3.1. A simple example is provided by the permutation

$$c_0(1, 2, \dots, d, d+1, \dots, 2d) = (d+1, d+2, \dots, 2d, 1, 2, \dots, d). \tag{2.23}$$

If we choose

$$p = p_1 = p_2 = \cdots = p_d, \quad q = q_1 = q_2 = \cdots = q_d,$$

then $M^{p,q}(c_0)$ is the space of tempered distributions such that

$$\|f\|_{M^{p,q}(c_0)} = \left(\int_{\mathbb{R}^d} \left(\int_{\mathbb{R}^d} |V_g f(x,\xi)|^p \, d\xi\right)^{\frac{q}{p}} dx\right)^{\frac{1}{q}} \asymp \|\hat{f}\|_{M^{p,q}} = \|f\|_{\mathcal{F}M^{p,q}}. \tag{2.24}$$

The second to last equivalence follows by using the fundamental identity of time–frequency analysis: $V_g f(x,\xi) = e^{-2\pi i x\cdot\xi} V_{\hat{g}} \hat{f}(\xi, -x)$, hence $|V_g f(x,\xi)| = |V_{\hat{g}} \hat{f}(\xi, -x)|$. Moreover, when computing the $M^{p,q}$-norm of $\hat{f}$ we used the window $\hat{g} \in \mathcal{S}(\mathbb{R}^d)$ in place of g. This does not matter in view of Proposition 2.3.2.

(iv) The definition of mixed modulation spaces does not depend on the choice of the window $g \in \mathcal{S}(\mathbb{R}^d)$. The proof follows a similar pattern as for the standard modulation spaces.

2.3.1 Properties of modulation spaces

In this section we shall recall the main properties of modulation spaces. We shall prove that they are Banach spaces. Moreover, we shall exhibit inclusion relations, duality and density properties.

For sake of simplicity, we limit our exposition to the original modulation spaces $M^{p,q}_m(\mathbb{R}^{2d})$ in [126]. The general case of Definition 2.3.1 enjoys analogous properties.

Let us first recall the following convolution property for $L^\infty_{v_s}(\mathbb{R}^{2d})$.

Lemma 2.3.5. *Consider the weight function v_s in (2.5). If $s > 2d$ and $f, g \in L^\infty_{v_s}(\mathbb{R}^{2d})$, then there exists $C_s > 0$ such that*

$$\|f * g\|_{L^\infty_{v_s}} \le C_s \|f\|_{L^\infty_{v_s}} \|g\|_{L^\infty_{v_s}}. \tag{2.25}$$

Proof. If $f, g \in L^\infty_{v_s}(\mathbb{R}^{2d})$, then

$$|f(z)| \le \|f\|_{L^\infty_{v_s}} v_{-s}(z), \quad z \in \mathbb{R}^{2d},$$

and the same estimate holds for g. Since the weight v_s is equivalent to $(1 + |z|)^s$, we have

$$\begin{aligned} |f * g(z)| &\le \|f\|_{L^\infty_{v_s}} \|g\|_{L^\infty_{v_s}} \int_{\mathbb{R}^{2d}} v_{-s}(w) v_{-s}(z - w)\, dw \\ &\le C_s \|f\|_{L^\infty_{v_s}} \|g\|_{L^\infty_{v_s}} \int_{\mathbb{R}^{2d}} (1 + |w|)^{-s} (1 + |z - w|)^{-s}\, dw \\ &\le C_s' \|f\|_{L^\infty_{v_s}} \|g\|_{L^\infty_{v_s}} (1 + |z|)^{-s}, \end{aligned}$$

and the last bound can be proved as follows. For a fixed $z \in \mathbb{R}^{2d}$, we can define the following partition of $\mathbb{R}^{2d}$:

$$N_z = \left\{ w \in \mathbb{R}^{2d} \,\Big|\, |w - z| \le \frac{|z|}{2} \right\}, \quad N_z^c = \left\{ w \in \mathbb{R}^{2d} \,\Big|\, |w - z| > \frac{|z|}{2} \right\}.$$

When $w \in N_z$, we have $|w| \ge |z|/2$ and

$$(1 + |w|)^{-s} \le \left(1 + \frac{|z|}{2}\right)^{-s} \le 2^s (1 + |z|)^{-s},$$

therefore

$$\int_{N_z} (1 + |w|)^{-s} (1 + |z - w|)^{-s}\, dw \le 2^s (1 + |z|)^{-s} \int_{N_z} (1 + |z - w|)^{-s}\, dw. \tag{2.26}$$

Analogously, when $w \in N_z^c$, since

$$(1 + |z - w|)^{-s} \le \left(1 + \frac{|z|}{2}\right)^{-s} \le 2^s (1 + |z|)^{-s},$$

one has

$$\int_{N_z^c} (1 + |w|)^{-s} (1 + |z - w|)^{-s}\, dw \le 2^s (1 + |z|)^{-s} \int_{N_z^c} (1 + |w|)^{-s}\, dw. \tag{2.27}$$

Now, since $s > 2d$, the integrals in the right-hand sides of (2.26) and (2.27) converge, and we can write

$$\int_{\mathbb{R}^{2d}} (1+|w|)^{-s}(1+|z-w|)^{-s}\,dw \le C_s(1+|z|)^{-s}$$

where

$$C_s = 2^{s+1}\int_{\mathbb{R}^{2d}} (1+|w|)^{-s}\,dw.$$

The proof is completed. □

Observe that the estimate (2.25) in particular proves that, for $s > 2d$,

$$\left(\frac{1}{v_s} * \frac{1}{v_s}\right)(z) \le C_s \frac{1}{v_s}(z), \quad z \in \mathbb{R}^{2d}.$$

2.3.1.1 Relations between modulation and other function spaces

We now present the first collection of useful results that specify the relations of some modulation spaces with other function spaces. We borrow many results from the textbook [160].

Proposition 2.3.6. *Fix a nonzero window $g \in \mathcal{S}(\mathbb{R}^d)$.*

(i) *If $|f(x)| \lesssim \langle x\rangle^{-s}$ and $s > d$, then $|V_g f(x,\xi)| \lesssim \langle x\rangle^{-s}$; analogously, if $|\hat{f}(\xi)| \lesssim \langle \xi\rangle^{-s}$ and $s > d$, then $|V_g f(x,\xi)| \lesssim \langle \xi\rangle^{-s}$.*

(ii) *If $m \in \mathcal{M}_v$ and $m(x,\xi) = m(x) = (m\otimes 1)(x,\xi)$, then $M^2_{m\otimes 1}(\mathbb{R}^d) = L^2_m(\mathbb{R}^d)$; analogously, if $m(x,\xi) = m(\xi) = (1\otimes m)(x,\xi)$, then $M^2_{1\otimes m}(\mathbb{R}^d) = \mathcal{F}L^2_m(\mathbb{R}^d)$.*

(iii) *If $v_s(\xi) = \langle\xi\rangle^s$ for some $s \in \mathbb{R}$, then $M^2_{1\otimes v_s}(\mathbb{R}^d) = H^s(\mathbb{R}^d)$ (the potential Sobolev space).*

(iv) *If $v_s(z) = \langle z\rangle^s$, $1 \le p,q \le \infty$, then*

$$\mathcal{S}(\mathbb{R}^d) = \bigcap_{s\ge 0} M^{p,q}_{v_s}(\mathbb{R}^d), \quad \mathcal{S}'(\mathbb{R}^d) = \bigcup_{s\ge 0} M^{p,q}_{1/v_s}(\mathbb{R}^d). \tag{2.28}$$

Proof. (i) We write $|V_g f(x,\xi)| \le \int_{\mathbb{R}^d} |f(t)||g(t-x)|\,dt = |f| * |\mathcal{I}g|(x)$, and apply the convolution properties (2.25) in dimension d to the functions f and g, which, by assumption, are in $L^\infty_{v_s}(\mathbb{R}^d)$

$$\big\| |f| * |\mathcal{I}g| \big\|_{L^\infty_{v_s}} \le C_s \|f\|_{L^\infty_{v_s}} \|g\|_{L^\infty_{v_s}}.$$

This implies

$$|V_g f(x,\xi)| \le C_s \|f\|_{L^\infty_{v_s}} \|g\|_{L^\infty_{v_s}} \langle x\rangle^{-s}, \quad \forall \xi \in \mathbb{R}^d.$$

The other estimate can be retrieved by the same argument applied to

$$|V_g f(x,\xi)| \le |\hat{f}| * |\mathcal{I}\hat{g}|(\xi), \quad \forall x \in \mathbb{R}^d,$$

cf. (1.29).
(ii) Since

$$\|f\|^2_{M^2_{m\otimes 1}} = \|V_g f(x,\xi)\cdot m(x)\|^2_{L^2(\mathbb{R}^{2d})} < \infty,$$

we have

$$V_g f(x,\xi) = \mathcal{F}[f \cdot T_x\overline{g}](\xi) \in L^2(\mathbb{R}^d_\xi), \quad \text{for a. e. } x \in \mathbb{R}^d.$$

By Plancherel's theorem,

$$\int_{\mathbb{R}^d} |\mathcal{F}[f\cdot T_x\overline{g}](\xi)|^2\, d\xi = \int_{\mathbb{R}^d} |f(t)|^2 |g(t-x)|^2\, dt,$$

and thus

$$\begin{aligned}\|f\|^2_{M^2_{m\otimes 1}} &= \int_{\mathbb{R}^d}\Big[\int_{\mathbb{R}^d} |f(t)|^2 |g(t-x)|^2 m(x)^2\, dt\Big]\, dx \\ &= \int_{\mathbb{R}^{2d}} |f(t)|^2 |g(u)|^2 m(t-u)^2\, dt\, du.\end{aligned}$$

Now, using $m \in \mathcal{M}_v$ and the weights' inequalities (2.6), we obtain

$$C^{-2}\|f\|^2_{L^2_m}\|g\|^2_{L^2_{1/v}} \le \|f\|^2_{M^2_{m\otimes 1}} \le C^2\|f\|^2_{L^2_m}\|g\|^2_{L^2_v}; \tag{2.29}$$

then $M^2_{m\otimes 1}(\mathbb{R}^d) = L^2_m(\mathbb{R}^d)$ with equivalent norms. Observe that for this argument it is enough to assume $g \in L^2_v(\mathbb{R}^d)$. If $m(x,\xi) = m(\xi)$, the fundamental identity of time–frequency analysis (1.31) yields

$$\int_{\mathbb{R}^{2d}} |V_g f(x,\xi)|^2 m(\xi)^2\, dx\, d\xi = \int_{\mathbb{R}^{2d}} |V_{\hat{g}}\hat{f}(\xi,-x)|^2 m(\xi)^2\, dx\, d\xi.$$

Then we apply the previous argument.
(iii) It follows from the preceding claim since $H^s(\mathbb{R}^d) = \mathcal{F}L^2_{v_s}(\mathbb{R}^d)$.
(iv) The left-hand side equality follows from Theorem 1.2.23, whereas the other from Proposition 1.2.20. □

Let us study the properties of V_g^* on modulation spaces.

Theorem 2.3.7. *Consider $m \in \mathcal{M}_v$ and $\gamma \in \mathcal{S}(\mathbb{R}^d)$. Then for $1 \le p, q \le \infty$,*

(i) $V_\gamma^* : L_m^{p,q}(\mathbb{R}^{2d}) \to M_m^{p,q}(\mathbb{R}^d)$ *and the following estimate holds:*

$$\|V_\gamma^* F\|_{M_m^{p,q}} = \|V_g(V_\gamma^* F)\|_{L_m^{p,q}} \lesssim \|V_g\gamma\|_{L_v^1} \|F\|_{L_m^{p,q}}. \tag{2.30}$$

(ii) *If* $F = V_g f$ *and* $\langle \gamma, g\rangle \neq 0$, *we have the inversion formula in* $M_m^{p,q}(\mathbb{R}^d)$

$$f = \frac{1}{\langle \gamma, g\rangle} \int_{\mathbb{R}^{2d}} V_g f(x,\xi) M_\xi T_x \gamma \, dx\, d\xi. \tag{2.31}$$

In short,

$$\mathrm{Id}_{M_m^{p,q}} = \langle \gamma, g\rangle^{-1} V_\gamma^* V_g.$$

Proof. (i) Consider $F \in L_m^{p,q}(\mathbb{R}^{2d})$. We first prove that $V_\gamma^* F \in \mathcal{S}'(\mathbb{R}^d)$. Taking $\phi \in \mathcal{S}(\mathbb{R}^d)$ and using Hölder's inequality for mixed norm spaces in (2.11), we majorize

$$\begin{aligned} |\langle V_\gamma^* F, \phi\rangle| &= |\langle F, V_\gamma \phi\rangle| \le \|F\|_{L_m^{p,q}} \|V_\gamma \phi\|_{L_{1/m}^{p',q'}} \\ &\le \|F\|_{L_m^{p,q}} \|\langle\cdot\rangle^n V_\gamma \phi\|_\infty \|\langle\cdot\rangle^{-n}\|_{L_{1/m}^{p',q'}}. \end{aligned}$$

For sufficiently large n, this expression is finite and, from Corollary 1.2.25, we obtain the claim. Therefore $V_\gamma^* F$ has a continuous STFT in view of Corollary 1.2.19, given by

$$\begin{aligned} V_g V_\gamma^* F(u,\eta) &= \langle V_\gamma^* F, M_\eta T_u g\rangle \\ &= \int_{\mathbb{R}^{2d}} F(x,\xi) \overline{V_\gamma(M_\eta T_u g)(x,\xi)} \, dx\, d\xi \\ &= \int_{\mathbb{R}^{2d}} F(x,\xi) V_g \gamma(u-x, \eta-\xi) e^{-2\pi i x\cdot(\eta-\xi)} \, dx\, d\xi. \end{aligned}$$

Taking the absolute values of the previous equality, we obtain the pointwise estimate

$$|V_g V_\gamma^* F(u,\eta)| \le (|F| * |V_g \gamma|)(u,\eta),$$

and Young's inequality (2.12) gives

$$\|V_g V_\gamma^* F\|_{L_m^{p,q}} \lesssim \|F\|_{L_m^{p,q}} \|V_g \gamma\|_{L_v^1}.$$

Since $g, \gamma \in \mathcal{S}(\mathbb{R}^d)$, by Theorem 1.2.23, we have $V_g\gamma \in \mathcal{S}(\mathbb{R}^{2d}) \hookrightarrow L_v^1(\mathbb{R}^{2d})$ and the right-hand side of the inequality above is finite. Hence the $M_m^{p,q}$-norm of $V_\gamma^* F$, related to the window g, is controlled by

$$\|V_\gamma^* F\|_{M_m^{p,q}} = \|V_g(V_\gamma^* F)\|_{L_m^{p,q}} \lesssim \|F\|_{L_m^{p,q}} \|V_g \gamma\|_{L_v^1} < \infty,$$

that is, $V_\gamma^* F \in M_m^{p,q}(\mathbb{R}^d)$.

(ii) If $V_g f \in L_m^{p,q}(\mathbb{R}^{2d})$, then $\tilde{f} = \langle \gamma, g\rangle^{-1} V_\gamma^* V_g f \in M_m^{p,q}(\mathbb{R}^d)$ by the preceding step. The equality $\tilde{f} = f$ follows from the inversion formula in $\mathcal{S}'(\mathbb{R}^d)$, cf. (1.68). □

Proposition 2.3.8 (Density). *If the weight m satisfies* (2.17) *and* $1 \leq p,q < \infty$, *then* $\mathcal{S}(\mathbb{R}^d)$ *is a dense subspace of* $M_m^{p,q}(\mathbb{R}^d)$.

Proof. Since by assumption there exist $C, s_0 > 0$ such that $m(z) \leq C v_{s_0}(z)$ for every $z \in \mathbb{R}^{2d}$, we can majorize the $M_m^{p,q}$-norm of $f \in \mathcal{S}(\mathbb{R}^d)$ as follows:

$$\|f\|_{M_m^{p,q}} = \|V_g f\|_{L_m^{p,q}} \leq C\|V_g f v_s\|_\infty \|v_s^{-1}\|_{L_m^{p,q}}.$$

If $f \in \mathcal{S}(\mathbb{R}^d)$ and $s \geq s_0$ is large enough, both terms on the right-hand side are finite because $V_g f \in \mathcal{S}(\mathbb{R}^{2d})$ by Theorem 1.2.23, and thus $\mathcal{S}(\mathbb{R}^d) \subseteq M_m^{p,q}(\mathbb{R}^d)$. Now choose, for example, the exhausting sequence of compact sets $K_n = \{z \mid |z| \leq n\}$ (closed balls of radius n) and a window $g \in \mathcal{S}(\mathbb{R}^d)$ with $\|g\|_2 = 1$. For $f \in M_m^{p,q}(\mathbb{R}^d)$, set $F_n = (V_g f)\chi_{K_n}$ and

$$f_n = V_g^* F_n = \int_{\mathbb{R}^{2d}} F_n(x,\xi) M_\xi T_x g \, dx \, d\xi = \int_{\mathbb{R}^{2d}} V_g f(x,\xi)\chi_{K_n}(x,\xi) M_\xi T_x g \, dx \, d\xi.$$

Since F_n is of rapid decay, we see that $f_n \in \mathcal{S}(\mathbb{R}^d)$ by Theorem 1.2.22. Using the inversion formula and the continuity of the mapping $V_g^* : L_m^{p,q}(\mathbb{R}^{2d}) \to M_m^{p,q}(\mathbb{R}^d)$ (cf. Theorem 2.3.7), we can write

$$\|f - f_n\|_{M_m^{p,q}} = \left\|V_g^*(V_g f - F_n)\right\|_{M_m^{p,q}} \lesssim \|V_g f - F_n\|_{L_m^{p,q}}.$$

If $p,q < \infty$, we have $\|V_g f - F_n\|_{L_m^{p,q}} \to 0$ as $n \to \infty$ and then $\|f - f_n\|_{M_m^{p,q}} \to 0$, hence $\mathcal{S}(\mathbb{R}^d) \hookrightarrow M_m^{p,q}(\mathbb{R}^{2d})$, as desired. □

Theorem 2.3.9. *Consider $m \in \mathcal{M}_v$, then*
(i) *If $p,q \in [1,\infty]$, the space $M_m^{p,q}(\mathbb{R}^d)$ is a Banach space.*
(ii) *The space $M_m^{p,q}(\mathbb{R}^d)$ is invariant under time–frequency shifts, and there exists a constant $C > 0$ such that*

$$\left\|\pi(z)f\right\|_{M_m^{p,q}} \leq C v(z)\|f\|_{M_m^{p,q}}, \quad z \in \mathbb{R}^{2d}. \tag{2.32}$$

(iii) *If $p = q$ and there exists $C > 0$ such that $m(Jz) \leq Cm(z)$, for every $z \in \mathbb{R}^{2d}$, then $M_m^p(\mathbb{R}^d)$ is invariant under Fourier transform.*

Proof. (i) Observe that $M_m^{p,q}(\mathbb{R}^d)$ is a normed space. Let us show that it is complete. For a fixed window function $g \in \mathcal{S}(\mathbb{R}^d) \setminus \{0\}$, set

$$V(g) = \{F \in L_m^{p,q}(\mathbb{R}^{2d}) \mid F = V_g f,\ f \in \mathcal{S}'(\mathbb{R}^d)\}.$$

By definition $V(g)$ is a linear subspace of $L_m^{p,q}(\mathbb{R}^{2d})$ isometrically isomorphic to $M_m^{p,q}(\mathbb{R}^d)$. Let us now show that $M_m^{p,q}(\mathbb{R}^d)$ is complete.

Consider a Cauchy sequence $\{f_n\} \subset M_m^{p,q}(\mathbb{R}^d)$, then $\{V_g f_n\}$ is a Cauchy sequence in $L_m^{p,q}(\mathbb{R}^{2d})$ which is complete, and thus there exists $F \in L_m^{p,q}(\mathbb{R}^{2d})$ such that $\|V_g f_n -$

$F\|_{L^{p,q}_m} \to 0$, as $n \to \infty$. Now define $f = \|g\|_{L^2}^{-2} V_g^* F$. Theorem 2.3.7 gives $f \in M^{p,q}_m(\mathbb{R}^d)$ and $f_n = \|g\|_{L^2}^{-2} V_g^* V_g f_n$. By the same result, we have the estimate

$$\|f - f_n\|_{M^{p,q}_m} = \|g\|_{L^2}^{-2} \|V_g^*(F - V_g f_n)\|_{M^{p,q}_m} \lesssim \|F - V_g f_n\|_{L^{p,q}_m} \to 0$$

as $n \to \infty$, which gives the claim.
(ii) We use the covariance formula (1.51), $|V_g(\pi(z)f)(w)| = |T_z V_g f(w)|$, so that

$$\begin{aligned}\|\pi(z)f\|_{M^{p,q}_m} &= \|V_g(\pi(z)f)\|_{L^{p,q}_m} = \|V_g f(\cdot - z)m\|_{L^{p,q}} = \|V_g f(\cdot)m(\cdot + z)\|_{L^{p,q}} \\ &\le Cv(z)\|V_g(f)m\|_{L^{p,q}} = Cv(z)\|f\|_{M^{p,q}_m},\end{aligned}$$

as desired.
(iii) It follows by an explicit computation involving the fundamental identity (1.31) and the norm equivalence in Proposition 2.3.2:

$$\begin{aligned}\|\hat f\|^p_{M^p_m} &= \|V_g \hat f\|^p_{L^p_m} \lesssim \|V_{\hat g} \hat f\|^p_{L^p_m} = \int_{\mathbb{R}^{2d}} |V_{\hat g} \hat f(z)|^p m(z)^p \, dz \\ &= \int_{\mathbb{R}^{2d}} |V_g f(-Jz)|^p m(z)^p \, dz = \int_{\mathbb{R}^{2d}} |V_g f(z)|^p m(Jz)^p \, dz \\ &\lesssim \|f\|^p_{M^p_m},\end{aligned}$$

as desired, since $m(Jz) \lesssim m(z)$ for $z \in \mathbb{R}^{2d}$. □

Theorem 2.3.10 (Duality). *If $1 \le p, q < \infty$ then $(M^{p,q}_m(\mathbb{R}^d))^* \simeq M^{p',q'}_{1/m}(\mathbb{R}^d)$ and the duality between $f \in M^{p,q}_m$ and $h \in M^{p',q'}_{1/m}$ is concretely given by the continuous sesquilinear form (which extends the inner product of $L^2(\mathbb{R}^d)$, see Theorem 1.2.11)*

$$\langle f, h\rangle = \int_{\mathbb{R}^{2d}} V_g f(z)\overline{V_g h(z)}\, dz, \tag{2.33}$$

where $g \in \mathcal{S}(\mathbb{R}^d) \setminus \{0\}$ is a fixed window function.

Proof. Suppose that $h \in M^{p',q'}_{1/m}(\mathbb{R}^d)$. Then $\ell_h(f) := \int V_g f(z)\overline{V_g h(z)}\, dz$ defines a bounded linear functional over $M^{p,q}_m(\mathbb{R}^d)$. From Hölder's inequality (2.11), we have

$$|\ell_h(f)| \le \|V_g f\|_{L^{p,q}_m} \|V_g h\|_{L^{p',q'}_{1/m}} = \|f\|_{M^{p,q}_m} \|h\|_{M^{p',q'}_{1/m}}.$$

Conversely, if we assume that $\ell \in (M^{p,q}_m(\mathbb{R}^d))^*$, since $M^{p,q}_m(\mathbb{R}^d)$ is isometrically isomorphic to the close subspace $V(g) = \{F \in L^{p,q}_m(\mathbb{R}^{2d}) \mid F = V_g f, f \in \mathcal{S}'(\mathbb{R}^d)\}$ of $L^{p,q}_m(\mathbb{R}^{2d})$, ℓ provides a linear functional $\tilde\ell$ on $V(g)$ by setting $\ell(f) = \tilde\ell(V_g f)$. Hahn–Banach theorem allows extending $\tilde\ell$ to a bounded functional on $L^{p,q}_m(\mathbb{R}^{2d})$ and Theorem 2.2.3 (iii) shows that there is a function $H \in L^{p',q'}_{1/m}(\mathbb{R}^{2d})$ such that

$$\tilde\ell(V_g f) = \int_{\mathbb{R}^{2d}} V_g f(z)\overline{H(z)}\, dz.$$

Now set $h = V_g^* H$, therefore $h \in M_{1/m}^{p',q'}(\mathbb{R}^d)$ by Theorem 2.3.7 and

$$\langle f, h\rangle = \int_{\mathbb{R}^{2d}} V_g f(z)\overline{H(z)}\,dz = \int_{\mathbb{R}^{2d}} V_g f(z)\overline{V_g h(z)}\,dz = \ell(f),$$

since $V_g^*(H - V_g h) = 0$ if $\|g\|_2 = 1$ and

$$\int_{\mathbb{R}^{2d}} V_g f(z)\overline{H(z)}\,dz = \int_{\mathbb{R}^d} f(x)\overline{V_g^* H(x)}\,dx,$$

cf. Remark 1.2.27 (ii), and argue by density. Hence every linear functional has the prescribed form and this gives the claim. □

By the previous result, we can compute the $M^{p,q}$-norm by duality. In fact, we have

Corollary 2.3.11. *Assume* $1 < p, q \le \infty$ *and* $f \in M_m^{p,q}(\mathbb{R}^d)$*. Then*

$$\|f\|_{M_m^{p,q}} = \sup_{\|g\|_{M_{1/m}^{p',q'}} \le 1} |\langle f, g\rangle|. \tag{2.34}$$

Notice that (2.34) *still holds true whenever* $p = 1$ *or* $q = 1$ *and* $f \in \mathcal{S}(\mathbb{R}^d)$*, simply by extending Theorem* 2.3.10 *to the duality* ${}_{\mathcal{S}'}\langle\cdot,\cdot\rangle_{\mathcal{S}}$.

We now widen the space of admissible windows from $\mathcal{S}(\mathbb{R}^d)$ to $M_v^1(\mathbb{R}^d)$.

Theorem 2.3.12. *Assume that* $m \in \mathcal{M}_v(\mathbb{R}^{2d})$ *and let* $g, \gamma \in M_v^1(\mathbb{R}^d)\setminus\{0\}$*. Then*
(i) *The mapping* $V_\gamma^* : L_m^{p,q}(\mathbb{R}^{2d}) \to M_m^{p,q}(\mathbb{R}^d)$ *is bounded and the estimate* (2.30) *still holds.*
(ii) *For* $f \in M_m^{p,q}(\mathbb{R}^d)$*, we have* $V_g f \in L_m^{p,q}(\mathbb{R}^{2d})$*, and, if* $\langle g, \gamma\rangle \neq 0$*, the inversion formula* (2.31) *remains valid.*
(iii) $\|V_g f\|_{L_m^{p,q}}$ *is an equivalent norm on* $M_m^{p,q}(\mathbb{R}^d)$.

Proof. (i) Given $F \in L_m^{p,q}(\mathbb{R}^{2d})$, consider the mapping $\gamma \mapsto V_\gamma^* F$. Then by Theorem 2.3.7, it maps $\mathcal{S}(\mathbb{R}^d)$ into $M_m^{p,q}(\mathbb{R}^d)$ and

$$\|V_\gamma^* F\|_{M_m^{p,q}} \lesssim \|F\|_{L_m^{p,q}} \|V_g \gamma\|_{L_v^1} = \|F\|_{L_m^{p,q}} \|\gamma\|_{M_v^1}.$$

Moreover, since $\mathcal{S}(\mathbb{R}^d)$ is dense in $M_v^1(\mathbb{R}^d)$, $\gamma \mapsto V_\gamma^* F$ extends to a bounded map from $M_v^1(\mathbb{R}^d)$ to $M_m^{p,q}(\mathbb{R}^d)$ by the density principle (cf. Theorem A.0.9).

(ii) (*First step*) Let us show that for $f \in M_m^{p,q}(\mathbb{R}^d)$ and $g \in M_v^1(\mathbb{R}^d)$ the linear mapping $g \mapsto V_g f$ is bounded from $M_v^1(\mathbb{R}^d)$ to $L_m^{p,q}(\mathbb{R}^{2d})$. Using Lemma 1.2.29 for $g, h = \gamma = g_0 \in \mathcal{S}(\mathbb{R}^d)$, we have the pointwise estimate

$$|V_g f(x,\xi)| \le \|g_0\|_2^{-2} (|V_{g_0} f| * |V_g g_0|)(x,\xi).$$

Young's inequality yields

$$\|V_g f\|_{L^{p,q}_m} \lesssim \|g_0\|_2^{-2}\|V_{g_0} f\|_{L^{p,q}_m}\|V_g g_0\|_{L^1_v} \lesssim \|g_0\|_2^{-2}\|f\|_{M^{p,q}_m}\|g\|_{M^1_v}. \tag{2.35}$$

By density of $\mathcal{S}(\mathbb{R}^d)$ in M^1_v, this estimate implies that for fixed $f \in M^{p,q}_m$ the map $g \mapsto V_g f$ is bounded from M^1_v into $L^{p,q}_m$.
(*Second step*) Consider $g, \gamma \in M^1_v(\mathbb{R}^d)\backslash\{0\}$ as in the assumptions. By the density of $\mathcal{S}(\mathbb{R}^d)$ in $M^1_v(\mathbb{R}^d)$, we can find two sequences $\{g_n\}, \{\gamma_n\} \subset \mathcal{S}(\mathbb{R}^d)$ such that $\|g_n - g\|_{M^1_v} \to 0$, $\|\gamma_n - \gamma\|_{M^1_v} \to 0$ and $\langle \gamma_n, g_n\rangle \neq 0$. Using the estimates in (2.35) and (i), we obtain, for $f \in M^{p,q}_m(\mathbb{R}^d)$ and $F \in L^{p,q}_m(\mathbb{R}^{2d})$,

$$\|V_{g_n} f - V_g f\|_{L^{p,q}_m} \lesssim \|g_n - g\|_{M^1_v} \to 0, \quad \left\|V^*_{\gamma_n} F - V^*_\gamma F\right\|_{M^{p,q}_m} \lesssim \|\gamma_n - \gamma\|_{M^1_v} \to 0.$$

Finally, writing the inversion formula (2.31) for γ_n, g_n and taking the limit $n \to +\infty$,

$$\langle \gamma, g\rangle^{-1} V^*_\gamma V_g f = \lim_{n\to\infty} \langle \gamma_n, g_n\rangle^{-1} V^*_{\gamma_n} V_{g_n} f = f,$$

we obtain the result.
(iii) The proof of the norm equivalence is similar to the proof of Proposition 2.3.2. □

We now study the action of the multiplication operator $f \mapsto \langle\cdot\rangle^s f$ and of the Fourier multiplier

$$\langle D\rangle^s f = \mathcal{F}^{-1}\langle\cdot\rangle^s \mathcal{F} f \tag{2.36}$$

on modulation spaces, cf. [287]. We need the following preliminary result.

Lemma 2.3.13. *For $s \in \mathbb{R}$, take $m = s + d + 1$, and define*

$$\Phi(x,t) = \frac{v_s(t)}{v_s(x) v_m(t-x)}, \quad x, t \in \mathbb{R}^d \tag{2.37}$$

and

$$\Phi_2(x,\xi) = [\mathcal{F}_2 \Phi(x,\cdot)](\xi), \quad x, \xi \in \mathbb{R}^d, \tag{2.38}$$

the partial Fourier transform of Φ with respect to the t-variable. Then for any $r \geq 0$, we have

$$\int_{\mathbb{R}^d} \langle\xi\rangle^r \sup_{x\in\mathbb{R}^{2d}} |\Phi_2(x,\xi)|\, d\xi < \infty. \tag{2.39}$$

Proof. Observe that $\Phi \in C^\infty(\mathbb{R}^{2d})$, moreover,

$$0 \leq \Phi(x,t) = \frac{v_s(t-x+x)}{v_s(x) v_m(t-x)} \leq C v_{-d-1}(t-x),$$

so that

$$\int_{\mathbb{R}^d} \Phi(x,t)\,dt \le C \int_{\mathbb{R}^d} v_{-d-1}(t-x)\,dt = C \int_{\mathbb{R}^d} v_{-d-1}(y)\,dy < \infty$$

and

$$R_\Phi := \sup_{x\in\mathbb{R}^d} \int_{\mathbb{R}^d} \Phi(x,t)\,dt < \infty.$$

Using (0.3), one easily verifies that $\forall \alpha \in \mathbb{N}^d$,

$$|\partial_t^\alpha \Phi(x,t)| \le C_\alpha \Phi(x,t).$$

By applying the partial Fourier transform with respect to the t-variable and using the Riemann–Lebesgue lemma,

$$\sup_{x,\xi} |\xi^\alpha \Phi_2(x,\xi)| \le C \sup_{x,\xi} |\mathcal{F}_2[\partial_t^\alpha \Phi(x,t)](\xi)| \le C \sup_{x\in\mathbb{R}^d} \int_{\mathbb{R}^d} |\partial_t^\alpha \Phi(x,t)|\,dt \le C' R_\Phi,$$

for every $\alpha \in \mathbb{N}^d$. The previous estimate easily implies (2.39). □

Theorem 2.3.14. *Consider* $1 \le p,q \le \infty$, $m \in \mathcal{M}_v(\mathbb{R}^{2d})$, $s \in \mathbb{R}$.

(i) *For* $(v_s \otimes 1)(x,\xi) = v_s(x)$, *the mapping* $f \mapsto v_s f$ *is a topological isomorphism from* $M^{p,q}_{m(v_s\otimes 1)}(\mathbb{R}^d)$ *to* $M^{p,q}_m(\mathbb{R}^d)$.

(ii) *For* $(1\otimes v_s)(x,\xi) = v_s(\xi)$, *the Fourier multiplier* $\langle D\rangle^s$ *defined in* (2.36) *is a topological isomorphism from* $M^{p,q}_{m(1\otimes v_s)}(\mathbb{R}^d)$ *to* $M^{p,q}_m(\mathbb{R}^d)$.

Proof. Consider a real-valued window function $g_0 \in \mathcal{S}(\mathbb{R}^d)\setminus\{0\}$ and set $m = s+d+1$, $g = v_{-m} g_0 \in \mathcal{S}(\mathbb{R}^d)\setminus\{0\}$. Then for $f \in M^{p,q}_{m(v_s\otimes 1)}(\mathbb{R}^d)$,

$$\begin{aligned} V_g(v_s f)(x,\xi) &= \mathcal{F}(v_s f T_x g)(\xi) = \mathcal{F}(v_s f v_{-m}(\cdot - x) T_x g_0)(\xi) \\ &= v_s(x)\mathcal{F}(f T_x g_0 \Phi(x,\cdot))(\xi) \\ &= v_s(x)\mathcal{F}(f T_x g_0) * \mathcal{F}[\Phi(x,\cdot)](\xi) \\ &= v_s(x)\mathcal{F}(f T_x g_0) * \Phi_2(x,\cdot)(\xi) \end{aligned}$$

with Φ, Φ_2 defined in (2.37) and (2.38), respectively, and $\mathcal{F} = \mathcal{F}_2$. Consider the case $p,q < \infty$, then

$$\begin{aligned} \|v_s f\|_{M^{p,q}_m} &\asymp \|V_g(v_s f)\|_{L^{p,q}_m} \\ &= \Big(\int_{\mathbb{R}^d}\Big(\int_{\mathbb{R}^d} |\mathcal{F}(f T_x g_0) * \Phi_2(x,\cdot)(\xi)|^p m^p(x,\xi) v_s^p(x)\,dx\Big)^{\frac{q}{p}} d\xi\Big)^{\frac{1}{q}} \\ &= \Big(\int_{\mathbb{R}^d}\Big(\int_{\mathbb{R}^d}\Big|\int_{\mathbb{R}^d} \mathcal{F}(f T_x g_0)(\xi-\eta)\Phi_2(x,\eta)\,d\eta\Big|^p m^p(x,\xi) v_s^p(x)\,dx\Big)^{\frac{q}{p}} d\xi\Big)^{\frac{1}{q}}. \end{aligned}$$

Define $h(\xi) = \sup_{x\in\mathbb{R}^d} |\Phi_2(x,\xi)|$ and observe that the estimate (2.39) implies $h \in L^1_{v_r}(\mathbb{R}^d)$, for every $r \geq 0$. Using Minkowski's integral inequality (cf. Appendix A, Theorem A.0.1),

$$
\begin{aligned}
&\Big(\int_{\mathbb{R}^d}\Big|\int_{\mathbb{R}^d} \mathcal{F}(fT_xg_0)(\xi-\eta)\Phi_2(x,\eta)\,d\eta\Big|^p m^p(x,\xi)v_s^p(x)\,dx\Big)^{\frac{1}{p}}\\
&\leq \int_{\mathbb{R}^d}\Big(\int_{\mathbb{R}^d} |\mathcal{F}(fT_xg_0)(\xi-\eta)|^p|\Phi_2(x,\eta)|^p m^p(x,\xi)v_s^p(x)\,dx\Big)^{\frac{1}{p}} d\eta\\
&\leq \int_{\mathbb{R}^d}\Big(\int_{\mathbb{R}^d} |\mathcal{F}(fT_xg_0)(\xi-\eta)|^p m^p(x,\xi)v_s^p(x)\,dx\Big)^{\frac{1}{p}} h(\eta)\,d\eta\\
&\leq C\int_{\mathbb{R}^d}\Big(\int_{\mathbb{R}^d} |\mathcal{F}(fT_xg_0)(\xi-\eta)|^p m^p(x,\xi-\eta)v^p(0,\eta)v_s^p(x)\,dx\Big)^{\frac{1}{p}} h(\eta)\,d\eta\\
&\leq C'\int_{\mathbb{R}^d}\Big(\int_{\mathbb{R}^d} |\mathcal{F}(fT_xg_0)(\xi-\eta)|^p m^p(x,\xi-\eta)v_{s_0}^p(\eta)v_s^p(x)\,dx\Big)^{\frac{1}{p}} h(\eta)\,d\eta\\
&= C'\int_{\mathbb{R}^d}\Big(\int_{\mathbb{R}^d} |\mathcal{F}(fT_xg_0)(\xi-\eta)|^p m^p(x,\xi-\eta)v_s^p(x)\,dx\Big)^{\frac{1}{p}} v_{s_0}(\eta)h(\eta)\,d\eta
\end{aligned}
$$

where we used $m \in \mathcal{M}_v$ and the polynomial growth in (2.17) (observe that $v_{s_0}(0,\eta) = v_{s_0}(\eta)$). Coming back to the computation of $\|v_sf\|_{M^{p,q}_m}$, we apply Minkowski's integral inequality again, so that

$$
\begin{aligned}
\|v_sf\|_{M^{p,q}_m} &\leq C'\Big\|\int_{\mathbb{R}^d}\Big(\int_{\mathbb{R}^d} |\mathcal{F}(fT_xg_0)(\cdot-\eta)|^p m^p(x,\cdot-\eta)v_s^p(x)\,dx\Big)^{\frac{1}{p}} v_{s_0}(\eta)h(\eta)\,d\eta\Big\|_q\\
&\leq C'\int_{\mathbb{R}^d}\Big\|\Big(\int_{\mathbb{R}^d} |\mathcal{F}(fT_xg_0)(\cdot-\eta)|^p m^p(x,\cdot-\eta)v_s^p(x)\,dx\Big)^{\frac{1}{p}}\Big\|_q v_{s_0}(\eta)h(\eta)\,d\eta\\
&= C'\int_{\mathbb{R}^d}\Big\|\Big(\int_{\mathbb{R}^d} |\mathcal{F}(fT_xg_0)(\cdot)|^p m^p(x,\cdot)v_s^p(x)\,dx\Big)^{\frac{1}{p}}\Big\|_q v_{s_0}(\eta)h(\eta)\,d\eta\\
&= C'\|f\|_{M^{p,q}_{m(v_s\otimes 1)}}\|hv_{s_0}\|_1 < \infty,
\end{aligned}
$$

since $\|hv_{s_0}\|_1 < \infty$ because $h \in L^1_{v_r}(\mathbb{R}^d)$, for every $r \geq 0$. The cases $p = \infty$ or $q = \infty$ follow by using the same arguments above.

Finally, using $v_sv_{-s} = 1$ and the continuity of the mapping $f \mapsto v_sf$ from $M^{p,q}_{m(v_s\otimes 1)}(\mathbb{R}^d)$ to $M^{p,q}_m$ shown above, we have

$$
\|f\|_{M^{p,q}_{m(v_s\otimes 1)}} = \big\|v_{-s}(v_sf)\big\|_{M^{p,q}_{m(v_s\otimes 1)}} \leq C\|v_sf\|_{M^{p,q}_m}.
$$

Hence the mapping $f \mapsto v_s f$ is a linear continuous bijection from $M^{p,q}_{m(v_s \otimes 1)}$ to $M^{p,q}_m$. This gives the claim.

(ii) Consider $g \in \mathcal{S}(\mathbb{R}^d) \setminus \{0\}$ and define $g_0 = \hat{g} v_m$, with $m = s + d + 1$ as in the previous part. Using the fundamental identity of time–frequency analysis (1.31), we can write

$$
\begin{aligned}
|V_g(\langle D\rangle^s f)(x,\xi)| &= |V_{\hat{g}}(v_s \hat{f})(\xi,-x)| = |\mathcal{F}(v_s \hat{f} T_\xi \bar{\hat{g}})(-x)| \\
&= v_s(\xi) |\mathcal{F}(\hat{f} T_\xi \overline{g_0} \Phi(\xi,\cdot))(-x)| \\
&= v_s(\xi) \Big| \int_{\mathbb{R}^d} \mathcal{F}(\hat{f} T_\xi \overline{g_0})(-x-y) \Phi_2(\xi,y)\, dy \Big| \\
&\le v_s(\xi) \int_{\mathbb{R}^d} |\mathcal{F}(\hat{f} T_\xi \overline{g_0})(-x-y) \Phi_2(\xi,y)|\, dy
\end{aligned}
$$

where, as in the previous step, Φ and Φ_2 are defined in (2.37) and (2.38), respectively. Setting $\chi_0 = \mathcal{F}^{-1} g_0$ and using (1.26),

$$
|\mathcal{F}(\hat{f} T_\xi \overline{g_0})(-x-y)| = |V_{\chi_0} f(x+y,\xi)|,
$$

so that, using Minkowski's integral inequality and proceeding as in the previous part,

$$
\begin{aligned}
\|\langle D\rangle^s f\|_{M^{p,q}_m} &\le \Big(\int_{\mathbb{R}^d} \Big(\int_{\mathbb{R}^d} \Big(\int_{\mathbb{R}^d} |V_{\chi_0} f(x+y,\xi) \Phi_2(\xi,y)|\, dy \Big)^p m^p(x,\xi) v_s^p(\xi)\, dx \Big)^{\frac{q}{p}} d\xi \Big)^{\frac{1}{q}} \\
&\le C \Big(\int_{\mathbb{R}^d} \Big(\int_{\mathbb{R}^d} |V_{\chi_0} f(x,\xi)|^p m^p(x,\xi) v_s^p(\xi)\, dx \Big)^{\frac{q}{p}} d\xi \Big)^{\frac{1}{q}} \|v_{s_0} h\|_1 \\
&= C'' \|f\|_{M^{p,q}_{m(1\otimes v_s)}}
\end{aligned}
$$

where $h(y) = \sup_\xi |\Phi_2(\xi,y)| \in L^1_{v_{s_0}}(\mathbb{R}^d)$ by (2.39), with s_0 coming from (2.17). So we obtain the result. □

In the sequel we shall often use the closure of the Schwartz class with respect to the modulation norm.

Definition 2.3.15. For $m \in \mathcal{M}_v(\mathbb{R}^{2d})$, $1 \le p,q \le \infty$, we define by $\mathcal{M}^{p,q}_m(\mathbb{R}^d)$ the closure of $\mathcal{S}(\mathbb{R}^d)$ in the $M^{p,q}_m$-norm. Observe that $\mathcal{M}^{p,q}_m(\mathbb{R}^d) = M^{p,q}_m(\mathbb{R}^d)$, whenever the indices p and q are finite.

Notice that these spaces enjoy the duality property $(\mathcal{M}^{p,q}_m)^* = \mathcal{M}^{p',q'}_{1/m}$, with $1 \le p,q \le \infty$, and p',q' being the conjugate exponents.

Finally, we recall the behavior of modulation spaces with respect to complex interpolation. For the proof, we refer to the original literature [131, 132] and [126, Corollary 2.3].

Proposition 2.3.16. *Let* $0 < \theta < 1$, $p_j, q_j \in [1, \infty]$ *and* $m_j \in \mathcal{M}_v(\mathbb{R}^{2d})$, $j = 1, 2$. *Set*

$$\frac{1}{p} = \frac{1-\theta}{p_1} + \frac{\theta}{p_2}, \quad \frac{1}{q} = \frac{1-\theta}{q_1} + \frac{\theta}{q_2}, \quad m = m_1^{1-\theta} m_2^{\theta},$$

then

$$(\mathcal{M}_{m_1}^{p_1,q_1}(\mathbb{R}^d), \mathcal{M}_{m_2}^{p_2,q_2}(\mathbb{R}^d))_{[\theta]} = \mathcal{M}_m^{p,q}(\mathbb{R}^d).$$

Proposition 2.3.17. *Let* $1 \le p_1, p_2, q_1, q_2 \le \infty$, *with* $q_2 < \infty$. *If* T *is a linear operator such that, for* $i = 1, 2$,

$$\|Tf\|_{M^{p_i,q_i}} \le A_i \|f\|_{M^{p_i,q_i}} \quad \forall f \in M^{p_i,q_i},$$

then

$$\|Tf\|_{M^{p,q}} \le C A_1^{1-\theta} A_2^{\theta} \|f\|_{M^{p,q}} \quad \forall f \in M^{p,q},$$

where $1/p = (1-\theta)/p_1 + \theta/p_2$, $1/q = (1-\theta)/q_1 + \theta/q_2$, $0 < \theta < 1$, *and* C *is independent of* T.

We shall also use the following interpolation result.

Proposition 2.3.18. *Let* $0 < \theta < 1$, $p_j, q_j \in [1, \infty]$ *and* $m_j \in \mathcal{M}_v(\mathbb{R}^{2d})$, $j = 1, 2$. *Then*

$$(M_{m_1\otimes 1}^{\infty,1}(\mathbb{R}^d), M_{m_2\otimes 1}^{\infty,1}(\mathbb{R}^d))_{[\theta]} = M_{m\otimes 1}^{\infty,1}(\mathbb{R}^d),$$

with $m = m_1^{1-\theta} m_2^{\theta}$.

Proof. We limit to an outline. By [132], this amounts to computing $[\ell^1(L_{m_1}^{\infty}), \ell^1(L_{m_2}^{\infty})]_{[\theta]} = \ell^1([L_{m_1}^{\infty}, L_{m_2}^{\infty}]_{[\theta]})$, where the equality is due to [17, Theorem 5.1.2]. Using the connection between complex and real interpolation given by [17, Theorem 4.7.2], we have

$$[L_{m_1}^{\infty}, L_{m_2}^{\infty}]_{[\theta]} = (L_{m_1}^{\infty}, L_{m_2}^{\infty})_{\theta,\infty}.$$

Finally, the interpolation theorem by Stein–Weiss [17, Theorem 5.4.1] gives

$$(L_{m_1}^{\infty}, L_{m_2}^{\infty})_{\theta,\infty} = L_m^{\infty}, \quad m = m_1^{1-\theta} m_2^{\theta},$$

as desired. □

2.3.1.2 Potential Sobolev spaces

As a consequence of the previous fact, we can establish relations between modulation spaces and potential Sobolev spaces. We have already seen in Proposition 2.3.6 (iii) that the Sobolev spaces $H^s(\mathbb{R}^d)$, $s \in \mathbb{R}$, are basic examples of modulation spaces. We now study the more general case $W_s^p(\mathbb{R}^d)$, $1 \le p \le \infty$.

Definition 2.3.19. Given $s \in \mathbb{R}$, the Fourier multiplier $\langle D\rangle^s$ in (2.36) is also called the Bessel potential operator. The potential Sobolev space $W_s^p(\mathbb{R}^d)$ is defined by

$$W_s^p(\mathbb{R}^d) = \{f \in \mathcal{S}'(\mathbb{R}^d), \|\langle D\rangle^s f\|_p < \infty\}, \quad 1 \le p \le \infty. \tag{2.40}$$

For $s \in \mathbb{R}$ the Bessel kernel is

$$G_s = \mathcal{F}^{-1}\{(1 + |\cdot|^2)^{-s/2}\} = \mathcal{F}^{-1}(\langle\cdot\rangle^{-s}), \tag{2.41}$$

so that the potential Sobolev space can be equivalently defined by

$$W_s^p(\mathbb{R}^d) = G_s * L^p(\mathbb{R}^d) = \{f \in \mathcal{S}'(\mathbb{R}^d), f = G_s * g,\ g \in L^p(\mathbb{R}^d)\}. \tag{2.42}$$

Of course, if $s = 0$, $W_0^p(\mathbb{R}^d) = L^p(\mathbb{R}^d)$. Moreover, $W_s^2(\mathbb{R}^d) = H^s(\mathbb{R}^d)$.

Let us first report the following isomorphism property for Bessel potentials.

Theorem 2.3.20. *For $1 \le p, q \le \infty$, $s_1, s_2 \in \mathbb{R}$ given, the Bessel potential*

$$Tf = G_s * f, \quad f \in \mathcal{S}(\mathbb{R}^d)$$

with kernel G_s defined in (2.41) and $s = s_2 - s_1$, gives rise to an isomorphism between $M^{p,q}_{1\otimes v_{s_1}}(\mathbb{R}^d)$ and $M^{p,q}_{1\otimes v_{s_2}}(\mathbb{R}^d)$. More precisely,

$$M^{p,q}_{1\otimes v_{s_2}}(\mathbb{R}^d) = G_s * M^{p,q}_{1\otimes v_{s_1}}(\mathbb{R}^d). \tag{2.43}$$

Proof. It follows immediately from Theorem 2.3.14 (ii) since $Tf = G_s * f = \langle D\rangle^{-s} f$. □

According to the previous isomorphism, one can also define the weighted modulation spaces $M^{p,q}_{1\otimes v_s}(\mathbb{R}^d)$ as images of the unweighted ones via Bessel potentials

$$M^{p,q}_{1\otimes v_s}(\mathbb{R}^d) = G_s * M^{p,q}(\mathbb{R}^d), \quad 1 \le p, q \le \infty,\ s \in \mathbb{R}. \tag{2.44}$$

Proposition 2.3.21. *For $1 \le p \le \infty$ and $s \in \mathbb{R}$*

$$W_s^p(\mathbb{R}^d) \hookrightarrow M^{p,\infty}_{1\otimes v_s}(\mathbb{R}^d).$$

Proof. Consider $g \in \mathcal{S}(\mathbb{R}^d)$. Assume $s = 0$ first. Using formula (1.28) to express the STFT and Young's inequality, we find that

$$\begin{aligned}\|f\|_{M^{p,\infty}} \asymp \|V_g f\|_{L^{p,\infty}} &= \sup_{\xi\in\mathbb{R}^d}\Big(\int_{\mathbb{R}^d} |V_g f(x,\xi)|^p\, dx\Big)^{1/p} \\ &= \sup_{\xi\in\mathbb{R}^d}\|(f * M_\xi g^*)\|_p \le \sup_{\xi\in\mathbb{R}^d}\|f\|_p\|M_\xi g^*\|_1 \lesssim \|f\|_p.\end{aligned}$$

Consequently, $L^p(\mathbb{R}^d) \subseteq M^{p,\infty}(\mathbb{R}^d)$. For arbitrary $s \in \mathbb{R}$, we observe that $W_s^p = G_s * L^p$ by definition, and $M^{p,\infty}_{1\otimes v_s}(\mathbb{R}^d) = G_s * M^{p,\infty}(\mathbb{R}^d)$ by (2.44). So the embedding follows for all $s \in \mathbb{R}$. □

Concerning weighted modulation spaces $L^p_m(\mathbb{R}^d)$, with $m \in \mathcal{M}_v(\mathbb{R}^d)$, they are included in the modulation spaces $M^{p,\infty}_{m\otimes 1}(\mathbb{R}^d)$ as follows.

Proposition 2.3.22. *For $1 \le p \le \infty$ and $m \in \mathcal{M}_v(\mathbb{R}^d)$,*

$$L^p_m(\mathbb{R}^d) \hookrightarrow M^{p,\infty}_{m\otimes 1}(\mathbb{R}^d).$$

Proof. Consider $g \in \mathcal{S}(\mathbb{R}^d)$. Using formula (1.28) to express the STFT, we can write

$$\begin{aligned}\|f\|_{M^{p,\infty}_{m\otimes 1}} &\asymp \|V_g f\|_{L^{p,\infty}_{m\otimes 1}} = \sup_{\xi\in\mathbb{R}^d}\Big(\int_{\mathbb{R}^d} |V_g f(x,\xi)|^p m(x)^p\,dx\Big)^{1/p}\\ &= \sup_{\xi\in\mathbb{R}^d}\|(f * M_\xi g^*)\|_{L^p_m} \le \sup_{\xi\in\mathbb{R}^d}\|f\|_{L^p_m}\|M_\xi g^*\|_{L^1_v} \lesssim \|f\|_{L^p_m},\end{aligned}$$

where in the last row we used the weighted Young's inequality in (2.12) in the form $L^p_m * L^1_v \hookrightarrow L^p_m$. Also, $\|M_\xi g^*\|_{L^1_v} = \|g^*\|_{L^1_v} < \infty$, since $g \in \mathcal{S}(\mathbb{R}^d)$ by assumption. □

2.3.2 Alternative definition of modulation spaces using frequency-uniform localization techniques

Frequency-uniform localization techniques can be applied to discretize modulation spaces, see [298, 303, 176].

For $k \in \mathbb{Z}^d$, we denote by $\mathcal{Q}_k$ the unit closed cube centered at k. The family $\{Q_k\}_{k\in\mathbb{Z}^d}$ is a covering of $\mathbb{R}^d$. We define $|\xi|_\infty := \max_{i=1,\dots,d}|\xi_i|$. Consider now a smooth function $\rho : \mathbb{R}^d \to [0,1]$ satisfying $\rho(\xi) = 1$ for $|\xi|_\infty \le 1/2$ and $\rho(\xi) = 0$ for $|\xi|_\infty \ge 3/4$. Define

$$\rho_k(\xi) = T_k\rho(\xi) = \rho(\xi - k), \quad k \in \mathbb{Z}^d, \tag{2.45}$$

that is, ρ_k is the translation of ρ at k. By the assumption on ρ, we infer that $\rho_k(\xi) = 1$ for $\xi \in \mathcal{Q}_k$ and

$$\sum_{k\in\mathbb{Z}^d} \rho_k(\xi) \ge 1, \quad \forall \xi \in \mathbb{R}^d.$$

Denote by

$$\sigma_k(\xi) = \frac{\rho_k(\xi)}{\sum_{l\in\mathbb{Z}^d}\rho_l(\xi)}, \quad \xi \in \mathbb{R}^d,\ k \in \mathbb{Z}^d. \tag{2.46}$$

Observe that $\sigma_k(\xi) = \sigma_0(\xi - k) \in \mathcal{D}(\mathbb{R}^d)$ and the sequence $\{\sigma_k\}_{k\in\mathbb{Z}^d}$ is a smooth partition of unity

$$\sum_{k\in\mathbb{Z}^d} \sigma_k(\xi) = 1, \quad \forall \xi \in \mathbb{R}^d.$$

Definition 2.3.23. For $k \in \mathbb{Z}^d$, we define the frequency-uniform decomposition operator by

$$\Box_k := \mathcal{F}^{-1}\sigma_k\mathcal{F}. \tag{2.47}$$

The previous operators allow introducing an alternative norm on the weighted modulation spaces $M^{p,q}_{u\otimes w}(\mathbb{R}^d)$ as follows.

Definition 2.3.24. For $1 \le p, q \le \infty$, $u, w \in \mathcal{M}_v(\mathbb{R}^d)$ (that is, v-moderate weights on $\mathbb{R}^d$), we define the norm

$$\|f\|_{\tilde{M}^{p,q}_{u\otimes w}(\mathbb{R}^d)} = \Big(\sum_{k\in\mathbb{Z}^d} \|\Box_k f\|^q_{L^p_u} w(k)^q\Big)^{\frac{1}{q}}, \quad f \in \mathcal{S}'(\mathbb{R}^d), \tag{2.48}$$

with an obvious modification for $q = \infty$.

Let us show that $\|f\|_{\tilde{M}^{p,q}_{u\otimes w}(\mathbb{R}^d)}$ is an equivalent norm for $M^{p,q}_{u\otimes w}(\mathbb{R}^d)$.

Proposition 2.3.25. *For $1 \le p, q \le \infty$, $u, w \in \mathcal{M}_v(\mathbb{R}^d)$, the norm $\|\cdot\|_{\tilde{M}^{p,q}_{u\otimes w}}$ is an equivalent norm on $M^{p,q}_{u\otimes w}(\mathbb{R}^d)$.*

Proof. We prove the case $q < \infty$; the case $q = \infty$ can be obtained similarly. Consider a window function $\phi \in \mathcal{S}(\mathbb{R}^d)$ such that $\operatorname{supp}\hat{\phi} \subset B_{10\sqrt{d}}(0)$, satisfying $\hat{\phi}(\xi) = 1$ in $B_{5\sqrt{d}}(0)$. For any $f \in \mathcal{S}'(\mathbb{R}^d)$, we reckon

$$\Box_k f = \mathcal{F}^{-1}\sigma_k\mathcal{F}f = \mathcal{F}^{-1}\sigma_k T_\xi \bar{\hat{\phi}}\mathcal{F}f, \quad \text{for } \xi \in \mathcal{Q}_k,$$

since $T_\xi\bar{\hat{\phi}} = 1$ in $\operatorname{supp}\sigma_k$ for $\xi \in \mathcal{Q}_k$. Using Young's inequality, for $\xi \in \mathcal{Q}_k$, we obtain

$$\|\Box_k f\|_{L^p_u} \lesssim \|\mathcal{F}^{-1}\sigma_k\|_{L^1_v}\|\mathcal{F}^{-1}T_\xi\bar{\hat{\phi}}\mathcal{F}f\|_{L^p_u} \lesssim \|\mathcal{F}^{-1}T_\xi\bar{\hat{\phi}}\mathcal{F}f\|_{L^p_u}.$$

Observing that $w(\xi) \asymp w(k)$ for $\xi \in \mathcal{Q}_k$, cf (2.7) in Lemma 2.1.5, we infer

$$\|\Box_k f\|^q_{L^p_u} w(k)^q \lesssim \|\mathcal{F}^{-1}T_\xi\bar{\hat{\phi}}\mathcal{F}f\|^q_{L^p_u} w(\xi)^q \lesssim \int_{\mathcal{Q}_k} \|\mathcal{F}^{-1}T_\xi\bar{\hat{\phi}}\mathcal{F}f\|^q_{L^p_u} w(\xi)^q\, d\xi.$$

Hence using the equivalent formula (1.26) for the STFT, we obtain

$$\begin{aligned}\sum_{k\in\mathbb{Z}^d} \|\Box_k f\|^q_{L^p_u} w(k)^q &\lesssim \sum_{k\in\mathbb{Z}^d} \int_{\mathcal{Q}_k} \|\mathcal{F}^{-1}T_\xi\bar{\hat{\phi}}\mathcal{F}f\|^q_{L^p_u} w(\xi)^q\, d\xi \\ &= \int_{\mathbb{R}^d} \|V_\phi f\|^q_{L^p_u} w(\xi)^q\, d\xi,\end{aligned}$$

which gives $\|f\|_{\tilde{M}^{p,q}_{u\otimes w}} \lesssim \|f\|_{M^{p,q}_{u\otimes w}}$.

Vice versa, for $\xi \in \mathcal{Q}_k$, we can write

$$|V_\phi f(x,\xi)| = |\mathcal{F}^{-1}T_\xi \bar{\hat{\phi}}\mathcal{F}f(x)| = \Big|\sum_{|l-k|\le C} \mathcal{F}^{-1}\sigma_l T_\xi \bar{\hat{\phi}}\mathcal{F}f(x)\Big|.$$

Thus, for $\xi \in \mathcal{Q}_k$, using Young's inequality again,

$$\begin{aligned}\|V_\phi f(\cdot,\xi)\|_{L^p_u} &= \Big\|\sum_{|l-k|\le C} \mathcal{F}^{-1}\sigma_l T_\xi \bar{\hat{\phi}}\mathcal{F}f\Big\|_{L^p_u} \\ &\lesssim \sum_{|l-k|\le C} \|\mathcal{F}^{-1}[T_\xi \bar{\hat{\phi}}]\|_{L^p_v}\|\Box_l f\|_{L^p_u} \\ &= \sum_{|l-k|\le C} \|M_\xi \mathcal{F}^{-1}[\bar{\hat{\phi}}]\|_{L^p_v}\|\Box_l f\|_{L^p_u} \\ &\lesssim \sum_{|l-k|\le C} \|\Box_l f\|_{L^p_u}.\end{aligned}$$

Taking the q-power integration over $\xi \in \mathcal{Q}_k$ with weight w yields

$$\int_{\mathcal{Q}_k} \|V_\phi f(\cdot,\xi)\|_{L^p_u} w(\xi)^q \, d\xi \lesssim \sum_{|l-k|\le C} \|\Box_l f\|^q_{L^p_u} w(k)^q.$$

Summing over k,

$$\|f\|_{M^{p,q}_{u\otimes w}} \lesssim \|f\|_{\tilde{M}^{p,q}_{u\otimes w}}.$$

This concludes the proof. □

The previous proposition is an ingredient for the following characterization.

For tempered distributions compactly supported either in time or in frequency, the $M^{p,q}$-norm is equivalent to the $\mathcal{F}L^q$-norm or L^p-norm, respectively, cf. [244, 255].

Proposition 2.3.26. *Let* $1 \le p,q \le \infty$.

(i) *For every* $f \in \mathcal{S}'(\mathbb{R}^d)$, *supported in a compact set* $K \subset \mathbb{R}^d$, *we have* $f \in M^{p,q} \Longleftrightarrow f \in \mathcal{F}L^q$, *and*

$$C_K^{-1}\|f\|_{M^{p,q}} \le \|f\|_{\mathcal{F}L^q} \le C_K\|f\|_{M^{p,q}}, \tag{2.49}$$

where $C_K > 0$ *depends only on* K.

(ii) *For every* $f \in \mathcal{S}'(\mathbb{R}^d)$, *whose Fourier transform is supported in a compact set* $K \subset \mathbb{R}^d$, *we have* $f \in M^{p,q} \Longleftrightarrow f \in L^p$, *and*

$$C_K^{-1}\|f\|_{M^{p,q}} \le \|f\|_{L^p} \le C_K\|f\|_{M^{p,q}}, \tag{2.50}$$

where $C_K > 0$ *depends only on* K.

Proof. (i) Consider $f \in \mathcal{S}'(\mathbb{R}^d)$ supported in a compact set $K \subset \mathbb{R}^d$ and $R > 0$ such that $K \subset B_R(0)$.

Assume first $f \in \mathcal{F}L^q(\mathbb{R}^d)$. Let $g \in \mathcal{D}(\mathbb{R}^d)$ with $\operatorname{supp} g \subset B_R(0)$. Then for each $\omega \in \mathbb{R}^d$, $V_g f(\cdot,\omega)$ is supported in $B_{2R}(0)$. Thus, using (1.26),

$$|V_g f(x,\xi)| = |\mathcal{F}^{-1}(\hat{f} T_\xi \bar{\hat{g}})(x)|,$$

and setting $\operatorname{meas}(B_{2R}(0))^{1/p} = C_{R,p}$, we have the following estimates:

$$\begin{aligned}\|V_g f(\cdot,\xi)\|_{L^p} &\le \operatorname{meas}(B_{2R}(0))^{1/p}\|V_g f(\cdot,\xi)\|_{L^\infty} = C_{R,p}\|\mathcal{F}^{-1}(\hat{f} T_\xi \bar{\hat{g}})\|_{L^\infty}\\ &\le C_{R,p}\|\hat{f} T_\xi \bar{\hat{g}}\|_{L^1} \le C_{R,p}|\hat{f}| * |\check{g}|(\xi).\end{aligned}$$

Finally, taking the L^q-norm in the above inequalities,

$$\|\|V_g f(x,\xi)\|_{L^p_x}\|_{L^q_\xi} \le C_{R,p}\||\hat{f}| * |\check{g}|\|_{L^q} \le C_{R,p}\|\hat{f}\|_{L^q}\|\check{g}\|_{L^1},$$

i. e., $\|f\|_{M^{p,q}} \lesssim \|f\|_{\mathcal{F}L^q}$. Vice versa, consider $f \in M^{p,q}(\mathbb{R}^d)$, $g \in \mathcal{D}(\mathbb{R}^d)$ with $g \equiv 1$ on $B_{2R}(0)$. Observe that

$$g(t-x) = 1, \quad \forall t,x \in B_R(0).$$

Hence for every $\xi \in \mathbb{R}^d$

$$\hat{f}(\xi)\chi_{B_R(0)}(x) = V_g f(x,\xi)\chi_{B_R(0)}(x)$$

and, taking the L^p-norm with respect to the x-variable,

$$\operatorname{meas}(B_R(0))^{1/p}|\hat{f}(\xi)| = \|V_g f(\cdot,\xi)\chi_{B_R(0)}(\cdot)\|_{L^p} \le \|V_g f(\cdot,\xi)\|_{L^p}.$$

This yields

$$\|\hat{f}\|_{L^q} \le \operatorname{meas}(B_R(0))^{-1/p}\|\|V_g f(x,\xi)\|_{L^p_x}\|_{L^q_\xi},$$

as desired.

(ii) Using the equivalent $M^{p,q}$-norm in (2.48),

$$\|f\|_{M^{p,q}} \asymp \Big(\sum_{k\in\mathbb{Z}^d} \|\square_k f\|_{L^p}^q\Big)^{1/q}.$$

Now, if $\hat{f}$ has compact support, the above sum is finite, and one deduces at once the first estimate in (2.50), since the multipliers $\square_k$ are (uniformly) bounded on L^p. To obtain the second estimate in (2.50), we write $f = \sum_{k\in\mathbb{Z}^d} \square_k f$, then apply the triangle inequality and the finiteness of the sum over k again. □

For another characterization of modulation spaces using symplectic rotations, we refer the reader to the recent contribution [64].

2.3.2.1 Action of the Fourier transform on modulation spaces

We now consider the action of the Fourier transform on modulation spaces.

Theorem 2.3.27. *For $1 \le p \le \infty$, u, w even weights on $\mathbb{R}^d$, the Fourier transform $\mathcal{F}$ is a topological isomorphism from $M^p_{u\otimes w}(\mathbb{R}^d)$ onto $M^p_{w\otimes u}(\mathbb{R}^d)$ with inverse $\mathcal{F}^{-1}$. In particular, if $u = w$, $\mathcal{F}$ is an automorphism on $M^p_{u\otimes u}(\mathbb{R}^d)$.*

Proof. The key idea is to use the fundamental identity of time–frequency analysis (1.31), which gives

$$V_{\hat{g}}\hat{f}(x,\xi) = V_g f(-\xi, x), \quad (x,\xi) \in \mathbb{R}^{2d}.$$

In fact, consider $p < \infty$, let $f \in M^p_{u\otimes w}(\mathbb{R}^d)$, and a window function $g \in \mathcal{S}(\mathbb{R}^d)$. Then $\hat{g} \in \mathcal{S}(\mathbb{R}^d)$ and can be chosen as window function for $\hat{f}$, so that, using Tonelli theorem,

$$\begin{aligned}
\|\hat{f}\|^p_{M^p_{w\otimes u}} &\asymp \|V_{\hat{g}}\hat{f}\|^p_{L^p_{w\otimes u}} = \int_{\mathbb{R}^d}\Big(\int_{\mathbb{R}^d} |V_{\hat{g}}\hat{f}(x,\xi)|^p w^p(x)\,dx\Big)u^p(\xi)\,d\xi \\
&= \int_{\mathbb{R}^d}\Big(\int_{\mathbb{R}^d} |V_g f(-\xi,x)|^p u^p(\xi)\,d\xi\Big)w^p(x)\,dx \\
&= \|V_g f\|^p_{L^p_{u\otimes w}} \asymp \|f\|^p_{M^p_{u\otimes w}}.
\end{aligned}$$

The case $p = \infty$ is analogous.

Similarly one can show that the inverse Fourier transform $\mathcal{F}^{-1}$ is bounded from $M^p_{w\otimes u}(\mathbb{R}^d)$ to $M^p_{u\otimes w}(\mathbb{R}^d)$, and the claim follows. □

Using the same pattern, we obtain the following result.

Proposition 2.3.28. *Consider $1 \le p \le \infty$, m a weight function on $\mathbb{R}^{2d}$ such that $m(x,\xi) = m(\xi, x)$ for every $(x,\xi) \in \mathbb{R}^{2d}$ and even with respect to x and ξ,*

$$m(-x,\xi) = m(x,\xi) = m(x,-\xi) = m(-x,-\xi), \quad \forall (x,\xi) \in \mathbb{R}^{2d}.$$

Then the Fourier transform $\mathcal{F}$ is an automorphism on $M^p_m(\mathbb{R}^d)$ with inverse $\mathcal{F}^{-1}$.

Of course, the previous arguments do not apply for the more general case of modulation spaces with two different indices $p \ne q$. In this case, we enter the realm of Wiener amalgam spaces, see the next section.

Using the fundamental identity of time–frequency analysis (1.31), we write the STFT as follows:

$$|V_g f(x,\omega)| = |V_{\hat{g}}\hat{f}(\omega,-x)| = |\mathcal{F}[\hat{f}\,T_\omega\overline{\hat{g}}\,](-x)|. \tag{2.51}$$

Since the weights are even, for $1 < p, q < \infty$, we compute

$$\|f\|_{M^{p,q}_{u\otimes w}} = \Big(\int_{\mathbb{R}^d} \|\hat{f}\,T_\omega\overline{\hat{g}}(x,\omega)\|^q_{\mathcal{F}L^p_u} w^q(\omega)\,d\omega\Big)^{\frac{1}{q}}. \tag{2.52}$$

(The cases $p = \infty$ and $q = \infty$ are similar.) This allows, via the Fourier isomorphism, giving a Banach space structure to $\mathcal{F}M^{p,q}_{u\otimes w}(\mathbb{R}^d)$ with norm

$$\|f\|_{\mathcal{F}M^{p,q}_{u\otimes w}} = \left(\int_{\mathbb{R}^d}\left(\int_{\mathbb{R}^d}|V_g f(x,\omega)|^p u^p(\omega)\,d\omega\right)^{\frac{q}{p}} w^q(x)\,dx\right)^{\frac{1}{q}}, \tag{2.53}$$

for $1 \leq p, q < \infty$. If $p = \infty$ or $q = \infty$, the integral on $\mathbb{R}^d$ is replaced by the supremum on $\mathbb{R}^d$.

As observed by H. G. Feichtinger in his insightful retrospective survey [127], it would be natural to include such spaces into a more general class of modulation spaces, where a switch in the order of integration is considered as well in Definition 2.3.1. In the sequel, following a standard procedure, we shall rather regard $\mathcal{F}M^{p,q}_{u\otimes w}(\mathbb{R}^d)$ as the special case of the Wiener amalgam spaces with local components $\mathcal{F}L^p_u$ and global component L^q_w, $1 \leq p, q \leq \infty$, cf. (2.52). Namely, with the notation of the next section,

$$\mathcal{F}M^{p,q}_{u\otimes w}(\mathbb{R}^d) = W(\mathcal{F}L^p_u, L^q_w)(\mathbb{R}^d). \tag{2.54}$$

2.4 Wiener amalgam spaces

We now introduce a class of spaces of functions (or distributions) defined by a norm which "amalgamates" (i. e., mixes) a local criterion for membership with a global one. More precisely, the norm provides a global criterion for a local property, in a sense that will be specified in the sequel. The first appearance of these spaces traces back to several works by N. Wiener (cf. [310, 311, 313]), and we refer to the standard papers [38, 124, 148, 181, 189] for a systematic review of the subject.

A remark on the name is needed as the current formulation of the theory for these function spaces is due to H. G. Feichtinger [120, 121, 122] who, following a suggestion of J. J. Benedetto, named them "Wiener amalgam spaces."

For the sake of clarity, let us first introduce a particular kind of amalgam space, modeled on the L^p spaces, but much more flexible since it is possible to control the local regularity of a function and its decay at infinity separately.

Definition 2.4.1 (Weighted Wiener amalgams $W(L^p, L^q_m)$). Consider $p, q \in [1, \infty]$, a weight function $m \in \mathcal{M}_v(\mathbb{R}^d)$, and a compact set $Q \subset \mathbb{R}^d$ with nonempty interior. The Wiener amalgam space $W(L^p, L^q_m)(\mathbb{R}^d)$ consists of the functions $f : \mathbb{R}^d \to \mathbb{C}$ such that $f \in L^p_{loc}(\mathbb{R}^d)$ and for which the *control function*

$$F_Q(x) := \|f \cdot T_x \chi_Q\|_{L^p} \in L^q_m(\mathbb{R}^d), \quad x \in \mathbb{R}^d. \tag{2.55}$$

The norm on $W(L^p, L^q_m)$ is given by

$$\begin{aligned}\|f\|_{W(L^p,L^q_m)} &:= \|F_Q(x)\|_{L^q_m}\\ &= \big\|\|f\cdot T_x\chi_Q\|_{L^p}\big\|_{L^q_m}\\ &= \left(\int_{\mathbb{R}^d}\left(\int_{\mathbb{R}^d}|f(t)|^p\chi_Q(t-x)\,dt\right)^{\frac{q}{p}} m^q(x)\,dx\right)^{\frac{1}{q}},\end{aligned} \tag{2.56}$$

with suitable adjustments for the cases $p, q = \infty$.

Remark 2.4.2.

(i) This special definition allows us to grasp the sense of the amalgam; we first view f "locally", through translations $T_x\chi_Q$ of the sharp cut-off function χ_Q, and measure those local pieces in the L^p-norm, then we measure the global behavior of those local pieces according to the L^q_m-norm. The "window", through which we view f locally need not be a unit d-dimensional cube.

(ii) There is no gain in generality by allowing the local component to be a weighted L^p space. It can be shown that $W(L^p_v, L^q_m) = W(L^p, L^q_{vm})$; so it suffices to only consider the case where the global component is a weighted space.

Generalizing the previous argument, we now want to consider a space of suitable functions with global behavior (decay and/or summability) expressed by the global component C and with local properties given by the local component B, for certain Banach spaces B, C. Even if the construction can be carried out in a more abstract context (cf. [124]) we will restrict ourselves to Euclidean spaces $\mathbb{R}^d$ and to subspaces of temperate distributions, following [125], and we will provide technical conditions on B and C in order to properly define $W(B, C)$. We observe that general Wiener amalgams $W(B, C)$ are not simply "gluing together disjoint local pieces." In particular, overlaps are essential when moving to more general amalgams determined by continuity, smoothness, or other criteria. In this case the simple cut-off function χ_Q of Definition 2.4.1 is clearly inadequate. To define general Wiener amalgams $W(B, C)$, an appropriate smooth cut-off function needs to be employed in place of the sharp cut-off function χ_Q above. For local components, we require

Definition 2.4.3 (Local component space). A Banach space $(B, \|\cdot\|_B)$ is a local component space if

(i) $\mathcal{S}(\mathbb{R}^d) \hookrightarrow B \hookrightarrow \mathcal{S}'(\mathbb{R}^d)$ are continuous embeddings;

(ii) For every test function $g \in \mathcal{D}(\mathbb{R}^d)$ and $f \in B$ one has $gf \in B$. Moreover, there exists a constant $C = C(g) > 0$ such that

$$\|fT_x g\|_B \le C\|f\|_B \quad \forall x \in \mathbb{R}^d,\ \forall f \in B.$$

In view of the previous condition, the following space is well defined:

$$B_{\mathrm{loc}} := \{f \in \mathcal{S}'(\mathbb{R}^d) : gf \in B,\ \forall g \in \mathcal{D}(\mathbb{R}^d)\}.$$

Following the standard terminology of time–frequency analysis, any given nonzero $g \in \mathcal{D}(\mathbb{R}^d)$ will be called a *window function* (windows usually can be taken from a much larger space), and we can define the related *control function* as

$$F_g(x) := \|fT_x g\|_B. \tag{2.57}$$

The role of the control function is to localize f by means of the window g, measured in the B-norm.

Example 2.4.4. Examples of local component Banach spaces are the following:
(i) The ordinary $L^p(\mathbb{R}^d)$ spaces, $p \in [1,\infty]$;
(ii) For $m \in \mathcal{M}_v(\mathbb{R}^d)$, the Fourier–Lebesgue spaces $\mathcal{F}L^p_m(\mathbb{R}^d)$, $1 \le p \le \infty$. These are local component spaces, since the embeddings $\mathcal{S} \hookrightarrow \mathcal{F}L^1_m \hookrightarrow \mathcal{S}'$ and Young's inequality $L^1_v * L^p_m \subseteq L^p_m$ imply that the elements of $\mathcal{F}L^1_v$ define bounded multipliers on $\mathcal{F}L^p_m$.
(iii) The mixed norm spaces $L^{p,q}(\mathbb{R}^{2d})$, $p,q \in [1,\infty]$.

For the global behavior, we introduce the following conditions.

Definition 2.4.5 (Global component space). A Banach space $(C, \|\cdot\|_C)$ can serve as a global component space if
(i) $(C, \|\cdot\|_C)$ is a solid BF-space, i. e., a Banach space of locally integrable functions, continuously embedded into $L^1_{\mathrm{loc}}(\mathbb{R}^d)$, such that for every $f \in C$, $g \in L^1_{\mathrm{loc}}(\mathbb{R}^d)$,

$$|g(x)| \le |f(x)| \quad \text{for a. e. } x \in \mathbb{R}^d \Longrightarrow g \in C \text{ and } \|g\|_C \le \|f\|_C. \tag{2.58}$$

(ii) $(C, \|\cdot\|_C)$ is with tempered translation, that is, it satisfies the following estimate for some $s \ge 0$:

$$\|T_x f\|_C \lesssim v_s(x)\|f\|_C, \quad v_s(x) = \langle x\rangle^s. \tag{2.59}$$

(iii) $(C, \|\cdot\|_C)$ is a Banach convolution module with respect to the Beurling algebra $L^1_{v_s}(\mathbb{R}^d)$, $s \ge 0$, i. e.,

$$f * g \in C \quad \text{for } f \in C,\ g \in L^1_{v_s}(\mathbb{R}^d),$$

and we have the norm estimate

$$\|f * g\|_C \le \|g\|_{L^1_{v_s}} \|f\|_C. \tag{2.60}$$

Example 2.4.6. Examples of global component spaces are the following:

(i) The ordinary L^p spaces, $p \in [1, \infty]$;
(ii) Weighted $L^p_m(\mathbb{R}^d)$ spaces, where $m \in \mathcal{M}_{v_s}(\mathbb{R}^d)$, $s \geq 0$;
(iii) The mixed norm spaces $L^{p,q}_m(\mathbb{R}^{2d})$, $p, q \in [1, \infty]$, $m \in \mathcal{M}_{v_s}(\mathbb{R}^{2d})$, $s \geq 0$;
(iv) Orlicz spaces, as described in a recent paper [293].

Definition 2.4.7 (Wiener amalgam spaces $W(B, C)$). Let $(B, \|\cdot\|_B)$ be a local component space and $(C, \|\cdot\|_C)$ a global component space. We define the Wiener amalgam space $W(B, C)$ with local component B and global component C as

$$W(B, C)(\mathbb{R}^d) = \{f \in B_{\mathrm{loc}} : F_g \in C\}, \tag{2.61}$$

where $g \in \mathcal{D}(\mathbb{R}^d)\setminus\{0\}$ is a fixed window function and F_g is the associated control function. The natural norm on $W(B, C)$ is

$$\|f\|_{W(B,C)} := \|F_g\|_C.$$

Example 2.4.8. Combining Examples 2.4.4 and 2.4.6, we obtain some Wiener amalgam spaces, relevant for the applications in the sequel.
(i) $W(L^p, L^q_m)(\mathbb{R}^d)$ are the spaces in Definition 2.4.1. The use of the window χ_Q does not produce more generality than $g \in \mathcal{D}(\mathbb{R}^d)$.
(ii) $W(\mathcal{F}L^p_u, L^q_w)(\mathbb{R}^d) = \mathcal{F}M^{p,q}_{u\otimes w}(\mathbb{R}^d)$, according to (2.54).
(iii) For $p = q$ and under the assumptions of Theorem 2.3.27, we infer

$$W(\mathcal{F}L^p_u, L^p_w)(\mathbb{R}^d) = M^p_{u\otimes w}(\mathbb{R}^d). \tag{2.62}$$

For conditions on weights u and w which guarantee invariance under Fourier transform, we refer to [124]. Notice that, when $p = q = 1$, the space

$$W(\mathcal{F}L^1, L^1)(\mathbb{R}^d) = M^1(\mathbb{R}^d) \tag{2.63}$$

is the well-known Feichtinger algebra, also denoted by $S_0(\mathbb{R}^d)$, see [202]. It can be shown that it is the smallest Banach algebra invariant with respect to pointwise and convolution products. The algebra's property follows from the convolution and multiplication relations for modulation spaces shown in Propositions 2.4.19 and 2.4.23.
(iv) The space $W(\mathcal{F}L^1, L^{p,q}_m)(\mathbb{R}^{2d})$ will describe the regularity of $V_g f$ for $f \in M^{p,q}_m(\mathbb{R}^d)$ ($g \in M^1(\mathbb{R}^d)$); see the sequel.

We now state the main properties of $W(B, C)$ in the general setting, for proofs we refer to the original literature [130], see also the master thesis of T. Dobler [112].

First, we recall that a Banach space $(Y, \|\cdot\|_Y)$ has an absolutely continuous norm if the Banach dual Y' coincides with the Köthe-dual $Y^\alpha := \{H \in L^1_{\mathrm{loc}}(\mathbb{R}^d) : HF \in L^1(\mathbb{R}^d), \forall F \in Y\}$. For example, for $1 \leq p < \infty$, the Lebesgue space $L^p(\mathbb{R}^d)$ has an absolutely continuous norm, but $L^\infty(\mathbb{R}^d)$ does not.

Theorem 2.4.9 (Properties of Wiener amalgam spaces $W(B,C)$). *Let B, B_i local component spaces and C, C_i global component spaces ($i = 1,2,3$).*

(i) (Banach space property) *$(W(B,C)(\mathbb{R}^d), \|\cdot\|_{W(B,C)})$ is a Banach space continuously embedded into $\mathcal{S}'(\mathbb{R}^d)$; different choices of window function in $\mathcal{D}(\mathbb{R}^d)$ provide equivalent norms.*

(ii) (Convolution) *If $B_1 * B_2 \hookrightarrow B_3$ and $C_1 * C_2 \hookrightarrow C_3$, then*

$$W(B_1, C_1) * W(B_2, C_2) \hookrightarrow W(B_3, C_3).$$

If, moreover, (B_1, B_2, B_3) and (C_1, C_2, C_3) are Banach convolution triples, i. e.,

$$\|f * g\|_{B_3} \lesssim \|f\|_{B_1} \|g\|_{B_2} \quad \|f * g\|_{C_3} \lesssim \|f\|_{C_1} \|g\|_{C_2},$$

then $(W(B_1, C_1), W(B_2, C_2), W(B_3, C_3))$ is a Banach convolution triple

$$\|f * g\|_{W(B_3,C_3)} \lesssim \|f\|_{W(B_1,C_1)} \|g\|_{W(B_2,C_2)}.$$

(iii) (Inclusions) *If $B_1 \hookrightarrow B_2$ and $C_1 \hookrightarrow C_2$, then $W(B_1, C_1) \hookrightarrow W(B_2, C_2)$.*

(iv) (Complex interpolation) *For $0 < \theta < 1$, we have*

$$[W(B_1, C_1), W(B_2, C_2)]_{[\theta]} = W([B_1, B_2]_{[\theta]}, [C_1, C_2]_{[\theta]}) \tag{2.64}$$

if C_1 or C_2 has absolutely continuous norm. The same holds if every Wiener amalgam space is replaced by the closure of the Schwartz space into itself.

(v) (Duality) *If B', C' are the topological dual spaces of B, C respectively, and the space of test functions $\mathcal{D}$ is dense in both B and C, then*

$$W(B, C)' = W(B', C').$$

(vi) (Pointwise products) *If $B_1 \cdot B_2 \hookrightarrow B_3$ and $C_1 \cdot C_2 \hookrightarrow C_3$, we have*

$$W(B_1, C_1) \cdot W(B_2, C_2) \hookrightarrow W(B_3, C_3).$$

One class of Wiener amalgam spaces we are going to work with is given by $W(L^p, L^q_m)(\mathbb{R}^{2d})$, $1 \le p, q \le \infty$. Let us show that they are translation-invariant.

Lemma 2.4.10. *If $m \in \mathcal{M}_v(\mathbb{R}^d)$, then $W(L^p, L^q_m)(\mathbb{R}^d)$ is translation invariant, with*

$$\|T_y f\|_{W(L^p,L^q_m)} \le Cv(y) \|f\|_{W(L^p,L^q_m)}, \quad \forall y \in \mathbb{R}^d.$$

In particular, if $m \equiv 1$, the translation T_y is an isometry on $W(L^p, L^q_m)(\mathbb{R}^d)$.

Proof. Fix a window function g in $\mathcal{D}(\mathbb{R}^d)$ and consider $f \in W(L^p, L^q_m)$. Since L^p is translation-invariant, we have $T_y f \in L^p$ and

$$\|T_y f T_x g\|_{L^p} = \|f T_{x-y} g\|_{L^p} = F_g(x-y) = T_y F_g(x).$$

Since weighted Lebesgue space L^q_m is translation invariant as well, by using (2.3) as standard, we have

$$\|T_y f\|_{W(L^p,L^q_m)} = \|T_y F_g\|_{L^q_m} \leq Cv(y)\|f\|_{W(L^p,L^q_m)}.$$

If $m \equiv 1$, we use the fact that translation is an isometry on L^q. □

In the next result, we show that if $p = q$ then the Wiener amalgam space $W(L^p, L^p_m)$ turns into the Lebesgue space L^p_m and in particular $W(L^p, L^p) = L^p$. To this end, we need a lemma which introduces an important property of weight functions.

Lemma 2.4.11. *Let m be a v-moderate weight. Then for any compact set $K \subset \mathbb{R}^d$ with positive (Lebesgue) measure, there exist constants $C_1(K) > 0$ and $C_2(K) > 0$ such that*

$$C_1(K)m(y) \leq \int_{K+x} m(t)\,dt \leq C_2(K)m(y), \quad x \in \mathbb{R}^d \tag{2.65}$$

for all $y \in K + x$.

Proof. First, we check that given a compact set $K \subset \mathbb{R}^d$ with positive measure, there exists a constant $C(K) > 0$ such that

$$\sup_{t\in K+x} m(t) \leq C(K) \inf_{t\in K+x} m(t), \quad \forall x \in \mathbb{R}^d. \tag{2.66}$$

Fix any compact set $K \subset \mathbb{R}^d$ and any $x \in \mathbb{R}^d$. Let $t = r + x$ and $u = s + x$ be any points in $K + x$. Then

$$m(t) = m(r + x) \leq Am(x)v(r) \leq A \sup_{r\in K} v(r)m(x) \leq A_1(K)m(x)$$

and

$$m(x) = m(u - s) \leq Bm(u)v(-s) \leq \sup_{s\in -K} v(s)m(u) \leq B_1(K)m(u).$$

If we set $C(K) = A_1(K)B_1(K)$, we get $m(t) \leq C(K)m(u)$ for all $t, u \in K + x$ and consequently (2.66).

Next, given any $y \in K + x$, we deduce

$$m(y) \leq \sup_{t\in K+x} m(t) \leq C(K) \inf_{t\in K+x} m(t) \leq \frac{C(K)}{|K|} \int_{K+x} m(t)\,dt,$$

and

$$\int_{K+x} m(t)\,dt \leq |K| \sup_{t\in K+x} m(t) \leq |K|C(K) \inf_{t\in K+x} m(t) \leq |K|C(K)m(y).$$

Therefore we obtain (2.65), with $C_1(K) = |K|/C(K)$ and $C_2(K) = |K|C(K)$. □

Proposition 2.4.12. *Let m be a v-moderate weight. We have*

$$W(L^p, L^p_m)(\mathbb{R}^d) = L^p_m(\mathbb{R}^d). \tag{2.67}$$

Proof. Suppose that $1 \le p < \infty$, take a window function $g \in \mathcal{D}(\mathbb{R}^d)$, then

$$\begin{aligned}
\|f\|^p_{W(L^p,L^p_m)} &= \int_{\mathbb{R}^d}\int_{\mathbb{R}^d} |f(y)|^p |g(x-y)|^p m^p(x)\,dx\,dy \\
&= \int_{\mathbb{R}^d} |f(y)|^p \Big(\int_{\mathbb{R}^d} |g(x-y)|^p m^p(x)\,dx\Big)\,dy.
\end{aligned}$$

We can take as compact set $K = \operatorname{supp} g$. Consequently,

$$\int_{\mathbb{R}^d} |g(x-y)|^p m^p(x)\,dx = \int_{K+y} |g(x-y)|^p m^p(x)\,dx.$$

Lemma 2.4.11 yields

$$Am^p(y) \le \int_{K+y} m^p(x)\,dx \le Bm^p(y), \tag{2.68}$$

so that, using the left-hand side inequality above,

$$\begin{aligned}
A\|f\|^p_{L^p_m} &= A\int_{\mathbb{R}^d} |f(y)|^p m^p(y)\,dy \\
&\le \int_{\mathbb{R}^d} |f(y)|^p \Big(\int_{K+y} m^p(x)\,dx\Big)\,dy \\
&= \int_{\mathbb{R}^d} |f(y)|^p \Big(\int_{\mathbb{R}^d} \chi_K(x-y) m(x)^p\,dx\Big)\,dy \\
&\asymp \|f\|^p_{W(L^p,L^p_m)}.
\end{aligned}$$

On the other hand, using the right-hand side inequality in (2.68),

$$\begin{aligned}
\|f\|^p_{W(L^p,L^p_m)} &\le \sup_{x\in K}|g(x)|^p \int_{\mathbb{R}^d} |f(y)|^p \Big(\int_{K+y} m^p(x)\,dx\Big)\,dy \\
&\le B\sup_{x\in K}|g(x)|^p \|f\|^p_{L^p_m}.
\end{aligned}$$

Hence $W(L^p, L^p_m)$ is equal to L^p_m with equivalent norm. The remaining case $p = \infty$ is similar. □

The same arguments show that, for $m \in \mathcal{M}_v(\mathbb{R}^{2d})$,

$$W(L^{p,q}, L^{p,q}_m)(\mathbb{R}^{2d}) = L^{p,q}_m(\mathbb{R}^{2d}). \tag{2.69}$$

For further references, we recall the action of the Fourier transform on the Wiener amalgam spaces $W(\mathcal{F}L^q, L^p)$ (cf. [124]).

Proposition 2.4.13. *For every $1 \le p \le q \le \infty$, the Fourier transform $\mathcal{F}$ maps $W(\mathcal{F}L^q, L^p)$ in $W(\mathcal{F}L^p, L^q)$ continuously.*

For $p = q$, this was already proved in Theorem 2.3.27, cf. (ii) in Example 2.4.8. Note also that in view of Proposition 2.4.13, exchanging the role of p, q and using (2.54), (ii) in Example 2.4.8, we have, for $p \ge q$,

$$M^{p,q} = \mathcal{F}W(\mathcal{F}L^p, L^q) \hookrightarrow W(\mathcal{F}L^q, L^p)$$

and, in particular,

$$M^{\infty,1} \hookrightarrow W(\mathcal{F}L^1, L^\infty). \tag{2.70}$$

The above inclusion is strict, this will play a crucial role in Chapter 4.

2.4.1 Inclusion, convolution, and multiplication relations

We now mainly focus on modulation spaces and present their main properties in terms of inclusion, convolution, and multiplication relations. For this reason, a finer estimate in terms of Wiener amalgam spaces of the STFT is needed, see Lemma 2.4.15.

Lemma 2.4.14. *Fix $g \in M^1_v(\mathbb{R}^{2d}) \setminus \{0\}$. For $f \in W(\mathcal{F}L^1, L^{p,q}_m)(\mathbb{R}^{2d})$, we can express the norm by means of the STFT as follows:*

$$\|f\|_{W(\mathcal{F}L^1, L^{p,q}_m)} = \left(\int_{\mathbb{R}^d}\left(\int_{\mathbb{R}^d}\left(\int_{\mathbb{R}^{2d}} |V_g f(z_1, z_2, \zeta)|\, d\zeta\right)^p m(z_1, z_2)^p\, dz_1\right)^{q/p} dz_2\right)^{1/q}. \tag{2.71}$$

Proof. It is a simple computation. In fact, consider first $g \in \mathcal{S}(\mathbb{R}^{2d}) \setminus \{0\}$, $f \in \mathcal{S}(\mathbb{R}^{2d})$,

$$\begin{aligned}
\|f\|_{W(\mathcal{F}L^1, L^{p,q}_m)} &= \left(\int_{\mathbb{R}^d}\left(\int_{\mathbb{R}^d} \|f \cdot T_{(z_1,z_2)}\bar{g}\|^p_{\mathcal{F}L^1} m(z_1, z_2)^p\, dz_1\right)^{q/p} dz_2\right)^{1/q} \\
&= \left(\int_{\mathbb{R}^d}\left(\int_{\mathbb{R}^d}\left(\int_{\mathbb{R}^{2d}} |(f \cdot T_{(z_1,z_2)}\bar{g})\widehat{\ }(\zeta)|\, d\zeta\right)^p m(z_1, z_2)^p\, dz_1\right)^{q/p} dz_2\right)^{1/q} \\
&= \left(\int_{\mathbb{R}^d}\left(\int_{\mathbb{R}^d}\left(\int_{\mathbb{R}^{2d}} |V_g f(z_1, z_2, \zeta)|\, d\zeta\right)^p m(z_1, z_2)^p\, dz_1\right)^{q/p} dz_2\right)^{1/q}.
\end{aligned}$$

The extension to arbitrary f, g is done by approximation (see, e. g., the proof of Theorem 2.3.12). □

Lemma 2.4.15. *Let $1 \le p, q \le \infty$. If $f \in M_m^{p,q}(\mathbb{R}^d)$ and $g \in M_v^1(\mathbb{R}^d)$, then $V_g f \in W(\mathcal{F}L^1, L_m^{p,q})(\mathbb{R}^{2d})$ with norm estimate*

$$\|V_g f\|_{W(\mathcal{F}L^1, L_m^{p,q})} \lesssim \|f\|_{M_m^{p,q}} \|g\|_{M_v^1}. \tag{2.72}$$

Proof. Let $\varphi \in \mathcal{S}(\mathbb{R}^d) \setminus \{0\}$ and set $\Phi = V_\varphi \varphi \in \mathcal{S}(\mathbb{R}^{2d})$. We only consider $f, g \in \mathcal{S}(\mathbb{R}^d)$, the extension to arbitrary f, g can be obtained by approximation.

We use (2.71) and then Lemma 1.3.37 (i) to evaluate the $W(\mathcal{F}L^1, L_m^{p,q})$-norm of $V_g f$.

$$\begin{aligned}
&\|V_g f\|_{W(\mathcal{F}L^1, L_m^{p,q})} \\
&\asymp \bigg(\int_{\mathbb{R}^d}\bigg(\int_{\mathbb{R}^d}\bigg(\int_{\mathbb{R}^{2d}} |V_\Phi(V_g f)(z_1, z_2, \zeta_1, \zeta_2)|\, d\zeta_1\, d\zeta_2\bigg)^p m(z_1, z_2)^p\, dz_1\bigg)^{q/p} dz_2\bigg)^{1/q} \\
&= \bigg(\int_{\mathbb{R}^d}\bigg(\int_{\mathbb{R}^d}\bigg(\int_{\mathbb{R}^{2d}} |V_\varphi g(-z_1 - \zeta_2, \zeta_1)||V_\varphi f(-\zeta_2, z_2 + \zeta_1)|\, d\zeta_1\, d\zeta_2\bigg)^p \\
&\qquad \times m(z_1, z_2)^p\, dz_1\bigg)^{q/p} dz_2\bigg)^{1/q} \\
&= \bigg(\int_{\mathbb{R}^d}\bigg(\int_{\mathbb{R}^d}\bigg(\int_{\mathbb{R}^{2d}} |V_\varphi f(w_1, w_2)||V_\varphi g(w_1 - z_1, w_2 - z_2)|\, dw_1\, dw_2\bigg)^p \\
&\qquad \times m(z_1, z_2)^p\, dz_1\bigg)^{q/p} dz_2\bigg)^{1/q} \\
&= \bigg(\int_{\mathbb{R}^d}\bigg(\int_{\mathbb{R}^d} (|V_\varphi f| * |V_\varphi g|^*)^p (z_1, z_2) m(z_1, z_1)^p\, dz_1\bigg)^{q/p} dz_2\bigg)^{1/q} \\
&= \big\||V_\varphi f| * |V_\varphi g|^*\big\|_{L_m^{p,q}}.
\end{aligned}$$

Now Young's inequality for mixed norm spaces (cf. Theorem 2.2.3 (v)) yields the desired estimate

$$\|V_g f\|_{W(\mathcal{F}L^1, L_m^{p,q})} \lesssim \|V_\varphi f\|_{L_m^{p,q}} \|V_\varphi g\|_{L_v^1} \asymp \|f\|_{M_m^{p,q}} \|g\|_{M_v^1}.$$ □

Corollary 2.4.16. *If $g \in M_v^1(\mathbb{R}^d) \setminus \{0\}$ and $f \in M_m^{p,q}(\mathbb{R}^d)$, then*

$$V_g f \in W(L^\infty, L_m^{p,q})(\mathbb{R}^{2d})$$

with

$$\|V_g f\|_{W(C_0, L_m^{p,q})} \le C \|V_g g\|_{W(L^\infty, L_v^1)} \|f\|_{M_m^{p,q}}. \tag{2.73}$$

Proof. It follows from Lemma 2.4.15 and the inclusion relations for Wiener amalgam spaces since, by Riemann–Lebesgue lemma, $\mathcal{F}L^1(\mathbb{R}^d) \hookrightarrow C_0(\mathbb{R}^d) \hookrightarrow L^\infty(\mathbb{R}^d)$. □

We have now all the instruments to prove the inclusion relations for modulation spaces.

Theorem 2.4.17 (Inclusion relations). *If $p_1 \le p_2$, $q_1 \le q_2$ and $m_2 \lesssim m_1$, then*

$$M^{p_1,q_1}_{m_1}(\mathbb{R}^d) \hookrightarrow M^{p_2,q_2}_{m_2}(\mathbb{R}^d). \tag{2.74}$$

Proof. The following inclusion relation between $\ell^p(\mathbb{Z}^d)$ spaces is well known:

$$\ell^{p_1}(\mathbb{Z}^d) \subseteq \ell^{p_2}(\mathbb{Z}^d) \quad \text{for } p_1 \le p_2.$$

Such relation extends easily to the discrete spaces $\ell^{p,q}(\mathbb{Z}^{2d})$. In fact, let us show

$$\ell^{p_1,q_1}(\mathbb{Z}^{2d}) \subseteq \ell^{p_2,q_2}(\mathbb{Z}^{2d}) \quad \text{for } p_1 \le p_2,\ q_1 \le q_2.$$

Given $a \in \ell^{p_1,q_1}(\mathbb{Z}^{2d})$, define the sequences $c = (c_n)$, $d = (d_n)$ by

$$c_n = \Big(\sum_{k\in\mathbb{Z}^d} |a_{k,n}|^{p_1}\Big)^{\frac{1}{p_1}}, \quad d_n = \Big(\sum_{k\in\mathbb{Z}^d} |a_{k,n}|^{p_2}\Big)^{\frac{1}{p_2}}.$$

Then $d_n \le c_n$ because $p_1 \le p_2$ and

$$\|a\|_{\ell^{p_2,q_2}} = \|d\|_{q_2} \le \|d\|_{q_1} \le \|c\|_{q_1} = \|a\|_{\ell^{p_1,q_1}}.$$

The inclusion

$$\ell^{p_1,q_1}_{m_1}(\mathbb{Z}^{2d}) \subseteq \ell^{p_2,q_2}_{m_2}(\mathbb{Z}^{2d}) \quad \text{for } p_1 \le p_2, q_1 \le q_2, m_2 \le Cm_1, \tag{2.75}$$

follows then from

$$\|a\|_{\ell^{p_2,q_2}_{m_2}} = \|am_2\|_{\ell^{p_2,q_2}} \le \|am_2\|_{\ell^{p_1,q_1}} \le C\|am_1\|_{\ell^{p_1,q_1}} = C\|a\|_{\ell^{p_1,q_1}_{m_1}}.$$

Fix a window $g \in \mathcal{S}(\mathbb{R}^d)\setminus\{0\}$. Given $f \in M^{p,q}_m(\mathbb{R}^d)$, we define the sequence $a = (a_{k,n})$ by

$$a_{k,n} = \sup_{(x,\xi)\in[0,1]^{2d}} |V_g f(x+k, \xi+n)|.$$

Hence we may write $\|a\|_{\ell^{p,q}_m} = \|V_g f\|_{W(L^\infty, L^{p,q}_m)}$. Finally, by Proposition 2.4.12 and (2.75),

$$\begin{aligned}
\|f\|_{M^{p_2,q_2}_{m_2}} &= \|V_g f\|_{L^{p_2,q_2}_{m_2}} = \|V_g f\|_{W(L^{p_2,q_2}, L^{p_2,q_2}_{m_2})} \\
&\le \|V_g f\|_{W(L^\infty, L^{p_2,q_2}_{m_2})} = \|a\|_{\ell^{p_2,q_2}_{m_2}} \le C\|a\|_{\ell^{p_1,q_1}_{m_1}} \\
&= C\|V_g f\|_{W(L^\infty, L^{p_1,q_1}_{m_1})} \le C'\|f\|_{M^{p_1,q_1}_{m_1}},
\end{aligned}$$

where the last estimate follows from Corollary 2.4.16. □

Other useful embeddings are as follows, cf. [126].

Proposition 2.4.18. *Given* $1 \le p_1, p_2, q_1, q_2 \le \infty$, *the weight* v_s *defined in* (2.5), *with* $s = s_1$ *or* $s = s_2$ *in* $\mathbb{R}$, *one has*

$$M^{p_1,q_1}_{1\otimes v_{s_1}}(\mathbb{R}^d) \hookrightarrow M^{p_2,q_2}_{1\otimes v_{s_2}}(\mathbb{R}^d) \tag{2.76}$$

if and only if

$$p_1 \le p_2 \tag{2.77}$$

and

$$q_1 \le q_2, \quad s_1 \ge s_2 \quad or \quad q_1 > q_2, \quad \frac{s_1}{d} + \frac{1}{q_1} > \frac{s_2}{d} + \frac{1}{q_2}. \tag{2.78}$$

Proof. We use the discrete modulation norm defined in (2.48), so that

$$\|f\|_{M^{p,q}_{1\otimes v_s}} \asymp \Big(\sum_{k\in\mathbb{Z}^d} \|\Box_k f\|^q_{L^p} \langle k\rangle^{sq} \Big)^{\frac{1}{q}}.$$

The necessity of (2.77) follows from the fact that $\mathcal{F}L^{p_1}$ is locally contained in $\mathcal{F}L^{p_2}$ if and only if $p_1 \le p_2$ (with strict inclusion if $p_1 < p_2$), cf. [20, 147]. The set of conditions in (2.78) in turn describes the inclusions between weighted ℓ^q spaces. Namely, recall that

$$\ell^q_s = \{a = (a_k)_{k\in\mathbb{Z}^d} : \|\langle k\rangle^s a_k\|_{\ell^q} < \infty\}$$

and

$$\ell^{q_1}_{s_1} \subset \ell^{q_2}_{s_2}$$

if and only if the indices' relations in (2.78) are satisfied; cf., for instance, [177, Lemma 2.10]. □

It is important to notice that the spaces above are equal if and only if all the parameters are equal. We now present the convolution and multiplication properties for modulation spaces.

Proposition 2.4.19. *Consider* $m \in \mathcal{M}_v(\mathbb{R}^{2d})$. *We write* $m_1(x) = m(x,0)$ *and* $m_2(\xi) = m(0,\xi)$ *for the restrictions to* $\mathbb{R}^d \times \{0\}$ *and* $\{0\} \times \mathbb{R}^d$, *and likewise for* v.

Let v *be an arbitrary weight function on* $\mathbb{R}^d$ *satisfying* (2.17) *and* $1 \le p, q, p_1, p_2, q_1, q_2 \le \infty$. *If*

$$\frac{1}{p} + 1 \le \frac{1}{p_1} + \frac{1}{p_2} \tag{2.79}$$

and

$$\frac{1}{q} \leq \frac{1}{q_1} + \frac{1}{q_2}, \tag{2.80}$$

then

$$M^{p_1,q_1}_{m_1\otimes v}(\mathbb{R}^d) * M^{p_2,q_2}_{v_1\otimes v_2 v^{-1}}(\mathbb{R}^d) \hookrightarrow M^{p,q}_m(\mathbb{R}^d) \tag{2.81}$$

with norm inequality

$$\|f * h\|_{M^{p,q}_m} \lesssim \|f\|_{M^{p_1,q_1}_{m_1\otimes v}} \|h\|_{M^{p_2,q_2}_{v_1\otimes v_2 v^{-1}}}. \tag{2.82}$$

Proof. We measure the modulation space norm with respect to the Gaussian windows $g_0(x) = e^{-\pi x^2}$ and $g(x) = 2^{-d/2}e^{-\pi x^2/2} = (g_0 * g_0)(x) \in \mathcal{S}(\mathbb{R}^d)$. Recall that different windows yield equivalent norms for the modulation spaces, specifically,

$$\|f\|_{M^{p,q}_m} \asymp \|V_{g_0}f\|_{L^{p,q}_m} \asymp \|V_g f\|_{L^{p,q}_m}.$$

Consider $f \in M^{p_1,q_1}_{m_1\otimes v}(\mathbb{R}^d)$, $h \in M^{p_2,q_2}_{v_1\otimes v_2 v^{-1}}(\mathbb{R}^d)$ such that

$$\frac{1}{p} + 1 = \frac{1}{p_1} + \frac{1}{p_2}, \qquad \frac{1}{q} = \frac{1}{q_1} + \frac{1}{q_2}.$$

Formula (1.28) for the STFT $V_g(f * h)(x,\xi)$ and the identity

$$M_\xi(g_0^* * g_0^*) = M_\xi g_0^* * M_\xi g_0^*,$$

allow expressing the STFT of $f * h$ as follows:

$$V_g(f * h)(x,\xi) = e^{-2\pi i x\cdot\xi}((f * h) * M_\xi g^*)(x) = e^{-2\pi i x\cdot\xi}((f * M_\xi g_0^*) * (h * M_\xi g_0^*))(x).$$

To estimate the $M^{p,q}$-norm of $f * h$, we first majorize m by $m(x,\xi) \lesssim m(x,0)v(0,\xi) = m_1(x)v_2(\xi)$ and then we use Young's inequality with $1/p_1+1/p_2 = 1+1/p$ in the x-variable and Hölder's inequality with $1/q_1 + 1/q_2 = 1/q$ in the ξ-variable, with weights as in Theorem 2.2.3. We obtain that

$$\begin{aligned}
\|f * h\|_{M^{p,q}_m} &\asymp \|V_g(f * h)\|_{L^{p,q}_m} \\
&\lesssim \left(\int_{\mathbb{R}^d}\left(\int_{\mathbb{R}^d} |(f * M_\xi g_0^*) * (h * M_\xi g_0^*)(x)|^p m_1(x)^p \,dx\right)^{q/p} v_2(\xi)^q \,d\xi\right)^{1/s} \\
&= \left(\int_{\mathbb{R}^d} \|(f * M_\xi g_0^*) * (h * M_\xi g_0^*)\|^q_{L^p_{m_1}} v_2(\xi)^q \,d\xi\right)^{1/q} \\
&\lesssim \left(\int_{\mathbb{R}^d} \|f * M_\xi g_0^*\|^q_{L^{p_1}_{m_1}} \|h * M_\xi g_0^*\|^q_{L^{p_2}_{v_1}} v_2(\xi)^q \,d\xi\right)^{1/q}
\end{aligned}$$

$$\lesssim \Big(\int_{\mathbb{R}^d} \|f * M_\xi g_0^*\|_{L^{p_1}_{m_1}}^{q_1} v(\xi)^{q_1}\,d\xi\Big)^{\frac{1}{q_1}} \Big(\int_{\mathbb{R}^d} \|h * M_\xi g_0^*\|_{L^{p_2}_{v_1}}^{q_2} \frac{v_2(\xi)^{q_2}}{v(\xi)^{q_2}}\,d\xi\Big)^{\frac{1}{q_2}}$$
$$\asymp \|f\|_{M^{p_1,q_1}_{m_1\otimes v}} \|h\|_{M^{p_2,q_2}_{v_1\otimes v_2 v^{-1}}}$$

The estimate (2.82) is proved when the indices satisfy (2.79) and (2.80) with the equalities. The case of the inequalities follows by the inclusion relations (2.74). □

Remark 2.4.20. Despite the large number of indices, the statement of this proposition has some intuitive meaning: a function $f \in M^{p,q}$ behaves like $f \in L^p$ and $\hat{f} \in L^q$; so the parameters related to the x-variable behave like those in Young's theorem for convolution, whereas the parameters related to ξ behave like in Hölder's inequality for pointwise multiplication.

For Wiener amalgam spaces, we have the following multiplication relations.

Proposition 2.4.21. *For every* $1 \le p, q \le \infty$, *we have*

$$\|fu\|_{W(\mathcal{F}L^p,L^q)} \le \|f\|_{W(\mathcal{F}L^1,L^\infty)} \|u\|_{W(\mathcal{F}L^p,L^q)}.$$

Proof. From the norm equivalence (2.54), the estimate to prove is equivalent to

$$\|\hat{f} * \hat{u}\|_{M^{p,q}} \le \|\hat{f}\|_{M^{1,\infty}} \|\hat{u}\|_{M^{p,q}},$$

but this a special case of Proposition 2.4.19. □

Convolution relations for $W(\mathcal{F}L^p, L^q)$ spaces follow immediately from Theorem 2.4.9 (ii). Precisely,

Proposition 2.4.22. *For every* $1 \le p, q \le \infty$, *we have*

$$\|f * u\|_{W(\mathcal{F}L^p,L^q)} \le \|f\|_{W(\mathcal{F}L^\infty,L^1)} \|u\|_{W(\mathcal{F}L^p,L^q)}. \tag{2.83}$$

Finally, returning to modulation spaces, we have the following multiplication properties.

Proposition 2.4.23. *Consider* $m \in \mathcal{M}_v(\mathbb{R}^{2d})$. *We write* $m_1(x) = m(x,0)$ *and* $m_2(\xi) = m(0,\xi)$ *for the restrictions to* $\mathbb{R}^d \times \{0\}$ *and* $\{0\} \times \mathbb{R}^d$, *and likewise for* v. *Let* v *be an arbitrary weight function on* $\mathbb{R}^d$ *satisfying (2.17). Consider* $1 \le p, q, p_1, p_2, q_1, q_2 \le \infty$ *such that*

$$\frac{1}{p} \le \frac{1}{p_1} + \frac{1}{p_2} \tag{2.84}$$

and

$$\frac{1}{q} + 1 \le \frac{1}{q_1} + \frac{1}{q_2}. \tag{2.85}$$

If $f \in M^{p_1,q_1}_{v_1\otimes m_2}(\mathbb{R}^d)$, $h \in M^{p_2,q_2}_{v^{-1}\otimes v_2}(\mathbb{R}^d)$, *the product* fh *is in* $M^{p,q}_m(\mathbb{R}^d)$, *with*

$$\|fh\|_{M^{p,q}_m} \lesssim \|f\|_{M^{p_1,q_1}_{v_1\otimes m_2}} \|h\|_{M^{p_2,q_2}_{v^{-1}\otimes v_2}}. \tag{2.86}$$

Proof. We measure the modulation space norm with respect to the Gaussian windows $g(x) = g_0^2(x)$, with $g_0(x) = e^{-\pi x^2}$. Observe that $m(x,\xi) \lesssim v(x,0)m(0,\xi) = v_1(x)m_2(\xi)$ and

$$\|fh\|_{M_m^{p,q}} \asymp \|V_g(fh)\|_{L_m^{p,q}} \lesssim \|V_g(fh)\|_{L^{p,q}_{v_1\otimes m_2}}.$$

Using (1.24), we write

$$V_g(fh)(x,\xi) = \widehat{fhT_x\bar{g}}(\xi) = \widehat{fT_xg_0} * \widehat{hT_xg_0}(\xi).$$

Consider $f \in M^{p_1,q_1}_{v_1v\otimes m_2}(\mathbb{R}^d)$, $h \in M^{p_2,q_2}_{v^{-1}\otimes v_2}(\mathbb{R}^d)$ such that relations (2.84) and (2.85) are satisfied with the equality signs. Using Minkowskii's integral inequality first, Hölder's inequality with $1/p_1 + 1/p_2 = 1/p$ in the x-variable, and finally Young's inequality with $1/q_1 + 1/q_2 = 1 + 1/q$ in the ξ-variable, we reckon

$$\begin{aligned}
\|fh\|_{M_m^{p,q}} &\asymp \|V_g(fh)\|_{L_m^{p,q}} \lesssim \|V_g(fh)\|_{L^{p,q}_{v_1\otimes m_2}}\\
&= \big\|\|(\widehat{fT_xg_0} * \widehat{hT_xg_0})(\xi)\|_{L^p_{v_1}(\mathbb{R}^d_x)}\big\|_{L^q_{m_2}(\mathbb{R}^d_\xi)}\\
&\le \Big\|\int_{\mathbb{R}^d} \|\widehat{fT_xg_0}(\xi-\eta)\widehat{hT_xg_0}(\eta)\|_{L^p_{v_1}(\mathbb{R}^d_x)}\,d\eta\Big\|_{L^q_{m_2}(\mathbb{R}^d_\xi)}\\
&\le \Big\|\int_{\mathbb{R}^d} \|\widehat{fT_xg_0}(\xi-\eta)\|_{L^{p_1}_{v_1v}(\mathbb{R}^d_x)}\|\widehat{hT_xg_0}(\eta)\|_{L^{p_2}_{v^{-1}}(\mathbb{R}^d_x)}\,d\eta\Big\|_{L^q_{m_2}(\mathbb{R}^d_\xi)}\\
&= \big\|(\|\widehat{fT_xg_0}\|_{L^{p_1}_{v_1v}(\mathbb{R}^d_x)} * \|\widehat{hT_xg_0}\|_{L^{p_2}_{v^{-1}}(\mathbb{R}^d_x)})(\xi)\big\|_{L^q_{m_2}(\mathbb{R}^d_\xi)}\\
&\le \big\|\|\widehat{fT_xg_0}\|_{L^{p_1}_{v_1v}(\mathbb{R}^d_x)}\big\|_{L^q_{m_2}(\mathbb{R}^d_\xi)}\big\|\|\widehat{hT_xg_0}\|_{L^{p_2}_{v^{-1}}(\mathbb{R}^d_x)}\big\|_{L^q_{m_2}(\mathbb{R}^d_\xi)}\\
&\asymp \|f\|_{M^{p_1,q_1}_{v_1v\otimes m_2}}\|h\|_{M^{p_2,q_2}_{v^{-1}\otimes v_2}}.
\end{aligned}$$

The inequalities in (2.84) and (2.85) follow by the inclusion relations (2.74). This concludes the proof. □

In a recent contribution [176], W. Guo, J. Chen, D. Fan, and G. Zhao characterize some properties of weighted modulation and Wiener amalgam spaces by the corresponding properties on weighted Lebesgue spaces. As applications, they obtain sharp conditions for product and convolution inequalities, and embedding on weighted modulation and Wiener amalgam spaces. The weights considered there are of the type $m(x,\xi) = m(x)\mu(\xi)$. We refer the interested reader to the original paper, see also [292].

2.5 Dilation properties for Wiener amalgam spaces $W(L^p, L^q)$

In this section we study the dilation properties of Wiener amalgam spaces $W(L^p, L^q)$, $1 \le p, q \le \infty$.

For $\lambda > 0$, we set $f_\lambda(x) = f(\lambda x)$. Then Wiener amalgam spaces fulfill the following estimates.

Proposition 2.5.1. *For* $1 \le p, q \le \infty$,

$$\|f_\lambda\|_{W(L^p,L^q)} \lesssim \lambda^{-d\max\{\frac{1}{p},\frac{1}{q}\}}\|f\|_{W(L^p,L^q)}, \quad \forall\lambda,\ 0 < \lambda \le 1, \tag{2.87}$$

and

$$\|f_\lambda\|_{W(L^p,L^q)} \lesssim \lambda^{-d\min\{\frac{1}{p},\frac{1}{q}\}}\|f\|_{W(L^p,L^q)}, \quad \forall\lambda \ge 1. \tag{2.88}$$

Also, we have

$$\|f_\lambda\|_{W(L^p,L^q)} \gtrsim \lambda^{-d\min\{\frac{1}{p},\frac{1}{q}\}}\|f\|_{W(L^p,L^q)}, \quad \forall\lambda,\ 0 < \lambda \le 1, \tag{2.89}$$

and

$$\|f_\lambda\|_{W(L^p,L^q)} \gtrsim \lambda^{-d\max\{\frac{1}{p},\frac{1}{q}\}}\|f\|_{W(L^p,L^q)}, \quad \forall\lambda \ge 1. \tag{2.90}$$

We first prove the following weaker estimates.

Lemma 2.5.2. *For* $1 \le p, q \le \infty$,

$$\|f_\lambda\|_{W(L^p,L^q)} \lesssim \lambda^{-d(\frac{1}{p}+\frac{1}{q})}\|f\|_{W(L^p,L^q)}, \quad \forall\lambda,\ 0 < \lambda \le 1, \tag{2.91}$$

and

$$\|f_\lambda\|_{W(L^p,L^q)} \lesssim \lambda^{d(1-\frac{1}{p}-\frac{1}{q})}\|f\|_{W(L^p,L^q)}, \quad \forall\lambda \ge 1. \tag{2.92}$$

Proof. To compute the Wiener norm, we choose the window function $g = \chi_{[0,1]^d}$, the characteristic function of the box $[0,1]^d$. Then

$$\begin{aligned}\|f_\lambda\|_{W(L^p,L^q)} &\asymp \big\|\|f(\lambda t)g(t-x)\|_{L^p_t}\big\|_{L^q_x}\\ &= \lambda^{-\frac{d}{p}}\big\|\|f(t)g_{1/\lambda}(t-\lambda x)\|_{L^p_t}\big\|_{L^q_x}\\ &= \lambda^{-d(\frac{1}{p}+\frac{1}{q})}\big\|\|f(t)g_{1/\lambda}(t-x)\|_{L^p_t}\big\|_{L^q_x}.\end{aligned} \tag{2.93}$$

If $0 < \lambda \le 1$, the window function g fulfills $g_{1/\lambda}(y) \le g(y)$, and (2.91) follows.

Let now $\lambda \ge 1$. To prove (2.92), we observe that

$$g_{1/\lambda}(t) \le \sum_{j\in\mathbb{Z}^d\cap[0,\lambda]^d} g(t-j).$$

Notice that the above sum contains $N_\lambda = O(\lambda^d)$ terms. Using this inequality, we dominate the expression in (2.93) by

$$\lambda^{-d(\frac{1}{p}+\frac{1}{q})}\sum_{j\in\mathbb{Z}^d\cap[0,\lambda]^d}\big\|\|f(t)g(t-x-j)\|_{L^p_t}\big\|_{L^q_x} = \lambda^{-d(\frac{1}{p}+\frac{1}{q})}N_\lambda\|f\|_{W(L^p,L^q)},$$

where we also performed a change of variable in the integral with respect to x. This concludes the proof. □

The complex interpolation of Wiener amalgam spaces (having Lebesgue spaces as global components) in Theorem 2.4.9, can be rephrased as follows (see also [120]).

Corollary 2.5.3. *Let B_0, B_1, be local components of Wiener amalgam spaces, as in Definition 2.4.3. Then for $1 \le q_0, q_1 \le \infty$ with $q_0 < \infty$ or $q_1 < \infty$, and $0 < \theta < 1$ we have*

$$[W(B_0, L^{q_0}), W(B_1, L^{q_1})]_{[\theta]} = W([B_0, B_1]_{[\theta]}, L^{q_\theta}),$$

with $1/q_\theta = (1-\theta)/q_0 + \theta/q_1$.

Proof of Proposition 2.5.1. We first prove (2.87) and (2.88) when $p = \infty$. We see at once that (2.87) coincides with (2.91) when $p = \infty$. On the other hand, (2.88) for $p = \infty$ follows by complex interpolation (cf. Corollary 2.5.3) from (2.92) with $(p, q) = (\infty, 1)$, i. e.,

$$\|f_\lambda\|_{W(L^\infty,L^1)} \lesssim \|f\|_{W(L^\infty,L^1)}$$

and the trivial estimate

$$\|f_\lambda\|_{W(L^\infty,L^\infty)} \asymp \|f_\lambda\|_{L^\infty} = \|f\|_{L^\infty} \asymp \|f\|_{W(L^\infty,L^\infty)}.$$

Since the estimates (2.87) and (2.88) also hold for $p = q$ (because $W(L^p, L^p) = L^p$ with equivalent norms, in view of Proposition 2.4.12), by interpolation with the case $p = \infty$, $1 \le q \le \infty$, we see that they hold for any pair (p, q), with $1 \le q \le p \le \infty$.

When $1 < p < q \le \infty$ they follow by duality arguments as follows, see (v) in Theorem 2.4.9.

Suppose first $\lambda \ge 1$. Then relation (2.87), applied to the pair (p', q'), yields

$$\begin{aligned}
\|f_\lambda\|_{W(L^p,L^q)} &= \sup_{\|g\|_{W(L^{p'},L^{q'})}=1} |\langle f_\lambda, g\rangle| \\
&= \sup_{\|g\|_{W(L^{p'},L^{q'})}=1} \lambda^{-d} |\langle f, g_{1/\lambda}\rangle| \\
&\le \lambda^{-d} \sup_{\|g\|_{W(L^{p'},L^{q'})}=1} \|f\|_{W(L^p,L^q)} \|g_{1/\lambda}\|_{W(L^{p'},L^{q'})} \\
&\lesssim \lambda^{-d} \lambda^{d \max\{\frac{1}{p'}, \frac{1}{q'}\}} \|f\|_{W(L^p,L^q)},
\end{aligned}$$

which is (2.88). Similarly, one proves (2.87) in that case.

It remains to prove the estimates (2.87) and (2.88) for $p = 1$, $1 < q \le \infty$. They follow from interpolation from the case $(p, q) = (1, 1)$ and the case $(p, q) = (1, \infty)$, where (2.87) and (2.88) coincide with (2.91) and (2.92), respectively.

Finally, (2.89) and (2.90) follow at once from (2.88) and (2.87), respectively, applied to the function $f_{1/\lambda}$. □

We now show that the result above is sharp.

Proposition 2.5.4 (Sharpness of (2.87) and (2.88)).

(i) *Suppose that, for some $\alpha \in \mathbb{R}$,*

$$\|f_\lambda\|_{W(L^p,L^q)} \lesssim \lambda^\alpha \|f\|_{W(L^p,L^q)}, \quad \forall\,\lambda, 0 < \lambda \le 1,\ \forall f \in W(L^p,L^q). \tag{2.94}$$

Then

$$\alpha \le -d \max\left\{\frac{1}{p}, \frac{1}{q}\right\}. \tag{2.95}$$

(ii) *Suppose that, for some $\alpha \in \mathbb{R}$,*

$$\|f_\lambda\|_{W(L^p,L^q)} \lesssim \lambda^\alpha \|f\|_{W(L^p,L^q)}, \quad \forall \lambda \ge 1,\ \forall f \in W(L^p,L^q). \tag{2.96}$$

Then

$$\alpha \ge -d \min\left\{\frac{1}{p}, \frac{1}{q}\right\}. \tag{2.97}$$

This also shows the sharpness of the estimates (2.89) and (2.90), since they are equivalent to (2.88) and (2.87), respectively.

Proof. (i) First, consider the case $p \ge q$. We have $W(L^p,L^q) \hookrightarrow W(L^q,L^q) = L^q$. Hence

$$\lambda^{-\frac{d}{q}} \|f\|_{L^q} = \|f_\lambda\|_{L^q} \lesssim \|f_\lambda\|_{W(L^p,L^q)}.$$

Combining this estimate with (2.94) and letting $\lambda \to 0^+$, we obtain $\alpha \le -d/q$.

Assume now $p < q$. It suffices to verify that, for every $\epsilon > 0$, there exists $f \in W(L^p,L^q)$ such that

$$\|f_\lambda\|_{W(L^p,L^q)} \ge C\lambda^{-\frac{d}{p}+\epsilon}. \tag{2.98}$$

We study the case of dimension $d = 1$. The general case follows by tensor products of functions of one variable. To this end, we choose

$$f(t) = \begin{cases} |t|^{-\frac{1}{p}+\epsilon} & \text{for } |t| \le 1, \\ 0 & \text{for } |t| > 1. \end{cases}$$

Observe that $f \in W(L^p,L^1) \hookrightarrow W(L^p,L^q)$, for every $1 \le q \le \infty$, and

$$f_\lambda(t) = f(\lambda t) = \lambda^{-\frac{1}{p}+\epsilon}|t|^{-\frac{1}{p}+\epsilon}, \quad \text{for } |t| \le \frac{1}{\lambda}. \tag{2.99}$$

Now, take $g = \chi_{B_1(0)}$ as window function. Of course,

$$\begin{aligned} \|f_\lambda\|_{W(L^p,L^q)} &= \left(\int \|f_\lambda T_y g\|_{L^p}^q \, dy\right)^{1/q} \\ &\ge \left(\int_{B_1(0)} \|f_\lambda T_y g\|_{L^p}^q \, dy\right)^{1/q}. \end{aligned}$$

By using (2.99) and the choice $g = \chi_{B_1(0)}$, for $\lambda \le 1/2$ we have $f_\lambda(t) \ge \lambda^{-\frac{1}{p}+\epsilon} f(t)$ and the last expression is estimated from below by

$$\lambda^{-1/p+\epsilon} \Big(\int_{B_1(0)} \|fT_y g\|_{L^p}^q \, dy \Big)^{1/q},$$

that is, (2.98) holds.

(ii) Again, we first consider the case $p \ge q$. Then $W(L^p, L^q) \hookrightarrow W(L^p, L^p) = L^p$. Hence

$$\|f_\lambda\|_{W(L^p,L^q)} \gtrsim \|f_\lambda\|_{L^p} = \lambda^{-\frac{d}{p}} \|f\|_{L^p}.$$

Combining this estimate with (2.96) and letting $\lambda \to +\infty$, we obtain $\alpha \ge -\frac{d}{p}$.

Suppose now $p < q$. As before it suffices to prove, in dimension $d = 1$, that for every $\epsilon > 0$ there exists a function $f \in W(L^p, L^q)$ such that

$$\|f_\lambda\|_{W(L^p,L^q)} \ge C\lambda^{-\frac{1}{q}-\epsilon}.$$

Therefore choose

$$f(t) = \begin{cases} |t|^{-\frac{1}{q}-\epsilon} & \text{for } |t| \ge 1, \\ 0 & \text{for } |t| < 1. \end{cases}$$

Then $f \in W(L^\infty, L^q) \hookrightarrow W(L^p, L^q)$, for every $1 \le p \le \infty$, and

$$f(\lambda t) = \lambda^{-\frac{1}{q}-\epsilon} |t|^{-\frac{1}{q}-\epsilon}, \quad \text{for } |t| \ge \frac{1}{\lambda}. \tag{2.100}$$

Again, choose $g = \chi_{B_1(0)}$ as window function. We have

$$\|f_\lambda\|_{W(L^p,L^q)} \ge \Big(\int_{B_2(0)} \|f_\lambda T_y g\|_{L^p}^q \, dy \Big)^{1/q}.$$

By using (2.100) and the choice $g = \chi_{B_1(0)}$, for $\lambda \ge 1$ we have $f_\lambda(t) \ge \lambda^{-\frac{1}{q}-\epsilon} f(t)$ and the last expression can be estimated from below by

$$\lambda^{-1/q-\epsilon} \Big(\int_{B_2(0)} \|fT_y g\|_{L^p}^q \, dy \Big)^{1/q},$$

which concludes the proof of (ii). □

2.6 Dilation properties for modulation spaces

These properties were essentially proved in the paper [70], extending the case $A = \lambda I$, $\lambda > 0$, already treated in [274].

Given a function f on $\mathbb{R}^d$ and $A \in \mathrm{GL}(d, \mathbb{R})$, we set $f_A(t) = f(At)$. We also consider the unitary operator $\mathcal{U}_A$ on $L^2(\mathbb{R}^d)$ defined by

$$\mathcal{U}_A f(t) = |\det A|^{1/2} f(At) = |\det A|^{1/2} f_A(t). \tag{2.101}$$

In this section we study the boundedness of this operator on modulation and Wiener amalgam spaces. We need the following preliminary results.

Lemma 1.3.36 helps compute the modulation norm of the rescaled Gaussian φ_A.

Lemma 2.6.1. *Let $1 \le p, q \le \infty$, $A \in \mathrm{GL}(d, \mathbb{R})$ and $\varphi(t) = e^{-\pi t^2}$. Then*

$$\|\varphi_A\|_{M^{p,q}} = p^{-d/(2p)} q^{-d/(2q)} |\det A|^{-1/p} (\det(A^T A + I))^{-(1-1/q-1/p)/2}. \tag{2.102}$$

Proof. Since the modulation space norm is independent of the choice of the window function, we choose the Gaussian φ, so that $\|\varphi_A\|_{M^{p,q}} \asymp \|V_\varphi \varphi_A\|_{L^{p,q}}$. Since

$$\begin{aligned}\int_{\mathbb{R}^d} e^{-\pi p(I-(A^T A+I)^{-1})x\cdot x}\, dx &= (\det(I - (A^T A + I)^{-1}))^{-1/2} p^{-d/2} \\ &= p^{-d/2} |\det A|^{-1} (\det(A^T A + I))^{1/2}\end{aligned}$$

and, analogously, $\int_{\mathbb{R}^d} e^{-\pi q(A^T A+I)^{-1}\xi\cdot\xi}\, d\xi = (\det(A^T A+I))^{1/2} q^{-d/2}$, the result immediately follows from Lemma 1.3.36. □

As a consequence, if we consider the matrix $A = \sqrt{\lambda} I$, $\lambda > 0$, and the rescaled Gaussian functions $\varphi_{\sqrt{\lambda}}(t) = \varphi(\sqrt{\lambda} t) = e^{-\pi \lambda t^2}$, we obtain the following result.

Lemma 2.6.2. *For $1 \le p, q \le \infty$, we have*

$$\|\varphi_{\sqrt{\lambda}}\|_{M^{p,q}} \asymp \lambda^{-\frac{d}{2q'}} \quad \text{as } \lambda \to +\infty, \tag{2.103}$$

$$\|\varphi_{\sqrt{\lambda}}\|_{M^{p,q}} \asymp \lambda^{-\frac{d}{2p}} \quad \text{as } \lambda \to 0^+. \tag{2.104}$$

Proposition 2.6.3. *Let $1 \le p, q \le \infty$ and $A \in \mathrm{GL}(d, \mathbb{R})$. Then for every $f \in M^{p,q}(\mathbb{R}^d)$*

$$\|f_A\|_{M^{p,q}} \lesssim |\det A|^{-(1/p-1/q+1)} (\det(I + A^T A))^{1/2} \|f\|_{M^{p,q}}. \tag{2.105}$$

Proof. First, by a change of variable, the dilation is transferred from the function f to the window φ:

$$V_\varphi f_A(x, \xi) = |\det A|^{-1} V_{\varphi_{A^{-1}}} f(Ax, (A^T)^{-1}\xi).$$

Whence performing the change of variables $Ax = u$, $(A^T)^{-1}\xi = v$,

$$\begin{aligned}\|f_A\|_{M^{p,q}} &= |\det A|^{-1}\Big(\int_{\mathbb{R}^d}\Big(\int_{\mathbb{R}^d}|V_{\varphi_{A^{-1}}}f(Ax,(A^T)^{-1}\xi)|^p\,dx\Big)^{q/p}d\xi\Big)^{1/q}\\ &= |\det A|^{-(1/p-1/q+1)}\|V_{\varphi_{A^{-1}}}f\|_{L^{p,q}}.\end{aligned}$$

Note that the same arguments apply to φ_A, hence

$$\|\varphi_A\|_{M^{p,q}} = |\det A|^{-(\frac{1}{p}-\frac{1}{q}+1)}\|V_{\varphi_{A^{-1}}}\varphi\|_{L^{p,q}},$$

so that combining with Lemma 2.6.1 we obtain, in particular,

$$\|V_{\varphi_{A^{-1}}}\varphi\|_{L^1} = (\det(I+A^TA))^{1/2}. \tag{2.106}$$

Now, Lemma 1.2.29, written for $h(t)=\gamma(t)=\varphi(t)=e^{-\pi t^2}$, yields the following bound:

$$|V_{\varphi_{A^{-1}}}f(x,\xi)| \le \|\varphi\|_{L^2}^{-2}(|V_\varphi f| * |V_{\varphi_{A^{-1}}}\varphi|)(x,\xi).$$

Finally, Young's inequality and (2.106) provide the desired result

$$\begin{aligned}\|f_A\|_{M^{p,q}} &\lesssim |\det A|^{-(1/p-1/q+1)}\big\||V_\varphi f| * |V_{\varphi_{A^{-1}}}\varphi|\big\|_{L^{p,q}}\\ &\lesssim |\det A|^{-(1/p-1/q+1)}\|V_\varphi f\|_{L^{p,q}}\|V_{\varphi_{A^{-1}}}\varphi\|_{L^1}\\ &\asymp |\det A|^{-(1/p-1/q+1)}(\det(I+A^TA))^{1/2}\|f\|_{M^{p,q}}.\end{aligned}$$

□

Corollary 2.6.4. *Let $1 \le p,q \le \infty$ and $\lambda \in \mathbb{R}\setminus\{0\}$. Then there exists a constant $C > 0$ such that*

$$\|f_\lambda\|_{M^{p,q}} \le C|\lambda|^{-d(1/p-1/q+1)}(1+\lambda^2)^{d/2}\|f\|_{M^{p,q}}.$$

Proof. It follows by Proposition 2.6.3 with the matrix $A = \mathrm{diag}[\lambda,\dots,\lambda]$. □

Corollary 2.6.5. *Let $1 \le p,q \le \infty$ and $A \in \mathrm{GL}(d,\mathbb{R})$. Then for every $f \in W(\mathcal{F}L^p,L^q)(\mathbb{R}^d)$,*

$$\|f_A\|_{W(\mathcal{F}L^p,L^q)} \lesssim |\det A|^{(1/p-1/q-1)}(\det(I+A^TA))^{1/2}\|f\|_{W(\mathcal{F}L^p,L^q)}. \tag{2.107}$$

Proof. It follows immediately from the relation between Wiener amalgam spaces and modulation spaces given by $W(\mathcal{F}L^p,L^q) = \mathcal{F}M^{p,q}$ and by the relation $\widehat{(f_A)} = |\det A|^{-1}(\hat{f})_{(A^T)^{-1}}$. □

In the sequel we give a more precise result about the behavior of the dilation operator $M^{p,q} \to M^{p,q}$ in terms of A, when A is a symmetric matrix.

For $1 \le p \le \infty$, let p' be the conjugate exponent of p, $1/p + 1/p' = 1$. For $(1/p, 1/q) \in [0,1] \times [0,1]$, we define the subsets

$$I_1 = \max(1/p, 1/p') \le 1/q, \quad I_1^* = \min(1/p, 1/p') \ge 1/q,$$
$$I_2 = \max(1/q, 1/2) \le 1/p', \quad I_2^* = \min(1/q, 1/2) \ge 1/p',$$
$$I_3 = \max(1/q, 1/2) \le 1/p, \quad I_3^* = \min(1/q, 1/2) \ge 1/p,$$

as shown in Figure 2.1.

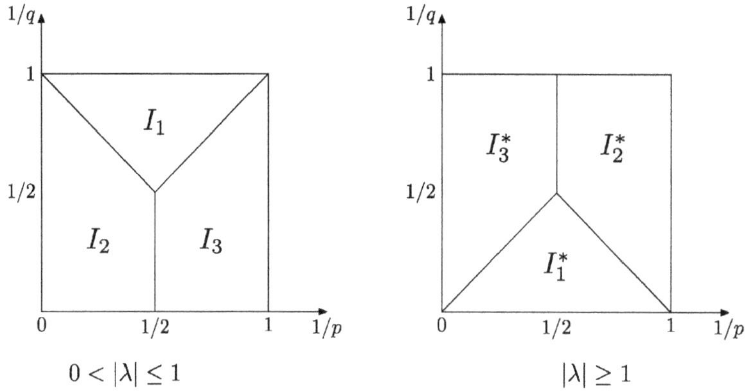

Figure 2.1: The index sets.

We introduce the indices

$$\mu_1(p,q) = \begin{cases} -1/p & \text{if } (1/p, 1/q) \in I_1^*, \\ 1/q - 1 & \text{if } (1/p, 1/q) \in I_2^*, \\ -2/p + 1/q & \text{if } (1/p, 1/q) \in I_3^*, \end{cases} \tag{2.108}$$

and

$$\mu_2(p,q) = \begin{cases} -1/p & \text{if } (1/p, 1/q) \in I_1, \\ 1/q - 1 & \text{if } (1/p, 1/q) \in I_2, \\ -2/p + 1/q & \text{if } (1/p, 1/q) \in I_3. \end{cases} \tag{2.109}$$

We are now able to present the dilation properties for modulation spaces. Let us first recall the basic result for $A = \lambda I$ in [274, Theorem 1.1].

Theorem 2.6.6. *Let $1 \le p, q \le \infty$, and $A = \lambda I$, $\lambda \neq 0$.*

(i) *We have*

$$\|f_A\|_{M^{p,q}} \lesssim |\lambda|^{d\mu_1(p,q)} \|f\|_{M^{p,q}}, \quad |\lambda| \ge 1, \ \forall f \in M^{p,q}(\mathbb{R}^d).$$

Conversely, if there exists $\alpha \in \mathbb{R}$ *such that*

$$\|f_A\|_{M^{p,q}} \lesssim |\lambda|^{\alpha} \|f\|_{M^{p,q}}, \quad |\lambda| \geq 1, \ \forall f \in M^{p,q}(\mathbb{R}^d),$$

then $\alpha \geq d\mu_1(p,q)$.

(ii) *We have*

$$\|f_A\|_{M^{p,q}} \lesssim |\lambda|^{d\mu_2(p,q)} \|f\|_{M^{p,q}}, \quad 0 < |\lambda| \leq 1, \ \forall f \in M^{p,q}(\mathbb{R}^d).$$

Conversely, if there exists $\beta \in \mathbb{R}$ *such that*

$$\|f_A\|_{M^{p,q}} \lesssim |\lambda|^{\beta} \|f\|_{M^{p,q}}, \quad 0 < |\lambda| \leq 1, \ \forall f \in M^{p,q}(\mathbb{R}^d),$$

then $\beta \leq d\mu_2(p,q)$.

For simplicity, from now on, we use the abuse of notation $f_\lambda = f_A$, with $A = \lambda I$, $\lambda \neq 0$.

Proof Theorem 2.6.6 (ii), *with* $(1/p, 1/q) \in I_1$ *and* (i), *with* $(1/p, 1/q) \in I_1^*$. *Case* (ii) *with* $(1/p, 1/q) \in I_1$. Then $\mu_2(p,q) = -1/p$. Let $1 \leq r \leq \infty$, by Corollary 2.6.4,

$$\|f_\lambda\|_{M^{r,1}} \leq C|\lambda|^{-d/r} \|f\|_{M^{r,1}}, \quad f \in M^{r,1}, \ 0 < |\lambda| \leq 1. \tag{2.110}$$

On the other hand, since $M^{2,2} = L^2$, we can write

$$\|f_\lambda\|_{M^{2,2}} \leq C|\lambda|^{-d/2} \|f\|_{M^{2,2}}, \quad f \in M^{2,2}, \ 0 < |\lambda| \leq 1. \tag{2.111}$$

For $1 \leq r \leq \infty$, $0 < \theta < 1$ such that $1/p = (1-\theta)/r + \theta/2$ and $1/q = (1-\theta)/1 + \theta/2$, by complex interpolation (Proposition 2.3.17), we obtain

$$\|f_\lambda\|_{M^{p,q}} \leq C(|\lambda|^{-d/r})^{1-\theta} (|\lambda|^{-d/2})^{\theta} \|f\|_{M^{p,q}}, \quad f \in M^{p,q}, \ 0 < |\lambda| \leq 1.$$

Since $(1-\theta)/r = 1/p + 1/q - 1$ and $\theta/2 = 1 - 1/q$, it follows that

$$\|f_\lambda\|_{M^{p,q}} \leq C|\lambda|^{-d/p} \|f\|_{M^{p,q}}, \quad f \in M^{p,q}, \ 0 < |\lambda| \leq 1,$$

which is the first part of the claim with $(1/p, 1/q) \in I_1$.

We next prove the necessary condition (second part of the claim) with $(1/p, 1/q) \in I_1$. Assume that there exists $\beta \in \mathbb{R}$ such that

$$\|f_\lambda\|_{M^{p,q}} \lesssim |\lambda|^{\beta} \|f\|_{M^{p,q}}, \quad 0 < |\lambda| \leq 1, \ \forall f \in M^{p,q}(\mathbb{R}^d).$$

Test the inequality on the Gaussian function φ. By Lemma 2.6.1 with $A = \lambda I$,

$$\|\varphi_\lambda\|_{M^{p,q}} = p^{-d/(2p)} q^{-d/(2q)} |\lambda|^{-d/p} (1+\lambda^2)^{-d(1-1/q-1/p)/2} \leq |\lambda|^{\beta} \|\varphi\|_{M^{p,q}},$$

for every $0 < |\lambda| \leq 1$. Letting $\lambda \to 0$, we get $\beta \leq -d/p$, as desired.

Case (i) *with* $(1/p, 1/q) \in I_1^*$. Note that, in this occurrence, $\mu_1(p,q) = -1/p$. Consider $1 \le p \le \infty$, $q \ge 2$ such that $(1/p, 1/q) \in I_1^*$. Then $(1/p', 1/q') \in I_1$. First, assume $p \ne 1$. Since $1 < p, q \le \infty$, by duality (cf. Eq. (2.34))

$$\begin{aligned}
\|f_\lambda\|_{M^{p,q}} &= \sup|\langle f_\lambda, g\rangle| = |\lambda|^{-d} \sup|\langle f, g_{1/\lambda}\rangle| \\
&= |\lambda|^{-d} \sup \|f\|_{M^{p,q}} \|g_{1/\lambda}\|_{M^{p',q'}} \\
&\le |\lambda|^{-d} \|f\|_{M^{p,q}} \sup(C(|\lambda|^{-1})^{-d/p'} \|g\|_{M^{p',q'}}) \\
&= C|\lambda|^{-d/p} \|f\|_{M^{p,q}},
\end{aligned}$$

for all $f \in M^{p,q}(\mathbb{R}^d)$, $|\lambda| \ge 1$, where the supremum is taken over all $g \in \mathcal{S}(\mathbb{R}^d)$ such that $\|g\|_{M^{p',q'}} = 1$. In the case $p = 1$, by Corollary 2.6.4,

$$\|f_\lambda\|_{M^{1,\infty}} \le C|\lambda|^{-d} \|f\|_{M^{1,\infty}}, \quad f \in M^{1,\infty}, \ |\lambda| \ge 1.$$

Hence we obtain the first part of case (i), with $(1/p, 1/q) \in I_1^*$.

Assume $1 < p < \infty$, $2 \le q < \infty$ such that $(1/p, 1/q) \in I_1^*$, and there exist $C > 0$ and $\alpha > 0$ such that

$$\|f_\lambda\|_{M^{p,q}} \le C|\lambda|^\alpha \|f\|_{M^{p,q}}, \quad \forall\, |\lambda| \ge 1, \ \forall f \in M^{p,q}(\mathbb{R}^d).$$

Then by duality and our assumption

$$\begin{aligned}
\|f_\lambda\|_{M^{p',q'}} &= \sup|\langle f_\lambda, g\rangle| = |\lambda|^{-d} \sup|\langle f, g_{1/\lambda}\rangle| \\
&= |\lambda|^{-d} \sup \|f\|_{M^{p',q'}} \|g_{1/\lambda}\|_{M^{p,q}} \\
&\le |\lambda|^{-d} \sup \|f\|_{M^{p',q'}} (C(|\lambda|^{-1})^\alpha \|g\|_{M^{p,q}}) \\
&= C|\lambda|^{-d-\alpha} \|f\|_{M^{p',q'}},
\end{aligned}$$

for all $f \in M^{p',q'}(\mathbb{R}^d)$, $0 < |\lambda| \le 1$, where the supremum is taken over all $g \in \mathcal{S}(\mathbb{R}^d)$ such that $\|g\|_{M^{p,q}} = 1$. Since $(1/p', 1/q') \in I_1$, by case (ii) with $(1/p', 1/q') \in I_1$, we get $-d - \alpha \le -d/p'$, that is $\alpha \ge -d/p$.

The duality argument extends to the case $p < \infty$, $q = \infty$. As for the case $p = \infty$ and $q = \infty$ we may observe that for $f(t) = 1$, $t \in \mathbb{R}^d$, we have $f_\lambda = f$, $\|f_\lambda\|_{M^\infty} = \|f\|_{M^\infty}$.

The proof is then completed. □

Next goal is to prove Case (i) with $(1/p, 1/q) \in I_2^*$ and Case (ii) with $(1/p, 1/q) \in I_2$. We need some preliminaries.

Lemma 2.6.7. *Let $1 \le p, q \le \infty$ such that $(1/p, 1/q) \in I_2^*$ and $1/p \ge 1/q$. Then there exists a constant $C > 0$ such that*

$$\|f_\lambda\|_{M^{p,q}} \le C|\lambda|^{-d(2/p-1/q)} (1 + \lambda^2)^{d(1/p-1/2)} \|f\|_{M^{p,q}},$$

for all $f \in M^{p,q}(\mathbb{R}^d)$, and $|\lambda| > 0$.

Proof. Consider $1 \le r \le \infty$. By Corollary 2.6.4, we have

$$\|f_\lambda\|_{M^{1,r}} \le C|\lambda|^{d(1/r-2)}(1+\lambda^2)^{d/2}\|f\|_{M^{1,r}},$$

for every $f \in M^{1,r}(\mathbb{R}^d)$ and $|\lambda| > 0$. Let $1 \le p, q \le \infty$ such that $(1/p, 1/q) \in I_2^*$ and $1/p \ge 1/q$. Take $1 \le r \le \infty$ and $0 < \theta < 1$ such that $1/p = (1-\theta)/1 + \theta/2$ and $1/q = (1-\theta)/r + \theta/2$. Since $M^{2,2} = L^2$, we can write

$$\|f_\lambda\|_{M^{2}2} \le C|\lambda|^{-d/2}\|f\|_{M^{2,2}}, \quad f \in M^{2,2},\ |\lambda| \ge 1. \tag{2.112}$$

Then by complex interpolation with (2.111) and (2.112),

$$\|f_\lambda\|_{M^{p,q}} \le C(|\lambda|^{d(1/r-2)}(1+\lambda^2)^{d/2})^{1-\theta}|\lambda|^{-d\theta/2}\|f\|_{M^{p,q}},$$

for all $f \in M^{p,q}$, and $|\lambda| > 0$. Using $(1-\theta)/r = 1/p + 1/q - 1$, $1-\theta = 2/p - 1$, $\theta/2 = -1/p + 1$, we get the desired estimate. □

Lemma 2.6.8. *There exists a constant $C > 0$ such that*

$$\|f_\lambda\|_{M^{2,\infty}} \le C|\lambda|^{-d}\|f\|_{M^{2,\infty}},$$

for all $f \in M^{2,\infty}(\mathbb{R}^d)$, and $0 < |\lambda| \le 1$.

Proof. We use the characterization of the $M^{2,\infty}$-norm in (2.48), that is,

$$\|f\|_{M^{2,\infty}(\mathbb{R}^d)} \asymp \sup_{k\in\mathbb{Z}^d} \|\Box_k f\|_{L^2}.$$

Hence

$$\begin{aligned}\|f_\lambda\|_{M^{2,\infty}} &\lesssim \sup_{k\in\mathbb{Z}^d}\|\sigma_0(\cdot - k)\hat{f}_\lambda\|_{L^2}\\ &= |\lambda|^{-d/2}\sup_{k\in\mathbb{Z}^d}\|\sigma_0(\lambda\cdot - k)\hat{f}\|_{L^2}\\ &= |\lambda|^{-d/2}\sup_{k\in\mathbb{Z}^d}\Big\|\sigma_0(\lambda\cdot - k)\Big(\sum_{l\in\mathbb{Z}^d}\sigma_0(\cdot - l)\Big)\hat{f}\Big\|_{L^2}.\end{aligned}$$

Observe that

$$\sum_{l\in\mathbb{Z}^d}|\sigma_0(\lambda t - k)\sigma_0(t-l)\hat{f}(t)|^2 \le C_d \sum_{\substack{|l_j - k_j/\lambda| \le 2/\lambda,\\ j=1,\dots,d}} |\sigma_0(\lambda t - k)\sigma_0(t-l)\hat{f}(t)|^2,$$

then we have

$$\begin{aligned}&\Big\|\sigma_0(\lambda\cdot - k)\sum_{l\in\mathbb{Z}^d}\sigma_0(\cdot - l)\hat{f}\Big\|_{L^2}\\ &\le C_d\Big(\sum_{\substack{|l_j - k_j/\lambda| \le 2/|\lambda|,\\ j=1,\dots,d}}\int_{\mathbb{R}^d}|\sigma_0(\lambda t - k)\sigma_0(t-l)\hat{f}(t)|^2 dt\Big)^{1/2}\end{aligned}$$

$$\le C_d\|\sigma_0\|_\infty\Big(\sum_{\substack{|l_j-k_j/\lambda|\le 2/|\lambda|,\\ j=1,\dots,d}}\|M_l\mathcal{F}^{-1}\sigma_0 * f\|_2^2\Big)^{1/2}$$

$$\le C_d\|\sigma_0\|_\infty \sup_{m\in\mathbb{Z}^d}\|M_m\mathcal{F}^{-1}\sigma_0 * f\|_{L^2}\Big(\sum_{\substack{|l_j-k_j/\lambda|\le 2/|\lambda|,\\ j=1,\dots,d}} 1\Big)^{1/2}$$

$$\le C'_d|\lambda|^{-d/2}\sup_{m\in\mathbb{Z}^d}\|\Box_m f\|_{L^2}.$$

Hence

$$\|f_\lambda\|_{M^{2,\infty}} \lesssim |\lambda|^{-d}\|f\|_{M^{2,\infty}},$$

as desired. □

Lemma 2.6.9. *Let $1\le p\le\infty$, $\lambda\in\mathbb{R}\setminus\{0\}$. Then the following statements are true:*
(i) *If $p\le 2$ then there exists a constant $C>0$ such that*

$$\|f_\lambda\|_{M^{p,1}} \le C\|f\|_{M^{p,1}}, \quad \forall f\in M^{p,1}(\mathbb{R}^d),\ |\lambda|\ge 1. \tag{2.113}$$

(ii) *If $p\ge 2$ then there exists a constant $C>0$ such that*

$$\|f_\lambda\|_{M^{p,1}} \le C|\lambda|^{-d(2/p-1)}\|f\|_{M^{p,1}}, \quad \forall f\in M^{p,1}(\mathbb{R}^d),\ |\lambda|\ge 1.$$

Proof. (i) Using Lemma 2.6.8, we can write

$$\begin{aligned}\|f_\lambda\|_{M^{2,1}} &= \sup|\langle f_\lambda, g\rangle| = |\lambda|^{-d}\sup|\langle f, g_{1/\lambda}\rangle|\\ &\le |\lambda|^{-d}\sup\|f\|_{M^{2,1}}\|g_{1/\lambda}\|_{M^{2,\infty}} \le |\lambda|^{-d}\sup\|f\|_{M^{2,1}}(C|1/\lambda|^{-d}\|g\|_{M^{2,\infty}})\\ &= C\|f\|_{M^{2,1}}\end{aligned}$$

for all $f\in\mathcal{S}(\mathbb{R}^d)$ and $|\lambda|\ge 1$, where the supremum is taken over all $g\in M^{2,\infty}(\mathbb{R}^d)$ such that $\|g\|_{M^{2,\infty}}=1$. Since $\mathcal{S}(\mathbb{R}^d)$ is dense in $M^{2,1}(\mathbb{R}^d)$, we get

$$\|f_\lambda\|_{M^{2,1}} \le C\|f\|_{M^{2,1}}, \quad \forall f\in M^{2,1}(\mathbb{R}^d),\ |\lambda|\ge 1. \tag{2.114}$$

On the other hand, by Corollary 2.6.4, we see

$$\|f_\lambda\|_{M^{1,1}} \le C\|f\|_{M^{1,1}}, \quad \forall f\in M^{1,1}(\mathbb{R}^d),\ |\lambda|\ge 1.$$

Hence by complex interpolation, we get (2.113).

(ii) By Corollary 2.6.4, we can write

$$\|f_\lambda\|_{M^{\infty,1}} \le C|\lambda|^{d}\|f\|_{M^{\infty,1}}, \quad \forall f\in M^{\infty,1}(\mathbb{R}^d),\ |\lambda|\ge 1. \tag{2.115}$$

Interpolating the latter inequality with (2.114), we obtain

$$\|f_\lambda\|_{M^{p,1}} \le C(|\lambda|^0)^{1-\theta}(|\lambda|^d)^\theta\|f\|_{M^{p,1}},$$

for all $f\in M^{p,1}(\mathbb{R}^d)$, $|\lambda|\ge 1$, with $1/p=(1-\theta)/2+\theta/\infty$, $1\le\theta\le 1$. Since $\theta=-2/p+1$, the claim is reached. □

Lemma 2.6.10. *Let $1 \le p \le \infty$, $\lambda \in \mathbb{R} \setminus \{0\}$. Then the following statements are true:*
(i) *If $p \le 2$ then there exists a constant $C > 0$ such that*

$$\|f_\lambda\|_{M^{p,\infty}} \le C|\lambda|^{-2d/p}\|f\|_{M^{p,\infty}}, \quad \forall f \in M^{p,\infty}(\mathbb{R}^d),\ 0 < |\lambda| \le 1. \tag{2.116}$$

(ii) *If $p \ge 2$ then there exists a constant $C > 0$ such that*

$$\|f_\lambda\|_{M^{p,\infty}} \le C|\lambda|^{-d}\|f\|_{M^{p,\infty}}, \quad \forall f \in M^{p,\infty}(\mathbb{R}^d),\ 0 < |\lambda| \le 1.$$

Proof. (i) Let $1 < p \le 2$. Using Lemma 2.6.9 (ii) and the duality argument,

$$\begin{aligned}\|f_\lambda\|_{M^{p,\infty}} &= \sup|\langle f_\lambda, g\rangle| = |\lambda|^{-d}\sup|\langle f, g_{1/\lambda}\rangle| \\ &\le |\lambda|^{-d}\sup\|f\|_{M^{p,\infty}}\|g_{1/\lambda}\|_{M^{p',1}} \\ &\le |\lambda|^{-d}\sup\|f\|_{M^{p,\infty}}(C|1/\lambda|^{-d(2/p'-1)}\|g\|_{M^{p',1}}) \\ &= C|\lambda|^{-2d/p}\|f\|_{M^{p,\infty}}\end{aligned}$$

for all $f \in M^{p,\infty}(\mathbb{R}^d)$ and $0 < |\lambda| \le 1$, where the supremum is taken over all $g \in \mathcal{S}(\mathbb{R}^d)$ such that $\|g\|_{M^{p',1}} = 1$. For $p = 1$, by Corollary 2.6.4,

$$\|f_\lambda\|_{M^{1,\infty}} \le C|\lambda|^{-2d}\|f\|_{M^{1,\infty}}, \quad \forall f \in M^{1,\infty}(\mathbb{R}^d),\ 0 < |\lambda| \le 1.$$

This gives the claim.

(ii) This case is obtained by using the same pattern as case (i). We leave the details to the interested reader. □

Lemma 2.6.11. *Let $1 \le p, q \le \infty$, $(p,q) \ne (1,\infty), (\infty,1)$, and $\epsilon > 0$. Set*

$$f = \sum_{k \in \mathbb{Z}^d \setminus \{0\}} |k|^{-d/q-\epsilon} M_k\varphi \quad \text{in } \mathcal{S}'(\mathbb{R}^d), \tag{2.117}$$

with φ being the standard Gaussian function. Then $f \in M^{p,q}(\mathbb{R}^d)$ and there exists a constant $C > 0$ such that

$$\|f_\lambda\|_{M^{p,q}} \ge C|\lambda|^{d(1/q-1)+\epsilon}, \quad 0 < |\lambda| \le 1. \tag{2.118}$$

Proof. First, we show $f \in M^{p,q}(\mathbb{R}^d)$. Using Leibniz' formula (0.5),

$$\begin{aligned}&\left|\int_{\mathbb{R}^d} e^{2\pi i k\cdot t}\varphi(t)\varphi(t-x)e^{-2\pi i\xi\cdot t}\,dt\right| \\ &= \left|\int_{\mathbb{R}^d} \varphi(t)\varphi(t-x)\{\langle 2\pi(\xi-k)\rangle^{-2d}(I-\Delta_t)^d e^{-2\pi i(\xi-k)\cdot t}\}\,dt\right| \\ &\lesssim \langle 2\pi(\xi-k)\rangle^{-2d}\left|\sum_{|\alpha|\le 2d}\sum_{\beta\le\alpha} C_{\alpha,\beta}\int_{\mathbb{R}^d} \partial^\beta\varphi(t)(\partial^{\alpha-\beta}\varphi)(t-x)e^{-2\pi i(\xi-k)\cdot t}\,dt\right| \\ &\lesssim \langle 2\pi(\xi-k)\rangle^{-2d}\sum_{|\alpha|\le 2d}\sum_{\beta\le\alpha}|\partial^\beta\varphi| * |\partial^{\alpha-\beta}\varphi|(x).\end{aligned}$$

Hence

$$\begin{aligned}\|f\|_{M^{p,q}} &\asymp \|V_\varphi f\|_{L^{p,q}}\\ &= \Bigg(\int_{\mathbb{R}^d}\Bigg(\int_{\mathbb{R}^d}\Bigg|\sum_{k\in\mathbb{Z}^d\setminus\{0\}}|k|^{-d/q-\epsilon}\int_{\mathbb{R}^d}M_k\varphi(t)\varphi(t-x)e^{-2\pi i\xi\cdot t}\,dt\Bigg|^p dx\Bigg)^{q/p}d\xi\Bigg)^{1/q}\\ &\le C\Bigg(\int_{\mathbb{R}^d}\Bigg(\sum_{k\in\mathbb{Z}^d\setminus\{0\}}|k|^{-d/q-\epsilon}\langle 2\pi(\xi-k)\rangle^{-2d}\Bigg)^q d\xi\Bigg)^{1/q}\\ &= C\Bigg(\sum_{l\in\mathbb{Z}^d}\int_{l+[-1/2,1/2]^d}\Bigg(\sum_{k\in\mathbb{Z}^d\setminus\{0\}}|k|^{-d/q-\epsilon}\langle 2\pi(\xi-k)\rangle^{-2d}\Bigg)^q d\xi\Bigg)^{1/q}\\ &\le C\Bigg(\sum_{l\in\mathbb{Z}^d}\Bigg(\sum_{k\in\mathbb{Z}^d\setminus\{0\}}|k|^{-d/q-\epsilon}\langle 2\pi(l-k)\rangle^{-2d}\Bigg)^q\Bigg)^{1/q}.\end{aligned}$$

Since $(|k|^{-d/q-\epsilon})_{k\in\mathbb{Z}^d\setminus\{0\}}\in\ell^q(\mathbb{Z}^d)$, by Young's inequality, we get $f\in M^{p,q}(\mathbb{R}^d)$. We next show (2.118). The Gaussian function φ is in $M^{p',q'}(\mathbb{R}^d)$, so by duality

$$\begin{aligned}\|f_\lambda\|_{M^{p,q}} &= \sup_{\|g\|_{M^{p',q'}}=1}|\langle f_\lambda,g\rangle| \ge |\langle f_\lambda,\varphi\rangle|\\ &= \Bigg|\sum_{k\in\mathbb{Z}^d\setminus\{0\}}|k|^{-d/q-\epsilon}\int_{\mathbb{R}^d}M_{\lambda k}\varphi(\lambda t)\varphi(t)\,dt\Bigg|\\ &= C(1+\lambda^2)^{-d/2}\sum_{k\in\mathbb{Z}^d\setminus\{0\}}|k|^{-d/q-\epsilon}e^{-\pi\frac{\lambda^2|k|^2}{4(1+\lambda^2)}}\\ &\ge C\sum_{\substack{0<|k_j|\le 1/|\lambda|,\\ j=1,\dots,d}}|k|^{-d/q-\epsilon}e^{-\pi\frac{\lambda^2|k|^2}{4(1+\lambda^2)}}\\ &\ge C|\lambda|^{d/q+\epsilon}\sum_{\substack{0<|k_j|\le 1/|\lambda|,\\ j=1,\dots,d}}1 \ge C|\lambda|^{d(1/q-1)+\epsilon},\end{aligned}$$

as desired. □

Proof Theorem 2.6.6 (ii), *with* $(1/p,1/q)\in I_2$. Note that $\mu_2(p,q)=1/q-1$. Let $p\ge 2$ and $1\le q\le\infty$ such that $(1/p,1/q)\in I_2$ and $1/p\le 1/q$. If $q=1$ then $p=\infty$ and we have already proved this case in Theorem 2.6.6 (ii), with $(1/p,1/q)\in I_1$. Hence we assume $1<q\le\infty$ and observe that $1\le p'\le 2$ and $1\le q'<\infty$. Since $(1/p',1/q')\in I_2^*$ and $1/p'\ge 1/q'$, by duality and Lemma 2.6.7,

$$\begin{aligned}\|f_\lambda\|_{M^{p,q}} &= \sup|\langle f_\lambda,g\rangle| = |\lambda|^{-d}\sup|\langle f,g_{1/\lambda}\rangle|\\ &\le |\lambda|^{-d}\|f\|_{M^{p,q}}\sup\|g_{1/\lambda}\|_{M^{p',q'}}\end{aligned}$$

$$\leq |\lambda|^{-d}\|f\|_{M^{p,q}} \sup(C(|\lambda|^{-1})^{-d(2/p'-1/q')}(1+\lambda^{-2})^{d(1/p'-1/2)}\|g\|_{M^{p',q'}})$$
$$\leq C|\lambda|^{d(1/q-1)}\|f\|_{M^{p,q}},$$

for all $f \in M^{p,q}(\mathbb{R}^d)$ and $0 < |\lambda| \leq 1$, where the supremum is taken over all $g \in \mathcal{S}(\mathbb{R}^d)$ such that $\|g\|_{M^{p',q'}} = 1$. This is the first part of Theorem 2.6.6 (ii), with $(1/p, 1/q) \in I_2$ and $1/p \leq 1/q$. Let $p \geq 2$ and $2 \leq q < \infty$ be such $(1/p, 1/q) \in I_2$ and $1/p \geq 1/q$. From the previous inequality, we obtain

$$\|f_\lambda\|_{M^{r,r}} \leq C|\lambda|^{d(1/r-1)}\|f\|_{M^{r,r}} \quad \forall f \in M^{r,r},\ 0 < |\lambda| \leq 1, \tag{2.119}$$

with $2 \leq r \leq \infty$. Take $2 \leq r \leq \infty$ and $0 < \theta < 1$ such that $1/p = (1-\theta)/2 + \theta/r$ and $1/q = (1-\theta)/\infty + \theta/r$. Since $q \neq \infty$ also $r \neq \infty$. Then by complex interpolation, Lemma 2.6.8, and (2.119),

$$\|f_\lambda\|_{M^{p,q}} \leq C(|\lambda|^{-d})^{1-\theta}(|\lambda|^{d(1/r-1)})^{\theta}\|f\|_{M^{p,q}} = C|\lambda|^{d(1/q-1)}\|f\|_{M^{p,q}},$$

for all $f \in M^{p,q}(\mathbb{R}^d)$ and $0 < |\lambda| \leq 1$. In the case $q = \infty$, by Lemma 2.6.10 (ii), we have nothing to prove. Hence we obtain the first part of Theorem 2.6.6 (ii), with $(1/p, 1/q) \in I_2$ and $1/p \geq 1/q$.

We now show the necessity of this part of the theorem. Let $p \geq 2$ and $1 \leq q \leq \infty$ such that $(1/p, 1/q) \in I_2$. Since $(1/\infty, 1/1) \in I_1$, we may assume $(p, q) \neq (\infty, 1)$. Suppose that there exist constants $C > 0$ and $\beta \in \mathbb{R}$ such that

$$\|f_\lambda\|_{M^{p,q}} \leq C|\lambda|^{\beta}\|f\|_{M^{p,q}}, \quad \forall f \in M^{p,q}(\mathbb{R}^d),\ 0 < |\lambda| \leq 1,$$

with $\beta > d(1/q-1)$. Then we can choose an $\epsilon > 0$ such that $d(1/q-1) + \epsilon < \beta$ and for this ϵ we consider the function f defined in (2.117). This function f is in $M^{p,q}(\mathbb{R}^d)$ by Lemma 2.6.11 and there exists a constant $C' > 0$ such that $\|f_\lambda\|_{M^{p,q}} \geq C'|\lambda|^{d(1/q-1)+\epsilon}$, so that

$$C'|\lambda|^{d(1/q-1)+\epsilon} \leq \|f_\lambda\|_{M^{p,q}} \leq C|\lambda|^{\beta}\|f\|_{M^{p,q}}, \quad 0 < |\lambda| \leq 1.$$

Though, since $d(1/q-1) + \epsilon < \beta$, this is a contradiction. Therefore β must satisfy $\beta \leq d(1/q-1)$, that is the claim. □

Proof Theorem 2.6.6 (i), *with* $(1/p, 1/q) \in I_2^*$. We note that $\mu_1(p, q) = 1/q - 1$ if $(1/p, 1/q) \in I_2^*$. In every case, except for $(p, q) \neq (1, \infty)$, the result follows from the preceding proofs, namely by duality, Theorem 2.6.6 (ii), with $(1/p', 1/q') \in I_2$, and the same argument as in the proof of Theorem 2.6.6 (i), with $(1/p, 1/q) \in I_1^*$. For the case $(p, q) = (1, \infty)$, the result was already proved in Theorem 2.6.6 (i), with $(1/p, 1/q) \in I_1^*$. □

The last goal is to prove Theorem 2.6.6 (i), with $(1/p, 1/q) \in I_3^*$, and (ii), with $(1/p, 1/q) \in I_3$. We use the fact that there exists $\gamma \in \mathcal{S}(\mathbb{R}^d)$ such that $\operatorname{supp} \gamma \subset [-1/8, 1/8]^d$ and $|\hat{\gamma}| \geq 1$ on $[-2, 2]^d$ (see, for example, the proof of [150, Theorem 2.6]).

Lemma 2.6.12. *Let $1 \le p \le \infty$, $1 \le q < \infty$ and $\epsilon > 0$. Suppose that $\gamma, \psi \in \mathcal{S}(\mathbb{R}^d)$ satisfy $\operatorname{supp}\gamma \subset [-1/8, 1/8]^d$, $\operatorname{supp}\psi \subset [-1/2, 1/2]^d$, $|\hat{\gamma}| \ge 1$ on $[-2, 2]^d$ and $\psi = 1$ on $[-1/4, 1/4]^d$. Set*

$$f = \sum_{k \in \mathbb{Z}^d \setminus \{0\}} |k|^{-d/q-\epsilon} M_k T_k \psi \quad \text{in } \mathcal{S}'(\mathbb{R}^d). \tag{2.120}$$

Then $f \in M^{p,q}(\mathbb{R}^d)$ and there exists a constant $C > 0$ such that

$$\|f_\lambda\|_{M^{p,q}} \asymp \|V_\gamma f_\lambda\|_{L^{p,q}} \ge C|\lambda|^{-d(2/p-1/q)+\epsilon}, \quad 0 < |\lambda| \le 1. \tag{2.121}$$

Proof. In the same way as in the proof of Lemma 2.6.11, we can prove that f is a function in $M^{p,q}(\mathbb{R}^d)$. Let us consider the second part. Since $\|V_\gamma f_\lambda\|_{L^{p,q}} = |\lambda|^{-d(1/p-1/q+1)}\|V_{\gamma_{1/\lambda}} f\|_{L^{p,q}}$, it is enough to show $\|V_{\gamma_{1/\lambda}} f\|_{L^{p,q}} \ge C|\lambda|^{-d/p+\epsilon+d}$ for every $0 < |\lambda| \le 1$. We note that $\operatorname{supp}\gamma((\cdot - x)/\lambda) \subset l + [-1/4, 1/4]^d$ for all $0 < \lambda \le 1$, $l \in \mathbb{Z}^d$ and $x \in l + [-1/8, 1, 8]^d$. Since $\operatorname{supp}\psi(\cdot - k) \subset k + [-1/2, 1/2]^d$ and $\psi(t-k) = 1$ if $t \in k + [-1/4, 1/4]^d$, it follows that

$$\begin{aligned}
&\Big(\int_{\mathbb{R}^d} |V_{\gamma_{1/\lambda}} f(x,\xi)|^p \, dx\Big)^{1/p} \\
&= \Big(\int_{\mathbb{R}^d}\Big|\sum_{k\in\mathbb{Z}^d\setminus\{0\}} |k|^{-d/q-\epsilon}\int_{\mathbb{R}^d} e^{-2\pi i(\xi-k)\cdot t}\psi(t-k)\overline{\gamma((t-x)/\lambda)}\,dt\Big|^p dx\Big)^{1/p} \\
&\ge \Big(\sum_{l\in\mathbb{Z}^d\setminus\{0\}}\int_{l+[-1/8,1/8]^d}\Big|\sum_{k\in\mathbb{Z}^d\setminus\{0\}} |k|^{-d/q-\epsilon}\int_{\mathbb{R}^d} e^{-2\pi i(\xi-k)\cdot t}\psi(t-k)\overline{\gamma((t-x)/\lambda)}\,dt\Big|^p dx\Big)^{1/p} \\
&= \Big(\sum_{l\in\mathbb{Z}^d\setminus\{0\}}\int_{l+[-1/8,1/8]^d}\Big||l|^{-d/q-\epsilon}\int_{\mathbb{R}^d} e^{-2\pi i(\xi-l)\cdot t}\overline{\gamma((t-x)/\lambda)}\,dt\Big|^p dx\Big)^{1/p} \\
&= 4^{-d/p}\Big(\sum_{l\in\mathbb{Z}^d\setminus\{0\}} ||l|^{-d/q-\epsilon}\lambda^d\hat{\gamma}(-\lambda(\xi-l))|^p\Big)^{1/p}.
\end{aligned}$$

Using $|\hat{\gamma}| \ge 1$ on $[-2, 2]^d$,

$$\begin{aligned}
\|V_{\gamma_{1/\lambda}} f\|_{L^{p,q}} &\ge 4^{-d/p}|\lambda|^d\Big(\int_{\mathbb{R}^d}\Big(\sum_{l\in\mathbb{Z}^d\setminus\{0\}} ||l|^{-d/q-\epsilon}\hat{\gamma}(-\lambda(\xi-l))|^p\Big)^{q/p} d\xi\Big)^{1/q} \\
&= 4^{-d/p}|\lambda|^{d-d/q}\Big(\int_{\mathbb{R}^d}\Big(\sum_{l\in\mathbb{Z}^d\setminus\{0\}} ||l|^{-d/q-\epsilon}\hat{\gamma}(\xi+\lambda l)|^p\Big)^{q/p} d\xi\Big)^{1/q} \\
&\ge 4^{-d/p}|\lambda|^{d-d/q}\Big(\int_{[-1,1]^d}\Big(\sum_{\substack{0<|l_j|\le 1/|\lambda|,\\ j=1,\dots,d}} ||l|^{-d/q-\epsilon}\hat{\gamma}(\xi+\lambda l)|^p\Big)^{q/p} d\xi\Big)^{1/q}
\end{aligned}$$

$$\geq C_d|\lambda|^{d-d/q}\Bigg(\sum_{\substack{0<|l_j|\leq 1/|\lambda|,\\ j=1,\dots,d}}|l|^{-p(d/q+\epsilon)}\Bigg)^{1/p}$$

$$\geq C_d|\lambda|^{d-d/q}|\lambda|^{d/q+\epsilon}\Bigg(\sum_{\substack{0<|l_j|\leq 1/|\lambda|,\\ j=1,\dots,d}}1\Bigg)^{1/p}$$

$$\geq C_d|\lambda|^{-d/p+d+\epsilon},$$

for every $0<|\lambda|\leq 1$, that is the claim. □

Lemma 2.6.13. *Let $1\leq p\leq\infty$ and $\gamma,\psi\in\mathcal{S}(\mathbb{R}^d)$ be as in Lemma 2.6.12. Set*

$$f=\sum_{k\in\mathbb{Z}^d}M_kT_k\psi \quad \text{in } \mathcal{S}'(\mathbb{R}^d). \tag{2.122}$$

Then $f\in M^{p,\infty}(\mathbb{R}^d)$ and there exists a constant $C>0$ such that

$$\|f_\lambda\|_{M^{p,\infty}}\asymp\|V_\gamma f_\lambda\|_{L^{p,\infty}}\geq C|\lambda|^{-2d/p},\quad 0<|\lambda|\leq 1. \tag{2.123}$$

In particular, for $1\leq p\leq 2$,

$$C'|\lambda|^{-2d/p}\leq\|f_\lambda\|_{M^{p,\infty}}\leq C|\lambda|^{-2d/p},\quad 0<|\lambda|\leq 1.$$

Proof. In the same way as in the proof of Lemma 2.6.11, we can prove

$$\Bigg|\int_{\mathbb{R}^d}e^{-2\pi i(\xi-k)\cdot t}\psi(t-k)\overline{\gamma(t-x)}\,dt\Bigg|\leq C\langle 2\pi(x-k)\rangle^{-2d}\langle 2\pi(\xi-k)\rangle^{-2d}.$$

This implies, for every $(x,\xi)\in\mathbb{R}^{2d}$,

$$\begin{aligned}|V_\gamma f(x,\xi)|&=\Bigg|\sum_{k\in\mathbb{Z}^d}\int_{\mathbb{R}^d}e^{-2\pi i(\xi-k)\cdot t}\psi(t-k)\overline{\gamma(t-x)}\,dt\Bigg|\\&\leq C\sum_{k\in\mathbb{Z}^d}\langle 2\pi(x-k)\rangle^{-2d}\langle 2\pi(\xi-k)\rangle^{-2d}\\&\leq C\langle 2\pi(x-\xi)\rangle^{-2d},\end{aligned}$$

so that $f\in M^{p,\infty}(\mathbb{R}^d)$. Consider now the second part. Since $\|V_{\gamma_{1/\lambda}}f(\cdot,\xi)\|_{L^p}$ is continuous with respect to $\xi\in\mathbb{R}^d$, we see that

$$\|V_{\gamma_{1/\lambda}}f\|_{L^{p,\infty}}=\sup_{\xi\in\mathbb{R}^d}\big\|V_{\gamma_{1/\lambda}}f(\cdot,\xi)\big\|_{L^p},$$

for each $0<|\lambda|\leq 1$. Hence by the same argument as in the proof of Lemma 2.6.12, we have

$$\begin{aligned}\|V_\gamma f_\lambda\|_{L^{p,\infty}}&=|\lambda|^{-d(1/p+1)}\|V_{\gamma_{1/\lambda}}f\|_{L^{p,\infty}}\\&\geq|\lambda|^{-d(1/p+1)}\big\|V_{\gamma_{1/\lambda}}f(\cdot,0)\big\|_{L^p}\end{aligned}$$

$$\begin{aligned}
&\geq C|\lambda|^{-d(1/p+1)}\Big(\sum_{l\in\mathbb{Z}^d}|\lambda^d\hat{\gamma}(\lambda l)|^p\Big)^{1/p}\\
&\geq C|\lambda|^{-d/p}\Big(\sum_{\substack{0<|l_j|\leq 1/|\lambda|,\\ j=1,\dots,d}}|\hat{\gamma}(\lambda l)|^p\Big)^{1/p}\\
&\geq C|\lambda|^{-2d/p},
\end{aligned}$$

for all $0 < |\lambda| \leq 1$. Combining Lemma 2.6.10 (i), we get $\|V_\gamma f_\lambda\|_{L^{p,\infty}} \asymp |\lambda|^{-2d/p}$, for all $0 < |\lambda| \leq 1$ and $1 \leq p \leq 2$, as desired. □

We are now ready to prove Theorem 2.6.6 (i), with $(1/p, 1/q) \in I_3^*$, and (ii) with $(1/p, 1/q) \in I_3$.

Proof Theorem 2.6.6 (ii), *with* $(1/p, 1/q) \in I_3$. We note that $\mu_2(p,q) = -2/p + 1/q$ if $(1/p, 1/q) \in I_3$. Let $1 \leq p \leq 2$ and $1 \leq q \leq \infty$ be such that $(1/p, 1/q) \in I_3$ and $1/p + 1/q \geq 1$. Note that the previous conditions imply $(1/p, 1/q) \in I_2^*$ and $1/p \geq 1/q$. Then by Lemma 2.6.7 there exists a constant $C > 0$ such that

$$\|f_\lambda\|_{M^{p,q}} \leq C|\lambda|^{-d(2/p-1/q)}\|f\|_{M^{p,q}}, \quad \forall f \in M^{p,q}(\mathbb{R}^d),\ 0 < |\lambda| \leq 1. \tag{2.124}$$

This is the first part of Theorem 2.6.6, (ii), with $(1/p, 1/q) \in I_3$ and $1/p+1/q \geq 1$. Consider $1 \leq p \leq 2$ and $2 \leq q < \infty$ be such that $(1/p, 1/q) \in I_3$ and $1/p + 1/q \leq 1$. The estimate in (2.124) implies

$$\|f_\lambda\|_{M^{r,r'}} \leq C|\lambda|^{-d(2/r-1/r')}\|f\|_{M^{r,r'}} = C|\lambda|^{-d(3/r-1)}\|f\|_{M^{r,r'}}, \tag{2.125}$$

for all $f \in M^{r,r'}(\mathbb{R}^d)$, $0 < |\lambda| \leq 1$, with $1 \leq r \leq 2$ and r' dual of r. Take $1 \leq r \leq 2$ and $0 < \theta < 1$ such that $1/p = (1-\theta)/2+\theta/r$, $1/q = (1-\theta)/\infty+\theta/r'$. Since $q \neq \infty$, we note that $r' \neq \infty$. Then by complex interpolation (Proposition 2.3.17), Lemma 2.6.8, and (2.125) give

$$\|f_\lambda\|_{M^{p,q}} \leq C(|\lambda|^{-d})^{1-\theta}(|\lambda|^{-d(3/r-1)})^{\theta}\|f\|_{M^{p,q}}, \quad \forall f \in M^{p,q}(\mathbb{R}^d),\ 0 < |\lambda| \leq 1.$$

Since $1-\theta = 2(1/p - \theta/r)$ and $\theta/r = \theta - 1/q$, we have

$$\|f_\lambda\|_{M^{p,q}} \leq C|\lambda|^{-d(2(1/p-\theta/r)+3\theta/r-\theta)}\|f\|_{M^{p,q}} = C|\lambda|^{-d(2/p-1/q)}\|f\|_{M^{p,q}},$$

for every $f \in M^{p,q}(\mathbb{R}^d)$, $0 < |\lambda| \leq 1$. The case $q = \infty$ is already proved in Lemma 2.6.10 (i). Hence we obtain the first part of Theorem 2.6.6 (ii), with $(1/p, 1/q) \in I_3$ and $1/p+1/q \leq 1$. By using Lemma 2.6.12 (or 2.6.13), we can prove the second part of Theorem 2.6.6 (ii), with $(1/p, 1/q) \in I_3$, in the same way as in the proof of the second part of Theorem 2.6.6 (ii), with $(1/p, 1/q) \in I_2$. We leave the details to the interested reader. □

Proof Theorem 2.6.6 (i), *with* $(1/p, 1/q) \in I_3^*$. We note that $\mu_1(p,q) = -2/p + 1/q$ if $(1/p, 1/q) \in I_3^*$. In every case, except for $(p,q) \neq (\infty, 1)$, by duality, Theorem 2.6.6 (ii), with $(1/p', 1/q') \in I_3$, and the same argument as in the proof of Theorem 2.6.6 (i), with $(1/p, 1/q) \in I_1^*$, we can prove Theorem 2.6.6 (i), with $(1/p, 1/q) \in I_3^*$. For the first part of Theorem 2.6.6 (i), with $(p,q) = (\infty, 1)$, by the estimate in (2.115), we have nothing to prove. By complex interpolation (Proposition 2.3.17), we can prove the second part in the same way as in the proof of Theorem 2.6.6 (i), with $q = \infty$. The proof of Theorem 2.6.6 is completed. □

The extension to a symmetric matrix $A \in \mathrm{GL}(d, \mathbb{R})$ reads as follows.

Theorem 2.6.14. *Let* $1 \leq p, q \leq \infty$. *There exists a constant* $C > 0$ *such that, for every symmetric matrix* $A \in \mathrm{GL}(d, \mathbb{R})$, *with eigenvalues* $\lambda_1, \ldots, \lambda_d$, *we have*

$$\|f_A\|_{M^{p,q}} \leq C \prod_{j=1}^{d} (\max\{1, |\lambda_j|\})^{\mu_1(p,q)} (\min\{1, |\lambda_j|\})^{\mu_2(p,q)} \|f\|_{M^{p,q}}, \tag{2.126}$$

for every $f \in M^{p,q}(\mathbb{R}^d)$.

Conversely, if there exist $\alpha_j > 0, \beta_j > 0$ *such that*

$$\|f_A\|_{M^{p,q}} \leq C \prod_{j=1}^{d} (\max\{1, |\lambda_j|\})^{\alpha_j} (\min\{1, |\lambda_j|\})^{\beta_j} \|f\|_{M^{p,q}}, \quad \forall f \in M^{p,q}(\mathbb{R}^d),$$

with $A = \mathrm{diag}[\lambda_1, \ldots, \lambda_d]$, *then* $\alpha_j \geq \mu_1(p,q)$ *and* $\beta_j \leq \mu_2(p,q)$.

Proof. The necessary conditions are an immediate consequence of the one dimensional case, already contained in Theorem 2.6.6. Indeed, it can be seen by taking f as tensor product of functions of one variable and by leaving free to vary just one eigenvalue, the remaining eigenvalues being all equal to one.

Let us come to the first part of the theorem. It suffices to prove it in the diagonal case $A = D = \mathrm{diag}[\lambda_1, \ldots, \lambda_d]$. Indeed, since A is symmetric, there exists an orthogonal matrix T such that $A = T^{-1}DT$, and D is a diagonal matrix. On the other hand, by Proposition 2.6.3, we have $\|f_A\|_{M^{p,q}} \lesssim \|f_{T^{-1}D}\|_{M^{p,q}} = \|(f_{T^{-1}})_D\|_{M^{p,q}}$ and $\|f_{T^{-1}}\|_{M^{p,q}} \lesssim \|f\|_{M^{p,q}}$; hence the general case in (2.126) follows from the diagonal case $A = D$, with f replaced by $f_{T^{-1}}$.

From now onward, $A = D = \mathrm{diag}[\lambda_1, \ldots, \lambda_d]$.

If the theorem holds true for a pair (p,q), with $(1/p, 1/q) \in [0,1] \times [0,1]$, then it is also true for their dual pair (p', q') (with $f \in \mathcal{S}$ if $p' = 1$ or $q' = 1$, see (2.34)). Indeed,

$$\begin{aligned}\|f_D\|_{M^{p',q'}} &= \sup_{\|g\|_{M^{p,q}} \leq 1} |\langle f_D, g\rangle| = |\det D|^{-1} \sup_{\|g\|_{M^{p,q}} \leq 1} |\langle f, g_{D^{-1}}\rangle| \\ &\leq |\det D|^{-1} \|f\|_{M^{p',q'}} \sup_{\|g\|_{M^{p,q}} \leq 1} \|g_{D^{-1}}\|_{M^{p,q}}\end{aligned}$$

$$\lesssim \prod_{j=1}^{d} |\lambda_j|^{-1} \prod_{j=1}^{d} (\max\{1, |\lambda_j|^{-1}\})^{\mu_1(p,q)} (\min\{1, |\lambda_j|^{-1}\})^{\mu_2(p,q)} \|f\|_{M^{p',q'}}$$

$$= \prod_{j=1}^{d} (\max\{1, |\lambda_j|\})^{\mu_1(p',q')} (\min\{1, |\lambda_j|\})^{\mu_2(p',q')} \|f\|_{M^{p',q'}},$$

for the index functions μ_1 and μ_2 fulfill

$$\mu_1(p',q') = -1 - \mu_2(p,q), \quad \mu_2(p',q') = -1 - \mu_1(p,q). \tag{2.127}$$

Hence it suffices to prove the estimates (2.126) for the case $p \geq q$. Notice that the estimate in $M^{1,q'}$, $q' > 1$, are proved for Schwartz functions only, but they extend to all functions in $M^{1,q'}$, $q' < \infty$, for $\mathcal{S}(\mathbb{R}^d)$ is dense in $M^{1,q'}$. The uncovered case $(1,\infty)$ will be verified directly at the end of the proof.

From Figure 2.1, it is clear that the estimate (2.126) follows by complex interpolation (Proposition 2.3.17) from the diagonal case $p = q$, and the three cases $(p,q) = (\infty, 1)$, $(p,q) = (1,\infty)$, and $(p,q) = (2,1)$. We shall then limit to consider such values of (p,q)

Case $p = q$. If $d = 1$ the claim is true by Theorem 2.6.6 in dimension $d = 1$. We then use the induction method. Namely, we assume that (2.126) is fulfilled in dimension $d - 1$ and prove that still holds in dimension d.

For $x, \xi \in \mathbb{R}^d$, we write $x = (x', x_d)$, $\xi = (\xi', \xi_d)$, with $x', \xi' \in \mathbb{R}^{d-1}$, $x_d, \xi_d \in \mathbb{R}$, $D' = \operatorname{diag}[\lambda_1, \dots, \lambda_{d-1}]$, and choose the Gaussian $\varphi(x) = e^{-\pi|x|^2} = e^{-\pi|x'|^2} e^{-\pi|x_d|^2} = \varphi'(x')\varphi_d(x_d)$ as window function. Observe that $V_\varphi f_D$ admits the two representations

$$\begin{aligned} V_\varphi f_D(x', x_d, \xi', \xi_d) &= \int_{\mathbb{R}^d} f(\lambda_1 t_1, \dots, \lambda_d t_d) \overline{M_{\xi'} T_{x'} \varphi'(t')}\, \overline{M_{\xi_d} T_{x_d} \varphi_d(t_d)}\, dt'\, dt_d \\ &= V_{\varphi'}((F_{x_d,\xi_d,\lambda_d})_{D'}) \\ &= V_{\varphi_d}((G_{x',\xi',D'})_{\lambda_d}) \end{aligned}$$

where

$$F_{x_d,\xi_d,\lambda_d}(t') = V_{\varphi_d}(f(t', \lambda_d \cdot))(x_d, \xi_d), \quad G_{x',\xi',D'}(t_d) = V_{\varphi'}(f(D' \cdot, t_d))(x', \xi').$$

By the inductive hypothesis, we have

$$\begin{aligned} \|f_D\|_{M^{p,p}(\mathbb{R}^d)} &= \|V_\varphi f_D\|_{L^p(\mathbb{R}^{2d})} \\ &= \Big(\int_{\mathbb{R}^2} \Big(\int_{\mathbb{R}^{2(d-1)}} |V_{\varphi'}((F_{x_d,\xi_d,\lambda_d})_{D'})(x', \xi')|^p\, dx'\, d\xi' \Big) dx_d\, d\xi_d \Big)^{\frac{1}{p}} \\ &\lesssim \prod_{j=1}^{d-1} (\max\{1, |\lambda_j|\})^{\mu_1(p,p)} (\min\{1, |\lambda_j|\})^{\mu_2(p,p)} \end{aligned}$$

$$
\begin{aligned}
&\times \Big(\int_{\mathbb{R}^{2d}} |V_{\varphi'}(F_{x_d,\xi_d,\lambda_d})(x',\xi')|^p \, dx \, d\xi \Big)^{1/p} \\
&= \prod_{j=1}^{d-1} (\max\{1,|\lambda_j|\})^{\mu_1(p,p)} (\min\{1,|\lambda_j|\})^{\mu_2(p,p)} \\
&\times \Big(\int_{\mathbb{R}^{2(d-1)}} \Big(\int_{\mathbb{R}^2} |V_{\varphi_d}((G_{x',\xi',I})_{\lambda_d})(x_d,\xi_d)|^p \, dx_d \, d\xi_d \Big) dx' \, d\xi' \Big)^{1/p} \\
&\lesssim \prod_{j=1}^{d} (\max\{1,|\lambda_j|\})^{\mu_1(p,p)} (\min\{1,|\lambda_j|\})^{\mu_2(p,p)} \|f\|_{M^{p,p}(\mathbb{R}^d)},
\end{aligned}
$$

where in the last raw we used Theorem 2.6.6 for $d = 1$.

Case $(p,q) = (2,1)$. First, we prove the case $(p,q) = (2,\infty)$ and then obtain the claim by duality as above, since $\mathcal{S}$ is dense in $M^{2,1}$. Namely, we want to show that

$$
\|f_D\|_{M^{2,\infty}} \lesssim \prod_{j=1}^{d} (\max\{1,|\lambda_j|\})^{-1/2} (\min\{1,|\lambda_j|\})^{-1} \|f\|_{M^{2,\infty}}, \quad \forall f \in M^{2,\infty}.
$$

We use the characterization of the $M^{2,\infty}$-norm in (2.48), that is,

$$
\|f\|_{M^{2,\infty}(\mathbb{R}^d)} \asymp \sup_{k\in\mathbb{Z}^d} \|\square_k f\|_{L^2}. \tag{2.128}
$$

Hence

$$
\begin{aligned}
\|f_D\|_{M^{2,\infty}} &\lesssim |\det D|^{-1/2} \sup_{k\in\mathbb{Z}^d} \|\sigma_0(D\cdot - k)\hat{f}\|_{L^2} \\
&= |\det D|^{-1/2} \sup_{k\in\mathbb{Z}^d} \Big\| \sigma_0(D\cdot - k) \Big(\sum_{l\in\mathbb{Z}^d} \sigma_0(\cdot - l) \Big) \hat{f} \Big\|_{L^2}.
\end{aligned} \tag{2.129}
$$

Observe that

$$
\begin{aligned}
\Big| \sigma_0(Dt-k) \Big(\sum_{l\in\mathbb{Z}^d} \sigma_0(t-l) \Big) \hat{f}(t) \Big|^2 &\le C_d \sum_{l\in\mathbb{Z}^d} |\sigma_0(Dt-k)\sigma_0(t-l)\hat{f}(t)|^2 \\
&= C_d \sum_{l\in\Lambda_k} |\sigma_0(Dt-k)\sigma_0(t-l)\hat{f}(t)|^2
\end{aligned}
$$

where

$$
\Lambda_k = \Big\{ l \in \mathbb{Z}^d : \Big| l_j - \frac{k_j}{\lambda_j} \Big| \le 1 + \frac{1}{|\lambda_j|} \Big\}
$$

and

$$
\#\Lambda_k \le C \prod_{j=1}^{d} \min\{1,|\lambda_j|\}^{-1}, \quad \forall k \in \mathbb{Z}^d
$$

(C being a constant depending on d only). Since $|\lambda_j| = \max\{1,|\lambda_j|\}\min\{1,|\lambda_j|\}$, the expression on the right-hand side of (2.129) is dominated by

$$C'\prod_{j=1}^{d}(\max\{1,|\lambda_j|\})^{-1/2}(\min\{1,|\lambda_j|\})^{-1}\sup_{k\in\mathbb{Z}^d}\|\Box_k f\|_{L^2}.$$

Thereby the norm equivalence (2.128) gives the desired estimate.

Case $(p,q)=(\infty,1)$. We have to prove that

$$\|f_D\|_{M^{\infty,1}}\lesssim\prod_{j=1}^{d}\max\{1,|\lambda_j|\}\|f\|_{M^{\infty,1}},\quad \forall f\in M^{\infty,1}.$$

This estimate immediately follows from (2.105), written for $A=D=\mathrm{diag}[\lambda_1,\dots,\lambda_d]$:

$$\|f_D\|_{M^{\infty,1}}\lesssim\prod_{j=1}^{d}(1+\lambda_j^2)^{1/2}\|f\|_{M^{\infty,1}}\lesssim\prod_{j=1}^{d}\max\{1,|\lambda_j|\}\|f\|_{M^{\infty,1}}.$$

Case $(p,q)=(1,\infty)$. We are left to prove that

$$\|f_D\|_{M^{1,\infty}}\lesssim\prod_{j=1}^{d}(\max\{1,|\lambda_j|\})^{-1}(\min\{1,|\lambda_j|\})^{-2}\|f\|_{M^{1,\infty}},\quad \forall f\in M^{1,\infty}.$$

This is again the estimate (2.105), written for $A=D=\mathrm{diag}[\lambda_1,\dots,\lambda_d]$,

$$\|f_D\|_{M^{1,\infty}}\lesssim\prod_{j=1}^{d}|\lambda_j|^{-2}\prod_{j=1}^{d}\max\{1,|\lambda_j|\}\|f\|_{M^{1,\infty}}.$$

□

Corollary 2.6.15. *Let* $1\le p,q\le\infty$. *There exists a constant* $C>0$ *such that, for every symmetric matrix* $A\in GL(d,\mathbb{R})$, *with eigenvalues* $\lambda_1,\dots,\lambda_d$, *we have*

$$\|f_A\|_{W(\mathcal{F}L^p,L^q)}\le C\prod_{j=1}^{d}(\max\{1,|\lambda_j|\})^{\mu_1(p',q')}(\min\{1,|\lambda_j|\})^{\mu_2(p',q')}\|f\|_{W(\mathcal{F}L^p,L^q)},\tag{2.130}$$

for every $f\in W(\mathcal{F}L^p,L^q)(\mathbb{R}^d)$.

Conversely, if there exist $\alpha_j>0,\beta_j>0$ *such that*

$$\|f_A\|_{W(\mathcal{F}L^p,L^q)}\le C\prod_{j=1}^{d}(\max\{1,|\lambda_j|\})^{\alpha_j}(\min\{1,|\lambda_j|\})^{\beta_j}\|f\|_{W(\mathcal{F}L^p,L^q)},$$

for every $f\in W(\mathcal{F}L^p,L^q)(\mathbb{R}^d)$, *with* $A=\mathrm{diag}[\lambda_1,\dots,\lambda_d]$, *then* $\alpha_j\ge\mu_1(p',q')$ *and* $\beta_j\le\mu_2(p',q')$.

Proof. It is a mere consequence of Theorem 2.6.14 and the index relation (2.127). Namely,

$$\begin{aligned}\|f_A\|_{W(\mathcal{F}L^p,L^q)} &= \|\widehat{f_A}\|_{M^{p,q}} = |\det A|^{-1}\|\hat{f}_{A^{-1}}\|_{M^{p,q}}\\ &\le C\prod_{j=1}^{d}|\lambda_j|^{-1}\prod_{j=1}^{d}(\max\{1,|\lambda_j|^{-1}\})^{\mu_1(p,q)}(\min\{1,|\lambda_j|^{-1}\})^{\mu_2(p,q)}\|\hat{f}\|_{M^{p,q}}\\ &= C\prod_{j=1}^{d}(\max\{1,|\lambda_j|\})^{\mu_1(p',q')}(\min\{1,|\lambda_j|\})^{\mu_2(p',q')}\|f\|_{W(\mathcal{F}L^p,L^q)}.\end{aligned}$$

The necessary conditions use the same argument. □

2.7 Sharpness of convolution, inclusion, and multiplication relations for modulation spaces

In this section we study the optimality of the convolution, inclusion, and pointwise multiplication relations for modulation spaces.

We need a few preliminary results concerning modulation space norms of Gaussians.

Lemma 2.7.1. *For $a, b \in \mathbb{R}$, $a > 0$, set*

$$\mathcal{G}_{(a+ib)}(t) = (a+ib)^{-d/2}e^{-\frac{\pi t^2}{a+ib}}, \quad t \in \mathbb{R}^d.$$

Then for every $1 \le p, q \le \infty$,

$$\|\mathcal{G}_{(a+ib)}\|_{W(\mathcal{F}L^p,L^q)} = \frac{((a+1)^2+b^2)^{\frac{d}{2}(\frac{1}{p}-\frac{1}{2})}}{p^{\frac{d}{2p}}(aq)^{\frac{d}{2q}}(a(a+1)+b^2)^{\frac{d}{2}(\frac{1}{p}-\frac{1}{q})}}. \tag{2.131}$$

Proof. We use the Gaussian $g(y) = e^{-\pi y^2}$ as window function, so that the norm $W(\mathcal{F}L^p, L^q)$ reads

$$\|\mathcal{G}_{(a+ib)}\|_{W(\mathcal{F}L^p,L^q)} \asymp \big\|\|\mathcal{G}_{(a+ib)}T_x g\|_{\mathcal{F}L^p}\big\|_{L^q_x}.$$

Now,

$$\begin{aligned}\widehat{\mathcal{G}_{(a+ib)}T_x g}(\xi) &= (\widehat{\mathcal{G}_{(a+ib)}} * M_{-x}g)(\xi)\\ &= \int e^{-\pi(a+ib)|\xi-y|^2}e^{-2\pi i x\cdot y}e^{-\pi y^2}\,dy\\ &= e^{-\pi(a+ib)\xi^2}\int e^{-\pi(a+1+ib)y^2-2\pi i(x+i(a+ib)\xi)\cdot y}\,dy\\ &= (a+1+ib)^{-d/2}e^{-\pi(a+ib)\xi^2}e^{-\pi\frac{(x+i(a+ib)\xi)\cdot(x+i(a+ib)\xi)}{a+1+ib}},\end{aligned} \tag{2.132}$$

where we used (0.27). Hence, after a simple computation,

$$|\widehat{\mathcal{G}_{(a+ib)}T_x g}(\xi)| = ((a+1)^2+b^2)^{-d/4} e^{-\frac{\pi}{(a+1)^2+b^2}[(a(a+1)+b^2)\xi^2+2bx\cdot\xi+(a+1)x^2]}.$$

It follows that

$$\begin{aligned}
\|\mathcal{G}_{(a+ib)}T_x g\|_{\mathcal{F}L^p} &= ((a+1)^2+b^2)^{-d/4} e^{-\frac{\pi(a+1)x^2}{(a+1)^2+b^2}} \\
&\quad\times\left(\int e^{-\frac{p\pi}{(a+1)^2+b^2}[(a^2+b^2+a)\xi^2+2bx\cdot\xi]}\,d\xi\right)^{\frac{1}{p}} \\
&= ((a+1)^2+b^2)^{-d/4} e^{-\frac{\pi a x^2}{a(a+1)+b^2}} \\
&\quad\times\left(\int e^{-\frac{p\pi}{(a+1)^2+b^2}\left|\sqrt{a(a+1)+b^2}\xi+\frac{b}{\sqrt{a(a+1)+b^2}}x\right|^2}\,d\xi\right)^{\frac{1}{p}} \\
&= ((a+1)^2+b^2)^{-d/4} e^{-\frac{\pi a x^2}{a(a+1)+b^2}}(a(a+1)+b^2)^{-\frac{d}{2p}} \\
&\quad\times\left(\frac{(a+1)^2+b^2}{p}\right)^{\frac{d}{2p}} \\
&= \frac{((a+1)^2+b^2)^{\frac{d}{2}(\frac{1}{p}-\frac{1}{2})}}{p^{\frac{d}{2q}}(a(a+1)+b^2)^{\frac{d}{2p}}} e^{-\frac{\pi a x^2}{a(a+1)+b^2}}.
\end{aligned}$$

By taking the L^q-norm of this expression, one obtains (2.131). □

Now, we turn our attention to the sharpness of the convolution properties for modulation spaces.

Proposition 2.7.2. *Let* $1 \le p, q, p_1, p_2, q_1, q_2 \le \infty$. *Then*

$$\|f * g\|_{M^{p,q}} \lesssim \|f\|_{M^{p_1,q_1}}\|g\|_{M^{p_2,q_2}} \tag{2.133}$$

if and only if the indices' relations (2.79) *and* (2.80) *hold true.*

Proof. The sufficiency is shown in Proposition 2.4.19. We prove the necessity. We consider the family of Gaussians $\varphi_{\sqrt{\lambda}}(x) = e^{-\pi\lambda x^2}$, for $\lambda > 0$. Obviously, $\varphi_{\sqrt{\lambda}} \in \mathcal{S}(\mathbb{R}^d) \subset M^{p,q}(\mathbb{R}^d)$, for every $1 \le p, q \le \infty$. Since $\|f\|_{M^{p,q}} \asymp \|\hat{f}\|_{W(\mathcal{F}L^p,L^q)}$ and $\widehat{\varphi_{\sqrt{\lambda}}} = \lambda^{-d/2}\varphi_{\sqrt{1/\lambda}}$, Lemma 2.7.1 yields

$$\begin{aligned}
\|\varphi_{\sqrt{\lambda}}\|_{M^{p,q}} &\asymp \lambda^{-d/2}\|\varphi_{\sqrt{1/\lambda}}\|_{W(\mathcal{F}L^p,L^q)} \\
&\asymp \|\mathcal{G}_{(\lambda)}\|_{W(\mathcal{F}L^p,L^q)} \asymp \frac{(\lambda+1)^{d(1/p-1/2)}}{\lambda^{d/(2q)}(\lambda^2+\lambda)^{(1/p-1/q)d/2}}.
\end{aligned} \tag{2.134}$$

A straightforward calculation shows that $(\varphi_{\sqrt{\lambda}} * \varphi_{\sqrt{\lambda}})(x) = (2\lambda)^{-\frac{d}{2}}\varphi_{\sqrt{\lambda/2}}(x)$. Hence using (2.134), we obtain

$$\|\varphi_{\sqrt{\lambda}} * \varphi_{\sqrt{\lambda}}\|_{M^{p,q}} \asymp \lambda^{-(1+1/p)d/2}, \quad \text{for } \lambda \to 0^+. \tag{2.135}$$

Using (2.134) again, we also obtain

$$\|\varphi_{\sqrt{\lambda}}\|_{M^{p_i,q_i}} \asymp \lambda^{-\frac{d}{2p_i}}, \quad i=1,2, \quad \text{for } \lambda \to 0^+. \tag{2.136}$$

Substituting in (2.82), we obtain (2.79). The relation (2.80) can be obtained similarly. In fact, the estimate (2.134) gives, for $\lambda \to +\infty$,

$$\|\varphi_{\sqrt{\lambda}} * \varphi_{\sqrt{\lambda}}\|_{M^{p,q}} \asymp \lambda^{-d(1-\frac{1}{2q})}, \quad \|\varphi_{\sqrt{\lambda}}\|_{M^{p_i,q_i}} \asymp \lambda^{-\frac{d}{2}(1-\frac{1}{q_i})}, \quad i=1,2,$$

and, using (2.82) again, the relation (2.80) follows.

An alternative proof of the necessary conditions (2.79) and (2.80) is provided by Lemma 2.3.26. Precisely, to prove (2.80), consider two compactly supported smooth functions f,g and their scaling $f_\lambda(x) = f(\lambda x)$, $g_\lambda(x) = g(\lambda x)$, with $\lambda \ge 1$. Since $\lambda \ge 1$, f_λ and g_λ (and therefore $f_\lambda * g_\lambda$) are all supported in a compact subset K, independent of λ. By Lemma 2.3.26 (i), the bilinear estimate (2.82) for f_λ and g_λ becomes

$$\|f_\lambda * g_\lambda\|_{\mathcal{F}L^q} \lesssim \|f_\lambda\|_{\mathcal{F}L^{q_1}} \|g_\lambda\|_{\mathcal{F}L^{q_2}}.$$

Using $f_\lambda * g_\lambda = \lambda^{-d}(f * g)_\lambda$, the dilation property for $\mathcal{F}L^q$ spaces

$$\|h(\lambda\cdot)\|_{\mathcal{F}L^q} = \lambda^{-\frac{d}{q'}} \|h\|_{\mathcal{F}L^q},$$

and letting $\lambda \to +\infty$, we obtain (2.80).

In order to prove (2.79), one argues similarly. Here the functions f,g have Fourier transforms $\hat{f},\hat{g}$ compactly supported and the scale λ satisfies $0 < \lambda \le 1$. By Lemma 2.3.26 (ii), the estimate (2.81) becomes

$$\|f_\lambda * g_\lambda\|_{L^p} \lesssim \|f_\lambda\|_{L^{p_1}} \|g_\lambda\|_{L^{p_2}}.$$

Using $f_\lambda * g_\lambda = \lambda^{-d}(f * g)_\lambda$, the dilation property $\|h(\lambda\cdot)\|_{L^p} = \lambda^{-d/p}\|h\|_{L^p}$, and letting $\lambda \to 0^+$, we prove (2.79). □

The family of Gaussians $\varphi_{\sqrt{\lambda}}$ also provides a proof for the sharpness of the inclusion relation for modulation spaces.

Proposition 2.7.3. *Let* $1 \le p_1, p_2, q_1, q_2 \le \infty$. *Then*

$$\|f\|_{M^{p_2,q_2}} \lesssim \|f\|_{M^{p_1,q_1}} \tag{2.137}$$

if and only if the following indices' relation holds:

$$p_1 \le p_2 \quad \text{and} \quad q_1 \le q_2. \tag{2.138}$$

Proof. The sufficiency is proved in Theorem 2.4.17. We show the necessity of (2.138). Let $\varphi_{\sqrt{\lambda}}(x) = e^{-\pi\lambda x^2}$, $\lambda > 0$. From the proof of Proposition 2.7.2,

$$\|\varphi_{\sqrt{\lambda}}\|_{M^{p_i,q_i}} \asymp \lambda^{-\frac{d}{2p_i}}, \quad \text{for } \lambda \to 0^+, \quad \text{and} \quad \|\varphi_{\sqrt{\lambda}}\|_{M^{p_i,q_i}} \asymp \lambda^{-\frac{d}{2}(1-\frac{1}{q_i})}, \quad \text{for } \lambda \to +\infty. \tag{2.139}$$

Hence for (2.137) to be satisfied, it must be that

$$\lambda^{-\frac{d}{2p_2}} \lesssim \lambda^{-\frac{d}{2p_1}}, \quad \text{for } \lambda \to 0^+, \quad \text{and} \quad \lambda^{-\frac{d}{2}(1-\frac{1}{q_2})} \lesssim \lambda^{-\frac{d}{2}(1-\frac{1}{q_1})}, \quad \text{for } \lambda \to +\infty,$$

which give the indices' relations in (2.138). □

In what follows we study the pointwise multiplication operator in modulation spaces (which is equivalent to studying the convolution operator for the Wiener amalgam spaces $W(\mathcal{F}L^p, L^q)$).

Proposition 2.7.4. *Let* $1 \le p, q, p_1, p_2, q_1, q_2 \le \infty$. *Then*

$$\|fg\|_{M^{p,q}} \lesssim \|f\|_{M^{p_1,q_1}} \|g\|_{M^{p_2,q_2}} \tag{2.140}$$

if and only if the indices' relations (2.84) *and* (2.85) *hold true.*

Proof. The sufficiency can be found in Proposition 2.4.23. For the necessity of the conditions (2.84) and (2.85), we test the estimate (2.140) on the Gaussians $f(x) = g(x) = \varphi_{\sqrt{\lambda}}(x) = e^{-\lambda\pi x^2}$. We observe that $\varphi_{\sqrt{\lambda}}\varphi_{\sqrt{\lambda}} = \varphi_{\sqrt{2\lambda}}$. Hence by applying (2.139) and substituting in (2.140), relation (2.85) follows by letting $\lambda \to 0^+$, whereas (2.84) follows by letting $\lambda \to +\infty$. □

2.7.1 Time-frequency tools

For further purpose we need to compute the modulation norm of the generalized Gaussian f defined in (1.126), namely for $a, b, c > 0$,

$$f(x,\xi) = f_{a,b,c}(x,\xi) = e^{-\pi a x^2} e^{-\pi b \xi^2} e^{2\pi i c x\cdot\xi}, \quad (x,\xi) \in \mathbb{R}^{2d}.$$

Proposition 2.7.5. *Consider the generalized Gaussian f defined in* (1.126) *and the window function* $\Phi(x,\xi) = e^{-\pi(x^2+\xi^2)}$. *Then for every* $1 \le p, q \le \infty$, *we have*

$$\begin{aligned} \|f\|_{M^{p,q}} &\asymp \|V_\Phi f\|_{L^{p,q}} \\ &\asymp \frac{[(a+1)(b+1)+c^2]^{\frac{d}{p}+\frac{d}{q}-\frac{d}{2}} [(c^2+ab+a)(c^2+ab+b)]^{\frac{d}{2q}-\frac{d}{2p}}}{[b^2(a+1)+b(c^2+a+1)]^{\frac{d}{2q}} [a^2(b+1)+a(c^2+b+1)]^{\frac{d}{2q}}}. \end{aligned} \tag{2.141}$$

The cases $p = \infty$ or $q = \infty$ can be obtained by using the rule $1/\infty = 0$ in formula (2.141).

Proof. By Proposition 1.3.33, we can write

$$|V_\Phi f(z,\zeta)| = \frac{1}{[(a+1)(b+1)+c^2]^{\frac{d}{2}}} \times e^{-\pi\frac{[a(b+1)+c^2]z_1^2+[(a+1)b+c^2]z_2^2+(b+1)\zeta_1^2+(a+1)\zeta_2^2-2c(z_1\cdot\zeta_2+z_2\cdot\zeta_1)}{(a+1)(b+1)+c^2}}.$$

It remains to compute the mixed $L^{p,q}$-norm of the previous function. We treat the cases $1 \le p,q < \infty$. The cases if either $p=\infty$ or $q=\infty$ are obtained with obvious modifications.

For simplicity, we set

$$\alpha = \frac{c^2+a(b+1)}{(a+1)(b+1)+c^2}, \quad \beta = \frac{c^2+(a+1)b}{(a+1)(b+1)+c^2}, \quad \gamma = \frac{(b+1)}{(a+1)(b+1)+c^2}, \tag{2.142}$$

$$\delta = \frac{(a+1)}{(a+1)(b+1)+c^2}, \quad \sigma = \frac{c}{(a+1)(b+1)+c^2}. \tag{2.143}$$

Hence

$$\frac{\|V_\Phi f\|_{L^{p,q}}}{[(a+1)(b+1)+c^2]^{-\frac{d}{2p}}} = \left(\int_{\mathbb{R}^{2d}} I^{\frac{q}{p}} e^{-\pi q(\gamma\zeta_1^2+\delta\zeta_2^2)}\, d\zeta_1\, d\zeta_2\right)^{\frac{1}{q}} =: A, \tag{2.144}$$

where $I := \int_{\mathbb{R}^{2d}} e^{-\pi p\alpha z_1^2-\pi p\beta z_2^2} e^{2\pi p\sigma(z_1\cdot\zeta_2+z_2\cdot\zeta_1)}\, dz_1\, dz_2$. Now straightforward computations and change of variables yield

$$\begin{aligned} I &= \int_{\mathbb{R}^d}\left(\int_{\mathbb{R}^d} e^{-\pi p(\alpha z_1^2-2\sigma\zeta_2\cdot z_1)}\, dz_1\right) e^{-\pi\beta z_2^2+2\pi p\sigma z_2\cdot\zeta_1}\, dz_2 \\ &= e^{\pi p\frac{\sigma^2}{\alpha}\zeta_2^2} e^{\pi p\frac{\sigma^2}{\beta}\zeta_1^2} \int_{\mathbb{R}^d} e^{-\pi p(\sqrt{\alpha}z_1-\frac{\sigma}{\sqrt{\alpha}}\zeta_2)^2}\, dz_1 \int_{\mathbb{R}^d} e^{-\pi p(\sqrt{\beta}z_2-\frac{\sigma}{\sqrt{\beta}}\zeta_1)^2}\, dz_2 \\ &= p^{-\frac{d}{2}}\alpha^{-\frac{d}{2}} p^{-\frac{d}{2}}\beta^{-\frac{d}{2}} e^{\pi p\frac{\sigma^2}{\alpha}\zeta_2^2} e^{\pi p\frac{\sigma^2}{\beta}\zeta_1^2}. \end{aligned}$$

Substituting the value of the integral I in (2.144), we obtain

$$\begin{aligned} A &= p^{-\frac{d}{p}}\alpha^{-\frac{d}{2p}}\beta^{-\frac{d}{2p}}\left(\int_{\mathbb{R}^d} e^{\pi q\frac{\sigma^2}{\beta}\zeta_1^2-\pi q\gamma\zeta_1^2}\, d\zeta_1 \int_{\mathbb{R}^d} e^{\pi q\frac{\sigma^2}{\alpha}\zeta_2^2-\pi q\delta\zeta_2^2}\, d\zeta_2\right)^{\frac{1}{q}} \\ &= p^{-\frac{d}{p}}\alpha^{-\frac{d}{2p}}\beta^{-\frac{d}{2p}}\left(\int_{\mathbb{R}^d} e^{-\pi q(\gamma-\frac{\sigma^2}{\beta})\zeta_1^2}\, d\zeta_1 \int_{\mathbb{R}^d} e^{-\pi q(\delta-\frac{\sigma^2}{\alpha})\zeta_2^2}\right)^{\frac{1}{q}} \\ &= p^{-\frac{d}{p}}\alpha^{-\frac{d}{2p}}\beta^{-\frac{d}{2p}} q^{-\frac{d}{q}}\left(\gamma-\frac{\sigma^2}{\beta}\right)^{-\frac{d}{2q}}\left(\gamma-\frac{\sigma^2}{\alpha}\right)^{-\frac{d}{2q}}. \end{aligned}$$

Finally, the goal is attained by substituting in A the values of the parameters $\alpha, \beta, \gamma, \delta, \sigma$ in (2.142) and (2.143) and observing that

$$\|f\|_{M^{p,q}} \asymp \|V_\Phi f\|_{L^{p,q}} = A\big[(a+1)(b+1)+c^2\big]^{-\frac{d}{2}}.$$

This concludes the proof. □

2.7.2 Further extensions of modulation spaces (quasi-Banach setting)

We end up this chapter recalling the extension of modulation spaces $M^{p,q}_m(\mathbb{R}^d)$ to the indices $0 < p, q \le \infty$ (quasi-Banach spaces) performed first by Y. V. Galperin and S. Samarah in [153].

Definition 2.7.6. Fix a nonzero window $g \in \mathcal{S}(\mathbb{R}^d)$, a weight $m \in \mathcal{M}_v(\mathbb{R}^{2d})$, and $0 < p, q \le \infty$. The modulation space $M^{p,q}_m(\mathbb{R}^d)$ consists of all tempered distributions $f \in \mathcal{S}'(\mathbb{R}^d)$ such that the (quasi-)norm

$$\|f\|_{M^{p,q}_m} = \|V_g f\|_{L^{p,q}_m} = \Bigg(\int_{\mathbb{R}^d}\Bigg(\int_{\mathbb{R}^d} |V_g f(x,\xi)|^p m(x,\xi)^p\, dx\Bigg)^{\frac{q}{p}} d\xi\Bigg)^{\frac{1}{q}} \tag{2.145}$$

(obvious changes with $p = \infty$ or $q = \infty$) is finite.

They were proved to be quasi-Banach spaces, whose definition does not depend on the window $g \in \mathcal{S}(\mathbb{R}^d)$. In this framework, it appears that the largest natural class of windows universally admissible for all spaces $M^{p,q}_m(\mathbb{R}^d)$, $0 < p, q \le \infty$ (with m having at most polynomial growth) is the Schwartz class $\mathcal{S}(\mathbb{R}^d)$.

They were further investigated in [211, 212, 248, 305] (see also references therein).

Actually, several papers involve modulation spaces with indices $1 \le p, q \le \infty$, whereas very few works deal with the quasi-Banach case $0 < p, q \le 1$. In fact, many properties related to the latter case are still unexplored. We recall here the convolution relations proved in the recent contribution [6], extending the Banach setting in Proposition 2.4.19.

Proposition 2.7.7. *Let $v(\xi) > 0$ be an arbitrary weight function on $\mathbb{R}^d$, $0 < p, q, r, t, u, \gamma \le \infty$, with*

$$\frac{1}{u} + \frac{1}{t} = \frac{1}{\gamma}$$

and

$$\frac{1}{p} + \frac{1}{q} = 1 + \frac{1}{r}, \quad \textit{for } 1 \le r \le \infty,$$

whereas

$$p = q = r, \quad \textit{for } 0 < r < 1.$$

For $m \in \mathcal{M}_v(\mathbb{R}^{2d})$, $m_1(x) = m(x, 0)$ *and* $m_2(\xi) = m(0, \xi)$ *are the restrictions to* $\mathbb{R}^d \times \{0\}$ *and* $\{0\} \times \mathbb{R}^d$, *and likewise for* v. *Then*

$$M^{p,u}_{m_1 \otimes v}(\mathbb{R}^d) * M^{q,t}_{v_1 \otimes v_2 v^{-1}}(\mathbb{R}^d) \hookrightarrow M^{r,y}_m(\mathbb{R}^d)$$

with norm inequality

$$\|f * h\|_{M^{r,y}_m} \lesssim \|f\|_{M^{p,u}_{m_1 \otimes v}} \|h\|_{M^{q,t}_{v_1 \otimes v_2 v^{-1}}}.$$

3 Gabor frames and linear operators

In the applications of time–frequency analysis, in particular signal theory, one refers to the discrete version of the short-time Fourier transform, and this will have also an important role in our book. Namely, one can consider the sampling of

$$V_g f(x,\xi) = \int_{\mathbb{R}^d} f(t)\overline{g(t-x)}e^{-2\pi i t\cdot\xi}\,dt$$

at the points $x = \alpha m$, $\xi = \beta n$ of the lattice $\Lambda = \alpha\mathbb{Z}^d \times \beta\mathbb{Z}^d$, for some fixed $\alpha > 0, \beta > 0$. Alternatively, one refers to the Gabor coefficients

$$c_{m,n} = \langle f, T_{\alpha m} M_{\beta n} g\rangle = e^{2\pi i \alpha\beta n\cdot m} V_g f(\alpha m, \beta n). \tag{3.1}$$

The basic issue is to discuss whether and how we may reconstruct f from the coefficients $c = (c_{m,n})$, possibly after normalization. Reminiscent of the classical orthonormal (o. n.) bases, we should try to recapture f by an expression of the type

$$\sum_{m,n\in\mathbb{Z}^d} c_{m,n} T_{\alpha m} M_{\beta n} g.$$

However, the Gabor system

$$\{T_{\alpha m} M_{\beta n} g,\ (m,n) \in \mathbb{Z}^{2d}\}$$

is not an o. n. basis, apart very particular cases. In fact, the so fruitful links with spectral theory, namely identification with eigenfunctions of a self-adjoint operator, are lost. The analysis of Gabor systems represents a challenging and fascinating field of the mathematical analysis. In our book we shall not discuss advanced topics in this context. Rather, aiming to the applications to partial differential equations, we shall limit ourselves to some basic notions and results necessary for proofs and numerical experiments in the sequel.

Starting from scratch, we first address nonexpert readers to Appendix A, Section A.1, concerning bases theory and series expansions in Hilbert and Banach spaces. A relevant point there, suggested by the previous expansion in terms of a Gabor system, is the unconditional convergence of a series. With respect to the classical theorems of Dirichlet and Riemann–Dini stating the equivalence in $\mathbb{R}$ with the absolute convergence, in Hilbert and Banach spaces absolute convergence implies unconditional convergence, but not vice versa, and a more precise analysis is needed, including in particular the concept of Riesz bases.

The first part of the present Chapter 3 has a preliminary character as well, being devoted to the theory of Frames in Hilbert spaces. Frames were proposed by Duffin and Schaeffer in 1952 [113] as a more flexible method of approximation, with respect to o. n. bases, by allowing nonorthogonal and redundant complete systems.

https://doi.org/10.1515/9783110532456-004

Even in problems where o. n. bases are available, frames are preferred because of their efficiency in computations. They represent also the correct setting for Gabor systems; let us describe in short the theory in this context.

A Gabor system is called a Gabor frame if it satisfies the frame condition

$$A\|f\|_2^2 \leq \sum_{m,n\in\mathbb{Z}^d} |\langle f, T_{\alpha m} M_{\beta n} g\rangle|^2 \leq B\|f\|_2^2, \quad \forall f \in L^2(\mathbb{R}^d),$$

where $0 < A \leq B$ and the Gabor coefficients $c_{m,n}$ are defined in (3.1). Under such a condition, the Gabor system is complete in $L^2(\mathbb{R}^d)$ and the series

$$Sf = \sum_{m,n\in\mathbb{Z}^d} c_{m,n} T_{\alpha m} M_{\beta n} g$$

is convergent. This is not a reproducing formula, that is, the operator S is not the identity in $L^2(\mathbb{R}^d)$. Nevertheless, $S : L^2(\mathbb{R}^d) \to L^2(\mathbb{R}^d)$ is invertible in $L^2(\mathbb{R}^d)$, and the application of its inverse S^{-1} provides the expected reconstruction of f. We may then reelaborate the formula by defining the dual window $\gamma = S^{-1}g$, and this provides after some computations

$$\begin{aligned} f &= \sum_{m,n\in\mathbb{Z}^d} c_{m,n} T_{\alpha m} M_{\beta n} \gamma \\ &= \sum_{m,n\in\mathbb{Z}^d} \langle f, T_{\alpha m} M_{\beta n} \gamma\rangle T_{\alpha m} M_{\beta n} g. \end{aligned}$$

So we have reduced the situation to the classical o. n. setting, apart the appearance of the dual window in the expression of the system or, alternatively, in the computation of the Gabor coefficients. Of course, it remains to check whether the initial frame condition is satisfied. This depends on the window g and the lattice parameters α, β. Intuitively, we may imagine that for sufficiently small α, β the frame condition is satisfied, but precise results are very difficult, and will be not given in this book, addressing the interested reader to the textbooks [50, 160, 183], see also the paper by I. Daubechies [101]. We shall limit to some examples, in particular for the Gaussian window $g(t) = e^{-\pi t^2}$, $t \in \mathbb{R}^d$, the Gabor system is a frame if and only if $\alpha\beta < 1$.

The role of Gabor frames in the sequel, apart numerical experiments, will be to provide an alternative definition of modulation spaces. Namely, we can use the sequence spaces $\ell_m^{p,q}$ in Definition 2.2.1 to state that $f \in M_m^{p,q}(\mathbb{R}^d)$ if and only if the Gabor coefficients $c = (c_{m,n})$ belong to $\ell_m^{p,q}$, with equivalence of norms

$$\|f\|_{M_m^{p,q}} \asymp \|c\|_{\ell_m^{p,q}}.$$

The last part of this chapter is devoted to integral expressions of operators, time-frequency counterpart of the classical Schwartz' kernel theorem for tempered distributions. There the material is preliminary to the next chapters, and the proofs use in an essential way a discretization of the problem by Gabor frames. Precisely, we characterize the continuity of a linear operator on the modulation space $M^p(\mathbb{R}^d)$, $1 \le p \le \infty$, by the membership of its kernel K in (mixed norm) modulation spaces. We recapture the result of H. Feichtinger announced in [126] and proved in [134], stating that linear continuous operators $A : M^1(\mathbb{R}^d) \to M^\infty(\mathbb{R}^d)$ are characterized by the condition $K \in M^\infty(\mathbb{R}^{2d})$. An obvious advantage, with respect to the Schwartz's kernel theorem in the topological vector spaces $(\mathcal{S}(\mathbb{R}^d), \mathcal{S}'(\mathbb{R}^d))$, is that $M^p(\mathbb{R}^d)$, $1 \le p \le \infty$, are Banach spaces. This reveals a superiority, in some respects, of the modulation space formalism, as emphasized in the Feichtinger's manifesto 2015 [128] for a post-modern harmonic analysis.

The contents of the chapter is organized in the following way. Section 3.1 is devoted to the main properties of frames in Hilbert spaces. Section 3.2 presents first the Gabor frames, and then treats the equivalent definition of modulation spaces. Finally, Section 3.3 concerns the kernel theorems for modulation spaces.

3.1 Frames in Hilbert spaces

Frames were originally introduced by Duffin and Schaeffer in 1952 [113], in the framework of nonharmonic Fourier series. In what follows we exhibit a brief survey on the topic, mainly borrowed from [183], to which we refer the interested reader.

Definition 3.1.1 (Frame). A sequence $\{x_n\}$ in a Hilbert space H is a frame if there exist constants $0 < A \le B$ such that

$$\forall x \in H, \quad A\|x\|_H^2 \le \sum_{n=1}^{\infty} |\langle x, x_n\rangle|^2 \le B\|x\|_H^2. \tag{3.2}$$

The constants A and B are called *lower* and *upper frame bound*, respectively. The largest possible lower frame bound is called *the optimal lower frame bound* and the smallest possible upper frame bound is called *the optimal upper frame bound.*

Thus if $\{x_n\}$ is a frame, $\|(\langle x, x_n\rangle)\|_{\ell^2}$ is an equivalent norm for H. In particular, if $A = B = 1$, then actually we have

$$\|x\|_H = \|(\langle x, x_n\rangle)\|_{\ell^2}.$$

Although Definition 3.1.1 says nothing explicitly about basis-like properties of $\{x_n\}$, we shall see that the equivalence (3.2) implies unconditional convergence, basis-like representations of vectors in H. We first exhibit special types of frames.

Definition 3.1.2. Let $\{x_n\}$ be a frame for a Hilbert space H.

(i) If $A = B$ (optimal frame bounds are equal) then $\{x_n\}$ is called a *tight frame*. We say in this case that $\{x_n\}$ is an A-tight frame.

(ii) If $A = B = 1$ then $\{x_n\}$ is called a *Parseval frame*.

(iii) A frame $\{x_n\}$ is called an *exact frame* if it ceases to be a frame whenever any single element is deleted from the sequence.

The easiest example of frame is provided by an orthonormal basis (o. n. basis, for short) $\{e_n\}$. In this case $A = B = 1$, so the frame is a Parseval frame and, moreover, it is exact.

Remark 3.1.3.

(i) In a frame a repetition of elements and the presence of the zero vector are allowed. For instance, for an o. n. basis $\{e_n\}$, the sequence $\{0, e_1, e_2, \dots, e_n, \dots\}$ is a Parseval frame but it is not exact. Another example is $\{e_1, e_1, e_2, \dots, e_n, \dots\}$ where we have a frame which is again not exact.

(ii) If $\{x_n\}$ is a frame, then $\sum_n |\langle x, x_n\rangle|^2$ is an absolutely convergent series, so by Lemma A.1.13, it is unconditionally convergent. So for any permutation σ of $\mathbb{N}$, we have that $\{x_{\sigma(n)}\}$ is still a frame, and therefore it does not matter what countable set we use to index a frame.

(iii) We will see that if a frame is exact then it is a basis for H, cf. Definition A.1.1. An inexact frame is redundant, or overcomplete, in the sense that a proper subset of it is still complete (in fact, still a frame).

(iv) Observe that, by (3.2), a frame is a complete sequence. Indeed, if we assume $\langle x, x_n\rangle = 0$, for every n, then by (3.2),

$$A\|x\|_H^2 \leq \sum_n |\langle x, x_n\rangle|^2 = 0 \Longrightarrow x = 0.$$

(v) Since any space that contains a countable complete set is separable, we conclude that a Hilbert space which has a frame is separable. Conversely, every separable Hilbert space contains an o. n. basis and hence has a frame.

Example 3.1.4. For $H = \mathbb{R}^2$, consider the vectors

$$x_1 = (1,0), \quad x_2 = (0,1), \quad x_3 = \left(\frac{1}{\sqrt{2}}, \frac{1}{\sqrt{2}}\right), \quad x_4 = \left(-\frac{1}{\sqrt{2}}, \frac{1}{\sqrt{2}}\right).$$

The sets $\{x_1, x_2\}$ and $\{x_3, x_4\}$ are orthonormal bases for $\mathbb{R}^2$. So

$$2\|x\|_H^2 = \sum_{n=1}^{4} |\langle x, x_n\rangle|^2,$$

that is, $\{x_1, \dots, x_4\}$ is an inexact tight frame with frame bounds $A = B = 2$. If we take $y_n = x_n/\sqrt{2}$, $n = 1, \dots, 4$, then the set $\{y_n\}$ is a Parseval frame, but observe that its elements are not orthogonal and are not a basis.

Example 3.1.5. Let $\{e_n\}$ be an orthonormal basis for a Hilbert space H. Then

(i) $\{e_1, e_1, e_2, e_2, \ldots, e_n, e_n, \ldots\}$ is a tight inexact frame with frame bounds $A = B = 2$. Observe that this frame is not orthogonal and is not a basis. Similarly, for another orthonormal basis $\{f_n\}$, the set $\{e_n\} \cup \{f_n\}$ is a tight inexact frame.
(ii) $\{e_1, e_2/2, e_3/3, \ldots\}$ is a complete orthogonal sequence and it is a basis for H, but it is not a frame, because it does not possess a lower bound.
(iii) $\{x_n\} = \{e_1, e_2/\sqrt{2}, e_2/\sqrt{2}, e_3/\sqrt{3}, e_3/\sqrt{3}, e_3/\sqrt{3}, \ldots\}$, for which we have $\|x\|_H^2 = \sum_n |\langle x, x_n\rangle|^2$, is an inexact Parseval frame.
(iv) $\{\sqrt{2}e_1, e_2, e_3, \ldots\}$ is a nontight exact frame with optimal frame bounds $A = 1$ and $B = 2$.

We now study more interesting examples of frames, coming from trigonometric systems.

Example 3.1.6 (Trigonometric systems). We consider the trigonometric system $\{e^{2\pi ibnt}\}_{n\in\mathbb{Z}}$, in $L^2(\mathbb{T})$, where $b > 0$ is fixed. Recall that

$$L^2(\mathbb{T}) = \{f : \mathbb{R} \to \mathbb{C},\ f \text{ measurable} : f(t+1) = f(t) \text{ a. e., and } \|f\|_2 < \infty\}/\sim,$$

where $\|f\|_2^2 = \int_0^1 |f(t)|^2\,dt < \infty$ and $f \sim g \Longleftrightarrow f = g$ a. e.

Since functions in $L^2(\mathbb{T})$ are 1-periodic, we are implicitly considering $e^{2\pi ibnt}$ to be defined on the interval $[0, 1)$ and then extended by 1-periodicity on $\mathbb{R}$.
(i) When $b = 1$, it is well known that $\{e^{2\pi int}\}_{n\in\mathbb{Z}}$ is an o. n. basis for $L^2(\mathbb{T})$.
(ii) Suppose $b > 1$. Then the function $e^{2\pi ibnt}$ is $\frac{1}{b}$-periodic on $\mathbb{R}$. Since $\frac{1}{b} < 1$ the interval $I_b := [0, 1 - \frac{1}{b}]$ is well defined and we have the inclusion $I_b \subset [0, 1]$. Moreover, for every $t \in I_b$, $t + \frac{1}{b} \in [0, 1]$ and

$$e^{2\pi ibn(t+\frac{1}{b})} = e^{2\pi ibnt},$$

so every function $f_n(t) := e^{2\pi ibnt}$ satisfies

$$f_n\left(t + \frac{1}{b}\right) = f_n(t), \quad t \in I_b. \tag{3.3}$$

This implies that $\operatorname{span}\{f_n\}$ satisfies (3.3). Moreover, $\overline{\operatorname{span}}\{f_n\}$ (in the $L^2(\mathbb{T})$-norm) satisfies (3.3) a. e. In fact, $f = \lim_n f_n$ in $L^2(\mathbb{T})$ implies pointwise convergence a. e. for a subsequence $\{f_{n_k}\}$. However, not every function $f \in L^2(\mathbb{T})$ fulfills (3.3). For example, the function $f(t) = t \in [0, 1]$ is one of these. Hence $\{e^{2\pi ibnt}\}_{n\in\mathbb{Z}}$ is incomplete and thus is not a frame.
(iii) Consider $0 < b < 1$. The function $e^{2\pi ibnt}$ is $\frac{1}{b}$-periodic but this time $\frac{1}{b} > 1$, so the previous argument does not apply to this case. For special choices of b, we can see that the system $\{e^{2\pi ibnt}\}_{n\in\mathbb{Z}}$ is a frame for $L^2(\mathbb{T})$. For example, for $b = 1/2$, setting $n = 2k$ for n even and $n = 2k + 1$ for n odd, we can write

$$\begin{aligned}\{e^{\pi int}\}_{n\in\mathbb{Z}} &= \{e^{2\pi ikt}\}_{k\in\mathbb{Z}} \cup \{e^{2\pi i(k+\frac{1}{2})t}\}_{k\in\mathbb{Z}} \\ &= \{e^{2\pi ikt}\}_{k\in\mathbb{Z}} \cup \{e^{\pi it}e^{2\pi ikt}\}_{k\in\mathbb{Z}}\end{aligned}$$

so the set is the union of two o. n. bases and it is an inexact tight frame with $A = B = 2$, that is,

$$2\|f\|_2^2 = \sum_{n\in\mathbb{Z}} |\langle f, e^{\pi int}\rangle|^2.$$

Similarly, if $b = 1/M$ with $M \in \mathbb{N}_+$, then

$$\{e^{2\pi ibnt}\}_{n\in\mathbb{Z}} = \bigcup_{j=0}^{M-1} \{e^{2\pi i\frac{Mk+j}{M}t}\}_{k\in\mathbb{Z}}$$

and $M\|f\|_2^2 = \sum_{n\in\mathbb{Z}} |\langle f, e^{2\pi ibnt}\rangle|^2$. This means that the system is an inexact M-tight frame (it has redundancy M).
We now show that $\{e^{2\pi ibnt}\}$ is a tight frame in $L^2(\mathbb{T})$ with $A = B = 1/b$, for all b with $0 < b < 1$. Observe that the sequence $\{b^{1/2}e^{2\pi ibnt}\}$ is an o. n. basis for $L^2([0, b^{-1}])$. The space $L^2(\mathbb{T}) = L^2([0, 1])$ can be viewed as a closed subspace of $L^2([0, b^{-1}])$ by extending a function $f \in L^2([0, 1])$ by zero on $(1, b^{-1}]$. Hence every function in $L^2(\mathbb{T})$ can be view as the orthogonal projection of a function on $[0, b^{-1}]$, where the projection $P : L^2([0, b^{-1}]) \to L^2(\mathbb{T})$ is defined by $Pf = f\chi_{[0,1]}$. Hence for every $f \in L^2([0, 1])$,

$$\begin{aligned}\|f\|_{L^2(\mathbb{T})}^2 = \|f\chi_{[0,1]}\|_{L^2([0,b^{-1}])}^2 &= b\sum_n \left|\langle f\chi_{[0,1)}, e^{2\pi ibnt}\rangle_{L^2([0,b^{-1}])}\right|^2 \\ &= b\sum_n \left|\langle f, e^{2\pi ibnt}\rangle_{L^2(\mathbb{T})}\right|^2.\end{aligned}$$

So the sequence $\{e^{2\pi ibnt}\}$ is a tight frame with $A = B = 1/b$. This frame is not a basis in the sense of Definition A.1.1, and so (as we will see in the sequel) it is not exact. For instance, we can represent the constant function $f(t) = 1$ in two different ways (with two different choices of coefficients). Indeed, $Pg = f$ for $g(t) = 1$ on $[0, b^{-1}]$, but also $g(t) = \chi_{[0,1]}$ (as function on $[0, b^{-1}]$) has the same projection. The former example $g = 1$ is an element of the sequence $\{e^{2\pi ibnt}\}$ $(n = 0)$ and so

$$1 = \sum_{n\in\mathbb{Z}} \delta_{0,n} e^{2\pi ibnt}. \tag{3.4}$$

For the latter one, we have $\chi_{[0,1]} \in L^2([0, b^{-1}])$ and $\chi_{[0,1]} \neq 1$ in $[0, b^{-1}]$. So there exists $(c_n) \in \ell^2$, $(c_n) \neq (\delta_{0,n})$ such that $\chi_{[0,1]} = b^{1/2}\sum_n c_n e^{2\pi ibnt}$, hence the representation of $1 \in L^2(\mathbb{T})$ in the series expansion (3.4) is not unique.

We now define Bessel sequences in Hilbert spaces.

Definition 3.1.7 (Bessel sequence). A sequence $\{x_n\}$ in a Hilbert space H is a Bessel sequence if

$$\forall x \in H, \quad \sum_{n=1}^{\infty} |\langle x, x_n\rangle|^2 < \infty. \tag{3.5}$$

Definition 3.1.8. Given a sequence $\{x_n\}$ in a Hilbert space H, the *coefficient operator*, also called *analysis operator* C, is defined to be

$$Cx = (\langle x, x_n \rangle), \quad x \in H. \tag{3.6}$$

Proposition 3.1.9. *If $\{x_n\}$ is a Bessel sequence, the coefficient operator C maps H into ℓ^2 and is a linear and continuous operator; that is, there exists a constant $B > 0$ such that*

$$\forall x \in H, \quad \|Cx\|_{\ell^2}^2 = \sum_{n=1}^{\infty} |\langle x, x_n \rangle|^2 \le B\|x\|_H^2. \tag{3.7}$$

The constant $B > 0$ in (3.7) *is called* Bessel constant. *Inequality* (3.7) *is called* Bessel's inequality.

Proof. We are going to present two different demonstrations.

First proof. If $\{x_n\}$ is a Bessel sequence, the coefficient operator C maps H into ℓ^2 by (3.5). Observe that C is a linear operator. We prove that C is bounded. We use the closed graph theorem (cf. Theorem A.0.2). Assume $\{y_k\} \subset H$ is such that $y_k \to y$ in H and $Cy_k \to c$ in ℓ^2. We shall show that $Cy = c$. We write $c = (c_n)$; since $Cy = (\langle y, x_n \rangle)$, to reach our goal we need to show that $c_n = \langle y, x_n \rangle$, for every $n \in \mathbb{N}_+$.

For every $m \in \mathbb{N}_+$, we need to prove that

$$\lim_{k \to \infty} \langle y_k, x_m \rangle = c_m, \tag{3.8}$$

or equivalently,

$$\lim_{k \to \infty} |\langle y_k, x_m \rangle - c_m|^2 = 0.$$

We have

$$\begin{aligned} |\langle y_k, x_m \rangle - c_m|^2 &\le \sum_{m=1}^{\infty} |\langle y_k, x_m \rangle - c_m|^2 \\ &= \|Cy_k - c\|_{\ell^2}^2 \to 0 \quad \text{for } k \to +\infty, \end{aligned}$$

and (3.8) is proved. Finally, by assumption that $y_k \to y$ in H and by the continuity of the inner product,

$$\lim_{k \to +\infty} \langle y_k, x_m \rangle = \langle y, x_m \rangle \quad \forall m$$

so that, by (3.8), we obtain the claim.

Second proof. Consider the sequence of operators $\{C_N\}_{N \in \mathbb{N}_+}$, with

$$C_N : x \mapsto (\langle x, x_1 \rangle, \langle x, x_2 \rangle, \ldots, \langle x, x_N \rangle, 0, 0, \ldots).$$

The operator C_N is trivially linear and maps H into ℓ^2, as $\sum_{n=1}^N |\langle x, x_n\rangle|^2 < \infty$. By using the Cauchy–Schwarz inequality, we infer that C_N is bounded with

$$\|C_N x\|_{\ell^2}^2 = \sum_{n=1}^N \left|\langle x, x_n\rangle\right|^2 \le \|x\|_H^2 \sum_{n=1}^N \|x_n\|_H^2 = B_N \|x\|_H^2.$$

Moreover, the sequence $(\|C_N x\|_{\ell^2})$ is an increasing sequence of nonnegative numbers indexed by N, so that, for every $x \in H$,

$$\sup_{N\in\mathbb{N}_+} \|C_N x\|_{\ell^2} = \lim_{N\to\infty} \|C_N x\|_{\ell^2} = \sqrt{\sum_{n=1}^\infty \left|\langle x, x_n\rangle\right|^2} = \|Cx\|_{\ell^2} < \infty.$$

We can apply the uniform boundedness principle (cf. Theorem A.0.3) and obtain

$$\sup_{N\in\mathbb{N}_+} \|C_N\|_{B(H,\ell^2)} =: \sqrt{B} < \infty,$$

for a suitable $B > 0$. Finally, the inequality

$$\|C_N x\|_{\ell^2} \le \|C_N\|_{B(H,\ell^2)} \|x\|_H \le \sqrt{B}\|x\|_H, \quad \forall x \in H$$

for $N \to \infty$ yields

$$\|Cx\|_{\ell^2} \le \sqrt{B}\|x\|_H, \quad \forall x \in H,$$

as desired. □

In particular, if $\{x_n\}$ is an o. n. basis of H, then by Plancherel theorem (see Theorem A.1.11 (iv)) we obtain that $C : H \to \ell^2$ is a (topological) isometric isomorphism.

Remark 3.1.10. We observe that a more well-known definition of Bessel sequence (which follows in our case from Proposition 3.1.9) is the following: A sequence $\{x_n\}$ in a Hilbert space H is a Bessel sequence if there exists $B > 0$ such that

$$\sum_{n=1}^\infty \left|\langle x, x_n\rangle\right|^2 \le B\|x\|_H^2, \quad \forall x \in H. \tag{3.9}$$

Remark 3.1.11. Let $\{x_n\}$ be a Bessel sequence in a Hilbert space H with Bessel bound $B > 0$. Then it is straightforward to prove the following inequalities:
(i) We have $\|x_n\|_H^2 \le B$ for every n.
(ii) If $\|x_m\|_H^2 = B$ for some m, then x_m is orthogonal to $\{x_n\}_{n\neq m}$.

Proposition 3.1.12. *Let $\{x_n\}$ be a Bessel sequence in a Hilbert space H with Bessel constant B. Then*

(i) *If* $c = (c_n) \in \ell^2$, *then the series* $\sum_n c_n x_n$ *converges unconditionally in* H *and*

$$D(c) = \sum_n c_n x_n$$

defines a linear and bounded operator from ℓ^2 *into* H.

(ii) $D^* = C$ *and* $\|D\| = \|C\| \leq \sqrt{B}$. *Consequently,*

$$\forall (c_n) \in \ell^2 \quad \left\| \sum_n c_n x_n \right\|_H^2 \leq B \sum_n |c_n|^2.$$

(iii) *If* $\{x_n\}$ *is a frame, then* C *is injective and* D *is surjective.*

Proof. (i) We use the equivalent formulation of unconditional convergence of Theorem A.1.21 (iii). Let $F \subset \mathbb{N}_+$ be a finite set. Then by the Cauchy–Schwarz inequality and using the assumption, we have

$$\begin{aligned}
\left\| \sum_{n\in F} c_n x_n \right\|_H &= \sup_{y\in H,\ \|y\|_H=1} \left| \left\langle \sum_{n\in F} c_n x_n, y \right\rangle \right| = \sup_{y\in H,\ \|y\|_H=1} \left| \sum_{n\in F} c_n \langle x_n, y\rangle \right| \\
&\leq \sup_{\|y\|_H=1} \left(\sum_{n\in F} |c_n|^2 \right)^{\frac{1}{2}} \left(\sum_{n\in F} |\langle x_n, y\rangle|^2 \right)^{\frac{1}{2}} \\
&\leq \sup_{\|y\|_H=1} \left(\sum_{n\in F} |c_n|^2 \right)^{\frac{1}{2}} \left(\sum_{n\in \mathbb{N}_+} |\langle x_n, y\rangle|^2 \right)^{\frac{1}{2}} \\
&\leq \sup_{\|y\|_H=1} \left(\sum_{n\in F} |c_n|^2 \right)^{\frac{1}{2}} (B\|y\|_H^2)^{\frac{1}{2}} \\
&= \sqrt{B} \left(\sum_{n\in F} |c_n|^2 \right)^{\frac{1}{2}}.
\end{aligned}$$

Now, $(c_n) \in \ell^2$, so $\sum_n |c_n|^2$ is unconditionally convergent. For any choice of $\epsilon > 0$, there exists $N_0 \in \mathbb{N}_+$ such that for every finite $F \subset \mathbb{N}_+$ with $\min(F) > N_0$, $\sum_{n\in F} |c_n|^2 < \epsilon^2/B$. Thus

$$\left\| \sum_{n\in F} c_n x_n \right\|_H^2 \leq B \sum_{n\in F} |c_n|^2 < B\epsilon^2/B = \epsilon^2,$$

proving our claim. Setting $F = \{1, \ldots, N\}$ and letting $N \to \infty$, we obtain

$$\left\| \sum_n c_n x_n \right\|_H^2 \leq B \sum_n |c_n|^2.$$

So the operator $Dc = \sum_n c_n x_n$ maps ℓ^2 into H and $\|D\| \leq \sqrt{B}$. The operator D is linear. In fact, for every $\alpha, \beta \in \mathbb{C}$ and every $c = (c_n), d = (d_n) \in \ell^2$,

$$D(\alpha(c_n) + \beta(d_n)) = \sum_n (\alpha c_n + \beta d_n) x_n = \alpha \sum_n c_n x_n + \beta \sum_n d_n x_n,$$

since the series is unconditionally convergent.

(ii) Observe that by (3.7) the operator $C : H \to \ell^2$ is linear and bounded with $\|C\| \le \sqrt{B}$. It remains to prove that $C^* = D$. The operator $C^* : \ell^2 \to H$ is linear and bounded; it remains to check that $C^*(c) = \sum_n c_n x_n$ for every $c = (c_n) \in \ell^2$. For every $x \in H$, $c = (c_n) \in \ell^2$,

$$\langle x, C^*(c)\rangle_H = \langle Cx, c\rangle_{\ell^2} = \langle ((\langle x, x_n\rangle)_n, (c_n)_n\rangle_{\ell^2}$$
$$= \sum_n \langle x, x_n\rangle \bar{c}_n = \sum_n \langle x, c_n x_n\rangle = \left\langle x, \sum_n c_n x_n \right\rangle$$

hence $C^*(c) = \sum_n c_n x_n$, that is, $C^* = D$.

(iii) If $\{x_n\}$ is also frame, then the inequality $A\|x\|_H^2 \le \sum_n |\langle x, x_n\rangle|^2$ holds for every $x \in H$. The claim follows by applying Theorem A.0.4 to the coefficient operator C. □

Definition 3.1.13. Consider a Bessel sequence $\{x_n\}$ in a Hilbert space H.
(i) The adjoint operator $C^* = D : \ell^2 \to H$ is called a *synthesis*, or *reconstruction*, *operator*.
(ii) The *frame operator* is defined as $S = DC : H \to H$.
(iii) The *Gram operator* or *Gram matrix* is $G = CD : \ell^2 \to \ell^2$.

Now consider the Gram operator $G : \ell^2 \to \ell^2$. It can be represented as multiplication by an infinite matrix. Indeed, it easy to see that the Gram operator can be represented as

$$G = [\langle x_n, x_m\rangle]_{m,n\in\mathbb{N}}. \tag{3.10}$$

That is, if we think of $c = (c_n) \in \ell^2$ as an infinite column vector, Gc is the product of the infinite matrix (3.10) with c. So the m-entry of Gc is

$$(Gc)_m = \sum_n c_n \langle x_n, x_m\rangle.$$

Remark 3.1.14. Since the analysis C and synthesis D operators associated to a Bessel sequence $\{x_n\}$ are bounded, so are the frame S and Gram G operators. Moreover, $S = DC = C^*C = DD^*$ and $G = CD = CC^* = D^*D$, so they are self-adjoint. We shall show that they are both positive. Recall that a self-adjoint operator P is positive if $\langle Px, x\rangle \ge 0$, for every $x \in H$.

First, by definition,

$$Sx = \sum_n \langle x, x_n\rangle x_n \quad \forall x \in H.$$

So

$$\langle Sx, x\rangle = \sum_n \left|\langle x, x_n\rangle\right|^2 \ge 0, \quad \forall x \in H. \tag{3.11}$$

Now, consider the Gram operator G. For every sequence $c \in \ell^2$, define $x = \sum_n c_n x_n$, then

$$\begin{aligned}\langle Gc, c\rangle &= \sum_m \sum_n c_n \langle x_n, x_m\rangle \bar{c}_m = \sum_m \sum_n \langle c_n x_n, c_m x_m\rangle \\ &= \left\langle \sum_n c_n x_n, \sum_m c_m x_m \right\rangle = \|x\|_H^2 \geq 0.\end{aligned}$$

3.1.1 Frame expansions and frame operator

Frames are Bessel sequences, so all the properties of Bessel sequences transfer to frames. For $U, V \in B(H)$, U, V being positive operators, we use the operator notation $U \leq B$ as follows:

$$U \leq V \Longleftrightarrow \langle Ux, x\rangle \leq \langle Vx, x\rangle \quad \forall x \in H.$$

Theorem 3.1.15 (Reproducing formulas for a frame). *Let $\{x_n\}$ be a frame for the Hilbert space H with frame bounds A, B. Then*

(i) *The frame operator S is a topological isomorphism of H onto itself, self-adjoint and positive, with*

$$AI \leq S \leq BI,$$

where I is the identity operator.

(ii) *S^{-1} is a topological isomorphism, self-adjoint and positive, and*

$$B^{-1}I \leq S^{-1} \leq A^{-1}I.$$

(iii) *$\{S^{-1}x_n\}$ is a frame for H, with frame bounds $0 < B^{-1} \leq A^{-1}$.*

(iv) *For each $x \in H$, we have the* reproducing formulae

$$x = \sum_n \langle x, S^{-1}x_n\rangle x_n, \quad x = \sum_n \langle x, x_n\rangle S^{-1}x_n, \tag{3.12}$$

and these series converge unconditionally in the norm of H.

(v) *If the frame is A-tight, then $S = AI$, $S^{-1} = A^{-1}I$, and*

$$\forall x \in H, \quad x = \frac{1}{A}\sum_n \langle x, x_n\rangle x_n.$$

Proof. (i) Since $\{x_n\}$ is a Bessel sequence, S is a continuous (self-adjoint) positive operator on H, and, by (3.11), $\langle Sx, x\rangle = \sum_n |\langle x, x_n\rangle|^2$. Since $\langle AIx, x\rangle = A\|x\|_H^2$, the frame definition can be rewritten as

$$\langle AIx, x\rangle \leq \langle Sx, x\rangle \leq \langle BIx, x\rangle \quad \forall x \in H,$$

which means $AI \leq S \leq BI$.

Applying the Cauchy–Schwarz inequality,

$$A\|x\|_H^2 \le \langle Sx, x\rangle \le \|Sx\|_H\|x\|_H,$$

so that $A\|x\|_H \le \|Sx\|_H$ for every $x \in H$. We then apply Theorem A.0.4 which gives that $\ker S = \{0\}$ and S^* is surjective. But $S^* = S$, so S is a topological isomorphism, by the Banach isomorphism theorem.

(ii) Since S is a self-adjoint positive topological isomorphism, the same holds for S^{-1} (left as an exercise). Since $AI \le S$, we have $\langle AIy, y\rangle \le \langle Sy, y\rangle$ for every $y \in H$. Choosing $y = S^{-1}x$, we can write

$$\begin{aligned} 0 \le A\|S^{-1}x\|_H^2 &\le \langle S(S^{-1}x), S^{-1}x\rangle \\ &= \langle x, S^{-1}x\rangle \\ &\le \|x\|_H\|S^{-1}x\|_H. \end{aligned}$$

Consequently, $\|S^{-1}x\|_H \le A^{-1}\|x\|_H$ and hence

$$\langle S^{-1}x, x\rangle \le \|S^{-1}x\|_H\|x\|_H \le A^{-1}\|x\|_H^2 = \langle A^{-1}Ix, x\rangle.$$

We now prove $S^{-1} \ge B^{-1}I$. We use the generalized Cauchy–Schwarz inequality (see (A.2) in Appendix A) applied to the semiinner product $(x, y) = \langle S^{-1}x, y\rangle$, which gives

$$|\langle S^{-1}u, v\rangle|^2 \le \langle S^{-1}u, u\rangle\langle S^{-1}v, v\rangle \quad u, v \in H.$$

Choosing $u = Sx$, $v = x$,

$$\|x\|_H^4 \le \langle x, Sx\rangle\langle S^{-1}x, x\rangle \le B\|x\|_H^2\langle S^{-1}x, x\rangle,$$

hence $\langle S^{-1}x, x\rangle \ge B^{-1}\|x\|_H^2$, that is, $S^{-1} \ge B^{-1}I$.

(iii) It is an easy exercise to show that if S is self-adjoint then S^{-1} is self-adjoint. Hence

$$\begin{aligned} \sum_n \langle x, S^{-1}x_n\rangle S^{-1}x_n &= S^{-1}\Big(\sum_n \langle x, S^{-1}x_n\rangle x_n\Big) \\ &= S^{-1}\Big(\sum_n \langle S^{-1}x, x_n\rangle x_n\Big) \\ &= S^{-1}S(S^{-1}x) = S^{-1}x. \end{aligned} \tag{3.13}$$

As a consequence,

$$\sum_n |\langle x, S^{-1}x_n\rangle|^2 = \sum_n \langle x, S^{-1}x_n\rangle\langle S^{-1}x_n, x\rangle = \Big\langle \sum_n \langle x, S^{-1}x_n\rangle S^{-1}x_n, x\Big\rangle = \langle S^{-1}x, x\rangle.$$

Using $B^{-1}I \le S^{-1} \le A^{-1}I$, we obtain

$$B^{-1}\|x\|_H^2 \le \sum_n |\langle x, S^{-1}x_n\rangle|^2 \le A^{-1}\|x\|_H^2.$$

This proves that $\{S^{-1}x_n\}$ is a frame with bounds B^{-1}, A^{-1}.

(iv) Since $\{x_n\}$ and $\{S^{-1}x_n\}$ are, in particular, Bessel sequences, for every $c = (c_n) \in \ell^2$, the series $\sum_n c_n x_n$ and $\sum_n c_n S^{-1}x_n$ converge unconditionally. Since the sequences $(\langle x, x_n\rangle)$ and $(\langle x, S^{-1}x_n\rangle)$ belong to ℓ^2 and S, S^{-1} are continuous operators on H, we can write

$$x = S(S^{-1}x) = \sum_n \langle S^{-1}x, x_n\rangle x_n = \sum_n \langle x, S^{-1}x_n\rangle x_n$$

and

$$x = S^{-1}(Sx) = \sum_n \langle Sx, S^{-1}x_n\rangle S^{-1}x_n = \sum_n \langle x, x_n\rangle S^{-1}x_n,$$

with unconditional convergence in H.

(v) This follows from the preceding statements. □

Let X be a dense subset of H. If $A \in B(H)$ satisfies $\langle Ax, x\rangle \ge 0$ for every $x \in X$, then $\langle Ax, x\rangle \ge 0$ for every $x \in H$. Applying this fact to the inequalities

$$\langle AIx, x\rangle \le \langle Sx, x\rangle \le \langle BIx, x\rangle, \quad x \in X,$$

it is enough to check the frame inequalities (3.2) for x in X.

Previous considerations yield the definition of a canonical dual frame.

Definition 3.1.16 (Canonical dual frame). Let $\{x_n\}$ be a frame for a Hilbert space H with frame operator S. Then the frame $\{S^{-1}x_n\}$ is called a *canonical dual frame* for $\{x_n\}$.

Observe that the frame operator of the canonical dual frame $\{S^{-1}x_n\}$ is the operator S^{-1}, see (3.13). The canonical dual of $\{S^{-1}x_n\}$ is the original frame $\{x_n\}$.

Definition 3.1.17 (Alternative duals). Let $\{x_n\}$ be a frame for a Hilbert space H. A sequence $\{y_n\} \subset H$ such that

$$x = \sum_n \langle x, y_n\rangle x_n, \quad \forall x \in H$$

with unconditional convergence is called an *alternative dual* for $\{x_n\}$. If $\{y_n\}$ is a frame, then it is called an *alternative dual frame*.

For example, if $\{e_n\}$ is an o. n. basis for H, then the frame $\{e_1, e_1, e_2, e_2, \dots\}$ has infinitely many duals. The canonical dual frame is $\{e_1/2, e_1/2, e_2/2, e_2/2, \dots\}$ but also $\{e_1, 0, e_2, 0, \dots\}$ and $\{0, e_1, 0, e_2 \dots\}$ are dual frames. The canonical dual frame is certainly the dual used more often. However, in applications, it may be useful to deal with a noncanonical dual frame. Indeed, sparse representations are the main goals in frame expansions. Just to give an idea of the problem, consider the previous example. If we choose alternative duals, we observe that half of the coefficients in the frame expansion are zero, whereas this is not the case for the canonical dual. The need of *well-concentrated* frame expansions is crucial in many applications.

Let us denote by $\{\tilde{x}_n\}$ the canonical dual frame of $\{x_n\}$, that is, $\tilde{x}_n = S^{-1}x_n$, where S is the frame operator for $\{x_n\}$.

We now show that among the choices of scalars $(c_n) \in \ell^2$ such that $\sum_n c_n x_n = x$, the coefficients $\langle x, \tilde{x}_n\rangle$, $n \in \mathbb{N}_+$, provided by the canonical dual frame, are those with minimal ℓ^2-norm.

Theorem 3.1.18. *Let $\{x_n\}$ be a frame for a Hilbert space H and fix $x \in H$. If $c = (c_n)$ is a sequence of scalars such that $x = \sum_n c_n x_n$, then*

$$\sum_n |c_n|^2 = \sum_n |\langle x, \tilde{x}_n\rangle|^2 + \sum_n |c_n - \langle x, \tilde{x}_n\rangle|^2. \tag{3.14}$$

In particular, the sequence $(\langle x, \tilde{x}_n\rangle)$ has the minimal ℓ^2-norm among all such sequences.

Proof. By Theorem 3.1.15, we have $x = \sum \langle x, \tilde{x}_n\rangle x_n$, with $(\langle x, \tilde{x}_n\rangle) \in \ell^2$. So if $(c_n) \notin \ell^2$, then $\sum_n |c_n|^2 = +\infty$ and equality (3.14) is trivially satisfied.

Assume now $(c_n) \in \ell^2$ and write $a_n = \langle x, \tilde{x}_n\rangle$ for short. We have

$$\begin{aligned}
\langle x, S^{-1}x\rangle &= \left\langle \sum_n a_n x_n, S^{-1}x \right\rangle \\
&= \sum_n a_n \langle \tilde{x}_n, x\rangle \\
&= \sum_n a_n \bar{a}_n \\
&= \langle (a_n), (a_n)\rangle_{\ell^2}
\end{aligned}$$

and

$$\begin{aligned}
\langle x, S^{-1}x\rangle &= \left\langle \sum_n c_n x_n, S^{-1}x \right\rangle \\
&= \sum_n c_n \langle \tilde{x}_n, x\rangle \\
&= \sum_n c_n \bar{a}_n \\
&= \langle (c_n), (a_n)\rangle_{\ell^2}.
\end{aligned}$$

So the sequence $(a_n - c_n)$ is orthogonal to (a_n) in ℓ^2 and, by the Pythagorean theorem, we can write

$$\|(c_n)\|_{\ell^2}^2 = \|(a_n) - (a_n - c_n)\|_{\ell^2}^2 = \|(a_n)\|_{\ell^2}^2 + \|(a_n - c_n)\|_{\ell^2}^2,$$

and the claim is proved. □

Theorem 3.1.19. *Let $\{x_n\}$ be a frame for a Hilbert space H.*

(i) *For every $m \in \mathbb{N}$,*

$$\sum_{n \neq m} \left|\langle x_m, \tilde{x}_n\rangle\right|^2 = \frac{1 - |\langle x_m, \tilde{x}_m\rangle|^2 - |1 - \langle x_m, \tilde{x}_m\rangle|^2}{2}. \tag{3.15}$$

(ii) *If $\langle x_m, \tilde{x}_m\rangle = 1$, then $\langle x_m, \tilde{x}_n\rangle = 0$ for every $n \neq m$.*

(iii) *The removal of an element x_m from a frame leaves either a frame or an incomplete set. Specifically,*

$$\langle x_m, \tilde{x}_m\rangle \neq 1 \Longrightarrow \{x_n\}_{n\neq m} \text{ is a frame,}$$
$$\langle x_m, \tilde{x}_m\rangle = 1 \Longrightarrow \{x_n\}_{n\neq m} \text{ is incomplete.}$$

Proof. (i) Set $a_n = \langle x_m, \tilde{x}_n\rangle$ for every n, then $x_m = \sum_n a_n x_n$. Observe that x_m can also be represented by $x_m = \sum_n \delta_{m,n} x_n$, so that using Theorem 3.1.18, formula (3.14) here becomes

$$\begin{aligned} 1 = \sum_n |\delta_{m,n}|^2 &= \sum_n |a_n|^2 + \sum_n |\delta_{m,n} - a_n|^2 \\ &= |a_m|^2 + \sum_{n\neq m} |a_n|^2 + |1 - a_m|^2 + \sum_{n\neq m} |a_n|^2. \end{aligned}$$

This yields

$$\sum_{n\neq m} |a_n|^2 = \frac{1 - |a_m|^2 - |1 - a_m|^2}{2}.$$

(ii) Suppose $\langle x_m, \tilde{x}_m\rangle = 1$, then by equation (3.15), $\sum_{n\neq m} |\langle x_m, \tilde{x}_n\rangle|^2 = 0$, which implies $\langle x_m, \tilde{x}_n\rangle = 0$, for every $n \neq m$.

(iii) Assume $\langle x_m, \tilde{x}_m\rangle = 1$, then by part (ii), $\langle x_m, \tilde{x}_n\rangle = 0$, for every $n \neq m$. Hence

$$0 = \langle x_m, \tilde{x}_n\rangle = \langle x_m, S^{-1}x_n\rangle = \langle S^{-1}x_m, x_n\rangle = \langle \tilde{x}_m, x_n\rangle.$$

This means that $\tilde{x}_m$ is orthogonal to the set $\{x_n\}_{n\neq m}$. However, $\tilde{x}_m \neq 0$ since $\langle x_m, \tilde{x}_m\rangle = 1$. So $\{x_n\}_{n\neq m}^{\perp} \neq \{0\}$ and this gives the incompleteness of the set $\{x_n\}_{n\neq m}$. Suppose now $\langle x_m, \tilde{x}_m\rangle \neq 1$, that is, in our notation $a_m \neq 1$. Then $x_m = \sum_n a_n x_n$ can be rewritten as

$$x_m = \frac{1}{1 - a_m} \sum_{n\neq m} a_n x_n.$$

For every $x \in H$, by the Cauchy–Schwarz inequality and the previous representation,

$$\begin{aligned} |\langle x, x_m\rangle|^2 &= \left|\left\langle x, \frac{1}{1 - a_m} \sum_{n\neq m} a_n x_n\right\rangle\right|^2 = \frac{1}{|1 - a_m|^2} \left|\sum_{n\neq m} \bar{a}_n \langle x, x_n\rangle\right|^2 \\ &\leq \frac{1}{|1 - a_m|^2} \Big(\sum_{n\neq m} |a_n|^2\Big)\Big(\sum_{n\neq m} |\langle x, x_n\rangle|^2\Big) \leq C \sum_{n\neq m} |\langle x, x_n\rangle|^2 \end{aligned}$$

where

$$C = \frac{1}{|1 - a_m|^2} \sum_{n \neq m} |a_n|^2 > 0.$$

Hence

$$\sum_n \left|\langle x, x_n\rangle\right|^2 = \left|\langle x, x_m\rangle\right|^2 + \sum_{n\neq m} \left|\langle x, x_n\rangle\right|^2 \leq (1 + C) \sum_{n\neq m} \left|\langle x, x_n\rangle\right|^2.$$

If the constants $A, B > 0$ are the frame bounds for $\{x_n\}$, then

$$\frac{A}{1 + C}\|x\|_H^2 \leq \frac{1}{1 + C} \sum_n \left|\langle x, x_n\rangle\right|^2 \leq \sum_{n\neq m} \left|\langle x, x_n\rangle\right|^2 \leq B\|x\|_H^2,$$

that is, $\{x_n\}_{n\neq m}$ is a frame. □

Let us recall the definition of biorthogonal sequences.

Definition 3.1.20. Two sequences $\{x_n\}$, $\{y_n\}$ in a Hilbert space H are called *biorthogonal* if

$$\langle x_m, y_n\rangle = \delta_{m,n}, \quad \forall m, n.$$

Corollary 3.1.21. *Let $\{x_n\}$ be a frame for a Hilbert space H. Then the following are equivalent:*
(i) *The sequence $\{x_n\}$ is an exact frame.*
(ii) *$\{x_n\}$ and $\{\tilde{x}_n\}$ are biorthogonal.*
(iii) *$\langle x_n, \tilde{x}_n\rangle = 1$ for all n.*

Consequently, if $\{x_n\}$ is a A-tight frame, then the following are equivalent:
(i) *The sequence $\{x_n\}$ is an exact frame.*
(ii) *$\{x_n\}$ is an orthogonal basis for H.*
(iii) *$\|x_n\|_H^2 = A$ for all n.*

Proof. (i) $\Longrightarrow$ (iii) If $\{x_n\}$ is exact, then for every m, the sequence $\{x_n\}_{n\neq m}$ is no more a frame, hence by Theorem 3.1.19, $\langle x_m, \tilde{x}_m\rangle = 1$.

(iii) $\Longrightarrow$ (i) For every fixed m, if $\langle x_m, \tilde{x}_m\rangle = 1$ then, by Theorem 3.1.19, $\{x_n\}_{n\neq m}$ is incomplete. This means that $\{x_n\}$ is exact.

(ii) $\Longrightarrow$ (iii) It follows immediately from Definition 3.1.20.

(iii) $\Longrightarrow$ (ii) It follows immediately from Theorem 3.1.19.

If $\{x_n\}$ is a A-tight frame, then $S = AI$ and $S^{-1} = A^{-1}I$, so that

$$\tilde{x}_n = S^{-1}x_n = \frac{1}{A}x_n.$$

Then the A-tight frame $\{x_n\}$ is exact iff $\langle x_n, A^{-1}x_m\rangle = \delta_{m,n}$, that is, $\{x_n\}$ is an orthogonal system and it is complete (because it is a frame), so it is a basis. □

We now present a characterization for Riesz bases, defined in Definition A.1.26, which we shall use in the sequel. For the proof we refer to [183, Theorem 7.13].

Theorem 3.1.22 (Characterization of Riesz bases). *Let $\{x_n\}$ be a sequence for a Hilbert space H. Then the following are equivalent:*
(i) *$\{x_n\}$ is a Riesz basis for H.*
(ii) *$\{x_n\}$ is a bounded unconditional basis for H.*
(iii) *$\{x_n\}$ is a basis for H and*

$$\sum_n c_n x_n \text{ converges} \iff \sum_n |c_n|^2 < \infty.$$

(iv) *$\{x_n\}$ is complete in H and there are constants $A, B > 0$ such that*

$$\forall c_1, \ldots, c_N, \quad A \sum_{n=1}^{N} |c_n|^2 \le \left\| \sum_{n=1}^{N} c_n x_n \right\|_H^2 \le B \sum_{n=1}^{N} |c_n|^2.$$

(v) *$\{x_n\}$ is a complete Bessel sequence and possesses a biorthogonal system $\{y_n\}$ that is also a complete Bessel sequence.*

The system

$$\{e_1, e_2/\sqrt{2}, e_2/\sqrt{2}, e_3/\sqrt{3}, e_3/\sqrt{3}, e_3/\sqrt{3}, \ldots\}$$

(cf. Example 3.1.5 (iii)) is an inexact Parseval frame which contains an orthogonal basis $\{e_1, e_2/\sqrt{2}, e_3/\sqrt{3}, \ldots\}$, not norm-bounded below, so it is not a Riesz basis.

Proposition 3.1.23. *Frames are preserved by topological isomorphisms. Specifically, if $\{x_n\}$ is a frame for a Hilbert space H and $T : H \to K$ is a topological isomorphism between the Hilbert spaces H and K, then $\{Tx_n\}$ is a frame for K. Moreover,*
(i) *If A, B are the frame bounds for $\{x_n\}$, then $\{Tx_n\}$ has frame bounds $A\|T^{-1}\|^{-2}$ and $B\|T\|^2$.*
(ii) *If S is the frame operator for $\{x_n\}$, then $\{Tx_n\}$ has frame operator TST^*.*
(iii) *$\{x_n\}$ is exact if and only if $\{Tx_n\}$ is exact.*

Proof. First, observe that T^* is a topological isomorphism. Indeed, $\ker T^* = R(T)^\perp = \{0\}$; using Theorem A.0.4, having $\ker T = \{0\}$ and $R(T)$ being closed is equivalent to saying that T^* is surjective. Since T being bounded implies T^* is bounded (see, e. g., [37]), a corollary of the open mapping theorem gives that T^* is a topological isomorphism. Moreover, observe that $TST^* : K \to K$; indeed, the following diagram is commutative:

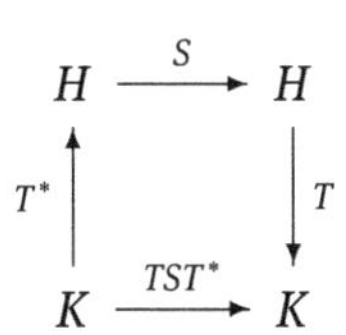

(the composition TST^* of topological isomorphisms is a topological isomorphism). For every $y \in K$, we have

$$TST^*y = T\Big(\sum_n \langle T^*y, x_n\rangle_H x_n\Big) = \sum_n \langle y, Tx_n\rangle_K Tx_n$$

with unconditional convergence because T is bounded. Let us prove that

$$A\|T^{-1}\|^{-2} I \le TST^* \le B\|T\|^2 I,$$

so that (i) and (ii) are satisfied. For every $y \in K$, $\langle TST^*y, y\rangle = \langle ST^*y, T^*y\rangle$ hence

$$A\|T^*y\|_H^2 \le \langle ST^*y, T^*y\rangle \le B\|T^*y\|_H^2. \tag{3.16}$$

Setting $y = (T^*)^{-1}T^*y$, so that $\|y\|_K \le \|(T^*)^{-1}\|\|T^*y\|_H = \|T^{-1}\|\|T^*y\|_H$, we get

$$\frac{\|y\|_K}{\|T^{-1}\|} \le \|T^*y\|_H \le \|T^*\|\|y\|_K = \|T\|\|y\|_K. \tag{3.17}$$

Combining (3.16) and (3.17) gives

$$A\frac{\|y\|_K^2}{\|T^{-1}\|^2} \le \langle TST^*y, y\rangle \le B\|T\|^2\|y\|_K^2$$

which yields the claim. Statement (iii) follows from the fact that a topological isomorphism preserves completeness. □

Corollary 3.1.24. *A sequence $\{x_n\}$ is an exact frame for a Hilbert space H if and only if it is a Riesz basis.*

Proof. Assume that sequence $\{x_n\}$ is an exact frame and so a complete Bessel sequence. By Theorem 3.1.15, also $\{S^{-1}x_n\}$ is a frame for H, so a complete Bessel sequence. Finally, by Corollary 3.1.21, $\{x_n\}$ and $\{S^{-1}x_n\}$ are biorthogonal systems. The claim follows by the characterization of Riesz bases in Theorem 3.1.22 (v).

Conversely, if $\{x_n\}$ is a Riesz basis, then by definition there exist a topological isomorphism T such that $Te_n = x_n$ with $\{e_n\}$ o. n. basis. Since o. n. bases are frames and, by Proposition 3.1.23, frames are preserved by topological isomorphisms, we obtain that $\{x_n\}$ is a frame. □

We can represent explicitly the isomorphism T giving an alternative to the first part of the proof of Corollary 3.1.24. Since S is a positive operator, which is a topological isomorphism from H onto itself, it has a square root $S^{1/2}$, which is a positive topological isomorphism (see Theorem A.0.6). Similarly, the inverse S^{-1} is a positive operator with square root $S^{-1/2} = (S^{1/2})^{-1}$. Since $\{x_n\}$ is exact, then $\{S^{-1}x_n\}$ and $\{x_n\}$ are biorthogonal, and we have

$$\langle S^{-1/2}x_m, S^{-1/2}x_n\rangle = \langle x_m, S^{-1/2}S^{-1/2}x_n\rangle = \langle x_m, S^{-1}x_n\rangle = \delta_{m,n}.$$

Thus $\{S^{-1/2}x_n\}$ is an o. n. sequence. Moreover, it is complete since topological isomorphisms preserve completeness. This implies that $\{S^{-1/2}x_n\}$ is an o. n. basis for H and the topological isomorphism $T = S^{1/2}$ maps this o. n. basis onto the frame $\{x_n\}$.

We can consider the sequence $\{S^{-1/2}x_n\}$ for any frame, not just exact. If this is the case, then $\{S^{-1/2}x_n\}$ will not be an o. n. basis for H, but it will be a Parseval frame.

Corollary 3.1.25. *Let $\{x_n\}$ be a frame for a Hilbert space H, with frame operator S. Then*
(i) *$\{S^{-1/2}x_n\}$ is a Parseval frame for H.*
(ii) *$\langle x_n, \tilde{x}_n\rangle = \|S^{-1/2}x_n\|_H^2$ and $0 \le \langle x_n, \tilde{x}_n\rangle \le 1$ for every n.*
(iii) *$\{x_n\}$ is an exact frame if and only if $\{S^{-1/2}x_n\}$ is an o. n. basis for H.*

Proof. (i) The operator $S^{-1/2}$ is a topological isomorphism because S is such. Since frames are preserved by topological isomorphisms, we have that $\{S^{-1/2}x_n\}$ is a frame. We prove that it is a Parseval frame. For every $x \in H$,

$$\sum_n \langle x, S^{-1/2}x_n\rangle S^{-1/2}x_n = S^{-1/2}\Big(\sum_n \langle S^{-1/2}x, x_n\rangle x_n\Big) = S^{-1/2}SS^{-1/2}x = Ix,$$

which is equivalent to saying that $\{S^{-1/2}x_n\}$ is a Parseval frame.

(ii) Using the self-adjointness of $S^{-1/2}$, we can write

$$\langle x_n, \tilde{x}_n\rangle = \langle x_n, S^{-1/2}S^{-1/2}x_n\rangle = \langle S^{-1/2}x_n, S^{-1/2}x_n\rangle = \|S^{-1/2}x_n\|_H^2.$$

Since $\{S^{-1/2}x_n\}$ is a 1-tight frame, we have $\|S^{-1/2}x_n\|_H \le 1$, for every n, by Remark 3.1.11.

(iii) The "$\Longrightarrow$" direction is already proved. Conversely, we have $x_n = S^{1/2}S^{-1/2}x_n$. Then x_n is the image of an o. n. basis by the topological isomorphism $S^{1/2}$, meaning that $\{x_n\}$ is a Riesz basis and the claim follows by Corollary 3.1.24. □

Let us end this general introduction to frames in Hilbert spaces by a few observations on the convergence of the frame expansions $\sum_n c_n x_n$, for arbitrary choice of scalars. By Proposition 3.1.12, if $c = (c_n) \in \ell^2$, then the frame expansion $\sum_n c_n x_n$ converges unconditionally. The converse is not true in general, as shown by the following example.

Example 3.1.26.
(i) Let $\{x_n\}$ be a frame with infinitely many zero elements. Consider the sequence $c = (c_n)$ defined as follows:

$$c_n = \begin{cases} 1, & \text{if } x_n = 0, \\ 0, & \text{if } x_n \neq 0. \end{cases}$$

Then $\sum_n c_n x_n = 0$ unconditionally, whereas $\sum_n |c_n|^2 = +\infty$.
(ii) Let $\{e_n\}_{n\in\mathbb{N}_+}$ be an o. n. basis for H and define

$$x_n = \frac{1}{n}e_n, \quad y_n = \left(1 - \frac{1}{n^2}\right)^{1/2} e_n.$$

Then $\{x_n\} \cup \{y_n\}$ is a Parseval frame for H (left as an exercise). Now consider the element $x = \sum_n n^{-1} e_n \in H$, since the sequence $(n^{-1})_{n\in\mathbb{N}_+}$ is in ℓ^2. We can write $x = \sum_n (1x_n + 0y_n)$ and $\sum_n (1^2 + 0^2) = +\infty$, so again the sequence of coefficients does not belong to ℓ^2.

Of course, in the previous example we have not considered the coefficient sequence $(\langle x_n, \tilde{x}_n\rangle)$, obtained by the canonical dual frame and which, as we already know, belongs to ℓ^2. (Recall $\sum_n |\langle x_n, \tilde{x}_n\rangle|^2 < \infty$.)

Let us emphasize that nonuniqueness is one of the major reasons why frames have been introduced as an alternative to bases. So it is important to consider alternative representation of elements with respect to a frame $\{x_n\}$.

For frames norm-bounded below, there is the following characterization, due to C. Heil.

Theorem 3.1.27. *Let $\{x_n\}$ be a frame for a Hilber space H which is norm-bounded below. Then*

$$\sum_n |c_n|^2 < \infty \Longleftrightarrow \sum_n c_n x_n \text{ converges unconditionally.}$$

Proof. The "$\Longrightarrow$" direction follows by Proposition 3.1.12.

On the other hand, assume that $\sum_n c_n x_n$ converges unconditionally. By Orlicz's theorem (cf. Theorem A.1.23), letting $0 < a = \inf_n \|x_n\|_H$, we obtain

$$a^2 \sum_n |c_n|^2 \leq \sum_n |c_n|^2 \|x_n\|_H^2 = \sum_n \|c_n x_n\|_H^2 < \infty,$$

and the claim follows. □

For exact frames the same result is given by Corollary 3.1.24 and Theorem 3.1.22 (iii). If a frame is not exact with $\inf_n \|x_n\|_H = 0$, the result fails, as shown by the preceding Example 3.1.26.

Regarding frames, we refer to an interesting paper by M. Frank, V. I. Paulsenand and T. Tiballi [149].

3.2 Gabor frames

The general theory seen so far finds an application for a particular type of frame, having time–frequency shifts as building blocks.

Definition 3.2.1 (Gabor frame). Given a nonzero window function $g \in L^2(\mathbb{R}^d)$ and lattice parameters $\alpha, \beta > 0$, the set of time–frequency shifts

$$\mathcal{G}(g, \alpha, \beta) = \{T_{\alpha m} M_{\beta n} g : m, n \in \mathbb{Z}^d\} \tag{3.18}$$

is called a *Gabor system*.

If $\mathcal{G}(g,\alpha,\beta)$ is a frame for $L^2(\mathbb{R}^d)$, it is called a *Gabor frame*. The associated frame operator, the Gabor frame operator S, has the form

$$\begin{aligned} Sf &= \sum_{m,n\in\mathbb{Z}^d} \langle f, T_{\alpha m}M_{\beta n}g\rangle T_{\alpha m}M_{\beta n}g \\ &= \sum_{m,n\in\mathbb{Z}^d} V_g f(\alpha m,\beta n) M_{\beta n}T_{\alpha m}g. \end{aligned} \tag{3.19}$$

We write $S_{g,g}^{\alpha,\beta}$ or $S_{g,g}$ whenever it is necessary to emphasize the dependence of the frame operator on g,α,β.

The series expansions above have coefficients that can be viewed as a sampling of the STFT on the lattice $\Lambda = \alpha\mathbb{Z}^d\times\beta\mathbb{Z}^d$. The order of translation and modulation in (3.19) is not important, due to the commutation relation (0.14). The order $M_\xi T_x$ is natural in the context of the STFT, whereas the order $T_x M_\xi$ is advantageous in connection with representation theory.

Gabor frames are named after the electrical engineer and physicist, most notable for inventing holography, Dennis Gabor (1900–1979). In his paper [152] Gabor conjectured that the Gabor system $\mathcal{G}(\varphi,1,1)$, where $\varphi(t) = e^{-\pi t^2}$ is the 1-dimensional Gaussian, is a basis for $L^2(\mathbb{R})$. In fact, he claimed that every function $f \in L^2(\mathbb{R})$ could be represented as

$$f = \sum_{m,n} c_{m,n}(f) M_n T_m \varphi,$$

for some scalars $c_{m,n}(f)$. This is the main reason why Gabor systems and Gabor frames are named in his honor. We shall see that his conjecture was false, but the previous expansions may make sense using frame theory.

The product $\alpha\beta$ of the lattice generators appears in many calculations involving Gabor systems. It is usually the product $\alpha\beta$ which is important, despite the single values of α and β. Indeed, by dilating g, we can change the value of α at the expense of a complementary change of β. This is contained in the following lemma.

Lemma 3.2.2. *Fix $g \in L^2(\mathbb{R}^d)$, $\alpha,\beta,\lambda > 0$ and consider the dilation operator D_λ defined in* (0.19). *Then the Gabor system $\mathcal{G}(g,\alpha,\beta)$ is a Gabor frame for $L^2(\mathbb{R}^d)$ iff $\mathcal{G}(D_\lambda g,\alpha/\lambda,\lambda\beta)$ is.*

Proof. We compute the action of the dilation operator D_λ on the sequence $\{T_{\alpha m}M_{\beta n}g\}$:

$$\begin{aligned} D_\lambda(T_{\alpha m}M_{\beta n}g)(x) &= \lambda^{d/2} e^{2\pi i\beta n\cdot(\lambda x-\alpha m)} g(\lambda x - \alpha m) \\ &= e^{2\pi i\lambda\beta n\cdot(x-\frac{\alpha}{\lambda}m)}\lambda^{d/2} g\Big(\lambda\Big(x-\frac{\alpha}{\lambda}m\Big)\Big) = T_{\alpha m/\lambda}M_{\lambda\beta n}D_\lambda g(x). \end{aligned}$$

So the Gabor system $\mathcal{G}(g,\alpha,\beta)$ is mapped by D_λ to the Gabor system $\mathcal{G}(D_\lambda g,\alpha/\lambda,\lambda\beta)$. Since D_λ is a bijective isometry of $L^2(\mathbb{R}^d)$, hence a topological isomorphism, by the general theory of frames (Proposition 3.1.23), we have that $\mathcal{G}(g,\alpha,\beta)$ is a frame if and only if $\mathcal{G}(D_\lambda g,\alpha/\lambda,\lambda\beta)$ is. □

The first task concerns the structure of the dual frame of a Gabor frame, first stated in the PhD thesis of O. Christensen [49], and for more general lattices in the paper by H. Feichtinger and W. Kozek [138].

Proposition 3.2.3. *If $\mathcal{G}(g,\alpha,\beta)$ is a frame for $L^2(\mathbb{R}^d)$, with frame bounds $0 < A \le B$, then there exists a window $\gamma \in L^2(\mathbb{R}^d)$ such that the canonical dual frame of $\mathcal{G}(g,\alpha,\beta)$ is $\mathcal{G}(\gamma,\alpha,\beta)$. Consequently, every $f \in L^2(\mathbb{R}^d)$ has the frame expansion*

$$\begin{aligned} f &= \sum_{m,n\in\mathbb{Z}^d} \langle f, T_{\alpha m}M_{\beta n}g\rangle T_{\alpha m}M_{\beta n}\gamma \\ &= \sum_{m,n\in\mathbb{Z}^d} \langle f, T_{\alpha m}M_{\beta n}\gamma\rangle T_{\alpha m}M_{\beta n}g, \end{aligned} \tag{3.20}$$

with unconditional convergence in $L^2(\mathbb{R}^d)$. Further, the following norm equivalences hold:

$$A\|f\|_2^2 \le \sum_{m,n\in\mathbb{Z}^d} |\langle f, T_{\alpha m}M_{\beta n}g\rangle|^2 \le B\|f\|_2^2,$$

$$B^{-1}\|f\|_2^2 \le \sum_{m,n\in\mathbb{Z}^d} |\langle f, T_{\alpha m}M_{\beta n}\gamma\rangle|^2 \le A^{-1}\|f\|_2^2.$$

Proof. Let us first show that the Gabor frame operator S in (3.19) commutes with the time–frequency shifts $T_{\alpha m}M_{\beta n}$. Given $f \in L^2(\mathbb{R}^d)$, $r,s \in \mathbb{Z}^d$,

$$\begin{aligned} (T_{\alpha r}M_{\beta s})^{-1}ST_{\alpha r}M_{\beta s}f &= (T_{\alpha r}M_{\beta s})^{-1} \sum_{m,n\in\mathbb{Z}^d} \langle T_{\alpha r}M_{\beta s}f, T_{\alpha m}M_{\beta n}g\rangle T_{\alpha m}M_{\beta n}g \\ &= \sum_{m,n\in\mathbb{Z}^d} \langle T_{\alpha r}M_{\beta s}f, T_{\alpha m}M_{\beta n}g\rangle (T_{\alpha r}M_{\beta s})^{-1}T_{\alpha m}M_{\beta n}g \\ &= \sum_{m,n\in\mathbb{Z}^d} \langle f, M_{-\beta s}T_{-\alpha r}T_{\alpha m}M_{\beta n}g\rangle M_{-\beta s}T_{-\alpha r}T_{\alpha m}M_{\beta n}g \end{aligned}$$

Observe that, by the commutation relations (0.14),

$$M_{-\beta s}T_{-\alpha r}T_{\alpha m} = M_{-\beta s}T_{\alpha(m-r)} = e^{-2\pi i\alpha\beta s\cdot(m-r)}T_{\alpha(m-r)}M_{-\beta s}.$$

The phase factor $e^{-2\pi i\alpha\beta s\cdot(m-r)}$ cancels, and we obtain

$$(T_{\alpha r}M_{\beta s})^{-1}ST_{\alpha r}M_{\beta s}f = \sum_{m,n\in\mathbb{Z}^d} \langle f, T_{\alpha(m-r)}M_{\beta(n-s)}g\rangle T_{\alpha(m-r)}M_{\beta(n-s)}g = Sf,$$

after renaming the indices.

As a consequence, also the inverse S^{-1} commutes with $T_{\alpha r}M_{\beta s}$ (left as an exercise) and the canonical dual frame consists of the functions

$$S^{-1}(T_{\alpha r}M_{\beta s}g) = T_{\alpha r}M_{\beta s}S^{-1}g, \quad r,s \in \mathbb{Z}^d.$$

We set $\gamma = S^{-1}g \in L^2(\mathbb{R}^d)$ and the claim follows by the general theory of frames, see Theorem 3.1.15. □

As a consequence of the previous proposition, we obtain the expression of the inverse frame operator.

Corollary 3.2.4. *Consider $\mathcal{G}(g,\alpha,\beta)$ to be a frame for $L^2(\mathbb{R}^d)$ with frame operator $S_{g,g}$. Setting $\gamma = S_{g,g}^{-1}g$, the inverse frame operator $S_{g,g}^{-1}$ is provided by*

$$S_{g,g}^{-1}f = S_{\gamma,\gamma}f = \sum_{m,n\in\mathbb{Z}^d} \langle f, T_{\alpha m}M_{\beta n}\gamma\rangle T_{\alpha m}M_{\beta n}\gamma.$$

The window function γ is called a *canonical dual window* of the window g.

Remark 3.2.5.

(i) Proposition 3.2.3 provides a discrete time–frequency representation of signals. If $\mathcal{G}(g,\alpha,\beta)$ is a frame for $L^2(\mathbb{R}^d)$, then (3.20) is a discrete version of the inversion formula (1.55) for the STFT. In addition, (3.20) provides a Gabor expansion of f with the canonical set of coefficients given by $c_{m,n} = \langle f, T_{\alpha m}M_{\beta n}\gamma\rangle$. The series expansion (3.20) can be rephrased in terms of the STFT as

$$f = \sum_{m,n\in\mathbb{Z}^d} V_g f(\alpha m, \beta n) M_{\beta n} T_{\alpha m}\gamma,$$

and it is a reconstruction of the signal f from the samples of its STFT.

(ii) The reconstruction formula (3.20) is based on the additional structure of Gabor frames. Since $S_{g,g}^{-1} = S_{\gamma,\gamma}$, the canonical dual window γ determines the inverse frame operator completely. So it suffices to solve a single linear equation $S\gamma = g$.

(iii) The Gabor frame operator $S_{g,g}^{\alpha,\beta}$ obviously depends on the window g and the lattice parameters α and β. It is proved in [140] that, when increasing the sampling rate, that is, $(\alpha\beta)^{-1} \to +\infty$, the dual window γ becomes similar to the original one, g. Precisely,

$$\lim_{(\alpha,\beta)\to(0,0)} (\alpha\beta)^{-1}(S_{g,g}^{\alpha,\beta})^{-1}g = g$$

in the L^2-norm.

(iv) Instead of separable lattices $\Lambda = \alpha\mathbb{Z}^d \times \beta\mathbb{Z}^d$ of the time–frequency plane, we can also consider more general regular lattices of the form $\Lambda = M\mathbb{Z}^{2d}$ where $M \in GL(2d,\mathbb{R})$, or even nonuniform discrete sets $X \subset \mathbb{R}^{2d}$. In this latter case, however, the structure of the dual frame is lost, and we only know that it is a frame, but not a Gabor frame anymore. For this subject, we refer to [133].

(v) The work by F. Weisz [308] is to be mentioned, where a signal f is approximated with the Riemannian sums

$$S_k f = \frac{1}{k^{2d}} \sum_{m,n\in\mathbb{Z}^d} V_g f\left(\frac{m}{k}, \frac{n}{k}\right) M_{\frac{n}{k}} T_{\frac{m}{k}} \gamma,$$

of the inverse short-time Fourier transform. For $f \in M^{p,q}(\mathbb{R}^d)$, $1 \le p, q < \infty$, it is shown that $S_k f \to f$ in the $M^{p,q}(\mathbb{R}^d)$-norm for $k \to \infty$, provided $g, \gamma \in M^1(\mathbb{R}^d)$.

3.2.1 Existence and density of Gabor frames

In what follows we shall provide a few examples of windows g and conditions on the lattice parameters so that the Gabor system $\mathcal{G}(g, \alpha, \beta)$ becomes a frame. We suggest the reader to look at the classical paper by C. Heil [182] for a detailed survey of this topic and for even more references.

We discuss first the following periodization trick.

Lemma 3.2.6. *If $f \in L^1(\mathbb{R}^d)$, then for all $\alpha > 0$,*

$$\int_{\mathbb{R}^d} f(x)\,dx = \int_{[0,\alpha]^d} \left(\sum_{k\in\mathbb{Z}^d} f(x+\alpha k)\right) dx.$$

Proof. Observe that the translated cubes $\alpha k + [0,\alpha]^d$ are disjoint except for an overlap of measure zero on the boundary, hence form a partition of $\mathbb{R}^d$. Using the σ-additivity of the Lebesgue integral and performing the change of variables $x \mapsto x + \alpha k$, we can write

$$\int_{\mathbb{R}^d} |f(x)|\,dx = \sum_{k\in\mathbb{Z}^d} \int_{\alpha k+[0,\alpha]^d} |f(x)|\,dx = \sum_{k\in\mathbb{Z}^d} \int_{[0,\alpha]^d} |f(x+\alpha k)|\,dx < \infty,$$

since $f \in L^1(\mathbb{R}^d)$ by assumption. This gives $T_{-\alpha k} f \in L^1([0,\alpha]^d \times \mathbb{Z}^d, dx \times \delta)$, where here δ is the discrete measure. An application of Fubini theorem yields

$$\begin{aligned}\int_{\mathbb{R}^d} f(x)\,dx &= \sum_{k\in\mathbb{Z}^d} \int_{\alpha k+[0,\alpha]^d} f(x)\,dx = \sum_{k\in\mathbb{Z}^d} \int_{[0,\alpha]^d} f(x+\alpha k)\,dx \\ &= \int_{[0,\alpha]^d} \left(\sum_{k\in\mathbb{Z}^d} f(x+\alpha k)\right) dx.\end{aligned}$$

This concludes the proof. □

Definition 3.2.7. Given $g, \gamma \in L^2(\mathbb{R}^d)$, $\alpha, \beta > 0$, we define the *correlation function* of the pair (g, γ) as

$$G_n(x) = \sum_{k\in\mathbb{Z}^d} \bar{g}\left(x - \frac{n}{\beta} - \alpha k\right) \gamma(x - \alpha k), \tag{3.21}$$

for $n \in \mathbb{Z}^d$.

Observe that, for any fixed n, the function G_n is a periodization of the function $(T_{n/\beta}\bar{g})\gamma \in L^1(\mathbb{R}^d)$ of period $\alpha\mathbb{Z}^d$. Setting

$$\mathcal{Q}_\alpha = [0,\alpha]^d, \tag{3.22}$$

we have $G_n \in L^1(\mathcal{Q}_\alpha)$ by Lemma 3.2.6. In particular, for $g = \gamma$, $n = 0$, we obtain

$$G_0(x) = \sum_{k\in\mathbb{Z}^d} |g(x-\alpha k)|^2. \tag{3.23}$$

Hence $G_0(x) \geq 0$ a. e. and, by Lemma 3.2.6,

$$\|G_0\|_{L^1(\mathcal{Q}_\alpha)} = \int_{\mathbb{R}^d} |g(x)|^2\,dx = \|g\|^2_{L^2(\mathbb{R}^d)}.$$

Example 3.2.8. The simplest example of Gabor frame, in dimension $d = 1$, is provided by

$$\mathcal{G}(\chi_{[0,1]}, 1, 1) = \{e^{2\pi inx}\chi_{[m,m+1]}\}, \quad m, n \in \mathbb{Z}.$$

For every fixed m, the sequence $\{e^{2\pi inx}\chi_{[m,m+1]}\}_{n\in\mathbb{Z}}$ is an o. n. basis for $L^2([m, m+1])$. Hence $\mathcal{G}(\chi_{[0,1]}, 1, 1)$ is simply the union of o. n. bases for $L^2([m, m+1])$ over all $m \in \mathbb{Z}$ and hence is an o. n. basis for $L^2(\mathbb{R})$. (Recall that this implies the frame property.)

Unfortunately, the previous example is not very useful in practice because of the lack of smoothness of the window $\chi_{[0,1]}$. Indeed, it is discontinuous and this implies slow convergence of functions in such o. n. basis. Such a basis will capture the decay property of a given f correctly, but it is impossible to distinguish a C^∞ function from a nonsmooth function, because of the artificial discontinuities introduced by the cut-off. On the frequency side, observe that

$$\mathcal{F}(\chi_{[0,1]})(\xi) = e^{-\pi i\xi}\frac{\sin(\pi\xi)}{\pi\xi}, \quad \xi \neq 0, \quad \mathcal{F}(\chi_{[0,1]})(0) = 1,$$

that is, $\mathcal{F}(\chi_{[0,1]})(\xi) = e^{-\pi i\xi}\operatorname{sinc}(\xi)$ and this function does not belong to $L^1(\mathbb{R})$, since its decay at infinity is of order $1/|\xi|$. For applications, we need Gabor frames which are generated by functions that are both smooth and well-localized.

Exercise 3.2.9. Fix $g \in L^2(\mathbb{R}^d)$ and $\alpha, \beta > 0$. Using Parseval formula, show that

$$\mathcal{G}(g,\alpha,\beta) \text{ is a frame} \Longleftrightarrow \mathcal{G}(\hat{g},\beta,\alpha) \text{ is a frame.}$$

Exercise 3.2.10. Assume $\mathcal{G}(g,\alpha,\beta)$ is a frame for $L^2(\mathbb{R})$. Show that $\mathcal{G}(g,\alpha,\beta)$ is a Riesz basis if and only if $\langle g,\gamma\rangle = 1$, where $\gamma = S^{-1}g$ is the canonical dual window. (Hint: use Corollary 3.1.24 and show that $\mathcal{G}(g,\alpha,\beta)$ is an exact frame. To prove this claim, use Corollary 3.1.21 (iii)).

If we consider dimension $d = 1$ and concentrate on functions g compactly supported in an interval of length $1/\beta$, then there exist Gabor frames $\mathcal{G}(g, \alpha, \beta)$ for $L^2(\mathbb{R})$ with smooth windows g, provided we choose the lattice parameters α, β properly. This was done first by Daubechies, Grossmann, Meyer in [102], and they were called *painless nonorthogonal expansions*, since they were easy to construct.

Theorem 3.2.11 (Painless nonorthogonal expansions). *Fix $\alpha, \beta > 0$ and $g \in L^2(\mathbb{R})$.*
(i) *If $\operatorname{supp} g \subseteq [0, \beta^{-1}]$ then $\mathcal{G}(g, \alpha, \beta)$ is a frame for $L^2(\mathbb{R})$ if and only if there exist constants A, B such that*

$$A\beta \leq \sum_{k \in \mathbb{Z}} |g(x - \alpha k)|^2 \leq B\beta \quad a.e.\ x \in \mathbb{R}, \tag{3.24}$$

that is, with the notation (3.23),

$$A\beta \leq G_0(x) \leq B\beta \quad a.e.\ x \in \mathbb{R}.$$

In this case, A and B are frame bounds for $\mathcal{G}(g, \alpha, \beta)$.
(ii) *If $0 < \alpha\beta < 1$, then there exists a g such that* $\operatorname{supp} g \subseteq [0, \beta^{-1}]$, *satisfying* (3.24) *and being as smooth as we like, even infinitely differentiable.*
(iii) *If $\alpha\beta = 1$, then any function that is supported in $[0, \beta^{-1}]$ and satisfies (3.24) is discontinuous.*
(iv) *If $\alpha\beta > 1$ and g is supported in $[0, \beta^{-1}]$, then equation (3.24) is not satisfied and $\mathcal{G}(g, \alpha, \beta)$ is incomplete in $L^2(\mathbb{R})$.*

Proof. (i) Assume $\operatorname{supp} g \subseteq [0, \beta^{-1}]$ and that (3.24) holds. We check the frame inequalities (3.2) for f in a dense subset of $L^2(\mathbb{R})$. This is enough to assure that $\mathcal{G}(g, \alpha, \beta)$ is a frame. We choose $C_c(\mathbb{R})$, the space of continuous functions with compact support, which is dense in $L^2(\mathbb{R})$. Since $g \in L^2(\mathbb{R})$ is supported in $[0, \beta^{-1}]$, the translates $T_{\alpha k} g$ are supported in $I_k = [\alpha k, \alpha k + \beta^{-1}]$ and belong to $L^2(I_k)$. Since f is bounded, $f T_{\alpha k} g \in L^2(I_k)$ as well. Now, $\{e^{2\pi i n x}\}_{n \in \mathbb{Z}}$ is an o. n. basis for $L^2([0, 1])$, so, by making a change of variables, we obtain that the sequence $\{\beta^{1/2} e^{2\pi i \beta n x}\}$ is an o. n. basis for $L^2(I_k)$. Applying Plancherel inequality and keeping in mind that g is supported in $[0, \beta^{-1}]$ (so that $T_{\alpha k} g$ is supported in I_k), we can write

$$\begin{aligned}
\int_{-\infty}^{+\infty} |f(x) g(x - \alpha k)|^2 \, dx &= \int_{\alpha k}^{\alpha k + \beta^{-1}} |f(x) T_{\alpha k} \overline{g}(x)|^2 \, dx \\
&= \|f T_{\alpha k} \overline{g}\|^2_{L^2(I_k)} \\
&= \sum_{n \in \mathbb{Z}} |\langle f T_{\alpha k} \overline{g}, \beta^{1/2} e^{2\pi i \beta n x} \rangle_{L^2(I_k)}|^2 \\
&= \beta \sum_{n \in \mathbb{Z}} \left| \int_{I_k} f(x) \overline{g}(x - \alpha k) e^{-2\pi i \beta n x} \, dx \right|^2
\end{aligned}$$

$$= \beta \sum_{n\in\mathbb{Z}} \left| \int_{-\infty}^{+\infty} f(x)\overline{g}(x-\alpha k) e^{-2\pi i\beta n x}\, dx \right|^2$$

$$= \beta \sum_{n\in\mathbb{Z}} \left| \int_{-\infty}^{+\infty} f(x)\overline{e^{2\pi i\beta n x} g(x-\alpha k)}\, dx \right|^2$$

$$= \beta \sum_{n\in\mathbb{Z}} |\langle f, M_{\beta n} T_{\alpha k} g\rangle|^2 .$$

Now, using Tonelli's theorem to interchange sum and integral,

$$\sum_{k\in\mathbb{Z}} \sum_{n\in\mathbb{Z}} |\langle f, M_{\beta n} T_{\alpha k} g\rangle|^2 = \frac{1}{\beta} \sum_{k\in\mathbb{Z}} \int_{-\infty}^{+\infty} |f(x) g(x-\alpha k)|^2\, dx$$

$$= \frac{1}{\beta} \int_{-\infty}^{+\infty} |f(x)|^2 \sum_{k\in\mathbb{Z}} |g(x-\alpha k)|^2\, dx$$

and, by the assumption (3.24),

$$\sum_{k,n\in\mathbb{Z}} |\langle f, M_{\beta n} T_{\alpha k} g\rangle|^2 = \frac{1}{\beta} \int_{-\infty}^{+\infty} |f(x)|^2 \sum_{k\in\mathbb{Z}} |g(x-\alpha k)|^2\, dx$$

$$\geq A \int_{-\infty}^{+\infty} |f(x)|^2\, dx = A\|f\|^2_{L^2(\mathbb{R})}.$$

The upper bound $\sum_{k,n\in\mathbb{Z}} |\langle f, M_{\beta n} T_{\alpha k} g\rangle|^2 \leq B\|f\|^2_{L^2(\mathbb{R})}$ is obtained analogously. Hence $\mathcal{G}(g,\alpha,\beta)$ is a frame. The converse will be proved in more generality in the subsequent Theorem 3.2.14.

(ii) Assume $0 < \alpha\beta < 1$ and let g be any continuous function such that $g(x) = 0$ if $x \notin [0,\beta^{-1}]$ and $g(x) > 0$ on $(0,\beta^{-1})$. For example, we can choose the smooth bell function centered at $1/(2\beta)$, that is,

$$g(x) = \begin{cases} e^{-\frac{1}{\frac{x}{\beta}-x^2}}, & 0 < x < \frac{1}{\beta}, \\ 0, & x \leq 0 \vee x \geq \frac{1}{\beta}. \end{cases}$$

Now, since $\alpha < \beta^{-1}$, the α-periodic function $G_0(x) = \sum_{k\in\mathbb{Z}} |g(x-\alpha k)|^2$ is strictly positive at every point. Further, at each point x, the function $G_0(x)$ is continuous (or in $\mathcal{C}^\infty(\mathbb{R})$ if we choose g properly; take, for instance, the bell function above) since it is a finite sum of continuous functions. So

$$0 < \inf_{x\in\mathbb{R}} G_0(x) = \min_{x\in[0,\alpha]} G_0(x) \leq \max_{x\in[0,\alpha]} G_0(x) = \sup_{x\in\mathbb{R}} G_0(x) < \infty,$$

that is, G_0 satisfies (3.24) and, using part (i), $\mathcal{G}(g,\alpha,\beta)$ is a frame.

(iii) If $\alpha\beta = 1$, then $\alpha = \beta^{-1}$. If $\operatorname{supp} g \subseteq [0, \beta^{-1}] = [0, \alpha]$, then $T_{\alpha k} g$ is supported in $[\alpha k, \alpha(k+1)]$. By contradiction, assume that g is continuous, then $g(0) = g(\alpha) = 0$. Since the intervals $[\alpha k, \alpha(k+1)]$ overlap at most one point, it follows that G_0 is continuous and $G_0(\alpha k) = 0$ for every $k \in \mathbb{Z}$, so it does not satisfy (3.24) and, by part (i), $\mathcal{G}(g, \alpha, \beta)$ is not a frame. Note that, arguing as in (ii), we may easily produce examples of discontinuous g with $\operatorname{supp} g \subset [0, \beta^{-1}]$, such that $\mathcal{G}(g, \beta^{-1}, \beta)$ is a frame, cf. Example 3.2.8.

(iv) If $\alpha\beta > 1$, then $\alpha > \beta^{-1}$. Hence the function $G_0(x)$ is equal to zero on the interval (β^{-1}, α) and cannot satisfy (3.24). By part (i), $\mathcal{G}(g, \alpha, \beta)$ is not a frame. We can also check directly that the Gabor system $\mathcal{G}(g, \alpha, \beta)$ is incomplete in $L^2(\mathbb{R})$: indeed, consider the characteristic function $\chi_{[\beta^{-1}, \alpha]}$, which is orthogonal to every element of $\mathcal{G}(g, \alpha, \beta)$. □

Remark 3.2.12. Note that it is the product $\alpha\beta$ which is important in the previous result. Indeed, by Lemma 3.2.2, one can change the value of α at the expense of a complementary change of β. Also the interval $[0, \beta^{-1}]$ can be replaced by any interval of length β^{-1}, simply by translating the function g.

Summarizing some important points of the painless nonorthogonal expansions:

- If $0 < \alpha\beta < 1$ then we can find a nice function g (even $g \in \mathcal{D}(\mathbb{R})$) such that $\mathcal{G}(g, \alpha, \beta)$ is a frame for $L^2(\mathbb{R})$.
- If $\alpha\beta = 1$ then there exist Gabor frames $\mathcal{G}(g, \alpha, \beta)$ for $L^2(\mathbb{R})$, with g discontinuous.
- If $\alpha\beta > 1$ and $\operatorname{supp} g \subseteq [0, \beta^{-1}]$ then the Gabor system $\mathcal{G}(g, \alpha, \beta)$ is incomplete and it is not a frame for $L^2(\mathbb{R})$.

Exercise 3.2.13. Assume the hypotheses of Theorem 3.2.11 is satisfied, that is, $\operatorname{supp} g \subseteq [0, \beta^{-1}]$ and (3.24) holds. Prove the following statements about the frame $\mathcal{G}(g, \alpha, \beta)$:

- The frame operator S is pointwise multiplication by $\beta^{-1} G_0$, that is,

$$Sf = \beta^{-1} G_0 f, \quad f \in L^2(\mathbb{R}).$$

  (Hint: use the definition of a frame operator.)
- The canonical dual window is $\gamma = (\beta g)/G_0$. (Hint: use the definition of a canonical dual window $\gamma = S^{-1} g$.)
- If $\alpha\beta = 1$, then $\mathcal{G}(g, \alpha, \beta)$ is a Riesz basis. (Hint: use Exercise 3.2.10.)

Theorem 3.2.11 gives necessary and sufficient conditions for the existence of Gabor frames $\mathcal{G}(g, \alpha, \beta)$ when the window g is supported in $[1, 1/\beta]$. This equivalence does not extend to general functions in $L^2(\mathbb{R})$, but the necessity still holds.

Theorem 3.2.14. *If $g \in L^2(\mathbb{R})$ and α, β are such that $\mathcal{G}(g, \alpha, \beta)$ is a frame for $L^2(\mathbb{R})$ with frame bounds $A, B > 0$, then we must have*

$$A\beta \leq G_0(x) \leq \beta B \quad a.e.,$$

that is, (3.24) *holds. In particular, g must be bounded.*

Proof. We use similar arguments as in the proof of Theorem 3.2.11. In this case we do not know if g is compactly supported and so we start with a function $f \in L^2(\mathbb{R})$ which is bounded and such that $\operatorname{supp} f$ is in an interval I of length $1/\beta$. So the product $fT_{\alpha k}\bar{g}$ is in $L^2(I)$, which has the sequence $\{\beta^{1/2}e^{2\pi i\beta nx}\}$ as o. n. basis, and again we can write

$$\int_{-\infty}^{+\infty} |fT_{\alpha k}\bar{g}(x)|^2\,dx = \beta \sum_{n\in\mathbb{Z}} |\langle f, M_{\beta n}T_{\alpha k}g\rangle|^2.$$

Using Tonelli's theorem and employing the lower frame bound A,

$$\begin{aligned}\beta A\|f\|^2_{L^2(\mathbb{R})} &\le \beta \sum_{k,n\in\mathbb{Z}} |\langle f, M_{\beta n}T_{\alpha k}g\rangle|^2 = \int_{-\infty}^{+\infty} |f(x)|^2 \sum_{k\in\mathbb{Z}} |T_{\alpha k}g(x)|^2\,dx \\ &= \int_{-\infty}^{+\infty} |f(x)|^2 G_0(x)\,dx.\end{aligned}$$

So every bounded $f \in L^2(I)$ fulfills

$$\int_{-\infty}^{+\infty} |f(x)|^2 (G_0(x) - \beta A)\,dx \ge 0. \tag{3.25}$$

Now, if $G_0 < \beta A$ on some subset E of I of positive measure, then we could take $f = \chi_E$ and obtain a contradiction to (3.25). Therefore, we must have $G_0 \ge \beta A$ a. e. on I, and a similar computation with the upper bound gives also $G_0 \le \beta B$. Since I is an arbitrary interval of length $1/\beta$ and the real line can be covered by countably many translates of I, we conclude $\beta A \le G_0 \le \beta B$ a. e. on $\mathbb{R}$. Moreover $\ell^2 \subset \ell^\infty$, so for a. e. x,

$$|g(x)| \le \sup_{k\in\mathbb{Z}} |g(x-\alpha k)| \le \Big(\sum_{k\in\mathbb{Z}} |g(x-\alpha k)|^2\Big)^{1/2} \le \sqrt{\beta B},$$

so that $\|g\|_\infty < \infty$. □

Exercise 3.2.15. Show that, if $\mathcal{G}(g,\alpha,\beta)$ is a Gabor frame with frame operator S and $g^\sharp = S^{-1/2}g$, then $\mathcal{G}(g^\sharp,\alpha,\beta)$ is a Parseval frame for $L^2(\mathbb{R})$. (Hint: use Corollary 3.1.25 and Theorem A.0.6 in Appendix A.)

The following corollary gives several features on Gabor frames on $L^2(\mathbb{R})$.

Corollary 3.2.16 (Density and frame bounds). *Consider $\alpha, \beta > 0$, $g \in L^2(\mathbb{R})$ such that $\mathcal{G}(g,\alpha,\beta)$ is a Gabor frame for $L^2(\mathbb{R})$ with frame bounds A and B. Then*
(i) $A\alpha\beta \le \|g\|_2^2 \le B\alpha\beta$.
(ii) *If $\mathcal{G}(g,\alpha,\beta)$ is a Parseval frame, then* $\|g\|_2^2 = \alpha\beta$.
(iii) $0 < \alpha\beta \le 1$.

(iv) $\langle g, \gamma\rangle = \alpha\beta$ *where* $\gamma = S^{-1}g$ *is the canonical dual window.*
(v) $\mathcal{G}(g, \alpha, \beta)$ *is a Riesz basis if and only if* $\alpha\beta = \langle g, \gamma\rangle = 1$.

Proof. (i) and (ii) Integrating (3.24) over the interval $[0, \alpha]$ and using Lemma 3.2.6 (observe that $|g|^2 \in L^1(\mathbb{R})$), we have

$$A\alpha\beta = \int_0^\alpha A\beta\,dx \le \int_0^\alpha \sum_{k\in\mathbb{Z}} |g(x-\alpha k)|^2\,dx = \int_{-\infty}^{+\infty} |g(x)|^2\,dx = \|g\|_2^2.$$

Similarly, one can prove $\|g\|_2^2 \le B\alpha\beta$. If the frame is a Parseval frame then $A = B = 1$, thus part (i) implies $\|g\|_2^2 = \alpha\beta$.

(iii) By Exercise 3.2.15, $\mathcal{G}(g^\sharp, \alpha, \beta)$, where $g^\sharp = S^{-1/2}g$, is a Parseval frame. Then by part (ii), $\|g^\sharp\|_2^2 = \alpha\beta$. By Remark 3.1.11, the elements $M_{\beta n}T_{\alpha k}g^\sharp$ must satisfy

$$\alpha\beta = \|g^\sharp\|_2^2 = \|M_{\beta n}T_{\alpha k}g^\sharp\|_2^2 \le 1,$$

so that $\alpha\beta \le 1$.

(iv) Using the previous step and the self-adjointness of $S^{-1/2}$,

$$\langle g, \gamma\rangle = \langle g, S^{-1/2}S^{-1/2}g\rangle = \langle S^{-1/2}g, S^{-1/2}g\rangle = \|g_\sharp\|_2^2 = \alpha\beta.$$

(v) It follows by Exercise 3.2.10. □

Summarizing, the Gabor systems $\mathcal{G}(g, \alpha, \beta)$ for $g \in L^2(\mathbb{R})$ can be divided into three parts:
- If $0 < \alpha\beta < 1$ then if $\mathcal{G}(g, \alpha, \beta)$ is a Gabor frame then it is a redundant frame.
- If $\alpha\beta = 1$ then if $\mathcal{G}(g, \alpha, \beta)$ is a Gabor frame then it is a Riesz basis.
- If $\alpha\beta > 1$ then the system $\mathcal{G}(g, \alpha, \beta)$ is not a frame.

The value $1/(\alpha\beta)$ is called the *density* of the Gabor system $\mathcal{G}(g, \alpha, \beta)$.

Remark 3.2.17. Corollary 3.2.16 and the necessity condition (3.24) is still true for $g \in L^2(\mathbb{R}^d)$ $(d \ge 1)$. For details on this subject we refer, for instance, to [160].

We end up this section by observing that there are window functions $\gamma \in L^2(\mathbb{R}^d)$ other than the canonical dual window $S^{-1}g$ that give rise to Gabor frames and are dual frames to the Gabor frame $\mathcal{G}(g, \alpha, \beta)$. Precisely, given a window $\gamma \in L^2(\mathbb{R}^d)$, we can consider the reconstruction operator D_γ, defined on finite sequences $c = (c_{m,n})$ by $D_\gamma c = \sum_{m,n} c_{m,n}M_{\beta n}T_{\alpha m}\gamma$. If D_γ is a bounded operator from $\ell^2(\mathbb{Z}^{2d})$ into $L^2(\mathbb{R}^d)$, with a slight abuse of language we can define the frame operator

$$S_{g,\gamma}f = D_\gamma C_g f = \sum_{m,n\in\mathbb{Z}^{2d}} \langle f, M_{\beta n}T_{\alpha m}g\rangle M_{\beta n}T_{\alpha m}\gamma, \tag{3.26}$$

which is a well defined operator $S_{g,\gamma} : L^2(\mathbb{R}^d) \to L^2(\mathbb{R}^d)$, since $C_g : L^2(\mathbb{R}^d) \to \ell^2(\mathbb{Z}^{2d})$ because the system $\mathcal{G}(g, \alpha, \beta)$ is a frame.

Definition 3.2.18. A function $\gamma \in L^2(\mathbb{R}^d)$ is called dual window for a Gabor frame $\mathcal{G}(g,\alpha,\beta)$ if the synthesis operator D_γ is a bounded operator from $\ell^2(\mathbb{Z}^{2d})$ to $L^2(\mathbb{R}^d)$ and if

$$S_{g,\gamma}f = I_{L^2(\mathbb{R}^d)},$$

that is, the frame operator is the identity operator on $L^2(\mathbb{R}^d)$.

The following condition characterizes all dual windows.

Theorem 3.2.19 (Wexler–Raz biorthogonality relations). *Let g,γ be in $L^2(\mathbb{R}^d)$ such that the synthesis operators D_g, D_γ are bounded. Then the following conditions are equivalent:*
(i) $S_{g,\gamma} = S_{\gamma,g} = I_{L^2(\mathbb{R}^d)}$.
(ii) $(\alpha\beta)^{-d}\langle M_{\frac{n}{\alpha}} T_{\frac{m}{\beta}}\gamma, M_{\frac{n'}{\alpha}} T_{\frac{m'}{\beta}} g\rangle = \delta_{m,m'}\delta_{n,n'}, m, m', n, n' \in \mathbb{Z}^d$.

Hence the Gabor systems $\mathcal{G}(g,\frac{1}{\beta},\frac{1}{\alpha})$ and $\mathcal{G}(\gamma,\frac{1}{\beta},\frac{1}{\alpha})$ are biorthogonal systems in $L^2(\mathbb{R}^d)$. The proof can be found in [160, Theorem 7.3.1].

One of the most used window functions for Gabor frames is represented by the Gaussian. A more general result on this topic is contained in [107, Proposition 10]. It extends to the multidimensional case the classical result of Lyubarskii [221] and Seip and Wallsten [262, 263]. We limit ourselves to the statement of such a result in our setting, with a more general lattice than the case $\alpha\mathbb{Z}^d \times \beta\mathbb{Z}^d$, $\alpha, \beta > 0$. First, we need a more general definition of Gabor systems.

Definition 3.2.20. For a general lattice $\Lambda \subset \mathbb{R}^{2d}$ and a nonzero window function $g \in L^2(\mathbb{R}^d)$, we define the Gabor system as

$$\mathcal{G}(g,\Lambda) = \{\pi(\lambda)g : \lambda \in \Lambda\}. \tag{3.27}$$

The Gaussian Gabor frames are then given by the following characterization.

Theorem 3.2.21. *Consider the lattice $\Lambda_{\alpha,\beta} = \alpha\mathbb{Z}^d \times \beta\mathbb{Z}^d$, with $\alpha = (\alpha_1,\dots,\alpha_d)$, $\beta = (\beta_1,\dots,\beta_d)$, and $\alpha_j, \beta_j > 0$, for every $j = 1,\dots,d$. Then the Gabor system $\mathcal{G}(\varphi,\Lambda_{\alpha,\beta})$, with the Gaussian function $\varphi(t) = e^{-\pi t^2}$, $t \in \mathbb{R}^d$, is a frame for $L^2(\mathbb{R}^d)$ if and only if $\alpha_j\beta_j < 1$, for every $j = 1,\dots,d$.*

Observe that in the critical case $\alpha_j\beta_j = 1$, for every $j = 1,\dots,d$, the system $\mathcal{G}(\varphi,1,1)$ is not a frame and, in particular, it is not a basis.

Another important tool for analyzing Gabor frames when $\Lambda = \alpha\mathbb{Z}^d \times \beta\mathbb{Z}^d$, $\alpha\beta = 1$, is the Zak transform. We recommend the article [204] for a survey of the properties of the Zak transform.

Definition 3.2.22. For $\alpha > 0$, the Zak transform of a function f on $\mathbb{R}^d$ is (formally)

$$Z_\alpha f(x,\xi) = \sum_{k\in\mathbb{Z}^d} f(x-\alpha k)e^{2\pi i\alpha k\cdot\xi} = \sum_{k\in\mathbb{Z}^d} M_{\alpha k}T_{\alpha k}f(x), \quad (x,\xi) \in \mathbb{R}^{2d}. \tag{3.28}$$

For example, the Zak transform Z_1 ($\alpha = 1$) of the Gaussian $\varphi(t) = e^{-\pi t^2}$ is given by $Z_1 f(x,\xi) = \sum_{k\in\mathbb{Z}^d} e^{-\pi(x-k)^2} e^{2\pi i k\cdot\xi}$.

Formally, we have the *quasiperiodicity relations*

$$Z_\alpha f(x+\alpha m,\xi) = e^{2\pi i \alpha m\cdot\xi} Z_\alpha(x,\xi) \quad \text{and} \quad Z_\alpha f\left(x,\xi+\frac{m}{\alpha}\right) = Z_\alpha f(x,\xi),$$

for $m \in \mathbb{Z}^d$. Thus $Z_\alpha f$ is completely determined by its values on

$$[0,\alpha)^d \times [0,1/\alpha)^d \subset \mathbb{R}^{2d}.$$

Recalling the cube $\mathcal{Q}_\alpha = [0,\alpha]^d$, $\alpha > 0$, the Plancherel's theorem for the Zak transform reads as follows [203] (see also [160, Theorem 8.2.3] or [186, Theorem 4.3.2]).

Theorem 3.2.23. *The Zak transform is a unitary map of $L^2(\mathbb{R}^d)$ onto $L^2(\mathcal{Q}_\alpha \times \mathcal{Q}_{1/\alpha})$.*

The unitary nature of the Zak transform allows us to translate conditions on frames for $L^2(\mathbb{R}^{2d})$ into those in $L^2(\mathcal{Q}_\alpha \times \mathcal{Q}_{1/\alpha})$, where things are frequently easier to deal with.

For $(u,\eta) \in \mathbb{R}^{2d}$, we easily compute the Zak transform of the time–frequency shifts $T_u M_\eta f$ of a function f in $L^2(\mathbb{R}^d)$. Namely,

$$\begin{aligned} Z_\alpha(T_u M_\eta f)(x,\xi) &= \sum_{k\in\mathbb{Z}^d} e^{2\pi i\eta\cdot(x-u-\alpha k)} f(x-u-\alpha k) e^{2\pi i\alpha k\cdot\xi} \\ &= e^{2\pi i\eta\cdot(x-u)} \sum_{k\in\mathbb{Z}^d} f(x-u-\alpha k) e^{2\pi i\alpha k\cdot(\xi-\eta)} \\ &= e^{2\pi i\eta\cdot(x-u)} Z_\alpha f(x-u,\xi-\eta). \end{aligned}$$

Hence $Z_\alpha(T_u M_\eta f)(x,\xi) = e^{2\pi i\eta\cdot(x-u)} T_{(u,\eta)} Z_\alpha f(x,\xi)$, which places great restrictions on the form $Z_\alpha g$ can take if $\mathcal{G}(g,\alpha,1/\alpha)$ is to generate a frame [100, 203] (see also [160, Theorem 8.2.3] or [186, Section 8.3]).

Theorem 3.2.24. *Suppose $\alpha\beta = 1$ and $g \in L^2(\mathbb{R}^d)$.*

(i) *The Gabor system $\mathcal{G}(g,\alpha,1/\alpha)$ is complete in $L^2(\mathbb{R}^d)$ if and only if $Z_\alpha f \neq 0$ a.e.*

(ii) *The following statements are equivalent:*
   (a) $0 < A \le |Z_\alpha f(x,\xi)|^2 \le B < \infty$ *a.e.*
   (b) *$\mathcal{G}(g,\alpha,1/\alpha)$ is a frame for $L^2(\mathbb{R}^d)$ with frame bounds A and B.*
   (c) *$\mathcal{G}(g,\alpha,1/\alpha)$ generates an exact frame for $L^2(\mathbb{R}^d)$ with frame bounds A, B.*

(iii) *$\mathcal{G}(g,\alpha,1/\alpha)$ generates an orthonormal basis for $L^2(\mathbb{R}^d)$ if and only if*

$$|Z_\alpha f(x,\xi)| = 1 \quad a.e.$$

As an example, in dimension $d = 1$, the Zak transform Z_α of the characteristic function $\chi_{[0,\alpha)}$ is

$$Z_\alpha \chi_{[0,\alpha)}(x,\xi) = \sum_{k\in\mathbb{Z}^d} \chi_{[0,\alpha)}(x-\alpha k) e^{2\pi i\alpha k\xi}$$

and $\chi_{[0,\alpha)}(x-\alpha k) \neq 0$ on $[0,\alpha)$ only if $x-\alpha k \in [0,\alpha)$ which forces $k=0$ and hence $Z_\alpha\chi_{[0,\alpha)}(x,\xi)=1$ on the rectangle $[0,\alpha)\times[0,1/\alpha)$. Similarly in higher dimensions the Zak transform of the characteristic function $\chi_{\mathcal{Q}_\alpha}$ is

$$Z_\alpha\chi_{\mathcal{Q}_\alpha}=1 \quad \text{for } (x,\xi)\in\mathcal{Q}_\alpha\times\mathcal{Q}_{1/\alpha}.$$

Using the previous theorem, we obtain that $\mathcal{G}(\chi_{\mathcal{Q}_\alpha},\alpha,1/\alpha)$ is an orthonormal basis for $L^2(\mathbb{R}^d)$, cf. Example 3.2.8.

Theorem 3.2.25 ([203]). *Let $f\in L^2(\mathbb{R}^d)$ be such that $Z_\alpha f$ is continuous on $\mathbb{R}^{2d}$. Then $Z_\alpha f$ has a zero.*

Example 3.2.26 ([102]). The Zak transform $Z_1\varphi$ of the Gaussian function $\varphi(t)=e^{-\pi t^2}$, $t\in\mathbb{R}$, is continuous and has a single zero in $[0,1)\times[0,1)$.

Therefore Theorem 3.2.24 states that $\mathcal{G}(\varphi,\alpha,1/\alpha)$ is complete but it is not a frame, cf. Theorem 3.2.21. While no function whose Zak transform is continuous can generate a Gabor frame when $\alpha\beta=1$, this does not exclude *smooth* window functions from generating Gabor frames. The following theorem, due to R. Balian [4] (and independently to F. Low [220]), shows that, in fact, if $\mathcal{G}(g,\alpha,\beta)$ generates a frame when $\alpha\beta=1$ then either g is not smooth or does not decay very fast. An elegant proof by G. Battle for the orthonormal basis case, based on the Heisenberg uncertainty principle, is in [7]. See also the results by Daubechies and Janssen in [104] and by Benedetto, Heil, and Walnut [11]. We state it in the original version in dimension $d=1$.

Theorem 3.2.27. *Consider $g\in L^2(\mathbb{R})$ and $\alpha,\beta>0$ with $\alpha\beta=1$. If $\mathcal{G}(g,\alpha,\beta)$ generates a frame, then either $tg(t)\notin L^2(\mathbb{R})$ or $\xi\hat{g}(\xi)\notin L^2(\mathbb{R})$.*

Example 3.2.28. With reference to the preceding Theorem 3.2.11 (iii), every function g generating a frame for $\alpha\beta=1$ and being compactly supported, hence $tg(t)\in L^2(\mathbb{R})$, is discontinuous, that is, consistent with the information $\xi\hat{g}(\xi)\notin L^2(\mathbb{R})$ from the Balian–Low theorem. In the opposite direction, consider the Fourier transform of such g, that is, $h=\hat{g}$. It generates a frame for $\alpha\beta=1$, see the previous Exercise 3.2.9. In this case we have obviously $\xi\hat{h}(\xi)\in L^2(\mathbb{R})$ but $th(t)\notin L^2(\mathbb{R})$. Note that h is an entire function, being the Fourier transform of a function with compact support. A simple example is $\mathcal{F}(\chi_{[0,1]})$, namely $h(t)=e^{-\pi i t}\operatorname{sinc}(t)$, see the computations before Exercise 3.2.9.

We conclude by presenting the so-called amalgam version of the Balian–Low theorem due to C. Heil [179]. Observe that Heil's formulation does not depend on the dimension d and uses the Wiener amalgam spaces $W(L^\infty,L^1)$ for the windows. Let us define

$$W_0(L^\infty,L^1)(\mathbb{R}^d):=W(L^\infty,L^1)(\mathbb{R}^d)\cap\mathcal{C}(\mathbb{R}^d).$$

Theorem 3.2.29. *If $\mathcal{G}(g,\alpha,1/\alpha)$, $\alpha>0$ is a frame (and hence a Riesz basis) for $L^2(\mathbb{R}^d)$, then both $g\notin W_0(L^\infty,L^1)(\mathbb{R}^d)$ and $\hat{g}\notin W_0(L^\infty,L^1)(\mathbb{R}^d)$.*

In summary, Gabor frames with $\alpha\beta = 1$ are bases for $L^2(\mathbb{R}^d)$ but have unpleasant window functions.

As a consequence we obtain a stronger version of the density theorem for *nice* windows.

Corollary 3.2.30. *If $\mathcal{G}(g,\alpha,\beta)$ is a frame for $L^2(\mathbb{R}^d)$ and if either $g \in W_0(L^\infty, L^1)(\mathbb{R}^d)$ or $\hat{g} \in W_0(L^\infty, L^1)(\mathbb{R}^d)$, then $\alpha\beta < 1$.*

For the Walnut representation of the Gabor frame operator, we refer the reader to Walnut's work [302], see also [160, Chapter 5].

Existence results for Gabor frames by pseudodifferential methods were obtained by P. Boggiatto and G. Garello in [25].

We end up this section by recalling the following conjecture known as the *HRT conjecture*, which is an open problem deeply rooted in time–frequency analysis. It was posed about 20 years ago by C. Heil, J. Ramanathan, and P. Topiwala in [184] as follows

Conjecture (The HRT conjecture). If $g \in L^2(\mathbb{R})$ and Λ is a finite subset of $\mathbb{R}^2$, then $\mathcal{G}(g,\Lambda)$ is linearly independent.

Although the HRT conjecture is still unresolved, there are quite a few results that might be regarded as evidence for an affirmative answer.

3.2.2 Gabor frames for modulation spaces

In this section we shall provide a discrete characterization of the modulation norm (2.20). Precisely, we shall extend Proposition 3.2.3 to the spaces $M^{p,q}_m$, showing that a function/distribution f is in $M^{p,q}_m$ if and only if the sequence of Gabor coefficients $\{\langle f, T_{\alpha m} M_{\beta n} g\rangle\}_{m,n}$ is in $\ell^{p,q}_m(\mathbb{Z}^{2d})$, cf. Definition 2.2.1.

Notice that the first connection between modulation spaces and atomic Gabor expansions in the spirit modern coorbit theory was given in [123].

We need some preliminaries.

Proposition 3.2.31. *Consider a weight function $m \in \mathcal{M}_v(\mathbb{R}^{2d})$, and a lattice $\Lambda = \alpha\mathbb{Z}^d \times \beta\mathbb{Z}^d$. For every $\alpha, \beta > 0$ there is a constant $C_{\alpha,\beta} > 0$ such that the estimate*

$$\|\{F(\alpha k, \beta n)\}_{k,n}\|_{\ell^{p,q}_m(\Lambda)} \le C_{\alpha,\beta}\|F\|_{W(L^\infty, L^{p,q}_m)}$$

is true for any function $F \in C(\mathbb{R}^{2d}) \cap W(L^\infty, L^{p,q}_m)(\mathbb{R}^{2d})$, for any $1 \le p, q \le \infty$.

Observe that this implies that the sequence $\{F(\alpha k, \beta n)\}_{k,n\in\mathbb{Z}^d}$ is in $\ell^{p,q}_m(\Lambda)$.

Proof. The continuity of F guarantees that $F(\alpha k, \beta n)$ is well defined, for every $k, n \in \mathbb{Z}^d$. Since the weight m is v-moderate, using (2.7), there exists a positive constant C satisfying

$$m(\alpha k, \beta n) \le Cm(r,s), \quad \text{for } (\alpha k, \beta n) \in (r,s) + [0,1]^{2d}.$$

This yields

$$|F(\alpha k, \beta n)|m(\alpha k, \beta n) \le C\|FT_{r,s}\chi\|_\infty m(r,s),$$

with χ characteristic function of $[0,1]^{2d}$. Since there are at most $C_\alpha = ([1/\alpha]+1)^d$ points $\alpha k \in r + [0,1]^d$, the ℓ^p-norm with respect to $k \in \mathbb{Z}^d$ can be controlled by

$$\Big(\sum_{k\in\mathbb{Z}^d} |F(\alpha k, \beta n)|^p m(\alpha k, \beta n)^p\Big)^{\frac{1}{p}} \le \Big(CC_\alpha \sum_{r\in\mathbb{Z}^d} \|FT_{r,s}\chi\|_\infty^p m(r,s)^p\Big)^{\frac{1}{p}},$$

for every $\beta n \in s + [0,1]^d$, of which there are at most $C_\beta = ([1/\beta]+1)^d$ points. Taking the ℓ^q-norm of the above inequality and estimating the right-hand side by the norm in $W(L^\infty, L^{p,q}_m)$, cf. Definition 2.4.1, we get the claim. □

We next show the boundness of the coefficient operator C_g from $M^{p,q}_m(\mathbb{R}^d)$ into $\ell^{p,q}_m(\Lambda)$, for every lattice $\Lambda = \alpha\mathbb{Z}^d \times \beta\mathbb{Z}^d$.

Theorem 3.2.32. *Consider a weight $m \in \mathcal{M}_v(\mathbb{R}^{2d})$, a nonzero window function $g \in M^1_v(\mathbb{R}^d)$, and a lattice $\Lambda = \alpha\mathbb{Z}^d \times \beta\mathbb{Z}^d$, for every $\alpha,\beta > 0$. Then the coefficient operator C_g is bounded from $M^{p,q}_m(\mathbb{R}^d)$ into $\ell^{p,q}_m(\Lambda)$, with*

$$\|C_g\|_{B(M^{p,q}_m, \ell^{p,q}_m)} \le C(v,\alpha,\beta)\|g\|_{M^1_v},$$

independently of p, q and m.

Proof. For $f \in M^{p,q}_m(\mathbb{R}^d)$ and the window $g \in M^1_v(\mathbb{R}^d)$, the STFT $V_g f$ is in the space $W(\mathcal{F}L^1, L^{p,q}_m)(\mathbb{R}^{2d})$, by Lemma 2.4.15. Since $\mathcal{F}L^1 \subset L^\infty$, by the inclusion relations for Wiener amalgam spaces, we infer that $V_g f \in W(L^\infty, L^{p,q}_m)(\mathbb{R}^{2d})$. The continuity of the STFT, together with Proposition 3.2.31 and estimate (2.72), give

$$\begin{aligned}\|C_g f\|_{\ell^{p,q}_m(\Lambda)} &= \|V_g f(\alpha k, \beta n)\|_{\ell^{p,q}_m(\Lambda)} \le C_{\alpha,\beta}\|V_g f\|_{W(L^\infty, L^{p,q}_m)} \\ &\le C_{\alpha,\beta}\|V_g f\|_{W(\mathcal{F}L^1, L^{p,q}_m)} \lesssim C_{\alpha,\beta}\|f\|_{M^{p,q}_m}\|g\|_{M^1_v}.\end{aligned}$$ □

We now present the continuity of the synthesis operator D_g from $\ell^{p,q}_m(\Lambda)$, for $\Lambda = \alpha\mathbb{Z}^d \times \beta\mathbb{Z}^d$, into $M^{p,q}_m(\mathbb{R}^d)$.

Theorem 3.2.33. *Consider a weight $m \in \mathcal{M}_v(\mathbb{R}^{2d})$, a nonzero window function $g \in M^1_v(\mathbb{R}^d)$, and a lattice $\Lambda = \alpha\mathbb{Z}^d \times \beta\mathbb{Z}^d$, for every $\alpha,\beta > 0$. Then the synthesis operator D_g is bounded from $\ell^{p,q}_m(\Lambda)$ into $M^{p,q}_m(\mathbb{R}^d)$, with*

$$\|D_g\|_{B(\ell^{p,q}_m, M^{p,q}_m)} \le C(v,\alpha,\beta)\|g\|_{M^1_v},$$

independently of p, q and m. If $p, q < \infty$, then

$$D_g c = \sum_{k,n\in\mathbb{Z}^d} c_{k,n} T_{\alpha k} M_{\beta n} g$$

converges unconditionally in $M^{p,q}_m(\mathbb{R}^d)$. Otherwise, $D_g c$ converges weak in $M^\infty_{1/v}(\mathbb{R}^d)$.*

Proof. Consider a sequence $c = (c_{k,n})$ in $\ell^{p,q}_m(\Lambda)$ and, for instance, the Gaussian $\varphi(t) = e^{-\pi t^2}$ as window function to measure the $M^{p,q}_m$-norm. We shall show that $V_\varphi(D_g c)$ is in $L^{p,q}_m(\mathbb{R}^{2d})$. Recalling the action of the STFT on the time–frequency shifts (1.51), we can write $V_\varphi(T_{\alpha k}M_{\beta n}g)(x,\xi) = T_{(\alpha k,\beta n)}V_\varphi g(x,\xi)$, so that

$$|V_\varphi(D_g c)| \le \sum_{k,n\in\mathbb{Z}^d} |c_{k,n}|T_{(\alpha k,\beta n)}|V_\varphi g|.$$

Since $\varphi, g \in M^1_v(\mathbb{R}^d)$, by Lemma 2.4.15 and the inclusion relations for Wiener amalgam spaces, we infer $V_\varphi g \in W(\mathcal{F}L^1, L^1_v)(\mathbb{R}^{2d}) \subset W(L^\infty, L^1_v)(\mathbb{R}^{2d})$, hence

$$|V_\varphi g|(x,\xi) \le \sum_{r,s\in\mathbb{Z}^d} a_{r,s}T_{(\alpha r,\beta s)}\chi_{[0,\alpha]^d\times[0,\beta]^d}(x,\xi),$$

with $a = (a_{r,s})$, $a_{r,s} \ge 0$ satisfying

$$\|a\|_{\ell^1_v} = \sum_{r,s\in\mathbb{Z}^d} a_{r,s}v(\alpha r,\beta s) \lesssim \|V_\varphi g\|_{W(\mathcal{F}L^1,L^1_v)} \lesssim \|g\|_{M^1_v}.$$

This implies, for every $(x,\xi) \in \mathbb{R}^{2d}$,

$$\begin{aligned}\sum_{k,n\in\mathbb{Z}^d} |c_{k,n}|T_{(\alpha k,\beta n)}|V_\varphi g| &\le \sum_{k,n,r,s\in\mathbb{Z}^d} |c_{k,n}|a_{r,s}T_{(\alpha(k+r),\beta(n+s))}\chi_{[0,\alpha]^d\times[0,\beta]^d}\\ &= \sum_{l,m\in\mathbb{Z}^d}\Big(\sum_{k,n\in\mathbb{Z}^d} |c_{k,n}|a_{l-k,m-n}\Big)T_{(\alpha l,\beta n)}\chi_{[0,\alpha]^d\times[0,\beta]^d}\\ &= \sum_{l,m\in\mathbb{Z}^d} (|c| * a)(l,m)T_{(\alpha l,\beta n)}\chi_{[0,\alpha]^d\times[0,\beta]^d}.\end{aligned}$$

Using now the norm equivalence (2.9),

$$\Big\|\sum_{k,n\in\mathbb{Z}^d} |c_{k,n}|T_{(\alpha k,\beta n)}|V_\varphi g|\Big\|_{L^{p,q}_m} \lesssim \||c| * a\|_{\ell^{p,q}_m} \lesssim \|c\|_{\ell^{p,q}_m}\|a\|_{\ell^1_v}.$$

To sum up,

$$\begin{aligned}\|D_g c\|_{M^{p,q}_m} &= \Big\|\sum_{k,n\in\mathbb{Z}^d} c_{k,n}T_{\alpha k,\beta n}g\Big\|_{M^{p,q}_m}\\ &\asymp \Big\|V_\varphi\Big(\sum_{k,n\in\mathbb{Z}^d} c_{k,n}T_{\alpha k,\beta n}g\Big)\Big\|_{L^{p,q}_m}\\ &\lesssim \|c\|_{\ell^{p,q}_m}.\end{aligned}$$

It remains to prove the unconditional convergence for $1 \le p,q < \infty$. Recall that in this case the finite sequences are dense in $\ell^{p,q}_m(\mathbb{Z}^{2d})$. Given $\epsilon > 0$, there is a finite set

$F_0 \in \mathbb{Z}^{2d}$ such that $\|c - c\chi_F\|_{\ell^{p,q}_m} < \epsilon$, for every $F \supseteq F_0$, cf. Theorem A.1.21 in Appendix A. Then

$$\Big\|D_g c - \sum_{(k,n)\in F} c_{k,n} T_{\alpha k} M_{\beta n} g\Big\|_{M^{p,q}_m} = \|D_g(c - c\cdot\chi_F)\|_{M^{p,q}_m}$$
$$\leq C\|c - c\cdot\chi_F\|_{\ell^{p,q}_m} < C\epsilon,$$

yielding the claim.

If $p = \infty$ or $q = \infty$, taking any $c = (c_{k,n}) \in \ell^{p,q}_m$, we have

$$\langle D_g c, f\rangle = \sum_{k,n\in\mathbb{Z}^d} \overline{\langle f, T_{\alpha k} M_{\beta n} g\rangle},$$

and, similarly as above, the series is unconditionally convergent for every $f \in M^1_v(\mathbb{R}^d)$, that is to say, $D_g c$ converges weak* in $(M^\infty_{1/v}, M^1_v)$. □

It can be shown that the constant $C(v,\alpha,\beta)$ can be chosen to be the same for compact ranges of lattice constants inside the (open) first quadrant of $\mathbb{R}^2$.

As a consequence of Theorem 3.2.33, we infer the boundedness of the frame operator on $M^{p,q}_m(\mathbb{R}^d)$.

Corollary 3.2.34. *If $m \in \mathcal{M}_v(\mathbb{R}^{2d})$, $g, \gamma \in M^1_v(\mathbb{R}^d)$, then the Gabor frame operator $S_{g,\gamma} = D_\gamma C_g$ is bounded on $M^{p,q}_m(\mathbb{R}^d)$, for every $1 \leq p, q \leq \infty$ and every $\alpha, \beta > 0$. The operator norm can be estimated uniformly by*

$$\|S_{g,\gamma}\|_{B(M^{p,q}_m)} \leq C(v,\alpha,\beta)\|g\|_{M^1_v}\|\gamma\|_{M^1_v},$$

with a constant $C(v,\alpha,\beta)$ depending only on v, α, β, but not on m, p, q.

It may be worthwhile to mention that the constants $C(v,\alpha,\beta)$ appearing in Theorems 3.2.32 and 3.2.33, as well as in Corollary 3.2.34, are in fact uniformly bounded for compact ranges of parameters (α,β) (assuming one has frames for the full range), for *good windows* (see [137], in contrast to [136], where it is demonstrated that Bessel bounds may vary like crazy for ordinary L^2-functions).

If, in addition, the function γ is a dual window for g, that is, $S_{g,\gamma} = I$, the identity operator on $L^2(\mathbb{R}^d)$, the modulation spaces can be characterized as follows.

Theorem 3.2.35. *If $m \in \mathcal{M}_v(\mathbb{R}^{2d})$, $g, \gamma \in M^1_v(\mathbb{R}^d)$ are such that $S_{g,\gamma} = I$ on $L^2(\mathbb{R}^d)$, then*

$$f = \sum_{k,n\in\mathbb{Z}^d} \langle f, T_{\alpha k} M_{\beta n} g\rangle T_{\alpha k} M_{\beta n}\gamma \tag{3.29}$$
$$= \sum_{k,n\in\mathbb{Z}^d} \langle f, T_{\alpha k} M_{\beta n}\gamma\rangle T_{\alpha k} M_{\beta n} g,$$

with unconditional convergence in $M^{p,q}_m$ for $p, q < \infty$ and weak convergence in $M^\infty_{1/v}$ otherwise. Further, the following norm equivalences hold:*

$$A\|f\|_{M^{p,q}_m} \leq \|\{\langle f, T_{\alpha k} M_{\beta n} g\rangle\}_{k,n}\|_{\ell^{p,q}_m} \leq B\|f\|_{M^{p,q}_m}, \tag{3.30}$$

$$B^{-1}\|f\|_{M_m^{p,q}} \le \|\{\langle f, T_{\alpha k}M_{\beta n}\gamma\rangle\}_{k,n}\|_{\ell_m^{p,q}} \le A^{-1}\|f\|_{M_m^{p,q}} \tag{3.31}$$

for suitable $0 < A \le B$.

Proof. Using the boundedness of the frame operator $S_{g,\gamma} = D_\gamma C_g$ on $M_m^{p,q}(\mathbb{R}^d)$, shown in Corollary 3.2.34, we obtain that the equality $S_{g,\gamma} = I$ holds for all $f \in M_m^{p,q}(\mathbb{R}^d)$. The left-hand side of the norm equivalence (3.30) is given by

$$\|f\|_{M_m^{p,q}} = \|D_\gamma C_g f\|_{M_m^{p,q}} \le \|D_\gamma\|_{B(\ell_m^{p,q}, M_m^{p,q})}\|C_g f\|_{\ell_m^{p,q}},$$

with $A = 1/\|D_\gamma\|_{B(\ell_m^{p,q}, M_m^{p,q})}$. The right-hand side of (3.30) follows by the continuity of the coefficient operator C_g:

$$\|C_g f\|_{\ell_m^{p,q}} \le \|C_g\|_{B(M_m^{p,q}, \ell_m^{p,q})}\|f\|_{M_m^{p,q}},$$

with $B = \|C_g\|_{B(M_m^{p,q}, \ell_m^{p,q})}$. The inequalities in (3.31) are obtained analogously, using the identity $D_g C_\gamma = I$. □

Corollary 3.2.36. *If* $g \in \mathcal{S}(\mathbb{R}^d)$, $\Lambda = \alpha\mathbb{Z}^d \times \beta\mathbb{Z}^d$ *and* $\mathcal{G}(g, \Lambda)$ *is a Gabor frame, then*

$$f \in \mathcal{S}(\mathbb{R}^d) \Longleftrightarrow \sup_{\lambda\in\Lambda} \langle\lambda\rangle^N |\langle f, \pi(\lambda)g\rangle| < \infty, \quad \forall N \in \mathbb{N}, \tag{3.32}$$

Proof. The characterization is a straightforward consequence of the norm equivalence in (3.30) (with $\lambda = (\alpha k, \beta n)$) and the representation

$$\mathcal{S}(\mathbb{R}^d) = \bigcap_{s\ge 0} M_{v_s}^\infty(\mathbb{R}^d)$$

obtained in (2.28) with $p = q = \infty$. □

An extension of Theorem 3.2.35 to the quasi-Banach setting is proved in [248, Theorem 8.3].

Theorem 3.2.37. *Let* $\mu \in \mathcal{M}_v$, $\mathcal{G}(g, \alpha, \beta)$ *be a frame for* $L^2(\mathbb{R}^d)$, *with lattice* $\Lambda = \alpha\mathbb{Z}^d \times \beta\mathbb{Z}^d$, *and* $g \in \mathcal{S}(\mathbb{R}^d)$.

(i) *For every* $0 < p, q \le \infty$, $C_g : M_\mu^{p,q} \to \ell_\mu^{p,q}$ *and* $D_g : \ell_\mu^{p,q} \to M_\mu^{p,q}$ *continuously. If* $f \in M_\mu^{p,q}(\mathbb{R}^d)$, *then the Gabor expansions (3.29) converge unconditionally in* $M_\mu^{p,q}$ *for* $0 < p, q < \infty$, *and weak*-*M_μ^∞ *unconditionally if* $p = \infty$ *or* $q = \infty$.

(ii) *The following (quasi-)norms are equivalent on* $M_\mu^{p,q}(\mathbb{R}^d)$:

$$\|f\|_{M_\mu^{p,q}} \asymp \|C_g f\|_{\ell_\mu^{p,q}}. \tag{3.33}$$

We end up this section by recalling an important result about the invertibility of the Gabor frame operator on $M_v^1(\mathbb{R}^d)$, with v being a submultiplicative weight. For generalizations and the proof of the result below, we refer to [159]; see also [160].

Theorem 3.2.38. *Assume that $m \in \mathcal{M}_v$. Consider a Gabor frame $\mathcal{G}(g,\alpha,\beta)$ with lattice $\Lambda = \alpha\mathbb{Z}^d \times \beta\mathbb{Z}^d$ satisfying $\alpha\beta \in \mathbb{Q}$ and window $g \in M^1_v$. Then the Gabor frame operator $S^{\alpha,\beta}_{g,g}$ is invertible on $M^1_v(\mathbb{R}^d)$. As a consequence, $S^{\alpha,\beta}_{g,g}$ is invertible on all modulation spaces $M^{p,q}_m(\mathbb{R}^d)$, $1 \le p, q \le \infty$, and the canonical dual window $\gamma = (S^{\alpha,\beta}_{g,g})^{-1} g$ belongs to $M^1_v(\mathbb{R}^d)$.*

It is noteworthy to mention the characterization of modulation spaces via Wilson bases in [36, 54, 103, 135, 315].

3.3 Kernel theorems for modulation spaces

Schwartz' kernel theorem certainly is one of the most important results in modern functional analysis. It states, in the framework of tempered distributions, that every linear continuous operator $A : \mathcal{S}(\mathbb{R}^d) \to \mathcal{S}'(\mathbb{R}^d)$ can be regarded as an integral operator in a generalized sense, namely

$$\langle Af, g\rangle = \langle K, g \otimes \bar{f}\rangle, \quad f, g \in \mathcal{S}(\mathbb{R}^d),$$

in the context of distributions, for some kernel $K \in \mathcal{S}'(\mathbb{R}^{2d})$, and vice versa [194]. We write, formally,

$$Af(x) = \int_{\mathbb{R}^d} K(x,y) f(y)\, dy, \quad f \in \mathcal{S}(\mathbb{R}^d). \tag{3.34}$$

This result provides a general framework that includes most of the operators of interest in harmonic analysis. However, for many applications in time–frequency analysis and mathematical signal processing, the spaces $\mathcal{S}(\mathbb{R}^d)$ and $\mathcal{S}'(\mathbb{R}^d)$ can be fruitfully replaced by the Feichtinger's algebra $M^1(\mathbb{R}^d)$ and its dual $M^\infty(\mathbb{R}^d)$. One of the obvious advantages is that $M^1(\mathbb{R}^d)$ and $M^\infty(\mathbb{R}^d)$ are Banach spaces; on the other hand, their norm gives a direct and transparent information on the time–frequency concentration of a function or tempered distribution. The same holds, more generally, for the whole scale of intermediate spaces $M^{p,q}(\mathbb{R}^d)$, $1 \le p, q \le \infty$.

A kernel theorem in the framework of modulation spaces was announced by H. Feichtinger in [119] and proved in [134]; see also [160, Theorem 14.4.1]. It states that linear continuous operators $A : M^1(\mathbb{R}^d) \to M^\infty(\mathbb{R}^d)$ are characterized by the membership of their distributional kernel K to $M^\infty(\mathbb{R}^{2d})$.

Theorem 3.3.1.

(i) *Every distribution $K \in M^\infty(\mathbb{R}^{2d})$ defines a bounded linear operator $A : M^1(\mathbb{R}^d) \to M^\infty(\mathbb{R}^d)$ according to*

$$\langle Af, g\rangle = \langle K, g \otimes \bar{f}\rangle, \quad \forall f, g \in M^1(\mathbb{R}^d),$$

with $\|A\|_{M^1 \to M^\infty} \le \|K\|_{M^\infty}$.

(ii) *For any bounded operator* $A : M^1(\mathbb{R}^d) \to M^\infty(\mathbb{R}^d)$ *there exists a unique kernel* $K \in M^\infty(\mathbb{R}^{2d})$ *such that*

$$\langle Af, g\rangle = \langle K, g \otimes \bar{f}\rangle, \quad \forall f, g \in M^1(\mathbb{R}^d).$$

In this section we provide a similar characterization of linear continuous operators acting on the following spaces:

- $M^1(\mathbb{R}^d) \to M^p(\mathbb{R}^d)$, for a fixed $1 \le p \le \infty$;
- $M^p(\mathbb{R}^d) \to M^\infty(\mathbb{R}^d)$, for a fixed $1 \le p \le \infty$;
- $M^p(\mathbb{R}^d) \to M^p(\mathbb{R}^d)$, for every $1 \le p \le \infty$.

In all cases the characterization is given in terms of the membership of their distributional kernels to certain mixed modulation spaces. In this case we shall use their wider definition contained in (2.3.3). These results are contained in [76].

From now on we assume that c is a permutation of the set $\{1, \dots, 2d\}$. We identify c with the linear bijection $\tilde{c}$ defined in (2.21).

From now on we shall work with the quadratic lattice $\Lambda = \alpha\mathbb{Z}^{2d}$, for suitable $\alpha > 0$ and use the time–frequency shift

$$\pi(\lambda) = M_{\lambda_1} T_{\lambda_2}, \quad \lambda = (\lambda_1, \lambda_2) \in \Lambda.$$

If the window function belongs to the Feichtinger's algebra $M^1(\mathbb{R}^d)$, then a Gabor frame for $L^2(\mathbb{R}^d)$ is a frame also for mixed modulation spaces. The arguments are similar to the classical case contained in Theorem 3.2.35 (cf. [19, Theorem 4.6]).

Theorem 3.3.2. *Let g be a window function in $M^1(\mathbb{R}^d)$ such that $\mathcal{G}(g, \Lambda)$ is a frame for $L^2(\mathbb{R}^d)$ with a dual frame $\mathcal{G}(\gamma, \Lambda)$. Consider $p_1, \dots, p_{2d} \in [1, \infty]$.*

(i) *There exist constants $0 < A \le B$ independent of $p_1, \dots, p_{2d}$ such that*

$$A\|f\|_{M(c)^{p_1,\dots,p_{2d}}} \le \|V_g f \circ \tilde{c}|_\Lambda\|_{\ell^{p_1,\dots,p_{2d}}} \le B\|f\|_{M(c)^{p_1,\dots,p_{2d}}}, \quad \forall f \in M(c)^{p_1,\dots,p_{2d}}.$$

(ii) *We have*

$$f = \sum_{\lambda \in \Lambda} \langle f, \pi(\lambda) g\rangle \pi(\lambda)\gamma = \sum_{\lambda \in \Lambda} \langle f, \pi(\lambda)\gamma\rangle \pi(\lambda) g, \tag{3.35}$$

with unconditional convergence in $M(c)^{p_1,\dots,p_{2d}}$ if $p_1, \dots, p_{2d} \in [1, \infty)$, and weak convergence in $M^\infty(\mathbb{R}^d)$ if $p_1, \dots, p_{2d} \in [1, \infty]$*

Indeed, this generalizes the basic fact, valid for the identity permutation, that the coefficient operator C_g and the reconstruction operator D_γ, here represented by

$$(C_g f)_\lambda = \langle f, \pi(\lambda) g\rangle, \quad D_\gamma a = \sum_{\lambda \in \Lambda} a_\lambda \pi(\lambda)\gamma,$$

satisfy

$$C_g : M^{p,q}(\mathbb{R}^d) \to \ell^{p,q}(\Lambda), \quad D_\gamma : \ell^{p,q}(\Lambda) \to M^{p,q}(\mathbb{R}^d), \tag{3.36}$$

with $D_\gamma C_g = I$, the identity on $M^{p,q}(\mathbb{R}^d)$, cf. (3.35) (similarly, $D_g C_\gamma = I$).

Concerning Theorem 3.3.2, we remark that even when $p_k = \infty$ for some or several indices $k \in \{1, \dots, 2d\}$ one still has unconditional weak* convergence.

Definition 3.3.3. Let g be a window function in $\mathcal{S}(\mathbb{R}^d)$ such that $\mathcal{G}(g, \Lambda)$ is a frame for $L^2(\mathbb{R}^d)$ with a dual frame $\mathcal{G}(\gamma, \Lambda)$, $\gamma \in \mathcal{S}(\mathbb{R}^d)$. Given a linear continuous operator $A : \mathcal{S}(\mathbb{R}^d) \to \mathcal{S}'(\mathbb{R}^d)$, we define the Gabor matrix related to A by

$$K_{\lambda,\mu} = \langle A\pi(\mu)\gamma, \pi(\lambda)g\rangle, \quad \lambda, \mu \in \Lambda. \tag{3.37}$$

For heuristic purposes at least, one can view $K_{\lambda,\mu}$ as an infinite matrix, with indices $\lambda, \mu \in \Lambda$. Properties of the operator A can be inferred from the related Gabor matrix $K_{\lambda,\mu}$, as we shall show in the sequel. Notice that this matrix turns out to be the kernel of the operator $\tilde{A} := C_g A D_\gamma$, acting on sequences on Λ, where C_g and D_γ are the coefficient and synthesis operator, respectively.

First, we need some preliminaries coming from the general theory of operators, see, for instance, Tao's lecture notes [278, Proposition 5.2].

Proposition 3.3.4. *Assume that (X, ν_X), (Y, μ_Y) are measure spaces and $p \in [1, \infty]$. Suppose that $K : Y \times X \to \mathbb{C}$, and $\|K(\cdot, x)\|_{L^p(Y)}$ is uniformly bounded on X. Then the operator A with kernel K maps $L^1(X)$ into $L^p(Y)$ and*

$$\|A\|_{L^1(X)\to L^p(Y)} = \operatorname{ess\,sup}_{x\in X} \|K(\cdot, x)\|_{L^p(Y)}. \tag{3.38}$$

Proof. For any $f \in L^1(X)$, Minkowski's inequality gives

$$\begin{aligned}\Big\| \int_X |K(\cdot, x)f(x)|\, d\nu_X \Big\|_{L^p(Y)} &\le \int_X \big\| |K(\cdot, x)|\, |f(x)| \big\|_{L^p(Y)} d\nu_X \\ &\le \int_X |f(x)| \operatorname{ess\,sup}_{x\in X} \|K(\cdot, x)\|_{L^p(Y)} d\nu_X \\ &= \|f\|_{L^1(X)} \operatorname{ess\,sup}_{x\in X} \|K(\cdot, x)\|_{L^p(Y)} < \infty,\end{aligned}$$

so that we infer $\|A\|_{L^1(X)\to L^p(Y)} \le \operatorname{ess\,sup}_{x\in X} \|K(\cdot, x)\|_{L^p(Y)}$.

To get the converse inequality, we observe from limiting arguments that we may take the kernel K to be a product of simple functions, i. e., a linear combination of tensor products of indicator functions on X and Y, separately. In such a case, we can perform a partition of X into finitely many sets of positive measure, on each of which K is independent of x up to sets of measure zero. On one of these sets, call it B, the supremum of $\|K(\cdot, x)\|_{L^p(Y)}$ is attained. If we then test A against $\chi_{B'}$, for some $B' \subset B$ of finite measure, we obtain the claim. □

Similarly, we have a dual statement (see, e. g., [278, Proposition 5.4]).

Proposition 3.3.5. *Assume that (X, ν_X), (Y, μ_Y) are measure spaces and $p \in [1, \infty]$. Suppose that $K : Y \times X \to \mathbb{C}$, and $\|K(\cdot, x)\|_{L^{p'}(Y)}$ is uniformly bounded on X. Then the operator A with kernel K maps $L^p(X)$ into $L^\infty(Y)$ and*

$$\|A\|_{L^p(X)\to L^\infty(Y)} = \operatorname{ess\,sup}_{x\in X}\|K(\cdot, x)\|_{L^{p'}(Y)}, \quad \frac{1}{p} + \frac{1}{p'} = 1. \tag{3.39}$$

We shall need the following Schur-type test.

Proposition 3.3.6. *Consider an operator defined on sequences on the lattice $\Lambda = \alpha\mathbb{Z}^d \times \beta\mathbb{Z}^d$ by*

$$(Kc)_{m',n'} = \sum_{m,n} K_{m',n',m,n} c_{m,n}.$$

(i) *If $K \in \ell_n^\infty \ell_{n'}^1, \ell_{m'}^\infty \ell_m^1$, K is continuous on $\ell_n^1 \ell_m^\infty$.*
(ii) *If $K \in \ell_{n'}^\infty \ell_n^1 \ell_m^\infty \ell_{m'}^1$, K is continuous on $\ell_n^\infty \ell_m^1$.*
(iii) *If $K \in \ell_n^\infty \ell_{n'}^1, \ell_{m'}^\infty \ell_m^1 \cap \ell_{n'}^\infty \ell_n^1 \ell_m^\infty \ell_{m'}^1$, and moreover $K \in \ell_{m',n'}^\infty \ell_{m,n}^1 \cap \ell_{m,n}^\infty \ell_{m',n'}^1$, the operator K is continuous on $\ell^{p,q} = \ell_n^q \ell_m^p$ for every $1 \le p, q \le \infty$.*
(iv) *Assume the hypotheses in (iii). Then K is continuous on all $\tilde{\ell}^{p,q}$, $1 \le p, q \le \infty$, where $\tilde{\ell}^{p,q}$ is the closure of the space of eventually zero sequences in $\ell^{p,q}$.*

Proof. (i) We have

$$\begin{aligned}
\|Kc\|_{\ell_{n'}^1, \ell_{m'}^\infty} &\le \sum_{n'} \sup_{m'} \sum_{m,n} |K_{m',n',m,n}||c_{m,n}| \\
&\le \sum_n \Big(\sum_{n'} \sup_{m'} \sum_m |K_{m',n',m,n}| \Big) \sup_m |c_{m,n}| \\
&\le \|K\|_{\ell_n^\infty \ell_{n'}^1, \ell_{m'}^\infty \ell_m^1} \|c\|_{\ell^1 \ell^\infty}.
\end{aligned}$$

(ii) It turns out that

$$\begin{aligned}
\|Kc\|_{\ell_{n'}^\infty \ell_{m'}^1} &\le \sup_{n'} \sum_{m'} \Big(\sum_{m,n} |K_{m',n',m,n}||c_{m,n}| \Big) \\
&\le \sup_{n'} \sum_n \Big(\sup_m \sum_{m'} |K_{m',n',m,n}| \sum_m |c_{m,n}| \Big) \\
&\le \|K\|_{\ell_{n'}^\infty \ell_n^1 \ell_m^\infty \ell_{m'}^1} \|c\|_{\ell^\infty \ell^1}.
\end{aligned}$$

(iii) Since the statement holds for $p = q$ by the classical Schur's test, and for $(p, q) = (1, \infty)$ and $(p, q) = (\infty, 1)$, by parts (i) and (ii), it follows by complex interpolation (see (3) on page 128 and (15) on page 134 of [296]) that it holds for all (p, q), except possibly in the cases $q = \infty$, $1 < p < \infty$. For these cases we argue by duality as follows.

In order to prove the continuity of K on $\ell^\infty \ell^p$, it suffices to verify that for any sequences $c = (c_{m,n}) \in \ell_n^\infty \ell_m^p$ and $d = (d_{m',n'}) \in \ell_{n'}^1 l_{m'}^{p'}$, with $d_{m',n'} \ge 0$, we have

$$\sum_{m',n'} |(Kc)_{m',n'}| d_{m',n'} \lesssim \|c\|_{\ell_n^\infty \ell_m^p} \|d\|_{\ell_n^1 \ell_m^{p'}}. \tag{3.40}$$

Now

$$\begin{aligned}\sum_{m',n'} |(Kc)_{m',n'}|d_{m',n'} &\le \sum_{m',n'}\sum_{m,n} |K_{m',n',m,n}||c_{m,n}|d_{m',n'}\\ &= \sum_{m,n}\Big(\sum_{m',n'} |K_{m',n',m,n}|d_{m',n'}\Big)|c_{m,n}|\\ &\le \|c\|_{\ell^\infty\ell^p}\|\tilde{K}d\|_{\ell^1\ell^{p'}},\end{aligned}$$

where $\tilde{K}$ is the operator with matrix kernel $\tilde{K}_{m,n,m',n'} = |K_{m',n',m,n}|$. Since it satisfies the same assumptions as K, it is continuous on $\ell^1\ell^{p'}$, which gives (3.40).

(iv) Since K is continuous on $\ell^{p,q}$ and by the definition of $\tilde{\ell}^{p,q}$, it suffices to verify that K maps every eventually zero sequence in $\tilde{\ell}^{p,q}$. This follows from the fact that K maps every eventually zero sequence in $\ell^1 \hookrightarrow \tilde{\ell}^{p,q}$, because K is bounded on ℓ^1. □

Let us introduce the permutation c_1 of the set $\{1,\dots,4d\}$ with related $\tilde{c}_1$ given by, for $x_i \in \mathbb{R}^d$, $i = 1,\dots,4$,

$$\tilde{c}_1(x_1,x_2,x_3,x_4) = \begin{pmatrix} I_d & 0_d & 0_d & 0_d\\ 0_d & 0_d & I_d & 0_d\\ 0_d & I_d & 0_d & 0_d\\ 0_d & 0_d & 0_d & I_d\end{pmatrix}\begin{pmatrix}x_1\\x_2\\x_3\\x_4\end{pmatrix} = (x_1,x_3,x_2,x_4). \tag{3.41}$$

We consider again lattices of the form $\Lambda = \alpha\mathbb{Z}^{2d}$.

Theorem 3.3.7. *Suppose $p \in [1,\infty]$. Let g be a window function in $\mathcal{S}(\mathbb{R}^d)$ such that $\mathcal{G}(g,\Lambda)$ is a frame for $L^2(\mathbb{R}^d)$ with a dual frame $\mathcal{G}(\gamma,\Lambda)$. A linear continuous operator $A : \mathcal{S}(\mathbb{R}^d) \to \mathcal{S}'(\mathbb{R}^d)$ is bounded $M^1(\mathbb{R}^d) \to M^p(\mathbb{R}^d)$ if and only if its distributional kernel K satisfies*

$$V_G K \circ \tilde{c}_1|_{\Lambda\times\Lambda} \in \ell^{p,\infty}(\Lambda\times\Lambda), \tag{3.42}$$

where $G = g \otimes \bar{\gamma}$. Moreover,

$$\|A\|_{M^1\to M^p} \asymp \|V_G K \circ \tilde{c}_1\|_{\ell^{p,\infty}(\Lambda\times\Lambda)}.$$

Proof. Assume first that (3.42) holds. For $\lambda = (\lambda_1,\lambda_2), \mu = (\mu_1,\mu_2) \in \Lambda$, observe that the Gabor matrix of the operator A in (3.37) is given by

$$\begin{aligned}K_{\lambda,\mu} &= \langle A\pi(\mu)\gamma, \pi(\lambda)g\rangle\\ &= \langle K, \pi(\lambda_1,\lambda_2)g \otimes \pi(\mu_1,-\mu_2)\bar{\gamma}\rangle\\ &= V_G K \circ \tilde{c}_1(\lambda_1,\lambda_2,\mu_1,-\mu_2).\end{aligned} \tag{3.43}$$

The operator $\tilde{A} := C_g A D_\gamma$ with kernel $K_{\lambda,\mu}$ acts on sequences on Λ. By (3.42) and Proposition 3.3.4, $\tilde{A}$ is a bounded operator $\ell^1(\Lambda) \to \ell^p(\Lambda)$. Since $C_g : M^p(\mathbb{R}^d) \to \ell^p(\Lambda)$ and $D_\gamma : \ell^p(\Lambda) \to M^p(\mathbb{R}^d)$ are continuous, we have $A = D_\gamma \tilde{A} C_g : M^1(\mathbb{R}^d) \to M^p(\mathbb{R}^d)$, with

$$\|A\|_{M^1 \to M^p} \le C \sup_{\mu \in \Lambda} \|K_{\cdot,\mu}\|_{\ell^p(\Lambda)}.$$

Conversely, consider the finite dimensional space

$$E_N = \{(a_\lambda)_{\lambda \in \Lambda} : a_\lambda = 0 \quad \text{for } |\lambda| > N\}$$

and the natural projection $P_N : \ell^\infty(\Lambda) \to E_N$. The composition operator

$$P_N C_g A D_\gamma|_{E_N} : (E_N, \|\cdot\|_{\ell^1}) \to (E_N, \|\cdot\|_{\ell^p})$$

is represented by the (finite) matrix

$$\{K_{\lambda,\mu}\}_{\lambda,\mu \in \Lambda, |\lambda| \le N, |\mu| \le N}.$$

Using formula (3.38) in Proposition 3.3.4 (we work on finite sets, so that the assumptions are trivially satisfied), we have

$$\sup_{\mu \in \Lambda, |\mu| \le N} \|K_{\lambda,\mu}\|_{\ell^p(\Lambda, |\lambda| \le N)} = \|P_N C_g A D_\gamma|_{E_N}\|_{\ell^1 \to \ell^p} \le \|C_g\| \|D_\gamma\| \|A\|_{M^1 \to M^p}.$$

Since the right-hand side is independent of N, taking the supremum with respect to N, we obtain

$$\sup_{\mu \in \Lambda} \|K_{\cdot,\mu}\|_{\ell^p(\Lambda)} \le C \|A\|_{M^1 \to M^p}.$$

This concludes the proof. □

Observe that, given a window function γ such that the Gabor system $\mathcal{G}(\gamma, \Lambda)$ is frame for $L^2(\mathbb{R}^d)$, and replacing γ by its complex conjugate $\bar{\gamma}$, the Gabor system $\mathcal{G}(\bar{\gamma}, \Lambda)$ is still a frame. We choose now a frame $\mathcal{G}(g, \Lambda)$ for $L^2(\mathbb{R}^d)$, with dual frame $\mathcal{G}(\gamma, \Lambda)$. Then $\mathcal{G}(g \otimes \bar{\gamma}, \Lambda \times \Lambda)$ is a frame for $L^2(\mathbb{R}^{2d})$.

Using Theorem 3.3.2, we reformulate the previous result in terms of mixed modulation spaces.

Corollary 3.3.8. *Suppose $p \in [1, \infty]$. A linear continuous operator $A : \mathcal{S}(\mathbb{R}^d) \to \mathcal{S}'(\mathbb{R}^d)$ is bounded from $M^1(\mathbb{R}^d)$ into $M^p(\mathbb{R}^d)$ if and only if its distributional kernel K satisfies*

$$K \in M(c_1)^{p,\infty}, \tag{3.44}$$

where c_1 is identified with (3.41).

Moreover

$$\|A\|_{M^1 \to M^p} \asymp \|K\|_{M(c_1)^{p,\infty}}.$$

Remark 3.3.9. When $p = \infty$, condition (3.44) becomes $K \in M^\infty(\mathbb{R}^{2d})$, and we recapture Feichtinger's kernel Theorem (Theorem 3.3.1).

A similar characterization of bounded operators $A : M^p(\mathbb{R}^d) \to M^\infty(\mathbb{R}^d)$ can be obtained arguing as above, using Proposition 3.3.5, (3.43) and Theorem 3.3.2. Precisely, define the permutation c_2 of the set $\{1, \dots, 4d\}$ with related $\tilde{c}_2$ by

$$\tilde{c}_2(x_1, x_2, x_3, x_4) = \begin{pmatrix} 0_d & 0_d & I_d & 0_d \\ I_d & 0_d & 0_d & 0_d \\ 0_d & 0_d & 0_d & I_d \\ 0_d & I_d & 0_d & 0_d \end{pmatrix} \begin{pmatrix} x_1 \\ x_2 \\ x_3 \\ x_4 \end{pmatrix} = (x_3, x_1, x_4, x_2), \tag{3.45}$$

for $x_i \in \mathbb{R}^d$, $i = 1, \dots, 4$.

Theorem 3.3.10. *Suppose $p \in [1, \infty]$. A linear continuous operator $A : \mathcal{S}(\mathbb{R}^d) \to \mathcal{S}'(\mathbb{R}^d)$ is bounded from $M^p(\mathbb{R}^d)$ into $M^\infty(\mathbb{R}^d)$ if and only if its distributional kernel K satisfies*

$$K \in M(c_2)^{p', \infty},$$

where $1/p + 1/p' = 1$.

Moreover,

$$\|A\|_{M^p \to M^\infty} \asymp \|K\|_{M(c_2)^{p',\infty}}.$$

As a consequence of Theorem 3.3.7 and Corollary 3.3.8, we have the following result.

Corollary 3.3.11. *Let $A : \mathcal{S}(\mathbb{R}^d) \to \mathcal{S}'(\mathbb{R}^d)$ be an operator with distributional kernel $K \in \mathcal{S}'(\mathbb{R}^{2d})$. The following conditions are equivalent:*

(i) *$A : M^p(\mathbb{R}^d) \to M^p(\mathbb{R}^d)$, bounded for every $1 \le p \le \infty$.*

(ii) *$A : M^1(\mathbb{R}^d) \to M^1(\mathbb{R}^d)$ and $A : M^\infty(\mathbb{R}^d) \to M^\infty(\mathbb{R}^d)$ bounded.*

(iii) *$K \in M(c_1)^{1,\infty} \cap M(c_2)^{1,\infty}$.*

Proof. All the equivalences follow trivially from Theorems 3.3.7 and 3.3.10 but (ii) $\Longrightarrow$ (i), which follows by complex interpolation for modulation spaces, see Proposition 2.3.17. □

3.3.1 Operators acting on $M^{p,q}$

Let us now consider the problem of boundedness of an integral operator on $M^{p,q}(\mathbb{R}^d)$, $1 \le p, q \le \infty$. We no longer expect a characterization as in the case $p = q$, studied in the previous section (see Remark 3.3.13 below). Nevertheless, sufficient conditions can be inferred.

Consider the permutations c_3 and c_4 of the set $\{1,\dots,4d\}$ with related $\tilde{c}_3, \tilde{c}_4$ defined by

$$\tilde{c}_3(x_1,x_2,x_3,x_4) = \begin{pmatrix} 0_d & I_d & 0_d & 0_d \\ 0_d & 0_d & I_d & 0_d \\ I_d & 0_d & 0_d & 0_d \\ 0_d & 0_d & 0_d & I_d \end{pmatrix} \begin{pmatrix} x_1 \\ x_2 \\ x_3 \\ x_4 \end{pmatrix} = (x_2,x_3,x_1,x_4) \tag{3.46}$$

and

$$\tilde{c}_4(x_1,x_2,x_3,x_4) = \begin{pmatrix} I_d & 0_d & 0_d & 0_d \\ 0_d & 0_d & 0_d & I_d \\ 0_d & I_d & 0_d & 0_d \\ 0_d & 0_d & I_d & 0_d \end{pmatrix} \begin{pmatrix} x_1 \\ x_2 \\ x_3 \\ x_4 \end{pmatrix} = (x_1,x_4,x_2,x_3), \tag{3.47}$$

for $x_i \in \mathbb{R}^d$, $i = 1,\dots,4$.

Proposition 3.3.6 can be rephrased in this framework as follows.

Proposition 3.3.12. *Consider an at most countable index set J and the operator defined on sequences $(a_{\lambda_1,\lambda_2})$ on the set $\Lambda = J \times J$ by*

$$(Aa)_{\lambda_1,\lambda_2} = \sum_{\mu_1,\mu_2 \in J} K_{\lambda_1,\lambda_2,\mu_1,\mu_2} a_{\mu_1,\mu_2}.$$

(i) *If $K \circ \tilde{c}_3 \in \ell^{1,\infty,1,\infty}(J^4)$, then A is continuous on $\ell^{\infty,1}(\Lambda)$. Moreover,*

$$\|A\|_{\ell^{\infty,1} \to \ell^{\infty,1}} \le \|K \circ \tilde{c}_3\|_{\ell^{1,\infty,1,\infty}}. \tag{3.48}$$

(ii) *If $K \circ \tilde{c}_4 \in \ell^{1,\infty,1,\infty}(J^4)$, then A is continuous on $\ell^{1,\infty}(\Lambda)$, with*

$$\|A\|_{\ell^{1,\infty} \to \ell^{1,\infty}} \le \|K \circ \tilde{c}_4\|_{\ell^{1,\infty,1,\infty}}. \tag{3.49}$$

(iii) *If K satisfies assumptions* (i) *and* (ii) *and, moreover, $K \circ c_0, K \in \ell^{1,\infty}(\Lambda \times \Lambda)$, where the permutation c_0 is defined in* (2.23), *then the operator A is continuous on $\ell^{p,q}(\Lambda)$, for every $1 \le p,q \le \infty$.*

Remark 3.3.13. Observe that the reverse inequalities in (3.48), (3.49) do not hold, even if one allows a multiplicative constant. Consider, for example, the estimate (3.48), with $J = \{0,1,\dots,N-1\}$, $N \in \mathbb{N}$, where

$$K_{\lambda_1,\lambda_2,\mu_1,\mu_2} = F_{\lambda_2,\mu_1} = \frac{1}{\sqrt{N}} e^{-2\pi i \lambda_2 \mu_1}$$

is the Fourier matrix. We have

$$(K \circ \tilde{c}_3)_{\mu_1,\lambda_1,\lambda_2,\mu_2} = K_{\lambda_1,\lambda_2,\mu_1,\mu_2}$$

so that

$$\|K \circ \tilde{c}_3\|_{\ell^{1,\infty,1,\infty}} = \|F\|_{\ell^1} = N^{3/2}$$

but

$$\|A\|_{\ell^{\infty,1} \to \ell^{\infty,1}} \leq N, \tag{3.50}$$

which blows up, as $N \to +\infty$, at a lower rate.

Let us verify (3.50). We have

$$\begin{aligned}\left\| \left\| \sum_{\mu_1,\mu_2 \in J} K_{\lambda_1,\lambda_2,\mu_1,\mu_2} a_{\mu_1,\mu_2} \right\|_{\ell^\infty_{\lambda_1}} \right\|_{\ell^1_{\lambda_2}} &= \left\| \sum_{\mu_1 \in J} F_{\lambda_2,\mu_1} \sum_{\mu_2 \in J} a_{\mu_1,\mu_2} \right\|_{\ell^1_{\lambda_2}} \\ &= \left\| \sum_{\mu_1 \in J} F_{\lambda_2,\mu_1} \tilde{a}_{\mu_1} \right\|_{\ell^1_{\lambda_2}}\end{aligned}$$

with $\tilde{a}_{\mu_1} := \sum_{\mu_2 \in J} a_{\mu_1,\mu_2}$. Now, using the embeddings

$$\|b\|_{\ell^1(J)} \leq \sqrt{N}\|b\|_{\ell^2(J)}, \quad \|b\|_{\ell^2(J)} \leq \sqrt{N}\|b\|_{\ell^\infty(J)}$$

and the fact that F is a unitary transformation of $\ell^2(J)$, we get (3.50).

This Schur-type test yields the following boundedness result.

Consider the permutations c_5 and c_6 of the set $\{1, \dots, 4d\}$ with related $\tilde{c}_5$, $\tilde{c}_6$ defined by

$$\tilde{c}_5 = \tilde{c}_1\tilde{c}_3 = \begin{pmatrix} 0_d & I_d & 0_d & 0_d \\ I_d & 0_d & 0_d & 0_d \\ 0_d & 0_d & I_d & 0_d \\ 0_d & 0_d & 0_d & I_d \end{pmatrix} \tag{3.51}$$

and

$$\tilde{c}_6 = \tilde{c}_1\tilde{c}_4 = \begin{pmatrix} I_d & 0_d & 0_d & 0_d \\ 0_d & I_d & 0_d & 0_d \\ 0_d & 0_d & 0_d & I_d \\ 0_d & 0_d & I_d & 0_d \end{pmatrix}. \tag{3.52}$$

Theorem 3.3.14.

(i) *A linear continuous operator $A : \mathcal{S}(\mathbb{R}^d) \to \mathcal{S}'(\mathbb{R}^d)$ is bounded on $M^{\infty,1}(\mathbb{R}^d)$ if its distributional kernel K satisfies*

$$K \in M(c_5)^{1,\infty,1,\infty}.$$

(ii) *A linear continuous operator $A : \mathcal{S}(\mathbb{R}^d) \to \mathcal{S}'(\mathbb{R}^d)$ is bounded on $M^{1,\infty}(\mathbb{R}^d)$ if its distributional kernel K satisfies*

$$K \in M(c_6)^{1,\infty,1,\infty}.$$

Proof. The result is a consequence of (3.43), Proposition 3.3.12 and Theorem 3.3.2, by using the same arguments as in the first part of the proof of Theorem 3.3.7. □

By interpolation we also obtain the following result.

Corollary 3.3.15. *Let $A : \mathcal{S}(\mathbb{R}^d) \to \mathcal{S}'(\mathbb{R}^d)$ be an operator with distributional kernel*

$$K \in M(c_5)^{1,\infty,1,\infty} \cap M(c_6)^{1,\infty,1,\infty}.$$

Then $A : M^{p',p}(\mathbb{R}^d) \to M^{p',p}(\mathbb{R}^d)$ continuously, for every $1 \le p \le \infty$, with $1/p + 1/p' = 1$.

By combining Corollaries 3.3.11 and 3.3.15, we finally obtain the following result.

Corollary 3.3.16. *Let $A : \mathcal{S}(\mathbb{R}^d) \to \mathcal{S}'(\mathbb{R}^d)$ be an operator with distributional kernel*

$$K \in M(c_1)^{1,\infty} \cap M(c_2)^{1,\infty} \cap M(c_5)^{1,\infty,1,\infty} \cap M(c_6)^{1,\infty,1,\infty}.$$

Then $A : M^{p,q}(\mathbb{R}^d) \to M^{p,q}(\mathbb{R}^d)$ continuously, for every $1 \le p, q \le \infty$.

4 Pseudodifferential operators

Let us first introduce pseudodifferential operators in the form arising as a generalization of the variable-coefficient differential operators

$$\sum_{|\alpha|\leq N} c_\alpha(x)\partial^\alpha,$$

for some $N \in \mathbb{N}_+$, where we recall that $\alpha = (\alpha_1,\dots,\alpha_d)$ is a multiindex in $\mathbb{N}^d$ with length $|\alpha| = \alpha_1 + \cdots + \alpha_d \leq N$. The coefficients $c_\alpha(x)$ belong to suitable function spaces. We start with the Fourier inversion formula for a suitable function f, say, in $\mathcal{S}(\mathbb{R}^d)$:

$$f(x) = \int_{\mathbb{R}^d} e^{2\pi i x\cdot\xi}\hat{f}(\xi)\,d\xi$$

and apply the monomial differential operator ∂_x^α

$$\partial_x^\alpha f(x) = \int_{\mathbb{R}^d} (2\pi i\xi)^\alpha e^{2\pi i x\cdot\xi}\hat{f}(\xi)\,d\xi,$$

where we recall the notation $\xi^\alpha = \xi_1^{\alpha_1}\cdots\xi_d^{\alpha_d}$, for any $\xi = (\xi_1,\dots,\xi_d) \in \mathbb{R}^d$. Thus we can write

$$\sum_{|\alpha|\leq N} c_\alpha(x)\partial^\alpha f(x) = \int_{\mathbb{R}^d} a(x,\xi)e^{2\pi i x\cdot\xi}\hat{f}(\xi)\,d\xi,$$

where the function

$$a(x,\xi) = \sum_{|\alpha|\leq N} c_\alpha(x)(2\pi i\xi)^\alpha$$

is called the *symbol* of the previous operator. Inspired by this, we give the following definition.

Definition 4.0.1 (Kohn–Nirenberg operator). Given a function or distribution a in $\mathbb{R}^{2d}$, let us first assume $a \in \mathcal{S}(\mathbb{R}^{2d})$ and define the pseudodifferential operator $a(x,D)$, acting initially on $\mathcal{S}(\mathbb{R}^d)$, as follows:

$$a(x,D)f(x) = \int_{\mathbb{R}^d} e^{2\pi i x\cdot\xi}a(x,\xi)\hat{f}(\xi)\,d\xi, \quad f \in \mathcal{S}(\mathbb{R}^d).$$

This kind of pseudodifferential operator is called *Kohn–Nirenberg operator* with *Kohn–Nirenberg symbol a*.

We are going to exhibit the connection between a pseudodifferential operator and its related time–frequency distribution. First, recall the Rihaczek distribution defined in (1.118):

$$W_0(f,g)(x,\xi) = R(f,g)(x,\xi) = e^{-2\pi i x\cdot\xi}f(x)\overline{\hat{g}(\xi)}, \quad f,g \in \mathcal{S}(\mathbb{R}^d).$$

https://doi.org/10.1515/9783110532456-005

Then it is straightforward to see that

$$\langle a, R(f,g)\rangle = \langle a(x,D)f, g\rangle, \tag{4.1}$$

for a symbol $a \in \mathcal{S}(\mathbb{R}^{2d})$ and functions $f, g \in \mathcal{S}(\mathbb{R}^d)$, where the brackets $\langle\cdot,\cdot\rangle$ denote possibly the extension to $\mathcal{S}' \times \mathcal{S}$ of the inner product on L^2. The proof is just a matter of changing variables and interchanging the integration order.

Recall that if $f, g \in \mathcal{S}(\mathbb{R}^d)$, the Rihaczek distribution $R(f,g)$ is in $\mathcal{S}(\mathbb{R}^{2d})$ (cf. Corollary 1.3.26), this implies that the left-hand side in (4.1) is well defined for any $a \in \mathcal{S}'(\mathbb{R}^{2d})$ and therefore $a(x,D)f \in \mathcal{S}'(\mathbb{R}^d)$, that is, $a(x,D)$ is a linear operator mapping $\mathcal{S}(\mathbb{R}^d)$ into $\mathcal{S}'(\mathbb{R}^d)$.

This shows that we can always study the action of a pseudodifferential operator by means of its symbol and a related time–frequency representation. We shall see in the sequel that the time–frequency distributions studied in Chapter 1 give rise to other particular types of pseudodifferential operators.

Remark 4.0.2. Formally, if we apply $a(x,D)$ to a plane wave $f(x) := e^{2\pi i x\cdot\eta}$, we see that

$$a(x,D)f(x) = a(x,\eta)f(x).$$

Thus we can (in principle) reconstruct the symbol a from the operator $a(x,D)$ by testing it against plane waves.

If the symbol depends only on x, that is, $a(x,\xi) = a(x)$, then

$$a(x,D)f(x) = \int_{\mathbb{R}^d} e^{2\pi i x\cdot\xi} a(x)\hat{f}(\xi)\,d\xi = a(x)f(x), \quad f \in \mathcal{S}(\mathbb{R}^d),$$

that is a multiplication operator. In particular, if $a(x,\xi) \equiv C$, for a suitable complex constant, then $a(x,D) = CI$, a multiple of the identity operator.

If the symbol depends only on ξ, that is, $a(x,\xi) = a(\xi)$, then

$$a(x,D)f(x) = \int_{\mathbb{R}^d} e^{2\pi i x\cdot\xi} a(\xi)\hat{f}(\xi)\,d\xi = \mathcal{F}^{-1}(a\hat{f})(\xi), \quad f \in \mathcal{S}(\mathbb{R}^d),$$

which is a *Fourier multiplier*. In other words, it can be regarded as a convolution operator, $a(x,D)f = h * f$, with $h = \mathcal{F}^{-1}a$.

Let us now write a pseudodifferential operator as an integral operator with kernel K. We first introduce the linear change of coordinates for measurable functions F on $\mathbb{R}^{2d}$:

$$\mathfrak{T}_b F(x,y) = F(x, y-x) \quad x,y \in \mathbb{R}^d. \tag{4.2}$$

Let $\mathcal{F}_2$ be the partial Fourier transform with respect to the second variable, already defined in (1.42); then for $f \in \mathcal{S}(\mathbb{R}^d)$ and $a \in \mathcal{S}(\mathbb{R}^{2d})$,

$$
\begin{aligned}
a(x,D)f(x) &= \int_{\mathbb{R}^d} e^{2\pi i x\cdot\xi} a(x,\xi)\hat{f}(\xi)\,d\xi \\
&= \int_{\mathbb{R}^d}\Big(\int_{\mathbb{R}^d} e^{-2\pi i(y-x)\cdot\xi} a(x,\xi)d\xi\Big) f(y)\,dy \\
&= \int_{\mathbb{R}^d} K(x,y)f(y)\,dy
\end{aligned}
$$

where we used $\hat{f}(\xi) = \int_{\mathbb{R}^d} f(y)e^{-2\pi i y\cdot\xi}\,dy$ and interchanged the order of integration. Hence the Kohn–Nirenberg operator $a(x,D)$ is an integral operator with kernel

$$K(x,y) = \mathcal{F}_2 a(x,y-x) = \mathfrak{T}_b\mathcal{F}_2 a(x,y). \tag{4.3}$$

Notice that $\mathfrak{T}_b$ and $\mathcal{F}_2$ are topological isomorphisms on $\mathcal{S}(\mathbb{R}^{2d})$, $\mathcal{S}'(\mathbb{R}^{2d})$, and $L^2(\mathbb{R}^{2d})$, as well as on many other function spaces (see in the sequel). The use of the integral expression of $a(x,D)$, as an alternative to (4.1), depends on the problem we are dealing with.

Let us give a short historical survey, with particular reference to boundedness problems. Pseudodifferential operators were first introduced (in the so-called Weyl form) in the framework of quantum mechanics. In fact, in 1931 this form was introduced by Hermann Weyl [309] as a *quatization rule*, that is, roughly speaking, a mapping which to any observable $a(x,\xi)$ on the phase space $\mathbb{R}^{2d}$ assigns an operator $\mathrm{Op}_w(a)$ on a Hilbert space, see the next Definition 4.0.3.

Besides quantum mechanics, they arise in the framework of the partial differential equations (PDEs) and engineering. In the context of PDEs they were introduced independently in [193] and [213]. Since then many symbol classes have been considered, according to several applications to PDEs. In particular, a deep analysis has been carried on for Hörmander's classes $S^m_{\rho,\delta}$, $m \in \mathbb{R}$, $0 \le \delta \le \rho \le 1$, of smooth functions $a(x,\xi)$ satisfying the estimates $|\partial_x^\alpha\partial_\xi^\beta a(x,\xi)| \le C_{\alpha,\beta}(1+|\xi|)^{m+\delta|\alpha|-\rho|\beta|}$.

Boundedness results on L^p-based Sobolev spaces for the corresponding pseudodifferential operators are of special interest because they imply regularity results for the solutions of the related PDEs.

The basic result is the boundedness on L^2 of operators in the above classes, with $\delta < \rho$, $m = 0$, which can be achieved by means of the symbolic calculus. In fact, L^2-boundedness still holds for $0 \le \delta = \rho < 1$, which is the classical Calderón–Vaillancourt theorem [40]. It is actually sufficient to suppose that the estimates on a are valid for a bounded number of derivatives, in particular, for $\rho = \delta = 0$, the assumption $a \in C_b^{2d+1}(\mathbb{R}^{2d})$ grants L^2-continuity.

Boundedness on L^p, $1 < p < \infty$, holds for symbols in $S^0_{1,\delta}$, $0 \le \delta < 1$, but generally fails for $\rho < 1$ and a loss of derivatives may then occur. We have no time to enter this topic and refer the interested reader to the works [172, 192, 197] and textbooks [193, 265].

There are also many results for symbols which are smooth and behave as usual with respect to ξ, but less regular with respect to x, e. g., just belonging to some Hölder class. In this connection see the books [280, 281], where important applications to nonlinear equations are presented as well.

Hence, the smoothness of the symbol or the boundedness of all derivatives of the symbol is not necessary for the boundedness of pseudodifferential operators on $L^2(\mathbb{R}^d)$. Being motivated by this argument, many authors (see, e. g., [55, 93, 207, 231]) have investigated the minimal assumption on the regularity of symbols for the corresponding operators to be bounded on L^2. In particular, Sugimoto [273] showed that symbols in the Besov space $B^{(\infty,\infty),(1,1)}_{d/2,d/2}$ imply L^2-boundedness (see also [272] and the references therein for extensions to the L^p framework). In 1994–1995 Sjöstrand introduced a new symbol class, larger than $S^0_{0,0}$, which was then recognized to be the modulation space $M^{\infty,1}(\mathbb{R}^d)$, first introduced in time–frequency analysis by Feichtinger [122, 120, 126]. In [266, 267] Sjöstrand proved that symbols in $M^{\infty,1}$ give rise to L^2-bounded operators. In view of the inclusion $C_b^{2d+1} \subset M^{\infty,1}$, this result represented an important generalization of the classical Calderón–Vaillancourt theorem. Since then several extensions appeared. In particular, in [160, 163], symbols in $M^{\infty,1}$ were proved to produce bounded operators on all $M^{p,q}$. A nice survey of results on the boundedness and the Schatten-class property of pseudodifferential operators using a time–frequency approach is for sure the chapter by Gröchenig [161]. Soon after Gröchenig and Strohmer [168] introduced the $M^{\infty,1}$ class of symbols of Sjöstrand type for pseudodifferential operators on a locally compact Abelian (LCA) group $\mathcal{G}$ and proved that such operators are continuous on $L^2(\mathcal{G})$, extending the $\mathbb{R}^d$ case to LCA groups.

Further refinements appeared in [162, 216, 286]. The case of weighted modulation spaces and Schatten–von Neumann properties were treated in [180, 287, 288]. Here we present more general results for boundedness on L^p spaces that were shown in [74]. The pseudodifferential operators considered there are in the Weyl form.

Definition 4.0.3 (Weyl operators). For $a \in \mathcal{S}(\mathbb{R}^{2d})$, we define the Weyl operator $\mathrm{Op}_w(a)$ with Weyl symbol a as

$$\mathrm{Op}_w(a)f(x) = \int_{\mathbb{R}^{2d}} e^{2\pi i(x-y)\cdot\xi} a\left(\frac{x+y}{2}, \xi\right) f(y)\,dy\,d\xi, \quad f \in \mathcal{S}(\mathbb{R}^d). \tag{4.4}$$

Such an operator is also known as Weyl transform [317]. For $f \in \mathcal{S}(\mathbb{R}^d)$ and $a \in \mathcal{S}(\mathbb{R}^{2d})$ (or in other suitable function spaces),

$$\begin{aligned}\mathrm{Op}_w(a)f(x) &= \int_{\mathbb{R}^{2d}} e^{2\pi i(x-y)\cdot\xi} a\left(\frac{x+y}{2}, \xi\right) f(y)\,dy\,d\xi \\ &= \int_{\mathbb{R}^d}\left(\int_{\mathbb{R}^d} e^{-2\pi i(y-x)\cdot\xi} a\left(\frac{x+y}{2}, \xi\right) d\xi\right) f(y)\,dy \\ &= \int_{\mathbb{R}^d} K(x,y) f(y)\,dy\end{aligned}$$

where

$$K = \mathfrak{T}_s^{-1}\mathcal{F}_2^{-1}a \tag{4.5}$$

and $\mathfrak{T}_s^{-1}$ is the isomorphism on $\mathcal{S}(\mathbb{R}^{2d})$ (or $L^2(\mathbb{R}^{2d})$, $\mathcal{S}'(\mathbb{R}^{2d})$ or other suitable space), defined in (1.88). Hence a Weyl operator is an integral operator with kernel K in (4.5). An easy consequence of a change of variables and an interchange of integration order in the integrals gives

$$\langle \mathrm{Op}_w(a)f, g\rangle = \langle a, W(g,f)\rangle, \quad f,g \in \mathcal{S}(\mathbb{R}^d). \tag{4.6}$$

Consequently, the time–frequency representation related to a Weyl operator is the Wigner distribution (cf. (1.79)).

Observe that from (4.3) and (4.5) we infer the equality $a(x,D) = \mathrm{Op}_w(\sigma)$, where the relations between σ, a and the kernel K of $a(x,D)$ are given by

$$\sigma = \mathcal{F}_2\mathfrak{T}_s K, \quad a = \mathcal{U}\sigma, \quad \sigma = \mathcal{U}^{-1}a, \tag{4.7}$$

with $\mathcal{F}_2$ being the partial Fourier transform in the second variable, $\mathfrak{T}_s$ the symmetric coordinate transformation defined in (1.87), and $\mathcal{U}$ defined by

$$\widehat{(\mathcal{U}\sigma)}(\xi_1,\xi_2) = e^{\pi i \xi_1\cdot\xi_2}\hat{\sigma}(\xi_1,\xi_2). \tag{4.8}$$

Let us describe the contents of the next sections. We first discuss some generalizations of the results by Sjöstrand.

A natural question is which modulation spaces $M^{p,q}$ give rise to symbols of bounded operators on L^p, $p \neq 2$. We provide in Section 4.1 a complete answer to this problem in the more general framework of the Wiener amalgam spaces $W(L^r, L^s)$ (see Corollary 4.1.6 and Proposition 4.1.15 below).

Section 4.2 is concerned with the following related question. Examples show that symbols merely on L^∞ generally do not produce bounded operators in L^2, but some additional regularity condition should be assumed. The above Sjöstrand's result is just an instance of this. There is a space larger than $M^{\infty,1}$ which still consists of bounded functions having locally the same regularity as a function whose Fourier transform is integrable. It is the modulation space $W(\mathcal{F}L^1, L^\infty)$, the subspace of tempered distributions f such that the function $\mathcal{F}(\overline{g(\cdot - x)}f)(\xi)$ belongs to $L_x^\infty L_\xi^1$, see (2.70). A natural question which arises is whether pseudodifferential operators with symbols in $W(\mathcal{F}L^1, L^\infty)$ are L^2-bounded. Fourier multipliers with symbols in $W(\mathcal{F}L^1, L^\infty)$ are, in fact, bounded on L^2 and M^p, and the same holds (more generally) for symbols in $W(\mathcal{F}L^1, L^\infty)$ of the type $a(x,\xi) = a_1(x)a_2(\xi)$ (see Proposition 4.2.2 below). However, contrary to what these special cases could suggest, we shall show in Proposition 4.2.4 that, for more general symbols in that class, boundedness on L^2 may fail.

Sections 4.3 and 4.4 are the core of the chapter. They concern boundedness of pseudodifferential operators on $M^{p,q}$ spaces. The symbols are also assumed to belong to modulation spaces. We want to present a generalization to the fundamental result by Gröchenig and Heil [163] 1999 (and its earlier precursor, the Heil–Ramanathan–Topiwala paper [185]), stating that symbols in $M^{\infty,1}(\mathbb{R}^{2d})$, in particular in $S^0_{0,0}$, produce bounded operators on modulation spaces. Such a property is extremely important in the applications to partial differential equations. In fact, the optimal regularity behavior of the solutions arises from the stability (i. e., no loss of derivatives) of the function spaces appearing in the a priori estimates. In turn, this corresponds to stable continuity estimates for the pseudodifferential operators expressing the solutions. As we observed, as soon as the symbol class $S^0_{0,0}$ and classes of nonsmooth symbols are involved, stable estimates in the Lebesgue L^p spaces fail. Modulation spaces M^p are then excellent substitutes of L^p in such situations.

In Section 4.3 we shall first introduce the τ-quantization $\mathrm{Op}_\tau(a)$ of a symbol a, defined formally by

$$\langle \mathrm{Op}_\tau(a)f, g\rangle = \langle a, W_\tau(g,f)\rangle$$

where $W_\tau(g,f)$ is the (cross)-τ-Wigner distribution defined in (1.109), see Chapter 1. Note that $\mathrm{Op}_1(a) = a(x,D)$, $\mathrm{Op}_{1/2}(a) = \mathrm{Op}_w(a)$. We shall prove that results for modulation symbols do not depend on the choice of the τ-quantization, $0 \le \tau \le 1$. Our strategy in Section 4.3 will be then to refer to the Weyl quantization, defined by $\langle \mathrm{Op}_w(a)f, g\rangle = \langle a, W(g,f)\rangle$ in (4.6). Hence, we shall discuss the validity of estimates of the type

$$\|W(f,g)\|_{M^{p,q}} \lesssim \|f\|_{M^{p_1,q_1}} \|g\|_{M^{p_2,q_2}}$$

for suitable indices p, q, p_1, q_1, p_2, q_2. Using these inequalities, it will be easy to deduce Theorem 4.4.15, extending the most part of the results in the literature about boundedness of $\mathrm{Op}_w(a)$ on modulation spaces, cf. Toft [286, 287]. In Section 4.4 we shall also discuss shortly localization operators; they provide an alternative method of quantization, and are important mathematical tools in signal theory and quantum mechanics.

Section 4.5 is devoted to the study of operators with symbols a in the Wiener amalgam spaces $W(\mathcal{F}L^p, L^q)$. Here we continue the analysis of Section 4.2, where boundedness of $a(x,D)$ and $\mathrm{Op}_w(a)$ was considered on $W(L^r, L^s)$. Attention is now focused on the τ-quantization $\mathrm{Op}_\tau(a)$ acting on modulation spaces. The strategy will be the same as in Section 4.4, namely taking as definition $\langle \mathrm{Op}_\tau(a)f, g\rangle = \langle a, W_\tau(g,f)\rangle$, we have reduced the situation to the study of estimates of the type

$$\|W_\tau(f,g)\|_{W(\mathcal{F}L^r, L^q)} \le C(\tau)\|f\|_{M^{p_1,p_2}} \|f\|_{M^{q_1,q_2}}$$

where the constant $C(\tau)$ depends on τ. The corresponding results for $\mathrm{Op}_\tau(a)$ turn out to be surprising, namely boundedness on $M^{p,q}$ holds for $0 < \tau < 1$, whereas may fail for $\tau = 0$, $\tau = 1$. This is in particular the case for symbols a in $W(\mathcal{F}L^1, L^\infty)$.

Section 4.6 is devoted to Born–Jordan quantization and related pseudodifferential operators. Born and Jordan [33] 1925 provided the first mathematical basis to the

Heisenberg's matrix mechanics, and recently we may note a strong regain in interest on it. Our rather long presentation is intended to give a detailed and rigorous treatment of Born–Jordan pseudodifferential operators $\mathrm{Op}_{BJ}(a)$. Continuity properties on modulation spaces are discussed in Section 4.7. Following our time–frequency approach, we shall use the definition

$$\langle \mathrm{Op}_{BJ}(a)f, g\rangle = \langle a, Q(g,f)\rangle,$$

where $Q(g,f)$ is the cross-Born–Jordan distribution. We address to Sections 4.6 and 4.7 for a precise setting and historical motivations.

We do not report here on other recent contributions to the theory of the pseudodifferential operators; nevertheless, we would like to mention the works of Ruzhansky and collaborators [46, 98, 109, 144, 154, 230, 256, 257, 258] concerning operators on Lie groups and applications. It would be interesting to reconsider this setting from the point of view of the time–frequency analysis.

4.1 Boundedness on L^p and Wiener amalgam spaces

We first give a proof of Sjöstrand's results [266, 267]. In the following remark we anticipate the contents of the discussion in Section 4.3, concerning different types of quantization.

Remark 4.1.1. Since $a(x,D) = \mathrm{Op}_w(\mathcal{U}^{-1}a)$, with $\mathcal{U}$ in (4.8), we shall show in the subsequent Proposition 4.3.6 that the modulation spaces $M^{p,q}$, $1 \le p,q \le \infty$, are invariant under $\mathcal{U}$, so that boundedness results for pseudodifferential operators with symbols in modulation spaces can be obtained using either the Weyl or the Kohn–Nirenberg form (for further references, see [286, 287, 288]). In the sequel, we shall adopt the operator form which is more convenient.

We also need the following useful remark.

Remark 4.1.2. Denote by $\overline{\mathcal{S}}^{W(L^r,L^\infty)}$ the closure of the Schwartz class in $W(L^r, L^\infty)$. If $1 \le r < \infty$, its dual space is given by

$$(\overline{\mathcal{S}}^{W(L^r,L^\infty)})' = W(L^{r'}, L^1),$$

see, e. g., [130, Theorem 2.8].

Theorem 4.1.3. *If $\sigma \in M^{\infty,1}(\mathbb{R}^{2d})$, then the Weyl operator $\mathrm{Op}_w(\sigma)$ is bounded on $W(L^2, L^s)(\mathbb{R}^d)$, for every $1 \le s \le \infty$, with the uniform estimate*

$$\|\mathrm{Op}_w(\sigma)f\|_{W(L^2,L^s)} \lesssim \|\sigma\|_{M^{\infty,1}} \|f\|_{W(L^2,L^s)}.$$

In view of Proposition 2.4.12, we have $W(L^2, L^2) = L^2$, hence the estimate above implies the L^2-boundedness of $\mathrm{Op}_w(\sigma)$.

Proof of Theorem 4.1.3. Let us show the estimate

$$|\langle \mathrm{Op}_w(\sigma)f, g\rangle| \lesssim \|\sigma\|_{M^{\infty,1}} \|f\|_{W(L^2,L^s)} \|g\|_{W(L^2,L^{s'})}, \quad \forall f, g \in \mathcal{S}(\mathbb{R}^d),$$

where $1/s + 1/s' = 1$. This will give at once the desired result if $s > 1$, whereas the case $s = 1$ follows by Remark 4.1.2.

Let $\varphi \in \mathcal{D}(\mathbb{R}^d)$ and set $\Phi = W(\varphi, \varphi) \in \mathcal{S}(\mathbb{R}^{2d})$. By the definition of the Weyl operator via cross-Wigner distribution in (4.6), the orthogonality relations in (1.44) and Hölder's inequality,

$$\begin{aligned} |\langle \mathrm{Op}_w(\sigma)f, g\rangle| &= |\langle \sigma, W(g,f)\rangle| \asymp |\langle V_\Phi \sigma, V_\Phi W(g,f)\rangle| \le \|V_\Phi \sigma\|_{L^{\infty,1}} \|V_\Phi W(g,f)\|_{L^{1,\infty}} \\ &\asymp \|\sigma\|_{M^{\infty,1}} \|W(g,f)\|_{M^{1,\infty}}. \end{aligned}$$

Then the result is proved if we show that $\|W(g,f)\|_{M^{1,\infty}} \lesssim \|f\|_{W(L^2,L^s)} \|g\|_{W(L^2,L^{s'})}$. If $\zeta = (\zeta_1, \zeta_2) \in \mathbb{R}^{2d}$, we write $\tilde{\zeta} = (\zeta_2, -\zeta_1)$. Then Lemma 1.3.37 says that

$$|V_\Phi(W(g,f))(z,\zeta)| = \left|V_\varphi f\left(z + \frac{\tilde{\zeta}}{2}\right)\right| \left|V_\varphi g\left(z - \frac{\tilde{\zeta}}{2}\right)\right|.$$

Consequently,

$$\|W(g,f)\|_{M^{1,\infty}} \asymp \sup_{\zeta \in \mathbb{R}^{2d}} \int_{\mathbb{R}^{2d}} \left|V_\varphi f\left(z + \frac{\tilde{\zeta}}{2}\right)\right| \left|V_\varphi g\left(z - \frac{\tilde{\zeta}}{2}\right)\right| dz.$$

We set $\pi(\tilde{\zeta}) = M_{\tilde{\zeta}_2} T_{\tilde{\zeta}_1}$. In what follows, we make the change of variables $z \mapsto z - \tilde{\zeta}/2$, and use the relations in (1.24) and (1.47), the Cauchy–Schwarz's and Parseval's inequalities with respect to the z_2 variable, so that

$$\begin{aligned} \|W(g,f)\|_{M^{1,\infty}} &\asymp \sup_{\zeta \in \mathbb{R}^{2d}} \int_{\mathbb{R}^{2d}} |V_\varphi f(z)| |V_\varphi g(z - \tilde{\zeta})| \, dz \\ &= \sup_{\zeta \in \mathbb{R}^{2d}} \int_{\mathbb{R}^{2d}} |V_\varphi f(z)| |V_\varphi(\pi(\tilde{\zeta})g)(z)| \, dz \\ &= \sup_{\zeta \in \mathbb{R}^{2d}} \int_{\mathbb{R}^d} \int_{\mathbb{R}^d} |\mathcal{F}(f T_{z_1}\varphi)(z_2)| |\mathcal{F}(\pi(\tilde{\zeta}) g T_{z_1}\varphi)(z_2)| \, dz_1 \, dz_2 \\ &\lesssim \sup_{\zeta \in \mathbb{R}^{2d}} \int_{\mathbb{R}^d} \|f T_{z_1}\varphi\|_2 \|\pi(\tilde{\zeta}) g T_{z_1}\varphi\|_2 \, dz_1 \\ &\lesssim \sup_{\zeta \in \mathbb{R}^{2d}} \|f\|_{W(L^2,L^s)} \|\pi(\tilde{\zeta}) g\|_{W(L^2,L^{s'})} \\ &= \|f\|_{W(L^2,L^s)} \|g\|_{W(L^2,L^{s'})}, \end{aligned}$$

where we have used Hölder's inequality in the second to last step and the invariance of the $W(L^2, L^s)$ spaces under time–frequency shifts $\pi(\tilde{\zeta})$ in the last one. □

If we choose symbols with a stronger decay, namely symbols $a \in M^{p,1} \subset M^{\infty,1}$, $1 \le p \le 2$, then the corresponding pseudodifferential operators $a(x,D)$ are bounded on every Wiener amalgam spaces $W(L^r, L^s)$, as shown in the following result.

Theorem 4.1.4. *If $a \in M^{p,1}(\mathbb{R}^{2d})$, $1 \le p \le 2$, then the operator $a(x,D)$ is bounded on $W(L^r,L^s)(\mathbb{R}^d)$, for every $1 \le r,s \le \infty$, with the uniform estimate*

$$\|a(x,D)f\|_{W(L^r,L^s)} \lesssim \|a\|_{M^{p,1}} \|f\|_{W(L^r,L^s)}.$$

Proof. By the inclusion relations for modulation spaces, we can just consider the case $a \in M^{2,1}$. We shall show that the integral kernel

$$K(x,y) = (\mathcal{F}_2 a)(x, y-x)$$

($\mathcal{F}_2$ stands for the partial Fourier transform with respect to the second variable) of $a(x,D)$ can be controlled from above by $|K(x,y)| \le F(x-y)$, where F is a positive function in $L^1(\mathbb{R}^d)$ with norm estimated by $\|a\|_{M^{2,1}}$. If it is so, $|a(x,D)f(x)| \le (F * |f|)(x)$ the convolution relations for Wiener amalgam spaces in Lemma 2.4.9 give the desired result.

We use the inversion formula (1.2.16) for the symbol a. Namely, for any window $g \in \mathcal{S}(\mathbb{R}^{2d})$, with $\|g\|_2 = 1$, we have

$$a(x,\xi) = \int_{\mathbb{R}^{4d}} (V_g a)(\alpha,\beta)(M_\beta T_\alpha g)(x,\xi)\, d\alpha\, d\beta. \tag{4.9}$$

Hence, if $\alpha = (\alpha_1,\alpha_2)$, $\beta = (\beta_1,\beta_2)$, it turns out that

$$K(x,y) = \int_{\mathbb{R}^{4d}} e^{2\pi i \alpha_2\cdot\beta_2} (V_g a)(\alpha_1,\alpha_2,\beta_1,\beta_2) M_{(\beta_1,\alpha_2)} T_{(\alpha_1,-\beta_2)} (\mathcal{F}_2^{-1} g)(x, x-y)\, d\alpha_1\, d\alpha_2\, d\beta_1\, d\beta_2.$$

Setting

$$H(\alpha_1, t; \beta_1, \beta_2) = \int_{\mathbb{R}^d} (V_g a)(\alpha,\beta) e^{2\pi i (t-\beta_2)\cdot\alpha_2}\, d\alpha_2,$$

and using the Cauchy–Schwarz's inequality with respect to the α_2 variable, we obtain

$$\begin{aligned} |K(x,y)| &\le \int_{\mathbb{R}^{3d}} \big|H(\alpha_1, x-y+\beta_2; \beta_1,\beta_2)\mathcal{F}_2^{-1} g(x-\alpha_1, x-y+\beta_2)\big|\, d\alpha_1\, d\beta_1\, d\beta_2 \\ &\le \int_{\mathbb{R}^{2d}} \|H(\cdot, x-y+\beta_2;\beta_1,\beta_2)\|_2 \|T_{(0,-\beta_2)} \mathcal{F}_2^{-1} g(\cdot, x-y)\|_2\, d\beta_1\, d\beta_2. \end{aligned}$$

For simplicity, let us set

$$F(t) := \int_{\mathbb{R}^{2d}} \|H(\cdot, t+\beta_2;\beta_1,\beta_2)\|_2 \|T_{(0,-\beta_2)} \mathcal{F}_2^{-1} g(\cdot, t)\|_2\, d\beta_1\, d\beta_2,$$

so that $|K(x,y)| \le F(x-y)$.

We are left to estimate $\|F\|_1$. Using the Cauchy–Schwarz's inequality with respect to the t variable,

$$\begin{aligned}\|F\|_1 &= \int_{\mathbb{R}^d}\int_{\mathbb{R}^{2d}} \|H(\cdot,t+\beta_2;\beta_1,\beta_2)\|_2 \|T_{(0,-\beta_2)}\mathcal{F}_2^{-1}g(\cdot,t)\|_2\, d\beta_1\, d\beta_2\, dt \\ &\le \int_{\mathbb{R}^{2d}} \|H(\cdot,\cdot;\beta_1,\beta_2)\|_2 \|T_{(0,-\beta_2)}\mathcal{F}_2^{-1}g\|_2\, d\beta_1\, d\beta_2 \\ &= \|g\|_2 \int_{\mathbb{R}^{2d}} \|H(\cdot,\cdot;\beta_1,\beta_2)\|_2\, d\beta_1\, d\beta_2 \\ &= \|H\|_{L^{2,1}} = \|V_g a\|_{L^{2,1}} \asymp \|a\|_{M^{2,1}},\end{aligned}$$

where in the last row we used Parseval's formula and the assumption $\|g\|_2 = 1$. This concludes the proof. □

Theorem 4.1.5. *Let $1 \le p,q,r,s \le \infty$ such that*

$$\frac{1}{p} \ge \left|\frac{1}{r} - \frac{1}{2}\right| + \frac{1}{q'}, \quad q \le \min\{r, r', s, s'\}. \tag{4.10}$$

Then every symbol $a \in M^{p,q}$ gives rise to a bounded operator $a(x,D)$ on $W(L^r,L^s)$ with the uniform estimate

$$\|a(x,D)f\|_{W(L^r,L^s)} \lesssim \|a\|_{M^{p,q}} \|f\|_{W(L^r,L^s)}.$$

Proof. We first make the complex interpolation between the estimates of Theorems 4.1.3 and 4.1.4 (which deal with the cases in which $q = 1$). Using the interpolation relations for Wiener amalgam and modulation spaces of Theorem 2.4.9 and Proposition 2.3.16, we obtain, for every $1 \le s < \infty$,

$$\|a(x,D)f\|_{W(L^r,L^s)} \lesssim \|a\|_{M^{p,1}} \|f\|_{W(L^r,L^s)},$$

where

$$\frac{1}{p} \ge \left|\frac{1}{r} - \frac{1}{2}\right|.$$

The remaining cases, when $s = \infty$, $p > 2$ (and therefore $r > 1$), follow by duality, for $W(L^r,L^\infty) = W(L^{r'},L^1)'$ in view of Theorem 2.4.9 and, considering the Weyl form $\mathrm{Op}_w(\sigma)$ of $a(x,D)$, we have $(\mathrm{Op}_w(\sigma))^* = \mathrm{Op}_w(\bar{\sigma})$.

Finally, by interpolation between what we just proved and the obvious case $p = q = r = s = 2$ (Weyl pseudodifferential operators with symbols in $M^2(\mathbb{R}^{2d}) = L^2(\mathbb{R}^{2d})$ are bounded on $W(L^2,L^2)(\mathbb{R}^d) = L^2(\mathbb{R}^d)$ by (4.6) and Moyal's identity), we obtain the claim. □

Recalling that, for $s = r$, we have $W(L^r, L^r) = L^r$, the above boundedness result can be rephrased for pseudodifferential operators acting on L^p spaces as follows (see Figure 4.1, where we take into account the necessary conditions in the sequel).

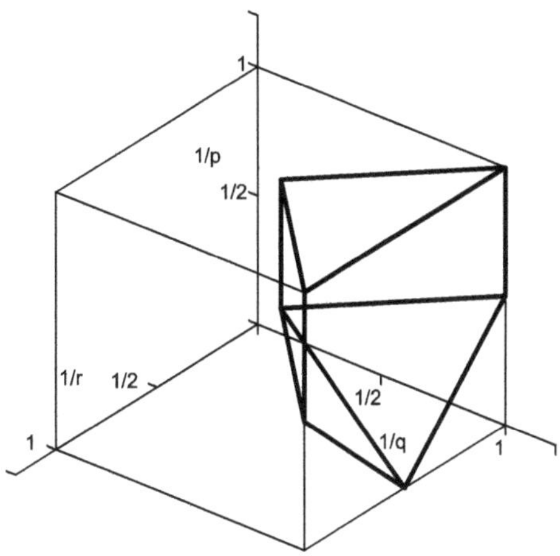

Figure 4.1: The triples $(1/r, 1/q, 1/p)$ inside the convex polyhedron are exactly those for which every symbol in $M^{p,q}$ produces a bounded operator on L^r.

Corollary 4.1.6. *Let $1 \le p, q, r \le \infty$ such that*

$$\frac{1}{p} \ge \left|\frac{1}{r} - \frac{1}{2}\right| + \frac{1}{q'}, \quad q \le \min\{r, r'\}. \tag{4.11}$$

Then every symbol in $a \in M^{p,q}$ gives rise to a bounded operator $a(x, D)$ on L^r with the uniform estimate

$$\|a(x, D)f\|_r \lesssim \|a\|_{M^{p,q}} \|f\|_r.$$

4.1.1 Boundedness on Wiener amalgam spaces: necessary conditions

In this section we show the optimality of Theorem 4.1.5 (and Corollary 4.1.6). We need the following auxiliary results.

Lemma 4.1.7. *Suppose that for some $1 \le p, q, r, s \le \infty$ the following estimate holds:*

$$\|\mathrm{Op_w}(\sigma)f\|_{W(L^r, L^s)} \le C\|\sigma\|_{M^{p,q}} \|f\|_{W(L^r, L^s)}, \quad \forall \sigma \in \mathcal{S}(\mathbb{R}^{2d}),\ \forall f \in \mathcal{S}(\mathbb{R}^d).$$

Then the same estimate is satisfied with r, s replaced by the dual indices r', s' (even if $r = \infty$ or $s = \infty$).

Proof. Indeed, observe that $\langle \mathrm{Op_w}(\sigma)f, g\rangle = \langle f, \mathrm{Op_w}(\overline{\sigma})g\rangle$, $\forall f, g \in \mathcal{S}(\mathbb{R}^d)$. Hence, by Theorem 2.4.9 and the assumptions rewritten for $\mathrm{Op_w}(\overline{\sigma})$ (observe that $\|\sigma\|_{M^{p,q}} = \|\bar{\sigma}\|_{M^{p,q}}$), we have

$$|\langle \mathrm{Op}_w(\sigma)f, g\rangle| \le C\|\sigma\|_{M^{p,q}}\|f\|_{W(L^{r'},L^{s'})}\|g\|_{W(L^r,L^s)}, \quad \forall f \in \mathcal{S}(\mathbb{R}^d),\ \forall g \in \mathcal{S}(\mathbb{R}^d). \tag{4.12}$$

Since $\mathrm{Op}_w(\sigma)f$ is a Schwartz function, it belongs to $W(L^{r'},L^{s'}) \subset W(L^r,L^s)'$, and $\|\mathrm{Op}_w(\sigma)f\|_{W(L^r,L^s)'} = \|\mathrm{Op}_w(\sigma)f\|_{W(L^{r'},L^{s'})}$, because $W(L^{r'},L^{s'})$ is isometrically embedded in $W(L^r,L^s)'$. Hence it suffices to prove that the estimate in (4.12) holds for every $g \in W(L^r,L^s)$. This follows by a density argument. Namely, consider, for a given $g \in W(L^r,L^s)$, a sequence g_n of Schwartz functions, with $g_n \to g$ in $\mathcal{S}'(\mathbb{R}^d)$ and $\|g_n\|_{W(L^r,L^s)} \le \|g\|_{W(L^r,L^s)}$ (1). Letting $n \to \infty$ in the above estimate (written with g_n in place of g) gives the desired conclusion. □

Lemma 4.1.8. *Let $h \in \mathcal{D}(\mathbb{R}^d)$, and consider the family of functions*

$$h_\lambda(x) = h(x)e^{-\pi i\lambda x^2}, \quad \lambda \ge 1.$$

Then for $1 \le q \le \infty$

$$\|\widehat{h_\lambda}\|_q \asymp \lambda^{\frac{d}{q}-\frac{d}{2}}. \tag{4.13}$$

Proof. The result is known and outlined, e. g., in [277, Exercise 2.34]. We report on a sketch of the proof for the sake of completeness.

Let $c > 0$ be such that $h(x)$ vanishes for $|x| > c$. First one shows the estimate $|\widehat{h_\lambda}(\xi)| \le C_N\langle\xi\rangle^{-N}\lambda^{-N}$, for every $N > 0$ and $\xi \in \mathbb{R}^d$ such that $|\xi| \ge 2c\lambda$. To this end, we observe that by rotational symmetry we can assume $\xi = (\xi_1, 0, \dots, 0)$. The claim then follows by applying the nonstationary phase theorem [269, Proposition 1, page 331] with the asymptotic parameter ξ_1 and the phase $\phi(x_1) := -2\pi x_1 - \pi\frac{\lambda}{\xi_1}x_1^2$ (the assumptions being satisfied for $|x_1| \le c$, uniformly with respect to the parameter λ/ξ_1).

In the region $|\xi| < 2c\lambda$ we have the estimate $|\widehat{h_\lambda}(\xi)| \le C\lambda^{-d/2}$, as a consequence of the stationary phase theorem (see [269, Section 5.13 (a), page 363]) with the phase given by the quadratic polynomial $\phi(x) := -\pi x^2$. One hence obtains the upper bound $\|\widehat{h_\lambda}\|_q \lesssim \lambda^{d(1/q-1/2)}$. Since

$$\|h\|_2^2 = \|\widehat{h_\lambda}\|_2^2 \le \|\widehat{h_\lambda}\|_q\|\widehat{h_\lambda}\|_{q'} \lesssim \lambda^{d(1/q'-1/2)}\|\widehat{h_\lambda}\|_q,$$

the lower bound follows as well. □

Finally, we establish a useful estimate for Wiener amalgam spaces, based on the classical Bernstein's inequality (see, e. g., [316]), that we are going to recall first. Let $B_R(x_0)$ be the ball of center $x_0 \in \mathbb{R}^d$ and radius $R > 0$ in $\mathbb{R}^d$.

1 For example, take $g_n(x) = n^d\varphi_1(x/n)(g * \varphi_2(n\cdot))(x)$, with $\varphi_1, \varphi_2 \in \mathcal{D}(\mathbb{R}^d)$, $\varphi_1(0) = 1$, $\|\varphi_2\|_1 = 1$.

Lemma 4.1.9 (Bernstein's inequality). *Let $f \in \mathcal{S}'(\mathbb{R}^d)$ be such that $\hat{f}$ is supported in $B_R(x_0)$, and let $1 \le p \le q \le \infty$. Then there exists a positive constant C independent of f, x_0, R, p, q, such that*

$$\|f\|_q \le CR^{d(\frac{1}{p}-\frac{1}{q})}\|f\|_p. \tag{4.14}$$

Lemma 4.1.10. *Let $1 \le p \le q \le \infty$. For every $R > 0$, there exists a constant $C_R > 0$ such that, for every $f \in \mathcal{S}'(\mathbb{R}^d)$ whose Fourier transform is supported in any ball of radius R, it turns out*

$$\|f\|_{W(L^q,L^p)} \le C_R\|f\|_p.$$

Proof. Choose a Schwartz function g whose Fourier transform $\hat{g}$ has compact support in $B_1(0)$, as window function arising in the definition of the norm in $W(L^q, L^p)$. Then the function $(T_x g)f$, $x \in \mathbb{R}^d$, has Fourier transform supported in a ball of radius $R + 1$. Therefore it follows from Bernstein's inequality (Lemma 4.1.9) that

$$\|(T_x g)f\|_q \le C_R\|(T_x g)f\|_p.$$

Taking the L^p-norm with respect to x gives the conclusion. □

We now establish a version of the upper bound in Lemma 4.1.8, for Wiener amalgam spaces.

Lemma 4.1.11. *With the notation of Lemma 4.1.8 we have, for $1 \le p, q \le \infty$,*

$$\|\widehat{h_\lambda}\|_{W(L^p,L^q)} \lesssim \lambda^{\frac{d}{q}-\frac{d}{2}}, \quad \lambda \ge 1.$$

Proof. When $p \ge q$ the desired result follows from Lemmata 4.1.10 and 4.1.8. When $p < q$ the result follows from the inclusion $L^q \hookrightarrow W(L^p, L^q)$ and Lemma 4.1.8. □

Proposition 4.1.12. *Let $\chi \in \mathcal{D}(\mathbb{R}^d)$, $\chi \ge 0$, $\chi(0) = 1$. Suppose that, for some $1 \le p, q, r, r_1, r_2 \le \infty$, $C > 0$, the estimate*

$$\|\chi a(x, D)f\|_r \le C\|a\|_{M^{p,q}}\|f\|_{W(L^{r_1},L^{r_2})}, \quad \forall a \in \mathcal{S}(\mathbb{R}^{2d}),\ f \in \mathcal{S}(\mathbb{R}^d), \tag{4.15}$$

holds. Then $q \le r_2'$ and

$$\frac{1}{p} \ge \frac{1}{2} - \frac{1}{r} + \frac{1}{q'}. \tag{4.16}$$

Proof. First we prove the constraint $q \le r_2'$. Let $h \in \mathcal{D}(\mathbb{R}^d)$, $h \ge 0$, $h(0) = 1$. Let h_λ be as in Lemma 4.1.8. We test (4.15) on the family of symbols

$$a_\lambda(x, \xi) = h(x)h_\lambda(\xi) = h(x)h(\xi)e^{-\pi i\lambda|\xi|^2},$$

and functions $f_\lambda = \mathcal{F}^{-1}(\overline{h_\lambda}) \in \mathcal{S}(\mathbb{R}^d)$. An explicit computation shows that

$$\chi(x)a_\lambda(x,D)f_\lambda(x) = \int e^{2\pi i x\cdot\xi}\chi(x)h(x)h^2(\xi)\,d\xi,$$

which is a nonzero Schwartz function independent of λ. On the other hand, by Lemma 4.1.8, we have

$$\|a_\lambda\|_{M^{p,q}} \asymp \|a_\lambda\|_{\mathcal{F}L^q} \lesssim \|h_\lambda\|_{\mathcal{F}L^q} \lesssim \lambda^{\frac{d}{q}-\frac{d}{2}}.$$

Similarly, by Lemma 4.1.11,

$$\|f_\lambda\|_{W(L^{r_1},L^{r_2})} \lesssim \lambda^{\frac{d}{r_2}-\frac{d}{2}}.$$

Taking into account these estimates and letting $\lambda \to +\infty$, (4.15) then gives $q \le r_2'$.

Let us now prove (4.16). Let h_λ be as above. We now test the estimate (4.15) on the family of symbols

$$a'_\lambda(x,\xi) = e^{-\pi\lambda|x|^2}\widehat{h_\lambda}(\xi),$$

and functions $f'_\lambda = \overline{h_\lambda}$. The operator $a'_\lambda(x,D)$ has integral kernel

$$K_\lambda(x,y) = e^{-\pi\lambda^2|x|^2}h_\lambda(x-y),$$

so that

$$\begin{aligned}\chi(x)\big|a'_\lambda(x,D)f'_\lambda(x)\big| &= \left|\int e^{-\pi\lambda^2|x|^2+2\pi i\lambda x\cdot y}h(x-y)\chi(x)h(y)\,dy\right|\\ &\ge \operatorname{Re}\int e^{-\pi\lambda^2|x|^2+2\pi i\lambda x\cdot y}h(x-y)\chi(x)h(y)\,dy.\end{aligned}$$

Now, $h(y)$ has compact support, say, in the ball $|y| \le C$. Moreover, if $|x| \le \lambda^{-1}$ for $\lambda \ge \lambda_0$ large enough, and $|y| \le C$ we have $\operatorname{Re}(e^{-\pi\lambda^2|x|^2+2\pi i\lambda x\cdot y}) \ge \frac{1}{2}$. Hence we deduce

$$\chi(x)\big|a'_\lambda(x,D)f'_\lambda(x)\big| \gtrsim 1, \quad \text{for } |x| \le \lambda^{-1},$$

which implies

$$\|\chi a'_\lambda(x,D)f'_\lambda\|_r \gtrsim \lambda^{-\frac{d}{r}}.$$

On the other hand, $\|f'_\lambda\|_{W(L^{r_1},L^{r_2})}$ is clearly independent of λ. Moreover, by Lemma 2.6.1,

$$\|e^{-\pi\lambda^2|\cdot|^2}\|_{M^{p,q}} \lesssim \lambda^{-\frac{d}{q'}}$$

and, by Lemmas 2.3.26 (ii) and 4.1.8, we have

$$\|\widehat{h_\lambda}\|_{M^{p,q}} \lesssim \|\widehat{h_\lambda}\|_p \lesssim \lambda^{\frac{d}{p}-\frac{d}{2}}.$$

Hence

$$\|a'_\lambda\|_{M^{p,q}} = \|e^{-\pi\lambda^2|\cdot|^2}\|_{M^{p,q}}\|\widehat{h_\lambda}\|_{M^{p,q}} \lesssim \lambda^{\frac{d}{p}-\frac{d}{2}-\frac{d}{q'}}.$$

Putting all together and letting $\lambda \to \infty$ we obtain (4.16). □

Proposition 4.1.13. *Suppose that, for some $1 \le p, q, r, s, r_1, r_2 \le \infty$, $C > 0$, the estimate*

$$\|a(x,D)f\|_{W(L^{r_1},L^{r_2})} \le C\|a\|_{M^{p,q}}\|f\|_{W(L^r,L^s)}, \quad \forall a \in \mathcal{S}(\mathbb{R}^{2d}),\ f \in \mathcal{S}(\mathbb{R}^d), \tag{4.17}$$

holds. Then $q \le r$.

Proof. Let h_1, h_2 be two Schwartz functions in $\mathbb{R}^d$ such that h_1 and $\widehat{h}_2$ are real-valued, with $h_1(0) = 1$, $\widehat{h_2}(0) = 1$, and satisfying

$$\operatorname{supp}\widehat{h_1} \subset B_1(0), \quad \operatorname{supp}\widehat{h_2} \subset B_1(0). \tag{4.18}$$

Consider, for every $N \ge 1$, the finite lattice

$$\Lambda_N = \{n = (n_1, \dots, n_d) \in 4\mathbb{Z}^d : 0 \le n_j \le 4(N-1),\ j = 1, \dots, d\}. \tag{4.19}$$

Observe that Λ_N has cardinality N^d, and

$$|n| \le d^{1/2}4(N-1),\ \forall n \in \Lambda_N \quad \text{and} \quad |n-m| \ge 4,\ \forall n, m \in \Lambda_N, n \ne m. \tag{4.20}$$

Moreover, let h be a smooth real-valued function, $h \ge 0$, $h(0) = 1$, supported in a ball $B_\epsilon(0)$, for a small ϵ to be chosen later. We test the estimate (4.17) on the family of functions $f_N(x) = h(Nx)$ and symbols

$$a_N(x,\xi) = \sum_{n\in\Lambda_N} b_n(x,\xi), \quad \text{where } b_n(x,\xi) = (M_{-n}h_1)(x)(T_n h_2)(\xi). \tag{4.21}$$

The integral kernel of the operator $a_N(x,D)$ is given by

$$K_N(x,y) = (\mathcal{F}_2^{-1}a_N)(x, x-y) = \sum_{n\in\Lambda_N} \mathcal{F}_2^{-1}(b_n)(x, x-y) = \sum_{n\in\Lambda_N} e^{-2\pi i n\cdot y} h_1(x)\widehat{h_2}(y-x).$$

We now show that, for a suitable $\delta > 0$,

$$|a_N(x,D)f_N(x)| \gtrsim 1, \quad \text{for } x \in B_\delta(0), \tag{4.22}$$

which implies

$$\|a_N(x,D)f_N\|_{W(L^{r_1},L^{r_2})} \gtrsim 1. \tag{4.23}$$

In order to prove (4.22), observe that, by the above computation,

$$a_N(x,D)f_N(x) = \int_{\mathbb{R}^d} \Big(\sum_{n\in\Lambda_N} e^{-2\pi i n\cdot y}\Big) h_1(x)\widehat{h_2}(y-x)h(Ny)\,dy.$$

Now, as a consequence of the first condition in (4.20), we see that on the support of $h(Ny)$, hence where $|y| \le \epsilon N^{-1}$, we have

$$\operatorname{Re}(e^{-2\pi i n\cdot y}) \ge \frac{1}{2}, \quad \forall n \in \Lambda_N,$$

if $\epsilon \leq d^{-1/2}/24$. This implies that

$$|a_N(x,D)f_N(x)| \geq \frac{N^d}{2} \int_{\mathbb{R}^d} h_1(x)\widehat{h_2}(y-x)h(Ny)\,dy.$$

Since $h_1(0) = \widehat{h_2}(0) = 1$, if δ and ϵ are small enough, so as $h_1(x) \geq 1/2$ and $\widehat{h_2}(y-x) \geq 1/2$ for $|y| \leq \epsilon$, $|x| \leq \delta$. It turns out that

$$|a_N(x,D)f_N(x)| \geq \frac{N^d}{8} \int_{\mathbb{R}^d} h(Ny)\,dy = \frac{\|h\|_1}{8}, \quad \text{for } |x| \leq \delta,$$

which implies (4.22).

We now prove that

$$\|a_N\|_{M^{p,q}} \lesssim N^{d/q}. \tag{4.24}$$

To see this, observe that

$$\widehat{b_n}(\zeta_1,\zeta_2) = (T_{-n}\widehat{h_1})(\zeta_1)(M_{-n}\widehat{h_2})(\zeta_2). \tag{4.25}$$

We choose a window $\Phi = \varphi \otimes \varphi$, where $\hat{\varphi}$ is a Schwartz function supported in $B_1(0)$. It follows from (4.25), the second condition in (4.20), and (4.18) that the functions $V_\Phi b_n(z_1,z_2,\omega_1,\omega_2) = (\widehat{b_n} * M_{-z}\widehat{\Phi}^*)(\omega)$ (with $z = (z_1,z_2)$, $\omega = (\omega_1,\omega_2)$, $\Phi^*(z) = \overline{\Phi(-z)}$), vanish unless $\omega_1 \in B_2(-n)$, $\omega_2 \in B_2(0)$. Hence, when n varies in Λ_N, they have pairwise disjoint supports, as well as the functions $\|V_\Phi b_n(\cdot,\cdot,\omega_1,\omega_2)\|_p$. We deduce that

$$\begin{aligned}
\|a_N\|_{M^{p,q}} &\asymp \left(\int_{\mathbb{R}^{2d}} \Big\|\sum_{n\in\Lambda_N} V_\Phi b_n(\cdot,\cdot,\omega_1,\omega_2)\Big\|_p^q\,d\omega_1\,d\omega_2\right)^{1/q}\\
&= \left(\int_{\mathbb{R}^d}\Big(\sum_{n\in\Lambda_N} \|V_\Phi b_n(\cdot,\cdot,\omega_1,\omega_2)\|_p^p\Big)^{q/p}\,d\omega_1\,d\omega_2\right)^{1/q}\\
&= \left(\int_{\mathbb{R}^{2d}} \sum_{n\in\Lambda_N} \|V_\Phi b_n(\cdot,\cdot,\omega_1,\omega_2)\|_p^q\,d\omega_1\,d\omega_2\right)^{1/q}\\
&= \left(\sum_{n\in\Lambda_N} \int_{B_2(-n)}\int_{B_2(0)} \|V_\Phi b_n(\cdot,\cdot,\omega_1,\omega_2)\|_p^q\,d\omega_1\,d\omega_2\right)^{1/q}.
\end{aligned} \tag{4.26}$$

On the other hand, since $V_\Phi b_n(z,\omega) = e^{-2\pi i z\cdot\omega}(b_n * M_\omega\Phi^*)(z)$, by Young's inequality, we have

$$\|V_\Phi b_n(\cdot,\cdot,\omega_1,\omega_2)\|_p \leq \|b_n\|_p,$$

and the expression for b_n in (4.21) shows that $\|b_n\|_p$ is, in fact, independent of n. Hence the expression in (4.26) is

$$\lesssim \Big(\sum_{n\in\Lambda_N} \int_{B_2(-n)} \int_{B_2(0)} dz_1\, dz_2 \Big)^{1/q} = C_d N^{d/q},$$

which gives (4.24).

Finally, since the functions f_N are supported in a fixed compact subset, we have

$$\|f_N\|_{W(L^r,L^s)} \asymp \|f_N\|_r = \|h\|_r N^{-d/r}. \tag{4.27}$$

Combining this estimate with (4.23), (4.24), and (4.17), and letting $N \to \infty$ yields the desired constraint $q \le r$. □

Theorem 4.1.14. *Suppose that, for some $1 \le p,q,r,s \le \infty$, the estimate*

$$\|a(x,D)f\|_{W(L^r,L^s)} \le C\|a\|_{M^{p,q}}\|f\|_{W(L^r,L^s)}, \quad \forall a \in \mathcal{S}(\mathbb{R}^{2d}),\ f \in \mathcal{S}(\mathbb{R}^d), \tag{4.28}$$

holds. Then $q \le \min\{r, r', s, s'\}$ and

$$\frac{1}{p} \ge \left|\frac{1}{2} - \frac{1}{r}\right| + \frac{1}{q'}. \tag{4.29}$$

Proof. We already know from Proposition 4.1.13 that $q \le r$. The constraints $q \le r'$ follows by duality arguments, namely by Lemma 4.1.7.

Now, we shall prove that

$$q \le s', \quad \frac{1}{p} \ge \frac{1}{2} - \frac{1}{r} + \frac{1}{q'}. \tag{4.30}$$

The remaining constraints will follow by duality as above.

Let $\chi \in \mathcal{D}(\mathbb{R}^d)$, with $0 \le \chi \le 1$, $\chi(0) = 1$. Then (4.28) implies

$$\|\chi a(x,D)f\|_{W(L^r,L^s)} \le C\|a\|_{M^{p,q}}\|f\|_{W(L^r,L^s)}, \quad \forall f \in \mathcal{S}(\mathbb{R}^d),\ \forall a \in \mathcal{S}(\mathbb{R}^{2d}).$$

Since for functions u supported in a fixed compact subset we have $\|u\|_{W(L^r,L^s)} \asymp \|u\|_r$, we deduce

$$\|\chi a(x,D)f\|_r \le C\|a\|_{M^{p,q}}\|f\|_{W(L^r,L^s)} \quad \forall f \in \mathcal{S}(\mathbb{R}^d),\ \forall a \in \mathcal{S}(\mathbb{R}^{2d}).$$

Then Proposition 4.1.12 implies (4.30). □

Proposition 4.1.15. *Suppose that, for some $1 \le p,q,r,s \le \infty$, every symbol $a \in M^{p,q}$ gives rise to an operator $a(x,D)$ bounded on $W(L^r,L^s)$. Then the constraints $q \le \min\{r, r', s, s'\}$ and (4.29) must hold.*

Proof. Let $\mathcal{W}(L^r, L^s)$ be the closure of $\mathcal{S}(\mathbb{R}^d)$ in $W(L^r, L^s)$. By assumption, the map

$$M^{p,q} \ni a \mapsto a(x,D) \in B(\mathcal{W}(L^r, L^s), W(L^r, L^s))$$

is well defined. By an application of the closed graph theorem and Theorem 4.1.14 we see that the desired conclusion follows if we prove that this map has a closed graph. To this end, let $a_n \to a$ in $M^{p,q}$, with $a_n(x,D) \to A$ in $B(\mathcal{W}(L^r, L^s), W(L^r, L^s))$. We have to prove that $A = a(x,D)$, i. e., $\langle Af, g\rangle = \langle a(x,D)f, g\rangle$ $\forall f, g \in \mathcal{S}(\mathbb{R}^d)$. Now, clearly, $\langle a_n(x,D)f, g\rangle \to \langle Af, g\rangle$. On the other hand, $\langle a_n(x,D)f, g\rangle = \langle a_n, G\rangle$, where $G(x,\omega) = \overline{e^{-2\pi i x\cdot\omega}\hat{f}(\omega)g(x)}$ is a fixed Schwartz function. Hence $\langle a_n(x,D)f, g\rangle$ tends to $\langle a, G\rangle = \langle a(x,D)f, g\rangle$, which concludes the proof. □

4.2 Symbols in Wiener amalgam spaces

As observed in the introduction to the chapter, a natural question which can be raised is whether the boundedness of pseudodifferential operators on L^2 or, more generally, on $M^{p,q}$, can be attained by widening the symbol Sjöstrand class $M^{\infty,1}$ to the Winer amalgam space $W(\mathcal{F}L^1, L^\infty)$. An example of a function belonging to $W(\mathcal{F}L^1, L^\infty)(\mathbb{R}^d) \setminus M^{\infty,1}(\mathbb{R}^d)$ is the chirp $\varphi(x) = e^{\pi i x^2}$, $x \in \mathbb{R}^d$, see Proposition 4.3.3 below. The multiplier operator $a(x,D)f(x) = e^{\pi i x^2} f(x)$ is a pseudodifferential operator with symbol $a(x,\xi) = (e^{\pi i|\cdot|^2} \otimes 1)(x,\xi) \in W(\mathcal{F}L^1, L^\infty)(\mathbb{R}^{2d}) \setminus M^{\infty,1}(\mathbb{R}^{2d})$, it is obviously bounded on $L^2(\mathbb{R}^d)$ and it is bounded on $M^{p,q}$ if and only if $p = q$, as shown below.

Proposition 4.2.1. *The multiplication $A : f \to e^{\pi i|\cdot|^2} f$ is unbounded on $M^{p,q}$, for every $1 \le p, q \le \infty$, with $p \ne q$.*

Proof. For $\lambda > 0$, we consider the one-parameter family of functions $f(x) = e^{-\pi\lambda x^2} \in M^{p,q}$, so that $\hat{f}(\xi) = \lambda^{-d/2} e^{-\pi(1/\lambda)\xi^2}$. For every $1 \le p, q \le \infty$, by (2.6.1), we have

$$\|f\|_{M^{p,q}} = \|\hat{f}\|_{W(\mathcal{F}L^p, L^q)} \asymp \frac{(\lambda+1)^{d(\frac{1}{p}-\frac{1}{2})}}{\lambda^{\frac{d}{2q}}(\lambda^2+\lambda)^{\frac{d}{2}(\frac{1}{p}-\frac{1}{q})}}.$$

Since $Af(x) = e^{-\pi(\lambda-i)x^2}$, its Fourier transform is given by

$$\widehat{Af}(\xi) = (\lambda - i)^{-d/2} e^{-\pi\frac{1}{\lambda-i}\xi^2},$$

so that the same formula as above yields

$$\|Af\|_{M^{p,q}} \asymp \frac{[(\lambda+1)^2+1]^{\frac{d}{2}(\frac{1}{p}-\frac{1}{2})}}{\lambda^{\frac{d}{2q}}(\lambda^2+\lambda+1)^{\frac{d}{2}(\frac{1}{p}-\frac{1}{q})}}.$$

As $\lambda \to 0$, we have

$$\|Af\|_{M^{p,q}} \asymp \lambda^{-\frac{d}{2q}}, \quad \|f\|_{M^{p,q}} \asymp \lambda^{-\frac{d}{2q}-\frac{d}{2}(\frac{1}{p}-\frac{1}{q})}$$

so that, if we assume $\|Af\|_{M^{p,q}} \le C\|f\|_{M^{p,q}}$, then $1/p - 1/q \ge 0$, that is, $p \le q$.

Moreover, the same argument applies to the adjoint operator $A^*f(x) = e^{-\pi i x^2} f(x)$.

Now we show that $p = q$. By contradiction, if A were bounded on $M^{p,q}$, with $p < q$, its adjoint A^* would satisfy

$$\|A^* f\|_{M^{p',q'}} \le C\|f\|_{M^{p',q'}}, \quad \forall f \in \mathcal{S}(\mathbb{R}^d),$$

with $q' < p'$, which is a contradiction to what has just been proved. □

The subsequent Proposition 4.2.4 shows that generally symbols in $W(\mathcal{F}L^1, L^\infty)$ do not produce bounded operators even on $L^2(\mathbb{R}^d)$. However, symbols expressed as tensor products $a(x,\xi) = a_1(x)a_2(\xi)$ are bounded on $M^p(\mathbb{R}^d)$, $1 \le p \le \infty$ (and hence on $L^2(\mathbb{R}^d)$), as shown in the next result.

Proposition 4.2.2. *If $a(x,\xi) = a_1(x)a_2(\xi)$, $a_i \in W(\mathcal{F}L^1, L^\infty)$, $i = 1, 2$, then $a(x,D)$ is bounded on $M^p(\mathbb{R}^d)$, $1 \le p \le \infty$.*

Proof. We have

$$a(x,D)f(x) = a_1(x)\mathcal{F}^{-1}(a_2\hat{f})(x) = a_1(x)[\mathcal{F}^{-1}(a_2) * f](x).$$

If $a_2 \in W(\mathcal{F}L^1, L^\infty)(\mathbb{R}^d)$, then $\mathcal{F}^{-1}(a_2) \in M^{1,\infty}(\mathbb{R}^d)$ and $\mathcal{F}^{-1}(a_2) * f \in M^{1,\infty}(\mathbb{R}^d) * M^p(\mathbb{R}^d) \hookrightarrow M^p(\mathbb{R}^d)$; see Proposition 2.4.19. It remains to show that the multiplier $A_{a_1}f(x) = a_1(x)f(x)$ is bounded on M^p. This immediately follows by the pointwise products for Wiener amalgam spaces in Theorem 2.4.9. Indeed, $\mathcal{F}L^1 \cdot \mathcal{F}L^p = \mathcal{F}(L^1 * L^p) \hookrightarrow \mathcal{F}L^p$, $L^\infty \cdot L^p \hookrightarrow L^p$, so that, for $M^p = W(\mathcal{F}L^p, L^p)$, we have $W(\mathcal{F}L^1, L^\infty) \cdot W(\mathcal{F}L^p, L^p) \hookrightarrow W(\mathcal{F}L^p, L^p)$, as desired. □

In particular, we obtain that the multiplier operator $a(x,D)f(x) = e^{\pi i|x|^2} f(x)$, which is not bounded on $M^{p,q}$, $p \ne q$, as discussed above, is bounded on M^p.

We now come to a necessary condition.

Proposition 4.2.3. *Suppose that, for some $1 \le p, q, r, s, r_1, r_2 \le \infty$, $C > 0$, either the estimate*

$$\|a(x,D)f\|_{W(L^{r_1},L^{r_2})} \le C\|a\|_{W(\mathcal{F}L^p,L^q)}\|f\|_{W(L^r,L^s)}, \quad \forall a \in \mathcal{S}(\mathbb{R}^{2d}),\ f \in \mathcal{S}(\mathbb{R}^d), \tag{4.31}$$

or

$$\|\mathrm{Op_w}(\sigma)f\|_{W(L^{r_1},L^{r_2})} \le C\|\sigma\|_{W(\mathcal{F}L^p,L^q)}\|f\|_{W(L^r,L^s)}, \quad \forall a \in \mathcal{S}(\mathbb{R}^{2d}),\ f \in \mathcal{S}(\mathbb{R}^d), \tag{4.32}$$

holds. Then $q \le r$.

Proof. We first suppose that (4.31) holds true. We test that estimate on the same families of functions and symbols as in the proof of Proposition 4.1.13, but with (4.18) replaced by

$$\operatorname{supp} h_1 \subset B_1(0), \quad \operatorname{supp} h_2 \subset B_1(0), \tag{4.33}$$

(the other conditions being unchanged).

Then (4.23) and (4.27) still hold, because in their proof we did not use (4.18). On the other hand, we can prove that

$$\|a_N\|_{W(\mathcal{F}L^p,L^q)} \lesssim N^{d/q}, \tag{4.34}$$

by using the same arguments in the proof of (4.24), with the roles of h_1, h_2 replaced by $\widehat{h_2}$ and $\widehat{h_1}$, respectively. Precisely, we now choose a window $\Phi = \varphi \otimes \varphi$ with $\varphi \in \mathcal{S}(\mathbb{R}^d)$ supported in $B_1(0)$. Then the functions $V_\Phi b_n(z_1, z_2, \omega_1, \omega_2)$ vanish unless $z_1 \in B_2(-n)$, $z_2 \in B_2(0)$, so that they have disjoint supports, as well as $\|V_\Phi b_n(z_1, z_2, \cdot, \cdot)\|_p$. Hence one deduces that

$$\begin{aligned}\|a_n\|_{W(\mathcal{F}L^p,L^q)} &\asymp \left(\int_{\mathbb{R}^{2d}} \Big\|\sum_{n\in\Lambda_N} V_\Phi b_n(z_1, z_2, \cdot, \cdot)\Big\|_p^q \, dz_1\, dz_2\right)^{1/q} \\ &= \left(\sum_{n\in\Lambda_N} \int_{B_2(-n)} \int_{B_2(0)} \|V_\Phi b_n(z_1, z_2, \cdot, \cdot)\|_p^q \, dz_1\, dz_2\right)^{1/q}.\end{aligned}$$

On the other hand, by Young's inequality, we have

$$\|V_\Phi b_n(z_1, z_2, \cdot, \cdot)\|_p \le \|\widehat{b_n}\|_p$$

and the right-hand side of the above inequality is independent of n. Hence (4.34) follows.

We now suppose that the estimate (4.32) holds. Since the arguments are similar to those just used, we only sketch the main point of the proof. We test the estimate (4.32) on the family of functions $f_N(x) = h(Nx)$, where h is a smooth real-valued function, with $h \ge 0$, $h(0) = 1$, supported in a ball $B_\epsilon(0)$, for a sufficiently small ϵ, and symbols

$$\sigma_N(x, \xi) = \sum_{n\in\Lambda_N} b_n(x, \xi), \quad \text{where } b_n(x, \xi) = (M_{-2n}h_1)(x)(T_n h_2)(\xi).$$

The lattice Λ_N is defined in (4.19) and h_1, h_2 are two Schwartz functions in $\mathbb{R}^d$ such that h_1 and $\widehat{h_2}$ are real valued, with $h_1(0) = 1$, $\widehat{h_2}(0) = 1$, and satisfying (4.33). By using the definition (4.6), we see that $\mathrm{Op}_w(\sigma)$ has integral kernel

$$K(x, y) = \mathcal{F}_2^{-1}\sigma\left(\frac{x+y}{2}, x-y\right) = \sum_{n\in\Lambda_N} e^{-4\pi i n\cdot y} h_1\left(\frac{x+y}{2}\right)\widehat{h_2}(y-x).$$

Hence, by arguing as in the proof of (4.22), one obtains, for a suitable $\delta > 0$,

$$|\mathrm{Op}_w(\sigma_N) f_N(x)| \gtrsim 1, \quad \text{for } x \in B_\delta(0),$$

which implies

$$\|\mathrm{Op}_w(\sigma_N) f_N\|_{W(L^{r_1}, L^{r_2})} \gtrsim 1.$$

The arguments in the first part of the present proof, with essentially no changes, show that

$$\|\sigma_N\|_{W(\mathcal{F}L^p,L^q)} \lesssim N^{d/q}.$$

Combining these estimates with $\|f_N\|_{W(L^r,L^s)} \lesssim N^{-d/r}$ and letting $N \to +\infty$ gives the desired conclusion. □

Proposition 4.2.4. *Suppose that, for some* $1 \le p,q,r,s \le \infty$, *every symbol* $a \in W(\mathcal{F}L^p, L^q)$ *gives rise to a bounded operator on* $W(L^r, L^s)$. *Then* $q \le r$. *The same happens if one replaces the Kohn–Nirenberg operator* $a(x,D)$ *by the Weyl one* $\mathrm{Op}_w(a)$.

Proof. The result follows from Proposition 4.2.3 by arguing as in the proof of Proposition 4.1.15. □

In particular, we obtain that for generic symbols in $W(\mathcal{F}L^1, L^\infty)(\mathbb{R}^{2d})$ continuity on $L^2(\mathbb{R}^d)$ may fail. We also deduce that L^2-boundedness fails in general for $a \in W(\mathcal{F}L^2, L^q)(\mathbb{R}^{2d})$ if $q > 2$. This result is sharp for Weyl operators, since for $a \in W(\mathcal{F}L^2, L^2)(\mathbb{R}^{2d}) = L^2(\mathbb{R}^{2d})$ the operator $\mathrm{Op}_w(a)$ is L^2-continuous, whereas obviously there are symbols $a \in L^2(\mathbb{R}^{2d})$ for which $a(x,D)$ is not bounded.

We finally tackle the problem of the invariance of the Wiener amalgam spaces $W(\mathcal{F}L^p, L^q)$ under the action of the operator $\mathcal{U}$ in (4.8), which expresses the Kohn–Nirenberg symbol of an operator in terms of the Weyl one. The lack of invariance, expressed by the following result, justifies the fact that the necessary conditions in this section were proved for both Kohn–Nirenberg and Weyl quantizations.

Proposition 4.2.5. *Let* $1 \le p,q \le \infty$. *If* $p \ne q$, *the operator* $\mathcal{U}$ *in* (4.8) *does not map* $W(\mathcal{F}L^p, L^q)$ *into itself.*

The proof, limited to the relevant case $p = 1, q = \infty$, is given by the following remark. For the general proof, we refer the reader to Proposition 6.1.21 in Chapter 6.

Remark 4.2.6. A concrete example of a Weyl symbol $\sigma \in W(\mathcal{F}L^1, L^\infty)$ such that the corresponding Kohn–Nirenberg symbol $a = \mathcal{U}\sigma$ does not belong to $W(\mathcal{F}L^1, L^\infty)$ is provided by $\sigma = \mathcal{U}^{-1}\delta$; therefore $a = \delta$. It is clear that δ does not belong to $W(\mathcal{F}L^1, L^\infty)$ (which consists only of continuous functions), whereas from (4.8) we have $\hat{\delta} = \widehat{\mathcal{U}\sigma} = e^{\pi i \xi_1\cdot\xi_2}\hat{\sigma}(\xi_1,\xi_2)$. Hence $\sigma = \mathcal{F}^{-1}(e^{-\pi i \xi_1\cdot\xi_2})$, belonging to $W(\mathcal{F}L^1, L^\infty)$, see (4.40) and Proposition 4.3.3 in the next section for details.

4.3 Pseudodifferential operators on modulation spaces

If we consider $\tau \in [0,1]$, the Shubin τ-representation of a pseudodifferential operator with symbol $a \in \mathcal{S}(\mathbb{R}^{2d})$ is defined by

$$\mathrm{Op}_\tau(a)f(x) = \int_{\mathbb{R}^{2d}} e^{2\pi i(x-y)\cdot\xi} a((1-\tau)x + \tau y, \xi) f(y)\, dy\, d\xi. \tag{4.35}$$

For $\tau = 1/2$, we recapture the Weyl operator in (4.4), that is, $\mathrm{Op}_{1/2}(a) = \mathrm{Op}_w(a)$. The case $\tau = 0$ gives $\mathrm{Op}_0(a) = a(x, D)$, that is, the Kohn–Nirenberg operator with symbol a. Indeed, we can generalize the weak definition of a Kohn–Nirenberg operator in (4.1) as follows.

Proposition 4.3.1. *Consider $a \in \mathcal{S}(\mathbb{R}^{2d})$, $f, g \in \mathcal{S}(\mathbb{R}^d)$. Then*

$$\langle \mathrm{Op}_\tau(a)f, g\rangle = \langle a, W_\tau(g, f)\rangle, \quad f, g \in \mathcal{S}(\mathbb{R}^d), \tag{4.36}$$

where $W_\tau(g, f)$ is the (cross-)τ-Wigner distribution defined in (1.109).

As already said for Weyl operators, the result is an easy consequence of a change of variables and an interchange of integration order. The case $\tau = 1/2$ gives the equality for Weyl operators in (4.5).

Almost diagonalization of τ-pseudodifferential operators with symbols in Wiener amalgam and modulation spaces is shown in [86].

Extensions of τ-operators with $\tau \in \mathbb{R}$ are treated in [45]. More general matrix formulations can be found in [8, 90, 290].

The goal of this section is the study of the continuity properties of pseudodifferential operators on modulation spaces. Their symbol will belong to modulation spaces as well. A direct computation, cf. [193] (see also [286, Remark 1.5]), shows that for every choice $\tau_1, \tau_2 \in [0, 1]$, $a_1, a_2 \in \mathcal{S}'(\mathbb{R}^{2d})$,

$$\mathrm{Op}_{\tau_1}(a_1) = \mathrm{Op}_{\tau_2}(a_2) \Longleftrightarrow \widehat{a_2}(\xi_1, \xi_2) = e^{-2\pi i(\tau_2-\tau_1)\xi_1\cdot\xi_2}\widehat{a_1}(\xi_1, \xi_2), \tag{4.37}$$

so for $\tau_1 \neq \tau_2$, applying the Fourier transform to both sides of the equality on the right-hand side of (4.37), we obtain

$$a_2(x, \xi) = \frac{1}{|\tau_1 - \tau_2|^d} e^{2\pi i(\tau_2-\tau_1)\Phi} * a_1(x, \xi), \tag{4.38}$$

where $\Phi(x, \xi) = x \cdot \xi$, see the subsequent formula (4.40).

The first step will be to show that the mapping $a \mapsto T_\Phi a = e^{2\pi i \tau\Phi} * a$ is a homeomorphism on $M^{p,q}_{1\otimes v_s}(\mathbb{R}^{2d})$, $1 \le p, q \le \infty$.

As a consequence, the results for pseudodifferential operators in the Weyl form with symbols in the modulation spaces above are valid for any other τ-pseudodifferential operator.

Let us recall the distribution Fourier transform of the generalized chirp below (cf. [145, Appendix A, Theorem 2]):

Proposition 4.3.2. *Let B a real, invertible, symmetric $n\times n$ matrix, and let $F_B(x) = e^{\pi i x\cdot Bx}$. Then the distribution Fourier transform of F_B is given by*

$$\widehat{F_B}(\xi) = e^{\pi i \sharp(B)/4} |\det B| e^{-\pi i \xi\cdot B^{-1}\xi}, \tag{4.39}$$

where $\sharp(B)$ is the number of positive eigenvalues of B minus the number of negative eigenvalues.

If we choose

$$B = B^{-1} = \begin{pmatrix} 0_{d\times d} & I_{d\times d} \\ I_{d\times d} & 0_{d\times d} \end{pmatrix}$$

(hence $n = 2d$) then formula (4.39) becomes

$$\mathcal{F}(e^{2\pi i\zeta_1\cdot\zeta_2})(z_1,z_2) = \widehat{F_B}(z_1,z_2) = e^{-2\pi i z_1\cdot z_2}. \tag{4.40}$$

We now show that F_B belongs to $W(\mathcal{F}L^1,L^\infty)(\mathbb{R}^{2d})$.

Proposition 4.3.3. *The chirp function $\varphi(t) = e^{\pi i t^2}$, $t \in \mathbb{R}^d$, belongs to $W(\mathcal{F}L^1,L^\infty)(\mathbb{R}^d)$. The function $F(\zeta_1,\zeta_2) = e^{2\pi i\zeta_1\cdot\zeta_2}$ belongs to $W(\mathcal{F}L^1,L^\infty)(\mathbb{R}^{2d})$.*

Proof. We fix attention on F, the argument for φ is similar. Consider the Gaussian function $g(\zeta_1,\zeta_2) = e^{-\pi\zeta_1^2}e^{-\pi\zeta_2^2}$ as window function to compute the $W(\mathcal{F}L^1,L^\infty)$-norm. Then we have

$$\|F\|_{W(\mathcal{F}L^1,L^\infty)(\mathbb{R}^{2d})} = \sup_{u\in\mathbb{R}^{2d}} \|\mathcal{F}(FT_u g)\|_{L^1(\mathbb{R}^{2d})}.$$

Let us compute $\mathcal{F}(FT_u g)(z)$. For $z = (z_1,z_2)$ by using (4.40),

$$\begin{aligned}
\mathcal{F}(FT_u g)(z_1,z_2) &= (\mathcal{F}(F) * M_{-u}\hat{g})(z_1,z_2) \\
&= \int_{\mathbb{R}^{2d}} e^{-2\pi i(z_1-y_1)\cdot(z_2-y_2)} e^{-2\pi i(u_1,u_2)\cdot(y_1,y_2)} e^{-\pi y_1^2} e^{-\pi y_2^2}\, dy_1\, dy_2 \\
&= e^{-2\pi i z_1\cdot z_2} \int_{\mathbb{R}^{2d}} e^{-2\pi i y_1\cdot y_2 + 2\pi i(z_2\cdot y_1 + z_1\cdot y_2) - 2\pi i(u_1\cdot y_1 + u_2\cdot y_2)} e^{-\pi y_1^2} e^{-\pi y_2^2}\, dy_1\, dy_2 \\
&= e^{-2\pi i z_1\cdot z_2} \int_{\mathbb{R}^d} e^{2\pi i(z_1\cdot y_2 - u_2\cdot y_2)} e^{-\pi y_2^2} \Big(\int_{\mathbb{R}^d} e^{-2\pi i y_1\cdot(y_2 - z_2 + u_1)} e^{-\pi y_1^2}\, dy_1 \Big)\, dy_2 \\
&= e^{-2\pi i z_1\cdot z_2} \int_{\mathbb{R}^d} e^{-2\pi i y_2\cdot(u_2 - z_1)} e^{-\pi y_2^2} e^{-\pi(y_2 - z_2 + u_1)^2}\, dy_2 \\
&= e^{-2\pi i z_1\cdot z_2} e^{-\pi(u_1 - z_2)^2} \int_{\mathbb{R}^d} e^{-2\pi i y_2\cdot(u_2 - z_1)} e^{-2\pi(y_2^2 + (u_1 - z_2)\cdot y_2)}\, dy_2 \\
&= e^{-2\pi i z_1\cdot z_2} e^{-\pi(u_1 - z_2)^2 + \frac{\pi}{2}(u_1 - z_2)^2} \int_{\mathbb{R}^d} e^{-2\pi i y_2\cdot(u_2 - z_1)} e^{-2\pi(y_2 + \frac{u_1 - z_2}{2})^2}\, dy_2 \\
&= e^{-2\pi i z_1\cdot z_2} e^{-\frac{\pi}{2}(u_1 - z_2)^2} \mathcal{F}(T_{-\frac{u_1 - z_2}{2}} e^{-2\pi|\cdot|^2})(u_2 - z_1) \\
&= 2^{-d/2} e^{-2\pi i z_1\cdot z_2} e^{-\frac{\pi}{2}(u_1 - z_2)^2} e^{-\frac{\pi}{2}(u_2 - z_1)^2} e^{\pi i(u_1 - z_2)\cdot(u_2 - z_1)} \\
&= 2^{-d/2} e^{\pi i(u_1\cdot u_2 - z_1\cdot z_2 - u_1\cdot z_1 - u_2\cdot z_2)} e^{-\frac{\pi}{2}(u_1 - z_2)^2} e^{-\frac{\pi}{2}(u_2 - z_1)^2}.
\end{aligned}$$

Hence

$$\|\mathcal{F}(FT_u g)\|_{L^1} = 2^{-d/2} \|e^{-\frac{\pi}{2}|\cdot|^2}\|_{L^1} = C,$$

for a constant C independent of the variable u and, consequently,

$$\sup_{u\in\mathbb{R}^{2d}} \|\mathcal{F}(FT_u g)\|_{L^1} < \infty$$

as desired. □

Corollary 4.3.4. *For $\zeta = (\zeta_1, \zeta_2)$, consider the function $F_J(\zeta) = F(J\zeta) = e^{-2\pi i\zeta_1\cdot\zeta_2}$, where the symplectic matrix J is defined in* (1.1). *Then $F_J \in W(\mathcal{F}L^1, L^\infty)(\mathbb{R}^{2d})$.*

Proof. The result immediately follows by Proposition 4.3.3 and by the dilation properties for modulation spaces (2.107). □

Corollary 4.3.5. *For $0 < |\tau| \le 1$, the function $e^{2\pi i\tau\Phi}$, $\Phi(x,\xi) = x\cdot\xi$, is in $M^{1,\infty}(\mathbb{R}^{2d})$.*

Proof. By Proposition 4.3.3, the function $e^{2\pi i\Phi}$ is in $M^{1,\infty}(\mathbb{R}^{2d})$, being the Fourier transform of the function F_J defined in Corollary 4.3.4. The dilation properties for modulation spaces in Theorem 2.6.6 give the claim. □

Proposition 4.3.6. *For $0 < |\tau| \le 1$, the mapping $a \mapsto T_\Phi a = e^{2\pi i\tau\Phi} * a$ is a homeomorphism on $M^{p,q}_{1\otimes v_s}(\mathbb{R}^{2d})$, $1 \le p, q \le \infty$.*

Proof. Using the convolution properties for modulation spaces in (2.81) with $p_1 = p$, $q_1 = q$, $p_2 = 1$ and $q_2 = \infty$, we get

$$\|T_\Phi a\|_{M^{p,q}_{1\otimes v_s}} = \|e^{2\pi i\tau\Phi} * a\|_{M^{p,q}_{1\otimes v_s}} \lesssim \|e^{2\pi i\tau\Phi}\|_{M^{1,\infty}} \|a\|_{M^{p,q}_{1\otimes v_s}},$$

as desired. □

From now on, we shall focus only on boundedness properties for Weyl operators, but the same results hold for any τ-pseudodifferential operator by the previous observations.

The key idea is the study of boundedness properties for the Wigner distribution.

4.4 Continuity results for the (cross-)Wigner distribution

An important issue related to such a distribution is the continuity of the map $(f_1, f_2) \mapsto W(f_1, f_2)$ in the relevant Banach spaces. The basic result in this connection is the Moyal's identity (1.82)

$$\|W(f_1,f_2)\|_{L^2(\mathbb{R}^{2d})} = \|f_1\|_{L^2(\mathbb{R}^d)}\|f_2\|_{L^2(\mathbb{R}^d)}, \quad f_1, f_2 \in L^2(\mathbb{R}^d).$$

Beside L^2, the time–frequency concentration of signals is often measured by the modulation space norm $M^{p,q}_m$, $1 \le p, q \le \infty$, for a suitable weight function m. Although several particular sufficient conditions have appeared in this connection in the literature, the general problem was solved in the paper [75], by providing the full range of modulation spaces in which the continuity of the cross-Wigner distribution $W(f,g)$ holds, as a function of f, g. The consequences of these results are manifold: new bounds for the short-time Fourier transform and the ambiguity function, boundedness results for pseudodifferential (in particular, localization) operators, and properties of the Cohen class (see the next sections).

Theorem 4.4.1. *Assume $p_i, q_i, p, q \in [1,\infty]$, $s \in \mathbb{R}$, are such that*

$$p_i, q_i \le q, \quad i = 1, 2, \tag{4.41}$$

and

$$\frac{1}{p_1} + \frac{1}{p_2} \ge \frac{1}{p} + \frac{1}{q}, \qquad \frac{1}{q_1} + \frac{1}{q_2} \ge \frac{1}{p} + \frac{1}{q}. \tag{4.42}$$

Then if $f_1 \in M^{p_1,q_1}_{v_{|s|}}(\mathbb{R}^d)$ and $f_2 \in M^{p_2,q_2}_{v_s}(\mathbb{R}^d)$, we have $W(f_1,f_2) \in M^{p,q}_{1\otimes v_s}(\mathbb{R}^{2d})$ and

$$\|W(f_1,f_2)\|_{M^{p,q}_{1\otimes v_s}} \lesssim \|f_1\|_{M^{p_1,q_1}_{v_{|s|}}} \|f_2\|_{M^{p_2,q_2}_{v_s}}. \tag{4.43}$$

Conversely, assume that there exists a constant $C > 0$ such that

$$\|W(f_1,f_2)\|_{M^{p,q}} \le C\|f_1\|_{M^{p_1,q_1}} \|f_2\|_{M^{p_2,q_2}}, \quad \forall f_1, f_2 \in \mathcal{S}(\mathbb{R}^{2d}). \tag{4.44}$$

Then (4.41) *and* (4.42) *must hold.*

In this section we prove Theorem 4.4.1. We will focus separately on the sufficient and necessary part in the statement. The sufficient conditions were proved by Toft in [287, Theorem 4.2] (cf. also [286, Theorem 4.1] for modulation spaces without weights) under the conditions

$$p \le p_i, q_i \le q, \quad i = 1, 2,$$

and

$$\frac{1}{p_1} + \frac{1}{p_2} = \frac{1}{q_1} + \frac{1}{q_2} = \frac{1}{p} + \frac{1}{q}. \tag{4.45}$$

Theorem 4.4.1 shows that the sufficient conditions can be widened and such an extension is sharp.

Theorem 4.4.2 (Sufficient conditions). *If $p_1, q_1, p_2, q_2, p, q \in [1,\infty]$ are indices which satisfy* (4.41) *and* (4.42), *$s \in \mathbb{R}$, $f_1 \in M^{p_1,q_1}_{v_{|s|}}(\mathbb{R}^d)$ and $f_2 \in M^{p_2,q_2}_{v_s}(\mathbb{R}^d)$, then $W(f_1,f_2) \in M^{p,q}_{1\otimes v_s}(\mathbb{R}^{2d})$, and the estimate* (4.43) *holds true.*

Proof. We first study the case $p, q < \infty$. Let $g \in \mathcal{S}(\mathbb{R}^d)$ and set $\Phi = W(g,g) \in \mathcal{S}(\mathbb{R}^{2d})$. If $\zeta = (\zeta_1, \zeta_2) \in \mathbb{R}^{2d}$, we write $\tilde{\zeta} = (\zeta_2, -\zeta_1)$. Then from Lemma 1.3.37,

$$|V_\Phi(W(f_1,f_2))(z,\zeta)| = \left|V_g f_2\left(z + \frac{\tilde{\zeta}}{2}\right)\right| \left|V_g f_1\left(z - \frac{\tilde{\zeta}}{2}\right)\right|. \tag{4.46}$$

Consequently,

$$\|W(f_1,f_2)\|_{M^{p,q}_{1\otimes v_s}} \asymp \Big(\int_{\mathbb{R}^{2d}}\Big(\int_{\mathbb{R}^{2d}}\Big|V_gf_2\Big(z+\frac{\tilde{\zeta}}{2}\Big)\Big|^p\Big|V_gf_1\Big(z-\frac{\tilde{\zeta}}{2}\Big)\Big|^p dz\Big)^{\frac{q}{p}}\langle\zeta\rangle^{sq}\,d\zeta\Big)^{1/q}.$$

After the change of variables $z \mapsto z-\tilde{\zeta}/2$, the integral over z becomes the convolution $(|V_gf_2|^p * |(V_gf_1)^*|^p)(\tilde{\zeta})$, and observing that $(1\otimes v_s)(z,\zeta) = \langle\zeta\rangle^s = v_s(\zeta) = v_s(\tilde{\zeta})$, we obtain

$$\begin{aligned}\|W(f_1,f_2)\|_{M^{p,q}_{1\otimes v_s}} &\asymp \Big(\iint_{\mathbb{R}^{2d}}(|V_gf_2|^p * |(V_gf_1)^*|^p)^{\frac{q}{p}}(\tilde{\zeta})v_s(\tilde{\zeta})^q\,d\zeta\Big)^{1/p}\\ &= \big\||V_gf_2|^p * |(V_gf_1)^*|^p\big\|^{\frac{1}{p}}_{L^{\frac{q}{p}}_{v_{ps}}}.\end{aligned}$$

Hence

$$\|W(f_1,f_2)\|^p_{M^{p,q}_{1\otimes v_s}} \asymp \big\||V_gf_2|^p * |(V_gf_1)^*|^p\big\|_{L^{\frac{q}{p}}_{v_{ps}}}. \tag{4.47}$$

Case $p \le q < \infty$.

Step 1. Here we prove the desired result in the case $p \le p_i, q_i$, $i = 1,2$.

Suppose first that (4.45) are satisfied (and hence $p_i, q_i \le q$, $i = 1,2$). Since $q/p \ge 1$, we can apply Young's inequality for mixed-normed spaces and majorize (4.47) as follows:

$$\begin{aligned}\|W(f_1,f_2)\|^p_{M^{p,q}_{1\otimes v_s}} &\lesssim \||V_gf_2|^p\|_{L^{r_2,s_2}_{v_{p|s|}}}\||(V_gf_1)^*|^p\|_{L^{r_1,s_1}_{v_{ps}}}\\ &= \||V_gf_1|^p\|_{L^{r_1,s_1}_{v_{p|s|}}}\||V_gf_2|^p\|_{L^{r_2,s_2}_{v_{ps}}}\\ &= \|V_gf_1\|^p_{L^{pr_1,ps_1}_{v_{|s|}}}\|V_gf_2\|^p_{L^{pr_2,ps_2}_{v_s}},\end{aligned}$$

for every $1 \le r_1, r_2, s_1, s_2 \le \infty$ such that

$$\frac{1}{r_1}+\frac{1}{r_2} = \frac{1}{s_1}+\frac{1}{s_2} = 1+\frac{p}{q}. \tag{4.48}$$

Choosing $r_i = p_i/p \ge 1$, $s_i = q_i/p \ge 1$, $i = 1,2$, the indices' relation (4.48) becomes (4.45), and we obtain

$$\|W(f_1,f_2)\|_{M^{p,q}_{1\otimes v_s}} \lesssim \|V_gf_1\|_{L^{p_1,q_1}_{v_{|s|}}}\|V_gf_2\|_{L^{p_2,q_2}_{v_s}} \asymp \|f_1\|_{M^{p_1,q_1}_{v_{|s|}}}\|f_2\|_{M^{p_2,q_2}_{v_s}}.$$

Now, still assume $p \le p_i, q_i$, $i = 1,2$ but

$$\frac{1}{p_1}+\frac{1}{p_2} \ge \frac{1}{p}+\frac{1}{q}, \qquad \frac{1}{q_1}+\frac{1}{q_2} = \frac{1}{p}+\frac{1}{q},$$

(hence $p_i, q_i \le q$, $i = 1, 2$). We set $u_1 = tp_1$, and look for $t \ge 1$ (hence $u_1 \ge p_1$) such that

$$\frac{1}{u_1} + \frac{1}{p_2} = \frac{1}{p} + \frac{1}{q},$$

which gives

$$0 < \frac{1}{t} = \frac{p_1}{p} + \frac{p_1}{q} - \frac{p_1}{p_2} \le 1$$

because $p_1(1/p + 1/q) - p_1/p_2 \le p_1(1/p_1 + 1/p_2) - p_1/p_2 = 1$, whereas the lower bound of the previous estimate follows by $1/(tp_1) = 1/p + 1/q - 1/p_2 > 0$ since $p \le p_2$. Hence the previous part of the proof gives

$$\begin{aligned}\|W(f_1, f_2)\|_{M^{p,q}_{1\otimes v_s}} &\lesssim \|f_1\|_{M^{u_1,q_1}_{v_{|s|}}} \|f_2\|_{M^{p_2,q_2}_{v_s}} \\ &\lesssim \|f_1\|_{M^{p_1,q_1}_{v_{|s|}}} \|f_2\|_{M^{p_2,q_2}_{v_s}}\end{aligned}$$

where the last inequality follows by inclusion relations for modulations spaces $M^{p_1,q_1}_{v_s}(\mathbb{R}^d) \subseteq M^{u_1,q_1}_{v_s}(\mathbb{R}^d)$ for $p_1 \le u_1$.

The general case, when

$$\frac{1}{p_1} + \frac{1}{p_2} \ge \frac{1}{p} + \frac{1}{q}, \qquad \frac{1}{q_1} + \frac{1}{q_2} \ge \frac{1}{p} + \frac{1}{q},$$

can be treated analogously.

Step 2. Assume now that $p_i, q_i \le q$, $i = 1, 2$, and they satisfy relation (4.42). If at least one of the indices p_1, p_2 is less than p, assume, for instance, $p_1 \le p$, whereas $p \le q_1, q_2$, then we proceed as follows. We choose $u_1 = p$, $u_2 = q$, and deduce by the results in Step 1 (with $p_1 = u_1$ and $p_2 = u_2$) that

$$\|W(f_1, f_2)\|_{M^{p,q}_{1\otimes v_s}} \lesssim \|f_1\|_{M^{u_1,q_1}_{v_{|s|}}} \|f_2\|_{M^{u_2,q_2}_{v_s}} \lesssim \|f_1\|_{M^{p_1,q_1}_{v_{|s|}}} \|f_2\|_{M^{p_2,q_2}_{v_s}}$$

where the last inequality follows by inclusion relations for modulation spaces, since $p_1 \le u_1 = p$ and $p_2 \le u_2 = q$.

Similarly, we argue when at least one of the indices q_1, q_2 is less than p and $p \le p_1, p_2$ or when at least one of the indices q_1, q_2 is less than p and at least one of the indices p_1, p_2 is less than p. The remaining case $p \le p_i, q_i \le q$ is treated in Step 1.

Case $p < q = \infty$. The argument is similar to the case $p \le q < \infty$.

Case $p = q = \infty$. We use (4.46) and the submultiplicative property of the weight v_s to get

$$\begin{aligned}\|W(f_1, f_2)\|_{M^{\infty}_{1\otimes v_s}} &= \sup_{z,\zeta\in\mathbb{R}^{2d}} \left|V_g f_2\left(z + \frac{\tilde{\zeta}}{2}\right)\right| \left|V_g f_1\left(z - \frac{\tilde{\zeta}}{2}\right)\right| v_s(\zeta) \\ &= \sup_{z,\zeta\in\mathbb{R}^{2d}} |V_g f_2(z)| |(V_g f_1)^*(z - \tilde{\zeta})| v_s(\zeta)\end{aligned}$$

$$= \sup_{z,\zeta\in\mathbb{R}^{2d}} \big|V_gf_2(z)\big|\big|(V_gf_1)^*(z-\tilde{\zeta})\big|v_s(\tilde{\zeta})$$

$$\leq \sup_{z\in\mathbb{R}^{2d}} (\|V_gf_1v_{|s|}\|_\infty\big|V_gf_2(z)v_s(z)\big|) = \|V_gf_1v_{|s|}\|_\infty\|V_gf_2v_s\|_\infty$$

$$\asymp \|f\|_{M^\infty_{v_{|s|}}}\|g\|_{M^\infty_{v_s}} \lesssim \|f\|_{M^{p_1,q_1}_{v_{|s|}}}\|f\|_{M^{p_2,q_2}_{v_s}},$$

for every $1 \leq p_i, q_i \leq \infty$, $i = 1, 2$. Notice that in this case conditions (4.41) and (4.44) are trivially satisfied.

Case $p > q$. Using the inclusion relations for modulation spaces, we majorize

$$\|W(f_1,f_2)\|_{M^{p,q}_{1\otimes v_s}} \lesssim \|W(f_1,f_2)\|_{M^{q,q}_{1\otimes v_s}} \lesssim \|f_1\|_{M^{p_1,q_1}_{v_{|s|}}}\|f_2\|_{M^{p_2,q_2}_{v_s}}$$

for every $1 \leq p_i, q_i \leq q$, $i = 1, 2$. Here we have applied the case $p \leq q$ with $p = q$. Notice that in this case condition (4.44) is trivially satisfied, since from $p_1, q_i \leq q$ we infer $1/p_1 + 1/p_2 \geq 1/q + 1/q$, $1/q_1 + 1/q_2 \geq 1/q + 1/q$. This concludes the proof. □

Remark 4.4.3.

(i) The result of Theorem 4.4.2 can be extended to more general weights. In particular, it holds for polynomial weights satisfying relation (4.10) in [287]. Hence our result extends Toft's result [287, Theorem 4.2] (cf. also [286, Theorem 4.1] for modulation spaces without weights). Other examples of suitable weights are given by (sub-)exponential weights of the type $v(z) = e^{\alpha|z|^\beta}$ for $\alpha > 0$ and $0 < \beta \leq 1$.

(ii) The particular case $p = 1$, $1 \leq q \leq \infty$, $p_1 = q_1 = 1$, $p_2 = q_2 = q$, $s \geq 0$, was already proved in [66, Proposition 2.2].

(iii) Among modulation spaces, the following well-known function spaces occur: Weighted L^2-spaces: if $v_s(x) = \langle x\rangle^s$, then

$$M^2_{v_s\otimes 1}(\mathbb{R}^d) = L^2_s(\mathbb{R}^d) = \{f : f(x)\langle x\rangle^s \in L^2(\mathbb{R}^d)\};$$

Sobolev spaces: for $s \in \mathbb{R}$, $v_s(\xi) = \langle\xi\rangle^s$,

$$M^2_{1\otimes v_s}(\mathbb{R}^d) = H^s(\mathbb{R}^d) = \{f : \hat{f}(\xi)\langle\xi\rangle^s \in L^2(\mathbb{R}^d)\};$$

Shubin–Sobolev spaces [265] (see their definition in (4.66) below): if $v_s(x,\xi) = \langle(x,\xi)\rangle^s$, then

$$M^2_{v_s}(\mathbb{R}^d) = L^2_s(\mathbb{R}^d) \cap H^s(\mathbb{R}^d) = Q_s(\mathbb{R}^d).$$

For $p_i = q_i = p = q = 2$, $i = 1, 2$, we obtain the following continuity result for the cross-Wigner distribution acting between Shubin spaces and Sobolev spaces: For $s \geq 0$, $f_1, f_2 \in Q_s(\mathbb{R}^d)$, the cross-Wigner distribution $W(f_1,f_2)$ is in $H^s(\mathbb{R}^{2d})$ with

$$\|W(f_1,f_2)\|_{H^s(\mathbb{R}^{2d})} \lesssim \|f_1\|_{Q_s(\mathbb{R}^d)}\|f_2\|_{Q_s(\mathbb{R}^d)}.$$

(iv) Continuity properties of the cross-Wigner distribution on modulation spaces with different weight functions can be inferred using the techniques of Theorem 4.4.2 and the Young-type inequalities for weighted spaces shown by Toft et al. in [292, Theorems 2.2–2.5].

The estimate in (4.43) can be slightly improved if $s \geq 0$. Precisely, we have the following result.

Theorem 4.4.4. *If $p_1, q_1, p_2, q_2, p, q \in [1, \infty]$ are indices which satisfy (4.41) and (4.42), $s \geq 0$, $f_1 \in M^{p_1,q_1}_{v_s}(\mathbb{R}^d)$ and $f_2 \in M^{p_2,q_2}_{v_s}(\mathbb{R}^d)$, then $W(f_1, f_2) \in M^{p,q}_{1\otimes v_s}(\mathbb{R}^{2d})$, with*

$$\|W(f_1, f_2)\|_{M^{p,q}_{1\otimes v_s}} \lesssim \|f_1\|_{M^{p_1,q_1}} \|f_2\|_{M^{p_2,q_2}_{v_s}} + \|f_1\|_{M^{p_1,q_1}_{v_s}} \|f_2\|_{M^{p_2,q_2}}.$$

Proof. The proof is similar to that of Theorem 4.4.2, but in this case from the estimate (4.47) we proceed by using

$$v_s(z) \lesssim v_s(z - w) + v_s(w), \quad s \geq 0$$

(with sp in place of s) instead of $v_s(z) \lesssim v_{|s|}(z - w)v_s(w)$. □

If, in particular, we consider the Wigner distribution $W(f, f)$, then Theorem 4.4.4 can be rephrased as follows.

Corollary 4.4.5. *Assume $s \geq 0$, $p_1, q_1, p, q \in [1, \infty]$ are such that*

$$2 \min\left\{\frac{1}{p_1}, \frac{1}{q_1}\right\} \geq \frac{1}{p} + \frac{1}{q}.$$

If $f \in M^{p_1,q_1}_{v_s}(\mathbb{R}^d)$, then the Wigner distribution $W(f, f)$ is in $M^{p,q}_{1\otimes v_s}(\mathbb{R}^{2d})$, with

$$\|W(f, f)\|_{M^{p,q}_{1\otimes v_s}(\mathbb{R}^{2d})} \lesssim \|f\|_{M^{p_1,q_1}(\mathbb{R}^d)} \|f\|_{M^{p_1,q_1}_{v_s}(\mathbb{R}^d)}. \tag{4.49}$$

We now prove the sharpness of Theorem 4.4.2 (and Corollary 4.4.5) in the unweighted case $s = 0$.

We need some preliminaries, concerning the modulation norm of Gaussians. The following result is an easy consequence of Proposition 1.3.33, and extends Lemma 1.8 in [286].

Corollary 4.4.6. *Consider the generalized Gaussian f defined in (1.126) and the window function $\Phi(x, \xi) = e^{-\pi(x^2+\xi^2)}$. Then for every $1 \leq p, q \leq \infty$ we have*

$$\|f\|_{M^{p,q}} \asymp \|V_\Phi f\|_{L^{p,q}} \asymp \frac{[(a+1)(b+1)+c^2]^{\frac{d}{p}+\frac{d}{q}-\frac{d}{2}}[(c^2+ab+a)(c^2+ab+b)]^{\frac{d}{2q}-\frac{d}{2p}}}{[b^2(a+1)+b(c^2+a+1)]^{\frac{d}{2q}}[a^2(b+1)+a(c^2+b+1)]^{\frac{d}{2q}}}. \tag{4.50}$$

The cases when $p = \infty$ or $q = \infty$ can be obtained by using the rule $1/\infty = 0$ in formula (4.50).

Proof. By Proposition 1.3.33, we can write

$$|V_\Phi f(z,\zeta)| = \frac{1}{[(a+1)(b+1)+c^2]^{\frac{d}{2}}} \times e^{-\pi\frac{[a(b+1)+c^2]z_1^2+[(a+1)b+c^2]z_2^2+(b+1)\zeta_1^2+(a+1)\zeta_2^2-2c(z_1\cdot\zeta_2+z_2\cdot\zeta_1)}{(a+1)(b+1)+c^2}}.$$

It remains to compute the mixed $L^{p,q}$-norm of the previous function. We treat the cases $1 \le p,q < \infty$. The cases when either $p = \infty$ or $q = \infty$ are obtained with obvious modifications.

For simplicity, we set

$$\alpha = \frac{c^2 + a(b+1)}{(a+1)(b+1)+c^2}, \quad \beta = \frac{c^2+(a+1)b}{(a+1)(b+1)+c^2}, \quad \gamma = \frac{(b+1)}{(a+1)(b+1)+c^2}, \tag{4.51}$$

$$\delta = \frac{(a+1)}{(a+1)(b+1)+c^2}, \quad \sigma = \frac{c}{(a+1)(b+1)+c^2}. \tag{4.52}$$

Hence

$$\frac{\|V_\Phi f\|_{L^{p,q}}}{[(a+1)(b+1)+c^2]^{-\frac{d}{2p}}} = \left(\int_{\mathbb{R}^{2d}} I^{\frac{q}{p}} e^{-\pi q(\gamma\zeta_1^2+\delta\zeta_2^2)}\, d\zeta_1 d\zeta_2\right)^{\frac{1}{q}} =: A, \tag{4.53}$$

where $I := \int_{\mathbb{R}^{2d}} e^{-\pi p\alpha z_1^2 - \pi p\beta z_2^2} e^{2\pi p\sigma(z_1\cdot\zeta_2+z_2\cdot\zeta_1)}\, dz_1\, dz_2$. Now straightforward computations and change of variables yield

$$\begin{aligned}
I &= \int_{\mathbb{R}^d}\left(\int_{\mathbb{R}^d} e^{-\pi p(\alpha z_1^2 - 2\sigma\zeta_2\cdot z_1)}\, dz_1\right) e^{-\pi\beta z_2^2 + 2\pi p\sigma z_2\cdot\zeta_1}\, dz_2 \\
&= e^{\pi p\frac{\sigma^2}{\alpha}\zeta_2^2} e^{\pi p\frac{\sigma^2}{\beta}\zeta_1^2} \int_{\mathbb{R}^d} e^{-\pi p(\sqrt{\alpha}z_1 - \frac{\sigma}{\sqrt{\alpha}}\zeta_2)^2}\, dz_1 \int_{\mathbb{R}^d} e^{-\pi p(\sqrt{\beta}z_2 - \frac{\sigma}{\sqrt{\beta}}\zeta_1)^2}\, dz_2 \\
&= p^{-\frac{d}{2}}\alpha^{-\frac{d}{2}} p^{-\frac{d}{2}}\beta^{-\frac{d}{2}} e^{\pi p\frac{\sigma^2}{\alpha}\zeta_2^2} e^{\pi p\frac{\sigma^2}{\beta}\zeta_1^2}.
\end{aligned}$$

Substituting the value of the integral I in (4.53), we obtain

$$\begin{aligned}
A &= p^{-\frac{d}{p}}\alpha^{-\frac{d}{2p}}\beta^{-\frac{d}{2p}}\left(\int_{\mathbb{R}^d} e^{\pi q\frac{\sigma^2}{\beta}\zeta_1^2 - \pi q\gamma\zeta_1^2}\, d\zeta_1 \int_{\mathbb{R}^d} e^{\pi q\frac{\sigma^2}{\alpha}\zeta_2^2 - \pi q\delta\zeta_2^2}\, d\zeta_2\right)^{\frac{1}{q}} \\
&= p^{-\frac{d}{p}}\alpha^{-\frac{d}{2p}}\beta^{-\frac{d}{2p}}\left(\int_{\mathbb{R}^d} e^{-\pi q(\gamma - \frac{\sigma^2}{\beta})\zeta_1^2}\, d\zeta_1 \int_{\mathbb{R}^d} e^{-\pi q(\delta - \frac{\sigma^2}{\alpha})\zeta_2^2}\right)^{\frac{1}{q}} \\
&= p^{-\frac{d}{p}}\alpha^{-\frac{d}{2p}}\beta^{-\frac{d}{2p}} q^{-\frac{d}{q}}\left(\gamma - \frac{\sigma^2}{\beta}\right)^{-\frac{d}{2q}}\left(\gamma - \frac{\sigma^2}{\alpha}\right)^{-\frac{d}{2q}}.
\end{aligned}$$

Finally, the goal is attained by substituting in A the values of the parameters $\alpha, \beta, \gamma, \delta, \sigma$ in (4.51) and (4.52) and observing that

$$\|f\|_{M^{p,q}} \asymp \|V_\Phi f\|_{L^{p,q}} = A\big[(a+1)(b+1)+c^2\big]^{-\frac{d}{2}}.$$

This concludes the proof. □

We have now all the tools to compute the modulation norm of the (cross-)Wigner distribution $W(\varphi, \varphi_{\sqrt{\lambda}})$ in (1.129). Precisely, setting in formula (4.50) the values $a = a_\lambda$, $b = b_\lambda$, $c = c_\lambda$, where $a_\lambda, b_\lambda, c_\lambda$ are defined as

$$a_\lambda = \frac{4\lambda}{1+\lambda} \quad b_\lambda = \frac{4}{1+\lambda} \quad c_\lambda = \frac{2(1-\lambda)}{1+\lambda},$$

and making easy simplifications we attain the following result.

Corollary 4.4.7. *For $\lambda > 0$ consider the (cross-)Wigner distribution $W(\varphi, \varphi_{\sqrt{\lambda}})$ defined in* (1.129) *(cf. Lemma* 1.3.34*). Then*

$$\big\|W(\varphi, \varphi_{\sqrt{\lambda}})\big\|_{M^{p,q}} \asymp \frac{[(2\lambda+1)(\lambda+2)]^{\frac{d}{2q}-\frac{d}{2p}}}{\lambda^{\frac{d}{2q}}(1+\lambda)^{\frac{d}{2}-\frac{d}{p}}}. \tag{4.54}$$

The cases when $p = \infty$ or $q = \infty$ can be obtained by using the rule $1/\infty = 0$ in formula (4.54).

Theorem 4.4.8 (Necessary conditions). *Consider $p_1, p_2, q_1, q_2, p, q \in [1, \infty]$. Assume that there exists a constant $C > 0$ such that*

$$\big\|W(f_1, f_2)\big\|_{M^{p,q}} \le C\|f_1\|_{M^{p_1,q_1}}\|f_2\|_{M^{p_2,q_2}}, \quad \forall f_1, f_2 \in \mathcal{S}(\mathbb{R}^{2d}), \tag{4.55}$$

then (4.41) *and* (4.42) *must hold.*

Proof. Let us first demonstrate the necessity of (4.42). We consider the dilated Gaussians $\varphi(\sqrt{\lambda}x)$, with $\varphi(x) = e^{-\pi x^2}$. An easy computation (see also [160, formula (4.20)]) shows that

$$W(\varphi_{\sqrt{\lambda}}, \varphi_{\sqrt{\lambda}})(x, \xi) = 2^{\frac{d}{2}}\lambda^{-\frac{d}{2}}\varphi_{\sqrt{2\lambda}}(x)\varphi_{\sqrt{2/\lambda}}(\xi),$$

hence

$$\big\|W(\varphi_{\sqrt{\lambda}}, \varphi_{\sqrt{\lambda}})\big\|_{M^{p,q}(\mathbb{R}^{2d})} = 2^{\frac{d}{2}}\lambda^{-\frac{d}{2}}\|\varphi_{\sqrt{2\lambda}}\|_{M^{p,q}(\mathbb{R}^d)}\|\varphi_{\sqrt{2/\lambda}}\|_{M^{p,q}(\mathbb{R}^d)}.$$

Now, from Lemma 2.6.1,

$$\|\varphi_{\sqrt{\lambda}}\|_{M^{r,s}} \asymp \lambda^{-\frac{d}{2r}}(\lambda+1)^{-\frac{d}{2}(1-\frac{1}{q}-\frac{1}{p})}.$$

The assumption (4.55) in this case becomes

$$\lambda^{-\frac{d}{2}}(\lambda+1)^{-\frac{d}{2}(1-\frac{1}{q}-\frac{1}{p})}(\lambda^{-1}+1)^{-\frac{d}{2}(1-\frac{1}{q}-\frac{1}{p})}$$
$$\lesssim \lambda^{-\frac{d}{2p_1}}(1+\lambda)^{-\frac{d}{2}(1-\frac{1}{q_1}-\frac{1}{p_1})}\lambda^{-\frac{d}{2p_2}}(1+\lambda)^{-\frac{d}{2}(1-\frac{1}{q_2}-\frac{1}{p_2})},$$

and, letting $\lambda \to +\infty$, we obtain

$$\frac{1}{p}+\frac{1}{q} \le \frac{1}{q_1}+\frac{1}{q_2}$$

whereas, for $\lambda \to 0^+$,

$$\frac{1}{p}+\frac{1}{q} \le \frac{1}{p_1}+\frac{1}{p_2},$$

so that (4.42) must hold.

It remains to prove the sharpness of (4.41). We first show the conditions $p_2, q_2 \le q$. We test (4.55) on the (cross-)Wigner distribution $W(\varphi, \varphi_{\sqrt{\lambda}})$ defined in (1.129), that is,

$$\|W(\varphi, \varphi_{\sqrt{\lambda}})\|_{M^{p,q}(\mathbb{R}^{2d})} \lesssim \|\varphi\|_{M^{p_1,q_1}(\mathbb{R}^d)}\|\varphi_{\sqrt{\lambda}}\|_{M^{p_2,q_2}(\mathbb{R}^d)}.$$

Using Corollary 4.4.7, the previous estimate can be rephrased as

$$\frac{[(2\lambda+1)(\lambda+2)]^{\frac{d}{2q}-\frac{d}{2p}}}{\lambda^{\frac{d}{2q}}(1+\lambda)^{\frac{d}{2}-\frac{d}{p}}} \lesssim \lambda^{-\frac{d}{2p_2}}\lambda^{-\frac{d}{2}(1-\frac{1}{q_2}-\frac{1}{p_2})}, \quad \forall \lambda > 0.$$

Letting $\lambda \to +\infty$, we obtain

$$q_2 \le q$$

whereas, for $\lambda \to 0^+$,

$$p_2 \le q.$$

The conditions $p_1, q_1 \le q$ then follow by using the cross-Wigner property,

$$W(\varphi_{\sqrt{\lambda}}, \varphi)(x,\xi) = \overline{W(\varphi, \varphi_{\sqrt{\lambda}})(x,\xi)},$$

so that

$$\|W(\varphi_{\sqrt{\lambda}}, \varphi)\|_{M^{p,q}(\mathbb{R}^{2d})} = \|\overline{W(\varphi, \varphi_{\sqrt{\lambda}})}\|_{M^{p,q}(\mathbb{R}^{2d})} = \|W(\varphi, \varphi_{\sqrt{\lambda}})\|_{M^{p,q}(\mathbb{R}^{2d})},$$

and applying the same argument as before. □

4.4.1 Continuity results for the short-time Fourier transform and the ambiguity function

The bounds obtained for the Wigner distribution can be transferred into new estimates for other time–frequency representations such that the STFT or the ambiguity function. Precisely, given $f,g \in L^2(\mathbb{R}^d)$, we recall the definition of the (cross-)ambiguity function in Definition 1.3.1:

$$A(f_1,f_2)(x,\xi) = \int_{\mathbb{R}^d} e^{-2\pi i t\cdot\xi} f_1\Big(t+\frac{x}{2}\Big)\overline{f_2\Big(t-\frac{x}{2}\Big)}\,dt.$$

In Lemma 1.3.11 it was shown that the Wigner distribution is the symplectic Fourier transform of the ambiguity function, in other words, we can write

$$W(f_1,f_2)(x,\xi) = \mathcal{F}\mathcal{U}A(f_1,f_2)(x,\xi), \quad f_1,f_2 \in L^2(\mathbb{R}^d), \tag{4.56}$$

where the operator $\mathcal{U}$ is the rotation $\mathcal{U}F(x,\xi) = F(\xi,-x)$ of a function F on $\mathbb{R}^{2d}$. We need the following norm equivalence.

Lemma 4.4.9. *For $s \in \mathbb{R}$, $1 \le p,q \le \infty$, the following equivalence holds:*

$$\|W(f_1,f_2)\|_{M^{p,q}_{1\otimes v_s}} = \|A(f_1,f_2)\|_{W(\mathcal{F}L^p,L^q_{v_s})} \asymp \|V_{\mathcal{I}f_2}f_1\|_{W(\mathcal{F}L^p,L^q_{v_s})},$$

where $\mathcal{I}f_2(x) = f_2(-x)$.

Proof. Let us observe that the weight v_s, $s \in \mathbb{R}$, is symmetric in each coordinate:

$$v_s(x,\xi) = v_s(x,-\xi) = v_s(-x,\xi) = v_s(-x,-\xi).$$

Using (4.56), the connection between modulation spaces and the symmetry of the weights v_s, we can write

$$\begin{aligned}\|W(f_1,f_2)\|_{M^{p,q}_{1\otimes v_s}} &= \|\mathcal{F}\mathcal{U}A(f_1,f_2)\|_{M^{p,q}_{1\otimes v_s}} = \|\mathcal{U}A(f_1,f_2)\|_{W(\mathcal{F}L^p,L^q_{v_s})}\\ &= \|A(f_1,f_2)\|_{W(\mathcal{F}L^p,L^q_{v_s})}.\end{aligned}$$

Now, Lemma 1.3.5 lets us write

$$A(f_1,f_2)(x,\xi) = 2^d e^{\pi i x\cdot\xi} V_{\mathcal{I}f_2}f_1(x,\xi).$$

It was proved in Proposition 4.3.3 that the function $F(x,\xi) = e^{\pi i x\cdot\xi}$ is in the Wiener amalgam space $W(\mathcal{F}L^1,L^\infty)(\mathbb{R}^{2d})$. This means that, by the product properties for Wiener amalgam spaces, for every $s \in \mathbb{R}$,

$$\|A(f_1,f_2)\|_{W(\mathcal{F}L^p,L^q_{v_s})} \lesssim \|F\|_{W(\mathcal{F}L^1,L^\infty)}\|V_{\mathcal{I}f_2}f_1\|_{W(\mathcal{F}L^p,L^q_{v_s})}$$

and, since $\bar{F}(x,\xi) = e^{-\pi i x\cdot\xi} \in W(\mathcal{F}L^1, L^\infty)$ as well, with $\|\bar{F}\|_{W(\mathcal{F}L^1,L^\infty)} = \|F\|_{W(\mathcal{F}L^1,L^\infty)}$, we can analogously write

$$\|V_{\mathcal{I}f_2} f_1\|_{W(\mathcal{F}L^p, L^q_{v_s})} \lesssim \|F\|_{W(\mathcal{F}L^1,L^\infty)} \|A(f_1,f_2)\|_{W(\mathcal{F}L^p,L^q_{v_s})}.$$

This proves the desired result. □

These observations, together with the Wigner property $W(f_1,f_2)(x,\xi) = \overline{W(f_2,f_1)}$, let us translate the sufficient and necessary conditions for Wigner distributions in terms of STFT acting from modulation spaces to Wiener amalgam spaces. Notice that the following two corollaries also hold for the ambiguity function $A(f_1,f_2)$ in place of the STFT.

Corollary 4.4.10. *Consider $s \in \mathbb{R}$ and assume that $p_1, p_2, q_1, q_2, p, q \in [1,\infty]$ satisfy conditions (4.41) and (4.42). Then if $f_1 \in M^{p_1,q_1}_{v_{|s|}}(\mathbb{R}^d)$ and $f_2 \in M^{p_2,q_2}_{v_s}(\mathbb{R}^d)$, we have $V_{f_1}f_2 \in W(\mathcal{F}L^p, L^q_{v_s})(\mathbb{R}^{2d})$ with*

$$\|V_{f_1}f_2\|_{W(\mathcal{F}L^p,L^q_{v_s})(\mathbb{R}^{2d})} \lesssim \|f_1\|_{M^{p_1,q_1}_{v_s}(\mathbb{R}^d)} \|f_2\|_{M^{p_2,q_2}_{v_s}(\mathbb{R}^d)}.$$

Conversely, assume that there exists a constant $C > 0$ such that

$$\|V_{f_1}f_2\|_{W(\mathcal{F}L^p,L^q)(\mathbb{R}^{2d})} \le C\|f_1\|_{M^{p_1,q_1}(\mathbb{R}^d)} \|f_2\|_{M^{p_2,q_2}(\mathbb{R}^d)}, \quad \forall f_1, f_2 \in \mathcal{S}(\mathbb{R}^{2d}).$$

Then (4.41) and (4.42) must hold.

The previous result has many special and interesting cases. Let us just give a flavor of the main important ones.

Corollary 4.4.11. *Assume that $p_1, p_2, p, q \in [1,\infty]$ satisfy*

$$p_1, p_2 \le q, \quad \frac{1}{p_1} + \frac{1}{p_2} \ge \frac{1}{p} + \frac{1}{q}. \tag{4.57}$$

For $s \in \mathbb{R}$, if $f_1 \in M^{p_1}_{v_{|s|}}(\mathbb{R}^d)$ and $f_2 \in M^{p_2}_{v_s}(\mathbb{R}^d)$, we have $V_{f_1}f_2 \in W(\mathcal{F}L^p, L^q_{v_s})(\mathbb{R}^{2d})$ with

$$\|V_{f_1}f_2\|_{W(\mathcal{F}L^p,L^q_{v_s})(\mathbb{R}^{2d})} \lesssim \|f_1\|_{M^{p_1}_{v_{|s|}}(\mathbb{R}^d)} \|f_2\|_{M^{p_2}_{v_s}(\mathbb{R}^d)}.$$

Conversely, assume that there exists a constant $C > 0$ such that

$$\|V_{f_1}f_2\|_{W(\mathcal{F}L^p,L^q)(\mathbb{R}^{2d})} \le C\|f_1\|_{M^{p_1}(\mathbb{R}^d)} \|f_2\|_{M^{p_2}(\mathbb{R}^d)}, \quad \forall f_1, f_2 \in \mathcal{S}(\mathbb{R}^{2d}).$$

Then (4.57) must hold.

Remark 4.4.12. The previous result holds also for the cross-Wigner distribution if we replace $\|V_{f_1}f_2\|_{W(\mathcal{F}L^p,L^q_{v_s})(\mathbb{R}^{2d})}$ by $\|W(f_1,f_2)\|_{M^{p,q}_{1\otimes v_s}(\mathbb{R}^{2d})}$.

The unweighted case $s = 0$ leads to a surprising refinement of Lemma 4.4.9.

Lemma 4.4.13. *For $1 \le p,q \le \infty$, we have the equivalence*

$$\|W(f_1,f_2)\|_{M^{p,q}} \asymp \|W(f_1,f_2)\|_{W(\mathcal{F}L^p,L^q)},$$

for every $f_1,f_2 \in \mathcal{S}'(\mathbb{R}^d)$ such that the norms above are bounded.

Proof. We use again the connection between cross-Wigner distribution and STFT in Lemma 1.3.5. For $a > 0$, define the dilation operator D_a on $\mathbb{R}^d$ by $D_a f(x) = a^{d/2} f(ax)$, $x \in \mathbb{R}^d$, so that the cross-Wigner distribution $W(f_1,f_2)$ can be written as

$$W(f_1,f_2)(x,\xi) = D_2(e^{\pi i x\cdot\xi} V_{\mathcal{I}f_2} f_1)(x,\xi).$$

Using the dilation properties for modulation spaces in Corollary 2.6.4, we estimate, similarly as in the proof of Lemma 4.4.9,

$$\begin{aligned}\|W(f_1,f_2)\|_{W(\mathcal{F}L^p,L^q)} &\le 2^{2d(1/p+1/q-1)} \|e^{\pi i x\cdot\xi} V_{\mathcal{I}f_2} f_1\|_{W(\mathcal{F}L^p,L^q)}\\ &\lesssim \|e^{\pi i x\cdot\xi}\|_{W(\mathcal{F}L^1,L^\infty)} \|V_{\mathcal{I}f_2} f_1\|_{W(\mathcal{F}L^p,L^q)} \lesssim \|V_{\mathcal{I}f_2} f_1\|_{W(\mathcal{F}L^p,L^q)}.\end{aligned}$$

Conversely,

$$V_{\mathcal{I}f_2} f_1(x,\xi) = 2^{-d} e^{-\pi i x\cdot\xi} W(f_1,f_2)\left(\frac{x}{2},\frac{\xi}{2}\right),$$

and the same arguments as before with the dilation $D_{1/2}$ give

$$\|V_{\mathcal{I}f_2} f_1\|_{W(\mathcal{F}L^p,L^q)} \lesssim \|e^{-\pi i x\cdot\xi}\|_{W(\mathcal{F}L^1,L^\infty)} \|W(f_1,f_2)\|_{W(\mathcal{F}L^p,L^q)},$$

hence $\|V_{\mathcal{I}f_2} f_1\|_{W(\mathcal{F}L^p,L^q)} \asymp \|W(f_1,f_2)\|_{W(\mathcal{F}L^p,L^q)}$. Finally, it is immediate to check that $\|V_{\mathcal{I}f_2} f_1\|_{W(\mathcal{F}L^p,L^q)} = \|V_{f_2} f_1\|_{W(\mathcal{F}L^p,L^q)}$. The result is then a consequence of Lemma 4.4.9. □

If we choose $s = 0$, $p = q'$ in Lemma 4.4.9 and apply the previous result, we can refine some Lieb's integral bounds for the ambiguity function showed in [218].

Corollary 4.4.14. *Assume $1 \le p_1,p_2,q_1,q_2 \le q$ are such that*

$$\frac{1}{p_1} + \frac{1}{p_2} \ge 1 \quad \frac{1}{q_1} + \frac{1}{q_2} \ge 1. \tag{4.58}$$

If $f_i \in M^{p_i,q_i}(\mathbb{R}^d)$, $i = 1,2$, then the ambiguity function satisfies $A(f_1,f_2) \in L^q(\mathbb{R}^{2d})$, with

$$\|A(f_1,f_2)\|_{L^q(\mathbb{R}^{2d})} \lesssim \|f_1\|_{M^{p_1,q_1}(\mathbb{R}^d)} \|f_2\|_{M^{p_2,q_2}(\mathbb{R}^d)}.$$

Conversely, assume that there exists a constant $C > 0$ such that

$$\|A(f_1,f_2)\|_{L^q(\mathbb{R}^{2d})} \le C\|f_1\|_{M^{p_1,q_1}(\mathbb{R}^d)} \|f_2\|_{M^{p_2,q_2}(\mathbb{R}^d)}.$$

Then we must have $1 \le p_1,p_2,q_1,q_2 \le q$ and (4.58).

Proof. The results in Lemmas 4.4.9 and 4.4.13 imply

$$\|A(f_1,f_2)\|_{M^{q',q}(\mathbb{R}^{2d})} \asymp \|A(f_1,f_2)\|_{W(\mathcal{F}L^{q'},L^q)(\mathbb{R}^{2d})}. \tag{4.59}$$

Using the inclusion relations between Lebesgue and modulation spaces [286, Proposition 1.7], we have, for $q \le 2$, $M^{q',q}(\mathbb{R}^d) \subseteq L^q(\mathbb{R}^d)$ and for $q \ge 2$, $L^q(\mathbb{R}^d) \subseteq M^{q',q}(\mathbb{R}^d)$. For $q \le 2$, $L^q(\mathbb{R}^d) \subseteq \mathcal{F}L^{q'}(\mathbb{R}^d)$, and the inclusion relations for Wiener amalgam spaces give, for $q \le 2$, $L^q(\mathbb{R}^d) = W(L^q, L^q)(\mathbb{R}^d) \subseteq W(\mathcal{F}L^{q'}, L^q)(\mathbb{R}^d)$, so that

$$M^{q',q}(\mathbb{R}^d) \subseteq L^q(\mathbb{R}^d) \subseteq W(\mathcal{F}L^{q'}, L^q)(\mathbb{R}^d).$$

Since $\mathcal{F}L^{q'}(\mathbb{R}^d) \subseteq L^q(\mathbb{R}^d)$ if $q \ge 2$, the inclusion relations for Wiener amalgam spaces let us write $W(\mathcal{F}L^{q'}, L^q)(\mathbb{R}^{2d}) \subseteq W(L^q, L^q)(\mathbb{R}^{2d}) = L^q(\mathbb{R}^{2d})$. Hence, for $q \ge 2$,

$$W(\mathcal{F}L^{q'}, L^q)(\mathbb{R}^d) \subseteq L^q(\mathbb{R}^d) \subseteq M^{q',q}(\mathbb{R}^d).$$

The sufficient and necessary conditions then follow from the equivalence (4.59), Lemma 4.4.9 and Corollary 4.4.10. □

Observe that for $p_i = q_i = 2$, $i = 1,2$, $M^{p_i,q_i}(\mathbb{R}^d) = L^2(\mathbb{R}^d)$, and we recapture Lieb's bound, see [218, Theorem 1].

4.4.2 Pseudodifferential operators on modulation spaces, conclusions

In this section we apply Theorem 4.4.1 to the study of pseudodifferential operators on modulation spaces. The key tool is the weak definition of a Weyl operator (4.6).

Theorem 4.4.15 below extends [89, Theorem 1.1] to weighted modulation spaces, thus widening the sufficient boundedness conditions presented by Toft in [287, Theorem 4.3]. It is stated for pseudodifferential operators in the Weyl form, but the same result holds for any τ-pseudodifferential operator by the previous observations. Using Theorem 4.4.1, the proof of Theorem 4.4.15 is decidedly simple.

Theorem 4.4.15. *Assume $s \ge 0$, $p_i, q_i, p, q \in [1,\infty]$, $i = 1,2$, are such that*

$$\min\left\{\frac{1}{p_1} + \frac{1}{p_2'}, \frac{1}{q_1} + \frac{1}{q_2'}\right\} \ge \frac{1}{p'} + \frac{1}{q'} \tag{4.60}$$

and

$$q \le \min\{p_1', q_1', p_2, q_2\}. \tag{4.61}$$

Then the pseudodifferential operator $\mathrm{Op}_w(a)$, from $\mathcal{S}(\mathbb{R}^d)$ to $\mathcal{S}'(\mathbb{R}^d)$, having symbol $a \in M^{p,q}_{1\otimes v_s}(\mathbb{R}^{2d})$, extends uniquely to a bounded operator from $\mathcal{M}^{p_1,q_1}_{v_s}(\mathbb{R}^d)$ to $\mathcal{M}^{p_2,q_2}_{v_s}(\mathbb{R}^d)$, with the estimate

$$\|\mathrm{Op}_w(a)f\|_{\mathcal{M}^{p_2,q_2}_{v_s}} \lesssim \|a\|_{M^{p,q}_{1\otimes v_s}} \|f\|_{\mathcal{M}^{p_1,q_1}_{v_s}}. \tag{4.62}$$

Conversely, if (4.62) *holds for* $s = 0$ *and for every* $f \in \mathcal{S}(\mathbb{R}^d)$, $\sigma \in \mathcal{S}'(\mathbb{R}^{2d})$, *then* (4.60) *and* (4.61) *must be satisfied.*

Proof. Assume $a \in M^{p,q}_{1\otimes v_s}(\mathbb{R}^{2d})$, $f \in \mathcal{M}^{p_1,q_1}_{v_s}(\mathbb{R}^d)$ are such that (4.60) and (4.61) are satisfied. For $g \in \mathcal{M}^{p_2',q_2'}_{v_{-s}}(\mathbb{R}^d)$, Theorem 4.4.1 says that the cross-Wigner distribution is in $M^{p',q'}_{1\otimes v_{-s}}(\mathbb{R}^{2d})$, provided that $p_1, q_1, p_2', q_2' \le q'$ and

$$\min\{1/p_1 + 1/p_2', 1/q_1 + 1/q_1'\} \ge 1/p' + 1/q',$$

which are conditions (4.61) and (4.60), respectively. Thereby, there exists a constant $C > 0$ such that

$$\begin{aligned} |\langle a, W(g,f)\rangle| &\le \|a\|_{M^{p,q}_{1\otimes v_s}(\mathbb{R}^{2d})} \|W(g,f)\|_{M^{p',q'}_{1\otimes v_{-s}}(\mathbb{R}^{2d})} \\ &\le C\|f\|_{\mathcal{M}^{p_1,q_1}_{v_s}(\mathbb{R}^d)} \|g\|_{\mathcal{M}^{p_2',q_2'}_{v_{-s}}(\mathbb{R}^d)}. \end{aligned}$$

Since $|\langle \mathrm{Op}_w(a)f, g\rangle| = |\langle a, W(g,f)\rangle|$, this concludes the proof of the sufficient conditions. The arguments for the necessity are similar, by using Theorem 4.4.8. □

Taking $p = \infty$, $q = 1$, $p_1 = p_2$, $q_1 = q_2$, we deduce in particular that for symbols $a \in M^{\infty,1}(\mathbb{R}^{2d})$ we have boundedness of $\mathrm{Op}_w(a)$ on $M^{r,s}(\mathbb{R}^d)$ for any $r, s \in [1, \infty]$, and we recapture [163].

4.4.3 Localization operators on modulation spaces

Time–frequency localization operators are a mathematical instrument to define a restriction of functions to a region in time–frequency plane that is compatible with the uncertainty principle and to extract time–frequency features. In this sense they have been introduced and studied by Daubechies [99] and Ramanathan and Topiwala [246], and they have been extensively investigated as an important mathematical tool in signal analysis and other applications [23, 66, 75, 139, 141, 142, 143, 318, 319].

Localization operators in connection with the Donoho–Stark uncertainty principle were considered in [21, 22].

In other mathematical contexts, time–frequency localization operators have already been used as a quantization procedure ("anti-Wick operators") by Berezin [16, 265] or as an approximation of pseudodifferential operators ("wave packets") by Cordoba and Fefferman [94, 145].

Given a function or distribution a on $\mathbb{R}^{2d}$ and windows φ_1, φ_2, we define the generalized anti-Wick operator $A_a^{\varphi_1,\varphi_2}$ by the (formal) integral

$$A_a^{\varphi_1,\varphi_2} f := \int_{\mathbb{R}^{2d}} a(x,\xi) V_{\varphi_1} f(x,\xi) M_\xi T_x \varphi_2 \, dx \, d\xi, \tag{4.63}$$

whenever this vector-valued integral makes sense. In most cases it is preferable to interpret the integral in a weak sense as

$$\begin{aligned}\langle A_a^{\varphi_1,\varphi_2} f, g\rangle &= \int a(x,\xi) V_{\varphi_1} f(x,\xi) \langle M_\xi T_x \varphi_2, g\rangle \, dx \, d\xi \\ &= \langle a, \overline{V_{\varphi_1} f} V_{\varphi_2} g\rangle \quad \text{for } f, g \in \mathcal{S}(\mathbb{R}^d). \end{aligned} \tag{4.64}$$

Here the brackets $\langle \cdot, \cdot \rangle$ express the duality on a suitable pair of dual spaces $B' \times B$ and extend the inner product on $L^2(\mathbb{R}^{2d})$.

If $\varphi_1(t) = \varphi_2(t) = e^{-\pi t^2}$, then $A_a = A_a^{\varphi_1,\varphi_2}$ is the classical anti-Wick operator and the mapping $a \to A_a$ is interpreted as a quantization rule [16, 318]. In this case A_a makes sense as a continuous operator from $\mathcal{S}$ to $\mathcal{S}'$ whenever $a \in \mathcal{S}'(\mathbb{R}^{2d})$. For general windows, $A_a^{\varphi_1,\varphi_2}$ is often viewed as a localization operator. Taking a to be a characteristic function $a = \chi_\Omega$ for $\Omega \subseteq \mathbb{R}^{2d}$, then $A_a^{\varphi_1,\varphi_2}$ evaluates "essentially" the restriction of $V_{\varphi_1} f$ to Ω. See [99, 246, 318] for investigations in this direction.

In the literature only operators of the form $A_a^{\varphi,\varphi}$ are studied because they satisfy certain additional properties required by a quantization rule; for instance, if $a \geq 0$, then $A_a^{\varphi,\varphi}$ is a positive operator, and if a is real-valued, then $A_a^{\varphi,\varphi}$ is self-adjoint. On the other hand, if we want to study how $A_a^{\varphi,\varphi}$ depends on φ, then it seems more appropriate to deal with anti-Wick operators with two different windows. Specifically, if $\varphi = \varphi_1 + \varphi_2$, then $A_a^{\varphi,\varphi} = A_a^{\varphi_1,\varphi_1} + A_a^{\varphi_2,\varphi_2} + A_a^{\varphi_1,\varphi_2} + A_a^{\varphi_2,\varphi_1}$, and it is natural to study $A_a^{\varphi_1,\varphi_2}$. In addition, some application in signal analysis suggest the use of windows with different regularities for the analysis $f \to V_{\varphi_1} f$ and for the synthesis $V_{\varphi_1} f \to A_a^{\varphi_1,\varphi_2}$. In this section we will therefore always work with this simple and natural generalization of the classical anti-Wick operators, presenting some results in [23, 66, 75].

Lemma 4.4.16 (Connection between Weyl and localization operators). *If $a \in \mathcal{S}'(\mathbb{R}^{2d})$, $\varphi_1, \varphi_2 \in \mathcal{S}(\mathbb{R}^d)$, then $A_a^{\varphi_1,\varphi_2}$ possesses the Weyl symbol $\sigma = a * W(\varphi_2, \varphi_1)$, in other words,*

$$A_a^{\varphi_1,\varphi_2} = \mathrm{Op}_w(\sigma), \tag{4.65}$$

with

$$\sigma = a * W(\varphi_2, \varphi_1),$$

where $W(\varphi_2, \varphi_1)$ is the cross-Wigner distribution of φ_2, φ_1.

Proof. Let $f, g \in \mathcal{S}(\mathbb{R}^d)$, and $a \in \mathcal{S}(\mathbb{R}^{2d})$, then $A_a^{\varphi_1,\varphi_2}$ is the integral operator with kernel

$$K(x,y) = \int a(z) \big(\pi(z)\varphi_2(x) \overline{\pi(z)\varphi_1(y)}\big) \, dz = \int a(z) \big(\pi(z)\varphi_2 \otimes \overline{\pi(z)\varphi_1}\big)(x,y) \, dz.$$

The connection between the distributional (or integral) kernel K of an operator and its Weyl symbol σ is given by the formula $\sigma = \mathcal{F}_2 \mathcal{T}_s K$, see (4.5). Since the integrals above are absolutely convergent, Lemma 1.3.9 implies that

$$\begin{aligned}\sigma &= \mathcal{F}_2 \mathcal{T}_s \Big(\int a(z) \pi(z)\varphi_2 \otimes \overline{\pi(z)\varphi_1} \, dz \Big) \\ &= \int a(z) \mathcal{F}_2 \mathcal{T}_s \big(\pi(z)\varphi_2 \otimes \overline{\pi(z)\varphi_1}\big) \, dz \end{aligned}$$

$$= \int a(z) W(\pi(z)\varphi_2, \pi(z)\varphi_1)\, dz = \int a(z) W(\varphi_2, \varphi_1)(\cdot - z)\, dz$$
$$= a * W(\varphi_2, \varphi_1).$$

Using a standard approximation argument, we then obtain the equivalence of the operators for all $a \in \mathcal{S}'(\mathbb{R}^{2d})$. □

Lemma 4.4.16 allows us to study localization operators by means of their Weyl symbol.

We now present sharp boundedness results for localization operators.

Theorem 4.4.17. *Assume $s \geq 0$, the indices $p_i, q_i, p, q \in [1, \infty]$, $i = 1, 2$, fulfill relations (4.60) and (4.61), and consider $r \in [1, 2]$. If $a \in M^{p,q}_{1\otimes v_{-s}}(\mathbb{R}^{2d})$ and $\varphi_1, \varphi_2 \in M^r_{v_{2s}}(\mathbb{R}^d)$, then the localization operator $A_a^{\varphi_1,\varphi_2}$ is continuous from $\mathcal{M}^{p_1,q_1}_{v_s}(\mathbb{R}^d)$ to $\mathcal{M}^{p_2,q_2}_{v_s}(\mathbb{R}^d)$ with*

$$\|A_a^{\varphi_1,\varphi_2}\|_{op} \lesssim \|a\|_{M^{p,q}_{1\otimes v_{-s}}} \|\varphi_1\|_{M^r_{v_{2s}}} \|\varphi_2\|_{M^r_{v_{2s}}}.$$

Proof. Using Theorem 4.4.1 for $\varphi_1, \varphi_2 \in M^r_{v_{2s}}(\mathbb{R}^d)$, we obtain that $W(\varphi_2, \varphi_1) \in M^{1,\infty}_{1\otimes v_{2s}}$, for every $r \in [1, 2]$. Now the convolution relations in Proposition 2.4.19, in the form $M^{p,q}_{1\otimes v_{-s}} * M^{1,\infty}_{1\otimes v_{2s}} \subseteq M^{p,q}_{1\otimes v_s}$, yield that the Weyl symbol $\sigma = a * W(\varphi_2, \varphi_1)$ belongs to $M^{p,q}_{1\otimes v_s}$. The result now follows from Theorem 4.4.15. □

Remark 4.4.18. Using the same techniques as in the proof Theorem 4.4.17, one can study conditions on symbols and window functions such that the operator $A_a^{\varphi_1,\varphi_2}$ is in the Schatten class S^p, cf., e. g., [66, Theorem 3.4] and [283, Theorem 3.9].

4.4.3.1 Relations between modulation and Shubin–Sobolev spaces

Shubin–Sobolev spaces [265]. Let $\varphi(t) = 2^{n/4} e^{-\pi t^2}$ and $a(z) = \langle z \rangle^s$ for $s \in \mathbb{R}$, and set $A_s = A_a^{\varphi,\varphi}$. Then the Shubin–Sobolev space Q_s for $s \in \mathbb{R}$ is defined by

$$Q_s(\mathbb{R}^d) := \{f \in \mathcal{S}'(\mathbb{R}^d) : A_s f \in L^2(\mathbb{R}^d)\} = A_s^{-1} L^2(\mathbb{R}^d), \tag{4.66}$$

with norm $\|u\|_{Q_s} := \|A_s u\|_2$.

Lemma 4.4.19 (Characterization of Shubin–Sobolev spaces). *For all $s \in \mathbb{R}$, we have*

$$M^2_{v_s}(\mathbb{R}^d) = Q_s(\mathbb{R}^d) \tag{4.67}$$

with equivalent norms.

Proof. (a) We first show the inclusion $M^2_{v_s} \subseteq Q_s$ for *all* $s \in \mathbb{R}$. Recall that the adjoint of the STFT $f \to V_g f$ is the operator

$$V_g^* F = \int_{\mathbb{R}^{2d}} F(z)\pi(z) g\, dz,$$

which is a bounded mapping from $L^2(\mathbb{R}^{2d})$ to $L^2(\mathbb{R}^d)$ by Theorem 2.3.7 (i), with

$$\|V_g^* F\|_2 = \Big\| \int_{\mathbb{R}^{2d}} F(z)\pi(z)g\,dz \Big\|_2 \le \|g\|_2 \|F\|_2 \quad \text{for } F \in L^2(\mathbb{R}^{2d}). \tag{4.68}$$

Since A_s is defined explicitly by the vector-valued integral

$$A_s f = \int_{\mathbb{R}^{2d}} \langle z\rangle^s V_\varphi f(z)\pi(z)\varphi\,dz,$$

(4.68) implies that

$$\|A_s f\|_2 \le \|\varphi\|_2 \big\| \langle\cdot\rangle^s V_\varphi f \big\|_2 \asymp \|f\|_{M^2_{v_s}},$$

and thus $M^2_{v_s}$ is continuously embedded in Q_s for all $s \in \mathbb{R}$.

(b) To obtain the reverse inclusion, we argue by duality. Since $Q_s' = Q_{-s}$ [265] and $(M^2_{v_s})' = M^2_{v_{-s}}$ by Theorem 2.3.10, we have

$$Q_s = (Q_{-s})' \subseteq (M^2_{v_{-s}})' = M^2_{v_s}$$

for all $s \in \mathbb{R}$. The equivalence of the norms follows from the inverse mapping theorem. □

Remark 4.4.20. (i) In other words, the previous result says that the localization operator $A_a^{\varphi,\varphi}$ provides an isomorphism between L^2 and $Q_s = M^2_{v_s}$. The lifting property between Sobolev–Shubin spaces has been extended to more general modulation spaces. For this topic, we refer the interested reader to the contributions [26, 169, 170].
(ii) The equality in (4.67) was first observed in [180, Lemma 7.3.1].

4.5 $\mathrm{Op}_\tau(a)$ operators with symbols in Wiener amalgam spaces

First we focus on the members of the Cohen class, defined in (1.96) as

$$Q_\sigma f = W(f,f) * \sigma, \quad f \in \mathcal{S}(\mathbb{R}^d),$$

where $\sigma \in \mathcal{S}'(\mathbb{R}^{2d})$ is called the Cohen kernel. If $\sigma = \delta$, then $Q_\sigma f = W(f,f)$ and we come back to the Wigner distribution. More generally, if $\sigma = \sigma_\tau$ in (1.112), then $Q_{\sigma_\tau} f = W_\tau(f,f)$, the τ-Wigner distribution.

In what follows we study the properties of the Cohen kernels σ_τ in the realm of modulation spaces.

Proposition 4.5.1. *We have, for every $\tau \in [0,1] \setminus \{1/2\}$,*

$$\sigma_\tau \in W(\mathcal{F}L^1, L^\infty)(\mathbb{R}^{2d}) \cap M^{1,\infty}(\mathbb{R}^{2d}).$$

Proof. We exploit the dilation properties in Corollary 2.6.5: for $A = \lambda I$, $\lambda > 0$,

$$\|f(A\cdot)\|_{W(\mathcal{F}L^p,L^q)} \le C\lambda^{d(\frac{1}{p}-\frac{1}{q}-1)}(\lambda^2+1)^{d/2}\|f\|_{W(\mathcal{F}L^p,L^q)}.$$

Using them for $\lambda = \sqrt{t}$, $0 < t < 1/2$, $p = 1$, $q = \infty$, we obtain

$$\|e^{\pm 2\pi i\zeta_1\cdot\zeta_2 t}\|_{W(\mathcal{F}L^1,L^\infty)} \le C\|e^{\pm 2\pi i\zeta_1\cdot\zeta_2}\|_{W(\mathcal{F}L^1,L^\infty)}$$

with constant $C > 0$ independent on the parameter t. For $t = \frac{2}{2\tau-1}$, when $\tau > 1/2$ and $t = -\frac{2}{2\tau-1}$, when $0 \le \tau < 1/2$, we obtain that $\sigma_\tau \in W(\mathcal{F}L^1, L^\infty)(\mathbb{R}^{2d})$. Now, an easy computation gives

$$\mathcal{F}\sigma_\tau(\zeta_1,\zeta_2) = e^{-\pi i(2\tau-1)\zeta_1\cdot\zeta_2},$$

so that, using $\mathcal{F}M^{1,\infty}(\mathbb{R}^{2d}) = W(\mathcal{F}L^1, L^\infty)(\mathbb{R}^{2d})$ and repeating the previous argument, we obtain $\sigma_\tau \in M^{1,\infty}(\mathbb{R}^{2d})$ for every $\tau \in [0,1] \setminus \{1/2\}$. □

The case $\tau = 1/2$ is different. Indeed, $\sigma_{1/2} = \delta$ and, for any fixed $g \in \mathcal{S}(\mathbb{R}^{2d}) \setminus \{0\}$, the STFT $V_g\delta$ is given by

$$V_g\delta(z,\zeta) = \langle \delta, M_\zeta T_z g\rangle = \overline{g(-z)},$$

yielding $\delta \in M^{1,\infty}(\mathbb{R}^{2d}) \setminus W(\mathcal{F}L^1, L^\infty)(\mathbb{R}^{2d})$. As already mentioned before, the τ-Wigner distribution comes into play in the weak definition the pseudodifferential operator $\mathrm{Op}_\tau(a)$ in (4.35), since

$$\langle \mathrm{Op}_\tau(a)f, g\rangle = \langle a, W_\tau(f,g)\rangle,$$

cf. (4.36).

The continuity properties on modulation spaces $M^{p,q}$ for $\mathrm{Op}_\tau(a)$ operators with symbols $a \in M^{p,q}(\mathbb{R}^{2d})$ are the same as for Weyl operators, see Theorem 4.4.15.

In the following two subsections we exhibit the continuity properties on modulation spaces for τ-pseudodifferential operators with symbols a in Wiener amalgam spaces $W(\mathcal{F}L^p, L^q)(\mathbb{R}^{2d})$ contained in [65, 110]. We will obtain boundedness results for $\tau \in (0,1)$ whereas, at the end-points $\tau = 0$ and $\tau = 1$, we shall show that the corresponding operators are in general unbounded. Furthermore, for $\tau \in (0,1)$, we will exhibit a function of τ which is an upper bound for the operator norm.

For $1 \le r_1, r_2 \le \infty$, we introduce the function

$$\alpha_{(r_1,r_2)}(\tau) = \frac{1}{\tau^{d(\frac{1}{r_1'}+\frac{1}{r_2})}(1-\tau)^{d(\frac{1}{r_1}+\frac{1}{r_2'})}}, \quad \tau \in (0,1). \tag{4.69}$$

Observe that the function $\alpha_{(r_1,r_2)}(\tau)$ is unbounded on $(0,1)$. Indeed, for $(r_1, r_2) \notin \{(1,\infty), (\infty, 1)\}$,

$$\lim_{\tau\to 0^+} \alpha_{(r_1,r_2)}(\tau) = \lim_{\tau\to 1^-} \alpha_{(r_1,r_2)}(\tau) = +\infty.$$

For $(r_1, r_2) = (1, \infty)$, we have $\lim_{\tau \to 1^-} \alpha_{(1,\infty)}(\tau) = +\infty$ whereas, for $(r_1, r_2) = (\infty, 1)$, $\lim_{\tau \to 0^+} \alpha_{(\infty,1)}(\tau) = +\infty$.

In the sequel we shall heavily use the following symplectic matrix:

$$\mathcal{A} = \begin{pmatrix} 0_{d\times d} & (\frac{1-\tau}{\tau})^{1/2} I_{d\times d} \\ -(\frac{\tau}{1-\tau})^{1/2} I_{d\times d} & 0_{d\times d} \end{pmatrix}, \quad \tau \in (0,1). \tag{4.70}$$

The main properties of $\mathcal{A}$ are detailed below. Their proof is attained by easy computations.

Lemma 4.5.2. *For any $\tau \in (0,1)$, the matrix $\mathcal{A}$ in (4.70) enjoys the following properties:*
(i) *$\mathcal{A}_\tau \in Sp(d, \mathbb{R})$; in particular, $\mathcal{A}_{1/2} = J$.*
(ii) *$\mathcal{A}_\tau^\top = -\mathcal{A}_{1-\tau}$, $\mathcal{A}_\tau^{-1} = -\mathcal{A}_\tau$.*
(iii) *$\mathcal{A}_{1-\tau}\mathcal{A}_\tau = \mathcal{A}_\tau^\top \mathcal{A}_\tau^{-1} = I_{2d\times 2d} - \mathcal{B}_\tau$, where*

$$\mathcal{B}_\tau = \begin{pmatrix} \frac{1}{1-\tau} I_{d\times d} & 0_{d\times d} \\ 0_{d\times d} & \frac{1}{\tau} I_{d\times d} \end{pmatrix}. \tag{4.71}$$

(iv) *$\sqrt{\tau(1-\tau)}(\mathcal{A}_\tau + \mathcal{A}_{1-\tau}) = \sqrt{\tau(1-\tau)}\mathcal{B}_\tau \mathcal{A}_\tau = J$.*

4.5.1 Boundedness properties of τ-Wigner distributions

This section is devoted to investigating the continuity properties of τ-Wigner distributions in the realm of Wiener and modulation spaces. For a submultiplicative weight v, we set

$$v_J(z) = v(Jz), \tag{4.72}$$

where J denotes the canonical symplectic matrix (1.1). A particular instance of the Young's inequality for mixed L^P spaces in (2.12) reads as follows.

Lemma 4.5.3. *For $1 \le p_i, q_i, r, s \le \infty$, $i = 1, 2$, $m \in \mathcal{M}_v(\mathbb{R}^{2d})$, $F \in L_v^{p_1,q_1}(\mathbb{R}^{2d})$, $G \in L_m^{p_2,q_2}(\mathbb{R}^{2d})$, we have $F * G \in L_m^{r,s}(\mathbb{R}^{2d})$, with $1/p_1 + 1/p_2 = 1 + 1/r$, $1/q_1 + 1/q_2 = 1 + 1/s$, and*

$$\|F * G\|_{L_m^{r,s}} \le \|F\|_{L_v^{p_1,q_1}} \|G\|_{L_m^{p_2,q_2}}. \tag{4.73}$$

We say that a measurable function f on $\mathbb{R}^{4d}$ is in the space $L_z^\infty(L_{\zeta,m}^1)(\mathbb{R}^{4d})$, with m weight function on $\mathbb{R}^{2d}$, if

$$\|f\|_{L_z^\infty(L_{\zeta,m}^1)} = \sup_{z\in\mathbb{R}^{2d}} \int_{\mathbb{R}^{2d}} |f(z,\zeta)| m(\zeta)\, d\zeta < \infty. \tag{4.74}$$

When working on the STFT of τ-Wigner distribution, we will use the following Young-type inequality:

Lemma 4.5.4. *If $m \in \mathcal{M}_v(\mathbb{R}^{2d})$, $f \in L^1_{1\otimes v}(\mathbb{R}^{4d})$ and $g \in L^\infty_z(L^1_{\zeta,m})(\mathbb{R}^{4d})$, then $f * g \in L^\infty_z(L^1_{\zeta,m})(\mathbb{R}^{4d})$, with*

$$\|f * g\|_{L^\infty_z(L^1_{\zeta,m})} \le \|f\|_{L^1_{1\otimes v}} \|g\|_{L^\infty_z(L^1_{\zeta,m})}.$$

Proof. Using the definition of $L^\infty_z(L^1_{\zeta,m})$-norm in (4.74),

$$\begin{aligned} I := \|f * g\|_{L^\infty_z(L^1_{\zeta,m})} &= \sup_{z\in\mathbb{R}^{2d}} \int_{\mathbb{R}^{2d}} |f * g|(z,\zeta) m(\zeta)\, d\zeta \\ &= \sup_{z\in\mathbb{R}^{2d}} \int_{\mathbb{R}^{2d}} \left| \int_{\mathbb{R}^{4d}} f(y,\eta) g(z-y,\zeta-\eta)\, dy\, d\eta \right| m(\zeta)\, d\zeta \\ &\le \sup_{z\in\mathbb{R}^{2d}} \int_{\mathbb{R}^{2d}} \int_{\mathbb{R}^{2d}} \left(\int_{\mathbb{R}^{2d}} |f|(y,\eta)|g|(z-y,\zeta-\eta)\, d\eta \right) m(\zeta)\, dy\, d\zeta \\ &= \sup_{z\in\mathbb{R}^{2d}} \int_{\mathbb{R}^{2d}} \int_{\mathbb{R}^{2d}} (|f|(y,\cdot) * |g|(z-y,\cdot))(\zeta) m(\zeta)\, dy\, d\zeta. \end{aligned}$$

By Young's inequality (4.73),

$$\begin{aligned} I &= \sup_{z\in\mathbb{R}^{2d}} \int_{\mathbb{R}^{2d}} \big\| |f|(y,\cdot) \big\|_{L^1_v} \big\| |g|(z-y,\cdot) \big\|_{L^1_m}\, dy \\ &\le \int_{\mathbb{R}^{2d}} \big\| |f|(y,\cdot) \big\|_{L^1_v} \sup_{z\in\mathbb{R}^d} \big\| |g|(z-y,\cdot) \big\|_{L^1_m}\, dy \\ &= \|g\|_{L^\infty_z(L^1_{\zeta,m})} \|f\|_{L^1_{1\otimes v}}, \end{aligned}$$

as claimed. □

A particular case of Lemma 4.5.3 gives:

Lemma 4.5.5. *Suppose $m \in \mathcal{M}_v(\mathbb{R}^{2d})$, $f \in L^1_{1\otimes v}(\mathbb{R}^{4d})$ and $g \in L^2_{1\otimes m}(\mathbb{R}^{4d})$. Then $f * g \in L^2_{1\otimes m}(\mathbb{R}^{4d})$, with*

$$\|f * g\|_{L^2_{1\otimes m}} \le \|f\|_{L^1_{1\otimes v}} \|g\|_{L^2_{1\otimes m}}.$$

Lemma 4.5.6. *Assume that $m \in \mathcal{M}_v(\mathbb{R}^{2d})$, $1 \le p_1, p_2 \le \infty$, $f \in M^{p_1,p_2}_m(\mathbb{R}^d)$, $g \in M^{p_1',p_2'}_{1/m}(\mathbb{R}^d)$. Then for every $\tau \in (0,1)$, the τ-Wigner distribution $W_\tau(g,f)$ is in $W(\mathcal{F}L^1_{1/v_J}, L^\infty)(\mathbb{R}^{2d})$, with*

$$\big\| W_\tau(g,f) \big\|_{W(\mathcal{F}L^1_{1/v_J}, L^\infty)} \le C \alpha_{(p_1,p_2)}(\tau) \|f\|_{M^{p_1,p_2}_m} \|g\|_{M^{p_1',p_2'}_{1/m}}, \tag{4.75}$$

where the function $\alpha_{(p_1,p_2)}(\tau)$ is defined in (4.69) and $C > 0$ is independent of τ.

Proof. We compute the STFT of $W_\tau(g,f)$ with respect to the window function $\Phi_\tau \in \mathcal{S}(\mathbb{R}^{2d})$ defined in Lemma 1.3.38. Using that lemma and the properties of the matrix $\mathcal{A}$ in Lemma 4.5.2, by performing the change of variables $\sqrt{\tau(1-\tau)}\mathcal{A}\zeta = \eta$, we deduce

$$\begin{aligned}\int_{\mathbb{R}^{2d}} & |V_{\Phi_\tau} W_\tau(g,f)|(z,\zeta)\frac{1}{v(J\zeta)}\,d\zeta\\ &= \int_{\mathbb{R}^{2d}} |V_{\varphi_1} g(z+\sqrt{\tau(1-\tau)}\mathcal{A}^T\zeta)||V_{\varphi_2} f(z+\sqrt{\tau(1-\tau)}\mathcal{A}\zeta)|\frac{1}{v(\sqrt{\tau(1-\tau)}\mathcal{B}_\tau\mathcal{A}\zeta)}\,d\zeta\\ &= \frac{1}{[\tau(1-\tau)]^d}\int_{\mathbb{R}^{2d}} |V_{\varphi_1} g(z+\mathcal{A}_{1-\tau}\mathcal{A}\eta)||V_{\varphi_2} f(z+\eta)|\frac{1}{v(\mathcal{B}_\tau\eta)}\,d\eta.\end{aligned}$$

Since m is a v-moderate weight, we can find a positive constant C, independent of τ, such that

$$\frac{1}{v(\mathcal{B}_\tau\eta)} \le C\frac{m(z+\eta)}{m(z+\mathcal{A}_{1-\tau}\mathcal{A}\eta)}, \tag{4.76}$$

so

$$\begin{aligned}\int_{\mathbb{R}^{2d}} & |V_{\Phi_\tau} W_\tau(g,f)|(z,\zeta)\frac{1}{v(J\zeta)}\,d\zeta\\ &\le C\frac{1}{[\tau(1-\tau)]^d}\int_{\mathbb{R}^{2d}} |V_{\varphi_1} g(z+\mathcal{A}_{1-\tau}\mathcal{A}\eta)||V_{\varphi_2} f(z+\eta)|\frac{m(z+\eta)}{m(z+\mathcal{A}_{1-\tau}\mathcal{A}\eta)}\,d\eta.\end{aligned}$$

Consequently,

$$\begin{aligned}&\|W_\tau(g,f)\|_{W(\mathcal{F}L^1_{1/v_J},L^\infty)}\\ &\asymp \sup_{z\in\mathbb{R}^{2d}}\int_{\mathbb{R}^{2d}} |V_{\varphi_1} g(z+\sqrt{\tau(1-\tau)}\mathcal{A}^T\zeta)||V_{\varphi_2} f(z+\sqrt{\tau(1-\tau)}\mathcal{A}\zeta)|\frac{1}{v(\sqrt{\tau(1-\tau)}\mathcal{B}_\tau\mathcal{A}\zeta)}\,d\zeta\\ &\le C\frac{1}{[\tau(1-\tau)]^d}\sup_{z\in\mathbb{R}^{2d}}\int_{\mathbb{R}^{2d}} |V_{\varphi_1} g(z+\mathcal{A}_{1-\tau}\mathcal{A}\eta)||V_{\varphi_2} f(z+\eta)|\frac{m(z+\eta)}{m(z+\mathcal{A}_{1-\tau}\mathcal{A}\eta)}\,d\eta\\ &\le C\frac{1}{[\tau(1-\tau)]^d}\|V_{\varphi_1} fm\|_{L^{p_1,p_2}}\left\|V_{\varphi_2} g\frac{1}{m}(z+\mathcal{A}_{1-\tau}\mathcal{A}\cdot)\right\|_{L^{p_1',p_2'}}\\ &\lesssim \frac{1}{[\tau(1-\tau)]^d}\left(\frac{1-\tau}{\tau}\right)^{d(\frac{1}{p_2}-\frac{1}{p_1})}\|f\|_{M^{p_1,p_2}_m}\|g\|_{M^{p_1',p_2'}_{1/m}}.\end{aligned}$$

The claim is proved. □

The previous estimate is not uniform with respect to τ, in the sense that the $W(\mathcal{F}L^1_{1/v_J}, L^\infty)$-norm of the τ-Wigner distribution has been calculated by using a window function Φ_τ depending on τ. The next goal is to find an upper bound of this norm independent of τ. We will need some preliminaries. The τ-Wigner distribution of the Gaussian function $\varphi(t) = e^{-\pi t^2}$ is in turn a generalized Gaussian function, as showed in the next lemma.

Lemma 4.5.7. *Consider $\varphi_1(t) = \varphi_2(t) = \varphi(t) = e^{-\pi t^2}$, $t \in \mathbb{R}^d$, and $\tau \in [0,1]$. Then*

$$W_\tau\varphi(x,\xi) = \frac{1}{(2\tau^2 - 2\tau + 1)^{d/2}} e^{-\pi\frac{1}{2\tau^2-2\tau+1}x^2} e^{-\pi\frac{1}{2\tau^2-2\tau+1}\xi^2} e^{2\pi i\frac{2\tau-1}{2\tau^2-2\tau+1}x\cdot\xi},$$

for all $(x,\xi) \in \mathbb{R}^{2d}$.

Proof. Using the definition of the τ-Wigner distribution in (1.109),

$$\begin{aligned}
W_\tau\varphi(x,\xi) &= \int_{\mathbb{R}^d} e^{-2\pi i\xi\cdot t} e^{-\pi(x+\tau t)^2} e^{-\pi(x-(1-\tau)t)^2}\, dt \\
&= \int_{\mathbb{R}^d} e^{-2\pi i\xi\cdot t} e^{-2\pi x^2} e^{-\pi[(2\tau^2-2\tau+1)t^2+2(2\tau-1)x\cdot t]}\, dt \\
&= e^{-2\pi x^2+\pi(\frac{2\tau-1}{\sqrt{2\tau^2-2\tau+1}})^2x^2} \int_{\mathbb{R}^d} e^{-2\pi i\xi\cdot t} e^{-\pi(\sqrt{2\tau^2-2\tau+1}t+\frac{2\tau-1}{\sqrt{2\tau^2-2\tau+1}}x)^2}\, dt.
\end{aligned}$$

We perform the following change of variables:

$$\sqrt{2\tau^2-2\tau+1}t + \frac{2\tau-1}{\sqrt{2\tau^2-2\tau+1}}x = y,$$

so that, letting $c(\tau) = 2\tau^2 - 2\tau + 1 > 0$,

$$\begin{aligned}
W_\tau\varphi(x,\xi) &= \frac{1}{c(\tau)^{\frac{d}{2}}} e^{-\pi(2-\frac{(2\tau-1)^2}{c(\tau)})x^2} \int_{\mathbb{R}^d} e^{-2\pi i\xi\cdot\frac{y}{\sqrt{c(\tau)}}} e^{2\pi i\frac{2\tau-1}{c(\tau)}\xi\cdot x} e^{-\pi y^2}\, dy \\
&= \frac{1}{c(\tau)^{\frac{d}{2}}} e^{-\pi\frac{1}{c(\tau)}x^2} e^{2\pi i\frac{2\tau-1}{c(\tau)}\xi\cdot x} e^{-\pi\frac{1}{c(\tau)}\xi^2},
\end{aligned}$$

as desired. □

Lemma 4.5.8. *Consider $\Phi(x,\xi) = e^{-\pi(x^2+\xi^2)}$, $(x,\xi) \in \mathbb{R}^{2d}$, and $\Phi_\tau = W_\tau(\varphi,\varphi)$, where $\varphi(t) = e^{-\pi t^2}$, $t \in \mathbb{R}^d$. Then for v_J in (4.72) there exists a constant $C > 0$ such that*

$$\|V_\Phi\Phi_\tau\|_{L^1_{1\otimes v_J}} \le C, \quad \forall\tau \in [0,1]. \tag{4.77}$$

Consequently,

$$\|\Phi_\tau\|_{M^1_{1\otimes v_J}} \le C, \quad \forall\tau \in [0,1]. \tag{4.78}$$

Proof. Using Lemma 4.5.7 and formula (1.127), with $z = (z_1, z_2), \zeta = (\zeta_1, \zeta_2) \in \mathbb{R}^{2d}$, we compute

$$
\begin{aligned}
|V_\Phi \Phi_\tau|(z,\zeta) &= \frac{1}{(2\tau^2-2\tau+1)^{d/2}} \frac{(2\tau^2-2\tau+1)^{d/2}}{(2\tau^2-2\tau+5)^{d/2}} \\
&\quad \times e^{-\pi \frac{\frac{3}{2\tau^2-2\tau+1}(z_1^2+z_2^2)+\frac{2\tau^2-2\tau+2}{2\tau^2-2\tau+1}(\zeta_1^2+\zeta_2^2)-2\frac{2\tau-1}{2\tau^2-2\tau+1}(z_1\cdot\zeta_2+z_2\cdot\zeta_1)}{\frac{2\tau^2-2\tau+5}{2\tau^2-2\tau+1}}} \\
&= \frac{1}{(2\tau^2-2\tau+5)^{d/2}} e^{-\pi \frac{3(z_1^2+z_2^2)+(2\tau^2-2\tau+2)(\zeta_1^2+\zeta_2^2)+(2-4\tau)(z_1\cdot\zeta_2+z_2\cdot\zeta_1)}{2\tau^2-2\tau+5}}.
\end{aligned}
$$

Observing that

$$
\frac{1}{(2\tau^2-2\tau+5)^{d/2}} \le \max_{\tau\in(0,1)} \frac{1}{(2\tau^2-2\tau+5)^{d/2}} = \left(\frac{2}{9}\right)^{d/2},
$$

by Lemma 2.1.4, we have

$$
\begin{aligned}
\|V_\Phi \Phi_\tau\|_{L^1_{1\otimes v_J}} &\le \left(\frac{2}{9}\right)^{d/2} \\
&\quad \times \int_{\mathbb{R}^{2d}} \int_{\mathbb{R}^{2d}} e^{-\pi \frac{3(z_1^2+z_2^2)+(2\tau^2-2\tau+2)(\zeta_1^2+\zeta_2^2)+(2-4\tau)(z_1\cdot\zeta_2+z_2\cdot\zeta_1)}{2\tau^2-2\tau+5}} v_J(\zeta)\, d\zeta_1\, d\zeta_2\, dz_1\, dz_2 \\
&\le C \int_{\mathbb{R}^{2d}} \int_{\mathbb{R}^{2d}} e^{-\pi \frac{3(z_1^2+z_2^2)+(2\tau^2-2\tau+2)(\zeta_1^2+\zeta_2^2)+(2-4\tau)(z_1\cdot\zeta_2+z_2\cdot\zeta_1)}{2\tau^2-2\tau+5}} e^{a|J\zeta|}\, d\zeta_1\, d\zeta_2\, dz_1\, dz_2 \\
&= C \int_{\mathbb{R}^{2d}} e^{-\pi \frac{3(z_1^2+z_2^2)}{2\tau^2-2\tau+5}} I_1\, dz_1\, dz_2,
\end{aligned}
$$

where

$$
I_1 := \int_{\mathbb{R}^{2d}} e^{-\pi \frac{(2\tau^2-2\tau+2)(\zeta_1^2+\zeta_2^2)+(2-4\tau)(z_1\cdot\zeta_2+z_2\cdot\zeta_1)}{2\tau^2-2\tau+5}} e^{a|J\zeta|}\, d\zeta_1\, d\zeta_2.
$$

The integral I_1 can be computed as follows:

$$
\begin{aligned}
I_1 &= \int_{\mathbb{R}^{2d}} e^{-\pi \frac{(2\tau^2-2\tau+2)(\zeta_1^2+\zeta_2^2)+(2-4\tau)(z_1\cdot\zeta_2+z_2\cdot\zeta_1)}{2\tau^2-2\tau+5}} e^{a|J\zeta|}\, d\zeta_1\, d\zeta_2 \\
&\le \int_{\mathbb{R}^{2d}} e^{-\pi \frac{(2\tau^2-2\tau+2)(\zeta_1^2+\zeta_2^2)+(2-4\tau)(z_1\cdot\zeta_2+z_2\cdot\zeta_1)}{2\tau^2-2\tau+5}} e^{a(|\zeta_1|+|\zeta_2|)}\, d\zeta_1\, d\zeta_2 \\
&= \left(\int_{\mathbb{R}^d} e^{-\pi \frac{(2\tau^2-2\tau+2)\zeta_1^2+(2-4\tau)z_2\cdot\zeta_1}{2\tau^2-2\tau+5}} e^{a|\zeta_1|}\, d\zeta_1 \right) \left(\int_{\mathbb{R}^d} e^{-\pi \frac{(2\tau^2-2\tau+2)\zeta_2^2+(2-4\tau)z_1\cdot\zeta_2}{2\tau^2-2\tau+5}} e^{a|\zeta_2|}\, d\zeta_2 \right).
\end{aligned}
$$

We calculate the integral with respect to the variable ζ_1 (the other integral is analogous):

$$\begin{aligned}
&\int_{\mathbb{R}^d} e^{-\pi\frac{(2\tau^2-2\tau+2)\zeta_1^2+(2-4\tau)z_2\cdot\zeta_1}{2\tau^2-2\tau+5}} e^{a|\zeta_1|}\,d\zeta_1\\
&= \int_{\mathbb{R}^d} e^{-\pi\frac{(2\tau^2-2\tau+2)\zeta_1^2+(2-4\tau)z_2\cdot\zeta_1+\frac{(1-2\tau)^2z_2^2}{2\tau^2-2\tau+2}-\frac{(1-2\tau)^2z_2^2}{2\tau^2-2\tau+2}}{2\tau^2-2\tau+5}} e^{a|\zeta_1|}\,d\zeta_1\\
&= e^{\pi\frac{(1-2\tau)^2z_2^2}{(2\tau^2-2\tau+2)(2\tau^2-2\tau+5)}} \int_{\mathbb{R}^d} e^{-\pi\frac{(\sqrt{2\tau^2-2\tau+2}\zeta_1+\frac{1-2\tau}{\sqrt{2\tau^2-2\tau+2}}z_2)^2}{2\tau^2-2\tau+5}} e^{a|\zeta_1|}\,d\zeta_1\\
&= e^{\pi\frac{(1-2\tau)^2z_2^2}{(2\tau^2-2\tau+2)(2\tau^2-2\tau+5)}} \underbrace{\int_{\mathbb{R}^d} e^{-\pi\frac{((2\tau^2-2\tau+2)\zeta_1+(1-2\tau)z_2)^2}{(2\tau^2-2\tau+5)(2\tau^2-2\tau+2)}} e^{a|\zeta_1|}\,d\zeta_1}_{:=I_3}.
\end{aligned}$$

In I_3 we perform the following change of variables:

$$(2\tau^2-2\tau+2)\zeta_1+(1-2\tau)z_2=\eta_1,$$

so that

$$\begin{aligned}
I_3 &= \frac{1}{(2\tau^2-2\tau+2)^d}\int_{\mathbb{R}^d} e^{-\pi\frac{\eta_1^2}{(2\tau^2-2\tau+5)(2\tau^2-2\tau+2)}} e^{\frac{a}{2\tau^2-2\tau+2}|\eta_1-(1-2\tau)z_2|}\,d\eta_1\\
&\le C_1^d e^{\frac{a|1-2\tau|}{2\tau^2-2\tau+2}|z_2|}\int_{\mathbb{R}^d} e^{-\pi C_2\eta_1^2}e^{aC_1|\eta_1|}\,d\eta_1,
\end{aligned}$$

where

$$C_1=\max_{\tau\in[0,1]}\frac{1}{(2\tau^2-2\tau+2)}=\frac{2}{3},\quad C_2=\min_{\tau\in[0,1]}\frac{1}{(2\tau^2-2\tau+5)(2\tau^2-2\tau+2)}=\frac{1}{10}.$$

Using $\lim_{|\eta_1|\to\infty} e^{-\pi\frac{C_1}{2}\eta_1^2}e^{aC_2|\eta_1|}=0$, for every $\epsilon>0$ there exists $R>0$ such that $e^{-\pi\frac{C_1}{2}\eta_1^2}e^{aC_2|\eta_1|}\le\epsilon$, for all $|\eta_1|$ with $|\eta_1|>R$. Hence

$$\begin{aligned}
I_3 &\le C_1^d e^{\frac{a|1-2\tau|}{2\tau^2-2\tau+2}|z_2|}\int_{\mathbb{R}^d} e^{-\pi C_2\eta_1^2}e^{aC_1|\eta_1|}\,d\eta_1\\
&= C_1^d e^{\frac{a|1-2\tau|}{2\tau^2-2\tau+2}|z_2|}\Bigg(\int_{\{\eta_1\in\mathbb{R}^d:|\eta_1|\le R\}} e^{-\pi C_2\eta_1^2}e^{aC_1|\eta_1|}\,d\eta_1\\
&\quad+\int_{\{\eta_1\in\mathbb{R}^d:|\eta_1|>R\}} e^{-\pi C_2\eta_1^2}e^{aC_1|\eta_1|}\,d\eta_1\Bigg)
\end{aligned}$$

$$\leq C_1^d e^{\frac{a|1-2\tau|}{2\tau^2-2\tau+2}|z_2|}\Big(e^{aC_1R}\int_{\mathbb{R}^d} e^{-\pi C_2\eta_1^2}\,d\eta_1+\epsilon\int_{\mathbb{R}^d} e^{-\pi\frac{C_2}{2}\eta_1^2}\,d\eta_1\Big)$$
$$=\tilde{C}e^{\frac{a|1-2\tau|}{2\tau^2-2\tau+2}|z_2|}<\infty,$$

where $\tilde{C}$ is a constant independent of τ. In conclusion, the integral I_1 can be majorized as

$$I_1\leq 2\tilde{C}e^{\pi\frac{(1-2\tau)^2z_2^2}{(2\tau^2-2\tau+2)(2\tau^2-2\tau+5)}+\frac{a|1-2\tau|}{2\tau^2-2\tau+2}|z_2|}e^{\pi\frac{(1-2\tau)^2z_1^2}{(2\tau^2-2\tau+2)(2\tau^2-2\tau+5)}+\frac{a|1-2\tau|}{2\tau^2-2\tau+2}|z_1|}.$$

Thus, there exists a constant $M_1>0$ independent of τ such that

$$\begin{aligned}\|V_\Phi\Phi_\tau\|_{L^1_{1\otimes v_J}}&\leq M_1\int_{\mathbb{R}^d} e^{-\pi\frac{3z_1^2}{2\tau^2-2\tau+5}}e^{\pi\frac{(1-2\tau)^2z_1^2}{(2\tau^2-2\tau+2)(2\tau^2-2\tau+5)}+\frac{a|1-2\tau|}{2\tau^2-2\tau+2}|z_1|}\,dz_1\\&\quad\times\int_{\mathbb{R}^d} e^{-\pi\frac{3z_2^2}{2\tau^2-2\tau+5}}e^{\pi\frac{(1-2\tau)^2z_2^2}{(2\tau^2-2\tau+2)(2\tau^2-2\tau+5)}+\frac{a|1-2\tau|}{2\tau^2-2\tau+2}|z_2|}\,dz_2\\&=2M_1\int_{\mathbb{R}^d} e^{-\pi\frac{3z_1^2}{2\tau^2-2\tau+5}}e^{\pi\frac{(1-2\tau)^2z_1^2}{(2\tau^2-2\tau+2)(2\tau^2-2\tau+5)}+\frac{a|1-2\tau|}{2\tau^2-2\tau+2}|z_1|}\,dz_1.\end{aligned}$$

The integral with respect the variable z_1 is computed analogously to that for ζ_1 above. The estimate (4.78) follows by

$$\|\Phi_\tau\|_{M^1_{1\otimes v_J}}\asymp\|V_\Phi\Phi_\tau\|_{L^1_{1\otimes v_J}}\leq C,$$

as desired. ☐

Proposition 4.5.9. *Under the assumptions of Lemma 4.5.6, there exists a constant $C>0$ independent of τ such that*

$$\|W_\tau(g,f)\|_{W(\mathcal{F}L^1_{1/v_J},L^\infty)}\leq C\alpha_{(p_1,p_2)}(\tau)\|f\|_{M_m^{p_1,p_2}}\|g\|_{M_{1/m}^{p_1',p_2'}},\quad \tau\in(0,1).\tag{4.79}$$

Proof. Changing window in the computation of the STFT as in Lemma (1.2.29), using Lemmas 4.5.4, 4.5.8 and Moyal's formula (1.115), we have

$$\begin{aligned}\|V_\Phi W_\tau(g,f)\|_{L_z^\infty(L^1_{\zeta,1/v_J})}&\leq\frac{1}{|\langle\Phi_\tau,\Phi_\tau\rangle|}\big\||V_{\Phi_\tau}W_\tau(g,f)|*|V_\Phi\Phi_\tau|\big\|_{L_z^\infty(L^1_{\zeta,1/v_J})}\\&\leq\frac{1}{\|\varphi\|^2\|\varphi\|^2}\|V_{\Phi_\tau}W_\tau(g,f)\|_{L_z^\infty(L^1_{\zeta,1/v_J})}\|V_\Phi\Phi_\tau\|_{L^1_{1\otimes v_J}}\\&\leq C\alpha_{(p_1,p_2)}(\tau)\|f\|_{M_m^{p_1,p_2}}\|g\|_{M_{1/m}^{p_1',p_2'}}.\end{aligned}$$

This completes the proof. ☐

Repeating the pattern of Lemma 4.5.6 and Proposition 4.5.9 in the Wiener amalgam space $W(\mathcal{F}L^2_{1/v_J}, L^2)(\mathbb{R}^{2d})$, we can state the following.

Proposition 4.5.10. *Let $m \in \mathcal{M}_v(\mathbb{R}^{2d})$, $f \in M^2_m(\mathbb{R}^d)$ and $g \in M^2_{1/m}(\mathbb{R}^d)$. For $\tau \in (0,1)$, the τ-Wigner distribution $W_\tau(g,f)$ is in $W(\mathcal{F}L^2_{1/v_J}, L^2)(\mathbb{R}^{2d})$, with the uniform estimate*

$$\|W_\tau(g,f)\|_{W(\mathcal{F}L^2_{1/v_J},L^2)} \le C\|f\|_{M^2_m}\|g\|_{M^2_{1/m}}, \tag{4.80}$$

where the positive constant C is independent of τ.

Proof. First step. We use Lemma 1.3.38, Young's inequality $L^1 * L^1 \subset L^1$, and the change of variables $\mathcal{B}_\tau\eta \to \eta$, to compute

$$\begin{aligned}
&\|W_\tau(g,f)\|_{W(\mathcal{F}L^2_{1/v_J},L^2)}\\
&\asymp \Big(\int_{\mathbb{R}^{2d}}\int_{\mathbb{R}^{2d}} |V_{\varphi_1}g(z+\sqrt{\tau(1-\tau)}\mathcal{A}^T\zeta)|^2 |V_{\varphi_2}f(z+\sqrt{\tau(1-\tau)}\mathcal{A}\zeta)|^2 \frac{1}{v^2(J\zeta)}\,d\zeta\,dz\Big)^{\frac12}\\
&\le C\frac{1}{[\tau(1-\tau)]^d}\Big(\int_{\mathbb{R}^{2d}}\int_{\mathbb{R}^{2d}} |V_{\varphi_1}g(z+\eta-\mathcal{B}_\tau\eta)|^2|V_{\varphi_2}f(z+\eta)|^2\frac{m^2(z+\eta)}{m^2(z+\eta-\mathcal{B}_\tau\eta)}\,d\eta\,dz\Big)^{\frac12}\\
&= C\frac{1}{[\tau(1-\tau)]^d}\Big(\int_{\mathbb{R}^{2d}} (|V_{\varphi_2}f|^2m^2) * \Big(|V_{\varphi_1}g|^2\frac{1}{m^2}\Big)(\mathcal{B}_\tau\eta)\,d\eta\Big)^{\frac12}\\
&\lesssim \||V_\varphi f|^2m^2\|_1 \Big\||V_\varphi g|^2\frac{1}{m^2}\Big\|_1\\
&\lesssim \|f\|_{M^2_m}\|g\|_{M^2_{1/m}}.
\end{aligned}$$

Second step. Consider now $\Phi \in \mathcal{S}(\mathbb{R}^{2d})$. Then the same pattern as in the proof of Proposition 4.5.9, with Lemma 4.5.4 replaced by Lemma 4.5.5, gives the uniform estimate (4.80). □

The previous issue can be rephrased in terms of modulation spaces as follows.

Corollary 4.5.11. *For $\tau \in (0,1)$, $m \in \mathcal{M}_v(\mathbb{R}^{2d})$, $f \in M^2_m(\mathbb{R}^d)$, $g \in M^2_{1/m}(\mathbb{R}^d)$, the τ-Wigner distribution belongs to $M^2_{1/v_J\otimes 1}(\mathbb{R}^{2d})$ with*

$$\|W_\tau(g,f)\|_{M^2_{1/v_J\otimes 1}} \le C\|f\|_{M^2_m}\|g\|_{M^2_{1/m}},$$

with $C > 0$ independent of τ.

4.5.2 Continuity properties of $\mathrm{Op}_\tau(a)$ operators

This section is devoted to the proof of Theorem 4.5.14. We will start with two preliminary results about τ-pseudodifferential operators acting on modulation spaces and having symbols in $W(\mathcal{F}L^{\infty}_{1/v_J}, L^1)(\mathbb{R}^{2d})$ and $W(\mathcal{F}L^{2}_{1/v_J}, L^2)(\mathbb{R}^{2d})$, respectively. Then by means of complex interpolation between Wiener amalgam spaces, we shall reach our goal.

Proposition 4.5.12. *Suppose that $m \in \mathcal{M}_v(\mathbb{R}^{2d})$ and consider a symbol function $a \in W(\mathcal{F}L^{\infty}_{v_J}, L^1)(\mathbb{R}^{2d})$. Then for every $\tau \in (0,1)$, the τ-pseudodifferential operator $\mathrm{Op}_\tau(a)$ is bounded on $\mathcal{M}^{p_1,p_2}_m(\mathbb{R}^d)$, for every $1 \le p_1, p_2 \le \infty$, with*

$$\|\mathrm{Op}_\tau(a)f\|_{\mathcal{M}^{p_1,p_2}_m} \le C\alpha_{(p_1,p_2)}(\tau)\|a\|_{W(\mathcal{F}L^{\infty}_{v_J},L^1)}\|f\|_{\mathcal{M}^{p_1,p_2}_m} \tag{4.81}$$

($C > 0$ does not depend on τ).

Proof. For every $f \in \mathcal{M}^{p_1,p_2}_m(\mathbb{R}^d)$ and $g \in \mathcal{M}^{p_1',p_2'}_{1/m}(\mathbb{R}^d)$, we can write

$$|\langle \mathrm{Op}_\tau(a)f, g\rangle| = |\langle a, W_\tau(g,f)\rangle| \le \|V_\Phi a\|_{L^1_z(L^\infty_{v_J,\zeta})}\|V_\Phi W_\tau(g,f)\|_{L^\infty_z(L^1_{1/v_J,\zeta})}.$$

Observing that

$$\|W_\tau(g,f)\|_{W(\mathcal{F}L^1_{1/v_J},L^\infty)} \asymp \|V_\Phi W_\tau(g,f)\|_{L^\infty_z(L^1_{1/v_J,\zeta})}$$

and using Proposition 4.5.9, we conclude the proof. □

Proposition 4.5.13. *Let $m \in \mathcal{M}_v(\mathbb{R}^{2d})$, $a \in W(\mathcal{F}L^2_{v_J}, L^2)(\mathbb{R}^{2d})$ and $\tau \in (0,1)$. Then the operator $\mathrm{Op}_\tau(a)$ is bounded on M^2_m with*

$$\|\mathrm{Op}_\tau(a)f\|_{M^2_m} \le C\|a\|_{W(\mathcal{F}L^2_{v_J},L^2)}\|f\|_{M^2_m}, \tag{4.82}$$

where the constant $C > 0$ is independent of τ.

Proof. The proof is similar to that of Proposition 4.5.12, where Proposition 4.5.9 is replaced by Proposition 4.5.10. □

Propositions 4.5.12 and 4.5.13 are the main ingredients in the proof of Theorem 4.5.14.

Theorem 4.5.14. *Suppose that $1 \le p, q, r_1, r_2 \le \infty$ satisfy*

$$q \le p' \tag{4.83}$$

and

$$\max\{r_1, r_2, r_1', r_2'\} \le p. \tag{4.84}$$

Let $m \in \mathcal{M}_v(\mathbb{R}^{2d})$ and $a \in W(\mathcal{F}L^p_{v_J}, L^q)(\mathbb{R}^{2d})$. For $\tau \in (0,1)$, every τ-pseudodifferential operator $\mathrm{Op}_\tau(a)$ is a bounded operator on $\mathcal{M}^{r_1,r_2}_m(\mathbb{R}^d)$. Moreover, there exists a constant $C > 0$ independent of τ such that

$$\|\mathrm{Op}_\tau(a)f\|_{\mathcal{M}^{r_1,r_2}_m} \le C\alpha_{(r_1,r_2)}(\tau)\|a\|_{W(\mathcal{F}L^p_{v_J},L^q)}\|f\|_{\mathcal{M}^{r_1,r_2}_m}, \quad \tau \in (0,1). \tag{4.85}$$

Proof. The key tool is the complex interpolation between Wiener amalgam and modulation spaces. We regard Op_τ as the bilinear map $(a,f) \mapsto \mathrm{Op}_\tau(a)f$. Propositions 4.5.12 and 4.5.13 give the continuity of the τ-pseudodifferential operator Op_τ on the following function spaces:

$$W(\mathcal{F}L^\infty_{v_J}, L^1)(\mathbb{R}^{2d}) \times \mathcal{M}^{p_1,p_2}_m(\mathbb{R}^d) \to \mathcal{M}^{p_1,p_2}_m(\mathbb{R}^d),$$
$$W(\mathcal{F}L^2_{v_J}, L^2)(\mathbb{R}^{2d}) \times M^2_m(\mathbb{R}^d) \to M^2_m(\mathbb{R}^d),$$

for $1 \le p_1, p_2 \le \infty$. Using the complex interpolation between Wiener amalgam and modulation spaces, for $\theta \in [0,1]$, we have

$$[W(\mathcal{F}L^\infty_{v_J}, L^1), W(\mathcal{F}L^2_{v_J}, L^2)]_\theta = W(\mathcal{F}L^p_{v_J}, L^{p'}),$$

with $2 \le p \le \infty$, and $[\mathcal{M}^{p_1,p_2}_m, M^2_m]_\theta = \mathcal{M}^{r_1,r_2}_m$, with

$$\frac{1}{r_1} = \frac{1-\theta}{p_1} + \frac{\theta}{2} = \frac{1-\theta}{p_1} + \frac{1}{p} \tag{4.86}$$

and

$$\frac{1}{r_2} = \frac{1-\theta}{p_2} + \frac{\theta}{2} = \frac{1-\theta}{p_2} + \frac{1}{p}, \tag{4.87}$$

so that $r_1, r_2 \le p$. Similarly, we obtain $r_1', r_2' \le p$, and thus the relation (4.84). Due to inclusion relations for Wiener amalgam spaces, we relax the assumptions on symbols, so that the symbol a may belong to $W(\mathcal{F}L^p_{v_J}, L^q)(\mathbb{R}^{2d})$, with $q \le p'$, which gives (4.83). Finally, the norm is provided by

$$\begin{aligned}
\|\mathrm{Op}_\tau\|_{B(W(\mathcal{F}L^p_{v_J},L^q)\times\mathcal{M}^{r_1,r_2}_m,\mathcal{M}^{r_1,r_2}_m)} &\le \|\mathrm{Op}_\tau\|^{1-\theta}_{B(W(\mathcal{F}L^\infty_{v_J},L^1)\times\mathcal{M}^{p_1,p_2}_m,\mathcal{M}^{p_1,p_2}_m)} \\
&\quad \times \|\mathrm{Op}_\tau\|^{\theta}_{B(W(\mathcal{F}L^2_{v_J},L^2)\times M^2_m, M^2_m)} \\
&\le C \frac{1}{\tau^{d(1-\theta)(1-\frac{1}{p_1}+\frac{1}{p_2})}(1-\tau)^{d(1-\theta)(1+\frac{1}{p_1}-\frac{1}{p_2})}} \\
&\le C \frac{1}{\tau^{d(1-\frac{1}{p_1}+\frac{1}{p_2})}(1-\tau)^{d(1+\frac{1}{p_1}-\frac{1}{p_2})}},
\end{aligned}$$

since $1-\theta \le 1$. This concludes the proof. □

We finally consider the end-points $\tau = 0$ and $\tau = 1$, for which the boundedness results stated above do not hold, in general. We remark that the modulation space $M^2(\mathbb{R}^d)$ is simply the Lebesgue space $L^2(\mathbb{R}^d)$. Hence Proposition 4.5.13 matches with what we observed after Proposition 4.2.4, namely $\mathrm{Op}_w(a)$ is not L^2-bounded in general for $a \in W(\mathcal{F}L^2, L^q)$ if $q > 2$, and L^2-continuity fails also if $a \in W(\mathcal{F}L^2, L^2) = L^2$ for $\mathrm{Op}_1(a)$ and $\mathrm{Op}_0(a)$. Concerning Proposition 4.5.12, the following example generalizes a 1-dimensional example exhibited by Boulkhemair in [34].

Proposition 4.5.15. *There exists a symbol $a \in W(\mathcal{F}L^\infty, L^1)(\mathbb{R}^{2d})$ such that the corresponding Kohn–Nirenberg operator $a(x,D)$ and $\mathrm{Op}_1(a)$ (so-called anti-Kohn–Nirenberg operator) are not bounded on $L^2(\mathbb{R}^d)$.*

Proof. Consider the symbol function

$$a(x_1,\dots,x_d,\xi_1,\dots,\xi_d) = x_1^{-1/2}\cdots x_d^{-1/2}\chi_{(0,1]}(x_1)\cdots\chi_{(0,1]}(x_d)e^{-\pi\xi^2}, \tag{4.88}$$

with $\xi^2 = \xi_1^2 + \cdots + \xi_d^2$. An easy computation shows that

$$a \in L^1(\mathbb{R}^{2d}) = W(L^1, L^1)(\mathbb{R}^{2d}) \subset W(\mathcal{F}L^\infty, L^1)(\mathbb{R}^{2d}).$$

Let us show that the Kohn–Niremberg operator $a(x,D)$ is unbounded on $L^2(\mathbb{R}^d)$. Consider the Gaussian function $\varphi(t) = e^{\pi t^2} \in L^2(\mathbb{R}^d)$, then $a(x,D)f \notin L^2(\mathbb{R}^d)$. Indeed, by a tensor product argument, we reduce to compute the following one-dimensional integral:

$$\int_{\mathbb{R}} e^{2\pi i x\cdot\xi}x^{-1/2}\chi_{(0,1]}(x)e^{-\pi\xi^2}e^{-\pi\xi^2}\,d\xi = \frac{1}{\sqrt{2}}x^{-1/2}\chi_{(0,1]}(x)e^{-\pi\frac{x^2}{2}},$$

whose result is a function that does not belong to $L^2(\mathbb{R})$.

To prove that the anti-Kohn–Nirenberg operator $\mathrm{Op}_1(a)$, where a is defined in (4.88), is unbounded on $L^2(\mathbb{R}^d)$, it is sufficient to observe that its adjoint operator is the Kohn–Niremberg one: $(\mathrm{Op}_1(a))^* = a(x,D)$, as detailed below:

$$\langle \mathrm{Op}_1(a)f, g\rangle = \langle a, \mathcal{R}^*(g,f)\rangle = \langle a, \overline{\mathcal{R}(f,g)}\rangle = \langle \mathcal{R}(f,g), \overline{a}\rangle = \langle f, \mathrm{Op}_0(a)g\rangle.$$

This proves our claim. □

4.6 Born–Jordan quantization and related pseudodifferential calculus

Roughly speaking, quantization is the process of associating to a function or distribution defined on phase space an operator. Historically, this notion appears explicitly for the first time in Born and Jordan's foundational paper [33] where they set out to give a firm mathematical basis to Heisenberg's matrix mechanics. Born and Jordan's quantization scheme was strictly limited to polynomials in the variables x and p; it was soon superseded by another rule due to Weyl, and whose extension is nowadays the preferred quantization in physics. However, it turns out that there is a recent regain in interest in an extension of Born and Jordan's initial rule, both in quantum physics and time–frequency analysis. In fact, on the one hand, it is the correct rule if one wants matrix and wave mechanics to be equivalent quantum theories (see the discussion in [106]). On the other hand, as a time–frequency representation the Born–Jordan distribution has been proved to be surprisingly successful, because it allows damping very well the unwanted “ghost frequencies”, as shown in [24], see also [59].

The difference between Born–Jordan and Weyl quantization is most easily apprehended on the level of monomial quantization: in dimension $d = 1$, for any integers $r, s \geq 0$, we have

$$\mathrm{Op}_{\mathrm{W}}(x^r\xi^s) = \frac{1}{2^s}\sum_{\ell=0}^{s}\binom{s}{\ell}\widehat{\xi}^{s-\ell}\widehat{x}^r\widehat{\xi}^{\ell} = \frac{1}{2^r}\sum_{\ell=0}^{r}\binom{r}{\ell}\widehat{x}^{\ell}\widehat{\xi}^s\widehat{x}^{r-\ell} \tag{4.89}$$

(see [224]) and

$$\mathrm{Op}_{BJ}(x^r\xi^s) = \frac{1}{s+1}\sum_{\ell=0}^{s}\widehat{\xi}^{s-\ell}\widehat{x}^r\widehat{\xi}^{\ell} = \frac{1}{r+1}\sum_{\ell=0}^{r}\widehat{x}^{\ell}\widehat{\xi}^s\widehat{x}^{r-\ell} \tag{4.90}$$

(see [33]). As usual, here $\widehat{\xi} = -i\hbar\partial/\partial x$ and $\widehat{x}$ is the multiplication operator by x. More generally, in this section we shall use the notation $\widehat{x}_j$ for the operator of multiplication by x_j and $\widehat{\xi}_j = -i\hbar\partial/\partial x_j$. These operators satisfy Born’s canonical commutation relations $[\widehat{x}_j, \widehat{\xi}_j] = i\hbar$, where $\hbar$ denotes the Planck constant.

Here we shall treat the general case of quantization depending on $\hbar$. Notice that the particular choice $\hbar = 1/(2\pi)$ gives the quantization and related pseudodifferential calculus in the framework of time–frequency analysis, recapturing the definition of pseudodifferential operators considered before.

The Born–Jordan scheme appears as being an equally-weighted quantization, as opposed to the Weyl scheme: $\mathrm{Op}_{BJ}(x^r\xi^s)$ is the average of all possible permutations of the product $\widehat{x}^r\widehat{\xi}^s$.

One can extend the Weyl and Born–Jordan quantizations to arbitrary symbols $a \in \mathcal{S}'(\mathbb{R}^{2d})$ by defining the operators $\widehat{A}_{\mathrm{W}} = \mathrm{Op}_{\mathrm{W}}(a)$ and $\widehat{A}_{\mathrm{BJ}} = \mathrm{Op}_{BJ}(a)\colon \mathcal{S}(\mathbb{R}^d) \to \mathcal{S}'(\mathbb{R}^d)$ as

$$\widehat{A}_{\mathrm{W}}\psi = \left(\frac{1}{2\pi\hbar}\right)^d \int_{\mathbb{R}^{2d}} a_\sigma(z)\widehat{T}(z)\psi\,dz, \tag{4.91}$$

$$\widehat{A}_{\mathrm{BJ}}\psi = \left(\frac{1}{2\pi\hbar}\right)^d \int_{\mathbb{R}^{2d}} a_\sigma(z)\Theta(z)\widehat{T}(z)\psi\,dz$$

where $\psi \in \mathcal{S}(\mathbb{R}^d)$ and the integrals are to be understood in the distributional sense; here $\widehat{T}(z_0) = e^{-i(x_0\cdot\widehat{\xi}-\xi_0\cdot\widehat{x})/\hbar}$, $z_0 = (x_0, \xi_0)$, is the Heisenberg operator, $a_\sigma(z)$ is the symplectic Fourier transform of a,[2] and Θ is Cohen [52] kernel function, defined by

$$\Theta(z) = \frac{\sin(x\cdot\xi/2\hbar)}{x\cdot\xi/2\hbar}, \quad z = (x, \xi) \neq 0,$$

2 The $\hbar$ symplectic Fourier transform of a function $a(z)$ in phase space $\mathbb{R}^{2d}$ is

$$\mathcal{F}_\sigma a(z) = a_\sigma(z) = \left(\frac{1}{2\pi\hbar}\right)^n \int_{\mathbb{R}^{2d}} e^{-\frac{i}{\hbar}\sigma(z,z')}a(z')\,dz'.$$

$\Theta(z) = 1$ for $z = 0$, see [24, 108] and the references therein. The presence of the function Θ produces a smoothing effect (in comparison with the Weyl quantization) which is responsible of the superiority of the Born–Jordan quantization in several respects. However, although this effect is numerically evident, it has remained a challenging open problem to quantify it analytically till the recent works [63, 59], which provide a mathematical explanation of the smoothing effects.

Now, it readily follows from the Schwartz kernel theorem that for every linear continuous operator $\widehat{A} : \mathcal{S}(\mathbb{R}^d) \longrightarrow \mathcal{S}'(\mathbb{R}^d)$ there exists a unique $b \in \mathcal{S}'(\mathbb{R}^{2d})$ such that $\widehat{A} = \mathrm{Op}_w(b)$. In other terms, dequantization can always be performed, and in a unique way. Instead, the situation is more complicated for Born–Jordan operators. In fact, to prove that there exists $a \in \mathcal{S}'(\mathbb{R}^{2d})$ such that $\widehat{A} = \mathrm{Op}_{BJ}(a)$, one has to solve a division problem, namely to find a distribution a such that its symplectic Fourier transform a_σ satisfies

$$b_\sigma = \Theta a_\sigma.$$

The existence of such a symbol a is far from being obvious because of the zeroes of Θ. Moreover, it turns out, as we shall see, that the solution is not even unique. In the following we shall investigate these issues, contained in [60].

The problem of the division of temperate distributions by smooth functions is in general a very subtle one, even in the presence of simple zeros. The basic idea here is, of course, that the space $\mathcal{S}'(\mathbb{R}^d)$ contains (generalized) functions rough enough to absorb the singularities and the loss of decay arising in the division by Θ, and we have, in fact, the following result:

Theorem 4.6.1. *For every $b \in \mathcal{S}'(\mathbb{R}^{2d})$, there exists a symbol $a \in \mathcal{S}'(\mathbb{R}^{2d})$ such that $\mathrm{Op}_{BJ}(a) = \mathrm{Op}_w(b)$.*

Hence, every linear continuous operator $\widehat{A} : \mathcal{S}(\mathbb{R}^d) \to \mathcal{S}'(\mathbb{R}^d)$ can be written in Born–Jordan form, i. e., there exists a symbol $a \in \mathcal{S}'(\mathbb{R}^{2d})$ such that $\widehat{A} = \mathrm{Op}_{BJ}(a)$.

We provide two proofs of this result. One is completely elementary and constructive. The other is shorter and based on the machinery of a priori estimates developed in [190] to prove that the division by a (nonidentically zero) polynomial is always possible in $\mathcal{S}'(\mathbb{R}^d)$. Actually, the reader familiar with [190] will notice that the tools used there are excessively sophisticated for our purposes: after all the function $\Theta(z)$ is the composition of the harmless "sinc" function with the monomial $x\xi$, and, in fact, we will show that a suitable change of variables will reduce matters to the problem of division by the "sinc" function. One can also rephrase this result as follows.

The map $\mathcal{S}'(\mathbb{R}^{2d}) \to \mathcal{S}'(\mathbb{R}^{2d})$

$$a \mapsto a * \Theta_\sigma, \tag{4.92}$$

which gives the Weyl symbol of an operator with Born–Jordan symbol a, is surjective.

It is important to observe that the above representation of the operator $\widehat{A}$ is never unique: if $\widehat{A} = \mathrm{Op}_{BJ}(a)$ and $z_0 \in \mathbb{R}^{2d}$ verifies $\Theta(z_0) = 0$ then the symbol $a(z) + e^{\frac{i}{\hbar}\sigma(z_0,z)}$ gives rise to the *same* operator $\widehat{A}$; see Example 4.6.6 below.

As one may suspect, imaginary-exponential symbols play an important role in the discussion. In fact, the function $e^{\frac{i}{\hbar}\sigma(z_0,z)}$ turns out to be the Weyl symbol of the operator $\widehat{T}(z_0)$ and, in general, any operator can be regarded as a superposition of $\widehat{T}(z)$'s; cf. (4.91). This suggests the study of the map (4.92) in spaces of smooth temperate functions which extend to entire functions in $\mathbb{C}^{2n}$, with a growth at most exponential in the imaginary directions. To be precise, for $r \geq 0$, let $\mathcal{A}_r$ be the space of smooth functions a in $\mathbb{R}^{2n}$ that extend to entire functions $a(\zeta)$ in $\mathbb{C}^{2n}$ and satisfying the estimate

$$|a(\zeta)| \leq C(1 + |\zeta|)^N \exp\left(\frac{r}{\hbar}|\operatorname{Im}\zeta|\right), \quad \zeta \in \mathbb{C}^{2n},$$

for some $C, N > 0$. For example, the symbol $e^{\frac{i}{\hbar}\sigma(z_0,z)}$ belongs to $\mathcal{A}_r$ with $r = |z_0|$, whereas $\mathcal{A}_0$ is nothing but the space of polynomials in phase space. Then we have the following result (Proposition 4.6.9 and Theorem 4.6.10).

The map $\mathcal{A}_r \to \mathcal{A}_r$ in (4.92) is surjective for every $r \geq 0$. It is also one-to-one (and therefore a bijection) if and only if $0 \leq r < \sqrt{4\pi\hbar}$.

We conjecture that the above threshold $r = \sqrt{4\pi\hbar}$ should have an interesting physical interpretation in terms of the Heisenberg uncertainty principle/symplectic capacity.

In what follows we present the tools we need to prove Theorem 4.6.1.

4.6.1 Division of distributions

We recall the Paley–Wiener–Schwartz theorem (see, e. g., [194, Theorem 7.3.1]).

Theorem 4.6.2. *For $r \geq 0$, let B_r be the closed ball $|x| \leq r$ in $\mathbb{R}^d$. If u is a distribution with compact support in B_r then its (symplectic) Fourier transform $\mathcal{F}_\sigma u$ extends to an entire analytic function in $\mathbb{C}^d$ and satisfies*

$$|\mathcal{F}_\sigma u(\zeta)| \leq C(1 + |\zeta|)^N \exp\left(\frac{r}{\hbar}|\operatorname{Im}\zeta|\right)$$

for some $C, N > 0$.

Conversely, every entire analytic function satisfying an estimate of this type is the (symplectic) Fourier transform of a distribution supported in B_r.

We begin with a technical result, inspired by [261, Theorem VII, page 123], which will be used in the sequel. The Schwartz seminorms $\|\psi\|_N$, for $\psi \in \mathcal{S}(\mathbb{R}^d)$, we are going to use here are

$$\|\psi\|_N := \sup_{|\alpha|+|\beta|\leq N} \sup_{x\in\mathbb{R}^d} |x^\alpha \partial_x^\beta \psi(x)| < \infty, \tag{4.93}$$

for every $N \geq 0$. Such seminorms on $\mathcal{S}(\mathbb{R}^d)$ are equivalent to those recalled in (1.57) (Chapter 1).

Proposition 4.6.3. *Let $v \in \mathcal{S}'(\mathbb{R}^d)$ and $\chi \in \mathcal{D}(\mathbb{R})$. For every $t \in \mathbb{R}$, there exists a distribution $u_t \in \mathcal{S}'(\mathbb{R}^d)$ satisfying*

$$(x_d - t)u_t = \chi(x_d - t)v. \tag{4.94}$$

Moreover, u_t can be chosen so that

$$|u_t(\varphi)| \leq C\|\varphi\|_N, \quad \forall \varphi \in \mathcal{S}(\mathbb{R}^d) \tag{4.95}$$

for some constants $C, N > 0$ independent of t.

Proof. We define u_t as follows. Write $x = (x', x_d)$, and let $\varphi \in \mathcal{S}(\mathbb{R}^d)$. We have

$$\varphi(x) = \varphi(x', t) + (x_d - t)\tilde{\varphi}_t(x)$$

with

$$\tilde{\varphi}_t(x) = \int_0^1 \partial_{x_d}\varphi(x', t + \tau(x_d - t))\, d\tau.$$

Observe that $\chi(x_d - t)\tilde{\varphi}_t \in \mathcal{S}(\mathbb{R}^d)$ (whereas $\tilde{\varphi}_t$ is not Schwartz in the x_d direction, in general) and define

$$u_t(\varphi) = v(\chi(x_d - t)\tilde{\varphi}_t).$$

It is easy to see that u_t is, in fact, a temperate distribution: since $v \in \mathcal{S}'(\mathbb{R}^d)$, we have

$$|u_t(\varphi)| \leq C\|\chi(x_d - t)\tilde{\varphi}_t\|_N$$

for some $C, N > 0$ independent of t. On the other hand, on the support of $\chi(x_d - t)$, we have $|x_d - t| \leq C_1$ and

$$1 + |x_d| \leq C_2(1 + |t|) \leq C_3(1 + |t + \tau(x_d - t)|),$$

for every $\tau \in [0, 1]$, so that

$$\begin{aligned} |x_d^{\alpha_d}\partial_x^\beta \tilde{\varphi}_t(x)| &\leq \int_0^1 |x_d^{\alpha_d}||\partial_x^\beta[\partial_{x_d}\varphi(x', t + \tau(x_d - t))]|\, d\tau \\ &\leq C' \int_0^1 (1 + |t + \tau(x_d - t)|)^{\alpha_d} |\partial_x^\beta[\partial_{x_d}\varphi(x', t + \tau(x_d - t))]|\, d\tau, \end{aligned}$$

which gives

$$\|\chi(x_d - t)\tilde{\varphi}_t\|_N \leq C'' \|\varphi\|_{N+1}$$

with constants C'', N independent of t. This gives (4.95).

Formula (4.94) is easily verified: for $\varphi \in \mathcal{S}(\mathbb{R}^d)$,

$$(x_d - t)u_t(\varphi) = u_t((x_d - t)\varphi) = v(\chi(x_d - t)\varphi) = \chi(x_d - t)v(\varphi). \qquad \square$$

We emphasize that the point in the above result is the control of the constants C and N with respect to t; in fact, the existence of a temperate distribution solution u_t for every fixed t already follows from a variant of the arguments in [261, page 127].

We also need the following division result *with control of the support*. This result could probably be proved by extracting and combining several arguments disseminated in [219], but we prefer to provide a self-contained and more accessible proof. As above we split the variable x in $\mathbb{R}^d$ into $x = (x', x_d)$.

Proposition 4.6.4. *Let $B' = \{x' \in \mathbb{R}^{d-1} : |x'| < 1\}$, $B = B' \times \mathbb{R} \subset \mathbb{R}^d$ and $f : B' \to \mathbb{R}$ be a smooth function. Let $K := \{x = (x', x_d) \in B : \ x_d \geq f(x')\}$. Suppose that*

$$K_0 := \{x' \in B' : f(x') = 0\} = \{x' \in B' : x_1 = \cdots = x_k = 0\}$$

for some $1 \leq k \leq d-1$, and

$$|f(x')| \geq C_0 \operatorname{dist}(x', K_0)^N \quad x' \in B \tag{4.96}$$

for some $C_0, N > 0$.

Then for every $v \in \mathcal{E}'(B)$ with supp $v \subset K$ *the equation*

$$x_d u = v \tag{4.97}$$

admits a solution $u \in \mathcal{E}'(B)$ with supp $u \subset K$.

The condition (4.96) is known as Lojasiewicz' inequality and is automatically satisfied if f is real-analytic ([219, Theorem 17]).

Proof. As a preliminary remark, it is clear that we can limit ourselves to constructing a solution $u \in \mathcal{D}'(B)$ with supp $u \subset K$, because one can then multiple u by a cut-off function, equal to 1 in a neighborhood of the support of v, and get another solution in $\mathcal{E}'(B)$, still satisfying supp $u \subset K$.

Now, it is easy to see that a solution of (4.97) is given by the distribution

$$\mathcal{D}(B) \ni \varphi \mapsto v(\tilde{\varphi})$$

where

$$\tilde{\varphi}(x) = \int_0^1 \partial_{x_d}\varphi(x', \tau x_d)\, d\tau. \tag{4.98}$$

More generally, any distribution of the form

$$u(\varphi) = w \otimes \delta + v(\tilde{\varphi}), \tag{4.99}$$

where w is an arbitrary distribution in B' and $\delta = \delta(x_d)$, is solution of (4.97). Hence we have reduced the situation to prove that w can be chosen so that $\operatorname{supp} u \subset K$, i. e., $u(\varphi) = 0$ for every φ supported in the *open* set $\Omega := \{x \in B : x_d < f(x)\}$. This condition, as we will see, forces the values of w on the test functions $\varphi_1 \in \mathcal{D}(\Omega)$.

We construct w as follows. Fix once for all a function $\varphi_2(x_d)$ in $\mathcal{D}(\mathbb{R})$, with $\varphi_2(0) = 1$, supported in the interval $[-1, 1]$. Let $\Omega' = \Omega \cap \{x_d = 0\}$. Let $\varphi_1(x')$ be any function in $\mathcal{D}(\Omega')$ and $\epsilon > 0$ such that

$$\operatorname{dist}(\operatorname{supp} \varphi_1, K_0) > \epsilon.$$

Then we define

$$w(\varphi_1) := -v(\tilde{\varphi})$$

where

$$\varphi(x) = \varphi_1(x')\varphi_2(x_d/(C_0\epsilon^N)), \tag{4.100}$$

with the constants C_0, N appearing in (4.96), and $\tilde{\varphi}$ is constructed from φ as in (4.98):

$$\begin{aligned} \tilde{\varphi}(x) &= \frac{\phi_1(x')}{C_0\epsilon^N} \int_0^1 \varphi_2'(\tau x_d/(C_0\epsilon^N))\, d\tau \\ &= \frac{\phi_1(x')}{x_d} \int_0^{x_d/(C_0\epsilon^N)} \varphi_2'(\tau)\, d\tau; \end{aligned} \tag{4.101}$$

see Figure 4.2.

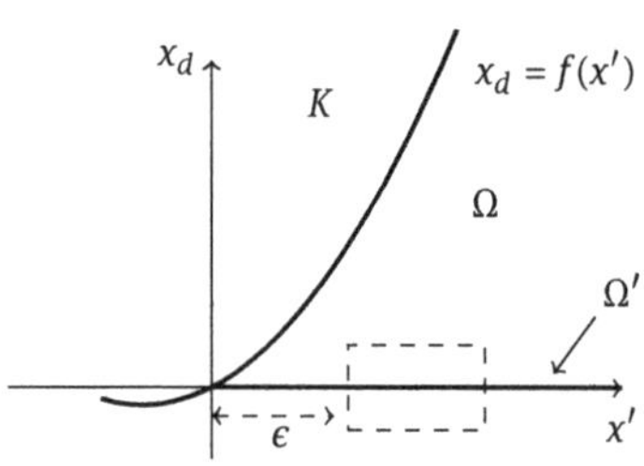

Figure 4.2: The box contains the support of φ in (4.100).

Observe that the function φ in (4.100) is supported in Ω by (4.96), but this is not the case for $\tilde{\varphi}$. We now prove the following facts.

(i) *w is well defined.* Let us verify that the definition of $w(\varphi_1)$ does not depend on the choice of ϵ. Let $\operatorname{dist}(\operatorname{supp}\varphi_1, K_0) > \epsilon > \epsilon' > 0$; then the difference function

$$\varphi(x) = \phi_1(x')[\varphi_2(x_d/(C_0\epsilon^N)) - \varphi_2(x_d/(C_0\epsilon'^N))]$$

is obviously still supported in Ω but, in addition, it vanishes at $x_d = 0$, so the corresponding function $\tilde{\varphi}$ in (4.98) has compact support contained in Ω. Since v is supported in K, we have $v(\tilde{\varphi}) = 0$.

(ii) $w \in \mathcal{D}'(\Omega')$. This is easy to verify and is also a consequence of the next point.

(iii) *w extends to a distribution in* $\mathcal{D}'(B')$. It is sufficient to prove an estimate of the type

$$|w(\varphi_1)| \le C \sup_{|\alpha|\le M} \sup_{x'\in B'} |\partial_{x'}^\alpha \varphi_1(x')|$$

for some constants $C, M > 0$ and every $\varphi_1 \in \mathcal{D}(\Omega')$. In fact, by the Hahn–Banach theorem, one can then extend the linear functional $w : \mathcal{D}(\Omega') \to \mathbb{C}$, which is continuous when $\mathcal{D}(\Omega')$ is endowed with the norm in the above right-hand side, to a functional on $\mathcal{D}(B')$, continuous with respect to the same norm and therefore a fortiori for the usual topology of this space.

Now, by the definition of w and since v has compact support in K, we have (cf. [194, Theorem 2.3.10])

$$|w(\varphi_1)| = |v(\tilde{\varphi})| \le C \sup_{|\beta|\le M} \sup_{\substack{x\in K\\ x_d\le C_1}} |\partial_x^\beta \tilde{\varphi}(x)| \tag{4.102}$$

for some $C, C_1, M > 0$. To estimate the last term, we observe that in the expression for $\tilde{\varphi}$ in (4.101) the integral is, in fact, constant if $x_d \ge C_0\epsilon^N$. In particular, this happens if $x \in K$ and $x' \in \operatorname{supp}\varphi_1$, because (4.96) implies that for such x it turns out

$$x_d \ge f(x) \ge C_0 \operatorname{dist}(x', K_0)^N \ge C_0\epsilon^N.$$

Hence for $x \in K$, $x_d \le C_1$, we have

$$\begin{aligned}\sup_{|\beta|\le M} |\partial_x^\beta \tilde{\varphi}(x)| &\le C \sup_{|\alpha|\le M} |\partial_{x'}^\alpha \varphi_1(x')| \cdot x_d^{-M}\\ &\le C \sup_{|\alpha|\le M} |\partial_{x'}^\alpha \varphi_1(x')| \cdot C_0^{-M} \operatorname{dist}(x', K_0)^{-NM}.\end{aligned}$$

On the other hand, we have

$$\operatorname{dist}(x', K_0) = (|x_1|^2 + \cdots + |x_k|^2)^{1/2},$$

so that a Taylor expansion (with remainder of order NM) of $\partial_{x'}^\alpha \varphi_1$ with respect to $x_1, \ldots, x_k$ (taking into account that φ_1 vanishes in a neighborhood of K_0) gives

$$\sup_{|\beta|\le M} \sup_{\substack{x\in K\\ x_d\le C_1}} |\partial_x^\beta \tilde{\varphi}(x)| \le C' \sup_{|\alpha|\le M+NM} \sup_{x'\in B'} |\partial_{x'}^\alpha \varphi_1(x')|.$$

This, together with (4.102), gives the desired conclusion.

(iv) *With the choice of w in* (4.99), *we have* $\operatorname{supp} u \subset K$. Let $\varphi \in \mathcal{D}(\Omega)$, and write

$$\varphi(x) = [\varphi(x) - \varphi(x',0)\varphi_2(x_d/(C_0\epsilon^N))] + \varphi(x',0)\varphi_2(x_d/(C_0\epsilon^N)),$$

where we choose $\epsilon < \operatorname{dist}(\operatorname{supp}\varphi(\cdot,0), K_0)$.

Now the distribution u vanishes when applied to the second term of this sum just by the definition of w (with $\varphi(x',0)$ playing the role of $\varphi_1(x')$). On the other hand, the function $\varphi(x) - \varphi(x',0)\varphi_2(x_d/(C_0\epsilon^N))$ is supported in Ω *and vanishes at* $x_d = 0$, so that one sees from the definition of u in (4.99) that its pairing with u is 0. □

4.6.2 Changes of coordinates for temperate distributions

In the sequel we will perform changes of coordinates which preserve Schwartz functions and temperate distributions in the following sense.

Let ϕ be a smooth diffeomorphism of the semispace $\{x_1 > 0\} \subset \mathbb{R}^d$ into itself. Suppose that ϕ is positively homogeneous for some *positive* order, say, $r > 0$, i. e., $\phi(\lambda x) = \lambda^r \phi(x)$ for every $x \in \mathbb{R}^d$ with $x_1 > 0$, $\lambda > 0$. It follows that the image of every truncated cone

$$U = \{x \in \mathbb{R}^d : x_1 \ge \epsilon|x|,\ |x| \ge \epsilon\},$$

with $0 < \epsilon \le 1$ (see Figure 4.3 below) is contained in another truncated cone of the same type. In fact, if $y = \phi(x)$, then for some $\epsilon' > 0$ we have $|y| \ge \epsilon'$ when $x \in U$ and $|x| = \epsilon$ by compactness, and therefore for every $x \in U$ by homogeneity. The same argument implies that $y_1/|y| \ge \epsilon' > 0$ for $x \in U$.

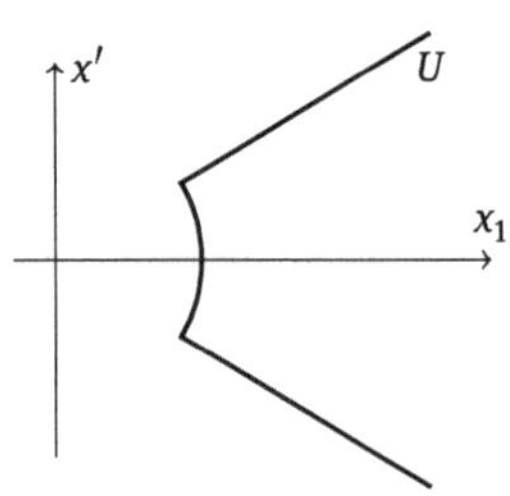

Figure 4.3: A truncated cone in the semispace $x_1 > 0$.

Moreover, the same applies to the inverse function $\phi^{-1}(x)$, which will be homogeneous of degree $1/r > 0$.

Now, let $\mathcal{S}(\mathbb{R}^d)_{\text{cone}}$, $\mathcal{S}'(\mathbb{R}^d)_{\text{cone}}$ be the spaces of Schwartz functions and temperate distributions in $\mathbb{R}^d$, respectively, with support contained in *some* truncated cone, as above. Then ϕ induces bijections

$$\phi_* : \mathcal{S}(\mathbb{R}^d)_{\text{cone}} \to \mathcal{S}(\mathbb{R}^d)_{\text{cone}}, \quad \phi_* : \mathcal{S}'(\mathbb{R}^d)_{\text{cone}} \to \mathcal{S}'(\mathbb{R}^d)_{\text{cone}}$$

defined as follows.

If $\psi \in \mathcal{S}(\mathbb{R}^d)_{\text{cone}}$ we define $\phi_*\psi(x) = \psi(\phi^{-1}(x))$ for $x_1 > 0$ and $= 0$ otherwise, and it is easy to see that $\phi_*\psi \in \mathcal{S}(\mathbb{R}^d)_{\text{cone}}$ using the homogeneity of ϕ^{-1} and the support condition on ψ.

If $u \in \mathcal{S}'(\mathbb{R}^d)_{\text{cone}}$ we define the distribution $\phi_* u$ by

$$\phi_* u(\varphi) = u(\chi \cdot \varphi \circ \phi |\det \phi'|),$$

for every $\varphi \in \mathcal{S}(\mathbb{R}^d)$, where $\chi : \mathbb{R}^d \to [0,1]$ is a smooth function, positively homogeneous of degree 0 for large $|x|$, $\chi(x) = 1$ on a truncated cone slightly larger than one containing the support of u and χ is supported in a truncated cone. It is easy to see that $\chi(x)\varphi(\phi(x))|\det \phi'(x)|$ is a function in $\mathcal{S}(\mathbb{R}^d)_{\text{cone}}$ because on every truncated cone the Jacobian determinant $|\det \phi'|$ is smooth and has an at most polynomial growth, together with its derivatives.

The maps $\phi_* : \mathcal{S}(\mathbb{R}^d)_{\text{cone}} \to \mathcal{S}(\mathbb{R}^d)_{\text{cone}}$ and $\phi_* : \mathcal{S}'(\mathbb{R}^d)_{\text{cone}} \to \mathcal{S}'(\mathbb{R}^d)_{\text{cone}}$ are bijections, the inverses being given by $(\phi^{-1})_*$ on $\mathcal{S}(\mathbb{R}^d)_{\text{cone}}$ and $\mathcal{S}'(\mathbb{R}^d)_{\text{cone}}$.

Observe that, more generally, if ψ is a smooth function with an at most polynomial growth together with its derivatives, we can similarly define $\phi_*\psi(x) = \psi(\phi^{-1}(x))$ for $x_1 > 0$ and have the formula

$$\phi_*(\psi u) = \phi_*\psi\phi_* u \tag{4.103}$$

if $u \in \mathcal{S}'(\mathbb{R}^d)_{\text{cone}}$ (the formula makes sense even if $\phi_*\psi$ is defined only for $x_1 > 0$, because u and therefore $\phi_* u$ is supported in a truncated cone).

4.6.3 Born–Jordan pseudodifferential operators

We recall the Grossmann–Royer operator, defined by, for $z_0 = (x_0, \xi_0)$,

$$\widehat{T}_{\text{GR}}(z_0)\psi(x) = e^{\frac{2i}{\hbar}(x-x_0)\cdot\xi_0}\psi(2x_0 - x).$$

It is a unitary operators on $L^2(\mathbb{R}^d)$ and an involution:

$$\widehat{T}_{\text{GR}}(z_0)\widehat{T}_{\text{GR}}(z_0) = I.$$

The two following important formulas hold:

$$\widehat{T}_{\mathrm{GR}}(z_0) = \widehat{T}(z_0)R^{\vee}\widehat{T}(z_0)^{-1} \tag{4.104}$$

where $\widehat{T}(z_0)$ is the Heisenberg operator defined above, $R^{\vee} = \widehat{T}_{\mathrm{GR}}(0)$ is the reflection operator, $R^{\vee}\psi(x) = \psi(-x)$, and

$$\widehat{T}_{\mathrm{GR}}(z_0)\psi(x) = 2^{-n}\mathcal{F}_{\sigma}[\widehat{T}(\cdot)\psi(x)](-z_0) \tag{4.105}$$

where recall $\mathcal{F}_{\sigma}$ is the symplectic Fourier transform.

Let $a \in \mathcal{S}'(\mathbb{R}^{2d})$. The Weyl operator $\widehat{A}_{\mathrm{W}} = \mathrm{Op}_{w}(a)$ can be defined as the image of the quantization mapping $a \mapsto \widehat{A}_{\mathrm{W}}$, with $\widehat{A}_{\mathrm{W}}$ defined by

$$\widehat{A}_{\mathrm{W}}\psi = \left(\frac{1}{\pi\hbar}\right)^{n} \int_{\mathbb{R}^{2d}} a(z_0)\widehat{T}_{\mathrm{GR}}(z_0)\psi\, dz_0, \tag{4.106}$$

which is equivalent, using (4.105), to

$$\widehat{A}_{\mathrm{W}}\psi = \left(\frac{1}{2\pi\hbar}\right)^{n} \int_{\mathbb{R}^{2d}} a_{\sigma}(z_0)\widehat{T}(z_0)\psi\, dz_0. \tag{4.107}$$

Notice that when, for instance, $a \in \mathcal{S}(\mathbb{R}^{2d})$, formula (4.106) can be rewritten in the familiar form

$$\widehat{A}_{\mathrm{W}}\psi(x) = \mathrm{Op}_{w}(a)\psi(x) = \left(\frac{1}{2\pi\hbar}\right)^{n} \int_{\mathbb{R}^{2d}} e^{\frac{i}{\hbar}(x-y)\cdot\xi} a\left(\frac{1}{2}(x+y), p\right)\psi(y)\, dp\, dy.$$

The Born–Jordan operator $\widehat{A}_{\mathrm{BJ}} = \mathrm{Op}_{BJ}(a)$ is constructed as follows: one first defines the Shubin τ-operator $\widehat{A}_{\tau} = \mathrm{Op}_{\tau}(a)$ by

$$\widehat{A}_{\tau}\psi = \int_{\mathbb{R}^{2d}} a_{\sigma}(z_0)\widehat{T}_{\tau}(z_0)\psi\, dz_0 \tag{4.108}$$

where $\widehat{T}_{\tau}(z)$ is the unitary operator on $L^2(\mathbb{R}^d)$ defined by

$$\widehat{T}_{\tau}(z_0) = e^{\frac{i}{\hbar}(\tau-\frac{1}{2})x_0\cdot\xi_0}\widehat{T}(z_0). \tag{4.109}$$

One thereafter defines

$$\widehat{A}_{\mathrm{BJ}}\psi = \int_0^1 \widehat{A}_{\tau}\psi\, d\tau.$$

Using the obvious formula

$$\Theta(z_0) := \int_0^1 e^{\frac{i}{\hbar}(\tau-\frac{1}{2})x_0\cdot\xi_0}\, d\tau = \begin{cases} \frac{\sin(x_0\cdot\xi_0/2\hbar)}{x_0\cdot\xi_0/2\hbar} & \text{for } x_0\cdot\xi_0 \neq 0, \\ 1 & \text{for } x_0\cdot\xi_0 = 0, \end{cases}$$

one thus has

$$\widehat{A}_{\mathrm{BJ}}\psi = \int a_\sigma(z_0)\widehat{T}_{\mathrm{BJ}}(z_0)\psi\, dz_0$$

where $\widehat{T}_{\mathrm{BJ}}(z_0) = \Theta(z_0)\widehat{T}(z_0)$.

It follows that $\widehat{A}_{\mathrm{BJ}}(a)$ is the Weyl operator with covariant symbol

$$(a_{\mathrm{W}})_\sigma = \Theta a_\sigma. \tag{4.110}$$

The Weyl symbol of $\widehat{A}_{\mathrm{BJ}}(a)$ is thus (taking the symplectic Fourier transform)

$$a_{\mathrm{W}} = \left(\frac{1}{2\pi\hbar}\right)^d a * \Theta_\sigma. \tag{4.111}$$

Conversely, assume that $\widehat{A} = \mathrm{Op}_w(a_{\mathrm{W}})$. Then $\widehat{A} = \mathrm{Op}_{BJ}(a)$, provided that a satisfies $(a_{\mathrm{W}})_\sigma = \Theta a_\sigma$.

4.6.4 Invertibility of the Born–Jordan quantization

In this section we investigate the injectivity and surjectivity of the map $\mathcal{S}'(\mathbb{R}^{2d}) \to \mathcal{S}'(\mathbb{R}^{2d})$ given by

$$a \longmapsto \left(\frac{1}{2\pi\hbar}\right)^d a * \Theta_\sigma, \tag{4.112}$$

namely the map which gives the Weyl symbol of an operator with Born–Jordan symbol a, cf. (4.111).

We have the following result.

Theorem 4.6.5. *The equation*

$$\left(\frac{1}{2\pi\hbar}\right)^d a * \Theta_\sigma = b$$

admits a solution $a \in \mathcal{S}'(\mathbb{R}^{2d})$, for every $b \in \mathcal{S}'(\mathbb{R}^{2d})$.

Proof. Taking the symplectic Fourier transform, we have reduced the situation to proving that the equation

$$\Theta a = b$$

admits at least a solution $a \in \mathcal{S}'(\mathbb{R}^{2d})$, for every $b \in \mathcal{S}'(\mathbb{R}^{2d})$. This is a problem of division of *temperate* distributions. We provide two proofs.

First proof. We localize the problem by considering a finite and smooth partition of unity $\psi_0(z)$, $\psi_j^{\pm}(z)$, $j = 1, \dots, 2d$, in phase space, where ψ_0 has support in a ball $|z| \le r$, with $r < \sqrt{4\pi\hbar}$, and ψ_j^+ for $j = 1, \dots, 2d$, is supported in a truncated cone (cf. Section 4.6.2) of the type

$$U_j^+ = \{z \in \mathbb{R}^{2d} : z_j \ge \epsilon|z|,\ |z| \ge \epsilon\}$$

contained in the semispace $z_j > 0$, and similarly ψ_j^- is supported in a similar truncated cone contained in the semispace $z_j < 0$, with $\psi_j^{\pm}$ homogeneous of degree 0 for large z (it is easy to see that such a partition of unity can be constructed if ϵ is small enough, e. g., if $\epsilon < r/2$ and $\epsilon < 1/(2\sqrt{d})$).

It is clear that if a_0 and $a_j^{\pm}$, $j = 1, \dots, 2d$, solve in $\mathcal{S}'(\mathbb{R}^{2d})$ the equations $\Theta a_0 = \psi_0 b$ and $\Theta a_j^{\pm} = \psi_j^{\pm} b$, then $a := a_0 + \sum_{j=1}^{d}(a_j^+ + a_j^-)$ solves $\Theta a = b$.

The equation $\Theta a_0 = \psi_0 b$ is easily solved as in the proof of Proposition 4.6.9, using the fact that $\psi_0 b$ is supported in a closed ball where $\Theta \ne 0$.

Let us now solve the equation

$$\Theta a = b$$

where $b \in \mathcal{S}'(\mathbb{R}^{2d})$ is supported in a truncated cone in the semispace, say, $z_1 > 0$ in phase space. We look for $a \in \mathcal{S}'(\mathbb{R}^{2d})$ supported in a truncated cone as well. We apply the following algebraic change of variables in phase space:

$$y_1 = z_1^2, \quad y_2 = z_1 z_2, \dots, \quad y_d = z_1 z_d,$$

$$y_{d+1} = z_1 z_{d+2}, \dots, \quad y_{2d-1} = z_1 z_{2d}, \quad y_{2d} = x \cdot \xi = \sum_{j=1}^{d} z_j z_{j+d}$$

where $z = (x, \xi)$.

It is easy to check that the map $z \mapsto y$ is a diffeomorphism of the semispace $z_1 > 0$ into itself ($y_1 > 0$),[3] and moreover it is homogeneous of degree 2. We now apply the remarks in Section 4.6.2 (where the dimension of the space is now $2d$), in particular (4.103), and we have reduced the situation to solve (in the new coordinates) the equation

$$\frac{\sin(y_{2d}/2\hbar)}{y_{2d}/2\hbar} a = b \tag{4.113}$$

3 The inverse change of variables is given by

$$z_1 = \sqrt{y_1},\ z_2 = y_2/\sqrt{y_1}, \dots,\ z_d = y_d/\sqrt{y_1},$$

$$z_{d+1} = \left(y_{2d} - \sum_{j=2}^{d} y_j y_{d+j}/y_1\right)/\sqrt{y_1},\ z_{d+2} = y_{d+1}/\sqrt{y_1}, \dots, z_{2d} = y_{2d-1}/\sqrt{y_1}.$$

where $b \in \mathcal{S}'(\mathbb{R}^{2d})$ is supported in a truncated cone in the semispace $y_1 > 0$, and we look for $a \in \mathcal{S}'(\mathbb{R}^{2d})$ similarly supported in a truncated cone in the same semispace.

We now consider a partition of unity in $\mathbb{R}$ obtained by translation of a fixed function, of the type $\chi(y_{2d} - 2\pi k\hbar)$, $k \in \mathbb{Z}$, where $\chi \in \mathcal{D}(\mathbb{R})$ is a fixed function supported in the interval $[-(3/2)\pi\hbar, (3/2)\pi\hbar]$. Observe that on the support of $\chi(y_{2d} - 2\pi k\hbar)$ the function $\frac{\sin(y_{2d}/2\hbar)}{y_{2d}/2\hbar}$ has only a simple zero at $y_{2d} = 2\pi k\hbar$, for $k \neq 0$, whereas it does not vanish on the support of χ (case $k = 0$).

We now solve, for every $k \in \mathbb{Z}$, the equation

$$\frac{\sin(y_{2d}/2\hbar)}{y_{2d}/2\hbar} a_k = \chi(y_{2d} - 2\pi k\hbar) b. \tag{4.114}$$

We suppose $k \neq 0$, the case $k = 0$ being easier. Since the function

$$\frac{\sin(y_{2d}/2\hbar)}{(y_{2d} - 2k\pi\hbar) y_{2d}/2\hbar}$$

is smooth and does not vanish on the support of $\chi(y_{2d} - 2\pi k\hbar)$, it is sufficient to solve the equation

$$(y_{2d} - 2k\pi\hbar) a_k = \chi(y_{2d} - 2\pi k\hbar) \left[\frac{(y_{2d} - 2k\pi\hbar) y_{2d}/2\hbar}{\sin(y_{2d}/2\hbar)} b \right].$$

Observe that the function $\frac{y_{2d} - 2k\pi\hbar}{\sin(y_{2d}/2\hbar)}$ has derivatives uniformly bounded with respect to k (in fact, we have $\sin(y_{2d}/2\hbar) = \pm \sin((y_{2d} - 2\pi k\hbar)/2\hbar)$, according to the parity of k, and the "sinc" function has bounded derivatives of any order). It follows from this remark and Proposition 4.6.3 that there exists a solution $a_k \in \mathcal{S}'(\mathbb{R}^{2d})$ of the above equation, satisfying the estimate

$$|a_k(\varphi)| \leq C \|\varphi\|_N \quad \forall \varphi \in \mathcal{S}(\mathbb{R}^{2d})$$

for some constants $C, N > 0$ independent of k.

Moreover, we can suppose that all the a_k's are supported in a fixed truncated cone (by multiplying by a cut-off function in phase space with the same property as $\psi_1^+(z)$ above, and $= 1$ on a truncated cone slightly larger than one containing the support of b). We can also multiply a_k by a cut-off function $\tilde{\chi}(y_{2d} - 2\pi k\hbar)$, where $\tilde{\chi} \in \mathcal{D}(\mathbb{R})$, and $\tilde{\chi} = 1$ in a neighborhood of the support of χ, and obtain new solutions

$$\tilde{a}_k := \tilde{\chi}(y_{2d} - 2\pi k\hbar) a_k$$

to (4.114), satisfying

$$|\tilde{a}_k(\varphi)| \leq C \|\tilde{\chi}(y_{2d} - 2\pi k\hbar)\varphi\|_N \tag{4.115}$$

for every $\varphi \in \mathcal{S}(\mathbb{R}^{2d})$; see Figure 4.4 below.

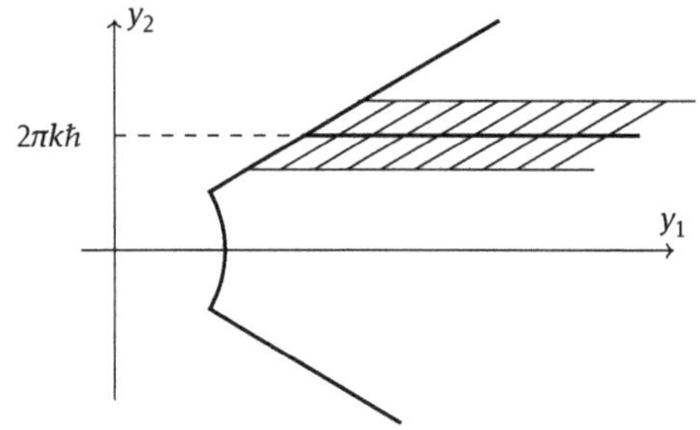

Figure 4.4: The distribution $\tilde{a}_k$ is supported in the strip ($d = 1$ in this figure).

Now we claim that $a := \sum_{k\in\mathbb{Z}} \tilde{a}_k$ solves (4.113). We have only to check that the series converges in $\mathcal{S}'(\mathbb{R}^{2d})$. Let us verify that, given $\varphi \in \mathcal{S}(\mathbb{R}^{2d})$, the series

$$\sum_{n\in\mathbb{Z}} \tilde{a}_k(\varphi)$$

converges absolutely.

On the support of $\tilde{\chi}(y_{2d}-2\pi k\hbar)$, we have $1+|y| \geq 1+|y_{2d}| \geq C(1+|k|)$, and $|y^\alpha \partial_y^\beta \varphi(y)| \leq C'_{N'}(1+|k|)^{-N'}$ *for every* N' and $\alpha, \beta \in \mathbb{N}^{2d}$. Hence by (4.115) we obtain

$$|\tilde{a}_k(\varphi)| \leq C''_{N'}(1+|k|)^{-N'},$$

and it is sufficient to take $N' = 2$ for the above series to converge absolutely.

Second proof. In [190] it was proved that the equation $Pu = v$ is always solvable in $\mathcal{S}'(\mathbb{R}^d)$ if P is a nonidentically zero polynomial. As observed there (page 556), that proof continues to hold if the polynomial P is replaced by a smooth function which satisfies the estimates (4.3) and (4.10) in that paper. Here we are interested in the division by the function $\Theta(z)$, which has only simple zeros, and those estimates read

$$|\Theta(z)| \geq C\,\mathrm{dist}(z, Z)^{\mu'}(1+|z|)^{-\mu''} \quad \forall z \in \mathbb{R}^{2d} \tag{4.116}$$

and

$$|\nabla\Theta(z)| \geq C(1+|z|)^{-\mu''} \quad \forall z \in Z \tag{4.117}$$

for some $C, \mu', \mu'' > 0$, where $Z = \{z \in \mathbb{R}^{2d} : \Theta(z) = 0\}$.

To check that these estimates are satisfied, let $\mathrm{sinc}(t) = \sin t/t$ ($\mathrm{sinc}(0) = 1$), so that $\Theta(z) = \Theta(x,\xi) = \mathrm{sinc}(x\cdot\xi/2\hbar)$, $z = (x,\xi)$. Observe that at points t where $\mathrm{sinc}(t/2\hbar) = 0$ we have

$$\left|\frac{d}{dt}\mathrm{sinc}(t/2\hbar)\right| = 1/|t|$$

so that

$$|\nabla\Theta(z)| = \frac{|(x,\xi)|}{|x\cdot\xi|} \geq \frac{2}{|z|}, \qquad \forall z \in Z,$$

which implies (4.117) with $\mu'' = 1$.

Concerning (4.116), observe first of all that, setting $Z_0 = \{2\pi k\hbar : k \in \mathbb{Z}, k \neq 0\} \subset \mathbb{R}$, we have, for $|t| > \pi\hbar$,

$$|\mathrm{sinc}(t/2\hbar)| = \frac{2\hbar}{|t|}|\sin(t/2\hbar)| \geq \frac{2\hbar}{|t|}\frac{1}{\pi\hbar}\,\mathrm{dist}(t,Z_0) = \frac{2}{\pi|t|}\,\mathrm{dist}(t,Z_0)$$

whereas if $|t| \leq \pi\hbar$,

$$|\mathrm{sinc}(t/2\hbar)| \geq \frac{2}{\pi} \geq \frac{1}{\pi^2\hbar}\,\mathrm{dist}(t,Z_0).$$

In both cases we have

$$|\mathrm{sinc}(t/2\hbar)| \geq C_0(1+|t|)^{-1}\,\mathrm{dist}(t,Z_0) \tag{4.118}$$

for some $C_0 > 0$.

Now, (4.116) is clearly satisfied in a neighborhood of 0, so that it is sufficient to prove it in any truncated cone contained in the semispaces $z_j > 0$ or $z_j < 0, j = 1,\dots,2d$. Consider for example a truncated cone U where $z_1 > 0$. We perform the change of coordinates $y = y(z)$ in this semispace, exactly as in the previous proof, and we observe that by (4.118) we have

$$|\mathrm{sinc}(y_{2d}/2\hbar)| \geq C_0(1+|y_{2d}|)^{-1}|y_{2d}-\overline{y}_{2d}| \geq C_0(1+|y|)^{-1}|y_{2d}-\overline{y}_{2d}|$$

where $\overline{y}_{2d} \in Z_0$ is such that $|y_{2d}-\overline{y}_{2d}| = \mathrm{dist}(y_{2d},Z_0)$. Now, for $z \in U$ we have $0 < \epsilon \leq |y| \leq C|z|^2$ and moreover the inverse map $z = z(y)$ is Lipschitz in any truncated cone $U' \supset z(U)$, because the derivatives $\partial z_j/\partial y_k$ are positively homogeneous of degree $-1/2 < 0$, and therefore bounded in U'. Using these facts and setting $\overline{y} = (y_1,\dots,y_{2d-1},\overline{y}_{2d})$, $\overline{z} = z(\overline{y}) \in Z$, we conclude that for every $z \in U$,

$$\begin{aligned}|\Theta(z)| = |\mathrm{sinc}(y_{2d}/2\hbar)| &\geq C_0(1+|y|)^{-1}|y-\overline{y}|\\ &\geq C(1+|z|)^{-2}|z-\overline{z}|,\\ &\geq C(1+|z|)^{-2}\,\mathrm{dist}(z,Z).\end{aligned}$$

This concludes the proof. □

From the previous theorem, we obtain at once the proof of Theorem 4.6.1 above.

Concerning the injectivity of the map (4.112), we begin with a simple example, which shows that the map (4.112) is not one-to-one, even when restricted to real analytic functions which extend to entire functions in $\mathbb{C}^{2d}$.

Example 4.6.6. Consider the Born–Jordan symbol

$$a(z) = e^{\frac{i}{\hbar}\sigma(z_0,z)} = e^{\frac{i}{\hbar}(x\cdot\xi_0 - x_0\cdot\xi)}, \tag{4.119}$$

where $z_0 = (x_0, \xi_0)$ is any point on the zero set $\Theta(z) = 0$.

The symplectic Fourier transform of a is

$$a_\sigma(z) = (2\pi\hbar)^d \delta(z - z_0)$$

and therefore $\Theta a_\sigma = 0$, because $\Theta(z_0) = 0$. Hence the corresponding Weyl symbol is $a_W = 0$ by (4.110).

Observe that the symbol a in (4.119) extends to an entire function $a(\zeta_1, \zeta_2) = e^{\frac{i}{\hbar}(\xi_0\cdot\zeta_1 - x_0\cdot\zeta_2)}$ in $\mathbb{C}^{2d}$, satisfying the estimate

$$|a(\zeta)| \le \exp\left(\frac{r}{\hbar}|\operatorname{Im}\zeta|\right), \quad \zeta \in \mathbb{C}^{2d},$$

where $r = |(\xi_0, -x_0)| = |z_0|$.

For future reference we observe that the minimum value of r is reached for the points z_0 at which the hypersurface $\Theta(z)$ has minimum distance from 0, which turns out to be $r = \sqrt{4\pi\hbar}$. For example, one can consider $x_0 = \xi_0 = (2\pi\hbar/d)^{1/2}(1, \dots, 1)$, so that $x_0 \cdot \xi_0 = 2\pi\hbar$ and $|x_0|^2 + |\xi_0|^2 = 4\pi\hbar$ (see Figure 4.5).

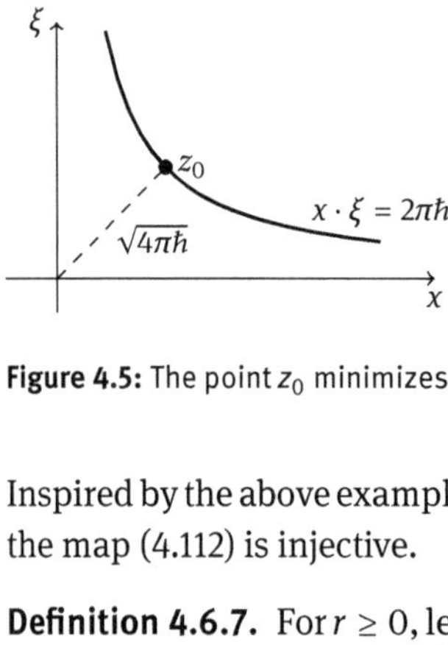

Figure 4.5: The point z_0 minimizes the distance of the zero set $\Theta(z) = 0$ from 0.

Inspired by the above example, we now exhibit a nontrivial class of functions on which the map (4.112) is injective.

Definition 4.6.7. For $r \ge 0$, let $\mathcal{A}_r$ be the space of smooth functions a in $\mathbb{R}^{2d}$ that extend to entire functions $a(\zeta)$ in $\mathbb{C}^{2d}$ and satisfying the estimate

$$|a(\zeta)| \le C(1 + |\zeta|)^N \exp\left(\frac{r}{\hbar}|\operatorname{Im}\zeta|\right), \quad \zeta \in \mathbb{C}^{2d},$$

for some $C, N > 0$.

Equivalently (by the Paley–Wiener–Schwartz theorem), $\mathcal{A}_r$ is the space of temperate distributions in $\mathbb{R}^{2d}$ whose (symplectic) Fourier transform is supported in the closed ball $|z| \le r$.

Remark 4.6.8. Observe that the space $\mathcal{A}_0$ is just the space of polynomials in phase space.

We have the following result.

Proposition 4.6.9. *The map* (4.112) *is a bijection* $\mathcal{A}_r \to \mathcal{A}_r$ *if and only if* $0 \le r < \sqrt{4\pi\hbar}$.

Proof. The "only if" part follows at once from the Remark 4.6.6 because the symbol $a(z)$ in (4.119) belongs to $\mathcal{A}_r$ for $r \ge \sqrt{4\pi\hbar}$ and is mapped to 0.

Consider now the "if" part. Taking the symplectic Fourier transform in (4.112) and by the Paley–Wiener–Schwartz theorem, we have reduced the situation to prove that, when $0 \le r < \sqrt{4\pi\hbar}$, the map

$$a \mapsto \Theta a$$

is a bijection $\mathcal{E}'(B_r) \to \mathcal{E}'(B_r)$, where $\mathcal{E}'(B_r)$ is the space of distributions on $\mathbb{R}^{2d}$ supported in the closed ball B_r given by $|z| \le r$.

Now, it is clear that if $a \in \mathcal{E}'(B_r)$ then $\Theta a \in \mathcal{E}'(B_r)$. On the other hand, since the function $\Theta(z)$ does not vanish for $|z| < \sqrt{4\pi\hbar}$, hence in a neighborhood of B_r (by assumption $r < \sqrt{4\pi\hbar}$), the equation $\Theta a = b$, for every $b \in \mathcal{E}'(B_r)$, has a unique solution $a \in \mathcal{E}'(B_r)$ obtained simply by multiplying by Θ^{-1}: $a = \Theta^{-1} b$. □

We now study the surjectivity of the map (4.112) on the spaces $\mathcal{A}_r$ when $r \ge \sqrt{4\pi\hbar}$.

Theorem 4.6.10. *Given* $r \ge 0$, *for every* $b \in \mathcal{A}_r$ *there exists* $a \in \mathcal{A}_r$ *such that*

$$\left(\frac{1}{2\pi\hbar}\right)^d a * \Theta_\sigma = b.$$

Proof. As above we have to prove that the equation

$$\Theta a = b$$

admits at least a solution $a \in \mathcal{E}'(B_r)$, for every $b \in \mathcal{E}'(B_r)$.

Since all the distributions here are compactly supported, the problem is local and we can solve the equation $\Theta a = b$ in $\mathcal{E}'(U_{z_0})$ for a sufficiently small open neighborhood U_{z_0} of any given point z_0 and conclude with a finite smooth partition of unity.

If $|z_0| > r$ and $U_{z_0} \subset \{|z| > r\}$ one can choose $a = 0$ in U_{z_0}.

When $|z_0| < r$ and $U_{z_0} \subset \{|z| < r\}$, we apply the classical division theorem valid for smooth functions with at most simple zeros [261, page 127]: for every $b \in \mathcal{E}'(U_{z_0})$ there therefore exists a solution $a \in \mathcal{E}'(U_{z_0})$.

Of course, if $\Theta(z_0) \ne 0$ the division is trivial, so that we now suppose that $z_0 = (x_0, \xi_0)$ belongs to both $|z| = r$ and $x \cdot \xi = 2\pi k\hbar$ for some $k \in \mathbb{Z}$, $k \ne 0$. Then necessarily we have $r \ge \sqrt{4\pi|k|\hbar}$, because this is the distance of the hypersurface $x \cdot \xi = 2\pi k\hbar$ from the origin. We therefore distinguish two cases.

First case: $r > \sqrt{4\pi|k|\hbar}$. Then the hypersurfaces $|z| = r$ and $x \cdot \xi = 2\pi k\hbar$ cut transversally at z_0, i. e., their normal vectors are linearly independent and the intersection Σ is therefore a submanifold of codimension 2. In fact, one sees easily that the vector normals to these two hypersurfaces at z_0 are linearly dependent if and only if $\xi_0 = \mathrm{sign}(k)x_0$ and $|x_0|^2 = 2\pi|k|\hbar$. In that case we must have $r = \sqrt{4\pi|k|\hbar}$.

Second case: $r = \sqrt{4\pi|k|\hbar}$. Then the hypersurfaces $|z| = r$ and $x \cdot \xi = 2\pi k\hbar$ touch along the submanifold Σ of codimension n having equations $\xi = \mathrm{sign}(k)x$, $|x|^2 = 2\pi|k|\hbar$.

In both cases by the implicit function theorem we can take analytic coordinates $y = (y', y_{2d})$ near z_0 so that z_0 has coordinates $y = 0$, the hypersurface $x \cdot \xi = 2\pi k\hbar$ is straightened to $y_{2d} = 0$ and moreover the above submanifold Σ has equations $y_1 = y_{2d} = 0$ (in the first case) or $y_1 = \cdots = y_{n-1} = y_{2n} = 0$ (in the second case). The portion of ball $|z| \le r$ near z_0 is defined now by the inequality $y_{2d} \ge f(y')$ for some real-analytic function $f(y')$ defined in a neighborhood of 0, and vanishing on $\Sigma \ni 0$.

We have reduced the situation to solving the equation

$$y_{2d}a = b$$

in a neighborhood of 0, where b is supported in the set $y_{2d} \ge f(y')$, and we look for a supported in the same set. This is exactly the situation of Proposition 4.6.4 (possibly after a rescaling). As already observed, condition (4.96) is satisfied by every real-analytic function and therefore Proposition 4.6.4 gives the desired conclusion. □

Example 4.6.11. We want to find a Born–Jordan symbol of the Heisenberg operator $\widehat{T}(z_0)$, $z_0 = (x_0, \xi_0) \in \mathbb{R}^{2d}$. First of all, we observe that $\widehat{T}(z_0)$ has Weyl symbol

$$b(z) = e^{\frac{i}{\hbar}\sigma(z_0,z)};$$

see [105, Proposition 198]. Hence we are looking for $a \in \mathcal{S}'(\mathbb{R}^{2d})$ such that $(\frac{1}{2\pi\hbar})^d a * \Theta_\sigma = b$, or equivalently, taking the symplectic Fourier transform, $\Theta a_\sigma = b_\sigma$, that is,

$$\Theta(z)a_\sigma(z) = (2\pi\hbar)^d\delta(z - z_0). \tag{4.120}$$

Now, if $\Theta(z_0) \neq 0$, we can take $a_\sigma(z) = \Theta(z_0)^{-1}(2\pi\hbar)^d\delta(z - z_0)$, namely $a(z) = \Theta(z_0)^{-1}e^{\frac{i}{\hbar}\sigma(z_0,z)}$.

If instead $\Theta(z_0) = 0$, we look for a_σ in the form

$$a_\sigma(z) = (2\pi\hbar)^d\sum_{j=1}^{2d} c_j\partial_j\delta(z - z_0) \tag{4.121}$$

for unknown $c_j \in \mathbb{C}$, $j = 1, \ldots, 2d$. Since $\Theta(z_0) = 0$, we have

$$\Theta(z)a_\sigma(z) = (2\pi\hbar)^d\left(-\sum_{j=1}^{2d} c_j\partial_j\Theta(z_0)\right)\delta(z - z_0),$$

so that equation (4.120) reduces to

$$-\sum_{j=1}^{2d} c_j \partial_j \Theta(z_0) = 1,$$

which has infinitely many solutions because $\nabla\Theta(z_0) \neq 0$ if $\Theta(z_0) = 0$ ($\Theta(z)$ has only simple zeros). For any solution $\mathbf{c} := (c_1, \dots, c_{2d})$, taking the inverse symplectic Fourier transform in (4.121) (using the formulas $(a_\sigma)_\sigma = a$, $a_\sigma(z) = \mathcal{F}a(Jz)$, $\mathcal{F}(\partial_j a) = \frac{i}{\hbar} z_j \mathcal{F}a)$, we find a Born–Jordan symbol

$$a(z) = \frac{i}{\hbar}\sigma(z, \mathbf{c}) e^{\frac{i}{\hbar}\sigma(z_0, z)}.$$

Observe that $b \in \mathcal{A}_r$ with $r = |z_0|$ and $a \in \mathcal{A}_r$ as well.

4.7 Born–Jordan pseudodifferential operators on modulation spaces

The main result concerning the sufficient boundedness conditions of Born–Jordan operators on modulation spaces shows that they behave similarly to Weyl pseudodifferential operators or any other τ-form of pseudodifferential operators. The necessary boundedness conditions still contain some open problems, as shown in the following Theorem 4.7.1. These results are contained in [62].

Since we apply the theory of the previous section in the framework of time–frequency analysis, we set

$$\hbar = \frac{1}{2\pi}$$

and consider the corresponding Cohen kernel

$$\Theta(z) := \operatorname{sinc}(x \cdot \xi) = \begin{cases} \frac{\sin(\pi x \cdot \xi)}{\pi x \cdot \xi} & \text{for } x \cdot \xi \neq 0, \\ 1 & \text{for } x \cdot \xi = 0. \end{cases} \tag{4.122}$$

The (cross-)Born–Jordan distribution $Q(f, g)$ is then defined by

$$Q(f, g) = W(f, g) * \Theta_\sigma, \quad f, g \in \mathcal{S}(\mathbb{R}^d), \tag{4.123}$$

where Θ_σ is the symplectic Fourier transform of Θ recalled in (1.17) below. Likewise, the Weyl operator, a Born–Jordan operator with symbol $a \in \mathcal{S}'(\mathbb{R}^{2d})$, can be defined as

$$\langle \mathrm{Op}_{BJ}(a) f, g\rangle = \langle a, Q(g, f)\rangle \quad f, g \in \mathcal{S}(\mathbb{R}^d). \tag{4.124}$$

Now, the first relevant feature of this section is to have the Cohen kernel Θ_σ computed explicitly (cf. the subsequent Proposition 4.7.5). Namely,

$$\Theta_\sigma(\zeta_1,\zeta_2) = \begin{cases} -2\mathrm{Ci}(4\pi|\zeta_1\zeta_2|), & (\zeta_1,\zeta_2) \in \mathbb{R}^2,\ d=1, \\ \mathcal{F}(\chi_{\{|s|\geq 2\}}|s|^{d-2})(\zeta_1\cdot\zeta_2), & (\zeta_1,\zeta_2)\in\mathbb{R}^{2d},\ d\geq 2, \end{cases}$$

where $\chi_{\{|s|\geq 2\}}$ is the characteristic function of the set $\{s \in \mathbb{R} : |s| \geq 2\}$ and where

$$\mathrm{Ci}(t) = -\int_t^{+\infty} \frac{\cos s}{s}\, ds, \quad t \in \mathbb{R}, \tag{4.125}$$

is the cosine integral function.

This expression of Θ_σ shows that this kernel behaves badly in general: it does not even belong to L^∞_{loc} (see Corollary 4.7.6) and has no decay at infinity (see Corollary 4.7.7). Despite these undesirable aspects, some directional smoothing effect is still present, as proved in [63], see the following section. Our main result below refers to continuity properties on the same modulation space $M^{r_1,r_2}(\mathbb{R}^d)$. Observe that a similar result for Weyl operators contained in Theorem 4.4.15 above studied the action of $\mathrm{Op}_w(a) : \mathcal{M}^{p_1,q_1}(\mathbb{R}^d) \to \mathcal{M}^{p_2,q_2}(\mathbb{R}^d)$, but this is only a straightforward generalization.

Theorem 4.7.1. *Consider $1 \leq p, q, r_1, r_2 \leq \infty$, such that*

$$p \leq q' \tag{4.126}$$

and

$$q \leq \min\{r_1, r_2, r_1', r_2'\}. \tag{4.127}$$

Then the Born–Jordan operator $\mathrm{Op}_{BJ}(a)$, from $\mathcal{S}(\mathbb{R}^d)$ to $\mathcal{S}'(\mathbb{R}^d)$, having symbol $a \in M^{p,q}(\mathbb{R}^{2d})$, extends uniquely to a bounded operator on $\mathcal{M}^{r_1,r_2}(\mathbb{R}^d)$, with the estimate

$$\|\mathrm{Op}_{BJ}(a)f\|_{\mathcal{M}^{r_1,r_2}} \lesssim \|a\|_{M^{p,q}}\|f\|_{\mathcal{M}^{r_1,r_2}} \quad f \in \mathcal{M}^{r_1,r_2}. \tag{4.128}$$

Conversely, if this conclusion holds true, the constraint (4.126) is satisfied and it must hold

$$\max\left\{\frac{1}{r_1}, \frac{1}{r_2}, \frac{1}{r_1'}, \frac{1}{r_2'}\right\} \leq \frac{1}{q} + \frac{1}{p}, \tag{4.129}$$

which is (4.127) when $p = \infty$.

Notice that condition (4.129) is weaker than (4.127) when $p < \infty$. Condition (4.129) is obtained by working with rescaled Gaussians which provide the best localization in terms of Wigner distribution (cf. [218]). On the Fourier side, the Born–Jordan distribution is the pointwise multiplication of the Wigner distribution with the kernel Θ. This reasoning conduces to conjecture that condition (4.129) should be optimal so that the sufficient boundedness conditions for Born–Jordan operators might be weaker than the corresponding ones for Weyl and τ-pseudodifferential operators. But the matter is really subtle and requires a new and most refined analysis of the kernel Θ. In particular, the zeroes of the Θ function should play a key role for a thorough understanding of such operators, which certainly deserve further study.

4.7.1 Analysis of the Cohen kernel Θ

Consider the Cohen kernel Θ defined in (4.122). Obviously, $\Theta \in \mathcal{C}^\infty(\mathbb{R}^{2d}) \cap L^\infty(\mathbb{R}^{2d})$, but it displays a vary bad decay at infinity, as clarified in what follows.

Proposition 4.7.2. *For $1 \le p < \infty$, the function $\Theta \notin L^p(\mathbb{R}^{2d})$.*

Proof. Observe that, for $t \in \mathbb{R}$, $|t| \le 1/2$, the function sinc$t$ satisfies $|\mathrm{sinc}(t)| \ge 2/\pi$. Hence, for any $1 \le p < \infty$,

$$\begin{aligned}\int_{\mathbb{R}^{2d}} |\Theta(x,\xi)|^p \, dx\, d\xi &= \int_{\mathbb{R}^{2d}} |\mathrm{sinc}(x\cdot\xi)|^p \, dx\, d\xi \\ &\ge \int_{|x\cdot\xi|\le 1/2} |\mathrm{sinc}(x\cdot\xi)|^p \, dx\, d\xi \\ &\ge \left(\frac{2}{\pi}\right)^p \int_{|x\cdot\xi|\le 1/2} dx\, d\xi \\ &= \left(\frac{2}{\pi}\right)^p \mathrm{meas}\{(x,\xi) : |x\cdot\xi| \le 1/2\} = +\infty.\end{aligned}$$

This concludes the proof. □

We continue our investigation of the function Θ by looking for the right Wiener amalgam and modulation spaces containing this function. For this reason, we first reckon explicitly the STFT of the Θ function, with respect to the Gaussian window $g(x,\xi) = e^{-\pi x^2} e^{-\pi\xi^2} \in \mathcal{S}(\mathbb{R}^{2d})$.

Proposition 4.7.3. *For $z_1, z_2, \zeta_1, \zeta_2 \in \mathbb{R}^d$,*

$$V_g\Theta(z_1,z_2,\zeta_1,\zeta_2) \tag{4.130}$$

$$= \int_{-1/2}^{1/2} \frac{1}{(t^2+1)^{d/2}} e^{-2\pi i[\frac{1}{t}\zeta_1\cdot\zeta_2 + \frac{t}{t^2+1}(z_1-\frac{1}{t}\zeta_2)\cdot(z_2-\frac{1}{t}\zeta_1)]} e^{-\pi\frac{t^2}{t^2+1}[(z_1-\frac{1}{t}\zeta_2)^2+(z_2-\frac{1}{t}\zeta_1)^2]} \, dt. \tag{4.131}$$

Proof. We write $\Theta(z_1,z_2) = F_1(z_1,z_2) + F_2(z_1,z_2)$, where $F_1(z_1,z_2) = \int_0^{1/2} e^{2\pi i z_1\cdot z_2 t}\, dt$ and $F_2(z) = F_1(Jz)$, $z = (z_1,z_2)$. Let us first reckon $V_gF_1(z,\zeta)$, $z = (z_1,z_2)$, $\zeta = (\zeta_1,\zeta_2) \in \mathbb{R}^{2d}$, where g is the Gaussian function above. For $t > 0$, we define the function $H_t(z_1, z_2) = e^{2\pi i t z_1\cdot z_2}$ and observe that

$$\mathcal{F}H_t(\zeta_1, \zeta_2) = \frac{1}{t^d} e^{-2\pi i \frac{1}{t}\zeta_1\cdot\zeta_2} \tag{4.132}$$

(cf. [145, Appendix A, Theorem 2]). By the dominated convergence theorem,

$$\begin{aligned}
V_gF_1(z,\zeta) &= \int_0^{1/2} \mathcal{F}(H_t T_z g)(\zeta)\,dt = \int_0^{1/2} (\mathcal{F}(H_t) * M_{-z}\hat{g})(\zeta_1,\zeta_2)\,dt \\
&= \int_0^{1/2} \frac{1}{t^d} \int_{\mathbb{R}^{2d}} e^{-2\pi i \frac{1}{t}(\zeta_1-y_1)\cdot(\zeta_2-y_2)} e^{-2\pi i (z_1,z_2)\cdot(y_1,y_2)} e^{-\pi y_1^2} e^{-\pi y_2^2}\,dy_1\,dy_2\,dt \\
&= \int_0^{1/2} \frac{1}{t^d} e^{-2\pi i \frac{1}{t}\zeta_1\cdot\zeta_2} \int_{\mathbb{R}^{2d}} e^{-2\pi i \frac{1}{t} y_1\cdot y_2 + 2\pi i \frac{1}{t}(\zeta_2\cdot y_1 + \zeta_1\cdot y_2) - 2\pi i (z_1\cdot y_1 + z_2\cdot y_2)} e^{-\pi(y_1^2+y_2^2)}\,dy_1\,dy_2\,dt \\
&= \int_0^{1/2} \frac{1}{t^d} e^{-2\pi i \frac{1}{t}\zeta_1\cdot\zeta_2} \int_{\mathbb{R}^d} e^{2\pi i (\frac{1}{t}\zeta_1\cdot y_2 - z_2\cdot y_2)} e^{-\pi y_2^2} \\
&\quad \times \Big(\int_{\mathbb{R}^d} e^{-2\pi i y_1\cdot(\frac{1}{t}y_2 - \frac{1}{t}\zeta_2 + z_1)} e^{-\pi y_1^2}\,dy_1 \Big)\,dy_2\,dt \\
&= \int_0^{1/2} \frac{1}{t^d} e^{-2\pi i \frac{1}{t}\zeta_1\cdot\zeta_2} \int_{\mathbb{R}^d} e^{-2\pi i y_2\cdot(z_2 - \frac{1}{t}\zeta_1)} e^{-\pi y_2^2} e^{-\pi(\frac{1}{t}y_2 - \frac{1}{t}\zeta_2 + z_1)^2}\,dy_2\,dt \\
&= \int_0^{1/2} \frac{1}{t^d} e^{-2\pi i \frac{1}{t}\zeta_1\cdot\zeta_2} e^{-\pi(z_1 - \frac{1}{t}\zeta_2)^2} \int_{\mathbb{R}^d} e^{-2\pi i y_2\cdot(z_2 - \frac{1}{t}\zeta_1)} e^{-\pi((1+\frac{1}{t^2})y_2^2 - 2(z_1 - \frac{1}{t}\zeta_2)\cdot\frac{1}{t}y_2)}\,dy_2\,dt \\
&= \int_0^{1/2} \frac{1}{t^d} e^{-2\pi i \frac{1}{t}\zeta_1\cdot\zeta_2} e^{-\pi\frac{t^2}{t^2+1}(u_1 - \frac{1}{t}\zeta_2)^2} \int_{\mathbb{R}^d} e^{-2\pi i y_2\cdot(u_2 - \frac{1}{t}\zeta_1)} \\
&\quad \times e^{-\pi(\frac{\sqrt{t^2+1}}{t}y_2 - \frac{t}{\sqrt{t^2+1}}(z_1 - \frac{1}{t}\zeta_2))^2}\,dy_2\,dt \\
&= \int_0^{1/2} \frac{1}{(t^2+1)^{d/2}} e^{-2\pi i \frac{1}{t}\zeta_1\cdot\zeta_2} e^{-\pi\frac{t^2}{t^2+1}(z_1 - \frac{1}{t}\zeta_2)^2} \\
&\quad \times \int_{\mathbb{R}^d} e^{-2\pi i (\frac{t}{\sqrt{t^2+1}}w + \frac{t}{t^2+1}(z_1 - \frac{1}{t}\zeta_2))\cdot(z_2 - \frac{1}{t}\zeta_1)} e^{-\pi w^2}\,dw\,dt \\
&= \int_0^{1/2} \frac{1}{(t^2+1)^{d/2}} e^{-2\pi i \frac{1}{t}\zeta_1\cdot\zeta_2} e^{-\pi\frac{t^2}{t^2+1}(z_1 - \frac{1}{t}\zeta_2)^2} e^{-\pi\frac{t^2}{t^2+1}(z_2 - \frac{1}{t}\zeta_1)^2} \\
&\quad \times e^{-2\pi i \frac{t}{t^2+1}(z_1 - \frac{1}{t}\zeta_2)\cdot(z_2 - \frac{1}{t}\zeta_1)}\,dt.
\end{aligned}$$

Now, an easy computation shows

$$V_gF_2(z,\zeta) = V_gF_1(Jz, J\zeta)$$

so that $V_g\Theta = V_gF_1 + V_gF_2$, and we obtain (4.131). □

Proposition 4.7.4. *The function* Θ *in* (4.122) *belongs to* $W(\mathcal{F}L^1, L^\infty)(\mathbb{R}^{2d})$.

Proof. We simply have to calculate

$$\sup_{z\in\mathbb{R}^{2d}} \int_{\mathbb{R}^{2d}} |V_g\Theta(z,\zeta)|\, d\zeta.$$

From (4.131) we observe that

$$\begin{aligned}
\|V_g\Theta(z,\cdot)\|_1 &\le \int_{-1/2}^{1/2}\int_{\mathbb{R}^{2d}} \frac{1}{(t^2+1)^{d/2}} e^{-\pi\frac{t^2}{t^2+1}(z_1-\frac{1}{t}\zeta_2)^2} e^{-\pi\frac{t^2}{t^2+1}(z_2-\frac{1}{t}\zeta_1)^2}\, d\zeta_1\, d\zeta_2\, dt \\
&= \int_{-1/2}^{1/2}\int_{\mathbb{R}^{2d}} \frac{1}{(t^2+1)^{d/2}} e^{-\pi\frac{1}{t^2+1}(tz_1-\zeta_2)^2} e^{-\pi\frac{1}{t^2+1}(tz_2-\zeta_1)^2}\, d\zeta_1\, d\zeta_2\, dt \\
&= \int_{-1/2}^{1/2}\int_{\mathbb{R}^{2d}} (t^2+1)^{d/2} e^{-\pi(v_1^2+v_2^2)}\, dv_1\, dv_2\, dt = C < \infty,
\end{aligned}$$

from which the claim follows. □

Using the STFT of the function Θ in (4.131), we observe that

$$\|V_g\Theta(\cdot,\zeta)\|_1 \le \int_{-1/2}^{1/2}\int_{\mathbb{R}^{2d}} \frac{1}{(t^2+1)^{d/2}} e^{-\pi\frac{t^2}{t^2+1}(u_1-\frac{1}{t}\zeta_2)^2} e^{-\pi\frac{t^2}{t^2+1}(u_2-\frac{1}{t}\zeta_1)^2}\, du_1\, du_2\, dt = +\infty$$

and conjecture that $\Theta \notin M^{1,\infty}(\mathbb{R}^{2d})$. The previous claim will follow if we prove that $\Theta_\sigma \notin W(\mathcal{F}L^1, L^\infty)(\mathbb{R}^{2d})$.

Note that $\Theta_\sigma(\zeta) = \mathcal{F}\Theta(J\zeta) = \mathcal{F}\Theta(\zeta)$. Furthermore, the distributional Fourier transform of Θ can be computed explicitly as follows. First, recall the definition of the cosine integral function (4.125).

Proposition 4.7.5. *For* $d \ge 1$, *the distribution symplectic Fourier transform* Θ_σ *of the function* Θ *is provided by*

$$\Theta_\sigma(\zeta_1,\zeta_2) = \begin{cases} -2\mathrm{Ci}(4\pi|\zeta_1\zeta_2|), & (\zeta_1,\zeta_2)\in\mathbb{R}^2,\ d=1, \\ \mathcal{F}(\chi_{\{|s|\ge 2\}}|s|^{d-2})(\zeta_1\cdot\zeta_2), & (\zeta_1,\zeta_2)\in\mathbb{R}^{2d},\ d\ge 2, \end{cases} \tag{4.133}$$

where $\chi_{\{|s|\ge 2\}}$ *is the characteristic function of the set* $\{s\in\mathbb{R} : |s|\ge 2\}$. *The case* $d=1$ *can be recaptured by the case* $d\ge 2$ *using* (4.125).

Proof. We carry out the computations of Θ_σ by studying first the case in dimension $d = 1$ and then, inspired by the former case, $d > 1$.

First step: $d = 1$. By Proposition 4.7.2, the function Θ is in

$$L^\infty(\mathbb{R}^2) \setminus L^p(\mathbb{R}^2) \subset \mathcal{S}'(\mathbb{R}^2), \quad 1 \le p < \infty,$$

so that the Fourier transform is meant in $\mathcal{S}'(\mathbb{R}^2)$. Observe that

$$\mathcal{F}\Theta(\zeta_1, \zeta_2) = \mathcal{F}_2\mathcal{F}_1\Theta(\zeta_1, \zeta_2),$$

where $\mathcal{F}_1$ (resp. $\mathcal{F}_2$) is the partial Fourier transform with respect to the first (resp. second) variable. Indeed, for every test function $\varphi \in \mathcal{S}(\mathbb{R}^2)$,

$$\langle \mathcal{F}\Theta, \varphi\rangle = \langle \Theta, \mathcal{F}^{-1}\varphi\rangle$$

and $\mathcal{F}^{-1}\varphi(x,\xi) = \mathcal{F}_1^{-1}\mathcal{F}_2^{-1}\varphi(x,\xi) = \mathcal{F}_2^{-1}\mathcal{F}_1^{-1}\varphi(x,\xi)$, by Fubini's theorem.

Using

$$\mathcal{F}_1\operatorname{sinc}(y_2\cdot)(\zeta_1) = \frac{1}{|y_2|}p_{1/2}(\zeta_1/y_2), \quad y_2 \neq 0,$$

where $p_{1/2}(t)$ is the box function defined by $p_{1/2}(t) = 1$ for $|t| \le 1/2$ and $p_{1/2}(t) = 0$ otherwise, we obtain, for $\zeta_1\zeta_2 \neq 0$ (hence in particular $|\zeta_1| > 0$),

$$\begin{aligned}
\mathcal{F}\Theta(\zeta_1,\zeta_2) &= \int_{\mathbb{R}} e^{-2\pi i\zeta_2 y_2}\frac{1}{|y_2|}p_{1/2}(\zeta_1/y_2)\,dy_2 = \int_{|y_2|\ge 2|\zeta_1|} e^{-2\pi i\zeta_2 y_2}\frac{1}{|y_2|}\,dy_2 \\
&= \int_{|s|\ge 2|\zeta_1\zeta_2|} e^{-2\pi i s}\frac{1}{|s|}\,ds \\
&= \int_{|s|\ge 2|\zeta_1\zeta_2|} \frac{\cos(2\pi s) - i\sin(2\pi s)}{|s|}\,ds \\
&= \int_{|s|\ge 2|\zeta_1\zeta_2|} \frac{\cos 2\pi s}{|s|}\,ds \\
&= 2\int_{2|\zeta_1\zeta_2|}^{+\infty} \frac{\cos 2\pi s}{s}\,ds = -2\operatorname{Ci}(4\pi|\zeta_1\zeta_2|),
\end{aligned}$$

by (4.125), so that, since $\zeta_1\zeta_2 = 0$ is a set of Lebesgue measure zero on $\mathbb{R}^2$, we can write

$$\Theta_\sigma(\zeta_1,\zeta_2) = -2\operatorname{Ci}(4\pi|\zeta_1\zeta_2|), \quad (\zeta_1,\zeta_2) \in \mathbb{R}^2. \tag{4.134}$$

Second step: $d > 1$. This is a simple generalization on the former step. For (z_1, z_2), $(\zeta_1,\zeta_2) \in \mathbb{R}^{2d}$, $d > 1$, we write

$$z_i = (z_i', z_{i,d}), \zeta_i = (\zeta_i', \zeta_{i,d}), \quad z_i', \zeta_i' \in \mathbb{R}^{d-1}, \quad z_{i,d}, \zeta_{i,d} \in \mathbb{R}, \quad i = 1,2. \tag{4.135}$$

We decompose $\mathcal{F}\Theta = \mathcal{F}_{2d}\mathcal{F}'\mathcal{F}_1\Theta$ where, for $\Theta = \Theta(z_1, z_2)$, $\mathcal{F}_1$ is the partial Fourier transform with respect to the variable $z_{1,d}$, $\mathcal{F}'$ is the partial Fourier transform with respect to the $2d-2$ variables $(z_1', z_2') \in \mathbb{R}^{2d-2}$, and $\mathcal{F}_{2d}$ is the partial Fourier transform with respect to the last variable $z_{2,d}$. We start by computing the partial Fourier transform $\mathcal{F}_1$:

$$\begin{aligned}\mathcal{F}_1\Theta(z_1', \cdot, z_2', z_{2,d})(\zeta_{1,d}) &= \mathcal{F}_1(T_{\frac{-z_1'\cdot z_2'}{z_{2,d}}} \operatorname{sinc}(z_{2,d}\cdot))(\zeta_{1,d}) \\ &= e^{2\pi i \frac{\zeta_{1,d}}{z_{2,d}} z_1'\cdot z_2'} \frac{1}{|z_{2,d}|} \mathcal{F}_1(\operatorname{sinc})\left(\frac{\zeta_{1,d}}{z_{2,d}}\right) \\ &= e^{2\pi i \frac{\zeta_{1,d}}{z_{2,d}} z_1'\cdot z_2'} \frac{1}{|z_{2,d}|} p_{1/2}\left(\frac{\zeta_{1,d}}{z_{2,d}}\right).\end{aligned}$$

Using the Gaussian integrals in [145, Appendix A, Theorem 2], we calculate

$$\mathcal{F}'(e^{2\pi i \frac{\zeta_{1,d}}{z_{2,d}} z_1'\cdot z_2'})(\zeta_1', \zeta_2') = \left|\frac{z_{2,d}}{\zeta_{1,d}}\right|^{d-1} e^{-2\pi i \frac{z_{2,d}}{\zeta_{1,d}} \zeta_1'\cdot\zeta_2'},$$

so that

$$\begin{aligned}\mathcal{F}\Theta(\zeta_1, \zeta_2) &= \mathcal{F}_{2d}\left(e^{-2\pi i \frac{z_{2,d}}{\zeta_{1,d}} \zeta_1'\cdot\zeta_2'} \left|\frac{z_{2,d}}{\zeta_{1,d}}\right|^{d-1} \frac{1}{|z_{2,d}|} p_{1/2}\left(\frac{\zeta_{1,d}}{z_{2,d}}\right)\right)(\zeta_{2,d}) \\ &= \int_{|\frac{\zeta_{1,d}}{z_{2,d}}|\le\frac{1}{2}} \left|\frac{z_{2,d}}{\zeta_{1,d}}\right|^{d-1} \frac{1}{|z_{2,d}|} e^{-2\pi i \frac{z_{2,d}}{\zeta_{1,d}} \zeta_1\cdot\zeta_2} \, dz_{2,d} \\ &= \int_{|s|\ge 2} e^{-2\pi i s(\zeta_1\cdot\zeta_2)} |s|^{d-2} \, ds,\end{aligned}$$

as claimed. □

Notice that the second equation (4.133) can be written as

$$\Theta_\sigma(\zeta_1, \zeta_2) = \int_{|s|\ge 2} e^{-2\pi i s(\zeta_1\cdot\zeta_2)} |s|^{d-2} \, ds.$$

Corollary 4.7.6. *We have*

$$\Theta_\sigma \notin L^\infty_{\mathrm{loc}}(\mathbb{R}^{2d}).$$

Proof. For the case $d = 1$, recall that the cosine integral $\operatorname{Ci}(x)$ has the series expansion

$$\operatorname{Ci}(x) = \gamma + \log x + \sum_{k=1}^{+\infty} \frac{(-x^2)^k}{2k(2k)!}, \quad x > 0$$

where γ is the Euler–Mascheroni constant, from which our claim easily follows.

For $d \ge 2$, Θ_σ is only defined as a tempered distribution. □

Corollary 4.7.7. *The function* $\Theta_\sigma \notin L^p(\mathbb{R}^{2d})$, *for any* $1 \le p \le \infty$.

Proof. The case $p = \infty$ is already treated in Corollary 4.7.6. For $d \ge 2$, again we observe that Θ_σ is not defined as a function but only as a tempered distribution. For $d = 1$, $1 \le p < \infty$, the claim follows by the expression (4.134). Indeed, choose $0 < \epsilon < \pi/2$, then $|\mathrm{Ci}(x)| \ge |\mathrm{Ci}(\epsilon)|$, for $0 < x < \epsilon$, so that

$$\int_{\mathbb{R}^2} |\Theta_\sigma(\zeta_1,\zeta_2)|^p \, d\zeta_1 \, d\zeta_2 \ge 2 \int_{|\zeta_1\zeta_2| < \frac{\epsilon}{4\pi}} |\mathrm{Ci}(4\pi|\zeta_1\zeta_2|)|^p \, d\zeta \, d\zeta_2$$

$$\ge C_p \mathrm{meas}\left\{(\zeta_1,\zeta_2) : |\zeta_1\zeta_2| < \frac{\epsilon}{4\pi}\right\} = +\infty,$$

for a suitable constant $C_p > 0$. □

Since $\mathcal{F}L^1 \subset L^\infty$, the Wiener amalgam space $W(\mathcal{F}L^1, L^\infty)$ is included in L^∞_{loc}. This proves our claim:

Corollary 4.7.8. *The function* $\Theta_\sigma \notin W(\mathcal{F}L^1, L^\infty)(\mathbb{R}^{2d})$ *or, equivalently,* $\Theta \notin M^{1,\infty}(\mathbb{R}^{2d})$.

4.7.2 Proof of Theorem 4.7.1

This section is focused on the proof of Theorem 4.7.1. This requires some preliminaries.

Lemma 4.7.9. *Let* $\chi \in \mathcal{D}(\mathbb{R})$. *Then for* $\zeta_1, \zeta_2 \in \mathbb{R}^d$ *the function* $\chi(\zeta_1\zeta_2)$ *belongs to* $W(\mathcal{F}L^1, L^\infty)(\mathbb{R}^{2d})$.

Proof. We write

$$\chi(\zeta_1\zeta_2) = \int_{\mathbb{R}} e^{2\pi i t \zeta_1 \cdot \zeta_2} \widehat{\chi}(t) \, dt,$$

and the desired result then follows as in proof of Proposition 4.7.4 because $\widehat{\chi}(t)$ lies in $\mathcal{S}(\mathbb{R}) \subset L^1(\mathbb{R})$. □

Lemma 4.7.10. *Let* $\psi \in \mathcal{D}(\mathbb{R}^d) \setminus \{0\}$ *and* $\lambda > 0$. *Then we have the following dilation properties:*

$$\|\psi(\lambda \cdot)\|_{W(\mathcal{F}L^p, L^q)} \asymp \lambda^{-d/p'} \quad \text{as } \lambda \to +\infty, \tag{4.136}$$

$$\|\psi(\lambda \cdot)\|_{W(\mathcal{F}L^p, L^q)} \asymp \lambda^{-d/q} \quad \text{as } \lambda \to 0. \tag{4.137}$$

Proof. Formula (4.136) follows by observing that for, say, $\lambda \ge 1$, the functions $\psi(\lambda \cdot)$ are supported in a fixed compact set, so that their $M^{p,q}$-norm is equivalent to their norm in $\mathcal{F}L^p$, which is easily estimated.

Let us now prove (4.137). Observe, first of all, that for $f \in \mathcal{S}(\mathbb{R}^d)$, $g \in \mathcal{D}(\mathbb{R}^d)$, $q \geq 1$, we have

$$\|fT_x g\|_{\mathcal{F}L^1} \lesssim \sum_{|\alpha|\leq d+1} \|\partial^\alpha (fT_x g)\|_{L^1} \lesssim \sum_{|\alpha|\leq d+1} \|\partial^\alpha (fT_x g)\|_{L^q},$$

so that

$$\|f\|_{W(\mathcal{F}L^p,L^q)} \lesssim \|f\|_{W(\mathcal{F}L^1,L^q)} \lesssim \sum_{|\alpha|\leq d+1} \|\partial^\alpha f\|_{L^q}.$$

Applying this formula with $f = \psi(\lambda x)$, $0 < \lambda \leq 1$, we obtain

$$\|\psi(\lambda\cdot)\|_{W(\mathcal{F}L^p,L^q)} \lesssim \lambda^{-d/q}.$$

To obtain the lower bound, we observe that

$$\begin{aligned}\lambda^{-d}\|\psi\|_{L^2}^2 &= \|\psi(\lambda\cdot)\|_{L^2}^2 \\ &\lesssim \|\psi(\lambda\cdot)\|_{W(\mathcal{F}L^{p'},L^{q'})}\|\psi(\lambda\cdot)\|_{W(\mathcal{F}L^p,L^q)} \\ &\lesssim \lambda^{-d/q'}\|\psi(\lambda\cdot)\|_{W(\mathcal{F}L^p,L^q)},\end{aligned}$$

which implies

$$\|\psi(\lambda\cdot)\|_{W(\mathcal{F}L^p,L^q)} \gtrsim \lambda^{-d/q}.$$

□

A straightforward generalization of the previous arguments is as follows:

Lemma 4.7.11. *Consider $1 \leq p,q \leq \infty$, $\psi \in \mathcal{D}(\mathbb{R}^d) \setminus \{0\}$ and $\lambda > 0$. Then*

$$\|\psi(\sqrt{\lambda}\cdot)\|_{W(\mathcal{F}L^p,L^q)} \asymp \lambda^{-\frac{d}{2p'}} \quad \text{as } \lambda \to +\infty, \tag{4.138}$$

$$\|\psi(\sqrt{\lambda}\cdot)\|_{W(\mathcal{F}L^p,L^q)} \asymp \lambda^{-\frac{d}{2q}} \quad \text{as } \lambda \to 0^+. \tag{4.139}$$

The same conclusion holds uniformly with respect to λ if ψ varies in bounded subsets of $\mathcal{D}(\mathbb{R}^d)$.

We first demonstrate the sufficient boundedness conditions.

Theorem 4.7.12. *Assume that $1 \leq p,q,r_1,r_2 \leq \infty$. Then the pseudodifferential operator $\mathrm{Op}_{BJ}(a)$, from $\mathcal{S}(\mathbb{R}^d)$ to $\mathcal{S}'(\mathbb{R}^d)$, having symbol $a \in M^{p,q}(\mathbb{R}^{2d})$, extends uniquely to a bounded operator on $M^{r_1,r_2}(\mathbb{R}^d)$, with the estimate* (4.128) *and the indices' conditions* (4.126) *and* (4.127).

The result relies on a thorough understanding of the action of the mapping

$$A : a \mapsto a * \Theta_\sigma, \tag{4.140}$$

which gives the Weyl symbol of an operator with Born–Jordan symbol a, on modulation spaces.

Proposition 4.7.13. *For every $1 \le p,q \le \infty$, the mapping A in (4.140), defined initially on $\mathcal{S}'(\mathbb{R}^{2d})$, restricts to a linear continuous map on $M^{p,q}(\mathbb{R}^{2d})$, i. e., there exists a constant $C > 0$ such that*

$$\|Aa\|_{M^{p,q}} \le C\|a\|_{M^{p,q}}. \tag{4.141}$$

Proof. By Proposition 4.7.4, the function $\Theta \in W(\mathcal{F}L^1, L^\infty)(\mathbb{R}^{2d})$. Its symplectic Fourier transform Θ_σ belongs to $\mathcal{F}_\sigma W(\mathcal{F}L^1, L^\infty)(\mathbb{R}^{2d}) = M^{1,\infty}(\mathbb{R}^{2d})$. Now, for every $1 \le p,q \le \infty$, the convolution relations for modulation spaces (2.81) give

$$M^{p,q}(\mathbb{R}^{2d}) * M^{1,\infty}(\mathbb{R}^{2d}) \hookrightarrow M^{p,q}(\mathbb{R}^{2d}),$$

and this shows the claim (4.141). □

Proof of Theorem 4.7.12. Assume $a \in M^{p,q}(\mathbb{R}^{2d})$, then Proposition 4.7.13 proves that $Aa = a * \Theta_\sigma \in M^{p,q}(\mathbb{R}^{2d})$ as well. We next write $\mathrm{Op}_{BJ}(a) = \mathrm{Op}_W(Aa)$ and use the sufficient conditions for Weyl operators in Theorem 4.7.1: if the Weyl symbol Aa is in $M^{p,q}(\mathbb{R}^{2d})$, then $\mathrm{Op}_W(Aa)$ extends to a bounded operator on $M^{r_1,r_2}(\mathbb{R}^d)$, with

$$\|\mathrm{Op}_{BJ}(a)f\|_{M^{r_1,r_2}} = \|\mathrm{Op}_W(Aa)f\|_{M^{r_1,r_2}} \lesssim \|Aa\|_{M^{p,q}}\|f\|_{M^{r_1,r_2}}$$

where the indices r_1, r_2, p, q satisfy (4.126) and (4.127). The inequality (4.141) then provides the claim. □

The necessary conditions of Theorem 4.7.1 require some preliminaries.

We reckon the adjoint operator $\mathrm{Op}_{BJ}(a)^*$ of a Born–Jordan operator $\mathrm{Op}_{BJ}(a)$ using the connection with Weyl operators. An easy computation shows

$$\mathrm{Op}_W(b)^* = \mathrm{Op}_W(\bar{b})$$

(see also [193]), so that

$$\mathrm{Op}_{BJ}(a)^* = \mathrm{Op}_W(a * \Theta_\sigma)^* = \mathrm{Op}_W(\overline{a * \Theta_\sigma}) = \mathrm{Op}_W(\bar{a} * \bar{\Theta}_\sigma) = \mathrm{Op}_W(\bar{a} * \Theta_\sigma) = \mathrm{Op}_{BJ}(\bar{a})$$

because Θ is an even real-valued function. Hence the adjoint of a Born–Jordan operator $\mathrm{Op}_{BJ}(a)$ with symbol a is the Born–Jordan operator having symbol $\bar{a}$ (the complex-conjugate of a). This nice property is the key argument for the following auxiliary result, already obtained for the case of Weyl operators in Lemma 4.1.7. The proof uses the same pattern as the former result and hence is omitted.

Lemma 4.7.14. *Suppose that, for some $1 \le p,q,r_1,r_2 \le \infty$, the following estimate holds:*

$$\|\mathrm{Op}_{BJ}(a)f\|_{M^{r_1,r_2}} \le C\|a\|_{M^{p,q}}\|f\|_{M^{r_1,r_2}}, \quad \forall a \in \mathcal{S}(\mathbb{R}^{2d}),\ \forall f \in \mathcal{S}(\mathbb{R}^d).$$

Then the same estimate is satisfied with r_1, r_2 replaced by r_1', r_2' (even if $r_1 = \infty$ or $r_2 = \infty$).

The above instruments let us show the necessity of (4.126) and (4.129). However, we first need a few preliminaries.

Proposition 4.7.15. *The function* $F(\zeta_1,\zeta_2) = e^{2\pi i\zeta_1\cdot\zeta_2}$ *belongs to* $W(\mathcal{F}L^1, L^\infty)(\mathbb{R}^{2d})$.

Proof. Consider the Gaussian function $g(\zeta_1,\zeta_2) = e^{-\pi\zeta_1^2}e^{-\pi\zeta_2^2}$ as a window function to compute the $W(\mathcal{F}L^1, L^\infty)$-norm. Then we have

$$\|F\|_{W(\mathcal{F}L^1,L^\infty)(\mathbb{R}^{2d})} = \sup_{u\in\mathbb{R}^{2d}} \|\mathcal{F}(FT_u g)\|_{L^1(\mathbb{R}^{2d})}.$$

Let us compute $\mathcal{F}(FT_u g)(z)$. For $z = (z_1, z_2)$,

$$\begin{aligned}
\mathcal{F}(FT_u g)(z_1,z_2) &= (\mathcal{F}(F) * M_{-u}\hat{g})(z_1,z_2)\\
&= \int_{\mathbb{R}^{2d}} e^{-2\pi i(z_1-y_1)\cdot(z_2-y_2)} e^{-2\pi i(u_1,u_2)\cdot(y_1,y_2)} e^{-\pi y_1^2} e^{-\pi y_2^2}\, dy_1\, dy_2\\
&= e^{-2\pi i z_1\cdot z_2} \int_{\mathbb{R}^{2d}} e^{-2\pi i y_1\cdot y_2 + 2\pi i(z_2\cdot y_1 + z_1\cdot y_2) - 2\pi i(u_1\cdot y_1 + u_2\cdot y_2)} e^{-\pi y_1^2} e^{-\pi y_2^2}\, dy_1\, dy_2\\
&= e^{-2\pi i z_1\cdot z_2} \int_{\mathbb{R}^d} e^{2\pi i(z_1\cdot y_2 - u_2\cdot y_2)} e^{-\pi y_2^2} \Big(\int_{\mathbb{R}^d} e^{-2\pi i y_1\cdot(y_2 - z_2 + u_1)} e^{-\pi y_1^2}\, dy_1 \Big)\, dy_2\\
&= e^{-2\pi i z_1\cdot z_2} \int_{\mathbb{R}^d} e^{-2\pi i y_2\cdot(u_2 - z_1)} e^{-\pi y_2^2} e^{-\pi(y_2 - z_2 + u_1)^2}\, dy_2\\
&= e^{-2\pi i z_1\cdot z_2} e^{-\pi(u_1 - z_2)^2} \int_{\mathbb{R}^d} e^{-2\pi i y_2\cdot(u_2 - z_1)} e^{-2\pi(y_2^2 + (u_1 - z_2)\cdot y_2)}\, dy_2\\
&= e^{-2\pi i z_1\cdot z_2} e^{-\pi(u_1 - z_2)^2 + \frac{\pi}{2}(u_1 - z_2)^2} \int_{\mathbb{R}^d} e^{-2\pi i y_2\cdot(u_2 - z_1)} e^{-2\pi(y_2 + \frac{u_1 - z_2}{2})^2}\, dy_2\\
&= e^{-2\pi i z_1\cdot z_2} e^{-\frac{\pi}{2}(u_1 - z_2)^2} \mathcal{F}(T_{-\frac{u_1 - z_2}{2}} e^{-2\pi|\cdot|^2})(u_2 - z_1)\\
&= 2^{-d/2} e^{-2\pi i z_1\cdot z_2} e^{-\frac{\pi}{2}(u_1 - z_2)^2} e^{-\frac{\pi}{2}(u_2 - z_1)^2} e^{\pi i(u_1 - z_2)\cdot(u_2 - z_1)}\\
&= 2^{-d/2} e^{\pi i(u_1\cdot u_2 - z_1\cdot z_2 - u_1\cdot z_1 - u_2\cdot z_2)} e^{-\frac{\pi}{2}(u_1 - z_2)^2} e^{-\frac{\pi}{2}(u_2 - z_1)^2}.
\end{aligned}$$

Hence

$$\|\mathcal{F}(FT_u g)\|_{L^1} = 2^{-d/2} \|e^{-\frac{\pi}{2}|\cdot|^2}\|_{L^1} = C,$$

for a constant C independent of the variable u, and, consequently,

$$\sup_{u\in\mathbb{R}^{2d}} \|\mathcal{F}(FT_u g)\|_{L^1} < \infty,$$

as desired. □

Corollary 4.7.16. *For $\zeta = (\zeta_1, \zeta_2)$, consider the function $F_J(\zeta) = F(J\zeta) = e^{-2\pi i \zeta_1 \cdot \zeta_2}$. Then $F_J \in W(\mathcal{F}L^1, L^\infty)(\mathbb{R}^{2d})$.*

Proof. The result immediately follows by Proposition 4.7.15 and by the dilation properties for Wiener amalgam spaces (2.107). □

Theorem 4.7.17. *Suppose that, for some $1 \le p, q, r_1, r_2 \le \infty$, $C > 0$, the estimate*

$$\|\mathrm{Op}_{BJ}(a)f\|_{M^{r_1,r_2}} \le C\|a\|_{M^{p,q}}\|f\|_{M^{r_1,r_2}} \quad \forall a \in \mathcal{S}(\mathbb{R}^{2d}),\ f \in \mathcal{S}(\mathbb{R}^d) \tag{4.142}$$

holds. Then the constraints in (4.126) *and* (4.129) *must hold.*

Proof. The estimate (4.142) can be written as

$$|\langle a, Q(f,g)\rangle| \le C\|a\|_{M^{p,q}}\|f\|_{M^{r_1,r_2}}\|g\|_{M^{r_1',r_2'}} \quad \forall a \in \mathcal{S}(\mathbb{R}^{2d}),\ f,g \in \mathcal{S}(\mathbb{R}^d),$$

which is equivalent to

$$\|Q(f,g)\|_{M^{p',q'}} \le C\|f\|_{M^{r_1,r_2}}\|g\|_{M^{r_1',r_2'}} \quad \forall f,g \in \mathcal{S}(\mathbb{R}^d).$$

Now, one should test this estimate on families of functions f_λ, g_λ such that $Q(f_\lambda, g_\lambda)$ is concentrated inside the hyperbola $|x \cdot \xi| < 1$ (say), see Figure 4.6, where $\theta \asymp 1$, so that the left-hand side is comparable to $\|W(f_\lambda, g_\lambda)\|_{M^{p',q'}}$ and can be estimated from below.

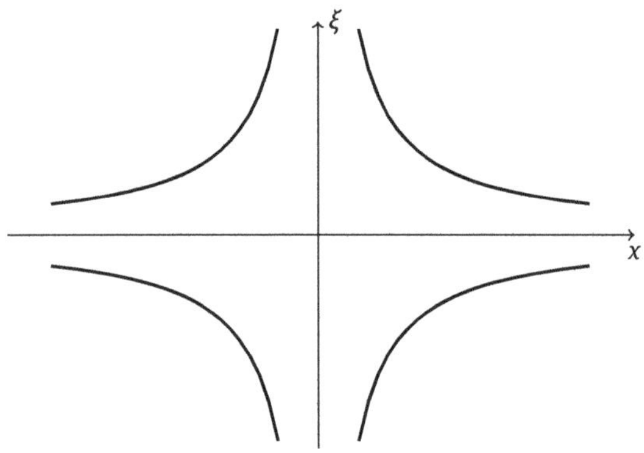

Figure 4.6: The region $|x \cdot \xi| < 1$ $(d = 1)$.

The choice $f_\lambda(x) = g_\lambda(x) = \varphi_{\sqrt{\lambda}}(x) = e^{-\pi\lambda x^2}$ provides the estimate (4.126) when $\lambda \to +\infty$. By (2.103), we obtain the estimate

$$\|\varphi_{\sqrt{\lambda}}\|_{M^{r_1,r_2}}\|\varphi_{\sqrt{\lambda}}\|_{M^{r_1',r_2'}} \lesssim \lambda^{-\frac{d}{2r_2'}}\lambda^{-\frac{d}{2r_2}}. \tag{4.143}$$

We gauge from below the norm $\|Q(\varphi_{\sqrt{\lambda}}, \varphi_{\sqrt{\lambda}})\|_{M^{p',q'}}$ as follows. By taking the symplectic Fourier transform and using Lemma 4.7.9 and the product property (4.103), we have

$$\begin{aligned}\|Q(\varphi_{\sqrt{\lambda}},\varphi_{\sqrt{\lambda}})\|_{M^{p',q'}} &= \|\Theta_\sigma * W(\varphi_{\sqrt{\lambda}},\varphi_{\sqrt{\lambda}})\|_{M^{p',q'}}\\ &\asymp \|\Theta\mathcal{F}_\sigma[W(\varphi_{\sqrt{\lambda}},\varphi_{\sqrt{\lambda}})]\|_{W(\mathcal{F}L^{p'},L^{q'})}\\ &\gtrsim \|\Theta(\zeta_1,\zeta_2)\chi(\zeta_1\cdot\zeta_2)\mathcal{F}_\sigma[W(\varphi_{\sqrt{\lambda}},\varphi_{\sqrt{\lambda}})]\|_{W(\mathcal{F}L^{p'},L^{q'})}\end{aligned}$$

for any $\chi\in\mathcal{D}(\mathbb{R})$. Choosing χ supported in the interval $[-1/4,1/4]$ and equal to 1 in the interval $[-1/8,1/8]$, we write

$$\chi(\zeta_1\cdot\zeta_2)=\chi(\zeta_1\cdot\zeta_2)\Theta(\zeta_1,\zeta_2)\Theta^{-1}(\zeta_1,\zeta_2)\tilde{\chi}(\zeta_1\cdot\zeta_2),$$

with $\tilde{\chi}\in\mathcal{D}(\mathbb{R})$ supported in $[-1/2,1/2]$ and $\tilde{\chi}=1$ on $[-1/4,1/4]$, therefore on the support of χ. Since, by Lemma 4.7.9, the function $\Theta^{-1}(\zeta_1,\zeta_2)\tilde{\chi}(\zeta_1\cdot\zeta_2)$ belongs to $W(\mathcal{F}L^1,L^\infty)$, again by the product property the last expression is estimated from below as

$$\gtrsim \|\chi(\zeta_1\cdot\zeta_2)\mathcal{F}_\sigma[W(\varphi_{\sqrt{\lambda}},\varphi_{\sqrt{\lambda}})]\|_{W(\mathcal{F}L^{p'},L^{q'})}.$$

Consider a function $\psi\in\mathcal{D}(\mathbb{R}^d)\setminus\{0\}$, supported in $[-1/4,1/4]$. Using

$$|\zeta_1\cdot\zeta_2|\le\frac{1}{2}(|\sqrt{\lambda}\zeta_1|^2+|\sqrt{\lambda}^{-1}\zeta_2|^2),$$

we see that $\chi(\zeta_1\cdot\zeta_2)=1$ on the support of $\psi(\sqrt{\lambda}\zeta_1)\psi(\sqrt{\lambda}^{-1}\zeta_2)$, for every $\lambda>0$.

Then we can write

$$\psi(\sqrt{\lambda}\zeta_1)\psi(\sqrt{\lambda}^{-1}\zeta_2)=\chi(\zeta_1\cdot\zeta_2)\psi(\sqrt{\lambda}\zeta_1)\psi(\sqrt{\lambda}^{-1}\zeta_2)$$

and, by Lemma 4.7.11, also infer

$$\|\psi(\sqrt{\lambda}\zeta_1)\psi(\sqrt{\lambda}^{-1}\zeta_2)\|_{W(\mathcal{F}L^1,L^\infty)}\lesssim 1,$$

so that we can continue the above estimate as

$$\gtrsim \|\psi(\sqrt{\lambda}\zeta_1)\psi(\sqrt{\lambda}^{-1}\zeta_2)\mathcal{F}_\sigma[W(\varphi_{\sqrt{\lambda}},\varphi_{\sqrt{\lambda}})]\|_{W(\mathcal{F}L^{p'},L^{q'})}.$$

Using (see, e. g., [160, formula (4.20)])

$$W(\varphi_{\sqrt{\lambda}},\varphi_{\sqrt{\lambda}})(x,\omega)=2^{\frac{d}{2}}\lambda^{-\frac{d}{2}}\varphi(\sqrt{2\lambda}x)\varphi\Big(\sqrt{\frac{2}{\lambda}}\omega\Big), \tag{4.144}$$

we calculate

$$\mathcal{F}_\sigma[W(\varphi_{\sqrt{\lambda}},\varphi_{\sqrt{\lambda}})](\zeta_1,\zeta_2)=2^{\frac{d}{2}}\lambda^{-\frac{d}{2}}\varphi((\sqrt{2\lambda})^{-1}\zeta_2)\varphi\Big(\sqrt{\frac{\lambda}{2}}\zeta_1\Big),$$

so that

$$\begin{aligned}&\|\psi(\sqrt{\lambda}\zeta_1)\psi(\sqrt{\lambda}^{-1}\zeta_2)\mathcal{F}_\sigma[W(\varphi_{\sqrt{\lambda}},\varphi_{\sqrt{\lambda}})]\|_{W(\mathcal{F}L^{p'},L^{q'})}\\&\quad=2^{d/2}\lambda^{-\frac{d}{2}}\|\psi(\sqrt{\lambda}\zeta_1)\varphi((1/\sqrt{2})\sqrt{\lambda}\zeta_1)\|_{W(\mathcal{F}L^{p'},L^{q'})}\\&\quad\times\|\psi(\sqrt{\lambda}^{-1}\zeta_2)\varphi((\sqrt{2\lambda})^{-1}\zeta_2)\|_{W(\mathcal{F}L^{p'},L^{q'})}.\end{aligned}$$

By Lemma 4.7.11, we can estimate the last expression and obtain

$$\|Q(\varphi_{\sqrt{\lambda}},\varphi_{\sqrt{\lambda}})\|_{M^{p',q'}}\gtrsim\lambda^{-d+\frac{d}{2p'}+\frac{d}{2q'}}\quad\text{as }\lambda\to+\infty.$$

Finally, using the estimate (4.143), we infer (4.126).

We now prove that $\max\{1/r_1,1/r_1'\}\le 1/q+1/p$. If we show the estimate $1/r_1\le 1/q+1/p$, then the constraint $1/r_1'\le 1/q+1/p$ follows by the duality argument of Lemma 4.7.14. To reach this goal, we consider $f_\lambda=\varphi$ (independent of the parameter λ) and $g=\varphi_{\sqrt{\lambda}}$ as before and use the previous pattern for these families of functions, in the case $\lambda\to 0^+$. By (2.103), the upper estimate becomes

$$\|\varphi\|_{M^{r_1,r_2}}\|\varphi_{\sqrt{\lambda}}\|_{M^{r_1',r_2'}}\lesssim\lambda^{-\frac{d}{2r_1'}}.\tag{4.145}$$

The same arguments as before let us write

$$\|Q(\varphi,\varphi_{\sqrt{\lambda}})\|_{M^{p',q'}}\gtrsim\|\psi(\sqrt{\lambda}\zeta_1)\psi(\sqrt{\lambda}^{-1}\zeta_2)\mathcal{F}_\sigma[W(\varphi,\varphi_{\sqrt{\lambda}})]\|_{W(\mathcal{F}L^{p'},L^{q'})},$$

where $\mathcal{F}_\sigma[W(\varphi,\varphi_{\sqrt{\lambda}})]$ is computed in (1.131). Observe that, given any $F\in W(\mathcal{F}L^{p'},L^{q'})$,

$$\begin{aligned}\|e^{\pi i\frac{\lambda-1}{\lambda+1}\zeta_1\cdot\zeta_2}F(\zeta_1,\zeta_2)\|_{W(\mathcal{F}L^{p'},L^{q'})}&\gtrsim\|e^{-\pi i\frac{\lambda-1}{\lambda+1}\zeta_1\cdot\zeta_2}e^{\pi i\frac{\lambda-1}{\lambda+1}\zeta_1\cdot\zeta_2}F(\zeta_1,\zeta_2)\|_{W(\mathcal{F}L^{p'},L^{q'})}\\&=\|F(\zeta_1,\zeta_2)\|_{W(\mathcal{F}L^{p'},L^{q'})},\end{aligned}$$

because $\|e^{-\pi i\frac{\lambda-1}{\lambda+1}\zeta_1\cdot\zeta_2}\|_{W(\mathcal{F}L^1,L^\infty)}\le C$, for every $\lambda>0$ by Corollary 4.7.16. So writing

$$c_\lambda=\frac{1}{(\lambda+1)^{\frac{d}{2}}}$$

(notice $c_\lambda\to 1$ for $\lambda\to 0^+$), we have reduced the situation to

$$\|Q(\varphi,\varphi_{\sqrt{\lambda}})\|_{M^{p',q'}}\gtrsim c_\lambda\|\psi(\sqrt{\lambda}\zeta_1)e^{-\frac{\pi\lambda}{\lambda+1}\zeta_1^2}\|_{W(\mathcal{F}L^{p'},L^{q'})}\|\psi(\sqrt{\lambda}^{-1}\zeta_2)e^{-\frac{\pi}{\lambda+1}\zeta_2^2}\|_{W(\mathcal{F}L^{p'},L^{q'})}.$$

By Lemma 4.7.11, we can estimate, for $\lambda\to 0^+$,

$$\|\psi(\sqrt{\lambda}\zeta_1)e^{-\frac{\pi\lambda}{\lambda+1}\zeta_1^2}\|_{W(\mathcal{F}L^{p'},L^{q'})}=\|\psi(\sqrt{\lambda}\zeta_1)e^{-\frac{\pi}{\lambda+1}(\sqrt{\lambda}\zeta_1)^2}\|_{W(\mathcal{F}L^{p'},L^{q'})}\asymp\lambda^{-\frac{d}{2q'}},$$

whereas

$$
\begin{aligned}
&\|\psi(\sqrt{\lambda}^{-1}\zeta_2)e^{-\frac{\pi}{\lambda+1}\zeta_2^2}\|_{W(\mathcal{F}L^{p'},L^{q'})} \\
&\gtrsim \lambda^{\frac{d}{2}}(\lambda+1)^{\frac{d}{2}}\left\| \int \hat{\psi}(\sqrt{\lambda}(\zeta_2-\eta))e^{-\pi(\lambda+1)\eta^2}\,d\eta\right\|_{L^{p'}} \\
&= \lambda^{\frac{d}{2}}(\lambda+1)^{\frac{d}{2}}\lambda^{-\frac{d}{2p'}}\left\| \int \hat{\psi}(\zeta_2-\sqrt{\lambda}\eta)e^{-\pi(\lambda+1)\eta^2}\,d\eta\right\|_{L^{p'}} \\
&= (\lambda+1)^{\frac{d}{2}}\lambda^{-\frac{d}{2p'}}\left\| \int \hat{\psi}(\zeta_2-t)e^{-\pi\frac{\lambda+1}{\lambda}t^2}\,dt\right\|_{L^{p'}} \\
&= \lambda^{\frac{d}{2}-\frac{d}{2p'}}\|\hat{\psi} * K_{1/\sqrt{\lambda}}\|_{L^{p'}} \\
&\sim \lambda^{\frac{d}{2}-\frac{d}{2p'}}\|\hat{\psi}\|_{p'}, \quad \text{as } \lambda \to 0^+
\end{aligned}
$$

where $K_{1/\sqrt{\lambda}}(\zeta_2) = \lambda^{-\frac{d}{2}}(\lambda+1)^{\frac{d}{2}}e^{-\frac{\pi(\lambda+1)}{\lambda}\zeta_2^2}$, $\lambda \to 0^+$, is an approximate identity. This yields

$$\lambda^{-\frac{d}{2r_1'}} \gtrsim \lambda^{-\frac{d}{2q'}}\lambda^{\frac{d}{2p}},$$

and, for $\lambda \to 0^+$, we obtain

$$\frac{1}{r_1} \le \frac{1}{q} + \frac{1}{p},$$

as desired.

It remains to prove that $\max\{1/r_2, 1/r_2'\} \le 1/q + 1/p$. Again, it is enough to show that $1/r_2 \le 1/q + 1/p$ and invoke Lemma 4.7.14 for $1/r_2' \le 1/q + 1/p$.

An explicit computation shows that

$$\mathcal{F}^{-1}\mathrm{Op}_W(\sigma)\mathcal{F} = \mathrm{Op}_W(\sigma \circ J), \tag{4.146}$$

where we recall $J(x,\omega) = (\omega, -x)$ as defined in (1.1) (this is also a consequence of the intertwining property of the metaplectic operator $\mathcal{F}$ with the Weyl operator $\mathrm{Op}_W(\sigma)$ [105, Corollary 221]).

Now, observing that $\Theta_\sigma \circ J = \Theta_\sigma$, we obtain

$$
\begin{aligned}
(a * \Theta_\sigma)(Jz) &= \int_{\mathbb{R}^{2d}} a(u)\Theta_\sigma(Jz-u)\,du = \int_{\mathbb{R}^{2d}} a(u)\Theta_\sigma(J(z-J^{-1}u))\,du \\
&= \int_{\mathbb{R}^{2d}} a(u)\Theta_\sigma(z-J^{-1}u)\,du = \int_{\mathbb{R}^{2d}} a(Ju)\Theta_\sigma(z-u)\,du \\
&= (a \circ J) * \Theta_\sigma(z).
\end{aligned}
$$

The previous computations, together with (4.146), give

$$\mathcal{F}^{-1}\mathrm{Op}_{BJ}(a)\mathcal{F} = \mathrm{Op}_{BJ}(a \circ J).$$

On the other hand, the map $a \mapsto a \circ J$ is an isomorphism of $M^{p,q}$, so that (4.142) is, in fact, equivalent to

$$\|\mathrm{Op}_{BJ}(a)f\|_{W(\mathcal{F}L^{r_1},L^{r_2})} \lesssim \|a\|_{M^{p,q}}\|f\|_{W(\mathcal{F}L^{r_1},L^{r_2})} \quad \forall a \in \mathcal{S}(\mathbb{R}^{2d}),\ f \in \mathcal{S}(\mathbb{R}^d). \tag{4.147}$$

The estimate (4.147) can be written as

$$|\langle a, Q(f,g)\rangle| \le C\|a\|_{M^{p,q}}\|f\|_{W(\mathcal{F}L^{r_1},L^{r_2})}\|g\|_{W(\mathcal{F}L^{r_1'},L^{r_2'})} \quad \forall a \in \mathcal{S}(\mathbb{R}^{2d}),\ f,g \in \mathcal{S}(\mathbb{R}^d),$$

which is equivalent to

$$\|Q(f,g)\|_{M^{p',q'}} \le C\|f\|_{W(\mathcal{F}L^{r_1},L^{r_2})}\|g\|_{W(\mathcal{F}L^{r_1'},L^{r_2'})} \quad \forall f,g \in \mathcal{S}(\mathbb{R}^d).$$

Now, taking $f = \varphi$ and $g = \varphi_{\sqrt{\lambda}}$ as before, we observe that, for $\lambda \to 0^+$, by (2.103),

$$\|\varphi_{\sqrt{\lambda}}\|_{W(\mathcal{F}L^{r_1'},L^{r_2'})} \asymp \lambda^{-\frac{d}{2}}\|\varphi_{1/\sqrt{\lambda}}\|_{M^{r_1',r_2'}} \asymp \lambda^{-\frac{d}{2}+\frac{d}{2r_2}} = \lambda^{-\frac{d}{2r_2'}}.$$

Arguing as in the previous case, we obtain $1/r_2 \le 1/q + 1/p$. This concludes the proof. □

We end up this chapter by observing that the Born–Jordan distribution can be generalized as follows.

Definition 4.7.18. For $n \in \mathbb{N}$, the nth Born–Jordan kernel is the function Θ^n on $\mathbb{R}^{2d}$ defined by

$$\Theta^n(x,\omega) = \mathrm{sinc}^n(x \cdot \omega), \quad (x,\omega) \in \mathbb{R}^{2d}. \tag{4.148}$$

The Born–Jordan distribution of order n (BJDn) is given by

$$Q^n f = Wf * \mathcal{F}_\sigma(\Theta^n), \quad f \in L^2(\mathbb{R}^d). \tag{4.149}$$

The cross-BJDn is given by

$$Q^n(f,g) = W(f,g) * \mathcal{F}_\sigma(\Theta^n), \quad f,g \in L^2(\mathbb{R}^d). \tag{4.150}$$

We write $Q^n(f,f) = Q^n f$ for every $f \in L^2(\mathbb{R}^d)$.

Note that $\Theta^0 \equiv 1$, hence $\mathcal{F}_\sigma(\Theta^0) = \delta$ and $Q^0 f = Wf$, the Wigner distribution of f.

The nth Born–Jordan pseudodifferential operator with symbol $a \in \mathcal{S}'(\mathbb{R}^d)$ is defined by

$$\langle \mathrm{Op}_{BJ,n}(a)f, g\rangle = \langle a, Q^n(g,f)\rangle, \quad f,g \in \mathcal{S}(\mathbb{R}^d). \tag{4.151}$$

(Observe that $n = 1$ is the standard Born–Jordan operator, whereas $n = 0$ gives the Weyl operator.) For a complete study of such operators, we refer to [59].

5 Time–frequency analysis of constant-coefficient partial differential equations

Here we apply time–frequency analysis tools in the study of constant-coefficient partial differential equations (PDEs).

In the first part of the chapter we study the dispersive properties of the constant-coefficient Schrödinger equation, presenting results contained in [70, 71, 72, 91]. Precisely, we look for estimates which give a control on the local regularity and decay at infinity separately. The Banach spaces that allow such a treatment are the Wiener amalgam spaces, which improve the classical Strichartz estimates in the case of large time. Attention is limited to the free particle equation $i\partial_t u + \Delta u = 0$ with initial datum $u(0) = u_0$. We prove in the homogeneous case that

$$\|e^{it\Delta}u_0\|_{W(L^{q_1},L^{q_2})_t W(L^{r_1},L^{r_2})_x} \lesssim \|u_0\|_{L^2_x},$$

under conditions on q_1, q_2, r_1, r_2 which are sharp, as obtained by testing the problem on Gaussian initial data. In the last part of the chapter we show that Gabor frames provide optimally sparse decompositions for representing solutions to hyperbolic and parabolic-type differential equations with constant coefficients. In fact, the Gabor matrix representation of the corresponding propagator displays super-exponential decay away from the diagonal. To achieve this goal, we first perform a time–frequency analysis of Fourier multipliers and, more generally, pseudodifferential operators with symbols of Gevrey, analytic and ultraanalytic type. These ideas stem from [83].

Let us remark a few aspects we shall use below and in the following chapter. Given a linear continuous operator $T : \mathcal{S}(\mathbb{R}^d) \to \mathcal{S}'(\mathbb{R}^d)$, providing the solution of a well-posed problem for a PDE, we expect T to be a pseudodifferential or a Fourier integral operator. For some problems, the space $\mathcal{S}(\mathbb{R}^d)$ can be replaced by other spaces of test functions, and $\mathcal{S}'(\mathbb{R}^d)$ by the respective duals. In any case we can compute the Gabor matrix of T. Consider a window function $g \in L^2(\mathbb{R}^d)$, a lattice $\Lambda = \alpha\mathbb{Z}^d \times \beta\mathbb{Z}^d$, with $\alpha, \beta > 0$ such that $\mathcal{G}(g, \alpha, \beta)$ (defined in (3.18)) is a frame for $L^2(\mathbb{R}^d)$ with dual window $\gamma \in L^2(\mathbb{R}^d)$. For simplicity, define the related Gabor atoms

$$g_{m,n} := M_{\beta n} T_{\alpha m} g, \quad (m, n) \in \Lambda, \tag{5.1}$$

and similarly $\gamma_{m,n}$. Any signal $f \in L^2(\mathbb{R}^d)$ admits the representations

$$f = \sum_{(m,n)\in\Lambda} \langle f, g_{m,n}\rangle \gamma_{m,n} = \sum_{(m,n)\in\Lambda} \langle f, \gamma_{m,n}\rangle g_{m,n}. \tag{5.2}$$

The analysis extends to f in $\mathcal{S}(\mathbb{R}^d)$, $\mathcal{S}'(\mathbb{R}^d)$ or other distributional setting, and we can choose its representation $f = \sum_{(m,n)\in\Lambda} \langle f, \gamma_{m,n}\rangle g_{m,n}$. Applying to f the operator T, we

https://doi.org/10.1515/9783110532456-006

can decompose Tf as

$$\begin{aligned} Tf(x) &= \sum_{(m',n')\in\Lambda} \langle Tf, g_{m',n'}\rangle \gamma_{m',n'} \\ &= \sum_{(m',n')\in\Lambda} \sum_{(m,n)\in\Lambda} \underbrace{\langle Tg_{m,n}, g_{m',n'}\rangle}_{T_{m',n',m,n}} c_{m,n}\gamma_{m',n'}, \end{aligned}$$

where $c_{m,n} = \langle f, \gamma_{m,n}\rangle$. The infinite Gabor matrix $T_{m',n',m,n}$ is defined by

$$T_{m',n',m,n} = \langle Tg_{m,n}, g_{m',n'}\rangle, \quad (m,n),(m',n') \in \Lambda \tag{5.3}$$

(by abuse of notation, we identify the matrix with its general entry) and the properties of T can be equivalently studied via those of the matrix above. We remark that this way of building the Gabor matrix involves only the function g. The dual window γ does not play any role. This has the advantage of working only with the Gabor frame $\{g_{m,n}\}_{(m,n)\in\Lambda}$, no information about its dual is needed. Of course, one can build up the Gabor matrix of an operator T considering both frames, as was already done in Chapter 3, see the Gabor matrix in (3.37). In what follows we will limit ourselves to the Gabor matrix defined in (5.3).

Hyperbolic problems were treated by many authors using frames. In particular, curvelet and shearlet representations of wave propagators were proved to be optimally sparse, cf. [41, 43, 173, 174]. Very recently new issues in shearlet theory have been shown in [1, 2, 5].

Our concern is limited to constant coefficient hyperbolic and evolution-type equations. The reason is simple: the example of the transport equation with nonconstant coefficients, standard test for numerical schemes, is somewhat discouraging. Let us denote by T the operator providing the solution of the related Cauchy problem. For a fixed time $t > 0$, the operator T turns out to be a change of variables, and Gabor frames are not suited to follow the corresponding nonlinear propagation of singularities. Heuristically this can be seen as follows. Consider a window function g such that the related Gabor system $\mathcal{G}(g, \alpha, \beta)$ is a frame. Decomposing a function or distribution f in terms of Gabor atoms $g_{m,n}$, that is,

$$f = \sum_{m,n\in\mathbb{Z}} c_{m,n}(f) g_{m,n},$$

corresponds geometrically to a uniform partition of the time–frequency plane (or phase plane, in PDEs terminology) into boxes, each atom occupying a box, loosely speaking; see Figure 5.1.

The correspondence principle in quantum mechanics suggests following the PDE's evolution in terms of its classical analog, namely the Hamiltonian flow in the phase plane. The propagator has therefore a sparse Gabor matrix if its Hamiltonian flow moves the above mentioned boxes, but introduces *only a controlled number of overlaps*. However, this is not the case for changes of variables, as one sees easily by

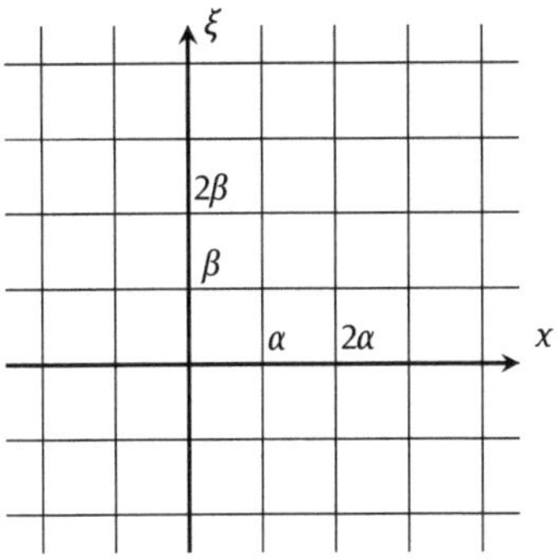

Figure 5.1: Decomposition of the time–frequency plane in uniform boxes.

direct inspection. More rigorously, it follows from the results in [78] that Hörmander's type Fourier integral operators, in particular, the changes of variables, do not have a sparse Gabor representation.

Because of this, the study of hyperbolic equations via Gabor frames will be limited to operators with constant coefficients. With respect to [41, 43], such results are stronger in the following aspects:

- Gabor frames allow treating hyperbolic equations, not necessarily strictly hyperbolic, of any order and in an arbitrary number of space variables;
- Gabor frames allow also considering parabolic equations;
- The decay of the Gabor matrix corresponding to the evolution operator is of superexponential type.

As a simple example, let us consider the Cauchy problem for the wave equation:

$$\begin{aligned} &\partial_t^2 u - \Delta_x u = 0, \quad (t,x) \in \mathbb{R} \times \mathbb{R}^d, \\ &u(0,x) = u_0(x), \quad \partial_t u(0,x) = u_1(x). \end{aligned} \tag{5.4}$$

We may express the solution in the form

$$u(t,x) = T_t u_1(x) + \partial_t T_t u_0(x), \tag{5.5}$$

where T_t is the Fourier multiplier

$$T_t f(x) = \int e^{2\pi i x\cdot\xi} \sigma_t(\xi) \widehat{f}(\xi)\, d\xi, \tag{5.6}$$

with symbol

$$\sigma_t(\xi) = \frac{\sin(2\pi|\xi|t)}{2\pi|\xi|}, \quad \xi \in \mathbb{R}^d. \tag{5.7}$$

The operator $T = T_t$ can be viewed in terms of its infinite Gabor matrix $T_{m',n',m,n}$ in (5.3).

Taking the Gaussian window function $g(x) = e^{-\pi x^2}$ and discretizing consequently T_t, we shall prove that the Gabor matrix satisfies

$$|T_{m',n',m,n}| \leq Ce^{-\epsilon(|m-m'|^2+|n-n'|^2)}. \tag{5.8}$$

We address to the sequel of the chapter for the dependence of C and ϵ on $t > 0$.

Let us describe in short how (5.8) can be obtained in the general context of symbols depending on $z = (x, \xi)$. The proof will be largely based on the properties of Gelfand–Shilov spaces introduced below. For $s \geq 0$, assume that the symbol σ satisfies

$$|\partial^\alpha \sigma(z)| \leq C^{|\alpha|+1}(\alpha!)^s, \quad z \in \mathbb{R}^{2d}, \tag{5.9}$$

for a constant $C > 0$ depending on σ. We say that the symbol σ is Gevrey when $s > 1$, analytic when $s = 1$, and ultraanalytic if $s < 1$. Let us observe that, when $s = 1$, σ extends to an analytic function in the strip $\{z + i\zeta \in \mathbb{C}^{2d} : |\zeta| < 1/C\}$, whereas when $s < 1$, σ extents to an entire function $\sigma(z + i\zeta)$ satisfying $|\sigma(z + i\zeta)| \leq \tilde{C}\exp(\delta|\zeta|^{1/(1-s)})$, for some constants $\tilde{C}, \delta > 0$ (see, e. g., [236, Section 6.1]).

As we will show, such estimates are satisfied by the symbols of the fundamental solutions for a large class of evolution operators with constant coefficients, with $0 \leq s < 1$. In particular, we have $s = 0$ for the wave equation (5.4).

Let T be a pseudodifferential operator with symbol σ satisfying (5.9) for some $s \geq 0$, and define $r = \min\{2, 1/s\}$. Fix as a window the Gaussian, or more generally any function g belonging to the Gelfand–Shilov classes $S_r^r(\mathbb{R}^d)$. Then we shall prove that

$$|T_{m',n',m,n}| \leq Ce^{-\epsilon(|m-m'|+|n-n'|)^r}. \tag{5.10}$$

Accordingly, every column or row of the Gabor matrix, rearranged in decreasing order, displays a similar decay, i. e., we obtain an *exponential-type sparsity*.

The contents of the chapter are organized in the following way. Section 5.1 is devoted to Strichartz estimates, in particular, the proof of the main result is given in Sections 5.1.1 and 5.1.2, the necessity of the conditions is proved in Section 5.1.3. Section 5.2 consists of six subsections. Namely, in Section 5.2.1 we recall the basics of Gelfand–Shilov spaces, in Section 5.2.2 we treat almost diagonalization of pseudodifferential operators in this setting, i. e., estimates (5.10) for operators with symbols satisfying (5.9). Sparsity of the corresponding Gabor matrix is proved in Section 5.2.3. Section 5.2.4 presents some applications to the boundedness of pseudodifferential operators on Gelfand–Shilov spaces. Finally, Sections 5.2.5 and 5.2.6 concern the applications to evolution partial differential equations, with numerics for wave equation, as mentioned before, and heat equation.

5.1 Strichartz estimates for the Schrödinger equation in Wiener amalgam spaces

Consider the Cauchy problem for the homogeneous Schrödinger equation (free particle)

$$\begin{cases} i\partial_t u + \Delta u = 0, \\ u(0,x) = u_0(x), \end{cases} \tag{5.11}$$

with $(t,x) \in \mathbb{R} \times \mathbb{R}^d$, $d \geq 1$. Applying the Fourier transform with respect to the space variable x, one easily sees that the solution $u(t,x)$ to the Cauchy problem in (5.11) can be formally written as

$$u(t,x) = \int_{\mathbb{R}^d} e^{2\pi i x\cdot\xi} e^{-4\pi^2 it\xi^2} \widehat{u_0}(\xi)\, d\xi,$$

that is, using the Fourier multiplier $e^{it\Delta}$ with symbol $a_t(\xi) = e^{-4\pi^2 it\xi^2}$,

$$u(t,x) = (e^{it\Delta}u_0)(x), \quad \text{for every fixed } t \in \mathbb{R}. \tag{5.12}$$

The study of space-time integrability properties for the solution of (5.11) has been pursued by many authors in the last 40 years. We recall the important fixed-time estimates

$$\|e^{it\Delta}u_0\|_{L^r(\mathbb{R}^d)} \lesssim |t|^{-d(\frac{1}{2}-\frac{1}{r})}\|u_0\|_{L^{r'}(\mathbb{R}^d)}, \quad 2 \leq r \leq \infty, \tag{5.13}$$

as well as the celebrated homogeneous Strichartz estimates [157, 206, 208, 322]:

$$\|e^{it\Delta}u_0\|_{L^q_t L^r_x} \lesssim \|u_0\|_{L^2_x}, \tag{5.14}$$

for $q \geq 2$, $r \geq 2$, with $2/q + d/r = d/2$, $(q,r,d) \neq (2,\infty,2)$, i. e., for (q,r) *Schrödinger admissible*. Here we set

$$\|F\|_{L^q_t L^r_x} = \left(\int \|F(t,\cdot)\|^q_{L^r_x}\, dt\right)^{1/q}.$$

These estimates express a gain of local x-regularity of the solution $u(t,\cdot)$, and a decay of its L^r_x-norm, both in some L^q_t-averaged sense. Recently several authors have turned their attention to fixed-time and space-time estimates of the Schrödinger propagator, see [306, 14, 15, 264, 303, 304, 305].

In what follows we study similar estimates in Wiener amalgam spaces. One of the main motivations for considering estimates in these spaces is that they control the local regularity of a function and its decay at infinity separately. Hence, they highlight and distinguish between the local properties and the global behavior of the solution

$u(t,x) = e^{it\Delta}u_0$ and therefore, as far as Strichartz estimates are concerned, they are natural candidates to perform an analysis of the solution which is finer than the classical one. Actually, Wiener amalgam spaces have already appeared as a technical tool in PDEs. In particular, the space $W(L^p, L^q)$ coincides with the space $X_0^{q,p}$ introduced in [276].

For comparison, the classical estimates (5.14) can be rephrased in terms of Wiener amalgam spaces as follows:

$$\|e^{it\Delta}u_0\|_{W(L^q,L^q)_t W(L^r,L^r)_x} \lesssim \|u_0\|_{L^2_x}. \tag{5.15}$$

Let us start with a brief discussion of the results. As usual we first establish an estimate of dispersive type. Namely, in our setting, we show that it is possible to move local regularity in (5.14) from the space variable to the time variable. As a result, we obtain new estimates involving the Wiener amalgam spaces $W(L^p, L^q)$ that generalize (5.14). This requires some preliminary steps.

First, we show the fixed-time estimates

$$\|e^{it\Delta}u_0\|_{W(L^2,L^r)} \lesssim (1+t^2)^{-\frac{d}{2}(\frac{1}{2}-\frac{1}{r})}\|u_0\|_{W(L^2,L^{r'})}, \quad 2 \le r \le \infty.$$

The related Strichartz estimates are achieved in Theorem 5.1.8 below. Finally, the complex interpolation with the classical estimates (5.14) yields our main result:

Theorem 5.1.1. *Let $1 \le q_1, r_1 \le \infty$, $2 \le q_2, r_2 \le \infty$ be such that $r_1 \le r_2$,*

$$\frac{2}{q_1} + \frac{d}{r_1} \ge \frac{d}{2}, \tag{5.16}$$

$$\frac{2}{q_2} + \frac{d}{r_2} \le \frac{d}{2}, \tag{5.17}$$

$(r_1, d) \ne (\infty, 2)$, $(r_2, d) \ne (\infty, 2)$, and, if $d \ge 3$, $r_1 \le 2d/(d-2)$. The same for $\tilde{q}_1, \tilde{q}_2, \tilde{r}_1, \tilde{r}_2$. Then, we have the homogeneous Strichartz estimates

$$\|e^{it\Delta}u_0\|_{W(L^{q_1},L^{q_2})_t W(L^{r_1},L^{r_2})_x} \lesssim \|u_0\|_{L^2_x}, \tag{5.18}$$

the dual homogeneous Strichartz estimates

$$\left\|\int e^{-is\Delta}F(s)\,ds\right\|_{L^2} \lesssim \|F\|_{W(L^{\tilde{q}_1'},L^{\tilde{q}_2'})_t W(L^{\tilde{r}_1'},L^{\tilde{r}_2'})_x}, \tag{5.19}$$

and the retarded Strichartz estimates

$$\left\|\int_{s<t} e^{i(t-s)\Delta}F(s)\,ds\right\|_{W(L^{q_1},L^{q_2})_t W(L^{r_1},L^{r_2})_x} \lesssim \|F\|_{W(L^{\tilde{q}_1'},L^{\tilde{q}_2'})_t W(L^{\tilde{r}_1'},L^{\tilde{r}_2'})_x}. \tag{5.20}$$

Figure 5.2 illustrates the range of exponents for the homogeneous estimates when $d \geq 3$. Notice that, if $q_1 \leq q_2$, these estimates follow immediately from (5.15) and the inclusion relations of Wiener amalgam spaces (cf. Theorem 2.4.9). So the issue consists in the case when $q_1 > q_2$.

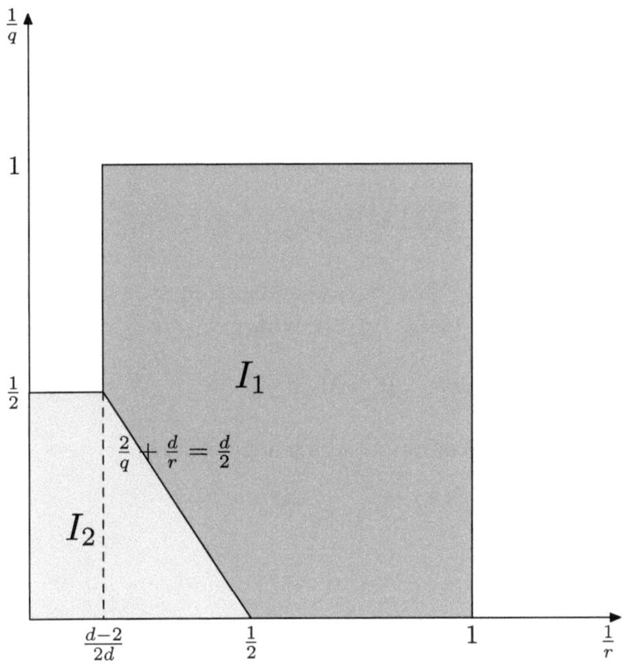

Figure 5.2: When $d \geq 3$, (5.18) holds for all pairs $(1/q_1, 1/r_1) \in I_1$, $(1/q_2, 1/r_2) \in I_2$, with $1/r_2 \leq 1/r_1$.

Since there are no relations between the pairs (q_1, r_1) and (q_2, r_2) other than $r_1 \leq r_2$, these estimates tell us that the analysis of the local regularity of the Schrödinger propagator is quite independent of its decay at infinity.

We will then discuss the sharpness of the results above. Let us start with the fixed-time estimates.

5.1.1 Fixed-time estimates

In this section we study estimates for the solution $u(t, x)$ to the Cauchy problem (5.11), for fixed t. We take advantage of the explicit formula for the solution

$$u(t, x) = (K_t * u_0)(x), \tag{5.21}$$

where

$$K_t(x) = \frac{1}{(4\pi i t)^{d/2}} e^{\frac{ix^2}{4t}}. \tag{5.22}$$

Precisely, we will first show that the function in (5.22) is in the Wiener amalgam space $W(\mathcal{F}L^1, L^\infty)$ and we shall compute its norm. This goal is attained thanks to the nice choice of the Gaussian function $e^{-\pi x^2} \in \mathcal{S}(\mathbb{R}^d)$, as a fitting window function.

Proposition 5.1.2. *For $a \in \mathbb{R}$, $a \neq 0$, let $f_a(x) = (ai)^{-d/2}e^{-\pi x^2/(ai)}$. Then $f_a \in W(\mathcal{F}L^1, L^\infty)$, with*

$$\|f_a\|_{W(\mathcal{F}L^1,L^\infty)} = \left(\frac{1+a^2}{a^4}\right)^{\frac{d}{4}}. \tag{5.23}$$

Proof. By definition,

$$\|f_a\|_{W(\mathcal{F}L^1,L^\infty)} = \sup_{x\in\mathbb{R}^d} \|f_a T_x g\|_{\mathcal{F}L^1},$$

for some nonzero window $g \in W(\mathcal{F}L^1, L^1)$ (different windows give equivalent norms). We then choose $g = e^{-\pi x^2}$, and, observing that $\hat{g} = g$, can write

$$\|f_a T_x g\|_{\mathcal{F}L^1} = \|\widehat{f_a T_x g}\|_{L^1} = \|\hat{f}_a * M_{-x} g\|_{L^1}.$$

Using (0.28), we compute the Fourier transform of f which happens to be

$$\hat{f}_a(\xi) = (ai)^{-d/2}(ai)^{d/2}e^{-\pi ai\xi^2} = e^{-\pi ai\xi^2}.$$

Thereby,

$$\begin{aligned}
\|\hat{f}_a * M_{-x} g\|_{L^1} &= \int_{\mathbb{R}^d}\left|\int_{\mathbb{R}^d} e^{-\pi ai(\xi-y)^2}e^{-2\pi ix\cdot y}e^{-\pi y^2}\,dy\right|d\xi \\
&= \int_{\mathbb{R}^d}\left|e^{-\pi ai\xi^2}\int_{\mathbb{R}^d} e^{-2\pi i(x-a\xi)\cdot y}e^{-\pi(1+ai)y^2}\,dy\right|d\xi \\
&= \int_{\mathbb{R}^d}\left|\mathcal{F}(e^{-\pi(1+ai)y^2})(x-a\xi)\right|d\xi \\
&= \int_{\mathbb{R}^d}\left|(1+ai)^{-d/2}e^{-\pi(x-a\xi)^2/(1+ai)}\right|d\xi,
\end{aligned}$$

where in the last equality we use (0.28). Performing the change of variables $x - a\xi = z$, hence $d\xi = a^{-d}dz$ and observing that $|(1+ai)^{-d/2}| = (1+a^2)^{-d/4}$, we can write

$$\begin{aligned}
\|\hat{f}_a * M_{-x} g\|_{L^1} &= (1+a^2)^{-d/4}a^{-d}\int_{\mathbb{R}^d}\left|e^{-\pi z^2/(1+ai)}\,dz\right| \\
&= (1+a^2)^{-d/4}a^{-d}\int_{\mathbb{R}^d} e^{-\pi z^2/(1+a^2)}\,dz
\end{aligned}$$

$$= (1+a^2)^{-d/4} a^{-d} (1+a^2)^{d/2}$$

$$= \left(\frac{1+a^2}{a^4}\right)^{d/4}.$$

Since the right-hand side does not depend on x, taking the supremum on $\mathbb{R}^d$ with respect to the x-variable, we attain the desired estimate. □

The space $W(\mathcal{F}L^1, L^\infty)$ is the finest Wiener amalgam space-norm for (5.22) which, consequently, gives the worst behavior in the time variable. We aim at improving the latter, at the expense of a rougher x-norm. This is achieved in the following result. Indeed, $W(\mathcal{F}L^1, L^\infty) \subset W(\mathcal{F}L^p, L^\infty)$ for $1 \le p \le \infty$ by the inclusion relations of these spaces (cf. Theorem 2.4.9).

Proposition 5.1.3. *For $a \in \mathbb{R}$, $a \neq 0$, let $f_{ai}(x) = (ai)^{-d/2} e^{-\pi x^2/(ai)}$. Then, for $1 \le p \le \infty$, $f_{ai} \in W(\mathcal{F}L^p, L^\infty)$ and*

$$\|f_{ai}\|_{W(\mathcal{F}L^p, L^\infty)} \asymp |a|^{-d/p} (1+a^2)^{(d/2)(1/p-1/2)}. \tag{5.24}$$

Proof. It follows the footsteps of Proposition 5.1.2. We consider the Gaussian $g(t) = e^{-\pi t^2}$ as a window function to compute

$$\|f_{ai}\|_{W(\mathcal{F}L^p, L^\infty)} \asymp \sup_{x \in \mathbb{R}^d} \|\hat{f_{ai}} * M_{-x} g\|_{L^p}.$$

Using $\widehat{f_{ai}}(\xi) = e^{-\pi a i \xi^2}$, we have, for $p < \infty$,

$$\begin{aligned}
\|\hat{f_a} * M_{-x} g\|_{L^p} &= \left(\int_{\mathbb{R}^d} \left| \int_{\mathbb{R}^d} e^{-\pi a i (\xi - y)^2} e^{-2\pi i x \cdot y} e^{-\pi^2 y^2} \, dy \right|^p d\xi \right)^{1/p} \\
&= (1+a^2)^{-d/4} |a|^{-d/p} \left(\int_{\mathbb{R}^d} e^{-\pi p z^2/(1+a^2)} \, dz \right)^{1/p} \\
&= (1+a^2)^{(d/2)(1/p-1/2)} |a|^{-d/p} p^{-d/(2p)}.
\end{aligned}$$

Since the right-hand side does not depend on x, taking the supremum on $\mathbb{R}^d$ with respect to the x-variable we attain the desired estimate.

If $p = \infty$,

$$\|f_a\|_{W(\mathcal{F}L^\infty, L^\infty)} \asymp \sup_{\xi \in \mathbb{R}^d} \left| (1+ai)^{-d/2} e^{-\pi (x - a\xi)^2/(1+ai)} \right| = (1+a^2)^{-d/4},$$

and we are done. □

For $a = 4\pi t$, $t \neq 0$, we infer:

Corollary 5.1.4. *Let $K_t(x)$ be the kernel in (5.22). Then, if $1 \le p \le \infty$,*

$$\|K_t\|_{W(\mathcal{F}L^p,L^\infty)} \asymp |t|^{-d/p}(1+t^2)^{(d/2)(1/p-1/2)}. \tag{5.25}$$

Lemma 5.1.5. *Let $1 \le p,q,r \le \infty$, with $1/r = 1/p + 1/q$, then*

$$W(\mathcal{F}L^p, L^\infty) * W(\mathcal{F}L^q, L^1) \hookrightarrow W(\mathcal{F}L^r, L^\infty). \tag{5.26}$$

Proof. This is a consequence of the convolution relations for Wiener amalgam spaces (cf. Theorem 2.4.9), namely $\mathcal{F}L^p * \mathcal{F}L^q = \mathcal{F}(L^p \cdot L^q) \hookrightarrow \mathcal{F}L^r$ by Hölder's inequality with $1/r = 1/p + 1/q$, and $L^\infty * L^1 \hookrightarrow L^\infty$. □

Proposition 5.1.6. *It turns out, for $2 \le q \le \infty$, that*

$$\|e^{it\Delta}u_0\|_{W(\mathcal{F}L^{q'},L^\infty)} \lesssim |t|^{d(2/q-1)}(1+t^2)^{d(1/4-1/q)}\|u_0\|_{W(\mathcal{F}L^q,L^1)}. \tag{5.27}$$

Proof. We use the explicit representation of the Schrödinger evolution operator $e^{it\Delta}u_0(x) = (K_t * u_0)(x)$. Let $1 \le p,q \le \infty$ satisfy

$$\frac{1}{p} + \frac{2}{q} = 1. \tag{5.28}$$

Then the kernel norm (5.25) and the convolution relations (5.26) yield the desired result. □

Theorem 5.1.7. *For $2 \le q,r,s \le \infty$ such that*

$$\frac{1}{s} = \frac{1}{r} + \frac{2}{q}\left(\frac{1}{2} - \frac{1}{r}\right), \tag{5.29}$$

we have

$$\|e^{it\Delta}u_0\|_{W(\mathcal{F}L^{s'},L^r)} \lesssim |t|^{d(\frac{2}{q}-1)(1-\frac{2}{r})}(1+t^2)^{d(\frac{1}{4}-\frac{1}{q})(1-\frac{2}{r})}\|u_0\|_{W(\mathcal{F}L^s,L^{r'})}.$$

In particular, for $2 \le r \le \infty$,

$$\|e^{it\Delta}u_0\|_{W(L^2,L^r)} \lesssim (1+t^2)^{-\frac{d}{2}(\frac{1}{2}-\frac{1}{r})}\|u_0\|_{W(L^2,L^{r'})}. \tag{5.30}$$

Proof. Estimates (5.30) follow by complex interpolation of (5.27), which corresponds to $r = \infty$, with the L^2 conservation law

$$\|e^{it\Delta}u_0\|_{L^2} = \|u_0\|_{L^2}, \tag{5.31}$$

which corresponds to $r = 2$.

Indeed, $L^2 = W(\mathcal{F}L^2, L^2) = W(L^2, L^2)$. By complex interpolation for Wiener amalgam spaces (2.64), for $0 < \theta = 2/r < 1$, and $1/s' = (1-2/r)/q' + 1/r$, so that relation (5.29) holds,

$$[W(\mathcal{F}L^{q'}, L^\infty), W(\mathcal{F}L^2, L^2)]_{[\theta]} = W([\mathcal{F}L^{q'}, \mathcal{F}L^2]_{[\theta]}, [L^\infty, L^2]_{[\theta]}) = W(\mathcal{F}L^{s'}, L^r)$$

and

$$[W(\mathcal{F}L^q, L^1), W(\mathcal{F}L^2, L^2)]_{[\theta]} = W([\mathcal{F}L^q, \mathcal{F}L^2]_{[\theta]}, [L^1, L^2]_{[\theta]}) = W(\mathcal{F}L^s, L^{r'}),$$

so that the estimate (5.30) is attained. □

5.1.2 Strichartz estimates

In this section we prove Theorem 5.1.1. Precisely, we first study the case $q_1 = \tilde{q}_1 = \infty$, $r_1 = \tilde{r}_1 = 2$. In view of the inclusion relation of Wiener amalgam spaces, it suffices to prove it for the pairs (q_2, r_2) which are scale invariant (i. e., satisfying (5.17) with equality), namely we have the following result.

Theorem 5.1.8. *Let $2 \le q \le \infty$, $2 \le r \le \infty$, be such that*

$$\frac{2}{q} + \frac{d}{r} = \frac{d}{2},$$

$(q, r, d) \neq (2, \infty, 2)$, *and similarly for $\tilde{q}, \tilde{r}$. Then we have the homogeneous Strichartz estimates*

$$\|e^{it\Delta} u_0\|_{W(L^\infty, L^q)_t W(L^2, L^r)_x} \lesssim \|u_0\|_{L^2_x}, \tag{5.32}$$

the dual homogeneous Strichartz estimates

$$\left\| \int e^{-is\Delta} F(s)\, ds \right\|_{L^2} \lesssim \|F\|_{W(L^1, L^{\tilde{q}'})_t W(L^2, L^{\tilde{r}'})_x}, \tag{5.33}$$

and the retarded Strichartz estimates

$$\left\| \int_{s<t} e^{i(t-s)\Delta} F(s)\, ds \right\|_{W(L^\infty, L^q)_t W(L^2, L^r)_x} \lesssim \|F\|_{W(L^1, L^{\tilde{q}'})_t W(L^2, L^{\tilde{r}'})_x}. \tag{5.34}$$

Proof in the nonendpoint case. Here we prove Theorem 5.1.8 in the nonendpoint case, namely for $q > 2, \tilde{q} > 2$. We first show the estimate (5.32). The case $q = \infty$, $r = 2$, follows at once from the conservation law (5.31). Indeed, $W(L^\infty, L^\infty)_t = L^\infty_t$ and $W(L^2, L^2)_x = L^2_x$, so that, taking the supremum over t in $\|e^{it\Delta} u_0\|_{L^2_x} = \|u_0\|_{L^2_x}$, we obtain the claim.

To prove the remaining cases, we can apply the usual TT^* method (or "orthogonality principle", see [157, Lemma 2.1] or [268, page 353]), because of Hölder-type inequality

$$|\langle F, G\rangle_{L^2_t L^2_x}| \le \|F\|_{W(L^\infty, L^q)_t W(L^2, L^r)_x} \|G\|_{W(L^1, L^{q'})_t W(L^2, L^{r'})_x}, \tag{5.35}$$

which can be proved directly from the definition of these spaces.

As a consequence, it suffices to prove the estimate

$$\left\| \int e^{i(t-s)\Delta} F(s)\, ds \right\|_{W(L^\infty, L^q)_t W(L^2, L^r)_x} \lesssim \|F\|_{W(L^1, L^{q'})_t W(L^2, L^{r'})_x}. \tag{5.36}$$

Recall the Hardy–Littlewood–Sobolev fractional integration theorem (see, e. g., [268], page 119) in dimension 1:

$$L^p(\mathbb{R}) * L^{1/\alpha, \infty}(\mathbb{R}) \hookrightarrow L^q(\mathbb{R}), \tag{5.37}$$

for $1 \le p < q < \infty$, $0 < \alpha < 1$, with $1/p = 1/q + 1 - \alpha$ (here $L^{1/\alpha,\infty}$ is the weak $L^{1/\alpha}$ space, see, e. g., [270]). Now, set $\alpha = d(1/2 - 1/r) = 2/q$ $(q > 2)$ so that, for $p = q'$, $L^{q'} *$

$L^{1/\alpha,\infty} \hookrightarrow L^q$. Moreover, observe that $(1+|t|)^{-\alpha} \in W(L^\infty, L^{1/\alpha,\infty})(\mathbb{R})$. The convolution relations for Wiener amalgam spaces in Theorem 2.4.9 then give

$$W(L^1, L^{q'})(\mathbb{R}) * W(L^\infty, L^{1/\alpha,\infty})(\mathbb{R}) \hookrightarrow W(L^\infty, L^q)(\mathbb{R}).$$

The preceding relations, together with the fixed-time estimates (5.30) and Minkowski's inequality, allow writing

$$\begin{aligned}
&\left\| \int e^{i(t-s)\Delta} F(s)\, ds \right\|_{W(L^\infty,L^q)_t, W(L^2,L^r)_x} \\
&\leq \left\| \int \|e^{i(t-s)\Delta} F(s) v\|_{W(L^2,L^r)_x}\, ds \right\|_{W(L^\infty,L^q)_t} \\
&\lesssim \left\| \int \|F(s)\|_{W(L^2,L^{r'})_x} (1+|t-s|)^{-\alpha} ds \right\|_{W(L^\infty,L^q)_t} \\
&\lesssim \|F\|_{W(L^1,L^{q'})_t, W(L^2,L^{r'})_x}.
\end{aligned}$$

The estimate (5.33) follows from (5.32) by duality.

Consider now the retarded Strichartz estimate (5.34). By complex interpolation, in order to get (5.34) with $(1/q, 1/r)$, $(1/\tilde{q}, 1/\tilde{r})$, $(1/\infty, 1/2)$ collinear, it suffices to prove (5.34) in the three cases $(\tilde{q}, \tilde{r}) = (q, r)$, $(q, r) = (\infty, 2)$, $(\tilde{q}, \tilde{r}) = (\infty, 2)$, as shown in Figure 5.3.

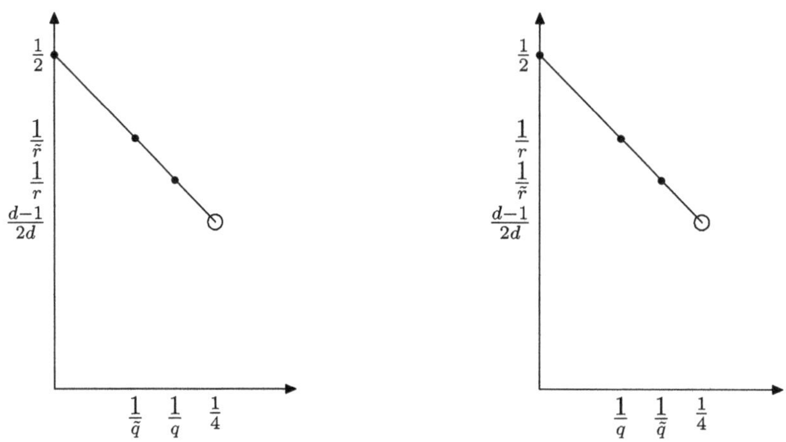

Figure 5.3: The complex interpolation method is applied in these two cases.

Now, the case $(\tilde{q}, \tilde{r}) = (q, r)$ follows from (5.36) with $\chi_{\{s<t\}} F$ in place of F.

The case $(q, r) = (\infty, 2)$ follows because

$$\left\| \int_{s<t} e^{i(t-s)\Delta} F(s)\, ds \right\|_{L^2_x} = \left\| \int_{s<t} e^{-is\Delta} F(s)\, ds \right\|_{L^2_x}$$
$$\lesssim \|F\|_{W(L^{(\tilde{q}/2)'}, L^{\tilde{q}'})_t, W(\mathcal{F}L^{\tilde{r}}, L^{\tilde{r}'})_x},$$

where we applied (5.31) and then (5.33) with $\chi_{\{s<t\}} F$ in place of F.

Finally, this latter argument applied to the adjoint operator

$$G \mapsto \int_{t>s} e^{-i(t-s)\Delta} G(t)\, dt$$

gives the case $(\tilde{q}, \tilde{r}) = (\infty, 2)$. □

Proof in the endpoint case. We are going to prove Theorem 5.1.8 in the endpoint case $(q, r) = P := (2, 2d/(d-2))$ or $(\tilde{q}, \tilde{r}) = P$, $d > 2$. That is, we prove the estimates

$$\|e^{it\Delta} u_0\|_{W(L^\infty, L^2)_t W(L^2, L^r)_x} \lesssim \|u_0\|_{L^2_x}, \quad r = 2d/(d-2), \tag{5.38}$$

$$\left\| \int e^{-is\Delta} F(s)\, ds \right\|_{L^2} \lesssim \|F\|_{W(L^1, L^{\tilde{q}'})_t W(L^2, L^{\tilde{r}'})_x}, \quad \tilde{r} = 2d/(d-2), \tag{5.39}$$

and

$$\left\| \int_{s<t} e^{i(t-s)\Delta} F(s)\, ds \right\|_{W(L^\infty, L^q)_t W(L^2, L^r)_x} \lesssim \|F\|_{W(L^1, L^{\tilde{q}'})_t W(L^2, L^{\tilde{r}'})_x}, \tag{5.40}$$

with (q, r), $(\tilde{q}, \tilde{r})$ Schrödinger admissible, and $(q, r) = P$ or $(\tilde{q}, \tilde{r}) = P$.

We study bilinear form estimates via a Whitney decomposition in time (see (5.42)). We estimate each dyadic contribution in (5.46) and (5.47). Finally, we conclude by a lemma of real interpolation theory to sum these estimates.

Precisely, by the same duality arguments as those used in the previous part of the proof, we observe that it suffices to prove (5.38). This is equivalent to the bilinear estimate

$$\left| \iint \langle e^{-is\Delta} F(s), e^{-it\Delta} G(t) \rangle\, ds\, dt \right| \lesssim \|F\|_{W(L^1, L^2)_t W(L^2, L^{r'})_x} \|G\|_{W(L^1, L^2)_t W(L^2, L^{r'})_x}.$$

By symmetry, it is enough to prove

$$|T(F, G)| \lesssim \|F\|_{W(L^1, L^2)_t W(L^2, L^{r'})_x} \|G\|_{W(L^1, L^2)_t W(L^2, L^{r'})_x}, \tag{5.41}$$

where

$$T(F, G) = \iint_{s<t} \langle e^{-is\Delta} F(s), e^{-it\Delta} G(t) \rangle\, ds\, dt.$$

Here the critical exponent $q = 2$ appears in the global component, which control the *decay in the t-variable at infinity*, hence the form $T(F, G)$ is decomposed dyadically as

$$T = \tilde{T} + \sum_{j \geq 0} T_j, \tag{5.42}$$

with

$$\tilde{T}(F, G) = \iint_{t-1<s<t} \langle e^{-is\Delta} F(s), e^{-it\Delta} G(t) \rangle\, ds\, dt \tag{5.43}$$

and

$$T_j(F,G) = \iint\limits_{t-2^{j+1}<s\leq t-2^j} \langle e^{-is\Delta}F(s), e^{-it\Delta}G(t)\rangle\, ds\, dt. \tag{5.44}$$

In the sequel we shall study the behavior of $\tilde{T}$ and T_j separately. We shall use repeatedly the following facts.

Lemma 5.1.9. *Let $1 \leq p, q, r \leq \infty$ be such that $1/p + 1/q = 1/r$. For every $f \in W(L^1, L^q)(\mathbb{R})$ supported in any interval I of length $l \geq 1$, it turns out that*

$$\|f\|_{W(L^1,L^r)} \leq C_p l^{1/p} \|f\|_{W(L^1,L^q)}. \tag{5.45}$$

Proof. To compute the $W(L^1, L^r)$-norm of f, we choose $g = \chi_{[0,1]}$, the characteristic function of the interval $[0, 1]$. Then, $\|f\|_{W(L^1,L^r)} \asymp \|\|fT_y g\|_{L^1}\|_{L^r}$. Since f is supported in an interval I of length l, the mapping

$$y \mapsto \|fT_y g\|_{L^1}$$

is supported in an interval $\tilde{I}$ of length $l+2$. Hence, for $1/p+1/q = 1/r$, Hölder's inequality yields

$$\begin{aligned}\big\|\|fT_y g\|_{L^1}\big\|_{L^r} &= \Big(\int_{\tilde{I}} \|fT_y g\|_{L^1}^r\, dy\Big)^{1/r} \leq (l+2)^{1/p}\big\|\|fT_y g\|_{L^1}\big\|_{L^q}\\ &\leq C_p l^{1/p}\|f\|_{W(L^1,L^q)},\end{aligned}$$

as desired. □

Lemma 5.1.10. *We have*

$$|\tilde{T}(F,G)| \lesssim \|F\|_{W(L^1,L^2)_t W(L^2,L^{r'})_x} \|G\|_{W(L^1,L^2)_t W(L^2,L^{r'})_x}. \tag{5.46}$$

Proof. We assume F and G compactly supported, with respect to the time variable, in intervals of length 1 (indeed, in (5.43), F and G can be replaced by $\chi_{[t-1,t]}(s)F(s)$ and $\chi_{[s,s+1]}(t)G(t)$, respectively).

Since $|t-s| \leq 1$, it follows from the duality properties in Theorem 2.4.9 and the fixed-time estimate (5.30) that

$$\begin{aligned}|\langle e^{-is\Delta}F(s), e^{-it\Delta}G(t)\rangle| &= |\langle F(s), e^{i(s-t)\Delta}G(t)\rangle|\\ &\lesssim \|F(s)\|_{W(L^2,L^{r'})_x}\|e^{i(s-t)\Delta}G(t)\|_{W(L^2,L^r)_x}\\ &\lesssim \|F(s)\|_{W(L^2,L^{r'})_x}(1+|s-t|)^{-d(\frac{1}{2}-\frac{1}{r})}\|G(t)\|_{W(L^2,L^{r'})_x}\\ &\lesssim \|F(s)\|_{W(L^2,L^{r'})_x}\|G(t)\|_{W(L^2,L^{r'})_x}.\end{aligned}$$

Integrating with respect to the variables s and t, we then obtain

$$|\tilde{T}(F,G)| \lesssim \|F\|_{L^1_t W(L^2,L^{r'})_x} \|G\|_{L^1_t W(L^2,L^{r'})_x}.$$

Lemma 5.1.9 with $p = q = 2$ and $r = 1$, applied to each function $\|F(t)\|_{W(L^2,L^{r'})_x}$ and $\|G(t)\|_{W(L^2,L^{r'})_x}$, gives the result. □

The estimates of the pieces $T_j(F,G)$ follow by the techniques of [208, Lemma 4.1].

Lemma 5.1.11. *We have*

$$|T_j(F,G)| \lesssim 2^{-j\beta(a,b)} \|F\|_{W(L^1,L^2)_t W(L^2,L^{a'})} \|G\|_{W(L^1,L^2)_t W(L^2,L^{b'})}, \tag{5.47}$$

for $(1/a, 1/b)$ *in a neighborhood of* $(1/r, 1/r)$.

Proof. Observe that here $d > 2$, hence $r < \infty$. Then, the result follows by complex interpolation from the following cases:
(i) $a = \infty$, $b = \infty$,
(ii) $2 \le a < r$, $b = 2$,
(iii) $a = 2$, $2 \le b < r$.

Case (i). We need to show the estimate

$$|T_j(F,G)| \lesssim 2^{-j(\frac{d}{2}-1)} \|F\|_{W(L^1,L^2)_t W(L^2,L^1)} \|G\|_{W(L^1,L^2)_t W(L^2,L^1)}. \tag{5.48}$$

Here the fixed-time estimate (5.30) and $|t-s| \lesssim 2^j$ yield

$$|\langle e^{-is\Delta}F(s), e^{-it\Delta}G(t)\rangle| \lesssim 2^{-j\frac{d}{2}} \|F(s)\|_{W(L^2,L^1)} \|G(t)\|_{W(L^2,L^1)}.$$

Integrating with respect to the variables s and t,

$$|T_j(F,G)| \lesssim 2^{-j\frac{d}{2}} \|F\|_{L^1_t(W(L^2,L^1)_x)} \|G\|_{L^1_t(W(L^2,L^1)_x)}.$$

Again, we can assume F and G compactly supported, with respect to the time variable, in intervals of length 2^j. Applying Lemma 5.1.9 with $p = q = 2$ and $r = 1$ to both functions $\|F(t)\|_{W(L^2,L^1)_x}$ and $\|G(t)\|_{W(L^2,L^1)_x}$, we obtain (5.48).

Case (ii). We have to show

$$|T_j(F,G)| \lesssim 2^{-j(\frac{d}{4}-1-\frac{d}{2a})} \|F\|_{W(L^1,L^2)_t W(L^2,L^{a'})_x} \|G\|_{W(L^1,L^2)_t L^2_x}. \tag{5.49}$$

Using similar arguments as in the previous case, we obtain

$$|T_j(F,G)| \lesssim \sup_t \left\| \int_{t-2^{j+1} < s \le t-2^j} e^{-is\Delta}F(s)\,ds \right\|_{L^2_x} \|G\|_{L^1_t L^2_x} \tag{5.50}$$

and

$$\|G\|_{L^1_t L^2_x} \lesssim 2^{j/2} \|G\|_{W(L^1,L^2)_t L^2_x}. \tag{5.51}$$

For $a \geq 2$, let now $\tilde{q} = \tilde{q}(a)$ be defined by

$$\frac{2}{\tilde{q}(a)} + \frac{d}{a} = \frac{d}{2}. \tag{5.52}$$

The nonendpoint case of (5.33), written for $\tilde{r} = a$ and $\tilde{q}$ above, gives

$$\begin{aligned}\sup_t \Big\| \int_{t-2^{j+1}<s\leq t-2^j} e^{-is\Delta} F(s)\, ds \Big\|_{L^2_x} &= \sup_t \Big\| \int_{\mathbb{R}} e^{-is\Delta} (T_{-t}\chi_{[-2^{j+1},-2^j]})(s) F(s)\, ds \Big\|_{L^2_x} \\ &\lesssim \|F\|_{W(L^1,L^{\tilde{q}'})W(L^2,L^{a'})},\end{aligned}$$

for every $2 \leq a < r$. Since the support of F with respect to the time is contained in an interval of length 2^j, Lemma 5.1.9 with $q = 2$, $r = \tilde{q}'$, and

$$\frac{1}{p} = \frac{1}{\tilde{q}'} - \frac{1}{2} = \frac{1}{2} - \frac{1}{\tilde{q}} = \frac{1}{2} - \frac{d}{4} + \frac{d}{2a},$$

gives

$$\|F\|_{W(L^1,L^{\tilde{q}'})W(L^2,L^{a'})} \lesssim 2^{j(\frac{1}{2}+\frac{d}{2a}-\frac{d}{4})} \|F\|_{W(L^1,L^2)W(L^2,L^{a'})}.$$

This estimate, together with (5.50) and (5.51), yields the estimate (5.49).
Case (iii). Use the same arguments as in case (ii). □

It remains to show

$$\sum_{j\geq 0} |T_j(F,G)| \lesssim \|F\|_{W(L^1,L^2)_t W(L^2,L^{r'})_x} \|G\|_{W(L^1,L^2)_t W(L^2,L^{r'})_x}. \tag{5.53}$$

Now, (5.53) can be achieved from (5.47) and some real interpolation results (collected in Appendix A below).

In details, we single out a_0, a_1, b_0, b_1 such that $(1/r, 1/r)$ is inside a small triangle with vertices $(1/a_0, 1/b_0)$, $(1/a_1, 1/b_0)$ and $(1/a_0, 1/b_1)$ (see Figure 5.4), so that

$$\beta(a_0, b_1) = \beta(a_1, b_0) \neq \beta(a_0, b_0).$$

Then, we apply Lemma A.2.5 with[1] $T = \{T_j\}$, (after setting $T_j = 0$ for $j < 0$) $C_0 = l_\infty^{\beta(a_0,b_0)}$, $C_1 = l_\infty^{\beta(a_0,b_1)}$ and, for $k = 0, 1$, we take

$$A_k = W(L^1(W(L^2, L^{a_k'})_x), L^2),\ B_k = W(L^1(W(L^2, L^{b_k'})_x), L^2).$$

1 We set $l_q^s = L^q(\mathbb{Z}, 2^{js} dj)$, where dj is the counting measure. Recall (see [17, Section 5.6]) that $(l_\infty^{s_0}, l_\infty^{s_1})_{\theta,1} = l_1^s$ whenever $s_0 \neq s_1$ and $s = (1-\theta)s_0 + \theta s_1$, $0 < \theta < 1$.

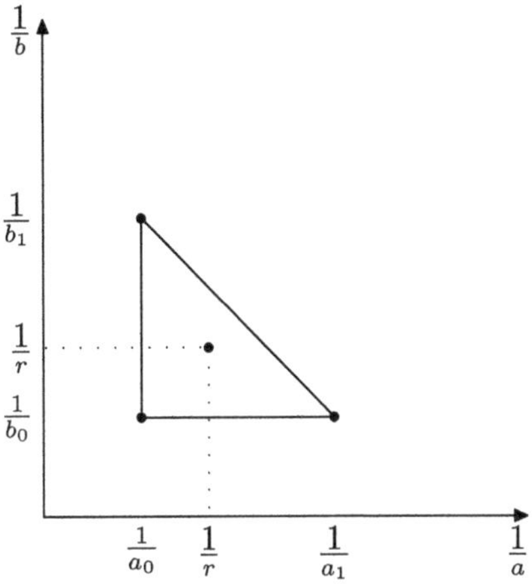

Figure 5.4: Triangle with vertices $(1/a_0, 1/b_0)$, $(1/a_1, 1/b_0)$ and $(1/a_0, 1/b_1)$.

Here we choose θ_0, θ_1, so that

$$1/r = (1-\theta_0)/a_0 + \theta_0/a_1, \quad 1/r = (1-\theta_1)/b_0 + \theta_1/b_1.$$

The assumptions are satisfied in view of (5.47). Moreover, with $\theta = \theta_0 + \theta_1$, we have

$$(1-\theta)\beta(a_0, b_0) + \theta\beta(a_0, b_1) = \beta(r, r) = 0.$$

Hence we attain the desired estimate (5.53) because

$$\begin{aligned}(A_0, A_1)_{\theta_0,2} &\hookrightarrow W((L^1(W(L^2, L^{a_0'})_x), L^1(W(L^2, L^{a_1'})_x))_{\theta_0,2}, L^2)\\ &\hookrightarrow W(L^1(W(L^2, L^{a_0'})_x, W(L^2, L^{a_1'})_x)_{\theta_0,2}, L^2)\\ &\hookrightarrow W(L^1(W(L^2, L^{r'})_x), L^2) = W(L^1, L^2)_t W(L^2, L^{r'})_x,\end{aligned}$$

where we used Proposition A.2.3 for the first and third embedding, and Proposition A.2.4 for the second one (the same holds for $(B_0, B_1)_{\theta_0,2}$).

The corresponding retarded estimates can be obtained as follows. The case $(\tilde{q}, \tilde{r}) = (q, r) = P$ is exactly (5.41). The case $(\tilde{q}, \tilde{r}) = P$, $(q, r) \neq P$, can be obtained by a repeated use of Hölder's inequality to interpolate from the case $(\tilde{q}, \tilde{r}) = (q, r) = P$, and the case $(\tilde{q}, \tilde{r}) = P$, $(q, r) = (\infty, 2)$ (which is clear from (5.39)). Precisely, we want to prove

$$\left\| \int_{s<t} e^{i(t-s)\Delta} F(s)\, ds \right\|_{W(L^\infty, L^q)_t W(L^2, L^r)_x} \lesssim \|F\|_{W(L^1, L^2)_t W(L^2, L^{\tilde{r}'})_x},$$

for $\tilde{r} = 2d/(d-1)$, $2/q + d/r = d/2$, $q > 2$.

We know that such an estimate holds for $(q,r) = (\infty,2)$, as well as $(q,r) = (\tilde{q},\tilde{r}) = P$, that is,

$$\Big\|\int_{s<t} e^{i(t-s)\Delta}F(s)\,ds\Big\|_{W(L^\infty,L^2)_t W(L^2,L^{\tilde{r}})_x} \lesssim \|F\|_{W(L^1,L^2)_t W(L^2,L^{\tilde{r}'})_x}.$$

Hence, upon setting $I = \int_{s<t} e^{i(t-s)\Delta}F(s)\,ds$, it suffices to prove that

$$\|I\|_{W(L^\infty,L^q)_t W(L^2,L^r)_x} \leq \|I\|^{1-\theta}_{W(L^\infty,L^2)_t W(L^2,L^{\tilde{r}})_x}\|I\|^{\theta}_{L^\infty_t L^2_x}, \tag{5.54}$$

with $1/r = (1-\theta)/\tilde{r} + \theta/2$, $1/q = (1-\theta)/2 + \theta/\infty$, $0 < \theta < 1$. To this end, we start with the well-known inequality

$$\|u\|_{L^r} \lesssim \|u\|^{1-\theta}_{L^{\tilde{r}}}\|u\|^{\theta}_{L^2}. \tag{5.55}$$

We apply (5.55) with $u = \|I(t)T_z g\|_{L^2_x}$, where $g \in \mathcal{D}(\mathbb{R}^d_x)$ is a nonzero window. We obtain

$$\begin{aligned}\|I(t)\|_{W(L^2,L^r)_x} &= \big\|\|I(t)T_z g\|_{L^2_x}\big\|_{L^r_z}\\ &\leq \big\|\|I(t)T_z g\|_{L^2_x}\big\|^{1-\theta}_{L^{\tilde{r}}_x}\big\|\|I(t)T_z g\|_{L^2_x}\big\|^{\theta}_{L^2_z}\\ &\leq \|I(t)\|^{1-\theta}_{W(L^2,L^r)_x}\|I(t)\|^{\theta}_{L^2_x}.\end{aligned} \tag{5.56}$$

Hence, given any nonzero window $h \in \mathcal{D}(\mathbb{R}_t)$, it follows that

$$\|I(t)\|_{W(L^2,L^r)_x}|(T_s h)(t)| \leq \big(\|I(t)\|_{W(L^2,L^r)_x}|(T_s h)(t)|\big)^{1-\theta}\big(\|I(t)\|_{L^2_x}|(T_s h)(t)|\big)^{\theta}. \tag{5.57}$$

We then take the $L^\infty_s L^q_t$-norm of the expression in the left-hand side of (5.57) and apply again Hölder's inequality. This gives the desired result (5.54).

Eventually, one can prove the retarded estimate in the case $(q,r) = P$, $(\tilde{q},\tilde{r}) \neq P$ showing, by the arguments above, that the dual inequality holds for the adjoint operator $G \mapsto \int_{t>s} e^{-i(t-s)\Delta}G(t)\,dt$.

This concludes the proof of Theorem 5.1.8. □

The Strichartz estimates in Theorem 5.1.8 can be combined with the classical ones (5.14) to obtain the estimates in Theorem 5.1.1.

Proof of Theorem 5.1.1. We prove the homogeneous estimates (5.18), the others follow by similar arguments. Recall that the classical homogeneous Strichartz estimates can be written as

$$\|e^{it\Delta}u_0\|_{W(L^{\tilde{q}},L^{\tilde{q}})_t W(L^{\tilde{r}},L^{\tilde{r}})_x} \lesssim \|u_0\|_{L^2_x}, \tag{5.58}$$

for every Schrödinger admissible pair $(\tilde{q},\tilde{r})$, i. e., $\tilde{q},\tilde{r} \geq 2$, with $2/\tilde{q}+d/\tilde{r} = d/2$, $(\tilde{q},\tilde{r},d) \neq (2,\infty,2)$. By complex interpolation between (5.58) and (5.32), one has

$$\|e^{it\Delta}u_0\|_{W(L^{q_1},L^{q_2})_t W(L^{r_1},L^{r_2})_x} \lesssim \|u_0\|_{L^2_x}, \tag{5.59}$$

with (q_i, r_i), $i = 1, 2$, Schrödinger admissible pairs.

Here

$$\frac{1}{q_1} = \frac{1-\theta}{\tilde{q}} + \frac{\theta}{\infty}, \quad \frac{1}{q_2} = \frac{1-\theta}{\tilde{q}} + \frac{\theta}{q},$$

so that

$$\frac{1}{q_2} = \frac{1}{q_1} + \frac{\theta}{q}.$$

Hence $q_1 \geq q_2$, i. e., $r_1 \leq r_2$. This shows (5.32) when (q_1, r_1) and (q_2, r_2) satisfy (5.16) and (5.17) with equality. The general case follows from the inclusion relations of Wiener amalgam spaces, which allow us to increase q_2, r_2, and diminish q_1, r_1 (see Figure 5.2). □

5.1.3 Sharpness of fixed-time and Strichartz estimates

In this section we prove the sharpness of the estimate in (5.18).

First, observe that the solution of the Cauchy problem (5.11), with initial datum $u_0(x) = e^{-\pi x^2}$, is given by the formula

$$u(t,x) = (1 + 4\pi i t)^{-d/2} e^{-\frac{\pi x^2}{1+4\pi i t}}. \tag{5.60}$$

This follows at once from Lemma 0.2.4, since the solution $u(t,x)$ can be represented as

$$u(t,x) = (K_t * u_0)(x) = \frac{1}{(4\pi i t)^{d/2}} (\phi^{((4\pi i t)^{-1})} * \phi^{(1)})(x).$$

The following two auxiliary results are needed.

Lemma 5.1.12. *Let $a > 0, b \in \mathbb{R}$. With the notation in Lemma 0.2.4, we have*

$$\|\phi^{(a+ib)}\|_{W(L^{r_1},L^{r_2})} \asymp a^{-\frac{d}{2r_2}} (a+1)^{\frac{d}{2}(\frac{1}{r_2} - \frac{1}{r_1})}. \tag{5.61}$$

Proof. We take the Gaussian function $g(x) = e^{-\pi x^2}$ as a window in the definition of $W(L^{r_1}, L^{r_2})$. Then

$$|\phi^{(a+ib)}(x) T_y g(x)| = e^{-\pi[(a+1)x^2 - 2x\cdot y + y^2]},$$

so that (0.27) gives

$$\|\phi^{(a+ib)} T_y g\|_{L^{r_1}} = [r_1(a+1)]^{-\frac{d}{2r_1}} e^{-\frac{\pi a}{a+1} y^2}.$$

Now we compute the L^{r_2}-norm of this expression (again by using (0.27)) and obtain (5.61). □

Lemma 5.1.13. *Let $u(t,x)$ be the solution of* (5.11) *with the initial datum $u_0(x) = e^{-\pi x^2}$. Then, for the solution $u(\lambda^2 t, \lambda x)$ corresponding to the initial datum $u_0(\lambda x), \lambda > 0$, we have*

$$\|u(\lambda^2 t, \lambda \cdot)\|_{W(L^{r_1},L^{r_2})} \asymp \lambda^{-\frac{d}{r_2}}(1 + (t\lambda^2)^2)^{\frac{d}{2}(\frac{1}{r_1}-\frac{1}{2})}(1 + \lambda^2 + (t\lambda^2)^2)^{\frac{d}{2}(\frac{1}{r_2}-\frac{1}{r_1})}. \tag{5.62}$$

Proof. We consider the one-parameter family of initial data $u_0(\lambda x) = e^{-\pi\lambda^2 x^2}$, $\lambda > 0$. If u is the function in (5.60), the corresponding solutions will be

$$\begin{aligned} u(\lambda^2 t, \lambda x) &= (1 + 4\pi i t\lambda^2)^{-d/2} e^{-\frac{\pi\lambda^2 x^2}{1+4\pi i t\lambda^2}} \\ &= (1 + 4\pi i t\lambda^2)^{-d/2} \phi^{(a+ib)}, \end{aligned}$$

with

$$a = \frac{\lambda^2}{1 + (4\pi t\lambda^2)^2}, \qquad b = -\frac{4\pi t\lambda^2}{1 + (4\pi t\lambda^2)^2}.$$

Hence (5.62) follows from the estimate (5.61). □

Proposition 5.1.14. *Let $u_0(x) = e^{-\pi x^2}$. Then*

$$\|e^{it\Delta} u_0\|_{W(L^{r_1},L^{r_2})} \sim C t^{d(\frac{1}{r_2}-\frac{1}{2})}, \quad \text{as } t \to +\infty, \tag{5.63}$$

for some $C > 0$.

Moreover, suppose that for some $t_0 \neq 0$, $1 \le r_1, r_2 \le \infty$, $C > 0$, the following estimate holds:

$$\|e^{it_0\Delta} u_0\|_{W(L^{r_1},L^{r_2})} \le C\|u_0\|_{W(L^{r_1'},L^{r_2'})}, \quad \forall u_0 \in \mathcal{S}(\mathbb{R}^d). \tag{5.64}$$

Then $r_1 \le r_2$.

Proof. Formula (5.63) is an immediate consequence of the norm (5.62), with $\lambda = 1$.

To prove the remaining part of the statement, we apply assumption (5.64) to the initial data $u_0(\lambda x) = e^{-\pi\lambda^2 x^2}$, $\lambda > 0$. As a consequence of the norm expressions computed in (5.61) and (5.62), we see that, when $\lambda \to +\infty$, the left- and right-hand sides of (5.64) are asymptotically equivalent to $c_1\lambda^{\frac{d}{r_2}-d}$, and $c_2\lambda^{-\frac{d}{r_1'}}$, respectively, for some $c_1, c_2 > 0$. This implies $\frac{d}{r_2} - d \le -\frac{d}{r_1'}$, namely $r_1 \le r_2$. □

Proposition 5.1.15 (Sharpness of (5.18)). *Suppose that, for some $1 \le q_1, q_2, r_1, r_2 \le \infty$, $C > 0$, the following estimate holds:*

$$\|e^{it\Delta} u_0\|_{W(L^{q_1},L^{q_2})_t W(L^{r_1},L^{r_2})_x} \le C\|u_0\|_{L^2}, \quad \forall u_0 \in \mathcal{S}(\mathbb{R}^d). \tag{5.65}$$

Then we have

$$\frac{2}{q_1} + \frac{d}{r_1} \ge \frac{d}{2}, \tag{5.66}$$

$$\frac{2}{q_2}+\frac{d}{r_2}\le\frac{d}{2}. \tag{5.67}$$

In particular (because of (5.67)), $r_2\ge 2$. *Finally, it must be* $q_2\ge 2$.

Proof. We apply the estimate (5.65) to the family of initial data $u_0(\lambda x)=e^{-\pi\lambda^2x^2}$, $\lambda>0$. Choose $g=\chi_{[-1,1]}$ as a window function in the definition of the space $W(L^{q_1},L^{q_2})_t$. Then, for $y\in\mathbb{R}$, $\lambda\ge 1$,

$$\begin{aligned}&\|u(\lambda^2t\lambda,\cdot)\|_{W(L^{r_1},L^{r_2})}T_yg(t)\|_{L^{q_1}}\\&\asymp\lambda^{\frac{d}{r_2}}\Bigg(\int\limits_{y-1}^{y+1}(\lambda^{-4}+t^2)^{\frac{q_1d}{2}(\frac{1}{r_1}-\frac{1}{2})}(\lambda^{-2}+t^2)^{\frac{q_1d}{2}(\frac{1}{r_2}-\frac{1}{r_1})}\,dt\Bigg)^{\frac{1}{q_1}}.\end{aligned} \tag{5.68}$$

To estimate this expression from below, we apply the easily verified formula

$$\int\limits_{-\frac{1}{2}}^{\frac{1}{2}}(\mu^2+t^2)^{\gamma_1}(\mu+t^2)^{\gamma_2}\,dt\gtrsim\mu^{1+2\gamma_1+\gamma_2},\quad 0<\mu\le 1,$$

with $\mu=\lambda^{-2}$. Thereby,

$$\|\|u(\lambda^2t,\lambda\cdot)\|_{W(L^{r_1},L^{r_2})}T_yg(t)\|_{L^{q_1}}\gtrsim\lambda^{-\frac{2}{q_1}-\frac{d}{r_1}}\quad\text{for } |y|\le\frac{1}{2},\ \lambda\ge 1.$$

Taking the L^{q_2}-norm of this expression yields

$$\|u(\lambda^2\cdot,\lambda\cdot)\|_{W(L^{r_1},L^{r_2})_t W(L^{r_1},L^{r_2})_x}\gtrsim\lambda^{-\frac{2}{q_1}-\frac{d}{r_1}},\quad\lambda\ge 1.$$

On the other hand, the right-hand side of (5.65) is equal to $\lambda^{-\frac{d}{2}}$; therefore, letting $\lambda\to+\infty$, we obtain the index relation (5.66).

We now prove (5.67). If $0<\lambda\le 1$, it follows from the norm estimate (5.62) that

$$\|u(\lambda^2t,\lambda\cdot)\|_{W(L^{r_1},L^{r_2})}\asymp\lambda^{\frac{d}{r_2}-d}(\lambda^{-4}+t^2)^{\frac{d}{2}(\frac{1}{r_2}-\frac{1}{2})}.$$

Hence, we have

$$\|\|u(\lambda^2t,\lambda\cdot)\|_{W(L^{r_1},L^{r_2})}T_yg\|_{L^{q_1}}\asymp\lambda^{\frac{d}{r_2}-d}(\lambda^{-4}+y^2)^{\frac{d}{2}(\frac{1}{r_2}-\frac{1}{2})},\quad 0<\lambda\le 1.$$

Taking the L^{q_2}-norm of this expression and applying the formula

$$\int\limits_{-\infty}^{+\infty}(\mu+y^2)^{\gamma}\,dy=c_\gamma\mu^{\frac{1}{2}+\gamma},\quad\mu>0,$$

(for every γ, and for a convenient $c_\gamma \in (0,\infty]$, independent of μ ($c_\gamma < \infty$ if $\gamma < -\frac{1}{2}$)) with $\mu = \lambda^{-4}$, we obtain the minorization

$$\|u(\lambda^2\cdot,\lambda\cdot)\|_{W(L^{q_1},L^{q_2})_t W(L^{r_1},L^{r_2})_x} \gtrsim \lambda^{-\frac{2}{q_2}-\frac{d}{r_2}}, \quad 0<\lambda\leq 1.$$

Since the right-hand side of (5.65) is equal to $\lambda^{-\frac{d}{2}}$, the estimate (5.67) follows by letting $\lambda \to 0^+$.

We are left to prove the condition $q_2 \geq 2$. We follow the pattern outlined in [277, Exercise 2.42]. Precisely, we take as initial datum

$$u_0(x) = \sum_{j=1}^N e^{-it_j\Delta} f,$$

where f is a fixed test function, normalized so that the function $v(t,x)$, defined by $v(t,x) = (e^{it\Delta}f)(x)$, satisfies

$$\|v\|_{W(L^{q_1},L^{q_2})_t W(L^{r_1},L^{r_2})_x} = 1. \tag{5.69}$$

Here $t_1,\dots,t_N$ are widely separated times that will be chosen later on. Notice that the corresponding solution will be

$$u(t,x) = (e^{it\Delta}u_0)(x) = \sum_{j=1}^N v(t-t_j,x).$$

We claim that

$$\|u_0\|_{L^2} = \left\|\sum_{j=1}^N e^{-it_j\Delta} f\right\|_{L^2} \leq (N+1)^{\frac{1}{2}}\|f\|_{L^2}, \tag{5.70}$$

if $t_1,\dots,t_N$ are suitable separated.

Indeed,

$$\begin{aligned}\left\|\sum_{j=1}^N e^{-it_j\Delta} f\right\|_{L^2}^2 &= \sum_{j=1}^N \|e^{-it_j\Delta}f\|_{L^2}^2 + \sum_{j\neq k}\langle e^{i(t_j-t_k)\Delta}f, f\rangle \\ &\leq N\|f\|_{L^2}^2 + C\sum_{j\neq k}|t_j-t_k|^{-\frac{d}{2}}\|f\|_{L^1}^2,\end{aligned}$$

where we used Cauchy–Schwarz inequality and the classical dispersive estimate $\|e^{it\Delta}f\|_{L^\infty} \leq C|t|^{-d/2}\|f\|_{L^1}$. Hence (5.70) follows if

$$|t_j-t_k|^{-d/2} \leq [C(N^2-N)\|f\|_{L^1}^2]^{-1}\|f\|_{L^2}^2.$$

We now estimate from below the left-hand side of (5.65). To this end, let $\tilde{v}(t,x) = v(t,x)\chi_R(t)$, where $\chi_R(t)$ is the characteristic function of the interval $[-R,R]$. Moreover, we assume $q_2 < \infty$ and choose R large enough so that

$$\|v - \tilde{v}\|_{W(L^{q_1},L^{q_2})_t W(L^{r_1},L^{r_2})_x} \le \frac{1}{N}. \tag{5.71}$$

We claim that, if $|t_j - t_k| \ge 2R + 2$, for every $j \ne k$, then

$$\begin{aligned}\|u(t,x)\|_{W(L^{q_1},L^{q_2})_t W(L^{r_1},L^{r_2})_x} &= \left\|\sum_{j=1}^N v(t-t_j,x)\right\|_{W(L^{q_1},L^{q_2})_t W(L^{r_1},L^{r_2})_x}\\ &\ge N^{\frac{1}{q_2}}\left(1-\frac{1}{N}\right) - 1.\end{aligned} \tag{5.72}$$

This, together with the assumption (5.65) and the L^2-estimate of the initial datum (5.70), gives the condition $q_2 \ge 2$, for N large enough.

In order to prove the minorization (5.72), observe that, by assumption (5.71),

$$\left\|\sum_{j=1}^N v(t-t_j,x) - \sum_{j=1}^N \tilde{v}(t-t_j,x)\right\|_{W(L^{q_1},L^{q_2})_t W(L^{r_1},L^{r_2})_x} \le 1.$$

Hence it suffices to prove

$$\left\|\sum_{j=1}^N \tilde{v}(t-t_j,x)\right\|_{W(L^{q_1},L^{q_2})_t W(L^{r_1},L^{r_2})_x} \ge N^{\frac{1}{q_2}}\left(1-\frac{1}{N}\right). \tag{5.73}$$

Now,

$$\left\|\sum_{j=1}^N \tilde{v}(t-t_j,\cdot)\right\|_{W(L^{r_1},L^{r_2})} = \sum_{j=1}^N \|\tilde{v}(t-t_j,\cdot)\|_{W(L^{r_1},L^{r_2})},$$

since, for every fixed t, there is at most one function in the sum which is not identically zero. Hence, upon setting $h_j(t) := \|\tilde{v}(t-t_j,\cdot)\|_{W(L^{r_1},L^{r_2})}$, we have

$$\left\|\sum_{j=1}^N \tilde{v}(t-t_j,\cdot)\right\|_{W(L^{q_1},L^{q_2})_t W(L^{r_1},L^{r_2})_x} = \left\|\left\|\sum_{j=1}^N h_j T_y g\right\|_{L^{q_1}}\right\|_{L^{q_2}_y}.$$

Choosing the window function g supported in $[0,1]$, since the h_j's are supported in intervals separated by a distance ≥ 2, we see that the last expression is equal to

$$\left\|\sum_{j=1}^N \|h_j T_y g\|_{L^{q_1}}\right\|_{L^{q_2}_y}.$$

In turn, since the functions $y \mapsto \|h_j T_y g\|_{L^{q_1}}$, $j = 1, \dots, N$, have disjoint supports, the norm above can be written as

$$\left(\sum_{j=1}^{N} \big\| \|h_j T_y g\|_{L^{q_1}} \big\|_{L_y^{q_2}}^{q_2} \right)^{\frac{1}{q_2}} = N^{\frac{1}{q_2}} \|\tilde{v}\|_{W(L^{q_1},L^{q_2})_t W(L^{r_1},L^{r_2})_x}.$$

Hence, the minorization (5.73) follows from the assumptions (5.69) and (5.71). □

One can also widen this theory to the study of partial differential equations *with variable coefficients*. Precisely, consider the Cauchy problem for the Schrödinger equation with a quadratic Hamiltonian, namely

$$\begin{cases} i\frac{\partial u}{\partial t} + Hu = 0, \\ u(0,x) = u_0(x), \end{cases} \tag{5.74}$$

where H is the Weyl quantization of a quadratic form on $\mathbb{R}^d \times \mathbb{R}^d$. The most interesting case is certainly the Schrödinger equation with a quadratic potential. In this case the solution $u(t,x)$ to (5.74) is given by

$$u(t,x) = e^{itH} u_0,$$

where the operator e^{itH} is a metaplectic operator, see Section (1.1.2) and the work [70], where the authors show fixed-time estimates for the solution $u(t,x)$ in terms of the initial datum u_0. An example is provided by the harmonic oscillator

$$H = -\frac{1}{4\pi}\Delta + \pi x^2 \tag{5.75}$$

(see, e. g., [145]), for which they deduce the dispersive estimate

$$\|e^{itH} u_0\|_{W(\mathcal{F}L^1,L^\infty)} \lesssim |\sin t|^{-d} \|u_0\|_{W(\mathcal{F}L^\infty,L^1)}.$$

Another Hamiltonian that has been considered is $H = -\frac{1}{4\pi}\Delta - \pi|x|^2$. In this case, the dispersive estimate reads

$$\|e^{itH} u_0\|_{W(\mathcal{F}L^1,L^\infty)} \lesssim \left(\frac{1 + |\sinh t|}{\sinh^2 t} \right)^{\frac{d}{2}} \|u_0\|_{W(\mathcal{F}L^\infty,L^1)}.$$

Using the fixed-time estimates above, one can infer suitable Strichartz estimates. For instance, the homogeneous Strichartz estimates achieved for the harmonic oscillator $H = -\frac{1}{4\pi}\Delta + \pi|x|^2$ read

$$\|e^{itH} u_0\|_{L^{q/2}([0,T])W(\mathcal{F}L^{r'},L^r)_x} \lesssim \|u_0\|_{L_x^2},$$

for every $T > 0$, $4 < q, \tilde{q} \leq \infty$, $2 \leq r, \tilde{r} \leq \infty$, such that $2/q + d/r = d/2$, and, similarly, for $\tilde{q}, \tilde{r}$. In the endpoint case $(q,r) = (4, 2d/(d-1))$, $d > 1$, the same estimate can be proved with $\mathcal{F}L^{r'}$ replaced by the slightly larger space $\mathcal{F}L^{r',2}$, where $L^{r',2}$ is a Lorentz space. We refer the interested reader to the original paper [70].

We end up this section by recalling that Strichartz estimates for Schrödinger equations can be found in the recent contributions [209, 210]. Similar estimates for the vibrating plate equation were studied in [92]. For the time–frequency analysis of the Dirac equation, we refer to the contribution [294].

5.2 Gabor representations of evolution operators

In this section we shall apply Gabor frames to the study of PDEs, following the plan in the introduction of the chapter. The functional frame is given by Gelfand–Shilov spaces. Modulation spaces are not abandoned in this presentation, but play a minor role. In fact, we have here to relax the assumption of algebraic growth (2.17) for the weight functions and use exponential weights

$$v(z) = e^{a|z|^b}, \quad a > 0, b > 0.$$

As we observed in Chapter 2, the main properties of $M^{p,q}_m$, $m \in \mathcal{M}_v$, extend to this case, provided $v(z)$ is submultiplicative. Recall that submultiplicativity is valid for exponential weights if and only if $b \leq 1$, cf. Lemma 2.1.4 in Chapter 2. This technical requirement restricts the application of modulation spaces, see the next Definition 5.2.6 and the subsequent Proposition 5.2.16, concerning boundedness of pseudodifferential operators on $M^{p,q}_m$. For recent results on the subject, let us refer to [45, 45, 82, 289].

5.2.1 Gelfand–Shilov spaces

The Schwartz class $\mathcal{S}(\mathbb{R}^d)$ does not give enough information about how fast a function $f \in \mathcal{S}(\mathbb{R}^d)$ and its derivatives decay at infinity. This is the main motivation to use subspaces of the Schwartz class, the so-called Gelfand–Shilov-type spaces, introduced in [155, 156] and now available also in a textbook [236]. Let us recall their definition and main properties; for proofs, we refer the reader to [155, 156, 236, 242].

Definition 5.2.1. Let there be given $s, r \geq 0$ and $A, B > 0$. The Gelfand–Shilov-type space $S^{s,A}_{r,B}(\mathbb{R}^d)$ is defined by

$$S^{s,A}_{r,B}(\mathbb{R}^d) = \{f \in \mathcal{S}(\mathbb{R}^d) : |x^\alpha \partial^\beta f(x)| \lesssim A^{|\alpha|} B^{|\beta|} (\alpha!)^r (\beta!)^s,\ \alpha, \beta \in \mathbb{N}^d\}. \tag{5.76}$$

Their projective and inductive limits are respectively denoted by

$$\Sigma^s_r = \operatorname{proj} \lim_{A>0,B>0} S^{s,A}_{r,B}; \quad S^s_r = \operatorname{ind} \lim_{A>0,B>0} S^{s,A}_{r,B}.$$

The space $S_r^s(\mathbb{R}^d)$ is nontrivial if and only if $r+s > 1$, or $r+s = 1$ and $r, s > 0$. So the smallest nontrivial space with $r = s$ is provided by $S_{1/2}^{1/2}(\mathbb{R}^d)$. Every function of the type $P(x)e^{-a|x|^2}$, with $a > 0$ and $P(x)$ polynomial on $\mathbb{R}^d$, is in the class $S_{1/2}^{1/2}(\mathbb{R}^d)$. We observe the trivial inclusions $S_{r_1}^{s_1}(\mathbb{R}^d) \subset S_{r_2}^{s_2}(\mathbb{R}^d)$ for $s_1 \le s_2$ and $r_1 \le r_2$. Moreover, if $f \in S_r^s(\mathbb{R}^d)$, also $x^\delta \partial^\gamma f$ belongs to the same space for every fixed δ, γ.

The action of the Fourier transform on $S_r^s(\mathbb{R}^d)$ interchanges the indices s and r, as explained in the following theorem.

Theorem 5.2.2. *For $f \in \mathcal{S}(\mathbb{R}^d)$, we have $f \in S_r^s(\mathbb{R}^d)$ if and only if $\hat{f} \in S_s^r(\mathbb{R}^d)$.*

Therefore, for $s = r$, the spaces $S_s^s(\mathbb{R}^d)$ are invariant under the action of the Fourier transform.

Theorem 5.2.3. *Assume $s > 0, r > 0, s + r \ge 1$. For $f \in \mathcal{S}(\mathbb{R}^d)$, the following conditions are equivalent:*
(a) $f \in S_r^s(\mathbb{R}^d)$.
(b) *There exist constants $A, B > 0$ such that*

$$\|x^\alpha f\|_{L^\infty} \lesssim A^{|\alpha|}(\alpha!)^r \quad \text{and} \quad \|\xi^\beta \hat{f}\|_{L^\infty} \lesssim B^{|\beta|}(\beta!)^s, \quad \alpha, \beta \in \mathbb{N}^d.$$

(c) *There exist constants $A, B > 0$ such that*

$$\|x^\alpha f\|_{L^\infty} \lesssim A^{|\alpha|}(\alpha!)^r \quad \text{and} \quad \|\partial^\beta f\|_{L^\infty} \lesssim B^{|\beta|}(\beta!)^s, \quad \alpha, \beta \in \mathbb{N}^d.$$

(d) *There exist constants $h, k > 0$ such that*

$$\|f e^{h|x|^{1/r}}\|_{L^\infty} < \infty \quad \text{and} \quad \|\hat{f} e^{k|\xi|^{1/s}}\|_{L^\infty} < \infty.$$

A suitable window class for weighted modulation spaces (see the subsequent Definition 5.2.6) is the Gelfand–Shilov-type space $\Sigma_1^1(\mathbb{R}^d)$, consisting of functions $f \in \mathcal{S}(\mathbb{R}^d)$ such that for *every* constant $A > 0$ and $B > 0$,

$$|x^\alpha \partial^\beta f(x)| \lesssim A^{|\alpha|} B^{|\beta|} \alpha! \beta!, \quad \alpha, \beta \in \mathbb{N}^d. \tag{5.77}$$

We have $S_s^s(\mathbb{R}^d) \subset \Sigma_1^1(\mathbb{R}^d) \subset S_1^1(\mathbb{R}^d)$ for every $s < 1$. Observe that the characterization of Theorem 5.2.3 can be adapted to $\Sigma_1^1(\mathbb{R}^d)$ by replacing the words "there exist" by "for every" and taking $r = s = 1$.

Let us underline the following property, which exhibits two equivalent ways of expressing the decay of a continuous function f on $\mathbb{R}^d$. This follows immediately from [236, Proposition 6.1.5], where the property was formulated for $f \in \mathcal{S}(\mathbb{R}^d)$. For the sake of clarity, we shall detail the proof showing the mutual dependence between the constants ϵ and C below.

Proposition 5.2.4 ([236, Proposition 6.1.5]). *Consider $r > 0$ and let h be a continuous function on $\mathbb{R}^d$. Then the following conditions are equivalent:*

(i) *There exists a constant* $\epsilon > 0$ *such that*

$$|h(x)| \lesssim e^{-\epsilon|x|^{\frac{1}{r}}}, \quad x \in \mathbb{R}^d, \tag{5.78}$$

(ii) *There exists a constant* $C > 0$ *such that*

$$|x^\alpha h(x)| \lesssim C^{|\alpha|}(\alpha!)^r, \quad x \in \mathbb{R}^d,\ \alpha \in \mathbb{N}^d. \tag{5.79}$$

Proof. We rewrite (5.78) in the form

$$|h(x)|^{\frac{1}{r}} \lesssim e^{-\frac{\epsilon}{r}|x|^{\frac{1}{r}}}, \quad x \in \mathbb{R}^d. \tag{5.80}$$

In turn, (5.80) can be rewritten as

$$\sup_{x\in\mathbb{R}^d} \sum_{n=0}^{\infty} \left(\frac{\epsilon}{r}\right)^n (n!)^{-1}|x|^{\frac{n}{r}}|h(x)|^{\frac{1}{r}} < \infty. \tag{5.81}$$

Hence the sequence of the terms of the series is uniformly bounded, as well as the sequence of the rth powers:

$$\frac{\epsilon^{rn}}{r^{rn}}(n!)^{-r}|x|^n|h(x)|, \quad n \in \mathbb{N},$$

and we obtain

$$|x|^n|h(x)| \lesssim \frac{r^{rn}}{\epsilon^{rn}}(n!)^r, \quad x \in \mathbb{R}^d,\ n = 0, 1, \ldots$$

Writing $|\alpha| = n$ and applying (0.2) gives

$$|x^\alpha h(x)| \lesssim \left(\frac{rd}{\epsilon}\right)^{r|\alpha|}(\alpha!)^r = C^{|\alpha|}(\alpha!)^r, \quad x \in \mathbb{R}^d,\ \alpha \in \mathbb{N}^d,$$

where $C = (\frac{rd}{\epsilon})^r$. Therefore (5.79) is proved.

Conversely, let (5.79) be satisfied. Using the following inequalities:

$$|x|^n \le \sum_{|\alpha|=n} \frac{n!}{\alpha!}|x^\alpha|, \quad \sum_{|\alpha|=n} \frac{n!}{\alpha!} = d^n, \quad \text{and} \quad \alpha! \le |\alpha|!,$$

and assumption (5.79), we obtain

$$|x|^n|h(x)| \le \sum_{|\alpha|=n} \frac{n!}{\alpha!}|x^\alpha h(x)| \lesssim \sum_{|\alpha|=n} \frac{n!}{\alpha!}C^{|\alpha|}(\alpha!)^r \le C^n(n!)^r d^n = (dC)^n(n!)^r,$$

for every $x \in \mathbb{R}^d$, $n \in \mathbb{N}$. Therefore the sequence

$$(dC)^{-n}(n!)^{-r}|x|^n|h(x)|, \quad n \in \mathbb{N},$$

is uniformly bounded for $x \in \mathbb{R}^d$, as well as the sequence

$$(dC)^{-\frac{n}{r}}(n!)^{-1}|x|^{\frac{n}{r}}|h(x)|^{\frac{1}{r}}, \quad n \in \mathbb{N}.$$

If we choose $\epsilon = q(dC)^{-\frac{1}{r}}$, for a fixed $q \in (0,1)$, we conclude

$$e^{\epsilon|x|^{\frac{1}{r}}}|h(x)|^{\frac{1}{r}} = \sum_{n=0}^{\infty} q^n (dC)^{-\frac{n}{r}}(n!)^{-1}|x|^{\frac{n}{r}}|h(x)|^{\frac{1}{r}} \lesssim \sum_{n=0}^{\infty} q^n.$$

This is (5.80), hence the proof is complete. □

The precise relation between the constants ϵ and C follows from this proof. Indeed, assuming (5.78), then (5.79) is satisfied with $C = (\frac{rd}{\epsilon})^r$. Conversely, (5.79) implies (5.78) for any $\epsilon < r(dC)^{-\frac{1}{r}}$. The bound is sharp for $d = 1$. Furthermore, it follows from the proof that the constant implicit in the notation $\lesssim$ in (5.78) depends only on the corresponding one in (5.79), and vice versa.

The strong dual spaces of $S_r^s(\mathbb{R}^d)$ and $\Sigma_1^1(\mathbb{R}^d)$ are called spaces of tempered ultradistributions and denoted by $(S_r^s)'(\mathbb{R}^d)$ and $(\Sigma_1^1)'(\mathbb{R}^d)$, respectively. Notice that they contain the space of tempered distributions $\mathcal{S}'(\mathbb{R}^d)$.

The spaces $S_r^s(\mathbb{R}^d)$ are nuclear spaces [226]. This property yields a kernel theorem for Gelfand–Shilov spaces, cf. [295].

Theorem 5.2.5. *There exists an isomorphism between the space of linear continuous maps T from $S_r^s(\mathbb{R}^d)$ to $(S_r^s)'(\mathbb{R}^d)$ and $(S_r^s)'(\mathbb{R}^{2d})$, which associates to every T a kernel $K_T \in (S_r^s)'(\mathbb{R}^{2d})$ such that*

$$\langle Tu, v\rangle = \langle K_T, v \otimes \bar{u}\rangle, \quad \forall u, v \in S_r^s(\mathbb{R}^d).$$

This K_T is called the kernel of T.

We now list some results about time–frequency representations of Gelfand–Shilov functions, cf. [58, 87, 171, 282]:

$$f, g \in S_s^s(\mathbb{R}^d), \quad s \geq 1/2 \Longrightarrow V_g f \in S_s^s(\mathbb{R}^{2d}), \tag{5.82}$$

$$f, g \in S_s^s(\mathbb{R}^d), \quad s \geq 1/2 \Longrightarrow W(f, g) \in S_s^s(\mathbb{R}^{2d}), \tag{5.83}$$

$$f, g \in \Sigma_1^1(\mathbb{R}^d) \Longrightarrow W(f, g) \in \Sigma_1^1(\mathbb{R}^{2d}). \tag{5.84}$$

If $g \in S_s^s(\mathbb{R}^d)$, $s \geq 1/2$, then

$$f \in S_s^s(\mathbb{R}^d) \Longleftrightarrow |V_g(f)(z)| \lesssim e^{-\epsilon|z|^{1/s}} \quad \text{for some } \epsilon > 0. \tag{5.85}$$

We can extend modulation spaces to the realm of ultradistributions as follows. Note that we permit now exponential submultiplicative weights v, and they represent, in fact, the more interesting case here.

Definition 5.2.6. Given $g \in \Sigma_1^1(\mathbb{R}^d)$, a weight function $m \in \mathcal{M}_v(\mathbb{R}^{2d})$, and $1 \le p,q \le \infty$, the *modulation space* $M_m^{p,q}(\mathbb{R}^d)$ consists of all tempered ultradistributions $f \in (\Sigma_1^1)'(\mathbb{R}^d)$ such that $V_g f \in L_m^{p,q}(\mathbb{R}^{2d})$. The norm on $M_m^{p,q}(\mathbb{R}^d)$ is as usual (cf. (2.20))

$$\|f\|_{M_m^{p,q}} = \|V_g f\|_{L_m^{p,q}} = \left(\int_{\mathbb{R}^d}\left(\int_{\mathbb{R}^d} |V_g f(x,\xi)|^p m(x,\xi)^p \, dx\right)^{q/p} d\xi\right)^{1/q} \tag{5.86}$$

(obvious changes if $p = \infty$ or $q = \infty$).

We observe that for $f, g \in \Sigma_1^1(\mathbb{R}^d)$ the above integral is convergent and thus $\Sigma_1^1(\mathbb{R}^d) \subset M_m^{p,q}(\mathbb{R}^d)$, $1 \le p,q \le \infty$, cf. [87], with dense inclusion when $p,q < \infty$, cf. [58]. When $p = q$, we simply write $M_m^p(\mathbb{R}^d)$ instead of $M_m^{p,p}(\mathbb{R}^d)$. The spaces $M_m^{p,q}(\mathbb{R}^d)$ are Banach spaces, and every nonzero $g \in M_v^1(\mathbb{R}^d)$ yields an equivalent norm in (5.86), so $M_m^{p,q}(\mathbb{R}^d)$ is independent of the choice of $g \in M_v^1(\mathbb{R}^d)$.

In this section we characterize the smoothness and growth of a function f on $\mathbb{R}^d$ in terms of the decay of its STFT $V_g f$, for a suitable window g. In the proofs we shall detail the relations among the constants C_f, C_g, $C_{f,g}$ and ϵ which come into play, since they could be useful for constructing efficient algorithms related to these objects.

Theorem 5.2.7. *Consider $s > 0$, $m \in \mathcal{M}_v(\mathbb{R}^d)$, $g \in M^1_{v\otimes 1}(\mathbb{R}^d) \setminus \{0\}$ such that there exists $C_g > 0$ satisfying*

$$\|\partial^\alpha g\|_{L_v^1(\mathbb{R}^d)} \lesssim C_g^{|\alpha|}(\alpha!)^s, \quad \alpha \in \mathbb{N}^d. \tag{5.87}$$

For $f \in C^\infty(\mathbb{R}^d)$ the following conditions are equivalent:

(i) *There exists a constant $C_f > 0$ such that*

$$|\partial^\alpha f(x)| \lesssim m(x) C_f^{|\alpha|}(\alpha!)^s, \quad x \in \mathbb{R}^d,\ \alpha \in \mathbb{N}^d. \tag{5.88}$$

(ii) *There exists a constant $C_{f,g} > 0$ such that*

$$|\xi^\alpha V_g f(x,\xi)| \lesssim m(x) C_{f,g}^{|\alpha|}(\alpha!)^s, \quad (x,\xi) \in \mathbb{R}^{2d},\ \alpha \in \mathbb{N}^d. \tag{5.89}$$

(iii) *There exists a constant $\epsilon > 0$ such that*

$$|V_g f(x,\xi)| \lesssim m(x) e^{-\epsilon|\xi|^{\frac{1}{s}}}, \quad (x,\xi) \in \mathbb{R}^{2d},\ \alpha \in \mathbb{N}^d. \tag{5.90}$$

If equivalent conditions (5.88), (5.89), and (5.90) are satisfied, we say that f is an m-Gevrey symbol when $s > 1$, an m-analytic symbol if $s = 1$, and an m-ultraanalytic symbol when $s < 1$.

Proof. (i) $\Longrightarrow$ (ii). Since $f, T_x g \in C^\infty(\mathbb{R}^d)$, for every $x \in \mathbb{R}^d$, we can use Leibniz formula and write

$$\partial^\alpha(fT_xg)(t) = \sum_{\beta\leq\alpha}\binom{\alpha}{\beta}\partial^{\alpha-\beta}f(t)T_x(\partial^\beta g)(t), \quad t \in \mathbb{R}^d,\ \alpha \in \mathbb{N}^d.$$

Let us estimate the $L^1(\mathbb{R}^d)$-norm of the products $\partial^{\alpha-\beta}fT_x(\partial^\beta g)$. Using the positivity and v-moderateness of the weight m, Hölder's inequality, and assumptions (5.88) and (5.87),

$$\begin{aligned}\|\partial^{\alpha-\beta}fT_x(\partial^\beta g)\|_{L^1} &\leq \|\partial^{\alpha-\beta}f\|_{L^\infty_{1/m}}\|mT_x(\partial^\beta g)\|_{L^1}\\ &\lesssim C_f^{|\alpha-\beta|}((\alpha-\beta)!)^s\|v(\cdot-x)\partial^\beta g(\cdot-x)\|_{L^1}m(x)\\ &\lesssim C_f^{|\alpha-\beta|}((\alpha-\beta)!)^s C_g^{|\beta|}(\beta!)^s m(x)\\ &\leq C^{|\alpha|}(\alpha!)^s m(x)\end{aligned}$$

where $C := \max\{C_f, C_g\}$ and we have used $(\alpha-\beta)!\beta! \leq \alpha!$. These estimates tell us, in particular, that $V_gf(x,\xi) = \mathcal{F}(fT_xg)(\xi)$ is well-defined. We can exchange partial derivatives and Fourier transform as follows:

$$\xi^\alpha V_gf(x,\xi) = \frac{1}{(2\pi i)^{|\alpha|}}\mathcal{F}(\partial^\alpha(fT_xg))(\xi) = \frac{1}{(2\pi i)^{|\alpha|}}\sum_{\beta\leq\alpha}\binom{\alpha}{\beta}\mathcal{F}(\partial^{\alpha-\beta}fT_x(\partial^\beta g))(\xi).$$

Using

$$|\mathcal{F}(\partial^{\alpha-\beta}fT_x(\partial^\beta g))(\xi)| \leq \|\mathcal{F}(\partial^{\alpha-\beta}fT_x(\partial^\beta g))\|_{L^\infty} \leq \|\partial^{\alpha-\beta}fT_x(\partial^\beta g)\|_{L^1},$$

the majorizations above, and (0.1),

$$|\xi^\alpha V_gf(x,\xi)| \lesssim \frac{m(x)}{(2\pi)^{|\alpha|}}\sum_{\beta\leq\alpha}\binom{\alpha}{\beta}C^{|\alpha|}(\alpha!)^s = \frac{m(x)}{\pi^{|\alpha|}}C^{|\alpha|}(\alpha!)^s.$$

This proves (5.89), with constant $C_{f,g} := \frac{\max\{C_f,C_g\}}{\pi}$.

(ii) $\Longrightarrow$ (i). We use the inversion formula (1.55), observing that (5.89) implies $f \in M^{\infty,1}_{m\otimes 1}(\mathbb{R}^d)$, so that the equality in (1.55) holds a. e. and we can assume that it holds everywhere since f is smooth. Let us consider the partial derivatives of f and exchange them with the integrals, the estimates below will provide a justification of this operation. So formally we can write

$$\partial^\alpha f(t) = \frac{1}{\|g\|_2^2}\int_{\mathbb{R}^{2d}} V_gf(x,\xi)\partial^\alpha(M_\xi T_xg)(t)\,dx\,d\xi, \quad t \in \mathbb{R}^d.$$

Using Leibniz formula $\partial^\alpha(M_\xi T_xg) = \sum_{\beta\leq\alpha}\binom{\alpha}{\beta}(2\pi i\xi)^\beta M_\xi T_x(\partial^{\alpha-\beta}g)$, we estimate

$$|\partial^\alpha f(t)| \leq \frac{1}{\|g\|_2^2}\sum_{\beta\leq\alpha}\binom{\alpha}{\beta}(2\pi)^{|\beta|}\int_{\mathbb{R}^{2d}}|V_gf(x,\xi)\xi^\beta|\cdot|T_x\partial^{\alpha-\beta}g(t)|\,dx\,d\xi, \quad t \in \mathbb{R}^d.$$

We set

$$I_{\alpha,\beta}(t) := \int_{\mathbb{R}^{2d}} |V_g f(x,\xi)\xi^\beta| \cdot |T_x \partial^{\alpha-\beta} g(t)| \, dx \, d\xi$$

and prove that, for every fixed $t \in \mathbb{R}^d$, $I_{\alpha,\beta}(t)$ are absolutely convergent integrals. Using $1 + |\xi|^{d+1} \leq c_d \sum_{|\gamma|\leq d+1} |\xi^\gamma|$, where c_d depends only on the dimension d, and assumption (5.89),

$$\begin{aligned}
I_{\alpha,\beta}(t) &= \int_{\mathbb{R}^{2d}} |V_g f(x,\xi)\xi^\beta| \frac{1 + |\xi|^{d+1}}{1 + |\xi|^{d+1}} |T_x \partial^{\alpha-\beta} g(t)| \, dx \, d\xi \\
&\lesssim \int_{\mathbb{R}^{2d}} \sum_{|\gamma|\leq d+1} |V_g f(x,\xi)\xi^{\beta+\gamma}| \frac{1}{1 + |\xi|^{d+1}} |T_x \partial^{\alpha-\beta} g(t)| \, dx \, d\xi \\
&\lesssim \sum_{|\gamma|\leq d+1} C_{f,g}^{|\beta+\gamma|} ((\beta+\gamma)!)^s \Big(\int_{\mathbb{R}^d} \frac{1}{1 + |\xi|^{d+1}} \, d\xi \Big) \Big(\int_{\mathbb{R}^d} m(x) |T_x \partial^{\alpha-\beta} g(t)| \, dx \Big).
\end{aligned}$$

Now, $(\beta+\gamma)! \leq 2^{|\beta|+|\gamma|}(\beta!)(\gamma!) \leq 2^{d+1}(d+1)!2^{|\beta|}\beta!$ and $C_{f,g}^{|\beta+\gamma|} = C_{f,g}^{d+1} C_{f,g}^{|\beta|}$ so forgetting about the constants depending only on d, f, and g,

$$\begin{aligned}
I_{\alpha,\beta}(t) &\lesssim C_{f,g}^{|\beta|} 2^{s\beta} (\beta!)^s \Big(\int_{\mathbb{R}^d} m(x) |T_x \partial^{\alpha-\beta} g(t)| \, dx \Big) \\
&\lesssim C_{f,g}^{|\beta|} 2^{s\beta} (\beta!)^s m(t) \Big(\int_{\mathbb{R}^d} v(x-t) |\partial^{\alpha-\beta} g(t-x)| \, dx \Big) \\
&= C_{f,g}^{|\beta|} 2^{s\beta} (\beta!)^s m(t) \|\partial^{\alpha-\beta} g\|_{L^1_v}.
\end{aligned}$$

The estimate above and assumption (5.87) on g allow the following majorization:

$$\begin{aligned}
|\partial^\alpha f(t)| &\lesssim \frac{m(t)}{\|g\|_2^2} \sum_{\beta\leq\alpha} \binom{\alpha}{\beta} (2\pi)^{|\beta|} (2^s C_{f,g})^{|\beta|} (\beta!)^s C_g^{|\alpha-\beta|} ((\alpha-\beta)!)^s \\
&\leq \frac{m(t)}{\|g\|_2^2} (\alpha!)^s \max\{C_{f,g}, C_g\}^{|\alpha|} 2^{s|\alpha|} (2\pi)^{|\alpha|} 2^{|\alpha|} \\
&\lesssim m(t) C_f^\alpha (\alpha!)^s,
\end{aligned}$$

where $C_f := 2^{s+2}\pi \max\{C_{f,g}, C_g\}$, and we used $(\beta)!(\alpha-\beta)! \leq \alpha!$ and (0.1).

(ii) $\Longleftrightarrow$ (iii). The equivalence is an immediate consequence of Proposition 5.2.4 and the subsequent remarks on the constants, where, for every fixed $x \in \mathbb{R}^d$, we choose $h(\xi) := V_g f(x,\xi)/m(x)$ and $r = s$. □

A natural question is whether we may find window functions satisfying (5.87). To this end, we recall the following characterization of Gelfand–Shilov spaces.

Proposition 5.2.8. *Let* $g \in \mathcal{S}(\mathbb{R}^d)$. *We have* $g \in S^s_r(\mathbb{R}^d)$, *with* $s, r > 0$, $r + s \geq 1$, *if and only if there exist constants* $A > 0$, $\epsilon > 0$ *such that*

$$|\partial^\alpha g(x)| \lesssim A^{|\alpha|}(\alpha!)^s e^{-\epsilon|x|^{\frac{1}{r}}}, \quad x \in \mathbb{R}^d,\ \alpha \in \mathbb{N}^d.$$

We have $g \in \Sigma^1_1(\mathbb{R}^d)$ *if and only if, for every* $A > 0$, $\epsilon > 0$,

$$|\partial^\alpha g(x)| \lesssim A^{|\alpha|}\alpha! e^{-\epsilon|x|}, \quad x \in \mathbb{R}^d,\ \alpha \in \mathbb{N}^d.$$

Proof. The first part of the statement is in [236, Proposition 6.1.7]. For the second part, the assumption $g \in \Sigma^1_1(\mathbb{R}^d)$ means that for every $A > 0, B > 0$,

$$|x^\beta \partial^\alpha g(x)| \lesssim A^{|\alpha|}B^{|\beta|}\alpha!\beta!, \quad x \in \mathbb{R}^d,\ \alpha,\ \beta \in \mathbb{N}^d.$$

Therefore the function $h = \frac{\partial^\alpha g}{A^{|\alpha|}\alpha!}$ satisfies (5.79) in Proposition 5.2.4 for every $C > 0$, and the estimate (5.78) is then satisfied for every $\epsilon > 0$ (see the observations after the proof of Proposition 5.2.4 for the uniformity of the constants). This gives the claim. □

Hence every $g \in S^s_r(\mathbb{R}^d)$, with $s > 0$, $0 < r < 1$, $s + r \geq 1$, satisfies (5.87) for every submultiplicative weight v. The same holds true if $g \in \Sigma^1_1(\mathbb{R}^d)$ and $s \geq 1$.

5.2.2 Almost diagonalization for pseudodifferential operators

Some results of [165] can be extended to the case of pseudodifferential operators having (ultra)analytic symbols. The results below cover also the Gevrey-analytic case which was already discussed in [165]. The almost diagonalization of different classes of pseudodifferential operators by using Gabor frames is given in [97, 243, 252, 275].

In view of the kernel theorem in $\mathcal{S} - \mathcal{S}'$, every linear continuous map from $\mathcal{S}(\mathbb{R}^d)$ to $\mathcal{S}'(\mathbb{R}^{2d})$ can be represented by means of a kernel $K \in \mathcal{S}'(\mathbb{R}^{2d})$ and, consequently, in the form of a Weyl pseudodifferential operator, cf. (4.5). Same arguments are valid in Gelfand–Shilov classes with $s = r \geq 1/2$ (see [226, 295]). Namely, considering $\sigma \in (S^s_s)'(\mathbb{R}^{2d})$, $f, g \in S^s_s(\mathbb{R}^d)$, we have $\mathrm{Op_w}(\sigma) : S^s_s(\mathbb{R}^d) \to (S^s_s)'(\mathbb{R}^d)$ continuously, in view of (5.83). The kernel K of $\mathrm{Op_w}(\sigma)$ belongs to $(S^s_s)'(\mathbb{R}^{2d})$. In the opposite direction, in view of Theorem 5.2.5, every linear continuous map from $S^s_s(\mathbb{R}^d)$ to $(S^s_s)'(\mathbb{R}^d)$ can be represented in the Weyl form $\mathrm{Op_w}(\sigma)$, with $\sigma \in (S^s_s)'(\mathbb{R}^{2d})$. The same holds for the couple of spaces Σ^1_1, $(\Sigma^1_1)'$.

The crucial relation between the action of the Weyl operator $\mathrm{Op_w}(\sigma)$ on time–frequency shifts and the short-time Fourier transform of its symbol, contained in [162, Lemma 3.1], can now be extended to Gelfand–Shilov spaces and their dual spaces as follows.

Lemma 5.2.9. *Consider $s \geq 1/2$, $g \in S_s^s(\mathbb{R}^d)$, $\Phi = W(g,g)$. Then, for $\sigma \in (S_s^s)'(\mathbb{R}^{2d})$,*

$$|\langle \mathrm{Op}_w(\sigma)\pi(z)g, \pi(w)g\rangle| = \left|V_\Phi\sigma\left(\frac{z+w}{2}, J(w-z)\right)\right| = |V_\Phi\sigma(u,v)| \tag{5.91}$$

and

$$|V_\Phi\sigma(u,v)| = \left|\left\langle \mathrm{Op}_w(\sigma)\pi\left(u - \frac{1}{2}J^{-1}(v)\right)g, \pi\left(u + \frac{1}{2}J^{-1}(v)\right)g\right\rangle\right|, \tag{5.92}$$

where we recall $J(z_1,z_2) = (z_2,-z_1)$. Moreover, the same results hold true if we replace the space $S_s^s(\mathbb{R}^d)$ with the space $\Sigma_1^1(\mathbb{R}^d)$.

Proof. Since $\Phi = W(g,g) \in S_s^s(\mathbb{R}^{2d})$ for $g \in S_s^s(\mathbb{R}^d)$ the duality $\langle\sigma, \pi(u,v)\Phi\rangle_{(S_s^s)'\times S_s^s}$ is well defined so that the short-time Fourier transform $V_\Phi\sigma(u,v)$ makes sense. The same pattern applies to the case $g \in \Sigma_1^1(\mathbb{R}^d)$. Using the weak definition of the Weyl operator in (4.6) and the intertwining property in Proposition 1.3.6 (iv), we can write

$$\begin{aligned}\langle \mathrm{Op}_w(\sigma)\pi(z)g, \pi(w)g\rangle &= \langle\sigma, W(\pi(w)g, \pi(z)g)\rangle\\ &= \langle\sigma, cM_{J(w-z)}T_{\frac{w+z}{2}}W(g,g)\rangle\\ &= \bar{c}V_{W(g,g)}\sigma\left(\frac{w+z}{2}, J(w-z)\right),\end{aligned} \tag{5.93}$$

with c being the phase factor ($|c| = 1$). Setting $u = (w+z)/2$, $v = J(w-z)$, we obtain (5.91). Furthermore, with $w = u + J^{-1}(v)/2$ and $z = u - J^{-1}(v)/2$, reading formula (5.93) backwards yields (5.92). □

We now exhibit a characterization of smooth symbols $\sigma \in \mathcal{C}^\infty(\mathbb{R}^{2d})$ for $\mathrm{Op}_w(\sigma)$ in terms of the continuous Gabor matrix.

Theorem 5.2.10. *Let $s \geq 1/2$ and $m \in \mathcal{M}_v(\mathbb{R}^{2d})$. If $1/2 \leq s < 1$, consider a window function $g \in S_s^s(\mathbb{R}^d)$; otherwise, if $s \geq 1$, assume either $g \in \Sigma_1^1(\mathbb{R}^d)$, or $g \in S_s^s(\mathbb{R}^d)$ and the following growth condition on the weight v:*

$$v(z) \lesssim e^{\epsilon|z|^{1/s}}, \quad z \in \mathbb{R}^{2d},$$

for every $\epsilon > 0$. Then the following properties are equivalent for $\sigma \in \mathcal{C}^\infty(\mathbb{R}^{2d})$:

(i) *The symbol σ satisfies*

$$|\partial^\alpha\sigma(z)| \lesssim m(z)C^{|\alpha|}(\alpha!)^s, \quad \forall z \in \mathbb{R}^{2d}, \ \forall\alpha \in \mathbb{N}^{2d}. \tag{5.94}$$

(ii) *There exists $\epsilon > 0$ such that*

$$|\langle \mathrm{Op}_w(\sigma)\pi(z)g, \pi(w)g\rangle| \lesssim m\left(\frac{w+z}{2}\right)e^{-\epsilon|w-z|^{\frac{1}{s}}}, \quad \forall z,w \in \mathbb{R}^{2d}. \tag{5.95}$$

Proof. (i) $\Longrightarrow$ (ii). Proposition 5.2.8 used to the window $\Phi = W(g,g)$ in Lemma 5.2.9, which lives in the space $S_s^s(\mathbb{R}^{2d})$ since $g \in S_s^s(\mathbb{R}^d)$ by (5.83), and the assumptions on v imply that Φ satisfies the assumptions of Theorem 5.2.7. Hence, using the equivalence (5.88) $\Longleftrightarrow$ (5.90), the assumption (5.94) is equivalent to the following decay estimate of the corresponding short-time Fourier transform:

$$|V_\Phi \sigma(u,v)| \lesssim m(u) e^{-\epsilon|v|^{\frac{1}{s}}}, \quad u, v \in \mathbb{R}^{2d},$$

for a suitable $\epsilon > 0$, which, combined with (5.91), yields

$$\left| V_\Phi \sigma\left(\frac{z+w}{2}, J(w-z)\right)\right| \lesssim m\left(\frac{w+z}{2}\right) e^{-\epsilon|J(w-z)|^{\frac{1}{s}}} = m\left(\frac{w+z}{2}\right) e^{-\epsilon|w-z|^{\frac{1}{s}}},$$

that is, (ii) holds.

(ii) $\Longrightarrow$ (i). Relation (5.92) and the decay assumption (5.95) give

$$\begin{aligned} |V_\Phi \sigma(u,v)| &= \left|\left\langle \mathrm{Op_w}(\sigma)\pi\left(u - \frac{1}{2}J^{-1}(v)\right)g, \pi\left(u + \frac{1}{2}J^{-1}(v)\right)g\right\rangle\right| \\ &\lesssim m(u) e^{-\epsilon|J^{-1}(v)|^{\frac{1}{s}}} = m(u) e^{-\epsilon|v|^{\frac{1}{s}}}, \end{aligned}$$

and, using the equivalence (5.88) $\Longleftrightarrow$ (5.90), we obtain the claim. □

Of course, from (5.95) we deduce the discrete Gabor matrix decay in (5.98) below. The converse requires some preliminaries. We apply a result obtained by Gröchenig and Lyubarskii in [164]. They find sufficient conditions on the lattice $\Lambda = A\mathbb{Z}^2$, $A \in GL(2, \mathbb{R})$, such that $g = \sum_{k=0}^{n} c_k H_k$, with H_k Hermite function, forms a so-called Gabor (super)frame $\mathcal{G}(g, \Lambda)$. Besides, they prove the existence of dual windows γ that belong to the space $S_{1/2}^{1/2}(\mathbb{R})$ (cf. [164, Lemma 4.4]). This theory transfers to the d-dimensional case by taking a tensor product $g = g_1 \otimes \cdots \otimes g_d \in S_{1/2}^{1/2}(\mathbb{R}^d)$ of windows as above, which defines a Gabor frame on the lattice $\Lambda_1 \times \cdots \times \Lambda_d$ and possesses a dual window $\gamma = \gamma_1 \otimes \cdots \otimes \gamma_d$ in the same space $\in S_{1/2}^{1/2}(\mathbb{R}^d)$. Let us simply call *Gabor superframe* $\mathcal{G}(g, \Lambda)$ for $L^2(\mathbb{R}^d)$ a Gabor frame with the above properties. Gabor superframes are the key for the discretization of the kernel in (5.95). First, we need the preliminary result below, which reflects the algebra property of Gelfand–Shilov spaces. For $s \geq 1/2$, $\epsilon > 0$, we define the following weight functions:

$$w_{s,\epsilon}(z) := e^{-\epsilon|z|^{\frac{1}{s}}}, \quad z \in \mathbb{R}^{2d}. \tag{5.96}$$

Lemma 5.2.11. *Let $\Lambda \subset \mathbb{R}^{2d}$ be a lattice of $\mathbb{R}^{2d}$. Then the sampling $\{w_{s,\epsilon}(\lambda)\}_{\lambda\in\Lambda}$ of (5.96) (defined on $\mathbb{R}^{2d}$) satisfies*

$$(w_{s,\epsilon} * w_{s,\epsilon})(\lambda) := \sum_{\nu\in\Lambda} w_{s,\epsilon}(\lambda - \nu) w_{s,\epsilon}(\nu) \lesssim w_{s,\epsilon 2^{-1/s}}(\lambda). \tag{5.97}$$

Proof. We use the arguments of [160, Lemma 11.1.1(c)]. For $\lambda \in \Lambda$, we divide the lattice Λ into the subsets $N_\lambda = \{\nu \in \Lambda : |\lambda - \nu| \le |\lambda|/2\}$ and $N_\lambda^c = \{\nu \in \Lambda : |\lambda - \nu| > |\lambda|/2\}$. For $\nu \in N_\lambda$, $|\nu| \ge |\lambda|/2$ and $|\nu|^{1/s} \ge (|\lambda|/2)^{1/s}$, so

$$(w_{s,\epsilon} * w_{s,\epsilon})(\lambda) \le e^{-(\epsilon 2^{-\frac{1}{s}})|\lambda|^{\frac{1}{s}}} \Big(\sum_{\nu \in N_\lambda} e^{-\epsilon|\lambda-\nu|^{\frac{1}{s}}} + \sum_{\nu \in N_\lambda^c} e^{-\epsilon|\nu|^{\frac{1}{s}}} \Big) \lesssim e^{-(\epsilon 2^{-\frac{1}{s}})|\lambda|^{\frac{1}{s}}}.$$

This concludes the proof. □

Theorem 5.2.12. *Let $\mathcal{G}(g,\Lambda)$ be a Gabor superframe for $L^2(\mathbb{R}^d)$. Consider $m \in \mathcal{M}_v(\mathbb{R}^{2d})$, $s \ge 1/2$, and a symbol $\sigma \in C^\infty(\mathbb{R}^{2d})$. Then the following properties are equivalent:*
(i) *There exists $\epsilon > 0$ such that the estimate* (5.95) *holds.*
(ii) *There exists $\epsilon > 0$ such that*

$$|\langle \mathrm{Op}_w(\sigma)\pi(\mu)g, \pi(\lambda)g\rangle| \lesssim m\Big(\frac{\lambda+\mu}{2}\Big) e^{-\epsilon|\lambda-\mu|^{\frac{1}{s}}}, \quad \forall \lambda, \mu \in \Lambda. \tag{5.98}$$

Proof. It remains to show that (ii) $\Longrightarrow$ (i). The arguments of [162, Theorem 3.2] can be adapted to this proof by using a Gabor superframe $\mathcal{G}(g,\Lambda)$, with a dual window $\gamma \in S^{1/2}_{1/2}(\mathbb{R}^d)$.

Let $\mathcal{Q}$ be a symmetric relatively compact fundamental domain of the lattice $\Lambda \subset \mathbb{R}^{2d}$. Given $w, z \in \mathbb{R}^{2d}$, we can write them uniquely as $w = \lambda + u$, $z = \mu + u'$, for $\lambda, \mu \in \Lambda$ and $u, u' \in \mathcal{Q}$. Using the Gabor reproducing formula for the time–frequency shift $\pi(u)g \in S^{1/2}_{1/2}(\mathbb{R}^d)$, we can write

$$\pi(u)g = \sum_{\nu \in \Lambda} \langle \pi(u)g, \pi(\nu)\gamma\rangle \pi(\nu)g.$$

Inserting the prior expansions in the assumption (5.98) gives

$$\begin{aligned} &|\langle \mathrm{Op}_w(\sigma)\pi(\mu+u')g, \pi(\lambda+u)g\rangle| \\ &\quad \le \sum_{\nu,\nu' \in \Lambda} |\langle \mathrm{Op}_w(\sigma)\pi(\mu+\nu')g, \pi(\lambda+\nu)g\rangle| |\langle \pi(u')g, \pi(\nu')\gamma\rangle| |\langle \pi(u)g, \pi(\nu)\gamma\rangle| \\ &\quad \lesssim \sum_{\nu,\nu' \in \Lambda} m\Big(\frac{\lambda+\mu+\nu+\nu'}{2}\Big) e^{-\epsilon|\lambda+\nu-\mu-\nu'|^{\frac{1}{s}}} |V_\gamma g(\nu'-u')| |V_\gamma g(\nu-u)|. \end{aligned} \tag{5.99}$$

Since the window functions g, γ are both in $S^{1/2}_{1/2}(\mathbb{R}^d)$, the STFT $V_\gamma g$ is in $S^{1/2}_{1/2}(\mathbb{R}^{2d})$ in view of (5.82). Thus there exists $h > 0$ such that $|V_\gamma g(z)| \lesssim e^{-h|z|^2}$, for every $z \in \mathbb{R}^{2d}$. In particular, being $\mathcal{Q}$ relatively compact and $u \in \mathcal{Q}$,

$$|V_\gamma g(\nu - u)| \lesssim e^{-h|\nu-u|^2} \le \sup_{u \in Q} e^{-h|\nu-u|^2} \lesssim e^{-h|\nu|^2}.$$

The assumption $m \in \mathcal{M}_v(\mathbb{R}^{2d})$ yields

$$m\Big(\frac{\lambda+\mu+\nu+\nu'}{2}\Big) \lesssim m\Big(\frac{\lambda+\mu}{2}\Big) v\Big(\frac{\nu}{2}\Big) v\Big(\frac{\nu'}{2}\Big)$$

and, for every $0 < \tilde{h} < h$,

$$v\left(\frac{\nu}{2}\right)e^{-h|\nu|^2} \lesssim e^{-\tilde{h}|\nu|^2}, \quad \forall \nu \in \Lambda.$$

Inserting these estimates in (5.99) yields

$$\begin{aligned}
&|\langle \mathrm{Op}_w(\sigma)\pi(\mu+u')g, \pi(\lambda+u)g\rangle| \\
&\quad \lesssim m\left(\frac{\lambda+\mu}{2}\right)\sum_{\nu,\nu'\in\Lambda} e^{-\epsilon|\lambda+\nu-\mu-\nu'|^{\frac{1}{s}}} v\left(\frac{\nu}{2}\right)e^{-h|\nu|^2} v\left(\frac{\nu'}{2}\right)e^{-h|\nu'|^2} \\
&\quad \lesssim m\left(\frac{\lambda+\mu}{2}\right)\sum_{\nu,\nu'\in\Lambda} e^{-\epsilon|\lambda+\nu-\mu-\nu'|^{\frac{1}{s}}} e^{-\tilde{h}|\nu|^2} e^{-\tilde{h}|\nu'|^2} \\
&\quad \lesssim m\left(\frac{\lambda+\mu}{2}\right)\sum_{\nu,\nu'\in\Lambda} e^{-b(|\lambda+\nu-\mu-\nu'|^{\frac{1}{s}}+|\nu|^{\frac{1}{s}}+|\nu'|^{\frac{1}{s}})},
\end{aligned} \tag{5.100}$$

with $b = \epsilon$ for $s > 1/2$ whereas $b = \min\{\epsilon, \tilde{h}\}$ for $s = 1/2$ (there may be a loss of decay). We observe that the row (5.100) can be rewritten as

$$m\left(\frac{\lambda+\mu}{2}\right)(w_{s,b} * w_{s,b} * w_{s,b})(\lambda-\mu),$$

where $w_{s,b}(\lambda) = e^{-b|\lambda|^{1/s}}$. Now we apply Lemma 5.2.11 twice and obtain

$$(5.100) \lesssim m\left(\frac{\lambda+\mu}{2}\right)e^{-\tilde{\epsilon}|\lambda-\mu|^{\frac{1}{s}}} \tag{5.101}$$

with $\tilde{\epsilon} = b2^{-2/s}$.

If $w, z \in \mathbb{R}^{2d}$ and $w = \lambda + u$, $z = \mu + u'$, $\lambda, \mu \in \Lambda$, $u, u' \in \mathcal{Q}$, then $\lambda - \mu = w - z + u' - u$ and $u' - u \in \mathcal{Q} - \mathcal{Q}$, which is a relatively compact set, thus

$$e^{-\tilde{\epsilon}|\lambda-\mu|^{\frac{1}{s}}} \lesssim \sup_{u\in\mathcal{Q}-\mathcal{Q}} e^{-\tilde{\epsilon}|w-z+u|^{\frac{1}{s}}} \lesssim e^{-\tilde{\epsilon}|w-z|^{\frac{1}{s}}}. \tag{5.102}$$

Finally, the v-moderateness of the weight $m \in \mathcal{M}_v(\mathbb{R}^{2d})$, together with the fact that v is continuous and the set $\mathcal{Q} + \mathcal{Q}$ is relatively compact, let us write

$$\begin{aligned}
m\left(\frac{\lambda+\nu}{2}\right) &= m\left(\frac{w+z}{2} - \frac{u+u'}{2}\right) \lesssim m\left(\frac{w+z}{2}\right)v\left(-\frac{u+u'}{2}\right) \\
&\lesssim m\left(\frac{w+z}{2}\right)\sup_{u\in\mathcal{Q}+\mathcal{Q}} v\left(\frac{u}{2}\right) \lesssim m\left(\frac{w+z}{2}\right).
\end{aligned} \tag{5.103}$$

Combining estimates (5.102) and (5.103) with (5.101), we obtain (5.95), with the parameter $\epsilon = \tilde{\epsilon}$ which appears in (5.101). □

5.2.3 Sparsity of the Gabor matrix

The operators $\mathrm{Op}_w(\sigma)$ which satisfy Theorem 5.2.10, say, with $m = \nu = 1$, enjoy a fundamental sparsity property. Let $\mathcal{G}(g, \Lambda)$ be a Gabor frame for $L^2(\mathbb{R}^d)$, with $g \in S^s_s(\mathbb{R}^d)$, $s \geq 1/2$. Then, as we saw,

$$|\langle \mathrm{Op}_w(\sigma)\pi(\mu)g, \pi(\lambda)g\rangle| \leq Ce^{-\epsilon|\lambda-\mu|^{\frac{1}{s}}}, \quad \forall \lambda, \mu \in \Lambda, \tag{5.104}$$

with suitable constants $C > 0$, $\epsilon > 0$. This gives at once an exponential-type sparsity, in the sense specified by Proposition 5.2.15 (cf. [41, 173] for the more standard notion of superpolynomial sparsity). We need first preliminary results on sequences.

Lemma 5.2.13. *Let $a = (a_n)_{n\in\mathbb{N}_+}$ be a nonincreasing sequence with $a_n \geq 0$, for every $n \in \mathbb{N}_+$. If $\|a\|_{\ell^1} < \infty$, then*

$$a_n \leq \frac{\|a\|_{\ell^1}}{n}, \quad n \in \mathbb{N}_+. \tag{5.105}$$

Proof. Since $a_n \geq 0$ for every $n \in \mathbb{N}_+$, for any $N \in \mathbb{N}$, we can write

$$\|a\|_{\ell^1} \geq \sum_{n=1}^{N} a_n \geq Na_N,$$

and the claim follows. □

Corollary 5.2.14. *If $a = (a_n)_{n\in\mathbb{N}_+} \in \ell^p$, for $0 < p < \infty$, then*

$$|a_n| \leq \frac{\|a\|_{\ell^p}}{n^{\frac{1}{p}}}, \quad n \in \mathbb{N}_+. \tag{5.106}$$

Proof. By assumption, $\|a\|^p_{\ell^p} = \sum_{n=1}^{\infty} |a_n|^p < \infty$. Since the series $\sum_{n=1}^{\infty} |a_n|^p$ is absolutely convergent, it is also unconditionally convergent, and we can rearrange the entries $|a_n|^p$ of the sequence $(|a_n|^p)_{n\in\mathbb{N}_+}$ in a decreasing order. The claim then follows by applying Lemma 5.2.13. □

Proposition 5.2.15. *Let the Gabor matrix $\langle \mathrm{Op}_w(\sigma)\pi(\mu)g, \pi(\lambda)g\rangle$ satisfy* (5.104). *Then it is sparse in the following sense. Let a be any column or row of the matrix, and let $|a|_n$ be the nth largest entry of the sequence a. Then $|a|_n$ satisfies*

$$|a|_n \leq Ce^{-\epsilon n^{\frac{1}{2ds}}}, \quad n \in \mathbb{N},$$

for some constants $C > 0, \epsilon > 0$.

Proof. By a discrete analog of Proposition 5.2.4, it suffices to prove that

$$n^{\alpha}|a|_n \leq C^{\alpha+1}(\alpha!)^{2ds}, \quad \alpha \in \mathbb{N}.$$

On the other hand, applying the estimate in (5.106) to the nth largest entry $|a|_n$, we have

$$n^{\frac{1}{p}} \cdot |a|_n \leq \|a\|_{\ell^p},$$

for every $0 < p < \infty$. Hence by (5.104) and setting $p = 1/\alpha$, we obtain

$$n^\alpha |a|_n \lesssim \Big(\sum_{\lambda \in \Lambda} e^{-\epsilon p|\lambda - \mu|^{\frac{1}{s}}} \Big)^{\frac{1}{p}} = \Big(\sum_{\lambda \in \Lambda} e^{-\epsilon p|\lambda|^{\frac{1}{s}}} \Big)^{\frac{1}{p}}.$$

Let $\mathcal{Q}$ be a fundamental domain of the lattice Λ. Then if $x \in \lambda + \mathcal{Q}$, $\lambda \in \Lambda$, we have $|x| \leq |\lambda| + C_0$, therefore $|x|^{1/s} \leq C_1(|\lambda|^{1/s} + 1)$. Hence

$$\begin{aligned}\sum_{\lambda \in \Lambda} e^{-\epsilon p|\lambda|^{\frac{1}{s}}} &\leq C_2 \int_{\mathbb{R}^{2d}} e^{-\epsilon p|x|^{\frac{1}{s}}}\, dx = \int_{\mathbb{S}^{2d-1}} d\sigma \int_0^{+\infty} e^{-\epsilon p \rho^{\frac{1}{s}}} \rho^{2d-1}\, d\rho \\ &= \frac{C_3 s}{(\epsilon p)^{2ds}} \int_0^{+\infty} e^{-t} t^{2ds-1}\, dt = \frac{C_3 s \Gamma(2ds)}{(\epsilon p)^{2ds}} = \frac{C_4}{p^{2ds}}.\end{aligned}$$

Finally, by Stirling's formula,

$$n^\alpha |a|_n \leq \frac{C_4^{1/p}}{p^{\frac{2ds}{p}}} \leq C_5^{\alpha+1} (\alpha!)^{2ds}. \qquad \square$$

5.2.4 Boundedness of pseudodifferential operators on modulation and Gelfand–Shilov spaces

Although not necessary for our numerical applications, we list some continuity results on modulation and Gelfand–Shilov spaces which follow easily from Theorem 5.2.10.

Proposition 5.2.16. *Let $s \geq 1/2$ and consider a symbol $\sigma \in \mathcal{C}^\infty(\mathbb{R}^{2d})$ satisfying the estimates*

$$|\partial^\alpha \sigma(z)| \lesssim C^{|\alpha|} (\alpha!)^s, \tag{5.107}$$

for some $C > 0$. Let $m \in \mathcal{M}_v(\mathbb{R}^{2d})$ and, if $s \geq 1$, assume the weight v satisfies

$$v(z) \lesssim e^{\epsilon |z|^{1/s}}, \quad z \in \mathbb{R}^{2d}$$

for every $\epsilon > 0$. Then the Weyl operator $\mathrm{Op}_w(\sigma)$ extends to a bounded operator on $M_m^{p,q}(\mathbb{R}^d)$.

Proof. Let $g \in \Sigma_1^1(\mathbb{R}^d)$ with $\|g\|_{L^2} = 1$. From the inversion formula (1.55),

$$V_g(\mathrm{Op}_w(\sigma)f)(u) = \int_{\mathbb{R}^{2d}} \langle \mathrm{Op}_w(\sigma)\pi(z)g, \pi(u)g\rangle V_g f(z)\, dz.$$

The desired result thus follows if we can prove that the map $M(\sigma)$ defined by

$$M(\sigma)G(u) = \int_{\mathbb{R}^{2d}} \langle \mathrm{Op}_w(\sigma)\pi(z)g, \pi(u)g\rangle G(z)\, dz \tag{5.108}$$

is continuous from $L^{p,q}_m(\mathbb{R}^{2d})$ into $L^{p,q}_m(\mathbb{R}^{2d})$. The characterization of Theorem 5.2.10 assures the existence of an $\epsilon > 0$ such that

$$|V_g(\mathrm{Op}_w(\sigma)f)| = |M(\sigma)V_g(f)| \lesssim (w_{s,\epsilon} * |V_g f|)(u), \quad u \in \mathbb{R}^{2d},$$

where the weight function $w_{s,\epsilon}$ on $\mathbb{R}^{2d}$ is defined in (5.96). The desired conclusion then follows from the relation $L^{p,q}_m * L^1_v \hookrightarrow L^{p,q}_m$, for $w_{s,\epsilon} \in L^1_v$ by the growth assumption on v. □

Proposition 5.2.17. *Let $s \geq 1/2$, and consider a symbol $\sigma \in C^\infty(\mathbb{R}^{2d})$ that satisfies* (5.107). *Then the Weyl operator* $\mathrm{Op}_w(\sigma)$ *is bounded on* $S^s_s(\mathbb{R}^d)$.

Proof. Fix a window function $g \in S^{1/2}_{1/2}(\mathbb{R}^d)$. For $f \in S^s_s(\mathbb{R}^d)$, we have $V_g f \in S^s_s(\mathbb{R}^{2d})$ in view of (5.82). Hence there exists a constant $h > 0$ such that $|V_g f|(z) \lesssim e^{-h|z|^{1/s}}$, for every $z \in \mathbb{R}^{2d}$. Thus, taking $\tilde{\epsilon} = \min\{\epsilon, h\}$ and using a continuous version of Lemma 5.2.11, for every $u \in \mathbb{R}^{2d}$,

$$|V_g(\mathrm{Op}_w(\sigma)f)(u)| \lesssim (w_{s,\epsilon} * w_{s,h})(u) \lesssim (w_{s,\tilde{\epsilon}} * w_{s,\tilde{\epsilon}})(u) \lesssim w_{s,\tilde{\epsilon}2^{-1/s}}(u).$$

Using (5.85), we obtain the claim. □

The reader interested in further boundedness results for pseudodifferential operators on Gelfand–Shilov spaces can look at the paper by J. Toft [289].

5.2.5 Evolution equations

In this section we apply the almost diagonalization result to the propagators of Cauchy problems for constant coefficient hyperbolic and parabolic-type equations. We consider operators

$$P(\partial_t, D_x) = \partial_t^m + \sum_{k=1}^m a_k(D_x)\partial_t^{m-k}, \quad t \in \mathbb{R},\ x \in \mathbb{R}^d, \tag{5.109}$$

where $a_k(\xi)$, $1 \leq k \leq m$, are polynomials. They may be nonhomogeneous, and their degree may be arbitrary. Here we set $D_{x_j} = \frac{1}{2\pi i}\partial_{x_j}$, $j = 1, \ldots, d$.

We are interested in the forward Cauchy problem

$$\begin{cases} P(\partial_t, D_x)u = 0, & (t,x) \in \mathbb{R}_+ \times \mathbb{R}^d, \\ \partial_t^k u(0,x) = u_k(x), & 0 \le k \le m-1, \end{cases}$$

where $u_k \in \mathcal{S}(\mathbb{R}^d)$, $0 \le k \le m-1$. Suppose that the *forward Hadamard–Petrowsky condition* is satisfied: *There exists a constant* $C > 0$ *such that*

$$(\tau, \xi) \in \mathbb{C} \times \mathbb{R}^d, \quad P(i\tau, \xi) = 0 \Longrightarrow \operatorname{Im} \tau \ge -C. \tag{5.110}$$

This is a sufficient and necessary condition for the above Cauchy problem with Schwartz data to be well posed [247, Section 3.10]. By taking the Fourier transform with respect to x and using classical results for the fundamental solution to ordinary differential operators [261, pp. 126–127], one sees that the solution is then given by

$$u(t,x) = \sum_{k=0}^{m-1} \partial_t^k E(t,\cdot) * \left(u_{m-1-k} + \sum_{j=1}^{m-k-1} a_j(D_x) u_{m-k-1-j} \right).$$

Here $E(t,x) = \mathcal{F}^{-1}_{\xi \to x} \sigma(t,\xi)$, where $\sigma(t,\xi)$ is the unique solution to

$$\left(\partial_t^m + \sum_{k=1}^{m} a_k(\xi) \partial_t^{m-k} \right) \sigma(t,\xi) = \delta(t)$$

supported in $[0,+\infty) \times \mathbb{R}^d$. The distribution $E(t,x)$ is therefore the fundamental solution of P supported in $[0,+\infty) \times \mathbb{R}^d$.

The study of the Cauchy problem is therefore reduced to that of the Fourier multiplier

$$\operatorname{Op}_w(\sigma) f = \sigma(t, D_x) f = \mathcal{F}^{-1} \sigma(t,\cdot) \mathcal{F} f = E(t,\cdot) * f. \tag{5.111}$$

(For Fourier multipliers the Weyl and Kohn–Nirenberg quantizations give the same operator.)

Example 5.2.18. Here are some classical operators and symbols $\sigma(t,\xi)$ of the corresponding propagators for $t \ge 0$ ($\sigma(t,\xi) = 0$ for $t < 0$):

- Wave operator $\partial_t^2 - \Delta$; $\sigma(t,\xi) = \frac{\sin(2\pi|\xi| t)}{2\pi|\xi|}$.
- Klein–Gordon operator $\partial_t^2 - \Delta + m^2$, with $m > 0$; $\sigma(t,\xi) = \frac{\sin(t\sqrt{4\pi^2|\xi|^2 + m^2})}{\sqrt{4\pi^2|\xi|^2 + m^2}}$.
- Heat operator $\partial_t - \Delta$; $\sigma(t,\xi) = e^{-4\pi^2|\xi|^2 t}$.

We now introduce an assumption under which the multiplier $\sigma(t, D_x)$ falls in the class of pseudodifferential operators considered above.

Assume that there are constants $C > 0$, $r \geq 1$ such that

$$(\tau,\zeta) \in \mathbb{C}\times\mathbb{C}^d, \quad P(i\tau,\zeta)=0 \Longrightarrow \operatorname{Im}\tau \geq -C(1+|\operatorname{Im}\zeta|)^r. \tag{5.112}$$

Clearly, this condition is stronger than the forward Hadamard–Petrowsky condition.

Theorem 5.2.19. *Assume P satisfies (5.112) for some $C > 0$, $r \geq 1$. Then the symbol $\sigma(t,\xi)$ of the corresponding propagator $\sigma(t,D_x)$ in (5.111) satisfies the following estimates:*

$$|\partial_\xi^\alpha \sigma(t,\xi)| \leq C^{(t+1)|\alpha|+t}(\alpha!)^\mu, \quad \xi \in \mathbb{R}^d,\ t \geq 0,\ \alpha \in \mathbb{N}^d, \tag{5.113}$$

with $\mu = 1 - 1/r$, for a new constant $C > 0$.

Proof. It is well known (see, e. g., [48, Proposition 1.3.2, Lemma 1.3.3], [247, Section 3.10]) that $\sigma(t,\xi)$ extends to an entire analytic function in the second variable, and that the estimates (5.112) imply the bound

$$|\sigma(t,\zeta)| \leq e^{Ct(1+|\operatorname{Im}\zeta|)^r}, \quad \zeta \in \mathbb{C}^d. \tag{5.114}$$

Now, given $\xi \in \mathbb{R}^d$, we consider the polydisk $B(\xi,R) = \prod_{j=1}^d B_j(\xi_j,R) = \{\zeta \in \mathbb{C}^d : |\zeta_j - \xi_j| \leq R, 1 \leq j \leq d\}$, with $R = (1+|\alpha|)^{1/r}$, $\alpha \in \mathbb{N}^d$. Observe that (5.114) implies

$$\sup_{\zeta \in B(\xi,R)} |\sigma(t,\zeta)| \leq e^{Ct(1+\sqrt{d}R)^r}. \tag{5.115}$$

The Cauchy's generalized integral formula

$$\partial_\xi^\alpha \sigma(t,\xi) = \frac{\alpha!}{(2\pi i)^d} \int \cdots \int_{\partial B_1(\xi_1,R)\times\cdots\times\partial B_d(\xi_d,R)} \frac{\sigma(t,\zeta_1,\ldots,\zeta_d)}{(\zeta_1-\xi_1)^{\alpha_1+1}\cdots(\zeta_d-\xi_d)^{\alpha_d+1}}\, d\zeta_1\cdots d\zeta_d$$

and estimate (5.115) yield

$$|\partial_\xi^\alpha \sigma(t,\xi)| \leq \frac{\alpha! e^{Ct(1+\sqrt{d}R)^r}}{R^{|\alpha|}} \leq C_1^{t(|\alpha|+1)} \frac{\alpha!}{(1+|\alpha|)^{|\alpha|/r}}. \tag{5.116}$$

Using Stirling formula and $1/r = 1-\mu$ gives

$$\frac{1}{(1+|\alpha|)^{|\alpha|/r}} \leq \frac{C_2^{|\alpha|}}{(\alpha!)^{1-\mu}},$$

which, combined with (5.116), provides the desired majorization (5.113). □

Observe that the assumption $r \geq 1$ in the above theorem implies $0 \leq \mu < 1$. Combining Theorems 5.2.19 and 5.2.10, we obtain at once our main application.

Theorem 5.2.20. *Assume P satisfies* (5.112) *for some $C > 0$, $r \geq 1$, and set $s = \max\{1/2, 1 - 1/r\}$. If $g \in S^s_s(\mathbb{R}^d)$ then $\sigma(t, D_x)$ in* (5.111) *satisfies*

$$|\langle \sigma(t, D_x)\pi(z)g, \pi(w)g\rangle| \leq Ce^{-\epsilon|w-z|^{\frac{1}{s}}}, \quad \forall z, w \in \mathbb{R}^{2d}, \tag{5.117}$$

for some $\epsilon > 0$ and for a new constant $C > 0$. The constants ϵ and C are uniform when t lies in bounded subsets of $[0, +\infty)$.

Notice that in the above theorem $1/2 \leq s < 1$, so that we always obtain superexponential decay, i. e., $1/s > 1$ in (5.117).

We now show that, if $P(\partial_t, D_x)$ is any hyperbolic operator, then (5.112) is satisfied with $r = 1$, and hence the above theorem applies with Gaussian decay ($s = 1/2$ in (5.117)), for windows $g \in S^{1/2}_{1/2}(\mathbb{R}^d)$.

We recall that the operator $P(\partial_t, D_x)$ is called hyperbolic with respect to t if the direction $N = (1, 0, \ldots, 0) \in \mathbb{R} \times \mathbb{R}^d$ is noncharacteristic for P (i. e., its principal symbol – the highest order homogeneous part in the symbol – does not vanish at N) and P satisfies the forward Hadamard–Petrowsky condition (5.110). It follows then that the operators $a_j(D_x)$ in (5.109) must have degree $\leq j$ and P has order m.

The wave and Klein–Gordon operators are, of course, the most important examples of hyperbolic operators. We emphasize, however, that P is *not* required to be strictly hyperbolic, namely the roots of the principal symbol are allowed to coincide.

For example, the operator $P = \partial_t^2 - \sum_{j,k=1}^d a_{j,k}\partial_{x_j}\partial_{x_k}$ is hyperbolic if the matrix $a_{j,k}$ is real, symmetric, and *positive semidefinite*.

Proposition 5.2.21. *Assume $P(\partial_t, D_x)$ is hyperbolic with respect to t. Then the condition* (5.112) *is satisfied with $r = 1$ for some $C > 0$, and hence*

$$|\langle \sigma(t, D_x)\pi(z)g, \pi(w)g\rangle| \leq Ce^{-\epsilon|w-z|^2}, \quad \forall z, w \in \mathbb{R}^{2d}, \tag{5.118}$$

if $g \in S^{1/2}_{1/2}(\mathbb{R}^d)$, for some $\epsilon > 0$ and for a new constant $C > 0$.

Proof. Denote by P_m the principal symbol of P and $\Gamma(P, N)$ the component of N in $\{\theta \in \mathbb{R} \times \mathbb{R}^d : P_m(\theta) \neq 0\}$. It follows from the hyperbolicity assumption (see, e. g., [193, Proposition 12.4.4]) that the symbol $Q(\tau, \xi) = P(i\tau, \xi)$ of $P(\partial_t, D_x)$ satisfies

$$Q(\Xi + \tau N + \sigma\theta) \neq 0 \quad \text{if } \Xi \in \mathbb{R}^{d+1},\ \operatorname{Im}\tau < -C,\ \operatorname{Im}\sigma \leq 0,\ \theta \in \Gamma(P, N),$$

for the same constant C which appears in (5.110). We then deduce that

$$(\tau, \zeta) \in \mathbb{C} \times \mathbb{C}^d, \quad Q(\tau, \zeta) = P(i\tau, \zeta) = 0 \Longrightarrow (\operatorname{Im}\tau, \operatorname{Im}\zeta) \notin -CN - \Gamma(P, N).$$

Indeed, if $(\tau, \zeta) = \Xi + i(-CN - \theta)$, with $\theta \in \Gamma(P, N)$, $\Xi \in \mathbb{R}^{d+1}$, then $\theta - \epsilon N \in \Gamma(P, N)$ is ϵ is small enough because $\Gamma(P, N)$ is open, and then

$$P(\tau, \zeta) = P(\Xi - i(C + \epsilon)N - i(\theta - \epsilon N)) \neq 0.$$

Hence $-(C + \operatorname{Im}\tau, \operatorname{Im}\zeta) \notin \Gamma(P,N)$, and since the cone $\Gamma(P,N)$ is open, this implies

$$\frac{-(C + \operatorname{Im}\tau, \operatorname{Im}\zeta)\cdot N}{|(C + \operatorname{Im}\tau, \operatorname{Im}\zeta)|} = \frac{-(C + \operatorname{Im}\tau)}{|(C + \operatorname{Im}\tau, \operatorname{Im}\zeta)|} \le C_1,$$

for some constant $C_1 < 1$, which gives

$$\operatorname{Im}\tau \ge -C - \frac{C_1}{1-C_1}|\operatorname{Im}\zeta|. \qquad \square$$

Example 5.2.22. Consider the wave operator $P = \partial_t^2 - \Delta$ in $\mathbb{R}\times\mathbb{R}^d$, with symbol $\sigma(t,\xi) = \frac{\sin(2\pi|\xi|t)}{2\pi|\xi|}$. Using some further information on the fundamental solution, we can give a very precise result on the matrix decay in dimension $d \le 3$. We take $g(x) = 2^{d/4}e^{-\pi x^2}$ as a window function, which is allowed because $g \in S^{1/2}_{1/2}(\mathbb{R}^d)$, and moreover $\|g\|_{L^2} = 1$ (Gaussian functions minimize the Heisenberg uncertainty so that they are, generally speaking, a natural choice for wave-packet decompositions). We claim that

$$|\langle\sigma(t,D_x)M_\xi T_x g, M_{\xi'}T_{x'}g\rangle| \le te^{-\frac{\pi}{2}[|\xi'-\xi|^2+(|x'-x|-t)_+^2]}, \quad x,x',\xi,\xi' \in \mathbb{R}^d,\ d\le 3,$$

where $(\cdot)_+$ denotes the positive part. This agrees with Proposition 5.2.21, with the constants made explicit.

In fact, in dimension $d \le 3$, we know that

$$\sigma(t,D_x)f(x) = \int f(x-y)\,d\mu_t(y)$$

where μ_t is a positive Borel measure, supported in the ball $B_t(0)$, with total mass $= t$ (see, e. g., [247, Chapter 4]). To be precise, we have $d\mu_t(y) = (1/2)\chi_{[-t,t]}dy$ in dimension 1, $d\mu_t(y) = (2\pi)^{-1}(t^2 - |y|^2)_+^{-1/2}dy$ in dimension 2, and $d\mu_t = (4\pi t)^{-1}d\sigma_{\partial B_t(0)}$ in dimension 3 (surface measure).

Hence, using $T_yM_\xi = e^{-2\pi i y\cdot\xi}M_\xi T_y$,

$$\langle\sigma(t,D_x)M_\xi T_x g, M_{\xi'}T_{x'}g\rangle = \int e^{-2\pi i y\cdot\xi}\langle M_\xi T_{x+y}g, M_{\xi'}T_{x'}g\rangle\,d\mu_t(y).$$

An explicit computation shows that

$$|\langle\pi(z)g,\pi(w)g\rangle| = e^{-\frac{\pi}{2}|w-z|^2}, \quad z,w \in \mathbb{R}^{2d},$$

so that

$$\begin{aligned}|\langle\sigma(t,D_x)M_\xi T_x g, M_{\xi'}T_{x'}g\rangle| &\le \int e^{-\frac{\pi}{2}[|\xi'-\xi|^2+|x'-x-y|^2]}\,d\mu_t(y)\\ &\le te^{-\frac{\pi}{2}[|\xi'-\xi|^2+(|x'-x|-t)_+^2]},\end{aligned}$$

which gives the claim.

Consider now the Gabor frame $\mathcal{G}(g,\Lambda)$, with $g(x) = 2^{d/4}e^{-\pi x^2}$, lattice $\Lambda = \mathbb{Z}^d \times (1/2)\mathbb{Z}^d$ ([160, Theorem 7.5.3]), and the Gabor matrix

$$T_{m',n',m,n} = \langle \sigma(t,D_x)M_nT_mg, M_{n'}T_{m'}g\rangle, \quad (m,n),\ (m',n') \in \Lambda.$$

We have

$$|T_{m',n',m,n}| \le \tilde{T}_{m',n',m,n} := te^{-\frac{\pi}{2}[|n'-n|^2+(|m'-m|-t)_+^2]}, \quad (m,n),\ (m',n') \in \Lambda,$$

with $d \le 3$. Figure 5.5 shows the magnitude of the entries, rearranged in decreasing order, of a generic column, e. g., $\tilde{T}_{m',n',0,0}$ (obtained for $m = n = 0$), at time $t = 0.75$, in dimension $d = 2$. In fact, the same figure applies to all columns, for $\tilde{T}_{m',n',m,n} = \tilde{T}_{m'-m,n'-n,0,0}$. This figure should be compared with [43, Figure 15], where a similar investigation was carried out for the curvelet matrix of the wave propagator on the unit square ($d = 2$) with periodic boundary conditions. It turns out that the Gabor decay is even better, despite the fact that we consider here the wave operator in the whole $\mathbb{R}^2$.

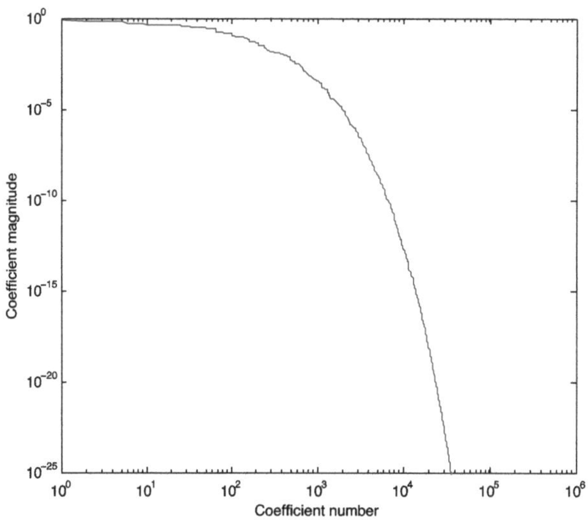

Figure 5.5: Decay of a generic column of the Gabor matrix for $\frac{\sin(2\pi|D|t)}{2\pi|D|}$ in dimension $d = 2$ and at time $t = 0.75$, with window $g(x) = \sqrt{2}e^{-\pi x^2}$ and lattice $\mathbb{Z}^2 \times (1/2)\mathbb{Z}^2$.

We now present a class of examples of operators which satisfy Theorem 5.2.20 and are not hyperbolic.

Example 5.2.23. Consider the operator

$$P(\partial_t, D_x) = \partial_t + (-\Delta)^k, \tag{5.119}$$

with $k \ge 1$ integer. When $k = 1$, we have the heat operator. Its symbol is the polynomial

$$P(i\tau,\zeta) = i\tau + (4\pi^2\zeta^2)^k.$$

We claim that it satisfies (5.112) with $r = 2k$.

If $\tau \in \mathbb{C}$, $\xi, \eta \in \mathbb{R}^d$, and $P(i\tau, \xi + i\eta) = 0$ then

$$\operatorname{Im}\tau = (2\pi)^{2k}\operatorname{Re}(|\xi|^2 + 2i\xi\cdot\eta - |\eta|^2)^k = (2\pi)^{2k}|\xi|^{2k} + Q(\xi,\eta),$$

where $Q(\xi,\eta)$ is a homogeneous polynomial of degree $2k$, with $Q(\xi,0) = 0$. Hence, for every $\epsilon > 0$ and some constant $C_\epsilon > 0$,

$$|Q(\xi,\eta)| \le \sum_{j+l=2k, j\ge 0, l\ge 1} |\xi|^j|\eta|^l \le \epsilon|\xi|^{2k} + C_\epsilon|\eta|^{2k},$$

where we applied the inequality $ab \le \frac{(\epsilon a)^p}{p} + \frac{(b/\epsilon)^q}{q}$ (with $a = |\xi|^j$, $b = |\eta|^l$, $p = 2k/j$ and $q = 2k/l$) to the terms of the sum with $j \ge 1$. Taking $\epsilon = (2\pi)^{2k}/2$, we get

$$\operatorname{Im}\tau \ge \frac{(2\pi)^{2k}}{2}|\xi|^{2k} - C|\eta|^{2k} \ge -C|\eta|^{2k},$$

which proves the claim.

Hence, for the operator P in (5.119), Theorem 5.2.20 applies with $r = 2k$ and $s = 1 - 1/(2k)$.

Remark 5.2.24. Theorems 5.2.10 and 5.2.12 allow also some applications to equations with variable coefficients. As an elementary example consider the transport equation

$$\begin{cases} \partial_t u - i(\sum_{j=1}^d a_j D_{x_j} + \sum_{j=1}^d b_j x_j)u = 0, \\ u(0,x) = u_0(x), \end{cases}$$

with $a_j, b_j \in \mathbb{R}$. The fundamental solution of the problem has symbol

$$\sigma(t,x,\xi) = \exp\left[it\sum_{j=1}^d (a_j\xi_j + b_j x_j)\right]$$

satisfying (5.94) with $m = 1$, $s = 0$, and the preceding arguments apply. More generally, we conjecture that pseudodifferential problems of the form

$$\partial_t - ia^w(x,D)u = 0, \quad u(0,x) = u_0(x),$$

where $a(x,\xi)$ is real-valued and $d_{x,\xi}a(x,\xi)$, as well as higher order derivatives, are bounded in $\mathbb{R}^{2d}$, can be treated similarly.

5.2.6 Some numerics for the heat equation

From Example 5.2.23 and Theorem 5.2.20, we infer that the heat propagator $\sigma(t, D_x) = e^{-4\pi^2 t|D|^2}$ satisfies the estimate

$$|\langle e^{-4\pi^2 t|D|^2}\pi(z)g, \pi(w)g\rangle| \le Ce^{-\epsilon|w-z|^2}, \quad \forall z, w \in \mathbb{R}^{2d},$$

for some $\epsilon > 0$, $C > 0$, if $g \in S^{1/2}_{1/2}(\mathbb{R}^d)$. Namely, the same decay as in the case of hyperbolic equations occurs.

In the following figures we summarized some numeric information about its Gabor discretization. Namely, Figure 5.6 shows the decay of a column of the Gabor matrix for the heat propagator, i. e.,

$$T_{m',n',0,0} = \langle e^{-4\pi^2 t|D|^2} g, M_{n'} T_{m'} g\rangle$$

for a Gaussian window, at different instant times t and in dimension $d = 2$. For $t = 0$, we get the identity operator, and therefore its matrix decay is optimal, compatibly with the uncertainty principle. As one sees from the other figures, the decay remains extremely good as time evolves. Also for $t = 0.75$, the decay matches that of the wave equation displayed in Figure 5.5, in spite that we no longer have here the finite speed propagation property.

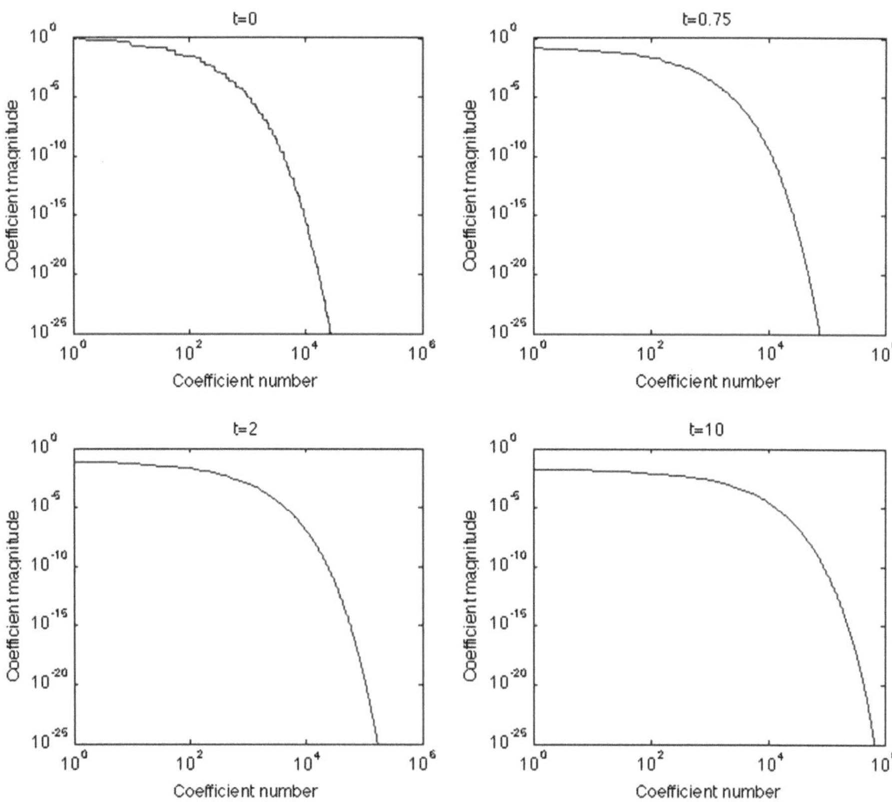

Figure 5.6: Decay of the column corresponding to $m = n = 0$, of the Gabor matrix for the heat propagator $e^{-4\pi^2 t|D|^2}$ in dimension $d = 2$ at different instant times, with window $g(x) = \sqrt{2}e^{-\pi x^2}$ and lattice $\mathbb{Z}^2 \times (1/2)\mathbb{Z}^2$.

Figure 5.7 instead shows, in dimension $d = 1$, the magnitude of the STFT of the solution $u(t,\cdot) = e^{-4\pi^2 t|D|^2}u_0$ of the heat equation, with initial datum $u_0(x) = \sqrt{2}e^{-\pi x^2}$ and window $g(x) = \sqrt{2}e^{-\pi x^2}$. The dissipative effect of the equation in the spatial domain is evident. A similar analysis for the Schrödinger equation will be carried on in the next chapter.

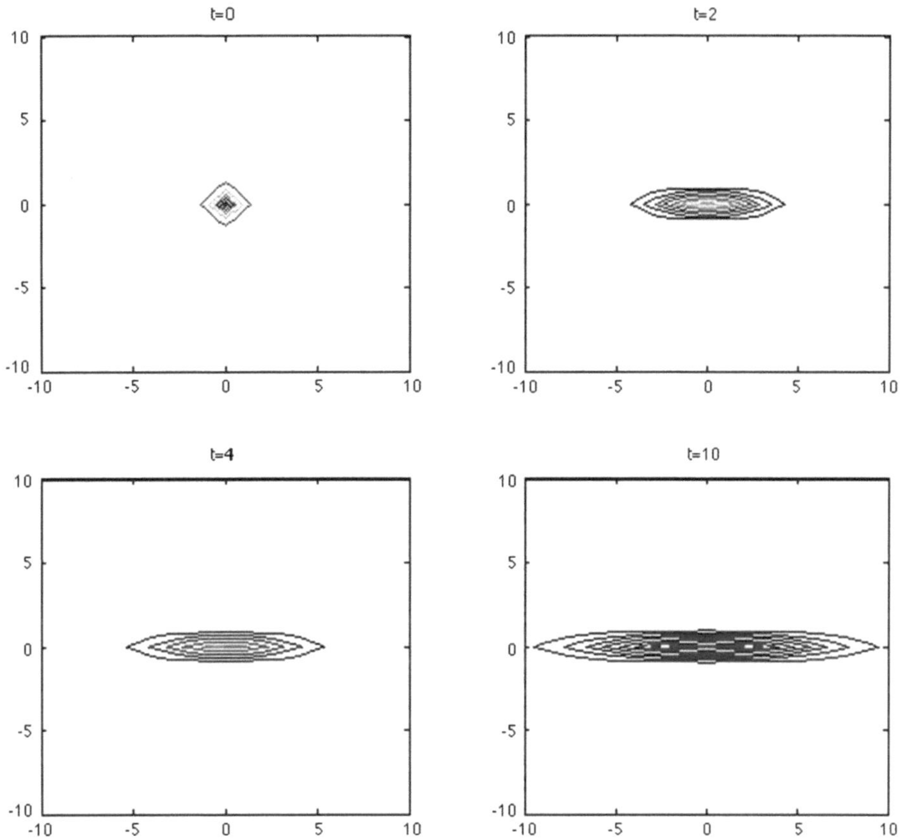

Figure 5.7: Magnitude of the STFT of the solution $u = e^{-4\pi^2 t|D|^2}u_0$ in dimension $d = 1$ and at different instant times, with initial datum and window $u_0(x) = g(x) = \sqrt{2}e^{-\pi x^2}$.

6 Fourier integral operators and applications to Schrödinger equations

In this chapter we define the Fourier integral operators, the object of our study, exhibiting their main features, mainly contained in [79]; see also [80, 88, 291]. Then we show that Gabor frames provide optimally sparse representations of such operators. Numerical examples for the Schrödinger case demonstrate the fast computation of these representations (cf. [77]). We address to the introduction of each section in the sequel for motivations and bibliographical remarks, and try to give here a flavor of the contents. The general idea is the same as in the second part of the preceding chapter, namely we express a linear continuous map $T : \mathcal{S}(\mathbb{R}^d) \to \mathcal{S}'(\mathbb{R}^d)$ in terms of the infinite Gabor matrix $T_{\mu,\lambda}$, $\lambda, \mu \in \Lambda$, defined by $T_{\mu,\lambda} = \langle Tg_\lambda, g_\mu \rangle$, with Gabor atoms $g_\lambda = \pi(\lambda)g$, see (0.15) for the notation and fix $\Lambda = \alpha\mathbb{Z}^d \times \beta\mathbb{Z}^d$, $\alpha, \beta > 0$ such that $\mathcal{G}(g, \alpha, \beta)$ is a frame. The operator T is understood to provide the solution of a well-posed problem for a PDE, and the window g is assumed to belong to a suitable function space. In the preceding chapter we proved for a class of problems, and very smooth g, that the Gabor matrix is almost diagonal, with exponential decay of $T_{\mu,\lambda}$ for $\langle \lambda - \mu \rangle \to \infty$. This provided exponential sparsity of the matrix with applications to numerics. In the present chapter we shall content ourselves with an algebraic off-diagonal decay, namely an estimate of the type $|T_{\mu,\lambda}| \leq C\langle \lambda - \mu \rangle^{-s}$, for some $C > 0$, $s \geq 0$. On the other hand, the analysis will be much more general, since the diagonal will be replaced by other submanifolds of $\Lambda \times \Lambda$. In fact, we shall consider the graph in $\Lambda \times \Lambda$ of a symplectic map $\chi : \mathbb{R}^{2d} \to \mathbb{R}^{2d}$. Assumptions on χ will be those natural in the global setting, including the linear symplectic map of Section 1.1 as basic examples. Consequently, we shall assume

$$|T_{\mu,\lambda}| \leq C\langle \lambda - \chi(\mu) \rangle^{-s}, \quad \lambda, \mu \in \Lambda. \tag{6.1}$$

Such estimates represent our very definition of Fourier integral operators, and grant sparsity properties and related numerics. Applications concern Schrödinger equations, under general assumptions on the Hamiltonians, including in particular quadratic Hamiltonians in the phase-space variables. In fact, the Gabor matrix of the corresponding Schrödinger propagator satisfies estimates of the form in (6.1), $\chi = \chi_t$ being the Hamiltonian flow at a given time $t \in \mathbb{R}$.

Let us describe in short the content of each section. Section 6.1 is devoted to the so-called naive theory of FIOs. It includes the case when the symplectomorphism χ is a small perturbation of the identity, as we have in the case of Schrödinger propagators $T = T_t$ for small values of the time t. We get then a simple expression for T of pseudodifferential type

$$Tf(x) = \int_{\mathbb{R}^d} e^{2\pi i \Phi(x,\eta)} \hat{f}(\eta)\, d\eta$$

https://doi.org/10.1515/9783110532456-007

which leads to estimates as in (6.1) and boundedness properties on modulation spaces. We shall refer to them as FIOs of type I.

In Section 6.2 we show that Gabor frames provide optimally sparse representations of such operators, with applications to numerics for the free particle Schrödinger equation. Section 6.3 presents our theory of FIOs in full generality, basing on (6.1). They form a Wiener algebra, containing as subset the FIOs of type I as shown in Section 6.4 (note that the space of the FIOs of type I is not closed under composition). In Section 6.5 we treat the relevant examples given by the generalized metaplectic operators, corresponding to the case when χ in (6.1) is a linear symplectic map, cf. Section 1.1. Section 6.6 is devoted to the applications to Schrödinger equations with Hamiltonians having principal part homogeneous of degree 2 in $z = (x, \xi)$. The propagator is expressed by the general FIOs of Section 6.3. With respect to the classical theory, novelty of our approach is represented by the fact that in terms of Gabor matrices we are able to provide a *global-in-time representation*. Besides, we introduce the notion of Gabor wave front set, which allows a precise description of propagation of microsingularities, see the conclusive examples, concerning in particular the quantum harmonic oscillator.

6.1 Fourier integral operators

Fourier integral operators (FIOs) are a mathematical tool to study a variety of problems arising in partial differential equations. Originally introduced by Lax [217] for the construction of parametrices in the Cauchy problem for hyperbolic equations, they have been widely employed to represent solutions in the framework of both pure and applied mathematics (see, e. g., the papers [34, 41, 114, 115, 175, 191], books [193, 269, 295] and references therein). In particular, a class of FIOs in $\mathbb{R}^d$ was applied by Helffer and Robert [187, 188] to study the spectral properties of a class of globally elliptic operators, generalizing the harmonic oscillator of the quantum mechanics. The Fourier integral operators we work with possess a phase function similar to those of [187, 188]. A simple example is the solution operator of the Cauchy problem for the Schrödinger equation with a quadratic Hamiltonian.

Following first the elementary presentation, given a function f on $\mathbb{R}^d$, the *Fourier integral operator* (FIO) T with symbol σ and phase Φ on $\mathbb{R}^{2d}$ can be formally defined by

$$Tf(x) = \int_{\mathbb{R}^d} e^{2\pi i \Phi(x,\eta)} \sigma(x,\eta) \hat{f}(\eta)\, d\eta.$$

Later we shall refer to such T as FIO of type I. The phase function $\Phi(x, \eta)$ is smooth on $\mathbb{R}^{2d}$, fulfills the estimates

$$|\partial_z^\alpha \Phi(z)| \le C_\alpha, \quad |\alpha| \ge 2,\ z \in \mathbb{R}^{2d}, \tag{6.2}$$

and the nondegeneracy condition

$$\left|\det \partial^2_{x,\eta}\Phi(x,\eta)\right| \geq \delta > 0, \quad (x,\xi) \in \mathbb{R}^{2d}. \tag{6.3}$$

The symbol σ on $\mathbb{R}^{2d}$ satisfies

$$\left|\partial_z^\alpha \sigma(z)\right| \leq C_\alpha, \quad |\alpha| \leq N, \text{ a. e. } z \in \mathbb{R}^{2d}, \tag{6.4}$$

for a fixed $N > 0$ (in the sequel we shall work also with rougher symbols). We denote as usual the Gabor frame $\mathcal{G}(g,\alpha,\beta) = \{g_{m,n}\}_{(m,n)}$, as defined in (5.1). Our goal will be to show that the Gabor matrix representation of an FIO T with respect to a Gabor frame with $g \in \mathcal{S}(\mathbb{R}^d)$ (defined in (5.3)) is well-organized (similarly to frames of curvelets and shearlets [41, 175]), provided that the symbol σ is smooth and satisfies the decay estimate (6.4) for every $N > 0$. Namely, we shall prove

For each $N > 0$, there exists a constant $C_N > 0$ such that

$$|T_{m',n',m,n}| \leq C_N \langle \chi(m,n) - (m',n')\rangle^{-2N}, \tag{6.5}$$

where χ is the canonical transformation generated by Φ (defined below).

In the special case of pseudodifferential operators, such an almost diagonalization was already obtained in [162, 252]. Indeed, pseudodifferential operators correspond to the phase $\Phi(x,\eta) = x\eta$ and canonical transformation $\chi(y,\eta) = (y,\eta)$.

One should mention that the use of almost diagonal estimates in proving continuity results goes back to the pioneering work [151], where the Calderón–Zygmund class of operators was studied via such a technique, and this was achieved by working with wavelets. Also observe that one simple case of a pseudodifferential operator is the Hilbert transform, and the usefulness of Gabor frames was first exhibited for this operator in [12].

A relevant byproduct of the results above are the boundedness properties of the operator T on modulation spaces. Indeed, the boundedness of the FIO T on $M^{p,q}(\mathbb{R}^d)$ is equivalent to that of the related Gabor matrix $\langle Tg_{m,n}, g_{m',n'}\rangle$ on the spaces of sequences $\ell^{p,q}$.

Hence, the estimates (6.5) readily give

For $1 \leq p < \infty$, the Fourier integral operator T, with symbol σ and phase Φ as above, extends to a continuous operator on $M^p(\mathbb{R}^d)$. (In the case $p = \infty$, the space $M^\infty(\mathbb{R}^d)$ is replaced by the closure of the Schwartz functions with respect to $\|\cdot\|_{M^\infty}$.)

The continuity property of an FIO T on $M^{p,q}$, with $p \neq q$, fails in general. Indeed, an example is provided by the operator $Tf(x) = e^{\pi i x^2} f(x)$, corresponding to $\Phi(x,\eta) = x\eta + x^2/2$, $\sigma \equiv 1$, which is bounded on $M^{p,q}$ if and only if $p = q$ (see Proposition 6.1.17 below).

To have a simple idea of the possible applications of the previous results, consider the Cauchy problem in (5.74) where H is the Weyl quantization of a quadratic form on $\mathbb{R}^d \times \mathbb{R}^d$. A simple example is the harmonic oscillator already defined in (5.75). The solution to (5.74) can be viewed as a one-parameter family of FIOs,

$$u(t,x) = e^{itH} u_0,$$

with symbol $\sigma \equiv 1$ and a phase given by a quadratic form $\Phi(x,\xi)$, satisfying trivially the preceding assumptions (6.2) and (6.3).

Our approach provides mathematical tools to solving Cauchy problems like (5.74) numerically, as we shall show in the sequel.

We remark that the L^2 boundedness of FIOs of the type above was first investigated Boulkhemair in [35], where he considered a more general situation compared to the condition (6.2). Namely, his assumption was

$$(\partial_z^\alpha \Phi) \in M^{\infty,1}(\mathbb{R}^{2d}), \quad |\alpha| = 2 \tag{6.6}$$

(observe that he did not assume that $|\alpha| \geq 2$, merely equality). In particular, no second order derivative of Φ needs to be smooth with this assumption. In [56] Concetti and Toft extend the results by Boulkhemair, using some basic arguments in [35] to show some continuity and Schatten properties of such FIOs. In particular, it is the first time where the authors prove that an FIO with symbol in $M^{\infty,1}(\mathbb{R}^{2d})$ is continuous on $M^p(\mathbb{R}^d)$. Finally, in [291], the authors extend the results in [56] to weighted modulation spaces.

6.1.1 Phase functions and canonical transformations

From now onward we shall work with a phase function satisfying the following assumptions.

Definition 6.1.1. A real phase function Φ on $\mathbb{R}^{2d}$ is called *tame* if the following three properties are satisfied:
A1. $\Phi \in C^\infty(\mathbb{R}^{2d})$;
A2. For $z = (x,\eta)$,

$$|\partial_z^\alpha \Phi(z)| \leq C_\alpha, \quad |\alpha| \geq 2; \tag{6.7}$$

A3. There exists $\delta > 0$ such that

$$|\det \partial_{x,\eta}^2 \Phi(x,\eta)| \geq \delta. \tag{6.8}$$

If we set

$$\begin{cases} y = \nabla_\eta \Phi(x,\eta), \\ \xi = \nabla_x \Phi(x,\eta), \end{cases} \tag{6.9}$$

we can solve with respect to (x,ξ) by the global inverse function theorem (see, e. g., [214]) and obtain a mapping χ defined by $(x,\xi) = \chi(y,\xi)$. The canonical transformation χ enjoys the following properties:

B1. $\chi : \mathbb{R}^{2d} \to \mathbb{R}^{2d}$ is smooth, invertible, and preserves the symplectic form in $\mathbb{R}^{2d}$, i. e., $dx \wedge d\xi = dy \wedge d\eta$; that is, χ is a *symplectomorphism*.

B2. For $z = (y,\eta)$,

$$|\partial_z^\alpha \chi(z)| \le C_\alpha, \quad |\partial_\zeta^\alpha \chi^{-1}(\zeta)| \le C_\alpha, \ |\alpha| \ge 1; \tag{6.10}$$

B3. There exists $\delta > 0$ such that, for $(x,\xi) = \chi(y,\eta)$,

$$\left|\det \frac{\partial x}{\partial y}(y,\eta)\right| \ge \delta. \tag{6.11}$$

Conversely, to every transformation χ satisfying *B1*, *B2*, *B3* (called *tame* canonical transformation in the sequel) corresponds a tame phase Φ, uniquely determined up to a constant. This can be easily proved by (6.11) and the global inverse function theorem [214].

When necessary and to avoid ambiguity, we shall denote by Φ_χ the phase function (up to constants) corresponding to the canonical transformation χ.

Observe that *B1* and *B2* imply that χ and χ^{-1} are globally Lipschitz. This property implies that

$$\langle w - \chi(z)\rangle \asymp \langle \chi^{-1}(w) - z\rangle \quad w,z \in \mathbb{R}^{2d},$$

which we will use frequently. Moreover, if χ and $\tilde{\chi}$ are two transformations satisfying *B1* and *B2*, the same is true for $\chi \circ \tilde{\chi}$, whereas the additional property *B3* is not necessarily preserved, even if χ and $\tilde{\chi}$ are linear. This reflects the lack of the algebra property of the corresponding FIOs; see Section 6.5 below.

6.1.2 Almost diagonalization of FIOs

For a given function f on $\mathbb{R}^d$, the FIO T with symbol σ and phase Φ can be formally defined by

$$Tf(x) = \int_{\mathbb{R}^d} e^{2\pi i \Phi(x,\eta)} \sigma(x,\eta) \hat{f}(\eta)\, d\eta. \tag{6.12}$$

To avoid technicalities, we initially take $f \in \mathcal{S}(\mathbb{R}^d)$ or, more generally, $f \in M^1(\mathbb{R}^d)$. If $\sigma \in L^\infty(\mathbb{R}^{2d})$ and the phase Φ is real then the integral converges absolutely and defines a function in $L^\infty(\mathbb{R}^d)$.

Assume that the phase function is tame as in Definition 6.1.1. We shall prove an almost diagonalization result for FIOs as above, with respect to a Gabor frame. Let us first consider the case of regular symbols. Precisely, for a given $N \in \mathbb{N}$, we consider symbols σ on $\mathbb{R}^{2d}$ satisfying, for $z = (x, \eta)$,

$$|\partial_z^\alpha \sigma(z)| \leq C_\alpha, \quad \text{a. e. } z \in \mathbb{R}^{2d}, \ |\alpha| \leq 2N \tag{6.13}$$

(here ∂_z^α denotes distributional derivatives).

Our goal is to study the decay properties of the matrix of the FIO T with respect to a Gabor frame $\{g_{m,n}\}_{(m,n)}$, with $g \in \mathcal{S}(\mathbb{R}^d)$.

Theorem 6.1.2. *Consider a tame phase function and a symbol satisfying* (6.13). *There exists $C_N > 0$ such that*

$$|\langle Tg_{m,n}, g_{m',n'}\rangle| \leq C_N \langle \nabla_z \Phi(m', n) - (n', m)\rangle^{-2N}, \tag{6.14}$$

with $(m, n), (m', n') \in \Lambda$.

Proof. Recall that the time-frequency shifts interchange under the action of the Fourier transform, $(T_x f)^\wedge = M_{-x}\hat{f}$ and $(M_\eta f)^\wedge = T_\eta \hat{f}$, besides they fulfill the commutation relations $T_x M_\eta = e^{-2\pi i x\cdot\eta} M_\eta T_x$. Using these properties, we compute the Gabor matrix entries as follows:

$$\begin{aligned}
&\langle Tg_{m,n}, g_{m',n'}\rangle \\
&= \int_{\mathbb{R}^d} Tg_{m,n}(x)\overline{M_{n'} T_{m'} g(x)}\, dx \\
&= \int_{\mathbb{R}^d}\int_{\mathbb{R}^d} e^{2\pi i\Phi(x,\eta)}\sigma(x,\eta) T_n M_{-m}\hat{g}(\eta) M_{-n'} T_{m'}\bar{g}(x)\, dx\, d\eta \\
&= \int_{\mathbb{R}^d}\int_{\mathbb{R}^d} M_{(0,-m)} T_{(0,-n)}(e^{2\pi i\Phi(x,\eta)}\sigma(x,\eta))\hat{g}(\eta) M_{-n'} T_{m'}\bar{g}(x)\, dx\, d\eta \\
&= \int_{\mathbb{R}^d}\int_{\mathbb{R}^d} T_{(-m',0)} M_{-(n',0)} M_{(0,-m)} T_{(0,-n)}(e^{2\pi i\Phi(x,\eta)}\sigma(x,\eta))\bar{g}(x)\hat{g}(\eta)\, dx\, d\eta \\
&= \int_{\mathbb{R}^d}\int_{\mathbb{R}^d} e^{2\pi i[\Phi(x+m',\eta+n)-(n',m)\cdot(x+m',\eta)]}\sigma(x+m',\eta+n)\bar{g}(x)\hat{g}(\eta)\, dx\, d\eta.
\end{aligned}$$

Since Φ is smooth, we may expand $\Phi(x, \eta)$ into a Taylor series around (m', n) and obtain

$$\Phi(x + m', \eta + n) = \Phi(m', n) + \nabla_z \Phi(m', n) \cdot (x, \eta) + \Phi_{2,(m',n)}(x, \eta)$$

where the remainder is given by

$$\Phi_{2,(m',n)}(x, \eta) = 2 \sum_{|\alpha|=2} \int_0^1 (1 - t)\partial^\alpha \Phi((m', n) + t(x, \eta))\, dt \frac{(x, \eta)^\alpha}{\alpha!}. \tag{6.15}$$

Whence, we can write

$$\begin{aligned}
&|\langle Tg_{m,n}, g_{m',n'}\rangle| \\
&= \left| \int_{\mathbb{R}^d}\int_{\mathbb{R}^d} e^{2\pi i\{[\nabla_z\Phi(m',n)-(n',m)]\cdot(x,\eta)\}} e^{2\pi i\Phi_{2,(m',n)}(x,\eta)} \sigma(x+m',\eta+n)\bar{g}(x)\hat{g}(\eta)\,dx\,d\eta \right|.
\end{aligned}$$

For $N \in \mathbb{N}$, using the identity

$$\begin{aligned}
&(1-\Delta_z)^N e^{2\pi i\{[\nabla_z\Phi(m',n)-(n',m)]\cdot(x,\eta)\}} \\
&\quad = \langle 2\pi(\nabla_z\Phi(m',n)-(n',m))\rangle^{2N} e^{2\pi i\{[\nabla_z\Phi(m',n)-(n',m)]\cdot(x,\eta)\}},
\end{aligned}$$

we integrate by parts and obtain

$$\begin{aligned}
&|\langle Tg_{m,n}, g_{m',n'}\rangle| \\
&= \frac{1}{\langle 2\pi(\nabla_z\Phi(m',n)-(n',m))\rangle^{2N}} \left| \int_{\mathbb{R}^d}\int_{\mathbb{R}^d} e^{2\pi i\{[\nabla_z\Phi(m',n)-(n',m)]\cdot(x,\eta)\}} \right. \\
&\quad \left. \times (1-\Delta_z)^N [e^{2\pi i\Phi_{2,(m',n)}(x,\eta)}\sigma(x+m',\eta+n)\bar{g}(x)\hat{g}(\eta)]\,dx\,d\eta \right|.
\end{aligned}$$

By means of Leibniz's formula, the factor

$$(1-\Delta_z)^N [e^{2\pi i\Phi_{2,(m',n)}(x,\eta)}\sigma(x+m',\eta+n)\bar{g}(x)\hat{g}(\eta)]$$

can be expressed as

$$e^{2\pi i\Phi_{2,(m',n)}(z)} \sum_{|\alpha|+|\beta|+|\gamma|\le 2N} C_{\alpha,\beta\gamma} p(\partial^{|\alpha|}\Phi_{2,(m',n)})(z)(\partial_z^\beta\sigma)(z+(m',n))\partial_z^\gamma(\bar{g}\otimes\hat{g})(z),$$

where $p(\partial^{|\alpha|}\Phi_{2,(m',n)})(z)$ is a polynomial made of derivatives of $\Phi_{2,(m',n)}$ of order at most $|\alpha|$.

As a consequence of (6.7), we have $\partial_z^\alpha\Phi_{2,(m',n)}(z) = O(\langle z\rangle^2)$, which combined with the assumption (6.13) and the hypothesis $g \in \mathcal{S}(\mathbb{R}^d)$ yields the desired estimate. □

Remark 6.1.3. More generally, one can consider symbols satisfying estimates of the form

$$|\partial_z^\alpha\sigma(z)| \le C_\alpha\mu(z), \quad \text{a. e. } z \in \mathbb{R}^{2d},\ |\alpha| \le 2N, \tag{6.16}$$

with $\mu \in \mathcal{M}_v$ and also more general windows g. Indeed, by arguing as above and using

$$|\partial_z^\beta\sigma(z+(m',n))| \le C_\beta\mu(z+(m',n)) \le C' C_\beta v(z)\mu(m',n),$$

one deduces the decay estimates

$$|\langle Tg_{m,n}, g_{m',n'}\rangle| \le C_N \frac{\mu(m',n)}{\langle\nabla_z\Phi(m',n)-(n',m)\rangle^{2N}},$$

provided the integral

$$\int_{\mathbb{R}^{2d}} \Big| \sum_{|\alpha|+|\beta|+|\gamma|\le 2N} C_{\alpha,\beta,\gamma} p(\partial^{|\alpha|}\Phi_{2,(m',n)})(z) v(z) \partial_z^\gamma (\bar{g}\otimes\hat{g})(z) \Big| \, dz \tag{6.17}$$

converges. This is guaranteed if, e. g., $\langle z\rangle^{2N} v(z)\partial_z^\gamma(\bar{g}\otimes\hat{g})(z) \in L^1$.

We are going to rewrite (6.14) in a form convenient for the applications to the continuity of FIOs in the next subsection. We need the following lemma.

Lemma 6.1.4. *Consider a tame phase function Φ. Then*

$$|\nabla_x\Phi(m',n) - n'| + |\nabla_\eta\Phi(m',n) - m| \gtrsim |x(m,n) - m'| + |\xi(m,n) - n'|, \tag{6.18}$$

where $(y,\eta)\mapsto(x,\xi)$ is the canonical transformation generated by Φ.

Proof. It suffices to prove the following inequalities:

$$|\nabla_\eta\Phi(m',n) - m| \gtrsim |x(m,n) - m'|, \tag{6.19}$$

$$|\nabla_x\Phi(m',n) - n'| \ge |\xi(m,n) - n'| - C|\nabla_\eta\Phi(m',n) - m|. \tag{6.20}$$

We observe that, by (6.9), we have

$$y = \nabla_\eta\Phi(x(y,\eta),\eta) \quad \forall (y,\eta)\in\mathbb{R}^{2d} \tag{6.21}$$

and

$$\nabla_x\Phi(x,\eta) = \xi(\nabla_\eta\Phi(x,\eta),\eta) \quad \forall (x,\eta)\in\mathbb{R}^{2d}. \tag{6.22}$$

Hence, we have $m = \nabla_\eta\Phi(x(m,n),n)$, so that

$$\begin{aligned} |\nabla_\eta\Phi(m',n) - m| &= |\nabla_\eta\Phi(m',n) - \nabla_\eta\Phi(x(m,n),n)| \\ &\gtrsim |x(m,n) - m'|, \end{aligned}$$

where the last inequality follows from the fact that, for every fixed η, the map $x\mapsto \nabla\Phi_\eta(x,\eta)$ has a Lipschitz inverse, with Lipschitz constant uniform with respect to η. This proves (6.19).

In order to prove (6.20), we observe that, in view of (6.22), one has

$$\begin{aligned} \nabla_x\Phi(m',n) - n' &= \xi(\nabla_\eta\Phi(m',n),n) - n' \\ &= \xi(m + \nabla_\eta\Phi(m',n) - m, n) - n' \\ &= \xi(m,n) - n' + O(\nabla_\eta\Phi(m',n) - m), \end{aligned}$$

where the last inequality follows from the Taylor formula for the function $y\mapsto\xi(y,n)$, taking into account that the function ξ has bounded derivatives. This proves (6.20). Hence the proof is completed. □

Combining the previous lemma with (6.14), we obtain the following result.

Theorem 6.1.5. *Consider a tame phase function* Φ *and a symbol satisfying* (6.13). *Let* $g \in \mathcal{S}(\mathbb{R}^d)$. *There exists a constant* $C_N > 0$ *such that*

$$|\langle Tg_{m,n}, g_{m',n'}\rangle| \leq C_N \langle \chi(m,n) - (m',n')\rangle^{-2N}, \tag{6.23}$$

where χ *is the canonical transformation generated by* Φ.

This result shows that the matrix representation of an FIO with respect a Gabor frame is well-organized, similarly to the results recently obtained by [41, 173] in terms of shearlets and curvelets frames. More precisely, if $\sigma \in S^0_{0,0}$, namely if (6.13) is satisfied for every $N \in \mathbb{N}$, then the Gabor matrix of T is highly concentrated along the graph of χ.

6.1.3 Continuity of FIOs on $M^p_\mu(\mathbb{R}^d)$

In this section we study the continuity of FIOs on the modulation spaces M^p_μ associated with a weight function $\mu \in \mathcal{M}_{v_s}(\mathbb{R}^{2d})$, $s \geq 0$. We first state some preliminaries.

Recall that $\mathcal{M}^{p,q}_\mu(\mathbb{R}^d)$ is the closure of the Schwartz class in $M^{p,q}_\mu(\mathbb{R}^d)$ (cf. Definition 2.3.15), so that $\mathcal{M}^{p,q}_\mu(\mathbb{R}^d) = M^{p,q}_\mu(\mathbb{R}^d)$ if $p < \infty$ and $q < \infty$. Also write $\tilde{\ell}^{p,q}_\mu$ for the closure of the space of eventually zero sequences in $\ell^{p,q}_\mu$. Similarly, $\tilde{\ell}^{p,q}_\mu = \ell^{p,q}_\mu$ if $p < \infty$ and $q < \infty$.

Theorem 6.1.6. *Let* $\mu \in \mathcal{M}_v(\mathbb{R}^{2d})$, $\mathcal{G}(g,\alpha,\beta)$ *be a Parseval Gabor frame for* $L^2(\mathbb{R}^d)$, *with lattice* $\Lambda = \alpha\mathbb{Z}^d \times \beta\mathbb{Z}^d$, *and* $g \in M^1_v(\mathbb{R}^d)$. *For every* $1 \leq p,q \leq \infty$, *the coefficient operator* C_g *is continuous from* $\mathcal{M}^{p,q}_\mu(\mathbb{R}^d)$ *into* $\tilde{\ell}^{p,q}_\mu(\Lambda)$, *whereas the synthesis operator* D_g *is continuous from* $\tilde{\ell}^{p,q}_\mu(\Lambda)$ *into* $\mathcal{M}^{p,q}_\mu(\mathbb{R}^d)$.

Proof. Since C_g is continuous from $M^{p,q}_\mu(\mathbb{R}^d)$ into $\ell^{p,q}_\mu(\Lambda)$ by Theorem 3.2.32, it suffices to verify that, if f is a Schwartz function then $C_g(f) \in \tilde{\ell}^{p,q}_\mu(\Lambda)$. This follows from the fact that $C_g(f) \in \ell^1_\mu(\Lambda)$. Similarly, by Theorem 3.2.33, the synthesis operator D_g is bounded from $\ell^{p,q}_\mu(\Lambda)$ into $M^{p,q}_\mu(\mathbb{R}^d)$. For D_g it suffices to verify that, if c is any eventually zero sequence, then $D_g(c) \in \mathcal{M}^{p,q}_\mu(\mathbb{R}^d)$. This is true because $D_g(c) \in M^1_\mu(\mathbb{R}^d)$. □

We need here the classical version of the so-called Schur's test (other versions of the Schur's test are given by Propositions 3.3.6 and 3.3.12 in Chapter 3).

Lemma 6.1.7. *Consider a lattice* Λ *and an operator* K *defined on sequences as*

$$(Kc)_\lambda = \sum_{\nu\in\Lambda} K_{\lambda,\nu} c_\nu.$$

(i) *For* $1 \le p \le \infty$, *assume*

$$\sup_{\nu\in\Lambda} \sum_{\lambda\in\Lambda} |K_{\lambda,\nu}| < \infty, \quad \sup_{\lambda\in\Lambda} \sum_{\nu\in\Lambda} |K_{\lambda,\nu}| < \infty.$$

Then K *is continuous on* $\ell^p(\Lambda)$ *and, moreover, maps the space* $c_0(\Lambda)$ *of sequences vanishing at infinity into itself.*

(ii) *If* $0 < p \le 1$ *and* $\sup_\nu \sum_\lambda |K_{\lambda,\nu}|^p < \infty$, *then*

$$\|Kc\|_{\ell^p} \le \Big(\sup_\nu \sum_\lambda |K_{\lambda,\nu}|^p\Big)^{1/p} \|c\|_{\ell^p},$$

that is, K *is continuous on* $\ell^p(\Lambda)$.

Proof. (i) Let $(c_\nu)_\nu$ be a sequence in $\ell^p(\Lambda)$, $1 \le p \le \infty$. Applying Hölder's inequality to

$$(Kc)_\lambda = \sum_{\nu\in\Lambda} K_{\lambda,\nu}^{\frac{1}{p'}} K_{\lambda,\nu}^{\frac{1}{p}} c_\nu,$$

we infer

$$|(Kc)_\lambda| \le \Big(\sum_{\nu\in\Lambda} |K_{\lambda,\nu}|\Big)^{\frac{1}{p'}} \Big(\sum_{\nu\in\Lambda} |K_{\lambda,\nu}||c_\nu|^p\Big)^{\frac{1}{p}}.$$

By summing over λ, we obtain the boundedness of the operator K

$$\begin{aligned}\|Kc\|_{\ell^p}^p &\le \sum_{\lambda\in\Lambda}\Big(\sum_{\nu\in\Lambda} |K_{\lambda,\nu}|\Big)^{\frac{p}{p'}} \Big(\sum_{\nu\in\Lambda} |K_{\lambda,\nu}||c_\nu|^p\Big)\\ &\lesssim \sum_{\nu\in\Lambda}\Big(\sum_{\lambda\in\Lambda} |K_{\lambda,\nu}|\Big)|c_\nu|^p\\ &\lesssim \|c\|_{\ell^p}^p.\end{aligned}$$

Let us show the second part. Since we know that K is continuous on ℓ^∞ and the space of eventually zero sequences is dense in c_0, it suffices to verify that K maps every eventually zero sequence in c_0. This follows from the fact that any eventually zero sequence belongs to ℓ^1 and therefore, since K is continuous on ℓ^1, is mapped in $\ell^1 \hookrightarrow c_0$.

(ii) Since, for any sequence $b = (b_\lambda)$, we have $\|b\|_{\ell^1} \le \|b\|_{\ell^p}$ if $0 < p \le 1$, it turns out that

$$\begin{aligned}\|Kc\|_{\ell^p}^p &\le \sum_\lambda\Big(\sum_\nu |K_{\lambda,\nu}||c_\nu|\Big)^p \le \sum_\lambda\sum_\nu |K_{\lambda,\nu}|^p|c_\nu|^p\\ &= \sum_\nu\sum_\lambda |K_{\lambda,\nu}|^p|c_\nu|^p \le \Big(\sup_\nu \sum_\lambda |K_{\lambda,\nu}|^p\Big)\|c\|_{\ell^p}^p.\end{aligned}$$

□

We can now state our result.

Theorem 6.1.8. *Consider a tame phase function and a symbol satisfying* (6.13). *Let* $0 \leq s < 2N - 2d$, *and* $\mu \in \mathcal{M}_{v_s}(\mathbb{R}^{2d})$. *For every* $1 \leq p < \infty$, T *extends to a continuous operator from* $M^p_{\mu\circ\chi}(\mathbb{R}^d)$ *into* $M^p_\mu(\mathbb{R}^d)$, *and for* $p = \infty$ *it extends to a continuous operator from* $\mathcal{M}^\infty_{\mu\circ\chi}(\mathbb{R}^d)$ *into* $\mathcal{M}^\infty_\mu(\mathbb{R}^d)$.

Moreover, observe that $\mu\circ\chi \in \mathcal{M}_{v_s}(\mathbb{R}^{2d})$. Indeed, $v_s \circ \chi \asymp v_s$, due to the bi-Lipschitz property of χ.

Proof. We first prove that

$$\|Tf\|_{M^p_\mu} \leq C\|f\|_{M^p_{\mu\circ\chi}},$$

for every $f \in \mathcal{S}(\mathbb{R}^d)$. This proves the theorem in the case $p < \infty$, since $\mathcal{S}(\mathbb{R}^d)$ is dense in $M^p_\mu(\mathbb{R}^d)$.

We see at once that, since $\sigma \in L^\infty(\mathbb{R}^{2d})$, T defines a bounded operator from $M^1(\mathbb{R}^d)$ into $L^\infty(\mathbb{R}^d) \hookrightarrow M^\infty(\mathbb{R}^d)$. Hence, for all $f \in \mathcal{S}(\mathbb{R}^d)$, we have $Tf \in M^\infty(\mathbb{R}^d)$, and Theorem 6.1.6 shows that $\|f\|_{M^p_{\mu\circ\chi}} \gtrsim \|C_g(f)\|_{\ell^p_{\mu\circ\chi}}$ and $\|Tf\|_{M^p_\mu} \lesssim \|C_g(Tf)\|_{\ell^p_\mu}$. On the other hand, the expansion (3.29) (with $\gamma = g$) holds for f with convergence in $M^1(\mathbb{R}^d)$. Therefore,

$$Tf = \sum_{m,n} \langle f, g_{m,n}\rangle Tg_{m,n}$$

with convergence in $M^\infty(\mathbb{R}^d)$. Hence,

$$\begin{aligned} C_g(Tf)_{m',n'} = \langle Tf, g_{m',n'}\rangle &= \sum_{m,n} \langle Tg_{m,n}, g_{m',n'}\rangle \langle f, g_{m,n}\rangle \\ &= \sum_{m,n} \langle Tg_{m,n}, g_{m',n'}\rangle C_g(f)_{m,n}. \end{aligned}$$

Therefore we are reduced to proving that the matrix operator

$$\{c_{m,n}\} \mapsto \sum_{m,n\in\mathbb{Z}^d} \langle Tg_{m,n}, g_{m',n'}\rangle c_{m,n} \tag{6.24}$$

is bounded from $\ell^p_{\mu\circ\chi}$ into ℓ^p_μ. This follows from Schur's test (Lemma 6.1.7) if we prove that, upon setting

$$K_{m',n',m,n} = \langle Tg_{m,n}, g_{m',n'}\rangle \mu(m',n')/\mu(\chi(m,n)),$$

we have

$$K_{m',n',m,n} \in \ell^\infty_{m,n}\ell^1_{m',n'} \tag{6.25}$$

and

$$K_{m',n',m,n} \in \ell^\infty_{m',n'}\ell^1_{m,n}. \tag{6.26}$$

In view of (6.23), we have

$$|K_{m',n',m,n}| \lesssim \langle \chi(m,n) - (m',n') \rangle^{-2N+s} \frac{\mu(m',n')}{\langle \chi(m,n) - (m',n') \rangle^{s} \mu(\chi(m,n))}. \tag{6.27}$$

Now, the last quotient in (6.27) is bounded because μ is v_s-moderate, so we deduce (6.25).

Finally, since χ is a bi-Lipschitz function, we have

$$|\chi(m,n) - (m',n')| \asymp |(m,n) - \chi^{-1}(m',n')|, \tag{6.28}$$

so that (6.26) follows as well.

The case $p = \infty$ is proved analogously by using Theorem 6.1.6 (with $p = q = \infty$), and the last part of the statement of Lemma 6.1.7. □

Remark 6.1.9. Theorem 6.1.8 with $\mu \equiv 1$ gives, in particular, continuity on the unweighted modulation spaces $M^p(\mathbb{R}^d)$. If, moreover, $p = 2$, we recapture the classical L^2-continuity result by Asada and Fujiwara [3].

Also Theorem 6.1.8 applies to $\mu = v_t$ with $|t| \leq s$. In that case we obtain continuity on $M^p_{v_t}(\mathbb{R}^d)$, because $v_t \circ \chi \asymp v_t$.

6.1.4 Continuity of FIOs on $M^{p,q}$

Now let us study the continuity of FIOs on modulation spaces $M^{p,q}(\mathbb{R}^d)$ possibly with $p \neq q$. As we shall see in Section 6.1.6, under the assumptions of Theorem 6.1.8 such operators may fail to be bounded when $p \neq q$. A counterexample is given by the phase $\Phi(x,\eta) = x\eta + x^2/2$, and symbol $\sigma = 1$, which does not yield a bounded operator on $M^{p,q}(\mathbb{R}^d)$, except for the case $p = q$. Here the obstruction is essentially due to the fact that the map $x \mapsto \nabla_x \Phi(x,\eta)$ has unbounded range. Indeed, we will show (for general phases) that if this map has range of finite diameter, uniformly with respect to η, then the corresponding operator is bounded on all $M^{p,q}(\mathbb{R}^d)$.

Theorem 6.1.10. *Consider a tame phase function Φ and a symbol satisfying* (6.13), *with $N > d$. Suppose, in addition, that*

$$\sup_{x,x',\eta} |\nabla_x \Phi(x,\eta) - \nabla_x \Phi(x',\eta)| < \infty. \tag{6.29}$$

Then the corresponding Fourier integral operator T extends to a bounded operator on $M^{p,q}(\mathbb{R}^d)$ for every $1 \leq p,q < \infty$ and on $\mathcal{M}^{p,q}(\mathbb{R}^d)$ if $p = \infty$ or $q = \infty$.

Proof. By arguing as in the proof of Theorem 6.1.8, it suffices to prove the continuity on $\ell^{p,q} = \ell^q_n \ell^p_m$ if $p < \infty$ and $q < \infty$, or $\tilde{\ell}^{p,q}$ if $p = \infty$ or $q = \infty$, of the operator

$$\{c_{m,n}\} \mapsto \sum_{m,n \in \mathbb{Z}^d} T_{m',n',m,n} c_{m,n},$$

where

$$T_{m',n',m,n} = \langle T g_{m,n}, g_{m',n'} \rangle.$$

By applying Proposition 3.3.6, it suffices to verify that

$$\{T_{m',n',m,n}\} \in \ell^\infty_n \ell^1_{n'} \ell^\infty_{m'} \ell^1_m, \tag{6.30}$$

$$\{T_{m',n',m,n}\} \in \ell^\infty_{n'} \ell^1_n \ell^\infty_m \ell^1_{m'}, \tag{6.31}$$

because we have already seen from (6.23) and (6.28) that

$$\{T_{m',n',m,n}\} \in \ell^\infty_{m,n} \ell^1_{m',n'} \cap \ell^\infty_{m',n'} \ell^1_{m,n}.$$

Let us now prove (6.30). It follows from (6.14) and (6.20) that

$$\begin{aligned}
|T_{m',n',m,n}| &\lesssim \left(1 + \left|\nabla_x \Phi(m',n) - n'\right|^2 + \left|\nabla_\eta \Phi(m',n) - m\right|^2\right)^{-N} \\
&\lesssim \left(1 + \left|\xi(m,n) - n'\right|^2 + \left|\nabla_\eta \Phi(m',n) - m\right|^2\right)^{-N} \\
&\lesssim \left(1 + \left|\xi(m,n) - n'\right|\right)^{-N} \left(1 + \left|\nabla_\eta \Phi(m',n) - m\right|\right)^{-N}.
\end{aligned}$$

By (6.9), we have

$$\xi(y,\eta) = \nabla_x \Phi(x(y,\eta),\eta), \quad \forall (y,\eta) \in \mathbb{R}^{2d},$$

so that the hypothesis (6.29) yields

$$\xi(m,n) = \xi(0,n) + O(1).$$

Hence (6.30) follows.

We now prove (6.31). As above, it follows from (6.23) and (6.19) that

$$\begin{aligned}
|T_{m',n',m,n}| &\lesssim \left(1 + \left|\nabla_x \Phi(m',n) - n'\right|^2 + \left|\nabla_\eta \Phi(m',n) - m\right|^2\right)^{-N} \\
&\lesssim \left(1 + \left|\nabla_x \Phi(m',n) - n'\right|^2 + \left|x(m,n) - m'\right|^2\right)^{-N} \\
&\lesssim \left(1 + \left|\nabla_x \Phi(m',n) - n'\right|\right)^{-N} \left(1 + \left|x(m,n) - m'\right|\right)^{-N}.
\end{aligned}$$

By (6.29), we have

$$\nabla_x \Phi(m',n) = \nabla_x \Phi(0,n) + O(1),$$

so that

$$\begin{aligned}
1 + \left|\nabla_x \Phi(m',n) - n'\right| &\gtrsim 1 + \left|\nabla_x \Phi(0,n) - n'\right| \\
&\gtrsim 1 + \left|n - \psi(n')\right|,
\end{aligned} \tag{6.32}$$

where ψ is the inverse function of the bi-Lipschitz function $\eta \mapsto \nabla_x \Phi(0,\eta)$. Therefore we obtain (6.31). This concludes the proof. □

Example 6.1.11. Theorem 6.1.10 applies, in particular, to phases of the type

$$\Phi(x,\eta) = x\cdot\eta + a(x,\eta) \quad \text{where } |\partial_x^\alpha\partial_\eta^\beta a(x,\eta)| \le C_{\alpha,\beta}, \quad 2|\alpha| + |\beta| \ge 2.$$

In the special case when $a(x,\eta) = a(\eta)$ is independent of x and the symbol $\sigma \equiv 1$, the FIO reduces to a Fourier multiplier

$$Tf(x) = \int_{\mathbb{R}^d} e^{2\pi i x\cdot\eta} e^{2\pi i a(\eta)} \hat{f}(\eta)\, d\eta$$

and we reobtain the result of [14, Theorem 1] on the continuity of T on all $\mathcal{M}^{p,q}(\mathbb{R}^d)$, $1 \le p, q \le \infty$.

6.1.5 Modulation spaces as symbol classes

In what follows we shall rephrase the quantity $|\langle Tg_{m,n}, g_{m',n'}\rangle|$ in terms of the STFT of the symbol σ, without assuming the existence of derivatives of σ. This will be applied to prove the continuity of FIOs with symbols in $M^{\infty,1}(\mathbb{R}^{2d})$ on modulation spaces $M^p(\mathbb{R}^d)$.

The same arguments as in the proof of Theorem 6.1.2 yield the equality

$$\langle Tg_{m,n}, g_{m',n'}\rangle = e^{2\pi i m\cdot n} \int_{\mathbb{R}^d} e^{2\pi i \Phi(x,\eta)} \sigma(x,\eta) M_{(-n',-m)} T_{(m',n)} (\bar{g}\otimes\hat{g})(x,\eta)\, dx\, d\eta.$$

Expanding the phase Φ into a Taylor series around (m',n), we obtain

$$\Phi(x,\eta) = \Phi(m',n) + \nabla_z\Phi(m',n)\cdot(x-m',\eta-n) + T_{(m',n)}\Phi_{2,(m',n)}(x,\eta)$$

where the remainder $\Phi_{2,(m',n)}$ is given by (6.15).

Inserting this expansion in the integrals above, we can write

$$\begin{aligned}\langle Tg_{m,n}, g_{m',n'}\rangle = e^{2\pi i (m\cdot n + \Phi(m',n) - \nabla_z\Phi(m',n)\cdot(m',n))} \int_{\mathbb{R}^d} e^{2\pi i \nabla_z\Phi(m',n)(x,\eta)} \sigma(x,\eta) \\ \times M_{(-n',-m)} T_{(m',n)} e^{2\pi i \Phi_{2,(m',n)}(x,\eta)} (\bar{g}\otimes\hat{g})(x,\eta)\, dx\, d\eta. \qquad (6.33)\end{aligned}$$

Defining

$$\Psi_{(m',n)}(x,\eta) := e^{2\pi i \Phi_{2,(m',n)}(x,\eta)} (\bar{g}\otimes\hat{g})(x,\eta), \qquad (6.34)$$

and computing the modulus of the left-hand side of (6.33), we are led to

$$|\langle Tg_{m,n}, g_{m',n'}\rangle| = |V_{\Psi_{(m',n)}} \sigma((m',n),(n' - \nabla_x\Phi(m',n), m - \nabla_\xi\Phi(m',n)))|. \qquad (6.35)$$

Observe that the window $\Psi_{(m',n)}$ of the STFT above depends on the pair (m',n).

We now study the continuity problem for T when the symbol σ is in the modulation space $M^{\infty,1}(\mathbb{R}^{2d})$.

Theorem 6.1.12. *Consider a tame phase function Φ and a symbol $\sigma \in M^{\infty,1}(\mathbb{R}^{2d})$. For every $1 \le p < \infty$, T extends to a continuous operator on $M^p(\mathbb{R}^d)$, for $p = \infty$ it extends to a continuous operator on $\mathcal{M}^\infty(\mathbb{R}^d)$.*

By arguing as in the proof of Theorem 6.1.8, it suffices to prove the continuity of the operator (6.24) on ℓ^p, if $1 \le p < \infty$, and on $\tilde{\ell}^\infty = c_0$. In view of Schur's test (Lemma 6.1.7) and (6.35), it suffices to prove the following result.

Proposition 6.1.13. *Consider a tame phase function Φ and a symbol $\sigma \in M^{\infty,1}(\mathbb{R}^{2d})$. If we set*

$$z_{m,n,m',n'} := ((m',n),(n' - \nabla_x\Phi(m',n), m - \nabla_\xi\Phi(m',n))), \quad m,m' \in \alpha\mathbb{Z}^d,\ n,n' \in \beta\mathbb{Z}^d, \tag{6.36}$$

then

$$\sup_{(m,n)\in\Lambda} \sum_{(m',n')\in\Lambda} |V_{\Psi_{(m',n)}}\sigma(z_{m,n,m',n'})| \lesssim \|\sigma\|_{M^{\infty,1}}, \tag{6.37}$$

$$\sup_{(m',n')\in\Lambda} \sum_{(m,n)\in\Lambda} |V_{\Psi_{(m',n)}}\sigma(z_{m,n,m',n'})| \lesssim \|\sigma\|_{M^{\infty,1}}. \tag{6.38}$$

We need the following lemma.

Lemma 6.1.14. *Let $\Psi_0 \in \mathcal{S}(\mathbb{R}^{2d})$ with $\|\Psi_0\|_{L^2} = 1$ and $\Psi_{(m',n)}$ be defined by (6.34), with $(m',n) \in \Lambda = \alpha\mathbb{Z}^d \times \beta\mathbb{Z}^d$, and $g \in \mathcal{S}(\mathbb{R}^d)$. Then*

$$\int_{\mathbb{R}^{4d}} \sup_{(m',n)\in\Lambda} |V_{\Psi_{(m',n)}}\Psi_0(w)|\, dw < \infty. \tag{6.39}$$

Proof of Lemma 6.1.14. We shall show that

$$|V_{\Psi_{(m',n)}}\Psi_0(w)| \le C\langle w\rangle^{-(4d+1)}, \quad \forall (m',n) \in \Lambda. \tag{6.40}$$

Using the switching property of the STFT

$$(V_f g)(x,\eta) = e^{-2\pi i\eta\cdot x}\overline{(V_g f)(-x,-\eta)},$$

we observe that $|V_{\Psi_{(m',n)}}\Psi_0(w_1,w_2)| = |V_{\Psi_0}\Psi_{(m',n)}|(-w_1,-w_2)$, and since the weight $\langle\cdot\rangle$ is even, relation (6.40) is equivalent to

$$|V_{\Psi_0}\Psi_{(m',n)}(w)| \le C\langle w\rangle^{-(4d+1)}, \quad \forall (m',n) \in \Lambda. \tag{6.41}$$

Now, the mapping V_{Ψ_0} is continuous from $\mathcal{S}(\mathbb{R}^{2d})$ to $\mathcal{S}(\mathbb{R}^{4d})$ (see [160, Chapter 11]). This means that there exist $M \in \mathbb{N}$ and $K > 0$ such that

$$\begin{aligned}|V_{\Psi_0}\Psi_{(m',n)}(w)| &= |V_{\Psi_0}\Psi_{(m',n)}(w)|\frac{\langle w\rangle^{4d+1}}{\langle w\rangle^{4d+1}}\\ &\leq \|V_{\Psi_0}\Psi_{(m',n)}\langle\cdot\rangle^{4d+1}\|_{L^\infty(\mathbb{R}^{4d})}\langle w\rangle^{-(4d+1)}\\ &\leq K\sum_{|\gamma|+|\delta|\leq M}\|D^\gamma X^\delta\Psi_{(m',n)}\|_{L^\infty(\mathbb{R}^{2d})}\langle w\rangle^{-(4d+1)},\end{aligned}$$

for every $(m',n) \in \Lambda$. We now claim that $\Psi_{(m',n)} \in \mathcal{S}(\mathbb{R}^{2d})$ uniformly with respect to (m',n). This is proved as follows: the function $e^{2\pi i\Phi_{2,(m',n)}(x,\eta)}$ is in $\mathcal{C}^\infty(\mathbb{R}^{2d})$ and possesses derivatives dominated by powers $\langle(x,\eta)\rangle^k$, $k \in \mathbb{N}$, uniformly with respect to (m',n), due to (6.7); since $(\bar{g}\otimes\hat{g}) \in \mathcal{S}(\mathbb{R}^{2d})$, it follows that $\Psi_{(m',n)} \in \mathcal{S}(\mathbb{R}^{2d})$, with seminorms uniformly bounded

$$p_M(\Psi_{(m',n)}) := \sum_{|\gamma|+|\delta|\leq M}\|D^\gamma X^\delta\Psi_{(m',n)}\|_{L^\infty} \leq C_M, \quad \forall (m',n) \in \Lambda.$$

Consequently,

$$\begin{aligned}|V_{\Psi_0}\Psi_{(m',n)}(w)| &\leq K\sum_{|\gamma|+|\delta|\leq M}\|D^\gamma X^\delta\Psi_{(m',n)}\|_{L^\infty(\mathbb{R}^{2d})}\langle w\rangle^{-(4d+1)}\\ &\leq KC_M\langle w\rangle^{-(4d+1)},\end{aligned}$$

for every $(m',n) \in \Lambda$, as desired. □

We exhibit a slight generalization of [160, Proposition 11.1.4] and its subsequent remark. This will be also essential in the following proof of Proposition 6.1.13.

Proposition 6.1.15. *Let $\mathcal{X}$ be a separated sampling set in $\mathbb{R}^{2d}$, that is, there exists $\delta > 0$ such that $\inf_{x,y\in\mathcal{X}:x\neq y}|x-y| \geq \delta$. Then there exists a constant $C > 0$ such that, if $F \in W(L^\infty, L^{p,q}_\mu)$ is any function everywhere defined on $\mathbb{R}^{2d}$ and lower semicontinuous, then the restriction $F_{|\mathcal{X}}$ is in $\ell^{p,q}_\mu$ with*

$$\|F_{|\mathcal{X}}\|_{\ell^{p,q}_\mu} \leq C\|F\|_{W(L^\infty,L^{p,q}_\mu)}.$$

Proof. One uses the arguments of [160, Proposition 11.1.4] and its subsequent remark. We just shall highlight the key points that make those arguments work under our assumptions.

First of all, if $(r,s) \in \mathbb{Z}^{2d}$ and $\mathcal{X}$ is separated, then the number of sampling points of $\mathcal{X}$ in $(r,s) + [0,1]^{2d}$ is bounded independently of (r,s).

Secondly, for $x \in \mathcal{X}$ such that $x \in (r,s) + [0,1]^{2d}$,

$$|F(x)|\mu(x) \leq C\|F\cdot T_{(r,s)}\chi_{[0,1]^{2d}}\|_{L^\infty}\mu(r,s).$$

Indeed, since F is everywhere defined on $\mathbb{R}^{2d}$ and lower semicontinuous, we have $\sup|f| = \operatorname{ess\,sup}|f|$ on every box, whereas $\mu(x) \leq C\mu(r,s)$ because μ is v-moderate and v is bounded on $[0,1]^{2d}$, see (2.7). □

Proof of Proposition 6.1.13. We shall prove (6.37). First, Lemma 1.2.29 for $h = \gamma = \Psi_0$, yields

$$|V_{\Psi_{(m',n)}}\sigma(z)| \leq (|V_{\Psi_0}\sigma| * |V_{\Psi_{(m',n)}}\Psi_0|)(z), \quad z \in \mathbb{R}^{4d},$$

so that

$$\begin{aligned}
&\sum_{(m',n')\in\Lambda} |V_{\Psi_{(m',n)}}\sigma(z_{m,n,m',n'})| \\
&\quad \leq \int_{\mathbb{R}^{4d}} \sum_{(m',n')\in\Lambda} |V_{\Psi_0}\sigma(z_{m,n,m',n'} - w)||V_{\Psi_{(m',n)}}\Psi_0(w)|\,dw \\
&\quad \leq \int_{\mathbb{R}^{4d}} \sum_{(m',n')\in\Lambda} |V_{\Psi_0}\sigma(z_{m,n,m',n'} - w)| \sup_{(m',n)\in\Lambda} |V_{\Psi_{(m',n)}}\Psi_0(w)|\,dw \\
&\quad \leq \sup_{w\in\mathbb{R}^{4d}} \sum_{(m',n')\in\Lambda} |V_{\Psi_0}\sigma(z_{m,n,m',n'} - w)| \int_{\mathbb{R}^{4d}} \sup_{(m',n)\in\Lambda} |V_{\Psi_{(m',n)}}\Psi_0(w)|\,dw \\
&\quad \leq C \sup_{w\in\mathbb{R}^{4d}} \sum_{(m',n')\in\Lambda} |V_{\Psi_0}\sigma(z_{m,n,m',n'} - w)|,
\end{aligned}$$

where the last majorization is due to Lemma 6.1.14. Since

$$\sum_{(m',n')\in\Lambda} |V_{\Psi_0}\sigma(z_{m,n,m',n'} - w)| \leq \sum_{(m',n')\in\Lambda} \sup_{u_1\in\mathbb{R}^{2d}} |V_{\Psi_0}\sigma(u_1, \tilde{z}_{m,n,m',n',w_2})|,$$

with

$$\tilde{z}_{m,n,m',n',w_2} := (n' - \nabla_x\Phi(m',n), m - \nabla_\xi\Phi(m',n)) - w_2, \quad w = (w_1, w_2) \in \mathbb{R}^{4d},$$

we shall prove that

$$\sum_{(m',n')\in\Lambda} \sup_{u_1\in\mathbb{R}^{2d}} |V_{\Psi_0}\sigma(u_1, \tilde{z}_{m,n,m',n',w_2})| \lesssim \|\sigma\|_{M^{\infty,1}}, \tag{6.42}$$

uniformly with respect to $(m,n) \in \Lambda$, $w_2 \in \mathbb{R}^{2d}$. For every fixed (m,n), the set $\mathcal{X} = \mathcal{X}_{m,n,w_2}$, given by

$$\mathcal{X}_{m,n,w_2} = \{\tilde{z}_{m,n,m',n',w_2};\ (m',n') \in \Lambda,\ w_2 \in \mathbb{R}^{2d}\},$$

is separated, uniformly with respect to (m,n), w_2. Indeed, given $(m'_1, n'_1) \neq (m'_2, n'_2)$, if $m'_1 \neq m'_2$,

$$|\tilde{z}_{m,n,m'_1,n'_1,w_2} - \tilde{z}_{m,n,m'_2,n'_2,w_2}| \geq |\nabla_\xi\Phi(m'_1,n) - \nabla_\xi\Phi(m'_2,n)| \geq C|m'_1 - m'_2| \geq \alpha C,$$

uniformly with respect to (m, n), w_2, because the mapping $x \mapsto \nabla_\xi \Phi(x, \xi)$ has an inverse that is Lipschitz continuous, thanks to (6.7) and (6.8). On the other hand, if $m'_1 = m'_2$,

$$|\tilde{z}_{m,n,m'_1,n'_1,w_2} - \tilde{z}_{m,n,m'_2,n'_2,w_2}| \geq |n'_1 - n'_2| \geq \beta, \quad \forall (m,n) \in \Lambda, w_2 \in \mathbb{R}^{2d}.$$

Hence, $\mathcal{X}$ is separated uniformly with respect to (m, n), w_2. Now, we apply Proposition 6.1.15 (with $p = q = 1$) to the function

$$F(u_2) = \sup_{u_1 \in \mathbb{R}^{2d}} |V_{\Psi_0}\sigma(u_1, u_2)|, \quad u_2 \in \mathbb{R}^{2d},$$

which is lower semicontinuous due to $V_{\Psi_0}\sigma$ being continuous. We obtain

$$\|F_{|\mathcal{X}}\|_{\ell^1} \leq C\|F\|_{W(L^\infty,L^1)} = \|V_{\Psi_0}\sigma\|_{W(C,L^{\infty,1})}. \tag{6.43}$$

If the symbol σ is in $M^{\infty,1}$, by Lemma 2.4.15, the STFT $V_{\Psi_0}\sigma$ belongs to the Wiener amalgam space $W(\mathcal{F}L^1, L^{\infty,1})$, and

$$\|V_{\Psi_0}\sigma\|_{W(C,L^{\infty,1})} \lesssim \|V_{\Psi_0}\sigma\|_{W(\mathcal{F}L^1,L^{\infty,1})} \lesssim \|\sigma\|_{M^{\infty,1}}\|\Psi_0\|_{M^1}.$$

The first inequality is due to $\mathcal{F}L^1 \hookrightarrow C$ and the inclusion relations between Wiener amalgam spaces. Combining this inequality with (6.43), we obtain (6.42), uniformly with respect to (m, n) and w_2, that is, (6.37) holds.

The estimate (6.38) is obtained by similar arguments. □

Remark 6.1.16. We observe that the continuity on $M^2(\mathbb{R}^d) = L^2(\mathbb{R}^d)$ of FIOs as above, with symbols in $M^{\infty,1}(\mathbb{R}^{2d})$, is also proved in [35] by other methods.

6.1.6 The case of quadratic phases: metaplectic operators

In this section we briefly discuss the particular case of quadratic phases, namely phases of the type

$$\Phi(x, \eta) = \frac{1}{2}Ax \cdot x + Bx \cdot \eta + \frac{1}{2}C\eta \cdot \eta + \eta_0 \cdot x - x_0 \cdot \eta, \tag{6.44}$$

where $x_0, \eta_0 \in \mathbb{R}^d$, A, C are real symmetric $d \times d$ matrices and B is a real $d \times d$ nondegenerate matrix.

It is easy to see that, if we take the symbol $\sigma \equiv 1$ and the phase (6.44), the corresponding FIO T is (up to a constant factor) a metaplectic operator (cf. Section 1.1.2). This can be seen by means of the easily verified factorization

$$T = M_{\eta_0} U_A D_B \mathcal{F}^{-1} U_C \mathcal{F} T_{x_0}, \tag{6.45}$$

where U_A and U_C are the multiplication operators by $e^{\pi i Ax\cdot x}$ and $e^{\pi i C\eta\cdot\eta}$, respectively, and D_B is the dilation operator $f \mapsto f(B\cdot)$. Each of the factors is (up to a constant factor) a metaplectic operator (see (1.15) and (1.16)), and so is T.

The corresponding canonical map, defined by (6.9), is now an affine symplectic map. For the benefit of the reader, some important special cases are detailed in the table below.

Operator	**Phase $\Phi(x,\eta)$**	**Canonical transformation**
T_{x_0}	$(x - x_0)\cdot\eta$	$\chi(y,\eta) = (y + x_0, \eta)$
M_{η_0}	$(\eta + \eta_0)\cdot x$	$\chi(y,\eta) = (y, \eta + \eta_0)$
D_B	$Bx\cdot\eta$	$\chi(y,\eta) = (B^{-1}y, {}^tB\eta)$
U_A	$x\cdot\eta + \frac{1}{2}Ax\cdot x$	$\chi(y,\eta) = (y, \eta + Ax)$

We observe that there are metaplectic operators, as the Fourier transform, which cannot be expressed as FIOs of the type (6.12).

Metaplectic operators are known to be bounded on $M^p_{v_s}(\mathbb{R}^d)$, see, e. g., [160, Proposition 12.1.3]. In our case, this also follows from Theorem 6.1.8. Indeed, since χ is a bi-Lipschitz function, we have $v_s \circ \chi \asymp v_s$.

Theorem 6.1.10 applies to quadratic phases whose affine symplectic map χ is (up to translations on the phase space) defined by an upper-triangular matrix, which happens precisely when $A = 0$. Indeed, we obtain the map χ by solving

$$\begin{cases} y = Bx + C\eta - x_0, \\ \xi = Ax + B\eta + \eta_0. \end{cases}$$

The phase condition (6.8) here becomes

$$\det B \neq 0,$$

so that B is an invertible matrix and $x = B^{-1}y - B^{-1}C\eta + B^{-1}x_0$. Whence the mapping $\chi : (y,\eta) \mapsto (x,\xi)$ is given by

$$\begin{pmatrix} x \\ \xi \end{pmatrix} = \begin{pmatrix} B^{-1} & -B^{-1}C \\ AB^{-1} & B - AB^{-1}C \end{pmatrix} \begin{pmatrix} y \\ \eta \end{pmatrix} + \begin{pmatrix} B^{-1}x_0 \\ AB^{-1}x_0 + \eta_0 \end{pmatrix}.$$

When $A = 0$, the phase Φ satisfies (6.29) and, consequently, the corresponding operators are bounded on all $M^{p,q}(\mathbb{R}^d)$. This can also be verified by means of the factorization (6.45) (with $A = 0$). Indeed, the continuity of the operators M_{η_0}, T_{x_0}, and D_B is easily seen, whereas that of the Fourier multiplier $\mathcal{F}^{-1}U_C\mathcal{F}$ was shown, e. g., in [163, Lemma 2.1].

On the other hand, generally the metaplectic operators are not bounded on $M^{p,q}(\mathbb{R}^d)$ if $p \neq q$. An example is given by the Fourier transform itself (see [120]). An instance which instead falls in the class of FIOs considered here is the following one.

Proposition 6.1.17. *The multiplication U_{I_d} is unbounded on $M^{p,q}(\mathbb{R}^d)$, for every $1 \leq p, q \leq \infty$, with $p \neq q$.*

Proof. We have $U_{I_d} f(x) = e^{\pi i x^2} f(x)$. For $\lambda > 0$, we consider the one-parameter family of Gaussian functions $\varphi_{\sqrt{\lambda}}(x) = e^{-\pi\lambda x^2} \in M^{p,q}(\mathbb{R}^d)$, so that $\widehat{\varphi_{\sqrt{\lambda}}}(\xi) = \lambda^{-d/2} e^{-\pi(1/\lambda)\xi^2}$. For every $1 \leq p, q \leq \infty$, by Lemma 2.7.1, we have

$$\|\varphi_{\sqrt{\lambda}}\|_{M^{p,q}} = \|\widehat{\varphi_{\sqrt{\lambda}}}\|_{W(\mathcal{F}L^p, L^q)} \asymp \frac{(\lambda+1)^{d(\frac{1}{p}-\frac{1}{2})}}{\lambda^{\frac{d}{2q}} (\lambda^2+\lambda)^{\frac{d}{2}(\frac{1}{p}-\frac{1}{q})}}.$$

Since $U\varphi_{\sqrt{\lambda}}(x) = e^{-\pi(\lambda-i)x^2}$ so that $\widehat{Uf}(\eta) = (\lambda - i)^{-d/2} e^{-(\pi/(\lambda-i))\xi^2}$, the same formula as above yields

$$\|U\varphi_{\sqrt{\lambda}}\|_{M^{p,q}} \asymp \frac{[(\lambda+1)^2+1]^{\frac{d}{2}(\frac{1}{p}-\frac{1}{2})}}{\lambda^{\frac{d}{2q}} (\lambda^2+\lambda+1)^{\frac{d}{2}(\frac{1}{p}-\frac{1}{q})}}.$$

As $\lambda \to 0$, we have

$$\|U\varphi_{\sqrt{\lambda}}\|_{M^{p,q}} \asymp \lambda^{-\frac{d}{2q}}, \quad \|f_{\sqrt{\lambda}}\|_{M^{p,q}} \asymp \lambda^{-\frac{d}{2q}-\frac{d}{2}(\frac{1}{p}-\frac{1}{q})}$$

so that, if we assume $\|U\varphi_{\sqrt{\lambda}}\|_{M^{p,q}} \leq C\|\varphi_{\sqrt{\lambda}}\|_{M^{p,q}}$, then $1/p - 1/q \geq 0$, that is, $p \leq q$.

Moreover, the same argument applies to the adjoint operator

$$U^* f(x) = e^{-\pi i x^2} f(x).$$

Now we show that $p = q$. By contradiction, if U were bounded on $M^{p,q}(\mathbb{R}^d)$, with $p < q$, its adjoint U^* would satisfy

$$\|U^* f\|_{M^{p',q'}} \leq C\|f\|_{M^{p',q'}}, \quad \forall f \in \mathcal{S}(\mathbb{R}^d),$$

with $q' < p'$, which is a contradiction to what we have just proved. □

6.1.6.1 Boundedness of metaplectic operators on $W(\mathcal{F}L^p, L^q)$ spaces

Let us continue the study of metaplectic operators, briefly defined in Section 1.1.2. We shall prove boundedness properties of the operator $\mu(\mathcal{A})$ (with $\mathcal{A}$ symplectic matrix, cf. (1.3)) on the Wiener amalgam spaces (or modulation spaces) $W(\mathcal{F}L^p, L^q)$, with $1 \leq p, q \leq \infty$. These issues were initially proved in the work [70].

Explicit integral representations for classical metaplectic operators, extending the results already contained in the literature [105, 145], were given by Morsche and Oonincx in [284] and applied to energy localization problems and to fractional Fourier transforms. The novelty of [284], with respect to the classical works [105, 145], is the explicit integral representation of metaplectic operators, covering all possible cases of symplectic matrices. Indeed, the integral representation of metaplectic operators in [105, 145] covers only the cases of nonsingular upper-left or upper-right component of the parameterizing matrix. For details, we refer the interested reader to the original work [284]. Here we also limit ourselves to metaplectic operators related to a symplectic matrix $\mathcal{A}$ having nonsingular upper-left or upper-right component, following the $d\times d$ block decomposition in (1.4). The subsequent formulae for metaplectic operators can be found in [145, Theorems 4.51 and 4.53].

Proposition 6.1.18. *Let $\mathcal{A}$ be a symplectic matrix with the $d \times d$ block decomposition in* (1.4)

$$\mathcal{A} = \begin{pmatrix} A & B \\ C & D \end{pmatrix}. \tag{6.46}$$

(i) *If* $\det B \neq 0$ *then*

$$\mu(\mathcal{A})f(x) = i^{d/2}(\det B)^{-1/2} \int e^{-\pi i x\cdot DB^{-1}x+2\pi i y\cdot B^{-1}x-\pi i y\cdot B^{-1}Ay} f(y)\,dy. \tag{6.47}$$

(ii) *If* $\det A \neq 0$,

$$\mu(\mathcal{A})f(x) = (\det A)^{-1/2} \int e^{-\pi i x\cdot CA^{-1}x+2\pi i\xi\cdot A^{-1}x+\pi i\xi\cdot A^{-1}B\xi} \hat{f}(\xi)\,d\xi. \tag{6.48}$$

The following hybrid formula will be also used in the sequel.

Proposition 6.1.19. *If $\mathcal{A}$ is in $Sp(d,\mathbb{R})$, having block decomposition in* (6.46) *with* $\det B \neq 0$ *and* $\det A \neq 0$, *then*

$$\mu(\mathcal{A})f(x) = (-i\det B)^{-1/2} e^{-\pi i x\cdot CA^{-1}x}(e^{-\pi i y\cdot B^{-1}Ay} * f)(A^{-1}x). \tag{6.49}$$

Proof. By (6.48), we can write

$$\begin{aligned}\mu(\mathcal{A})f(x) &= (\det A)^{-1/2} e^{-\pi i x\cdot CA^{-1}x} \int e^{2\pi i\xi\cdot A^{-1}x} \mathcal{F}(\mathcal{F}^{-1} e^{\pi i\xi\cdot A^{-1}B\xi})\hat{f}(\xi)\,d\xi \\ &= (-i\det B)^{-1/2} e^{-\pi i x\cdot CA^{-1}x} \int e^{2\pi i\xi\cdot A^{-1}x} \mathcal{F}(e^{-\pi i y\cdot B^{-1}Ay} * f)(\xi)\,d\xi,\end{aligned}$$

where we used the formula (see [145, Theorem 2, page 257])

$$\mathcal{F}^{-1}(e^{i\pi\xi\cdot A^{-1}B\xi})(y) = (-i\det A^{-1}B)^{-1/2} e^{-\pi i y\cdot B^{-1}Ay}.$$

Hence, from the Fourier inversion formula we obtain (6.49). □

The first continuity result is shown below.

Theorem 6.1.20. *Consider $\mathcal{A} \in Sp(d,\mathbb{R})$ with the block decomposition in* (6.46) *and* $1 \leq p \leq q \leq \infty$.

(i) *If* $\det B \neq 0$, *then*

$$\|\mu(\mathcal{A})f\|_{W(\mathcal{F}L^p,L^q)} \lesssim \alpha(\mathcal{A},p,q)\|f\|_{W(\mathcal{F}L^q,L^p)}, \tag{6.50}$$

where

$$\alpha(\mathcal{A},p,q) = |\det B|^{1/q-1/p-3/2} \left|\det(I+B^*B)(B+iA)(B+iD)\right|^{1/2}. \tag{6.51}$$

(ii) *If* $\det A \neq 0$ *and* $\det B \neq 0$, *then we have the estimate*

$$\|\mu(\mathcal{A})f\|_{W(\mathcal{F}L^1,L^\infty)} \lesssim \beta(\mathcal{A})\|f\|_{W(\mathcal{F}L^\infty,L^1)}, \tag{6.52}$$

with

$$\beta(\mathcal{A}) = |\det A|^{-3/2}|\det B|^{-1}\left|\det(I + A^*A)(B + iA)(A + iC)\right|^{1/2}. \tag{6.53}$$

Thanks to the previous theorem, we have all the tools to give the proof of Proposition 4.2.5 in Chapter 4. We recall its statement for the convenience of the reader.

Proposition 6.1.21 (Proposition 4.2.5). *Let* $1 \leq p, q \leq \infty$. *If* $p \neq q$, *the operator* $\mathcal{U}$ *in* (4.8) *does not map* $W(\mathcal{F}L^p, L^q)$ *into itself.*

Proof. Consider the symmetric matrix $B = \frac{1}{2}\begin{pmatrix} 0 & I_d \\ I_d & 0 \end{pmatrix}$, and the symplectic matrix $\mathcal{A} = \begin{pmatrix} I_{2d} & B \\ 0 & I_{2d} \end{pmatrix}$. It follows from the formula (6.48) that the operator $\mathcal{U}$ is exactly the metaplectic operator associated with the matrix $\mathcal{A}$. We write $\mathcal{U} = \mu(\mathcal{A})$. Hence, as a consequence of Theorem 6.1.20, we deduce that $\mathcal{U}$ and $\mathcal{U}^{-1} = \mu(\mathcal{A}^{-1})$ map $W(\mathcal{F}L^q, L^p)$ into $W(\mathcal{F}L^p, L^q)$ continuously for every $1 \leq p \leq q \leq \infty$.

Now, assume $p < q$. Let f be any distribution in $W(\mathcal{F}L^q, L^p)$; therefore by the boundedness result we have just recalled, $\mathcal{U}^{-1}f \in W(\mathcal{F}L^p, L^q)$. Suppose, by contradiction, that $\mathcal{U}$ maps $W(\mathcal{F}L^p, L^q)$ into itself. Then one would obtain $f = \mathcal{U}\mathcal{U}^{-1}f \in W(\mathcal{F}L^p, L^q)$, and therefore the inclusion $W(\mathcal{F}L^q, L^p) \subseteq W(\mathcal{F}L^p, L^q)$, which is false.

Suppose now $p > q$. Assume, by contradiction, that for every $f \in W(\mathcal{F}L^p, L^q)$ it turns out that $\mathcal{U}f \in W(\mathcal{F}L^p, L^q)$. Then we would have $f = \mathcal{U}^{-1}\mathcal{U}f \in W(\mathcal{F}L^q, L^p)$, and therefore the inclusion $W(\mathcal{F}L^p, L^q) \subseteq W(\mathcal{F}L^q, L^p)$, which is false. □

If the matrices A or B are symmetric, Theorem 6.1.20 can be sharpened as follows.

Theorem 6.1.22. *Consider* $\mathcal{A} \in Sp(d, \mathbb{R})$ *with the block decomposition in* (6.46).

(i) *If* $\det B \neq 0$, $B^* = B$, *with eigenvalues* $\lambda_1, \dots, \lambda_d$, $1 \leq p \leq q \leq \infty$, *then*

$$\|\mu(\mathcal{A})f\|_{W(\mathcal{F}L^p,L^q)} \lesssim \alpha'(\mathcal{A}, p, q)\|f\|_{W(\mathcal{F}L^q,L^p)}, \tag{6.54}$$

where

$$\begin{aligned} \alpha'(\mathcal{A}, p, q) &= \left|\det(B + iA)(B + iD)\right|^{1/2} \\ &\quad \times \prod_{j=1}^{d} (\max\{1, |\lambda_j|\})^{\mu_1(p,q)-1/2} (\min\{1, |\lambda_j|\})^{\mu_2(p,q)-1/2}, \end{aligned} \tag{6.55}$$

with μ_1 *and* μ_2 *defined in* (2.108) *and* (2.109), *respectively.*

(ii) *If* $\det A \neq 0$, $\det B \neq 0$, *and* $A^* = A$ *with eigenvalues* $\nu_1, \dots, \nu_d$, *then*

$$\|\mu(\mathcal{A})f\|_{W(\mathcal{F}L^1,L^\infty)} \lesssim \beta'(\mathcal{A})\|f\|_{W(\mathcal{F}L^\infty,L^1)}, \tag{6.56}$$

with

$$\beta'(\mathcal{A}) = |\det B|^{-1}\left|\det(B+iA)(A+iC)\right|^{1/2} \times \prod_{j=1}^{d}\left(\max\{1,|v_j|\}\right)^{-1/2}\left(\min\{1,|v_j|\}\right)^{-3/2}. \tag{6.57}$$

We now prove Theorems 6.1.20 and 6.1.22. We need the following preliminary result, improving Proposition 4.3.3.

Lemma 6.1.23. *Let R be a $d\times d$ real symmetric matrix, and $f(y)=e^{-\pi iRy\cdot y}$. Then*

$$\|f\|_{W(\mathcal{F}L^1,L^\infty)} = \left|\det(I+iR)\right|^{1/2}. \tag{6.58}$$

Proof. We first compute the short-time Fourier transform of f, with respect to the window $g(y)=e^{-\pi y^2}$. We have

$$\begin{aligned} V_g f(x,\xi) &= \int e^{-2\pi iy\cdot\xi}e^{-i\pi Ry\cdot y}e^{-\pi y^2}\,dy \\ &= e^{-\pi x^2}\int e^{-2\pi iy\cdot(\xi+ix)-\pi(I+iR)y\cdot y}\,dy \\ &= e^{-\pi x^2}\left(\det(I+iR)\right)^{-1/2}e^{-\pi(I+iR)^{-1}(\xi+ix)\cdot(\xi+ix)}, \end{aligned}$$

where we used [145, Theorem 1, page 256]. Hence

$$\left|V_g f(x,\xi)\right| = \left|\det(I+iR)\right|^{-1/2}e^{-\pi(I+R^2)^{-1}(\xi+Rx)\cdot(\xi+Rx)},$$

and, performing the change of variables

$$(I+R^2)^{-1/2}(\xi+Rx)=y,$$

with $d\xi = |\det(I+R^2)|^{1/2}dy$, we obtain

$$\int_{\mathbb{R}^d} V_g f(x,\xi)\,d\xi = \left|\det(I+iR)\right|^{-1/2}\left(\det(I+R^2)\right)^{1/2}\int_{\mathbb{R}^d} e^{-\pi y^2}\,dy = \left|\det(I+iR)\right|^{1/2}. \tag{6.59}$$

The last equality follows from $(I+iR) = (I+R^2)(I-iR)^{-1}$, so that $\det(I+iR)^{-1} = \det(I+R^2)^{-1}\det(I-iR)$. Now, relation (6.58) is proved by taking the supremum with the respect to $x\in\mathbb{R}^d$ in (6.59). □

Proof of Theorem 6.1.20. (i) We use the expression of $\mu(\mathcal{A})f$ in formula (6.47). The estimates below are obtained by using (in this order): Proposition 2.4.21 with Lemma 6.1.23, the estimate (2.107), Proposition 2.4.13, and, finally, Proposition 2.4.21 combined with Lemma 6.1.23 again:

$$\begin{aligned} \|\mu(\mathcal{A})f\|_{W(\mathcal{F}L^p,L^q)} &= |\det B|^{-1/2}\|e^{-\pi ix\cdot DB^{-1}x}\mathcal{F}^{-1}(e^{-\pi iy\cdot B^{-1}Ay}f)(B^{-1}x)\|_{W(\mathcal{F}L^p,L^q)} \\ &\le |\det B|^{-1/2}\|e^{-\pi ix\cdot DB^{-1}x}\|_{W(\mathcal{F}L^1,L^\infty)} \end{aligned}$$

$$
\begin{aligned}
&\times \|(\mathcal{F}^{-1}(e^{-\pi i y\cdot B^{-1}Ay}f))_{B^{-1}}\|_{W(\mathcal{F}L^p,L^q)} \\
&\lesssim |\det B|^{1/q-1/p-1/2}(\det(B^*B+I))^{1/2}|\det(I+iDB^{-1})|^{1/2} \\
&\times \|\mathcal{F}^{-1}(e^{-\pi i y\cdot B^{-1}Ay}f)\|_{W(\mathcal{F}L^p,L^q)} \\
&\lesssim |\det B|^{1/q-1/p-1/2}(\det(B^*B+I))^{1/2}\det(I+iDB^{-1})|^{1/2} \\
&\times \|e^{-\pi i y\cdot B^{-1}Ay}f\|_{W(\mathcal{F}L^q,L^p)} \\
&\lesssim \alpha(\mathcal{A},p,q)\|f\|_{W(\mathcal{F}L^q,L^p)}
\end{aligned}
$$

with $\alpha(\mathcal{A},p,q)$ given by (6.51).
(ii) In this case, we use formula (6.49). Then proceeding like in case (i), we majorize as follows:

$$
\begin{aligned}
\|\mu(\mathcal{A})f\|_{W(\mathcal{F}L^1,L^\infty)} &= |\det B|^{-1/2}\|e^{-\pi i x\cdot CA^{-1}x}(e^{-\pi i y\cdot B^{-1}Ay}*f)(A^{-1}x)\|_{W(\mathcal{F}L^1,L^\infty)} \\
&\lesssim |\det B|^{-1/2}\|e^{-\pi i x\cdot CA^{-1}x}\|_{W(\mathcal{F}L^1,L^\infty)} \\
&\quad \times \|(e^{-\pi i y\cdot B^{-1}Ay}*f)_{A^{-1}}\|_{W(\mathcal{F}L^1,L^\infty)} \\
&\lesssim |\det B|^{-1/2}|\det A|^{-1}(\det(A^*A+I))^{1/2}|\det(I+iCA^{-1})|^{1/2} \\
&\quad \times \|e^{-\pi i y\cdot B^{-1}Ay}*f\|_{W(\mathcal{F}L^1,L^\infty)} \\
&\lesssim \beta(\mathcal{A})\|f\|_{W(\mathcal{F}L^\infty,L^1)},
\end{aligned}
$$

where the last row is due to (2.83), with $\beta(\mathcal{A})$ defined in (6.53). □

Proof of Theorem 6.1.22. The proof uses the same arguments as in Theorem 6.1.20. Here, the estimate (2.107) is replaced by (2.130). Besides, the index relation (2.127) is applied in the final step. In details,

$$
\begin{aligned}
&\|\mu(\mathcal{A})f\|_{W(\mathcal{F}L^p,L^q)} \\
&\quad \le |\det B|^{-1/2}\|e^{-\pi i x\cdot DB^{-1}x}\|_{W(\mathcal{F}L^1,L^\infty)}\|(\mathcal{F}^{-1}(e^{-\pi i y\cdot B^{-1}Ay}f))_{B^{-1}}\|_{W(\mathcal{F}L^p,L^q)} \\
&\quad \lesssim \prod_{j=1}^{d}|\lambda_j|^{-1/2}|\det(I+iDB^{-1})(I+iB^{-1}A)|^{1/2} \\
&\quad\quad \times \prod_{j=1}^{d}(\max\{1,|\lambda_j|^{-1}\})^{\mu_1(p',q')}(\min\{1,|\lambda_j|^{-1}\})^{\mu_2(p',q')}\|f\|_{W(\mathcal{F}L^q,L^p)} \\
&\quad = |\det(B+iD)(B+iA)|^{1/2} \\
&\quad\quad \times \prod_{j=1}^{d}(\max\{1,|\lambda_j|\})^{\mu_1(p,q)-1/2}(\min\{1,|\lambda_j|\})^{\mu_2(p,q)-1/2}\|f\|_{W(\mathcal{F}L^q,L^p)},
\end{aligned}
$$

that is, case (i) is proved.

Case (ii) indeed is not an improvement of (6.52), but is just (6.52) rephrased in terms of the eigenvalues of A. □

Remark 6.1.24. The above theorems require the condition $\det B \neq 0$. However, in some special cases with $\det B = 0$, the previous results stay valid and can be used to obtain estimates between Wiener amalgam spaces. For example, if $\mathcal{A} = \left(\begin{smallmatrix} I & 0 \\ C & I \end{smallmatrix}\right)$, with $C = C^*$, then $\mu(\mathcal{A})f(x) = \pm e^{-\pi i Cx\cdot x}f(x)$ (see (1.16)), so that, for every $1 \leq p, q \leq \infty$, Proposition 2.4.21 and identity (6.58) give

$$\|\mu(\mathcal{A})f\|_{W(\mathcal{F}L^p,L^q)} \lesssim \prod_{j=1}^{d}(1+\lambda_j^2)^{1/4}\|f\|_{W(\mathcal{F}L^p,L^q)},$$

where the λ_j's are the eigenvalues of C. This estimate was also shown in [306, 14, 72].

6.2 Sparsity of Gabor representation of Schrödinger propagators

In this section we shall apply Fourier integral operators to study problems arising in partial differential equations. In this framework we list some of the many contributions [35, 41, 43, 95, 114, 115, 173, 187, 188, 190] (see also the references therein). Our attention will be initially fixed on numerical aspects. The Fourier integral operator T object of this study is of the type (6.12), with a tame phase function Φ. The symbol σ is a function on $\mathbb{R}^{2d}$ satisfying condition (6.13).

We recall that an important example of such an FIO T is the solution operator to the free particle Schrödinger equation

$$i\partial_t u + \Delta u = 0$$

with initial condition $u(0,x) = u_0(x)$, $x \in \mathbb{R}^d$, $d \geq 1$. The solution $u(t,x)$ is

$$u(t,x) = \int_{\mathbb{R}^d} e^{2\pi i(x\cdot\eta - 2\pi t\eta^2)}\widehat{u_0}(\eta)\,d\eta,$$

that is, $u(t,x) = T_t u_0(x)$, where $T = T_t$ is the FIO with phase $\Phi(x,\eta) = x\cdot\eta - 2\pi t\eta^2$ and symbol $\sigma \equiv 1$.

More generally, one can replace the Laplacian Δ by an operator H given by the Weyl quantization of a quadratic form on $\mathbb{R}^{2d}$. For instance, $H = -\frac{1}{4\pi}\Delta + \pi x^2$ or $H = -\frac{1}{4\pi}\Delta - \pi x^2$.

Let us emphasize that, as opposed to Hörmander's FIOs [114, 115, 190] where the phase function is assumed homogeneous of degree 1 with respect to the dual variables, these FIOs possess quite different properties of the phase, basic example being a quadratic form in x and ξ.

Consider a Gabor frame $\{g_{m,n}\}_{m,n}$, with $g_{m,n}(x) = e^{2\pi i n\cdot x}g(x-m)$, $(m,n) \in \Lambda = \alpha\mathbb{Z}^d \times \beta\mathbb{Z}^d$, $\alpha, \beta > 0$ and window $g \in \mathcal{S}(\mathbb{R}^d)$. In Theorem 6.1.2 it is shown that an FIO T with the features mentioned above has Gabor matrix satisfying

$$|T_{m',n',m,n}| = |\langle Tg_{m,n}, g_{m',n'}\rangle| \leq C_N\langle\chi(m,n) - (m',n')\rangle^{-2N}, \quad \forall N \in \mathbb{N},$$

where χ is the canonical transformation generated by the phase function Φ. Using this result, we shall exhibit the sparsity of the matrix $T_{m',n',m,n}$. Namely, let a be any column or row of the matrix, and let $|a|_n$ be the nth largest entry of the sequence a. Then for each $M > 0$, $|a|_n$ satisfies

$$|a|_n \leq C_M n^{-M}.$$

Notice that Hörmander's FIOs, arising as solution operators to Cauchy problems for hyperbolic differential equations, have been widely studied from the numerical point of view in the works [41, 43, 42, 173], where the authors prove that curvelet and shearlet frames almost diagonalize such operators and provide their sparsest possible representations. Our results show that

Gabor frames are the right choice of frames when studying Schrödinger-type operators.

Furthermore, differently from the frames studied in [41, 43, 42, 173], Gabor frames may be employed to characterize (quasi-)Banach and Hilbert spaces different from the L^2 case. Consequently, boundedness properties of FIOs acting on these spaces may be obtained via the continuity properties of the infinite matrix $T_{m',n',m,n}$. Before detailing these issues, we give a flavor of the related results.

Given $f \in M^p(\mathbb{R}^d)$, $0 < p \leq \infty$ (see Definition 2.7.6 for quasi-Banach modulation spaces), we infer a formula to compute Tf by nonlinear approximation, namely a formula which recovers Tf via a finite sum of Gabor atoms $g_{m,n}$ and an error term. Setting $\nu := (m,n)$, $b_\nu := \langle f, g_{m,n}\rangle$, we consider a non-increasing rearrangement $|b_{\nu_1}| \geq |b_{\nu_2}| \geq \cdots$. For $N \in \mathbb{N}$ and $B \geq 1$, introduce the index sets

$$J_N = \{\nu_1, \dots, \nu_N\}, \quad I_{N,B} = \{\lambda \in \Lambda : \exists \nu \in J_N, |\lambda - \chi(\nu)| \leq B\},$$

where χ is the canonical transformation generated by Φ. Then we show that (Theorem 6.2.5)

$$Tf = \sum_{\lambda \in I_{N,B}} \sum_{\nu \in J_N} T_{\lambda,\nu} \langle f, g_\nu \rangle g_\lambda + e_{N,B},$$

where the error $e_{N,B}$ satisfies the estimate

$$\|e_{N,B}\|_{M^q} \leq C B^{d \max\{1,1/q\} - N'} \|f\|_{M^q} + N^{1/q - 1/p} \|f\|_{M^p},$$

for every $0 < p < q \leq \infty$, $N' \in \mathbb{N}$, with $N' > d \max\{1, 1/q\}$ and the constant $C > 0$ is independent of f, B, N. Hence, the error term decays faster than any negative (fixed) power of N and B, as $N, B \to +\infty$. This is easily seen by fixing N' conveniently large and p small enough, provided that $f \in M^p(\mathbb{R}^d)$.

As a side preliminary result, boundedness properties of the FIO T on $M^p(\mathbb{R}^d)$, $0 < p < 1$ (Theorem 6.2.1) are obtained.

The last part of the section focuses on the solution $u(t,x)$ to the Schrödinger equation for the free particle case with initial datum given by $u_0(x) = e^{2\pi i m\cdot x}e^{-\pi|x-n|^2}$, where $(m,n) \in \alpha\mathbb{Z}^d \times \beta\mathbb{Z}^d$. Choosing a Gabor frame $\{g_{m,n}\}_{m,n}$, $(m,n) \in \alpha\mathbb{Z}^d \times \beta\mathbb{Z}^d$, $\alpha\beta < 1$, generated by the Gaussian $g(x) = e^{-\pi x^2}$, the entries of $\{T_{m',n',m,n}\}_{m',n',m,n}$ can be computed explicitly, see (6.67). Using Maple software, several numerical examples concerning the time–frequency concentration of the solution $u(t,x)$ varying in time are exhibited, according to the law (6.43). Indeed, they demonstrate the accuracy of (6.43). Moreover, numerical experiments on the coefficient decay, achieved in dimension $d = 1$ and $d = 2$, display the same results as for the curvelet case in [42].

We conclude with further open problems related to this theory. First, the off-diagonal decay of the matrix $T_{m',n',m,n}$ could be used in the finite section method [166] or in the adaptive application of operators [51]. Secondly, an interesting issue is the resolution of the wave front set of a distribution using Gabor frames (see [205]), instead of shearlets frames as done in [215].

We shall come back to Gabor wave front sets in Section 6.6.

6.2.1 Continuity of FIOs on $M^p_\mu(\mathbb{R}^d)$, $0 < p < 1$

The continuity of FIOs on the modulation spaces M^p_μ, $1 \le p \le \infty$, associated with a weight function $\mu \in \mathcal{M}_{v_s}(\mathbb{R}^{2d})$, $s \ge 0$, was studied in Section 6.1.3. Here we study the remaining cases of the quasi-Banach spaces $M^p_\mu(\mathbb{R}^d)$, $0 < p < 1$. The outcome will be used in the subsequent part to deal with the nonlinear approximation problem.

Theorem 6.2.1. *Consider a tame phase function Φ and a symbol satisfying* (6.13). *Let $0 \le s < 2N - 2d/p$, $0 < p < 1$, and $\mu \in \mathcal{M}_{v_s}(\mathbb{R}^{2d})$. Then T extends to a continuous operator from $M^p_{\mu\circ\chi}(\mathbb{R}^d)$ into $M^p_\mu(\mathbb{R}^d)$.*

Proof. We set $T_{m',n',m,n} = \langle Tg_{m,n}, g_{m',n'}\rangle$ and use Theorem 3.2.37. For $T = C_\gamma \circ T_{m',n',m,n} \circ D_\gamma$, the following diagram is commutative:

$$\begin{array}{ccc}
M^p_{\mu\circ\chi} & \xrightarrow{\quad T \quad} & M^p_\mu \\
{\scriptstyle C_\gamma}\big\downarrow & & \big\uparrow{\scriptstyle D_\gamma} \\
\ell^p_{\mu\circ\chi} & \xrightarrow{T_{m',n',m,n}} & \ell^p_\mu
\end{array}$$

where T is viewed as an operator with dense domain $\mathcal{S}(\mathbb{R}^d)$. Whence it is enough to prove the continuity of the infinite matrix $T_{m',n',m,n}$ from $\ell^p_{\mu\circ\chi}$ into ℓ^p_μ.

We apply Lemma 6.1.7 (ii) (with $\lambda = (m',n')$, $\nu = (m,n)$) to the matrix

$$K_{m',n',m,n} = T_{m',n',m,n}\frac{\mu(m',n')}{\mu(\chi(m,n))}.$$

That is, we need to prove

$$\sup_{m,n} \sum_{m',n'} |K_{m',n',m,n}|^p < \infty.$$

In view of (6.23), we have

$$\begin{aligned}|K_{m',n',m,n}|^p &\lesssim \langle \chi(m,n) - (m',n') \rangle^{(-2N+s)p} \frac{(\mu(m',n'))^p}{\langle \chi(m,n) - (m',n')\rangle^{sp} (\mu(\chi(m,n)))^p} \\ &\lesssim \langle \chi(m,n) - (m',n') \rangle^{(-2N+s)p},\end{aligned}$$

where the quotient in the second-to-last row above is bounded because μ is v_s-moderate. Since, by assumption, $(-2N + s)p < -2d$, the series

$$\sum_{m',n'} \langle \chi(m,n) - (m',n') \rangle^{(-2N+s)p}$$

converges, with sum uniformly bounded with respect to (m,n), yielding the desired result. □

6.2.2 Sparsity of the Gabor matrix and nonlinear approximation

Given an FIO T as above and $f \in M^q(\mathbb{R}^d)$, we are going to present a suitable formula to compute Tf by nonlinear approximation, namely we shall expresses Tf as a finite sum of Gabor atoms, modulo an error whose size is estimated in terms of its time–frequency concentration.

Consider $g \in \mathcal{S}$ and $\mathcal{G}(g,\alpha,\beta) = \{g_{m,n} = T_m M_n g, \quad (m,n) \in \Lambda = \alpha\mathbb{Z}^d \times \beta\mathbb{Z}^d\}$, a Gabor frame for $L^2(\mathbb{R}^d)$. Assume the symbol σ fulfills (6.13) for every $N \in \mathbb{N}$.

Set $\lambda = (m',n') \in \Lambda$ and $\nu = (m,n) \in \Lambda$. Define

$$T_{\lambda,\nu} = \langle Tg_\nu, g_\lambda \rangle.$$

Then Theorem 6.1.5 shows that, for every $N' \in \mathbb{N}$,

$$|T_{\lambda,\nu}| \le C_{N'} \langle \lambda - \chi(\nu) \rangle^{-N'}. \tag{6.60}$$

This gives at once that the Gabor matrix $T_{\lambda,\nu}$ is *sparse*, in the sense clarified by the following proposition

Proposition 6.2.2. *The Gabor matrix $T_{\lambda,\nu}$ is sparse. Namely, let a be any column or row of the matrix, and let $|a|_n$ be the nth largest entry of the sequence a. Then for each $M > 0$, $|a|_n$ satisfies*

$$|a|_n \le C_M n^{-M}.$$

Proof. We have

$$n^{1/p} \cdot |a|_n \leq \|a\|_{\ell^p},$$

for every $0 < p \leq \infty$. Hence it suffices to prove that every column or row is in ℓ^p for arbitrarily small p. This follows immediately from (6.60) (and was also already verified in the proof of Theorem 6.1.8). □

We now turn to the nonlinear approximation problem. First, some preliminary results are required.

Given a nonincreasing sequence b_j, $j = 1, 2, \ldots$, and $0 < p \leq \infty$, define

$$\sigma_{N,p}(b) = \Big(\sum_{j \geq N+1} |b_j|^p \Big)^{1/p}.$$

The quantity above satisfies the following estimate.

Lemma 6.2.3. *For every $0 < p < q \leq \infty$, it turns out that*

$$\sigma_{N,p}(b) \leq C N^{1/p - 1/q} \|b\|_{\ell^q}, \quad \forall b \in \ell^q,$$

for a constant $C = C_q$ depending only on q.

Proof. It follows from [111], see also [167, Lemma 5], that

$$\Big(\sum_{j=1}^{\infty} [j^{1/q - 1/p} \sigma_{j,p}(b)]^q \frac{1}{j} \Big)^{1/q} \leq C_q \|b\|_{\ell^q}.$$

Hence the desired estimate is a consequence of this elementary fact: if a numerical series $\sum_{j=1}^{\infty} a_j$ converges and the sequence a_j is nonincreasing and nonnegative, then $a_j \leq C/j$ for some constant $C > 0$. □

Let us show the easy estimate below.

Lemma 6.2.4. *For $\gamma > d$,*

$$\sum_{\lambda \in \Lambda : |\lambda - y| \geq B} \langle \lambda - y \rangle^{-\gamma} \lesssim (1 + B)^{d - \gamma},$$

uniformly with respect to $B \geq 0$, $y \in \mathbb{R}^d$. Precisely, the constant implicit in the notation $\lesssim$ depends only on d, γ, and the Gabor constants α, β.

Proof. By shifting the index in the summation, we can suppose that $y \in \alpha[0,1)^d \times \beta[0,1)^d$. Then $\langle \lambda - y \rangle \asymp \langle \lambda \rangle$, and it is enough to prove that

$$\sum_{\lambda \in \Lambda : |\lambda| \geq B} \langle \lambda \rangle^{-\gamma} \lesssim (1 + B)^{d - \gamma}.$$

Now, we can consider the partition of $\mathbb{R}^d$ defined by the boxes $\prod_{j=1}^d [k_j, k_j+1)$, $k_j \in \mathbb{Z}$, (having diameter $\sqrt{d}$). The number of points of Λ which fall into any such a box is uniformly bounded and we also have $\langle\lambda\rangle \asymp \langle x\rangle$, if λ, x belong to the same box. Hence, let $\{Q'\}$ be the boxes that intersect the set $|x| \geq R$. Then

$$\begin{aligned}\sum_{\lambda\in\Lambda:|\lambda|\geq B} \langle\lambda\rangle^{-\gamma} &= \sum_{Q'} \sum_{\lambda\in\Lambda\cap Q'} \langle\lambda\rangle^{-\gamma} \lesssim \sum_{Q'} \inf_{x\in Q'} \langle x\rangle^{-\gamma} \\ &\lesssim \sum_{Q'} \int_{Q'} \langle x\rangle^{-\gamma}\,dx \leq \int_{|x|\geq B-\sqrt{d}} \langle x\rangle^{-\gamma}\,dx \lesssim (1+B)^{d-\gamma},\end{aligned}$$

as an explicit computation shows. □

Given $N \in \mathbb{N}$ and $B \geq 1$, we associate to any function $f \in M^q_\mu(\mathbb{R}^d)$ two index sets J_N and $I_{N,B}$ defined as follows. Let b_ν be the sequence of weighted Gabor coefficients of f, namely $b_\nu = \langle f, g_\nu\rangle\mu(\nu)$, and let $|b_{\nu_1}| \geq |b_{\nu_2}| \geq \cdots$ be a nonincreasing rearrangement. We then define

$$J_N = \{\nu_1, \dots, \nu_N\}$$

and

$$I_{N,B} = \{\lambda \in \Lambda : \exists \nu \in J_N,\ |\lambda - \chi(\nu)| \leq B\},$$

where χ is the canonical transformation generated by Φ. The approximation result can be formulated as follows.

Theorem 6.2.5. *Let $0 < p < q \leq \infty$, and $f \in M^p_{\mu\circ\chi}(\mathbb{R}^d)$, with $\mu \in \mathcal{M}_{v_s}(\mathbb{R}^{2d})$, $s \geq 0$. Let J_N and $I_{N,B}$ be the index sets associated to f as above. Then for every $N', N \in \mathbb{N}$, with $N' > d\max\{1, 1/q\} + s$, and $B \geq 1$, we have*

$$Tf = \sum_{\lambda\in I_{N,B}} \sum_{\nu\in J_N} T_{\lambda,\nu}\langle f, g_\nu\rangle g_\lambda + e_{N,B},$$

where the error $e_{N,B}$ satisfies the estimate

$$\|e_{N,B}\|_{M^q_\mu} \lesssim B^{d\max\{1,1/q\}+s-N'} \|f\|_{M^q_{\mu\circ\chi}} + N^{1/q-1/p}\|f\|_{M^p_{\mu\circ\chi}}.$$

Here the constant implicit in the notation $\lesssim$ is independent of f, B and N.

Hence, given f, the error term decays faster of any negative fixed power of N and B, as $N, B \to +\infty$. This is seen by fixing N' conveniently large and p small enough.

Proof. Let $c_\nu = \langle f, g_\nu\rangle$. We have

$$e_{N,B} = e^{(1)}_{N,B} + e^{(2)}_{N,B}$$

where

$$e^{(1)}_{N,B} = \sum_{\lambda\notin I_{N,B}} \sum_{\nu\in J_N} T_{\lambda,\nu} c_\nu g_\lambda$$

and

$$e^{(2)}_{N,B} = \sum_{\lambda} \sum_{\nu \notin J_N} T_{\lambda,\nu} c_\nu g_\lambda.$$

We shall show the estimates

$$\|e^{(1)}_{N,B}\|_{M^q_\mu} \lesssim B^{d\max\{1,1/q\}+s-N'} \|f\|_{M^q_{\mu\circ\chi}} \tag{6.61}$$

and

$$\|e^{(2)}_{N,B}\|_{M^q_\mu} \lesssim N^{1/q-1/p} \|f\|_{M^p_{\mu\circ\chi}}, \tag{6.62}$$

where, as in the statement, the constant implicit in the notation $\lesssim$ is independent of f, B and N.

Let us prove (6.61). We have

$$\|e^{(1)}_{N,B}\|_{M^q_\mu} \lesssim \left(\sum_{\lambda \notin I_{N,B}} \left| \sum_{\nu \in J_N} T_{\lambda,\nu} c_\nu \right|^q \mu(\lambda)^q \right)^{1/q}.$$

Now, we want to apply Lemma 6.1.7 with $K_{\lambda,\nu} := T_{\lambda,\nu}\mu(\lambda)/\mu(\chi(\nu))$. To this end, we observe, as in the proof of Theorem 6.1.8, that

$$\frac{|T_{\lambda,\nu}|\mu(\lambda)}{\mu(\chi(\nu))} \lesssim \langle \lambda - \chi(\nu) \rangle^{-N'+s}.$$

Hence, Lemma 6.1.7, together with Lemma 6.2.4 (applied to $y = q(N' - s)$, if $0 < q \le 1$, and $y = N' - s$, if $1 \le q \le \infty$), and the equivalence $|\lambda - \chi(\nu)| \asymp |\chi^{-1}(\lambda) - \nu|$ (which follows from the bi-Lipschitz continuity of χ) give

$$\|e^{(1)}_{N,B}\|_{M^q_\mu} \lesssim \begin{cases} B^{d/q+s-N'} \|c\|_{\ell^q_{\mu\circ\chi}}, & 0 < q \le 1, \\ B^{d+s-N'} \|c\|_{\ell^q_{\mu\circ\chi}}, & 1 \le q \le \infty. \end{cases}$$

Since $\|c\|_{\ell^q_{\mu\circ\chi}} \lesssim \|f\|_{M^q_{\mu\circ\chi}}$, we deduce (6.61).

Let us now verify (6.62). From the continuity of the matrix $T_{\lambda,\nu}$ from $\ell^q_{\mu\circ\chi}$ into ℓ^q_μ it follows that

$$\begin{aligned} \|e^{(2)}_{N,B}\|_{M^q_\mu} &\lesssim \left(\sum_{\lambda} \left| \sum_{\nu \notin J_N} T_{\lambda,\nu} c_\nu \right|^q \mu(\lambda)^q \right)^{1/q} \\ &\lesssim \left(\sum_{\nu \notin J_N} |c_\nu|^q \mu(\chi(\nu))^q \right)^{1/q}. \end{aligned}$$

By Lemma 6.2.3, applied to the nonincreasing rearrangement of $\{c_\nu \mu(\chi(\nu))\}$, this last expression is

$$\lesssim N^{1/q-1/p} \|c\|_{\ell^p_{\mu\circ\chi}}.$$

Since $\|c\|_{\ell^p_{\mu\circ\chi}} \lesssim \|f\|_{M^p_{\mu\circ\chi}}$, we deduce (6.62). □

6.2.3 Application to the solution of the Schrödinger equation

We now turn our attention to some important examples of the class of FIOs studied here, represented by solution operators to the Cauchy problem for Schrödinger equations.

6.2.3.1 The free particle

Consider the Cauchy problem for the Schrödinger equation in (5.11) The explicit formula for the solution is given by (5.21) and (5.22).

The solution $u(t,x)$ can be expressed as an FIO

$$\begin{aligned} u(t,x) &= \int_{\mathbb{R}^d} e^{2\pi i x\cdot\eta}\mathcal{F}(K_t * u_0)(\eta)\,d\eta \\ &= \int_{\mathbb{R}^d} e^{2\pi i x\cdot\eta}\mathcal{F}(K_t)(\eta)\widehat{u_0}(\eta)\,d\eta \\ &= \int_{\mathbb{R}^d} e^{2\pi i(x\cdot\eta-2\pi t\eta^2)}\widehat{u_0}(\eta)\,d\eta, \end{aligned}$$

since, as an easy computation shows, $\mathcal{F}((ai)^{-d/2}e^{-\pi x^2/(ai)})(\eta) = e^{-\pi a i\eta^2}$, $a \in \mathbb{R}\setminus\{0\}$. Whence $u(t,x) = Tu_0(x)$, where T is the FIO with phase $\Phi(x,\xi) = x\cdot\eta - 2\pi t\eta^2$ and symbol $\sigma \equiv 1$.

The smooth bi-Lipschitz canonical transformation χ associated with the phase Φ is given by

$$\chi(y,\eta) = (y + 4\pi t\eta, \eta).$$

Hence, the matrix decay (6.60), for $\lambda = (m',n')$, $\nu = (m,n)$, is given by

$$|T_{\lambda,\nu}| \le C_{N'}\big(1 + \big|(m' - (m + 4\pi tn), n' - n)\big|\big)^{-N'}. \tag{6.63}$$

In what follows we shall compute the matrix entries explicitly.

Lemma 6.2.6. *For $c \in \mathbb{C}$, $c \ne 0$, consider the function $\phi^{(c)}(x) = e^{-\pi c x^2}$, $x \in \mathbb{R}^d$. For every $c_1, c_2 \in \mathbb{C}$, with $\operatorname{Re} c_1 \ge 0$, $\operatorname{Re} c_2 > 0$, $n \in \mathbb{Z}^d$, we have*

$$\phi^{(c_1)} * M_n\phi^{(c_2)}(x) = (c_1 + c_2)^{-d/2}\phi^{(\frac{1}{c_1+c_2})}(n)M_{\frac{c_1 n}{c_1+c_2}}\phi^{(\frac{c_1c_2}{c_1+c_2})}(x). \tag{6.64}$$

Proof. This is a straightforward computation. Set $z^2 = z_1^2 + \cdots + z_n^2$, for $z = (z_1,\ldots, z_n) \in \mathbb{C}^d$. Then

$$\begin{aligned} (\phi^{(c_1)} * M_n\phi^{(c_2)})(x) &= \int e^{-\pi c_1(x-t)^2+2\pi i t\cdot n-\pi c_2 t^2}\,dt \\ &= e^{-\pi c_1 x^2}\int e^{-\pi[(c_1+c_2)t^2-2(c_1x+in)\cdot t]}\,dt \end{aligned}$$

$$= e^{-\pi\left(c_1x^2 - \frac{(c_1x+in)^2}{c_1+c_2}\right)} \int e^{-\pi\left((c_1+c_2)^{1/2}t - \frac{c_1x+in}{(c_1+c_2)^{1/2}}\right)^2} dt$$

$$= (c_1+c_2)^{-d/2} e^{-\pi\frac{c_1c_2x^2+n^2}{c_1+c_2}} e^{2\pi i \frac{c_1}{c_1+c_2} n\cdot x},$$

as desired. □

Corollary 6.2.7. *Let $K_t(x)$ be the kernel in (5.22), then for $g_0 = e^{-\pi x^2}$, $n \in \mathbb{Z}^d$, we have*

$$(K_t * M_n g_0)(x) = \frac{1}{(1+4\pi i t)^{d/2}} e^{-\frac{4\pi^2 n^2 it}{1+4\pi it}} M_{\frac{n}{1+4\pi it}} e^{-\frac{\pi}{1+4\pi it}x^2}. \tag{6.65}$$

Proof. Since

$$(K_t * M_n g_0)(x) = \frac{1}{(4\pi i t)^{d/2}} (\phi^{((4\pi it)^{-1})} * M_n \phi^{(1)})(x),$$

the result follows at once from Lemma 6.2.6. □

Corollary 6.2.8. *Let $u(t,x)$ be the solution of the Cauchy problem (5.22), with $u_0(x) = M_n T_m e^{-\pi x^2}$. Then*

$$u(t,x) = (1+4\pi it)^{-d/2} e^{-\frac{4\pi^2 tn\cdot(2m+in)}{1+4\pi it}} M_{\frac{n}{1+4\pi it}} T_m e^{-\frac{\pi}{1+4\pi it}x^2}. \tag{6.66}$$

Proof. The convolution operator commutes with translations and $M_n T_m = e^{2\pi i m\cdot n} T_m M_n$, so that

$$u(t,x) = (K_t * M_n T_m g_0)(x) = e^{2\pi i m\cdot n} T_m (K_t * M_n g_0)(x).$$

Using (6.65),

$$\begin{aligned} u(t,x) &= (1+4\pi it)^{-d/2} e^{2\pi i m\cdot n} e^{-\frac{4\pi^2 n^2 it}{1+4\pi it}} T_m M_{\frac{n}{1+4\pi it}} e^{-\frac{\pi}{1+4\pi it}x^2} \\ &= (1+4\pi it)^{-d/2} e^{2\pi i m\cdot n} e^{-\frac{4\pi^2 n^2 it}{1+4\pi it}} e^{-2\pi i m\cdot \frac{n}{1+4\pi it}} M_{\frac{n}{1+4\pi it}} T_m e^{-\frac{\pi}{1+4\pi it}x^2} \\ &= (1+4\pi it)^{-d/2} e^{-\frac{4\pi^2 tn\cdot(2m+in)}{1+4\pi it}} M_{\frac{n}{1+4\pi it}} T_m e^{-\frac{\pi}{1+4\pi it}x^2}, \end{aligned}$$

that is, (6.66) holds. □

Hence, we have a more explicit formula for the matrix entries

$$\{T_{m',n',m,n}\}_{m',n',m,n} = \langle T(M_n T_m g_0), M_{n'} T_{m'} g_0 \rangle.$$

In fact,

$$T_{m',n',m,n} = (1+4\pi it)^{-d/2} e^{-\frac{4\pi^2 tn\cdot(2m+in)}{1+4\pi it}} \langle M_{\frac{n}{1+4\pi it}} T_m e^{-\frac{\pi}{1+4\pi it}|\cdot|^2}, M_{n'} T_{m'} g_0 \rangle. \tag{6.67}$$

We now present some numerical examples concerning (5.11). We choose the Gabor frame $\{M_n T_m g_0\}$, with $(m,n) \in \mathbb{Z}^d \times (1/2)\mathbb{Z}^d$ and $g_0(x) = e^{-\pi x^2}$.

First, we study the one-dimensional case ($d = 1$), taking the initial datum $u_0(x) = M_1 T_2 g_0(x) = e^{2\pi i x} e^{-\pi(x-2)^2}$. Figures 6.1, 6.2, 6.3, and 6.4 represent the magnitude of the STFT of the solution $u(t, x)$ for $t = 0, 1, 2, 10$, respectively. Figures 6.5, 6.6, 6.7, and 6.8 exhibit the contour plots of the STFT magnitude related to the same instant time as before.

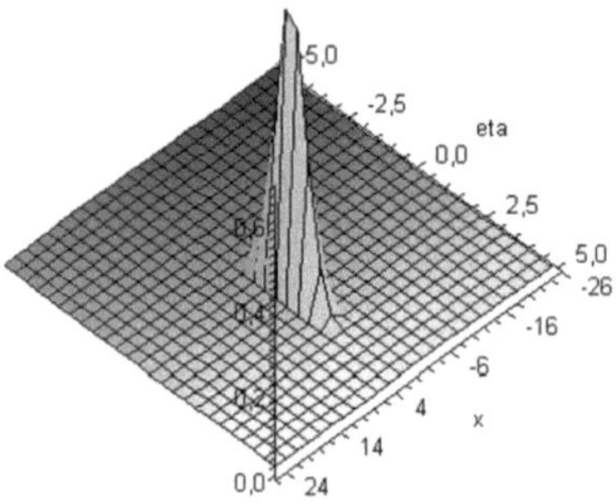

Figure 6.1: Time $t = 0$.

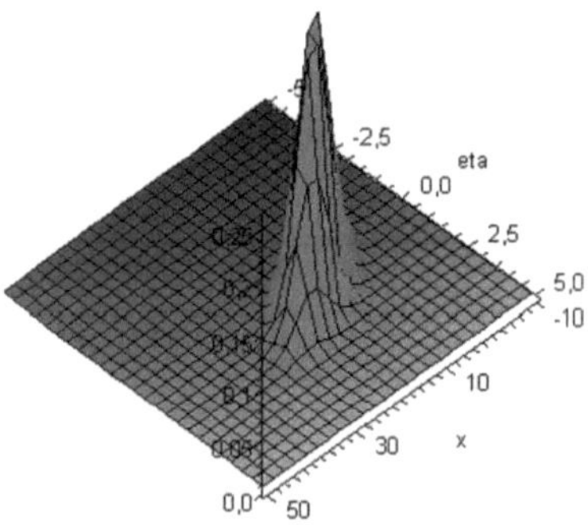

Figure 6.2: Time $t = 1$.

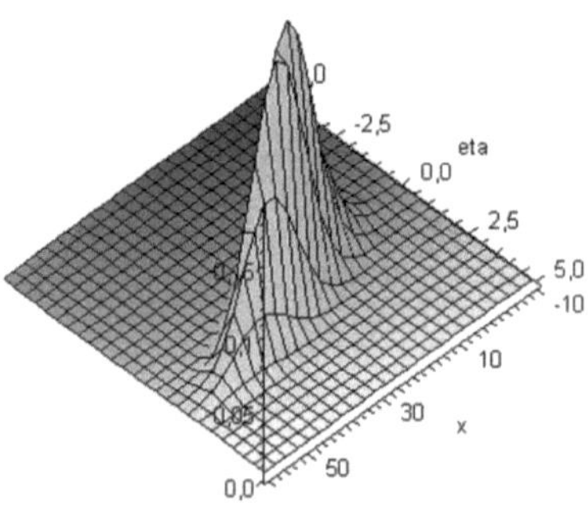

Figure 6.3: Time $t = 2$.

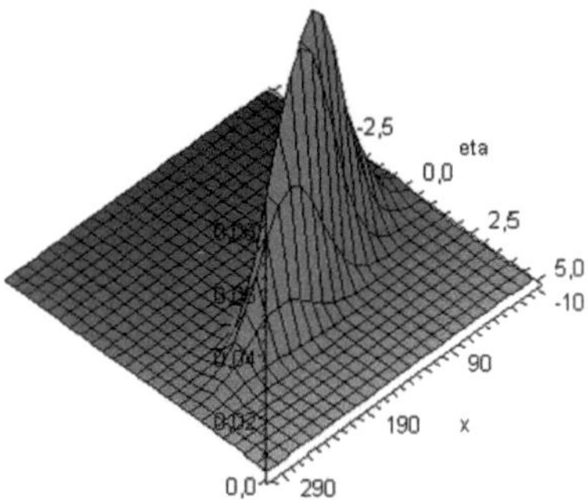

Figure 6.4: Time $t = 10$.

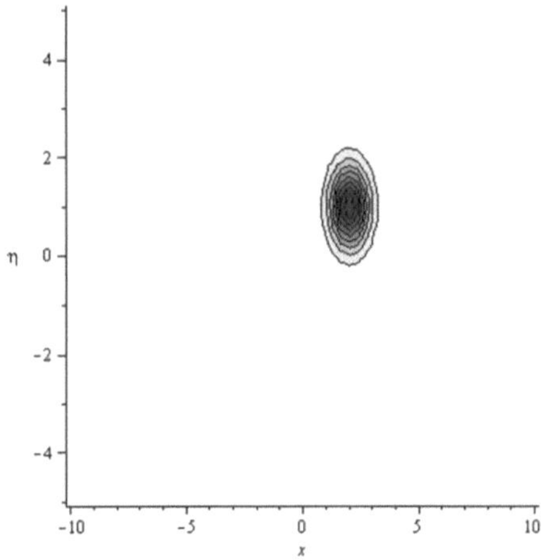

Figure 6.5: Time $t = 0$.

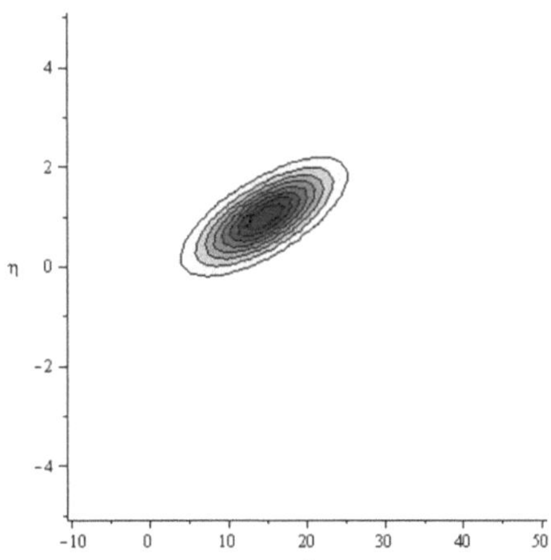

Figure 6.6: Time $t = 1$.

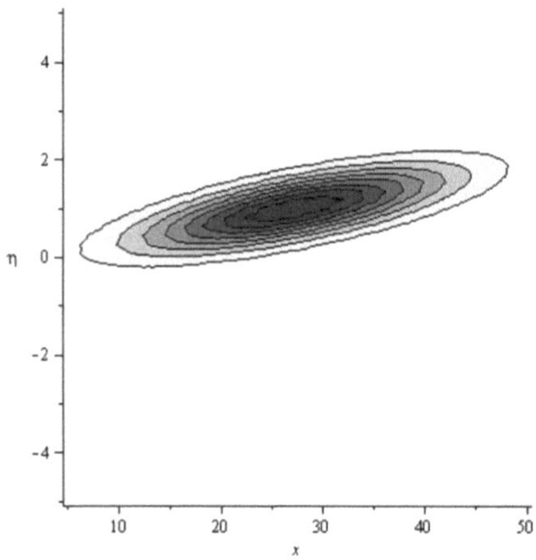

Figure 6.7: Time $t = 2$.

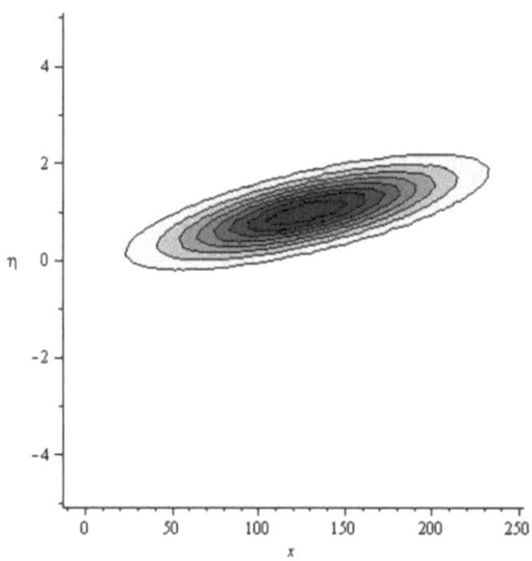

Figure 6.8: Time $t = 10$.

Indeed, the propagator $e^{it\Delta}$ moves the time–frequency concentration of the solution $u(t,x)$ when t increases, according to the law in (6.63) (with $\nu = (m,n) = (2,1)$).

Figures 6.9, 6.10, 6.11, and 6.12 represent the magnitude of the coefficients (sorted in decreasing order) of one column of the matrix $T_{m',n',m,n}$, namely that obtained by fixing $m = n = 0$. Notice that choosing the column corresponding to $m = n = 0$ does not represent a loss of generality. Indeed, as one can verify by a direct computation, $T_{m',n',m,n} = T_{m'-m-4\pi tn,n'-n,0,0}$, for every $(m,n),(m',n') \in \mathbb{Z}^d \times (1/2)\mathbb{Z}^d$, so that for all columns essentially the same figures as those presented can be drawn.

Finally, Figure 6.13 shows a similar analysis in dimension $d = 2$, for the column corresponding to $m = (0,0)$, $n = (0,0)$. Notice that we need about 10^6 coefficients to reach the threshold 10^{-25}. The same type of decay was obtained in [42, Figure 15] for the curvelet representation of the wave propagator.

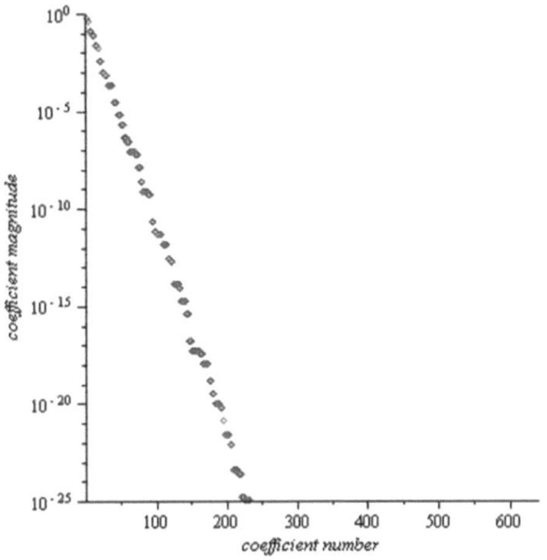

Figure 6.9: Time $t = 0$.

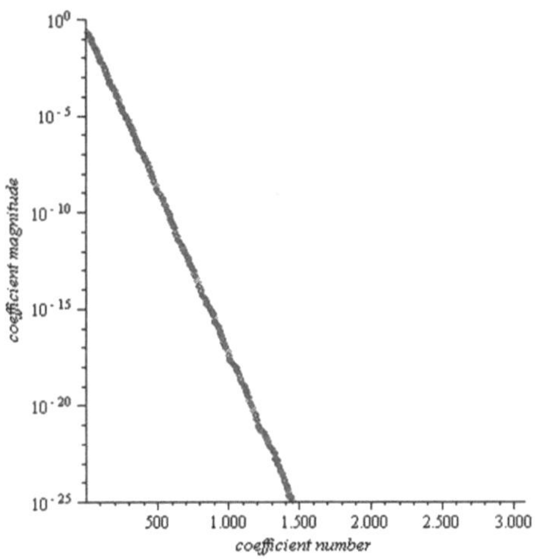

Figure 6.10: Time $t = 1$.

6.2.3.2 The harmonic oscillator

This is the case of the Cauchy problem for the Schrödinger equation below

$$\begin{cases} i\frac{\partial u}{\partial t} - \frac{1}{4\pi}\Delta u + \pi|x|^2 u = 0, \\ u(0,x) = u_0(x). \end{cases} \tag{6.68}$$

The solution is the metaplectic operator (hence FIO) given by [70, 145]

$$u(t,x) = (\cos t)^{-d/2} \int_{\mathbb{R}^d} e^{2\pi i[\frac{1}{\cos t} x\cdot\eta + \frac{\tan t}{2}(x^2+\eta^2)]} \hat{f}(\eta)\, d\eta, \quad t \neq \frac{\pi}{2} + k\pi,\ k \in \mathbb{Z}. \tag{6.69}$$

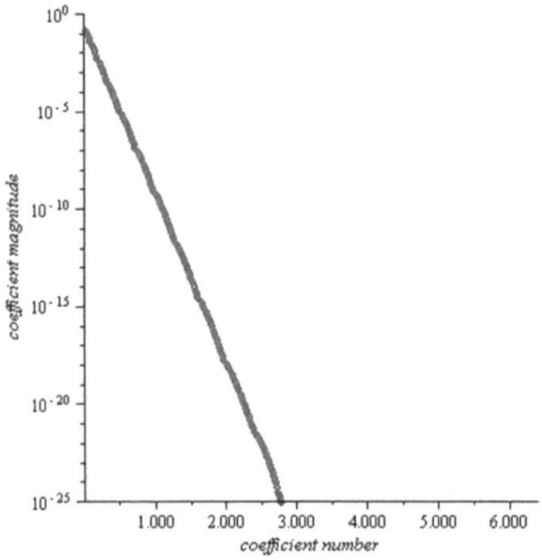

Figure 6.11: Time $t = 2$.

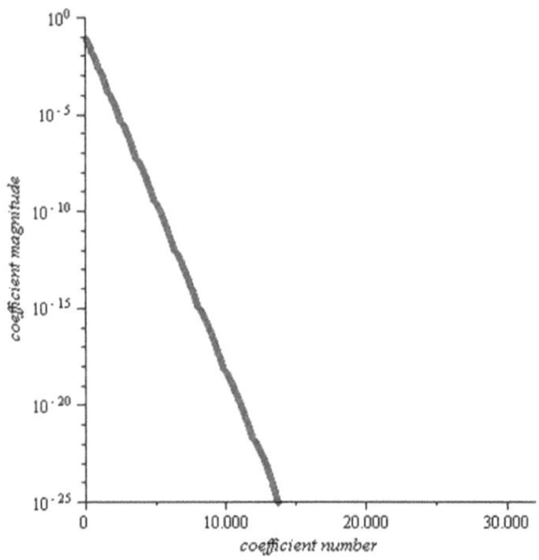

Figure 6.12: Time $t = 10$.

Here $u(t,x) = T_t$, where T_t is an FIO with symbol $\sigma \equiv 1$ and phase

$$\Phi_t(x,\eta) = \frac{1}{\cos t} x \cdot \eta + \frac{\tan t}{2}(x^2 + \eta^2).$$

The canonical transformation $\chi = \chi_t$ is then obtained by solving (6.9), so that

$$\chi_t(y,\eta) = \begin{pmatrix} (\cos t)I & (-\sin t)I \\ (\sin t)I & (\cos t)I \end{pmatrix} \begin{pmatrix} y \\ \eta \end{pmatrix}. \tag{6.70}$$

Since

$$\left|\det \partial^2_{x,\eta} \Phi_t(x,\eta)\right| = \frac{1}{|\cos t|^d} > 0, \quad t \neq \frac{\pi}{2} + k\pi,\ k \in \mathbb{Z},$$

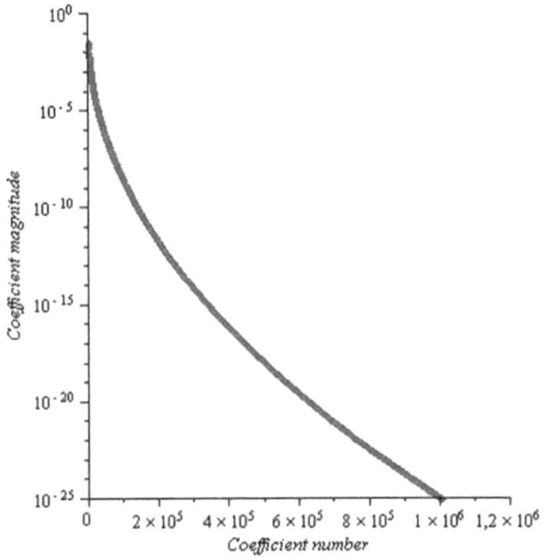

Figure 6.13: Coefficient magnitude at time $t = 1$ for $d = 2$.

the assumptions of Theorem 6.1.5 are fulfilled, and we get the estimate

$$|\langle Tg_{m,n}, g_{m',n'}\rangle| \leq C_N(1 + |(\cos t)m + (\sin t)n - m', -(\sin t)m + (\cos t)n - n'|)^{-N},$$

for every $N \in \mathbb{N}$, if $t \neq \frac{\pi}{2} + k\pi, k \in \mathbb{Z}$.

Remark 6.2.9. Gabor frames have also been successfully employed to study the semiclassical analysis of linear Schrödinger equations. In this context, time-dependent smooth Hamiltonians with at most quadratic growth have been considered and higher order parametrices for the corresponding Schrödinger equations have been constructed. Nonlinear parametrices, in the spirit of the nonlinear approximation, are presented and numerical experiments exhibited in [18]. A follow-up of the previous work is contained in [61]. Such papers represent an attempt to create a systematic common ground for semiclassical and time–frequency analysis. These two different areas combined together provide interesting outcomes for Schrödinger-type equations. In fact, continuity results of both Schrödinger propagators and their asymptotic solutions are obtained on $\hbar$-dependent Banach spaces, the semiclassical version of modulation spaces. Continuity properties of more general Fourier integral operators (FIOs) and their sparsity are also investigated. We refer the interested reader to [61].

Further applications of Gabor decompositions to Schrödinger and semilinear parabolic equations are contained in [85, 234].

6.3 Wiener algebras of Fourier integral operators

In this section we will consider algebras of Fourier integral operators and their properties, in particular Wiener algebras.

The original Wiener's lemma is a classical statement about absolutely convergent series, see [311] and [312]. Let us first recall this statement for the benefit of the reader. Let $A([0,1])$ denote the space of functions defined on the interval $[0,1]$ having absolutely convergent Fourier series expansion. That is, every f in $A([0,1])$ is given by

$$f(x) = \sum_{k\in\mathbb{Z}} c_k e^{2\pi ikx}$$

where

$$\sum_{k\in\mathbb{Z}} |c_k| < \infty.$$

Recall that every f in $A([0,1])$ is continuous. Moreover, the Fourier coefficients (c_k) corresponding to a given f are unique. This allows one to impose a norm on $A([0,1])$ by setting

$$\|f\|_{A([0,1])} = \sum_{k\in\mathbb{Z}} |c_k|.$$

It can be shown that $A([0,1])$ is complete with respect to this norm, making this a Banach space. Moreover, if f and g belong to $A([0,1])$, then so does fg, and

$$\|fg\|_{A([0,1])} \le \|f\|_{A([0,1])}\|g\|_{A([0,1])}.$$

This multiplicative structure, along with its usual interaction with addition and scalar multiplication, makes $A([0,1])$ a Banach algebra. Wiener's lemma states that the inverse of each nonzero element in $A([0,1])$ is again in $A([0,1])$.

Theorem 6.3.1 (Wiener's lemma). *Suppose that f is a function in $A([0,1])$ and is nonzero on $[0,1]$. Then $1/f$ belongs to $A([0,1])$.*

In a more general setting, Wiener's property represents nowadays one of the driving forces in the development of Banach algebra theory.

Let us first fix our attention on pseudodifferential operators, which we may express in the Kohn–Nirenberg form

$$\sigma(x,D)f(x) = \int e^{2\pi ix\cdot\eta}\sigma(x,\eta)\hat{f}(\eta)\,d\eta.$$

The best known result about Wiener's property for pseudodifferential operators is maybe that in [9], see also the subsequent contributions of [32, 299]. It concerns symbols σ in the Hörmander's class $S^0_{0,0}(\mathbb{R}^{2d})$, i. e., smooth functions on $\mathbb{R}^{2d}$ such that, for every multiindex α and every $z \in \mathbb{R}^{2d}$,

$$|\partial_z^\alpha \sigma(z)| \le C_\alpha. \tag{6.71}$$

The corresponding pseudodifferential operators form a subalgebra of $\mathcal{L}(L^2(\mathbb{R}^d))$, usually denoted by $L^0_{0,0}$. The standard symbolic calculus concerning the principal part of symbols of products does not hold, nevertheless, Wiener's lemma is still valid: *If $\sigma(x, D)$ is invertible in $\mathcal{L}(L^2(\mathbb{R}^d))$, then its inverse is again a pseudodifferential operator with symbol in $S^0_{0,0}(\mathbb{R}^{2d})$, hence belonging to $L^0_{0,0}$.*

In the absence of a symbolic calculus, such a version of Wiener's lemma seems to be the minimal property required of any reasonable algebra of pseudodifferential operators. A subalgebra $\mathcal{A}$ of operators in $\mathcal{L}(L^2(\mathbb{R}^d))$ that satisfies Wiener's lemma and is thus closed under inversion is usually called *spectrally invariant* or inverse-closed, or sometimes a Wiener algebra.

We take the point of view of time–frequency analysis and signal processing, for which Wiener's lemma provides an important justification of the engineering practice to model $\sigma(x, D)^{-1}$ as an almost diagonal matrix (this is a peculiar property of pseudodifferential operators, see Theorem 6.3.3 in the sequel).

Actually, in the applications to signal processing, the symbol $\sigma(x, \eta)$ is not always smooth, and it is convenient to use some generalized version of $S^0_{0,0}(\mathbb{R}^{2d})$ [271]. The spaces that do this job are the modulation classes

$$M^{\infty,1}(\mathbb{R}^{2d})$$

(the Sjöstrand class) and

$$M^{\infty}_{1\otimes v_s}(\mathbb{R}^{2d}) = M^{\infty,\infty}_{1\otimes v_s}(\mathbb{R}^{2d}), \quad v_s(z) = \langle z\rangle^s = \left(1 + |z|^2\right)^{s/2}, \; z \in \mathbb{R}^{2d},$$

with the parameter $s \in [0, \infty)$. The Hörmander symbol class $S^0_{0,0}(\mathbb{R}^{2d})$ can be characterized by means of modulation spaces as follows (see, for example, [165]):

$$S^0_{0,0} = \bigcap_{s\geq 0} M^{\infty}_{1\otimes v_s}(\mathbb{R}^{2d}). \tag{6.72}$$

As $s \to 2d$, the symbols in $M^{\infty}_{1\otimes v_s}(\mathbb{R}^{2d})$ have less regularity, until in the maximal space $M^{\infty,1}(\mathbb{R}^{2d})$ even differentiability is lost.

Theorem 6.3.2 ([266]). *The pseudodifferential operators with a symbol in $M^{\infty,1}(\mathbb{R}^{2d})$ form a Wiener subalgebra of $\mathcal{L}(L^2(\mathbb{R}^d))$, the so-called Sjöstrand algebra. The symbol classes $M^{\infty}_{1\otimes v_s}(\mathbb{R}^{2d})$ with $s > 2d$ provide a scale of Wiener subalgebras of the Sjöstrand algebra, and their intersection coincides with $L^0_{0,0}$.*

In this section we construct and investigate Wiener subalgebras consisting of Fourier integral operators and generalize Sjöstrand's theory in [266, 267] to FIOs. We will consider FIOs in (6.12), that we call *FIO of type I*. When necessary, we will also write the FIO T as

$$Tf(x) = T_{I,\Phi,\sigma}f(x) = \int_{\mathbb{R}^d} e^{2\pi i\Phi(x,\eta)}\sigma(x,\eta)\hat{f}(\eta)\, d\eta. \tag{6.73}$$

We first assume $\sigma \in S^0_{0,0}(\mathbb{R}^{2d})$ and consider a tame phase function Φ. If $\Phi(x,\eta) = x\eta$ then $\chi = \mathrm{Id}$, and we recapture the pseudodifferential operators in the Kohn–Nirenberg form. It is an easy exercise to show that the L^2-adjoint of an FIO of type I is an FIO of type II

$$Tf(x) = T_{II,\Phi,\tau}f(x) = \int_{\mathbb{R}^{2d}} e^{-2\pi i[\Phi(y,\eta)-x\cdot\eta]}\tau(y,\eta)f(y)\,dy\,d\eta. \tag{6.74}$$

For the L^2-boundedness of such FIOs of type I and II, see, for example, [3].

It is worth to observe that in contrast to the standard setting of Hörmander [191], we argue globally on $\mathbb{R}^d$, our basic examples being the propagators for Schrödinger-type equations. The second remark is that operators of the reduced form (6.73) or (6.74) do not form an algebra, quite in line with the calculus of Hörmander in [191]. The composition of FIOs requires heavier machinery and is addressed, for example, in [187], in the case when symbol and phase belong to the more restrictive Shubin class of [265].

As a minimal objective, we will present a cheap definition of a subalgebra of $\mathcal{L}(L^2(\mathbb{R}^d))$ containing FIOs of type I, II, and hence $L^0_{0,0}$, and prove the Wiener property for this class.

As a more ambitious objective, we will extend our analysis to the case when the symbol σ belongs to the symbol class $M^\infty_{1\otimes v_s}(\mathbb{R}^{2d})$, and we will define a corresponding scale of Wiener algebras of FIOs. The new algebras of FIOs will be constructed by means of Gabor frames and the decay properties of the corresponding Gabor matrix outside the graph of a symplectic map χ.

Let us recall that for the case of pseudodifferential operators the Gabor discretization provides an equivalent characterization of the Sjöstrand algebra in Theorem 6.3.2.

Theorem 6.3.3 ([162, 165]). *Assume that $\mathcal{G}(g,\Lambda)$ is a frame for $L^2(\mathbb{R}^d)$ with $g \in \mathcal{S}(\mathbb{R}^d)$ and fix $s > 2d$. Then the following statements are equivalent for a distribution $\sigma \in \mathcal{S}'(\mathbb{R}^d)$:*
(i) $\sigma \in M^\infty_{1\otimes v_s}(\mathbb{R}^{2d})$.
(ii) *There exists $C > 0$ such that*

$$|\langle\sigma(x,D)\pi(\lambda)g, \pi(\mu)g\rangle| \le C\langle\mu-\lambda\rangle^{-s}, \quad \forall\lambda,\mu \in \Lambda. \tag{6.75}$$

Hence, the assumption $\sigma \in S^0_{0,0}(\mathbb{R}^d)$ is equivalent to (6.75) being satisfied for all $s \ge 0$.

Moreover, $\sigma \in M^{\infty,1}(\mathbb{R}^{2d})$ if and only if there exists a sequence $h \in \ell^1(\Lambda)$ such that $|\langle\sigma(x,D)\pi(\lambda)g, \pi(\mu)g\rangle| \le h(\lambda-\mu)$.

The above theorem gives a precise meaning to the statement that the Gabor matrix of a pseudodifferential operators is almost diagonal, or that pseudodifferential operators are almost diagonalized by Gabor frames.

Definition 6.3.4. Let χ be a transformation satisfying *B1* and *B2*, and $s \geq 0$. Fix $g \in \mathcal{S}(\mathbb{R}^d) \setminus \{0\}$ and let $\mathcal{G}(g, \Lambda)$ be a Parseval frame for $L^2(\mathbb{R}^d)$. We say that a continuous linear operator $T : \mathcal{S}(\mathbb{R}^d) \to \mathcal{S}'(\mathbb{R}^d)$ is in the class $\mathrm{FIO}(\chi, s)$ if its Gabor matrix satisfies the decay condition

$$|\langle T\pi(\lambda)g, \pi(\mu)g\rangle| \leq C\langle \mu - \chi(\lambda)\rangle^{-s}, \quad \forall \lambda, \mu \in \Lambda. \tag{6.76}$$

The class $\mathrm{FIO}(\Xi, s) = \bigcup_\chi \mathrm{FIO}(\chi, s)$ is the union of these classes where χ runs over the set of all transformations satisfying *B1*, *B2*.

Note that we do not require assumption *B3*.

We shall show that this definition does not depend on the choice of the Gabor frame (cf. Lemma 6.3.6).

If $\chi = \mathrm{Id}$, the identity operator, then the corresponding Fourier integral operators are simply pseudodifferential operators.

The decomposition of an FIO with respect to a Gabor frame provides a technique to settle the following issues:

(i) *Boundedness of T on $L^2(\mathbb{R}^d)$* (Theorem 6.3.7).
If $s > 2d$ and $T \in \mathrm{FIO}(\chi, s)$, then T can be extended to a bounded operator on $L^2(\mathbb{R}^d)$.

(ii) *The algebra property* (Theorem 6.3.9). For $i = 1, 2$, $s_i > 2d$, $s = \min(s_1, s_2)$,

$$T^{(i)} \in \mathrm{FIO}(\chi_i, s_i) \Rightarrow T^{(1)}T^{(2)} \in \mathrm{FIO}(\chi_1 \circ \chi_2, s).$$

(iii) *Wiener property* (Theorem 6.3.10). If $s > 2d$, $T \in \mathrm{FIO}(\chi, s)$, and T is invertible on $L^2(\mathbb{R}^d)$, then $T^{-1} \in \mathrm{FIO}(\chi^{-1}, s)$.

These three properties can be summarized neatly by saying that the union

$$\bigcup_\chi \mathrm{FIO}(\chi, s)$$

is a Wiener subalgebra of $\mathcal{L}(L^2(\mathbb{R}^d))$ consisting of FIOs.

Although it is impossible to do justice to the vast literature on Fourier integral operators, let us mention some of the contributions that are most related to these ideas. From the formal point of view, this approach is very similar to that in [30, 31] and [279], where $\mathrm{FIO}(\chi, \infty)$ was treated. Instead of Gabor frames, in [30, 31] partitions of unity of the Weyl–Hörmander calculus are used, whereas in [279] the Bargmann transform is the main tool. The boundedness and composition of FIOs are treated in [35, 94, 291, 187]. Here we follow strictly [68].

6.3.1 Wiener algebra properties

We first present an equivalence between continuous decay conditions and the decay of the discrete Gabor matrix for a linear operator $\mathcal{S}(\mathbb{R}^d) \to \mathcal{S}'(\mathbb{R}^d)$.

Theorem 6.3.5. *Let T be a continuous linear operator $\mathcal{S}(\mathbb{R}^d) \to \mathcal{S}'(\mathbb{R}^d)$ and χ a canonical transformation which satisfies properties B1 and B2 following Definition* 6.1.1. *Let $\mathcal{G}(g, \Lambda)$ be a Parseval frame with $g \in \mathcal{S}(\mathbb{R}^d)$ and $s \geq 0$. Then the following properties are equivalent:*

(i) *There exists $C > 0$ such that*

$$|\langle T\pi(z)g, \pi(w)g\rangle| \leq C\langle w - \chi(z)\rangle^{-s}, \quad \forall z, w \in \mathbb{R}^{2d}. \tag{6.77}$$

(ii) *There exists $C > 0$ such that*

$$|\langle T\pi(\lambda)g, \pi(\mu)g\rangle| \leq C\langle \mu - \chi(\lambda)\rangle^{-s}, \quad \forall \lambda, \mu \in \Lambda. \tag{6.78}$$

Proof. The implication (i) $\Longrightarrow$ (ii) is obvious.

(ii) $\Longrightarrow$ (i). The argument is borrowed from the proof of a similar result for pseudodifferential operators in [162, Theorem 3.2]. Let C be a relatively compact fundamental domain of the lattice Λ. Given $z, w \in \mathbb{R}^{2d}$, we can write $w = \lambda + u$, $z = \mu + u'$ for unique $\lambda, \mu \in \Lambda$, $u, u' \in C$, and

$$|\langle T\pi(z)g, \pi(w)g\rangle| = |\langle T\pi(\mu)\pi(u')g, \pi(\lambda)\pi(u)g\rangle|.$$

Now we expand $\pi(u)g = \sum_{\nu\in\Lambda}\langle\pi(u)g, \pi(\nu)g\rangle\pi(\nu)g$, and likewise $\pi(u')g$. Since $V_g g \in \mathcal{S}(\mathbb{R}^{2d})$, the coefficients in this expansion satisfy

$$\sup_{u\in C}|\langle\pi(u)g, \pi(\nu)g\rangle| = \sup_{u\in C}|V_g g(\nu - u)| \lesssim \sup_{u\in C}\langle\nu - u\rangle^{-N} \lesssim \Big(\sup_{u\in C}\langle u\rangle^N\Big)\langle\nu\rangle^{-N} \lesssim \langle\nu\rangle^{-N}$$

for every N. Using (6.78), we now obtain that

$$\begin{aligned} |\langle T\pi(z)g, \pi(w)g\rangle| &\leq \sum_{\nu,\nu'\in\Lambda} |\langle T\pi(\mu + \nu')g, \pi(\lambda + \nu)g\rangle||\langle\pi(u')g, \pi(\nu')g\rangle||\langle\pi(u)g, \pi(\nu)g\rangle| \\ &\lesssim \sum_{\nu,\nu'\in\Lambda} \langle\lambda + \nu - \chi(\mu + \nu')\rangle^{-s}\langle\nu'\rangle^{-N}\langle\nu\rangle^{-N}. \end{aligned}$$

Since $v_s(z)^{-1} = \langle z\rangle^{-s}$ is v_s-moderate, we majorize the main term of the sum as

$$\begin{aligned} \langle\lambda + \nu - \chi(\mu + \nu')\rangle^{-s} &= \langle\lambda - \chi(\mu) + \nu - \chi(\mu + \nu') + \chi(\mu)\rangle^{-s} \\ &\leq \langle\lambda - \chi(\mu)\rangle^{-s}\langle\chi(\mu + \nu') - \chi(\mu) - \nu\rangle^{s} \lesssim \langle\lambda - \chi(\mu)\rangle^{-s}\langle\nu'\rangle^{s}\langle\nu\rangle^{s}, \end{aligned}$$

since $\chi(\mu + \nu') - \chi(\mu) = \mathcal{O}(\nu')$ by the Lipschitz property of χ.

Hence,

$$|\langle T\pi(z)g, \pi(w)g\rangle| \lesssim \langle\lambda - \chi(\mu)\rangle^{-s} \sum_{\nu,\nu'\in\Lambda} \langle\nu'\rangle^{s-N}\langle\nu\rangle^{s-N} \lesssim \langle\lambda - \chi(\mu)\rangle^{-s},$$

for $N \in \mathbb{N}$ large enough.

Finally, with $\lambda = w - u$, $\mu = z - u'$, we apply the above estimate again and obtain

$$\begin{aligned}\langle \lambda - \chi(\mu)\rangle^{-s} &= \langle w - u - \chi(z - u') - \chi(z) + \chi(z)\rangle^{-s} \\ &\le \langle w - \chi(z)\rangle^{-s} \langle u + \chi(z - u') - \chi(z)\rangle^{s} \\ &\lesssim \langle w - \chi(z)\rangle^{-s} \langle u\rangle^{s} \langle u'\rangle^{s} \lesssim \langle w - \chi(z)\rangle^{-s},\end{aligned}$$

since $\sup_{u\in C}\langle u\rangle^s < \infty$. Thus we have proved that $|\langle T\pi(z)g, \pi(w)g\rangle| \lesssim \langle w - \chi(z)\rangle^{-s}$. □

In view of the equivalence between (6.77) and (6.78), we observe that the class of FIOs in Definition 6.3.4 can be defined either using (6.78) or (6.77).

We first prove that Definition 6.3.4 does not depend on the choice of the Gabor frame.

Lemma 6.3.6. *The definition of* $\mathrm{FIO}(\chi, s)$ *is independent of the Gabor frame* $\mathcal{G}(g, \Lambda)$.

Proof. Let $\mathcal{G}(\varphi, \Lambda')$ be a Gabor frame with a window $\varphi \in \mathcal{S}(\mathbb{R}^d)$ and a possibly different lattice Λ'. As in the proof of Theorem 6.3.5, we expand $\pi(\lambda)\varphi = \sum_{\nu\in\Lambda}\langle \pi(\lambda)\varphi, \pi(\nu)g\rangle\pi(\nu)g$ with convergence in $\mathcal{S}(\mathbb{R}^d)$ and likewise $\pi(\mu)\varphi$, where $\lambda, \mu \in \Lambda'$. Consequently, $T\pi(\lambda)\varphi = \sum_{\nu\in\Lambda}\langle \pi(\lambda)\varphi, \pi(\nu)g\rangle T\pi(\nu)g$ converges weak* in $\mathcal{S}'(\mathbb{R}^d)$ and the following identity is well-defined:

$$\begin{aligned}\langle T\pi(\lambda)\varphi, \pi(\mu)\varphi\rangle &= \sum_{\nu\in\Lambda}\langle \pi(\lambda)\varphi, \pi(\nu)g\rangle\langle T\pi(\nu)g, \pi(\mu)\varphi\rangle \\ &= \sum_{\nu\in\Lambda}\sum_{\nu'\in\Lambda}\langle \pi(\lambda)\varphi, \pi(\nu)g\rangle\langle T\pi(\nu)g, \pi(\nu')g\rangle\langle \pi(\mu)\varphi, \pi(\nu')g\rangle.\end{aligned}$$

Since $\varphi, g \in \mathcal{S}(\mathbb{R}^d)$, the characterization (3.32) and the covariance property (1.50) imply that $|\langle \pi(\lambda)\varphi, \pi(\nu)g\rangle| = |\langle \varphi, \pi(\nu-\lambda)g\rangle| \lesssim \langle \nu-\lambda\rangle^{-N}$ for $\nu \in \Lambda, \lambda \in \Lambda'$ and every $N \ge 0$. After substituting these estimates and choosing N large enough, we obtain the majorization

$$\begin{aligned}|\langle T\pi(\lambda)\varphi, \pi(\mu)\varphi\rangle| &\lesssim \sum_{\nu\in\Lambda}\sum_{\nu'\in\Lambda}\langle \nu - \lambda\rangle^{-N}\langle \nu' - \chi(\nu)\rangle^{-s}\langle \nu' - \mu\rangle^{-N} \\ &\lesssim \sum_{\nu\in\Lambda}\langle \nu - \lambda\rangle^{-N}\langle \mu - \chi(\nu)\rangle^{-s} \asymp \sum_{\nu\in\Lambda}\langle \nu - \lambda\rangle^{-N}\langle \chi^{-1}(\mu) - \nu\rangle^{-s} \\ &\lesssim \langle \chi^{-1}(\mu) - \lambda\rangle^{-s} \asymp \langle \mu - \chi(\lambda)\rangle^{-s}.\end{aligned}$$ □

As in [165], the definition of classes of operators by their Gabor matrices facilitates the investigation of their basic properties. In line with Sjöstrand's original program, we next derive the boundedness, composition rules, and properties of the inverse operator in the classes $\mathrm{FIO}(\chi, s)$.

Theorem 6.3.7. *Let* $s > 2d$ *and* $T \in \mathrm{FIO}(\chi, s)$. *Then* T *extends to a bounded operator on* $\mathcal{M}^p(\mathbb{R}^d)$, $1 \le p \le \infty$, *and in particular on* $L^2(\mathbb{R}^d) = M^2(\mathbb{R}^d)$.

Proof. Let $\mathcal{G}(g, \Lambda)$ be a Parseval frame with $g \in \mathcal{S}(\mathbb{R}^d)$. Since the frame operator $S_g = D_g C_g$ is the identity operator, we can write T as $T = D_g C_g T D_g C_g$, where D_g and C_g

are the synthesis and coefficient operators, respectively. Since $\mathcal{G}(g,\Lambda)$ is a frame and $g \in \mathcal{S}(\mathbb{R}^d)$, C_g is bounded from $M^p(\mathbb{R}^d)$ to $\ell^p(\Lambda)$ (cf. Theorem 3.2.32) and $D_g = C_g^*$ is bounded from $\ell^p(\Lambda)$ to $\mathcal{M}^p(\mathbb{R}^d)$ (cf. Theorem 3.2.33).

The operator $C_g T D_g$ maps sequences to sequences, and its matrix K is precisely the Gabor matrix of T, namely, $K_{\mu,\lambda} = \langle T\pi(\lambda)g, \pi(\mu)g\rangle$. Since by assumption $|\langle T\pi(\lambda)g, \pi(\mu)g\rangle| \lesssim \langle \mu - \chi(\lambda)\rangle^{-s}$ and $s > 2d$, Schur's test implies that the matrix K representing $C_g T D_g$ is bounded on $\ell^p(\Lambda)$. Consequently, T is bounded on $\mathcal{M}^p(\mathbb{R}^d)$ for $1 \le p \le \infty$. □

Remark 6.3.8. The boundedness of the operator $T \in \mathrm{FIO}(\chi, s)$, $s > 2d$, fails in general on the modulation spaces $M^{p,q}(\mathbb{R}^{2d})$, with $p \neq q$. A concrete counterexample is the FIO of type I in (6.92) below, cf. Proposition 6.1.17.

Next we show that the class $\mathrm{FIO}(\Xi, s)$, for $s > 2d$, is an algebra.

Theorem 6.3.9. *If $T^{(i)} \in \mathrm{FIO}(\chi_i, s_i)$ with $s_i > 2d$, $i = 1,2$, then the composition $T^{(1)}T^{(2)}$ is in $\mathrm{FIO}(\chi_1 \circ \chi_2, s)$ with $s = \min(s_1, s_2)$. Consequently, the class $\mathrm{FIO}(\Xi, s) = \bigcup_\chi \mathrm{FIO}(\chi, s)$ is an algebra with respect to the composition of operators.*

Proof. We write the product $T^{(1)}T^{(2)}$ as

$$T^{(1)}T^{(2)} = D_g C_g T^{(1)} T^{(2)} D_g C_g = D_g (C_g T^{(1)} D_g)(C_g T^{(2)} D_g) C_g.$$

Then $C_g T^{(1)} T^{(2)} D_g$ is the Gabor matrix of $T^{(1)}T^{(2)}$ with entries

$$K_{\mu,\lambda} = \langle T^{(1)} T^{(2)} \pi(\lambda) g, \pi(\mu) g\rangle$$

and $C_g T^{(i)} D_g$, $i = 1,2$, is the Gabor matrix of $T^{(i)}$ with entries

$$K^{(i)}_{\mu,\lambda} = \langle T^{(i)} \pi(\lambda) g, \pi(\mu) g\rangle.$$

Thus the composition of operators corresponds to the multiplication of their Gabor matrices. Using the decay estimates for $K^{(i)}_{\mu,\lambda}$ and $s = \min(s_1, s_2)$, we estimate the size of the Gabor matrix of $T^{(1)}T^{(2)}$ as follows:

$$\begin{aligned} |K_{\mu,\lambda}| &= \sum_{\nu\in\Lambda} K^{(1)}_{\mu,\nu} K^{(2)}_{\nu,\lambda} \lesssim \sum_{\nu\in\Lambda} \langle \mu - \chi_1(\nu)\rangle^{-s_1} \langle \nu - \chi_2(\lambda)\rangle^{-s_2} \\ &\lesssim \sum_{\nu\in\Lambda} \langle \chi_1^{-1}(\mu) - \nu\rangle^{-s} \langle \nu - \chi_2(\lambda)\rangle^{-s}. \end{aligned}$$

Since $v_s^{-1} = \langle \nu\rangle^{-s} \in \ell^1(\Lambda)$ is subconvolutive for $s > 2d$, that is,

$$v_s^{-1} * v_s^{-1} \le C v_s^{-1},$$

cf. [160, Lemma 11.1.1 (d)], the last expression is dominated by $\langle \chi_1^{-1}(\mu) - \chi_2(\lambda)\rangle^{-s} \asymp \langle \mu - \chi_1(\chi_2(\lambda))\rangle^{-s}$.

By Proposition 6.3.7, $T^{(1)}$ and $T^{(2)}$ extend to bounded operators on $L^2(\mathbb{R}^d)$, so that the product $T^{(1)}T^{(2)}$ is well-defined and bounded on $L^2(\mathbb{R}^d)$. □

Finally, we consider the invertibility in the class $\mathrm{FIO}(\chi, s)$ and show that $\mathrm{FIO}(\Xi, s)$ is inverse-closed in $\mathcal{B}(L^2(\mathbb{R}^d))$ for $s > 2d$.

Theorem 6.3.10. *Let $T \in \mathrm{FIO}(\chi, s)$ with $s > 2d$. If T is invertible on $L^2(\mathbb{R}^d)$, then $T^{-1} \in \mathrm{FIO}(\chi^{-1}, s)$. Consequently, the algebra $\mathrm{FIO}(\Xi, s)$ is inverse-closed in $\mathcal{L}(L^2(\mathbb{R}^d))$.*

Proof. We first show that the adjoint operator T^* belongs to the class $\mathrm{FIO}(\chi^{-1}, s)$. Indeed, since χ is bi-Lipschitz, we have

$$\begin{aligned}|\langle T^*\pi(\lambda)g, \pi(\mu)g\rangle| &= |\langle \pi(\lambda)g, T(\pi(\mu)g)\rangle| = |\langle T(\pi(\mu)g, \pi(\lambda)g)\rangle| \\ &\lesssim \langle \lambda - \chi(\mu)\rangle^{-s} \asymp \langle \chi^{-1}(\lambda) - \mu\rangle^{-s}.\end{aligned}$$

Hence, by Theorem 6.3.9, the operator $P := T^*T$ is in $\mathrm{FIO}(\mathrm{Id}, s)$ and satisfies the estimate $|\langle P\pi(\lambda)g, \pi(\mu)g\rangle| \lesssim \langle \lambda - \mu\rangle^{-s}, \forall \lambda, \mu \in \Lambda$.

We now exploit the characterization for pseudodifferential operators contained in Theorem 6.3.3 and deduce that P is a pseudodifferential operator with a symbol in $M^{\infty}_{1\otimes v_s}(\mathbb{R}^{2d})$. Since T and therefore T^* are invertible on $L^2(\mathbb{R}^d)$, P is also invertible on $L^2(\mathbb{R}^d)$. Now we apply Theorem 6.3.2 and conclude that the inverse P^{-1} is again a pseudodifferential operator with a symbol in $M^{\infty}_{1\otimes v_s}(\mathbb{R}^{2d})$. Hence P^{-1} is in $\mathrm{FIO}(\mathrm{Id}, s)$. Finally, using the algebra property of Theorem 6.3.9 once more, we obtain that $T^{-1} = P^{-1}T^*$ is in $\mathrm{FIO}(\chi^{-1}, s)$ and thus satisfies the estimate $|\langle T^{-1}\pi(\lambda)g, \pi(\mu)g\rangle| \lesssim \langle \chi^{-1}(\lambda) - \mu\rangle^{-s}, \forall \lambda, \mu \in \Lambda$. □

Combining Theorems 6.3.7, 6.3.9, and 6.3.10, we see that $\mathrm{FIO}(\Xi, s)$ with $s > 2d$ is a Wiener subalgebra of $\mathcal{L}(L^2(\mathbb{R}^d))$ consisting of FIOs.

6.4 FIOs of type I

In this section we relate the Wiener algebra of FIOs in the preceding Section 6.3 to the FIOs of type I in Section 4.1. An evident connection is given by Theorem 6.1.5. Namely, assume T of the form (6.12) with symbol satisfying (6.13), for a given $N \in \mathbb{N}$, and a phase function Φ which fulfills *A1*, *A2*, and *A3* in Definition 6.1.1 (hence associated to a canonical transformation χ satisfying *B1*, *B2*, and *B3*). Then from (6.23) we have that $T \in \mathrm{FIO}(\chi, 2N)$; applying Theorem 6.3.7, we obtain that T is bounded on $\mathcal{M}^p(\mathbb{R}^d)$, $1 \le p \le \infty$, provided $s > 2N > 2d$, that is, $N > d$, and we recapture Theorem 6.1.8 in the case of weight $\mu = 1$.

A natural question is whether an operator $T \in \mathrm{FIO}(\chi, s)$, $s \ge 0$, with χ satisfying *B1*, *B2*, and *B3* can be written as an FIO of type I. A positive answer will be given in Theorem 6.4.10, where we shall characterize precisely the corresponding class of symbols, given by $M^{\infty}_{1\otimes v_s}(\mathbb{R}^{2d})$. When $s = 2N$, this includes strictly the class of symbols satisfying (6.13).

Note that by the Schwartz' kernel theorem every continuous linear operator $T : \mathcal{S}(\mathbb{R}^d) \to \mathcal{S}'(\mathbb{R}^d)$ can be written as an FIO of type I with a given phase $\Phi(x, \eta)$ for some symbol $\sigma(x,\eta)$ in $\mathcal{S}'(\mathbb{R}^{2d})$. Hence, if T is a continuous linear operator $\mathcal{S}(\mathbb{R}^d) \to \mathcal{S}'(\mathbb{R}^d)$ and χ satisfies *B1, B2, and B3*, then $T = T_{I,\Phi_\chi,\sigma}$ is an FIO of type I with symbol σ and phase Φ_χ.

As the first step we formulate the following result.

Proposition 6.4.1. *Let $T = T_{I,\Phi,\sigma}$ be an FIO of type I with symbol $\sigma \in \mathcal{S}'(\mathbb{R}^{2d})$ and a phase Φ satisfying A1 and A2. If the Gabor matrix of T satisfies*

$$|\langle T\pi(x,\eta)g, \pi(x',\eta')g\rangle| \le C\langle \nabla_x\Phi(x',\eta) - \eta', \nabla_\eta\Phi(x',\eta) - x\rangle^{-s}, \quad x,x',\eta,\eta' \in \mathbb{R}^d, \tag{6.79}$$

for some $C > 0$ and $s \ge 0$, then σ is in the generalized Sjöstrand class $M^{\infty}_{1\otimes v_s}(\mathbb{R}^{2d})$.

In particular, if $s > 2d$, then we have $\sigma \in M^{\infty,1}(\mathbb{R}^{2d})$.

Proof. To set up notation, let $\Phi_{2,z}$ be the remainder in the second order Taylor expansion of the phase Φ computed in (6.15) centered at z, that is,

$$\Phi_{2,z}(w) = 2\sum_{|\alpha|=2}\int_0^1 (1-t)\partial^\alpha\Phi(z+tw)\,dt\,\frac{w^\alpha}{\alpha!} \quad z,w \in \mathbb{R}^{2d},$$

and set

$$\Psi_z(w) = e^{2\pi i\Phi_{2,z}(w)}\overline{g}\otimes\hat{g}(w). \tag{6.80}$$

With the notation above, the fundamental relation between the Gabor matrix of an FIO and the STFT of its symbol in (6.35), for $g \in \mathcal{S}(\mathbb{R}^d)$, becomes

$$|\langle T\pi(x,\eta)g, \pi(x',\eta')g\rangle| = |V_{\Psi_{(x',\eta)}}\sigma((x',\eta),(\eta' - \nabla_x\Phi(x',\eta), x - \nabla_\eta\Phi(x',\eta)))|.$$

Writing $u = (x',\eta)$, $v = (\eta',x)$, (6.79) translates into

$$|V_{\Psi_u}\sigma(u, v - \nabla\Phi(u))| \le C\langle v - \nabla\Phi(u)\rangle^{-s},$$

and then into the estimate

$$\sup_{(u,w)\in\mathbb{R}^{2d}\times\mathbb{R}^{2d}} \langle w\rangle^s |V_{\Psi_u}\sigma(u,w)| < \infty. \tag{6.81}$$

Now, setting $G = \overline{g}\otimes\hat{g} \in \mathcal{S}(\mathbb{R}^{2d})$, we can write

$$\begin{aligned}
V_{G^2}\sigma(u,v) &= \int e^{-2\pi i t\cdot v}\sigma(t)\overline{G^2(t-u)}\,dt \\
&= \int e^{-2\pi i t\cdot v}\sigma(t)\overline{e^{-2\pi i\Phi_{2,u}(t-u)}G(t-u)}\,\overline{e^{2\pi i\Phi_{2,u}(t-u)}G(t-u)}\,dt \\
&= \int e^{-2\pi i t\cdot v}\sigma(t)\overline{\Psi_u(t-u)}\,\overline{e^{2\pi i\Phi_{2,u}(t-u)}G(t-u)}\,dt
\end{aligned}$$

$$= \mathcal{F}(\sigma T_u \overline{\Psi_u}) *_v \mathcal{F}(T_u(e^{2\pi i \Phi_{2,u}} \overline{G}))(v)$$
$$= V_{\Psi_u} \sigma(u, \cdot) * \mathcal{F}(T_u(e^{2\pi i \Phi_{2,u}} \overline{G}))(v). \tag{6.82}$$

Using (6.82) and the weighted Young's inequality $L^\infty_s(\mathbb{R}^{2d}) * L^1_s(\mathbb{R}^{2d}) \hookrightarrow L^\infty_s(\mathbb{R}^{2d})$, we get

$$\|\sigma\|_{M^\infty_{1\otimes v_s}(\mathbb{R}^{2d})} \asymp \sup_u \|V_{G^2} \sigma(u, \cdot)\|_{L^\infty_s} \lesssim \sup_u \|V_{\Psi_u} \sigma(u, \cdot)\|_{L^\infty_s} \sup_u \|\mathcal{F}(e^{2\pi i \Phi_{2,u}} \overline{G})\|_{L^1_s}.$$

The first factor on the right-hand side is finite by (6.81). The second is finite because the set $\{e^{2\pi i \Phi_{2,u}} \overline{G} : u \in \mathbb{R}^{2d}\}$ is bounded in $\mathcal{S}(\mathbb{R}^{2d})$, and the embedding $\mathcal{S} \hookrightarrow \mathcal{F}L^1_s$ is continuous. This gives $\sigma \in M^\infty_{1\otimes v_s}(\mathbb{R}^{2d})$.

The last statement follows from the inclusion relations for modulation spaces in Proposition 2.4.18. Namely, $M^{\infty,\infty}_{1\otimes v_s}(\mathbb{R}^{2d}) \hookrightarrow M^{\infty,1}(\mathbb{R}^{2d})$ if and only if $s > 2d$. □

The next lemma clarifies further the relation between the phase Φ and the canonical transformation χ.

Lemma 6.4.2. *Consider a phase function Φ satisfying A1, A2, and A3. Then*

$$|\nabla_x \Phi(x', \eta) - \eta'| + |\nabla_\eta \Phi(x', \eta) - x| \asymp |\chi_1(x, \eta) - x'| + |\chi_2(x, \eta) - \eta'|, \quad \forall x, x', \eta, \eta' \in \mathbb{R}^d. \tag{6.83}$$

Proof. The estimate $\gtrsim$ was already proved in Lemma 6.1.4.

For the converse estimate observe that $x = \nabla_\eta \Phi(\chi_1(x, \eta), \eta)$ by definition of χ_1, hence

$$|\nabla_\eta \Phi(x', \eta) - x| = |\nabla_\eta \Phi(x', \eta) - \nabla_\eta \Phi(\chi_1(x, \eta), \eta)| \leq C|x' - \chi_1(x, \eta)|, \tag{6.84}$$

because of assumption (A2) on Φ.

Since $\nabla_x \Phi(x', \eta) = \chi_2(\nabla_\eta \Phi(x', \eta), \eta)$, the first term on the left-hand side of (6.83) can be estimated as

$$|\nabla_x \Phi(x', \eta) - \eta'| \leq |\chi_2(\nabla_\eta \Phi(x', \eta), \eta) - \chi_2(x, \eta)| + |\chi_2(x, \eta) - \eta'|.$$

Finally, the Lipschitz continuity of χ and (6.84) imply that

$$|\nabla_x \Phi(x', \eta) - \eta'| \leq |\eta' - \chi_2(x, \eta)| + C|x' - \chi_1(x, \eta)|.$$ □

We shall identify the abstract class $\mathrm{FIO}(\chi, s)$ with a class of concrete Fourier integral operators. To chase this goal, we use an new characterization of the symbol classes $M^{\infty,1}_{1\otimes m}$ and $M^\infty_{1\otimes m}$ that is perfectly adapted to the investigation of Fourier integral operators with a tame phase Φ and was first introduced in [67]. The main technical work relies on showing that the set of windows Ψ_z in (6.80) possesses a joint time–frequency envelope as shown in Lemma 6.4.7 below. This property will allow us to replace the z-dependent family of windows Ψ_z by a single window in many estimates. Before proving the existence of a time–frequency envelope in Lemma 6.4.7 below, we first look at the phase factor $e^{2\pi i \Psi_{2,z}}$ occurring in (6.80).

Lemma 6.4.3. *For every $s \in \mathbb{R}$, $N \in \mathbb{N}$, $N > \frac{s}{2} + d$, and $\Psi \in \mathcal{S}(\mathbb{R}^{2d})$, we have*

$$\sup_{z\in\mathbb{R}^{2d}} |V_\Psi e^{2\pi i \Phi_{2,z}}| \in L^{\infty,1}_{v_{-4N}\otimes v_s}(\mathbb{R}^{4d}), \tag{6.85}$$

with v_{-4N} and v_s being weight functions on $\mathbb{R}^{2d}$, defined in (2.5).

Proof. Let $\Psi \in \mathcal{S}(\mathbb{R}^{2d})$, then

$$|V_\Psi e^{2\pi i \Phi_{2,z}}(u,w)| = \left| \int_{\mathbb{R}^{2d}} e^{2\pi i \Phi_{2,z}(\zeta)} T_u \bar{\Psi}(\zeta) e^{-2\pi i \zeta\cdot w}\, d\zeta \right|.$$

Using the identity

$$(1-\Delta_\zeta)^N e^{-2\pi i \zeta\cdot w} = \langle 2\pi w\rangle^{2N} e^{-2\pi i \zeta\cdot w},$$

we integrate by parts and obtain

$$|V_\Psi e^{2\pi i \Phi_{2,z}}(u,w)| = \frac{1}{\langle 2\pi w\rangle^{2N}} \left| \int_{\mathbb{R}^{2d}} (1-\Delta_\zeta)^N (e^{2\pi i \Phi_{2,z}(\zeta)} T_u \bar{\Psi}(\zeta)) e^{-2\pi i \zeta\cdot w}\, d\zeta \right|.$$

By means of Leibniz's formula, the factor $(1-\Delta_\zeta)^N (e^{2\pi i \Phi_{2,z}(\zeta)} T_u \bar{\Psi}(\zeta))$ can be expressed further as

$$e^{2\pi i \Phi_{2,z}(\zeta)} \sum_{|\alpha|+|\beta|\le 2N} p_\alpha(\partial \Phi_{2,z}(\zeta))(T_u \partial_\zeta^\beta \bar{\Psi})(\zeta),$$

where $p_\alpha(\partial\Phi_{2,z}(\zeta))$ is a polynomial of derivatives of $\Phi_{2,z}$ of degree at most $|\alpha|$.

As a consequence of condition (6.7), we have $|p_\alpha(\partial\Phi_{2,z}(\zeta))| \le C_\alpha \langle\zeta\rangle^{2|\alpha|}$ for every $z \in \mathbb{R}^{2d}$ with a constant independent of z. Moreover, the assumption $\Psi \in \mathcal{S}(\mathbb{R}^{2d})$ yields that

$$\sup_{|\beta|\le 2N} |T_u \partial_\zeta^\beta \bar{\Psi}| \le C_{N,s} \langle \zeta - u\rangle^{-\ell}$$

for every $\ell \ge 0$. Consequently,

$$\begin{aligned}
|V_\Psi e^{2\pi i \Phi_{2,z}}(u,w)| &\lesssim \frac{1}{\langle 2\pi w\rangle^{2N}} \int_{\mathbb{R}^{2d}} \sum_{|\alpha|+|\beta|\le 2N} \langle\zeta\rangle^{2|\alpha|} \langle \zeta - u\rangle^{-\ell}\, d\zeta \\
&\lesssim \frac{1}{\langle 2\pi w\rangle^{2N}} \int_{\mathbb{R}^{2d}} \langle\zeta\rangle^{4N} \langle \zeta - u\rangle^{-\ell}\, d\zeta \\
&\lesssim \frac{1}{\langle 2\pi w\rangle^{2N}} \langle u\rangle^{4N},
\end{aligned}$$

whenever $\ell > 4N + 2d$. Since $-2N + s < -2d$ by assumption, we obtain

$$\int_{\mathbb{R}^{2d}} \sup_{u\in\mathbb{R}^{2d}} \sup_{z\in\mathbb{R}^{2d}} |V_\Psi e^{2\pi i \Phi_{2,z}}|(u,w) \langle u\rangle^{-4N} \langle w\rangle^s\, dw < \infty,$$

whence (6.85) follows. □

Remark 6.4.4. Notice that the two weights v_{-4N} and v_s compensate each other in (6.85). It is easy to see that some form of compensation is necessary. For example, if $\Phi(\zeta) = \zeta^2/2$, then $\Phi_{2,z}(\zeta) = \zeta^2/2$ independent of z. A direct computation shows that the STFT of $e^{2\pi i \Phi_{2,z}(\zeta)}$ with Gaussian window $\Psi(\zeta) = e^{-\pi\zeta^2}$ has modulus, up to constants, $e^{-\pi(u_1-u_2)^2/2}$, $u_1, u_2 \in \mathbb{R}^{2d}$. This function belongs to $L^{\infty,1}_{v_{-4N}\otimes v_s}(\mathbb{R}^{4d})$ if and only if $-4N + s < -2d$. This is, in fact, better than the condition $-2N + s < -2d$ in the assumptions, due to the fact that here the derivatives of order ≥ 1 of $\Phi_{2,z}(\zeta)$ are bounded above by $\langle\zeta\rangle$, rather than $\langle\zeta\rangle^2$, as in the general case. This explains the presence of $2N$ instead of $4N$. In the general case the best upper bound for the derivatives of $\Phi_{2,z}(\zeta)$ is $\langle\zeta\rangle^2$, so that we conjecture the condition $-2N + s < -2d$ to be sharp.

Lemma 6.4.5. *Let $m \in \mathcal{M}_v(\mathbb{R}^d)$, v, v, w be weight functions on $\mathbb{R}^d$, and $\{f_z : z \in \mathbb{R}^{2d}\} \subseteq \mathcal{S}'(\mathbb{R}^d)$ be a set of distributions in $\mathcal{S}'$. If*

$$\sup_{z\in\mathbb{R}^d} |V_\varphi f_z| \in L^{\infty,1}_{v^{-1}\otimes m}(\mathbb{R}^{2d}), \tag{6.86}$$

for given $\varphi \in \mathcal{S}(\mathbb{R}^d)$, then for every $h \in M^1_{wv\otimes v}(\mathbb{R}^d)$

$$\sup_{z\in\mathbb{R}^d} \left|V_{\varphi^2}(f_z h)\right| \in L^1_{w\otimes m}(\mathbb{R}^{2d}). \tag{6.87}$$

Proof. Since $V_g f(x,\xi) = \widehat{(f \cdot T_x \bar{g})}(\xi)$ by (1.24), we obtain the identity

$$V_{\varphi^2}(f_z h)(x,\xi) = ((f_z \overline{T_x\varphi})(h\overline{T_x\varphi}))\hat{}(\xi) = (f_z\overline{T_x\varphi})\hat{} *_\xi (h\overline{T_x\varphi})\hat{}(\xi),$$

and consequently,

$$\left|V_{\varphi^2}(f_z h)(x,\xi)\right| \leq \left|\widehat{(f_z \cdot T_x\bar{\varphi})}\right| *_\xi \left|\widehat{(h \cdot T_x\bar{\varphi})}\right|(\xi),$$

where the convolution is in the second variable ξ. Now set

$$F(x,\xi) = \sup_{z\in\mathbb{R}^{2d}} \left|V_\varphi f_z(x,\xi)\right| = \sup_{z\in\mathbb{R}^{2d}} \left|(f_z \cdot T_x\bar{\varphi})\hat{}(\xi)\right|$$

and $H(x,\xi) = |V_\varphi h(x,\xi)|$. Then

$$\sup_{z\in\mathbb{R}^{2d}} \left|V_{\varphi^2}(f_z h)(x,\xi)\right| \leq (F *_\xi H)(x,\xi).$$

Finally,

$$\begin{aligned}
\|F *_\xi H\|_{L^1_{w\otimes m}} &= \int_{\mathbb{R}^d}\int_{\mathbb{R}^d}\int_{\mathbb{R}^d} F(x,\xi-\zeta)H(x,\zeta)\,d\zeta w(x)m(\eta)\,dx\,d\xi \\
&\leq \int_{\mathbb{R}^d}\int_{\mathbb{R}^d}\int_{\mathbb{R}^d} F(x,\xi-\zeta)v(x)^{-1}m(\xi-\zeta)H(x,\zeta)w(x)v(x)v(\zeta)\,dx\,d\xi\,d\zeta \\
&\leq \|F\|_{L^{\infty,1}_{v^{-1}\otimes m}} \|H\|_{L^1_{vw\otimes v}}.
\end{aligned}$$

In the last expression both norms are finite by assumption. □

Corollary 6.4.6. *Let $s \geq 0$, $N \in \mathbb{N}$, $N > \frac{s}{2} + d$ and $g \in M^1_{v_{4N}\otimes v_{4N}}(\mathbb{R}^d)$ and $\Psi \in \mathcal{S}(\mathbb{R}^{2d})$. Then the function $\Psi_z(w) = e^{2\pi i \Phi_{2,z}(w)}\bar{g} \otimes \hat{g}(w)$ defined as in (6.80) satisfies*

$$\sup_{z\in\mathbb{R}^{2d}} |V_{\Psi_z}\Psi^2| \in L^1_{1\otimes v_s}(\mathbb{R}^{4d}).$$

Proof. The symmetry property of the weight $\langle x\rangle^{4N}\langle\eta\rangle^{4N}$ implies that $M^1_{v_{4N}\otimes v_{4N}}(\mathbb{R}^d)$ is invariant under the Fourier transform, and thus $\hat{g} \in M^1_{v_{4N}\otimes v_{4N}}(\mathbb{R}^d)$. A tensor product argument then shows that $\bar{g}\otimes\hat{g} \in M^1_W(\mathbb{R}^{2d})$ with $W(x_1, x_2, \xi_1, \xi_2) = \langle x_1\rangle^{4N}\langle x_2\rangle^{4N}\langle \xi_1\rangle^{4N}\langle \xi_2\rangle^{4N}$. Since $W(x_1, x_2, \eta_1, \eta_2) \geq \langle (x_1, x_2)\rangle^{4N}\langle(\eta_1, \eta_2)\rangle^{4N}$, we also obtain that $\bar{g}\otimes\hat{g} \in M^1_{v_{4N}\otimes v_{4N}}(\mathbb{R}^{2d})$.

By Lemma 6.4.3, we have $\sup_{z\in\mathbb{R}^{2d}} |V_\Psi e^{2\pi i\Phi_{2,z}}| \in L^{\infty,1}_{v_{-4N}\otimes v_s}(\mathbb{R}^{4d})$.

We now apply Lemma 6.4.5 with $f_z = e^{2\pi i\Phi_{2,z}}$ and $h = \bar{g}\otimes\hat{g}$, $v = v_{4N}$, $m = v_s$, $w \equiv 1$, and $v = v_{4N}$ (observe that v_s is v_{4N}-moderate, since $s < 4N$). As a conclusion, we obtain that $\sup_{z\in\mathbb{R}^{2d}} |V_{\Psi^2}\Psi_z| \in L^1_{1\otimes v_s}(\mathbb{R}^{4d})$. The desired conclusion follows from (1.32). □

Lemma 6.4.7. *Under the assumptions of Corollary 6.4.6, we have*

$$\sup_{z\in\mathbb{R}^{2d}} |V_{\Psi_z}\Psi| \in W(L^\infty, L^1_{1\otimes v_s})(\mathbb{R}^{4d}). \tag{6.88}$$

Proof. Using Lemma 1.2.29 with $h = \gamma = \Psi^2$, $g = \Psi_z$, $f = \Psi$, we have

$$\begin{aligned}|V_{\Psi_z}\Psi|(u_1, u_2) &\leq \|\Psi^2\|_2^{-2}|V_{\Psi^2}\Psi| * |V_{\Psi_z}\Psi^2|(u_1, u_2)\\ &\leq \|\Psi^2\|_2^{-2}|V_{\Psi^2}\Psi| * \Big(\sup_{z\in\mathbb{R}^{2d}}|V_{\Psi_z}\Psi^2|\Big)(u_1, u_2).\end{aligned}$$

Since $V_{\Psi^2}\Psi \in \mathcal{S}(\mathbb{R}^{4d}) \subset W(L^\infty, L^1_{1\otimes v_s})$ and $\|\sup_{z\in\mathbb{R}^{2d}} |V_{\Psi_z}\Psi^2|\|_{L^1_{1\otimes v_s}} < \infty$ by Corollary 6.4.6, the convolution relations for Wiener amalgam spaces $L^1_{1\otimes v_s} * W(L^\infty, L^1_{1\otimes v_s}) \hookrightarrow W(L^\infty, L^1_{1\otimes v_s})$ (recall that $L^1_{1\otimes v_s} = W(L^1, L^1_{1\otimes v_s})$, cf. (2.67)) implies that

$$\Big\|\sup_{z\in\mathbb{R}^{2d}} |V_{\Psi_z}\Psi|\Big\|_{W(L^\infty, L^1_{1\otimes v_s})} \leq C\|V_{\Psi^2}\Psi\|_{W(L^\infty, L^1_{1\otimes v_s})}\Big\|\sup_{z\in\mathbb{R}^{2d}}|V_{\Psi_z}\Psi^2|\Big\|_{L^1_{1\otimes v_s}} < \infty,$$

and so $\sup_{z\in\mathbb{R}^{2d}} |V_{\Psi_z}\Psi| \in W(L^\infty, L^1_{1\otimes v_s})$. □

We can now formulate the new characterization of the symbol classes $M^{\infty,1}_{1\otimes m}$ and $M^\infty_{1\otimes m}$ mentioned above.

Proposition 6.4.8. *Let $s \geq 0$, $m \in \mathcal{M}_{v_s}(\mathbb{R}^{2d})$, $g \in \mathcal{S}(\mathbb{R}^d)$ and $\sigma \in \mathcal{S}'(\mathbb{R}^{2d})$.*

(i) *Then the symbol σ is in $M^{\infty,1}_{1\otimes m}(\mathbb{R}^{2d})$ if and only if*

$$\Big\|\sup_{z\in\mathbb{R}^{2d}} |V_{\Psi_z}\sigma|\Big\|_{L^{\infty,1}_{1\otimes m}(\mathbb{R}^{4d})} = \int_{\mathbb{R}^{2d}} \sup_{u_1\in\mathbb{R}^d}\sup_{z\in\mathbb{R}^{2d}} |V_{\Psi_z}\sigma(u_1, u_2)| m(u_2)\, du_2 < \infty, \tag{6.89}$$

with $\|\sigma\|_{M^{\infty,1}_{1\otimes m}} \asymp \|\sup_{z\in\mathbb{R}^{2d}} |V_{\Psi_z}\sigma|\|_{L^{\infty,1}_{1\otimes m}}$. In this case the function $H(u_2) = \sup_{u_1\in\mathbb{R}^d} \times \sup_{z\in\mathbb{R}^{2d}} |V_{\Psi_z}\sigma(u_1, u_2)|$ is in $W(L^\infty, L^1_m)(\mathbb{R}^{2d})$.

(ii) *Likewise,* $\sigma \in M^{\infty}_{1\otimes m}(\mathbb{R}^{2d})$ *if and only if* $\sup_{z\in\mathbb{R}^{2d}} |V_{\Psi_z}\sigma| \in L^{\infty}_{1\otimes m}(\mathbb{R}^{4d})$ *with* $\|\sigma\|_{M^{\infty}_{1\otimes m}} \asymp \|\sup_{z\in\mathbb{R}^{2d}} |V_{\Psi_z}\sigma|\|_{L^{\infty}_{1\otimes m}}$.

Proof. We detail the proof of case (i). Case (ii) is obtained similarly. Let $\Psi \in \mathcal{S}(\mathbb{R}^{2d})$ with $\|\Psi\|_2 = 1$.

Assume first that $\sigma \in M^{\infty,1}_{1\otimes m}(\mathbb{R}^{2d})$. Then, by Lemma 1.2.29, we have

$$|V_{\Psi_z}\sigma(u_1,u_2)| \le |V_\Psi\sigma| * |V_{\Psi_z}\Psi|(u_1,u_2) \le |V_\Psi\sigma| * \sup_{z\in\mathbb{R}^{2d}} |V_{\Psi_z}\Psi|(u_1,u_2).$$

Set $F(u_1,u_2) = \sup_{z\in\mathbb{R}^{2d}} |V_{\Psi_z}\Psi(u_1,u_2)|$, so that

$$|V_{\Psi_z}\sigma(u_1,u_2)| \le (|V_\Psi\sigma| * F)(u_1,u_2).$$

Finally,

$$\begin{aligned}\Big\|\sup_{z\in\mathbb{R}^{2d}} |V_{\Psi_z}\sigma|\Big\|_{L^{\infty,1}_{1\otimes m}} &= \int_{\mathbb{R}^{2d}} \sup_{u_1\in\mathbb{R}^d}\sup_{z\in\mathbb{R}^{2d}} \big|V_{\Psi_z}\sigma(u_1,u_2)\big| m(u_2)\,du_2 \\ &\le \big\||V_\Psi\sigma| * F\big\|_{L^{\infty,1}_{1\otimes m}} \\ &\le \|V_\Psi\sigma\|_{L^{\infty,1}_{1\otimes m}} \|F\|_{L^1_{1\otimes v_s}} \lesssim \|\sigma\|_{M^{\infty,1}_{1\otimes m}} \|F\|_{L^1_{1\otimes v_s}} < \infty,\end{aligned}$$

where in the last step we used the independence of the weighted $M^{\infty,1}$-norm of the window and Corollary 6.4.6.

To obtain a sharper estimate, set $H(u_2) = \sup_{u_1\in\mathbb{R}^d}\sup_{z\in\mathbb{R}^{2d}} |V_{\Psi_z}\sigma(u_1,u_2)|$, $F_1(u_2) = \int_{\mathbb{R}^{2d}} F(u_1,u_2)\,du_1$, and $G_1(u_2) = \sup_{u_1\in\mathbb{R}^{2d}} |V_\Psi\sigma(u_1,u_2)|$. Then the definition of $M^{\infty,1}_{1\otimes m}(\mathbb{R}^{2d})$ implies that $G_1 \in L^1_m(\mathbb{R}^{2d})$, and Lemma 6.4.7 implies that $F_1 \in W(L^\infty, L^1_{v_s})(\mathbb{R}^{2d})$. With these definitions, we obtain a pointwise estimate for H, namely

$$\begin{aligned}H(w) &= \sup_{u_1\in\mathbb{R}^d}\sup_{z\in\mathbb{R}^{2d}} \big|V_{\Psi_z}\sigma(u_1,w)\big| \\ &\le \int_{\mathbb{R}^{2d}} \sup_{u_1}\big|V_\Psi\sigma(u_1,u_2)\big|F(z-u_1,w-u_2)\,du_1\,du_2 \\ &= (G_1 * F_1)(w).\end{aligned}$$

The convolution relations for Wiener amalgam spaces (recall that $L^1_m = W(L^1, L^1_m)$ by (2.67)) now yield $H \le G_1 * F_1 \in L^1_m * W(L^\infty, L^1_{v_s}) \hookrightarrow W(L^\infty, L^1_{v_s})(\mathbb{R}^{2d})$, as claimed.

Conversely, assume (6.89). Using Lemma 1.2.29 again, we deduce that

$$|V_\Psi\sigma(u_1,u_2)| \le \frac{1}{\langle\Psi_z,\Psi_z\rangle}|V_{\Psi_z}\sigma| * |V_\Psi\Psi_z|(u_1,u_2),$$

and

$$\begin{aligned}\langle\Psi_z,\Psi_z\rangle &= \int_{\mathbb{R}^{2d}} \big|e^{2\pi i\Phi_{2,z}(w)}(\overline{g}\otimes\hat{g})(w)\big|^2\,dw \\ &= \int_{\mathbb{R}^{2d}} \big|(\overline{g}\otimes\hat{g})(w)\big|^2\,dw = \|\overline{g}\|_2^2\|\hat{g}\|_2^2 = \|g\|_2^4\end{aligned}$$

is, in fact, a constant (depending on g). Using the involution $f^*(z) = \overline{f(-z)}$, we continue with

$$\begin{aligned}|V_\Psi\sigma|(u_1,u_2) &\lesssim |V_{\Psi_z}\sigma| * |V_\Psi\Psi_z|(u_1,u_2)\\ &= |V_{\Psi_z}\sigma| * \big|(V_{\Psi_z}\Psi)^*\big|(u_1,u_2)\\ &\leq \Big(\sup_{z\in\mathbb{R}^{2d}}|V_{\Psi_z}\sigma|\Big) * \Big(\sup_{z\in\mathbb{R}^{2d}}\big|(V_{\Psi_z}\Psi)^*\big|\Big)(u_1,u_2).\end{aligned}$$

Since, by Corollary 6.4.6,

$$\sup_{z\in\mathbb{R}^{2d}}\big|(V_{\Psi_z}\Psi)^*\big| = \sup_{z\in\mathbb{R}^{2d}}|V_{\Psi_z}\Psi| \in L^1_{1\otimes v_s}(\mathbb{R}^{4d}),$$

we get

$$\|V_\Psi\sigma\|_{L^{\infty,1}_{1\otimes m}} \lesssim \Big\|\sup_{z\in\mathbb{R}^{2d}}|V_{\Psi_z}\sigma|\Big\|_{L^{\infty,1}_{1\otimes m}}\Big\|\sup_{z\in\mathbb{R}^{2d}}|V_{\Psi_z}\Psi|\Big\|_{L^1_{1\otimes v_s}} < \infty,$$

and the proof is complete. □

Theorem 6.4.9. *Let $s \geq 0$, N, $N > \frac{s}{2} + d$, $g \in M^1_{v_{4N}\otimes v_{4N}}(\mathbb{R}^d)$ and assume that $\mathcal{G}(g,\Lambda)$ is a Parseval frame for $L^2(\mathbb{R}^d)$. If the phase Φ is tame and $\sigma \in M^\infty_{1\otimes v_s}(\mathbb{R}^{2d})$, then there exists $C > 0$ such that*

$$\big|\langle T\pi(\mu)g,\pi(\lambda)g\rangle\big| \leq C\frac{\|\sigma\|_{M^\infty_{1\otimes v_s}}}{\langle\chi(\mu)-\lambda\rangle^s}, \quad \forall\lambda,\mu\in\mathbb{R}^{2d}. \tag{6.90}$$

Proof. Using again the fundamental relation between the Gabor matrix of an FIO and the STFT of its symbol in (6.35), for $g \in \mathcal{S}(\mathbb{R}^d)$, with the function Ψ_z defined in (6.80), we can write

$$\begin{aligned}\big|\langle T\pi(x,\xi)g,\pi(x',\xi')g\rangle\big| &= \big|V_{\Psi_{(x',\xi)}}\sigma((x',\xi),(\xi'-\nabla_x\Phi(x',\xi),x-\nabla_\eta\Phi(x',\xi)))\big|\\ &\leq \sup_{z\in\mathbb{R}^{2d}}\big|V_{\Psi_z}\sigma((x',\xi),(\xi'-\nabla_x\Phi(x',\xi),x-\nabla_\eta\Phi(x',\xi)))\big|.\end{aligned} \tag{6.91}$$

If $H(u_1,u_2) = \sup_{z\in\mathbb{R}^{2d}}|V_{\Psi_z}\sigma(u_1,u_2)|$, by Proposition 6.4.8 (ii), $H \in L^\infty_{1\otimes v_s}(\mathbb{R}^{4d})$ and $\|H\|_{L^\infty_{1\otimes v_s}} \lesssim \|\sigma\|_{M^\infty_{1\otimes v_s}}$. Now, using (6.91),

$$\begin{aligned}\big|\langle T\pi(x,\xi)g,\pi(x',\xi')g\rangle\big| &\leq \sup_{u_1\in\mathbb{R}^{2d}} H(u_1,\xi'-\nabla_x\Phi(x',\xi),x-\nabla_\eta\Phi(x',\xi))\\ &\lesssim \sup_{u_1\in\mathbb{R}^{2d}} H(u_1,\xi'-\nabla_x\Phi(x',\xi),x-\nabla_\eta\Phi(x',\xi))\\ &\quad\times\frac{\langle\xi'-\nabla_x\Phi(x',\xi),x-\nabla_\eta\Phi(x',\xi)\rangle^s}{\langle\xi'-\nabla_x\Phi(x',\xi),x-\nabla_\eta\Phi(x',\xi)\rangle^s}\\ &\lesssim \frac{\sup_{u_2}\sup_{u_1\in\mathbb{R}^{2d}}H(u_1,u_2)v_s(u_2)}{\langle\xi'-\nabla_x\Phi(x',\xi),x-\nabla_\eta\Phi(x',\xi)\rangle^s}\end{aligned}$$

$$= \frac{\|H\|_{L^\infty_{1\otimes v_s}}}{\langle \xi' - \nabla_x \Phi(x',\xi), x - \nabla_\eta \Phi(x',\xi)\rangle^s}$$
$$\lesssim \frac{\|\sigma\|_{M^\infty_{1\otimes v_s}}}{\langle \xi' - \chi_2(x,\xi), x' - \chi_1(x,\xi)\rangle^s},$$

where in the last inequality we have used Lemma 6.1.4. □

Theorem 6.4.10. *Fix $\mathcal{G}(g,\Lambda)$ be a Parseval frame with $g \in \mathcal{S}(\mathbb{R}^d)$ and let $s \geq 0$.*

Let T be a continuous linear operator $\mathcal{S}(\mathbb{R}^d) \to \mathcal{S}'(\mathbb{R}^d)$ and χ a canonical transformation which satisfies B1, B2, and B3. Then the following properties are equivalent:
(i) *$T = T_{I,\Phi_\chi,\sigma}$ is a FIO of type I for some $\sigma \in M^\infty_{1\otimes v_s}(\mathbb{R}^{2d})$.*
(ii) *$F \in FIO(\chi, s)$.*

Proof. The implication (i) $\Longrightarrow$ (ii) is proved in Theorem 6.4.9.

The implication (ii) $\Longrightarrow$ (i) follows immediately from Proposition 6.4.1 and Lemma 6.4.2, since $\langle (x,\eta') - \nabla\Phi(x,\eta)\rangle^{-s} \lesssim \langle (x',\eta') - \chi(x,\eta)\rangle^{-s}$. □

Corollary 6.4.11. *Under the same assumptions as in Theorem 6.4.10, the following statements are equivalent:*
(i) *$T = T_{I,\Phi_\chi,\sigma}$ is a FIO of type I for some $\sigma \in S^0_{0,0}$.*
(ii) *$T \in FIO(\chi,\infty) = \bigcap_{s\geq 0} FIO(\chi, s)$.*

Remark 6.4.12. The latter corollary should be juxtaposed to Tataru's characterization of $FIO(\chi,\infty)$ in [279, Theorem 4]. Tataru assumes only conditions *B1* and *B2*, but not *B3*, on the canonical transformation χ and therefore obtains a larger class of FIOs satisfying the decay condition (6.77). The new insight of Corollary 6.4.11 is that, under the additional assumption *B3*, *every* FIO admits the classical representation (6.12). As a byproduct, we see that the integral operator in *(14)* of [279] possesses the classical representation, provided that χ also satisfies *B3*. This observation might be useful for the solution of the Cauchy problem in [279, Section 5], where, for small time, χ is a small perturbation of the identity transformation and therefore certainly satisfies *B3*.

Thanks to Theorem 6.3.7, the FIOs of type I with symbol in $M^\infty_{1\otimes v_s}(\mathbb{R}^{2d})$, with $s > 2d$, are bounded on $M^p(\mathbb{R}^d)$, $1 \leq p \leq \infty$. Consider now the multiplication operator

$$Tf(x) = e^{\pi i x^2} f(x). \tag{6.92}$$

Then T is an FIO of type I having phase $\Phi(x,\xi) = x^2/2 + x\cdot\xi$ and symbol $\sigma(x,\xi) = 1$, for every $(x,\xi) \in \mathbb{R}^{2d}$. Observe that $\sigma = 1 \in S^0_{0,0}$, hence $\sigma \in M^{\infty,1}(\mathbb{R}^{2d})^s$, for every $s \geq 0$. However, the multiplication operator T is not bounded on $M^{p,q}$, when $p \neq q$, as proved in Proposition 6.1.17.

We now look at the Wiener property of the class of FIOs of type I with symbol in $M^\infty_{1\otimes v_s}(\mathbb{R}^{2d})$, with $s > 2d$. As we will see in Section 6.5, this class is not closed under composition, therefore the Wiener property must necessarily involve FIOs of type II (see (6.74)).

We first need to recall the following result, originally contained in [73, Lemma 2.10], though with different weights.

Lemma 6.4.13. *Let $\sigma \in M^{p,q}_{1\otimes v_s}(\mathbb{R}^{2d})$, $s \in \mathbb{R}$, and consider its adjoint σ^* defined by*

$$\sigma^*(x,\eta) = \overline{\sigma(\eta,x)}, \quad (x,\eta) \in \mathbb{R}^{2d}. \tag{6.93}$$

Then $\sigma^* \in M^{p,q}_{1\otimes v_s}(\mathbb{R}^{2d})$.

Proof. Fix a window function $g \in \mathcal{S}(\mathbb{R}^d)$. By a direct computation,

$$\begin{aligned} V_g\sigma^*(z_1,z_2;\zeta_1,\zeta_2) &= \int e^{-2\pi i(y_1\cdot\zeta_1+y_2\cdot\zeta_2)}\overline{\sigma(y_2,y_1)g(y_1-z_1,y_2-z_2)}\,dy_1\,dy_2 \\ &= \int e^{-2\pi i(v_2\cdot\zeta_1+v_1\cdot\zeta_2)}\overline{\sigma(v_1,v_2)g(v_2-z_1,v_1-\zeta_2)}\,dv_1\,dv_2 \\ &= V_{{}^tg}\overline{\sigma}(z_2,z_1;\zeta_2,\zeta_1) \end{aligned}$$

where we used the notation ${}^tg(y_1,y_2) := g(y_2,y_1)$. Since $\overline{\sigma} \in M^{p,q}_{1\otimes v_s}(\mathbb{R}^{2d})$, the result immediately follows by the independence of the window function for the computation of the modulation space norm and the symmetry of the weight v_s, that is, $v_s(\zeta_1,\zeta_2) = v_s(\zeta_2,\zeta_1)$. □

Theorem 6.4.14. *Let T be an FIO of type I with a tame phase Φ and a symbol $\sigma \in M^{\infty}_{1\otimes v_s}(\mathbb{R}^{2d})$, with $s > 2d$. If T is invertible on $L^2(\mathbb{R}^d)$, then T^{-1} is an FIO of type II with same phase Φ and a symbol $\tau \in M^{\infty}_{1\otimes v_s}(\mathbb{R}^{2d})$.*

Proof. Let χ be the canonical transformation associated to Φ. Then by Theorem 6.4.10 T belongs to $\mathrm{FIO}(\chi,s)$. As in the proof of Theorem 6.3.10, we consider $P = T^*T$ and write

$$T^{-1} = P^{-1}T^* = (T(P^{-1})^*)^* = (TP^{-1})^*.$$

We have already shown that P is in $\mathrm{FIO}(\mathrm{Id},s)$ and a pseudodifferential operator with a symbol in $M^{\infty}_{1\otimes v_s}(\mathbb{R}^{2d})$ and that also $P^{-1} \in \mathrm{FIO}(\mathrm{Id},s)$ by the spectral invariance of pseudodifferential operators of Theorem 6.3.2. Now Theorem 6.3.9 implies that $TP^{-1} \in \mathrm{FIO}(\chi,s)$ and Theorem 6.4.10 implies that TP^{-1} is an FIO of type I with tame phase Φ and a symbol $\rho \in M^{\infty}_{1\otimes v_s}(\mathbb{R}^{2d})$. Since $T^{-1}(TP^{-1})^*$, T^{-1} is an FIO of type II with phase Φ and the symbol $\tau(x,\eta) = \overline{\rho(\eta,x)}$. Lemma 6.4.13 shows that $\tau \in M^{\infty}_{1\otimes v_s}(\mathbb{R}^{2d})$. □

6.5 Generalized metaplectic operators

As an example, we will consider the class of $\mathrm{FIO}(\chi,s)$ whose phase is a linear transformation $\chi(z) = \mathcal{A}z$ for some invertible matrix $\mathcal{A} \in \mathrm{GL}(2d,\mathbb{R})$. Since χ must preserve the symplectic form (assumption *B2*), $\mathcal{A}$ must be a symplectic matrix, $\mathcal{A} \in Sp(d,\mathbb{R})$, cf. Definition 1.1.2.

Definition 6.5.1. Let $\mathcal{A} \in Sp(d,\mathbb{R})$ and $s \geq 0$. Fix a Parseval frame $\mathcal{G}(g,\Lambda)$ with $g \in \mathcal{S}(\mathbb{R}^d)$. We say that a continuous linear operator $T : \mathcal{S}(\mathbb{R}^d) \to \mathcal{S}'(\mathbb{R}^d)$ is a *generalized metaplectic operator*, in short, $T \in \mathrm{FIO}(\mathcal{A},s)$, if its Gabor matrix satisfies the decay condition

$$|\langle T\pi(\lambda)g, \pi(\mu)g\rangle| \leq C\langle \mu - \mathcal{A}\lambda\rangle^{-s}, \quad \forall \lambda,\mu \in \Lambda.$$

The union $\bigcup_{\mathcal{A}\in Sp(d,\mathbb{R})} \mathrm{FIO}(\mathcal{A},s)$ is called the class of generalized metaplectic operators and denoted by $\mathrm{FIO}(Sp,s)$.

Since $Sp(d,\mathbb{R})$ is a group, Theorems 6.3.9 and 6.3.10 imply the following statement.

Theorem 6.5.2. *For $s > 2d$, $\mathrm{FIO}(Sp,s)$ is a Wiener subalgebra of $\mathrm{FIO}(\Xi,s)$.*

The main examples in $\mathrm{FIO}(Sp,s)$ are the operators of the metaplectic representation of $Sp(d,\mathbb{R})$, see Section 1.1.2 for details. Let $\mathcal{A} = \begin{pmatrix} A & B \\ C & D\end{pmatrix} \in Sp(d,\mathbb{R})$ with $d\times d$ blocks A,B,C,D, cf. (1.4). Then condition *B3* of Definition 6.1.1 is equivalent to $\det A \neq 0$. In this case, $\mu(\mathcal{A})$ is explicitly given by the FIO of type I in (6.48). Namely,

$$\mu(\mathcal{A})(x) = (\det A)^{-1/2} \int_{\mathbb{R}^d} e^{2\pi i\Phi(x,\eta)}\hat{f}(\eta)\,d\eta$$

with the phase Φ given by

$$\Phi(x,\eta) = \frac{1}{2}x\cdot CA^{-1}x + \eta\cdot A^{-1}x - \frac{1}{2}\eta\cdot A^{-1}B\eta. \tag{6.94}$$

Even without the explicit form of $\mu(\mathcal{A})$, Definition 6.5.1 yields some interesting information about the metaplectic representation.

Proposition 6.5.3. *If $\mathcal{A} \in Sp(d,\mathbb{R})$, then $\mu(\mathcal{A}) \in \bigcap_{s\geq 0}\mathrm{FIO}(\mathcal{A},s)$.*

Proof. According to Theorem 6.3.5, it is enough to prove that $\mu(\mathcal{A})$ satisfies the continuous decay condition (6.77). Using the definition of $\mu(\mathcal{A})$ in (1.10) and the covariance property of the STFT (1.50), we write the Gabor matrix of $\mu(\mathcal{A})$ as

$$\begin{aligned}|\langle \mu(\mathcal{A})\pi(z)g, \pi(w)g\rangle| &= |\langle \pi(\mathcal{A}z)\mu(\mathcal{A})g, \pi(w)g\rangle| \\ &= |V_g(\pi(\mathcal{A}z)\mu(\mathcal{A})g)(w)| \\ &= |V_g(\mu(\mathcal{A})g)(w-\mathcal{A}z)|. \end{aligned} \tag{6.95}$$

Since both $g \in \mathcal{S}(\mathbb{R}^d)$ and $\mu(\mathcal{A})g \in \mathcal{S}(\mathbb{R}^d)$, we also have $V_g(\mu(\mathcal{A})g) \in \mathcal{S}(\mathbb{R}^{2d})$, by Corollary 1.2.24. This gives

$$|\langle \mu(\mathcal{A})\pi(z)g, \pi(w)g\rangle| \leq C\langle w - \mathcal{A}z\rangle^{-s}, \tag{6.96}$$

for every $s \geq 0$, as desired. □

The following theorem shows that every generalized metaplectic operator is a product of a metaplectic operator and a classical pseudodifferential operator.

Theorem 6.5.4. *Let $\mathcal{A} \in Sp(d,\mathbb{R})$ and $T \in \mathrm{FIO}(\mathcal{A},s)$ with $s > 2d$. Then there exist symbols $\sigma_1, \sigma_2 \in M^{\infty}_{1\otimes v_s}(\mathbb{R}^{2d})$ with the corresponding pseudodifferential operators $\sigma_1(x,D)$ and $\sigma_2(x,D)$, such that*

$$T = \sigma_1(x,D)\mu(\mathcal{A}) \quad \text{and} \quad T = \mu(\mathcal{A})\sigma_2(x,D). \tag{6.97}$$

Proof. We prove the factorization $T = \sigma_1(x,D)\mu(\mathcal{A})$, the other factorization is obtained analogously.

Since $\mu(\mathcal{A})^{-1} = \mu(\mathcal{A}^{-1})$ is in $\mathrm{FIO}(\mathcal{A}^{-1},s)$ by Proposition 6.5.3, the algebra property of Theorem 6.3.9 implies that $T\mu(\mathcal{A}^{-1}) \in \mathrm{FIO}(Id,s)$. The fundamental characterization of pseudodifferential operators of Theorem 6.3.3 implies that existence of a symbol $\sigma_1 \in M^{\infty}_{1\otimes v_s}(\mathbb{R}^{2d})$, such that $T\mu(\mathcal{A})^{-1} = \sigma_1(x,D)$, which is what we wanted to show. □

Finally, we check the counterpart of FIO I and II for generalized metaplectic operators.

Let $\mathcal{A} = \left(\begin{smallmatrix} A & B \\ C & D \end{smallmatrix}\right) \in Sp(d,\mathbb{R})$ with $\det A \neq 0$. As proved in Theorem 6.4.10, every generalized metaplectic operator $T \in \mathrm{FIO}(\mathcal{A},s)$ with $s > 2d$ is an FIO of type I,

$$Tf(x) = (\det \mathcal{A})^{-1/2} \int e^{2\pi i \Phi(x,\eta)} \sigma(x,\xi)\hat{f}(\eta)\,d\eta,$$

with a symbol $\sigma \in M^{\infty}_{1\otimes v_s}(\mathbb{R}^{2d})$ and phase $\Phi(x,\eta) = \frac{1}{2}x \cdot CA^{-1}x + \eta \cdot A^{-1}x - \frac{1}{2}\eta \cdot A^{-1}B\eta$.

We obtain examples of FIOs of type II by taking adjoints. If $T \in \mathrm{FIO}(\mathcal{A},s)$, then T^* is an FIO of type II.

As it is well known, the generalized metaplectic operators of type I do not enjoy the algebra property. In fact, consider, for instance, the operators $T_1 = \mu(\mathcal{A}_1)$ and $T_2 = \mu(\mathcal{A}_2)$, with

$$\mathcal{A}_1 = \begin{pmatrix} I_d & I_d \\ 0 & I_d \end{pmatrix}, \quad \mathcal{A}_2 = \begin{pmatrix} I_d & I_d \\ -I_d & 0 \end{pmatrix}.$$

Then both T_1 and T_2 are FIOs of type I but their product

$$T_1T_2 = \mu(\mathcal{A}_1)\mu(\mathcal{A}_2) = \mu(\mathcal{A}_1\mathcal{A}_2) = \mu(-J) = \mathcal{F}$$

cannot be an FIO of type I. Indeed, the Fourier transform $\mathcal{F} = \mu(-J)$ is an example of a metaplectic operator that is an FIO of neither type I nor of type II. Note that in this case assumption *B3* is not satisfied for $\mathcal{J}$.

As in Remark 6.4.12, we see again that there is a crucial difference between FIOs satisfying all axioms *B1*, *B2*, and *B3*, or only *B1* and *B2*.

6.6 Propagation of Gabor wave front set for Schrödinger equations

In the previous sections we proposed a new approach to the calculus of Fourier integral operators in terms of time–frequency localization, using *Gabor analysis*. The FIOs under consideration were of the type of those appearing in the study of the Schrödinger equations, typically the phase function being a homogeneous function of degree 2 in the whole of the phase space variables. With respect to the standard representations of FIOs, the time–frequency representation looks more involved, since old and new phase-space variables appear simultaneously, and everything depends on the choice of the window function. On the other hand, the problem of the caustics is automatically solved in this new setting and the expression provides an excellent tool for numerical analysis, as we have seen above.

In this last section we mainly report the outcomes of the paper [84], where the aforesaid results are applied to the analysis of Schrödinger equations. With respect to the enormous existing literature, such results are new in the following aspects. For a fixed a real-valued Hamiltonian which is homogeneous of degree 2, a pseudodifferential perturbation (also called potential in the following) is allowed with a bounded, complex-valued, nonsmooth symbol, for which even differentiability may be lost. A global-in-time propagator is constructed in the class of the FIOs in Definition 6.3.4, and well-posedness of the Cauchy problem is deduced in suitable modulation spaces. Regarding propagation of singularities, which is our main concern in this section, the known results do not apply to such a situation. We are then led to the new definition of the Gabor wave front set, which allows the expression of optimal results of propagation in our context.

To be precise, we aim at studying the representation in terms of time–frequency analysis of the propagator e^{itH},

$$H = a(x, D) + \sigma(x, D), \tag{6.98}$$

providing the solution to the Cauchy problem

$$\begin{cases} i\frac{\partial u}{\partial t} + a(x, D)u + \sigma(x, D)u = 0, \\ u(0, x) = u_0(x). \end{cases} \tag{6.99}$$

The Hamiltonian $a(x, D)$ is a pseudodifferential operator in the Kohn–Nirenberg form

$$Af(x) = a(x, D)f(x) = \int_{\mathbb{R}^d} e^{2\pi i x\cdot\xi} a(x, \xi)\hat{f}(\xi)\, d\xi, \tag{6.100}$$

where the symbol $a(z)$, $z = (x, \xi)$, is real-valued and positively homogeneous of degree 2, i. e., $a(\lambda z) = \lambda^2 a(z)$ for $\lambda > 0$, with $a \in C^\infty(\mathbb{R}^{2d} \setminus 0)$. This implies that $a(x, D)$ is formally self-adjoint modulo 0th order perturbations. Basic examples are real-valued quadratic forms $a(z)$, including the cases when $i\partial_t + a(x,D)$ is the free particle or the harmonic oscillator operator. When $a(z)$ is not a polynomial, we shall assume $a(z)$ to be modified in a bounded neighborhood of the origin in such a way that we have $a \in C^\infty(\mathbb{R}^{2d})$ keeping real values. As we shall see, cf. Example 6.6.28 below, the singularity at the origin of $a(z)$ can be admitted as well, by absorbing it in a nonsmooth potential. The pseudodifferential operator $a(x,D)$ enters the classes of [265], see also [187], to which we refer for the symbolic calculus and other properties, see also the next Section 6.6.1.

Concerning the potential $\sigma(x,D)$, the regularity assumptions will be expressed in terms of the modulation spaces $M^\infty_{1\otimes v_s}(\mathbb{R}^{2d})$. We shall not treat the case $\sigma \in M^{\infty,1}(\mathbb{R}^{2d})$, the interested reader can refer to [69], where quadratic Hamiltonians with a Sjöstrand potential are studied.

We want to study the Gabor matrix $k(t,w,z)$ of the propagator e^{itH}. Its structure will be linked, as expected, to the Hamiltonian field of $a(x,\xi)$. Namely, consider

$$\begin{cases} 2\pi\dot{x} = -\nabla_\xi a(x,\xi), \\ 2\pi\dot{\xi} = \nabla_x a(x,\xi), \\ x(0) = y, \quad \xi(0) = \eta, \end{cases}$$

(the factor 2π depends on our normalization of the STFT). Under our assumptions, the solution $\chi_t(y,\eta) = (x(t,y,\eta), \xi(t,y,\eta))$ exists for all $t \in \mathbb{R}$ and defines a symplectic diffeomorphism $\chi_t : \mathbb{R}^{2d}_{y,\eta} \to \mathbb{R}^{2d}_{x,\xi}$ which is homogeneous of degree 1 with respect to $w = (y,\eta)$ for large $|w|$, for every fixed $t \in \mathbb{R}$. We shall show the following result.

Theorem 6.6.1. *Let the preceding assumptions be satisfied, in particular, let* $\sigma \in M^\infty_{1\otimes v_s}(\mathbb{R}^{2d})$, $s > 2d$, *and* $k(t,w,z) = \langle e^{itH}\pi(z)g, \pi(w)g\rangle$ *be the Gabor matrix of the Schrödinger propagator* e^{itH}. *Then there exists* $C = C(t,s) > 0$ *such that*

$$|k(t,w,z)| \le C\langle z - \chi_t(w)\rangle^{-s}, \quad z = (x,\xi),\ w = (y,\eta) \in \mathbb{R}^{2d}. \tag{6.101}$$

According to the notations of Definition 6.3.4, this can be rephrased as $e^{itH} \in FIO(\chi_t, s)$. For t sufficiently small, our assumptions yield $\det \frac{\partial x}{\partial y}(t,y,\eta) \neq 0$ in the expression of χ_t, and (6.101) is then equivalent to

$$(e^{itH}u_0)(t,x) = \int_{\mathbb{R}^d} e^{2\pi i\Phi(t,x,\xi)} b(t,x,\xi)\hat{u}_0(\xi)\, d\xi, \tag{6.102}$$

with the phase Φ linked to χ_t as standard and $b(t,\cdot) \in M^\infty_{1\otimes v_s}(\mathbb{R}^{2d})$, see Theorem 6.4.10. In the classical approach, cf. [3], the occurrence of caustics makes the validity of (6.102) local in time. So for $t \in \mathbb{R}$, one is led to multiple compositions of local representations, with unbounded number of variables possibly appearing in the expression whereas $k(t,w,z)$ obviously remains defined for every $t \in \mathbb{R}$, and the estimates (6.101) hold for χ_t with $t \in \mathbb{R}$.

Let us now define the Gabor wave front set $WF^{p,r}_G(f)$ under our consideration.

Definition 6.6.2. Let $g \in \mathcal{S}(\mathbb{R}^d)$, $g \neq 0$, $r > 0$, $1 \leq p \leq \infty$. For $f \in M^p_{v_{-r}}(\mathbb{R}^d)$, $z_0 \in \mathbb{R}^{2d}$, $z_0 \neq 0$, we say that $z_0 \notin \mathrm{WF}^{p,r}_G(f)$ if there exists an open conic neighborhood $\Gamma_{z_0} \subset \mathbb{R}^{2d}$ of z_0 such that

$$\int_{\Gamma_{z_0}} |V_g f(z)|^p \langle z\rangle^{pr}\, dz < \infty \tag{6.103}$$

(with obvious changes for $p = \infty$).

Then $\mathrm{WF}^{p,r}_G(f)$ is well defined as a conic closed subset of $\mathbb{R}^{2d} \setminus \{0\}$. The main result of this last section can be summarized as follows.

Theorem 6.6.3. *Consider $\sigma \in M^\infty_{1\otimes v_s}(\mathbb{R}^{2d})$, $s > 2d$, $1 \leq p \leq \infty$. Then*

$$e^{itH} : M^p_{v_r}(\mathbb{R}^d) \to M^p_{v_r}(\mathbb{R}^d) \tag{6.104}$$

continuously, for $|r| < s - 2d$. Moreover, for $u_0 \in M^p_{v_{-r}}(\mathbb{R}^d)$,

$$\mathrm{WF}^{p,r}_G(e^{itH}u_0) = \chi_t(\mathrm{WF}^{p,r}_G(u_0)), \tag{6.105}$$

provided $0 < 2r < s - 2d$.

Observe the more restrictive assumption on r for (6.105) compared to that for (6.104).

As an elementary example, consider the perturbed harmonic oscillator (studied in Example 6.6.28 in the sequel)

$$\begin{cases} i\partial_t u - \frac{1}{4\pi}\partial_x^2 u + \pi x^2 u + |\sin x|^\mu u = 0, \\ u(0,x) = u_0(x), \end{cases} \tag{6.106}$$

with $\mu > 1$. We shall prove that $|\sin x|^\mu \in M^\infty_{1\otimes v_{\mu+1}}$ and then, from Theorem 6.6.3, we have that the Cauchy problem is well-posed for $u_0 \in M^p_{v_r}(\mathbb{R})$, $|r| < \mu - 2$ and the propagation of $\mathrm{WF}^{p,r}_G(u(t,\cdot))$ for $t \in \mathbb{R}$ takes place as in Theorem 6.6.3 for $0 < r < \mu/2 - 1$, where χ_t is defined in (6.70).

Using (6.72) and (6.105), we may recapture results for the propagation in the case of a smooth potential, i. e., $\sigma \in S^0_{0,0}$. We define the wave front set $\mathrm{WF}_G(f)$ by stating $z_0 \notin \mathrm{WF}_G(f)$ if there exists an open conic set $\Gamma_{z_0} \subset \mathbb{R}^{2d}$ containing z_0 such that for every $r > 0$,

$$|V_g f(z)| \leq C_r \langle z\rangle^{-r}, \quad z \in \Gamma_{z_0}, \tag{6.107}$$

for a suitable $C_r > 0$. Then the estimate (6.101) is satisfied for every s and, from Theorem 6.6.3, we recapture for $u_0 \in \mathcal{S}'(\mathbb{R}^d)$,

$$\mathrm{WF}_G(e^{itH}u_0) = \chi_t(\mathrm{WF}_G(u_0)). \tag{6.108}$$

Although it is impossible to do justice to the vast literature in this connection, let us mention some of the related contributions. The pioneering work is that of Hörmander [195], who defined the wave front set in (6.107), as well as its analytic version, and proved (6.108) in the case of the metaplectic operators. For subsequent results providing (6.108) and its analytic-Gevrey version for general smooth symbols, let us refer to [28, 178, 200, 201, 222, 223, 227, 232, 233, 321]. The wave front sets introduced there under different names actually coincide with those of Hörmander, cf. [253], see also [44, 47, 245, 259, 260, 301]. Still concerning propagation of singularities in the case of smooth or analytic symbols, we refer to [96, 225, 250, 251, 307]. Besides, concerning global-in-time representations of e^{itH}, solving the problem of the caustics for smooth symbols, see [3, 30, 31, 158, 279].

The general result in Theorem 6.6.3 is due to [84].

6.6.1 Preliminaries and Shubin classes

We shall use the following properties.

Lemma 6.6.4. *Consider $\mu > 0$. Then the function $f(x) = |\sin x|^{\mu}$ belongs to $M^{\infty}_{1\otimes v_{\mu+1}}(\mathbb{R})$.*

Proof. Consider a window function $g \in \mathcal{D}(\mathbb{R})$, with $\operatorname{supp} g \subset [-\pi/4, \pi/4]$ to compute the STFT $V_g f$ with $f(x) = |\sin x|^{\mu}$. Then $|V_g f(x,\xi)|$ is a periodic function of period π in the x variable. So

$$\|f\|_{M^{\infty}_{1\otimes v_{\mu+1}}} = \sup_{|x|\le\pi/2}\ \sup_{\xi\in\mathbb{R}} \langle\xi\rangle^{\mu+1}|V_g f|(x,\xi).$$

Now observe that $\operatorname{supp} T_x g \subset [-3\pi/4, 3\pi/4]$, for $x \in [-\pi/2, \pi/2]$, and on that interval $f(x) = |x|^{\mu}\varphi(x)$, with $\varphi \in \mathcal{D}(\mathbb{R})$. We can write

$$V_g f(x,\xi) = \int_0^{+\infty} e^{-2\pi it\xi} t^{\mu}\varphi(t)\overline{g(t-x)}\,dt + \int_{-\infty}^{0} e^{-2\pi it\xi}(-t)^{\mu}\varphi(t)\overline{g(t-x)}\,dt := A + B.$$

So it suffices to estimate integral A, the estimate of B is analogous. Setting $F_x(t) = e^t\varphi(t)\overline{g(t-x)} \in \mathcal{S}(\mathbb{R})$, we observe that the family $\{F_x\}_{x\in[-\pi/2,\pi/2]}$ belongs to a bounded subset of $\mathcal{S}(\mathbb{R})$. Now

$$\begin{aligned} A &= \int_0^{+\infty} e^{-2\pi it\xi} t^{\mu} e^{-t} e^{t}\varphi(t)\overline{g(t-x)}\,dt \\ &= \frac{\Gamma(\mu+1)}{(1+2\pi i\xi)^{\mu+1}} * \widehat{F_x}(\xi), \end{aligned}$$

and this yields

$$\langle\xi\rangle^{\mu+1}|A| \lesssim \frac{\langle\xi\rangle^{\mu+1}}{(1+2\pi i\xi)^{\mu+1}} * \left(\langle\xi\rangle^{\mu+1}\widehat{F_x}(\xi)\right) \in L^{\infty}(\mathbb{R}),$$

by Young's inequality, since the first factor of the convolution product is bounded and the second lies in a bounded subset of $\mathcal{S}(\mathbb{R}) \subset L^1(\mathbb{R})$. □

Corollary 6.6.5. *Consider the symbol* $\sigma(x,\xi) = |\sin x|^\mu$ *on* $\mathbb{R}^2$. *Then we have* $\sigma \in M^\infty_{1\otimes v_{\mu+1}}(\mathbb{R}^2)$.

Proof. It is an immediate consequence of Lemma 6.6.4. Indeed, taking $\psi(x,\xi) = g(x)\varphi(\xi)$, with g being the 1-dimensional window of the previous proof and $\varphi \in \mathcal{S}(\mathbb{R})$, we have $V_\psi\sigma((x_1,x_2),(\xi_1,\xi_2)) = V_g(|\sin(\cdot)|^\mu)(x_1,\xi_1)V_\varphi 1(x_2,\xi_2)$, and the claim follows since $\langle(\xi_1,\xi_2)\rangle \le \langle\xi_1\rangle\langle\xi_2\rangle$ and $1 \in S^0_{0,0} \subset M^\infty_{1\otimes v_s}(\mathbb{R})$, for every $s \ge 0$. □

Proposition 6.6.6. *Let* $h \in \mathcal{C}^\infty(\mathbb{R}^d \setminus \{0\})$ *be positively homogeneous of degree* $r > 0$, *i. e.,* $h(\lambda x) = \lambda^r h(x)$ *for* $x \neq 0$, $\lambda > 0$, *and* $\chi \in \mathcal{D}(\mathbb{R}^d)$. *Set* $f = h\chi$. *Then for* $\psi \in \mathcal{S}(\mathbb{R}^d)$ *there exists a constant* $C > 0$ *such that*

$$|V_\psi f(x,\xi)| \le C(1+|\xi|)^{-r-d}, \quad x,\xi \in \mathbb{R}^d.$$

Proof. We know that the Fourier transform of h is a homogeneous distribution of degree $-r-d$, smooth in $\mathbb{R}^d \setminus \{0\}$ [194, Theorems 7.1.16, 7.1.18]. Hence, if $\chi' \in \mathcal{D}(\mathbb{R}^d)$, $\chi' = 1$ in a neighborhood of the origin, we have

$$|(1-\chi'(\xi))\widehat{h}(\xi)| \le C(1+|\xi|)^{-r-d}, \quad \xi \in \mathbb{R}^d. \tag{6.109}$$

On the other hand, by the very definition of the STFT, we have

$$|V_\psi f(x,\xi)| \le |((\chi'\widehat{h}) *_\xi \widehat{\chi})| *_\xi |\widehat{\overline{\psi}}| + |((1-\chi')\widehat{h}) *_\xi \widehat{\chi}| *_\xi |\widehat{\overline{\psi}}|.$$

Since $\widehat{\overline{\psi}}, \widehat{\chi} \in \mathcal{S}(\mathbb{R}^d)$, the first term on the right-hand side has rapid decay because $\mathcal{E}' * \mathcal{S} \subset \mathcal{S}$, whereas the second term is estimated using (6.109), as at the end of the proof of Lemma 6.6.4. □

Here we are interested in operators with symbols in the Shubin classes (cf. [265], Helffer [187]); indeed, we shall use them as symbol and phase spaces for the unperturbed initial value problem for Schrödinger equations.

Definition 6.6.7. For $m \in \mathbb{R}$, the class $\Gamma^m(\mathbb{R}^{2d})$ is the set of functions $a \in \mathcal{C}^\infty(\mathbb{R}^{2d})$ such that for every $\alpha \in \mathbb{Z}^{2d}_+$ there exists a constant $C_\alpha > 0$ such that

$$|\partial^\alpha_z a(z)| \le C_\alpha v_{m-|\alpha|}(z), \quad z \in \mathbb{R}^{2d},$$

where we recall $v_s(z) = \langle z\rangle^s$ is defined in (2.5).

Consider $a_j \in \Gamma^{m_j}(\mathbb{R}^{2d})$ with m_j being a decreasing sequence tending to $-\infty$. Then a function $a \in \mathcal{C}^\infty(\mathbb{R}^{2d})$ satisfies

$$a \sim \sum_{j=1}^\infty a_j \tag{6.110}$$

if

$$\forall r \geq 2 \quad a - \sum_{j=1}^{r-1} a_j \in \Gamma^{m_r}(\mathbb{R}^{2d}).$$

Namely, our symbol class will be a subclass of $\Gamma^m(\mathbb{R}^{2d})$, defined as follows [187, Section 1.5].

Definition 6.6.8. A function a is in the class $\Gamma^{m,cl}(\mathbb{R}^{2d})$ if $a \in \Gamma^m(\mathbb{R}^{2d})$ and admits an asymptotic expansion

$$a \sim \sum_{j=0}^{\infty} a_{m-j}, \tag{6.111}$$

where $a_{m-j} \in C^\infty(\mathbb{R}^{2d})$ satisfies $a_{m-j}(\lambda z) = \lambda^{m-j} a_{m-j}(z)$, for $|z| \geq 1$, $\lambda \geq 1$. The function a_m corresponding to $j = 0$ in expansion (6.111) is called the *principal symbol* of the symbol a.

For $a \in \Gamma^m(\mathbb{R}^{2d})$, the corresponding pseudodifferential operator $a(x, D)$ is defined by the Kohn–Nirenberg form (6.100).

Definition 6.6.9. We say that $A \in G^m$ (resp. $A \in G^{m,cl}$) if its symbol satisfies $a \in \Gamma^m(\mathbb{R}^{2d})$ (resp. $a \in \Gamma^{m,cl}(\mathbb{R}^{2d})$).

A pseudodifferential operator $A \in G^{m,cl}$ is called globally elliptic if there exist $R > 0$ and $C > 0$ such that

$$|a_m(z)| \geq C\langle z\rangle^m, \quad \text{for } z \in \mathbb{R}^{2d},\ |z| \geq R, \tag{6.112}$$

where a_m is the principal symbol.

Let $a \in \Gamma^{2,cl}(\mathbb{R}^{2d})$ with real principal symbol a_2. The related classical evolution, given by the linear Hamilton–Jacobi system, following our normalization can be written as

$$\begin{cases} 2\pi\partial_t x(t,y,\eta) = -\nabla_\xi a_2(x(t,y,\eta), \xi(t,y,\eta)), \\ 2\pi\partial_t \xi(t,y,\eta) = \nabla_x a_2(x(t,y,\eta), \xi(t,y,\eta)), \\ x(0,y,\eta) = y, \\ \xi(0,y,\eta) = \eta. \end{cases} \tag{6.113}$$

The solution $(x(t,y,\eta), \xi(t,y,\eta))$ exists for every $t \in \mathbb{R}$. Indeed, setting $u := (x, \xi)$, $F(u) := (-\nabla_\xi a(u), \nabla_x a(u))$, the initial value problem (6.113) can be rephrased as

$$u'(t) = F(u(t)), \quad u(t_0) = u_0, \tag{6.114}$$

in the particular case $t_0 = 0$. Observe that $a \in \Gamma^{2,cl}(\mathbb{R}^{2d})$ implies $F_j \in \Gamma^{1,cl}(\mathbb{R}^{2d})$, for $j = 1, \dots, 2d$ and $\partial^\alpha F_j \in \Gamma^{0,cl}(\mathbb{R}^{2d})$, for every $|\alpha| > 0$, $j = 1, \dots, 2d$, hence in particular

$F : \mathbb{R}^{2d} \to \mathbb{R}^{2d}$ is a Lipschitz continuous mapping. Thus the previous ODE is an autonomous ODE with a mapping $F \in C^\infty(\mathbb{R}^{2d} \to \mathbb{R}^{2d})$ having at most linear growth, hence $\|F(u)\| \lesssim 1 + \|u\|$. This implies that for each $u_0 \in \mathbb{R}^{2d}$ and $t_0 \in \mathbb{R}$ there exists a unique classical global solution $u : \mathbb{R} \to \mathbb{R}^{2d}$ (in this case $u \in C^\infty(\mathbb{R} \to \mathbb{R}^{2d})$ since $F \in C^\infty(\mathbb{R}^{2d} \to \mathbb{R}^{2d})$) to (6.114). Moreover, the solution maps $S_{t_0}(t) : \mathbb{R}^{2d} \to C^\infty(\mathbb{R} \to \mathbb{R}^{2d})$, defined by $S_{t_0}(t)u_0 = u(t)$, and $S_{t_0}(t_0) = \mathrm{Id}$, the identity operator on $\mathbb{R}^{2d}$, are Lipschitz continuous mappings, obey the time translation invariance $S_{t_0}(t) = S_0(t - t_0)$ and the group laws

$$S_0(t)S_0(t') = S_0(t + t'), \quad S_0(0) = \mathrm{Id}. \tag{6.115}$$

Observe that $S_0(t)$ is a bi-Lipschitz diffeomorphism with $S_0^{-1}(t) = S_0(-t)$. To be consistent with our notations, we call the bi-Lipschitz diffeomorphism

$$\chi_t(y, \eta) := S_0(t)(y, \eta), \quad (y, \eta) \in \mathbb{R}^{2d}. \tag{6.116}$$

The theory of Hamilton–Jacobi allows finding a $T > 0$ such that for $t \in]-T, T[$ there exists a phase function $\Phi(t, x, \eta)$, solution of the eiconal equation (cf. [187, (3.2.12),(3.2.13)])

$$\begin{cases} 2\pi\partial_t\Phi + a_2(x, \nabla_x\Phi) = 0, \\ \Phi(0, x, \eta) = x\eta. \end{cases} \tag{6.117}$$

The phase $\Phi(t, x, \eta) \in C^\infty(]-T, T[, \Gamma^2(\mathbb{R}^{2d}))$ is real-valued since the principal symbol $a_2(x, \xi)$ is real-valued; moreover, Φ fulfills the condition of nondegeneracy

$$\left|\det \partial^2_{x,\eta}\Phi(t, x, \eta)\right| \geq c > 0, \quad (t, x, \eta) \in]-T, T[\times \mathbb{R}^{2d}, \tag{6.118}$$

after possibly shrinking $T > 0$ (cf. [187, pages 142–143] and [81]).

The relation between the phase Φ and the canonical transformation χ is given by

$$(x, \nabla_x\Phi(t, x, \eta)) = \chi_t(\nabla_\eta\Phi(t, x, \eta), \eta), \quad t \in]-T, T[. \tag{6.119}$$

In particular,

$$\begin{cases} y(t, x, \eta) = \nabla_\eta\Phi(t, x, \eta), \\ \xi(t, x, \eta) = \nabla_x\Phi(t, x, \eta), \end{cases} \tag{6.120}$$

and there exists $\delta > 0$ such that

$$\left|\det \frac{\partial x}{\partial y}(t, y, \eta)\right| \geq \delta, \quad t \in]-T, T[. \tag{6.121}$$

Observe that each component of χ_t is a function in $C^\infty(]-T, T[, \Gamma^1(\mathbb{R}^{2d}))$, positively homogeneous of degree 1 for (y, η) large. Moreover, using (6.115), we observe that the same holds, in fact, for every $t \in \mathbb{R}$.

For $t \in]-T, T[$, the phase function $\Phi(t, \cdot)$ above is a *tame* phase, and similarly for the canonical transformation χ_t, according to Definition 6.1.1.

Note that in this general context we have no assumption of homogeneity for large (x, η), nevertheless, the mapping χ is well defined by the global inverse function theorem; moreover, χ is a smooth bi-Lipschitz canonical transformation and satisfies, for $(x, \xi) = \chi(y, \xi)$,

$$|\partial_z^\alpha x_i(z)| + |\partial_z^\alpha \xi_i(z)| \le C_\alpha, \quad |\alpha| \ge 1,\ z = (y, \eta),\ i = 1, \dots, d. \tag{6.122}$$

Finally, the mapping χ fulfills condition (6.11), which allows us to uniquely determine (up to a constant) the related tame phase function Φ_χ (see the comments after Definition 6.1.1).

We shall refine and apply results for tame canonical transformations in Definition 6.1.1 to the special case of the canonical transformations coming from (6.113). First, we need to show a few properties of the class of FIOs introduced in Definition 6.3.4 which occur in this study.

Lemma 6.6.10. *For $s > 2d$, $T^{(i)} \in \mathrm{FIO}(\chi_i, s)$, $i = 1, 2$, the continuous Gabor matrix of the composition $T^{(1)}T^{(2)}$ is controlled by*

$$|\langle T^{(1)}T^{(2)}\pi(w)g, \pi(z)g\rangle| \le C_0 C_1 C_2 \langle z - \chi_1 \circ \chi_2(w)\rangle^{-s}, \quad w, z \in \mathbb{R}^{2d}, \tag{6.123}$$

where $C_i > 0$ is the constant of $T^{(i)}$ in (6.77), $i = 1, 2$, whereas $C_0 > 0$ depends only on s and on the Lipschitz constants of χ_1 and χ_1^{-1}.

Proof. Consider $g \in \mathcal{S}(\mathbb{R}^d)$ with $\|g\|_2 = 1$. We write the product $T^{(1)}T^{(2)}$ as

$$T^{(1)}T^{(2)} = V_g^* V_g T^{(1)}T^{(2)} V_g^* V_g = V_g^*(V_g T^{(1)} V_g^*)(V_g T^{(2)} V_g^*)V_g.$$

Thus the composition of operators corresponds to the multiplication of their (continuous) Gabor matrices. Using the decay estimates for the continuous Gabor matrices of $T^{(i)}$, $i = 1, 2$,

$$\begin{aligned}
|\langle T^{(1)}T^{(2)}\pi(w)g, \pi(z)g\rangle| &\le \int_{\mathbb{R}^{2d}} |\langle T^{(1)}\pi(w)g, \pi(y)g\rangle||\langle T^{(2)}\pi(y)g, \pi(z)g\rangle|\, dy \\
&\le C_1 C_2 \int_{\mathbb{R}^{2d}} \langle z - \chi_1(y)\rangle^{-s}\langle y - \chi_2(w)\rangle^{-s}\, dy \\
&\le C_1 C_2 C(\chi_1) \int_{\mathbb{R}^{2d}} \langle \chi_1^{-1}(z) - y\rangle^{-s}\langle y - \chi_2(w)\rangle^{-s}\, dy \\
&\le C_1 C_2 C(\chi_1) C_s \langle \chi_1^{-1}(z) - \chi_2(w)\rangle^{-s} \\
&\le C_1 C_2 C(\chi_1) C_s C(\chi_1^{-1}) \langle z - \chi_1 \circ \chi_2(w)\rangle^{-s}
\end{aligned} \tag{6.124}$$

for every $z, w \in \mathbb{R}^{2d}$, $s > 2d$, where C_1 and C_2 are the controlling constants in (6.77) of the operators $T^{(1)}$ and $T^{(2)}$, and the bi-Lipschitz property of χ_1 gives

$$\langle z - \chi_1(y)\rangle^{-s} \leq C(\chi_1)\langle \chi_1^{-1}(z) - y\rangle^{-s}, \quad \forall y, z \in \mathbb{R}^{2d},$$

and

$$\langle \chi_1^{-1}(y) - z\rangle^{-s} \leq C(\chi_1^{-1})\langle y - \chi_1(z)\rangle^{-s}, \quad \forall y, z \in \mathbb{R}^{2d}.$$

Furthermore, we used that v_{-s} is subconvolutive for $s > 2d$, i. e., $v_{-s} * v_{-s} \leq C_s v_{-s}$ [160, Lemma 11.1.1(d)]. If we call $C_0 = C(\chi_1)C_s C(\chi_1^{-1})$, the claim is proved. □

By induction, we immediately obtain

Corollary 6.6.11. *For $n \in \mathbb{N}$, $n \geq 2$, $s > 2d$, $T^{(i)} \in FIO(\chi_i, s)$, $i = 1, \ldots, n$, we have*

$$|\langle T^{(1)} T^{(2)} \cdots T^{(n)} \pi(w)g, \pi(z)g\rangle| \leq C_0 C_1 \cdots C_n \langle z - \chi_1 \circ \chi_2 \circ \cdots \circ \chi_n(w)\rangle^{-s}, \tag{6.125}$$

where C_0 depends on s and on the Lipschitz constants of the mappings

$$\chi_1, \chi_1^{-1}, \chi_1 \circ \chi_2, (\chi_1 \circ \chi_2)^{-1}, \ldots, \chi_1 \circ \chi_2 \circ \cdots \circ \chi_{n-1}, (\chi_1 \circ \chi_2 \circ \cdots \circ \chi_{n-1})^{-1}.$$

Observe that using Schur's test and the same techniques as in the proof of Theorem 6.3.7, it is straightforward to obtain the following weighted version of Theorem 6.3.7. We omit the proof.

Theorem 6.6.12. *Let $0 \leq r < s - 2d$, and $\mu \in \mathcal{M}_{v_r}(\mathbb{R}^{2d})$. For every $1 \leq p \leq \infty$, $T \in FIO(\chi, s)$ extends to a continuous operator from $M^p_{\mu\circ\chi}(\mathbb{R}^d)$ into $M^p_\mu(\mathbb{R}^d)$.*

Let us underline that $\mu \circ \chi \in \mathcal{M}_{v_r}(\mathbb{R}^{2d})$, since $v_r \circ \chi \asymp v_r$, due to the bi-Lipschitz property of χ.

The characterization below is written for pseudodifferential operators in the Kohn–Nirenberg form $\sigma(x, D)$, but it works the same for any τ-form (in particular, Weyl form $\sigma^w(x, D)$) in which a pseudodifferential operator is written.

Gluing together the results of Theorems 6.4.9 and 6.4.10, we observe that the constant C in (6.77) satisfies

$$C \asymp \|\sigma\|_{M^\infty_{1\otimes v_s}}. \tag{6.126}$$

For $\chi = \mathrm{Id}$, we recapture the characterization for pseudodifferential operators of Theorem 6.3.3.

Since we shall apply our results to the mappings $\chi_t(x, \eta)$ coming from the Hamilton–Jacobi system (6.113), we need to be more precise in the estimate (6.126): it is important to see how the constants involved in the equivalence depend on the time variable t. This amounts to rewriting the proofs of the results cited above for the special case of a phase function $\Phi \in C^\infty(]-T, T[, \Gamma^2(\mathbb{R}^{2d}))$ and following the time variable t. We state the result here, sketching the main points of the proofs, and leaving the details to the interested reader.

Theorem 6.6.13. *Consider $g \in \mathcal{S}(\mathbb{R}^d)$, $s \geq 0$, and $T > 0$ such that in $]-T, T[$ equation (6.117) is solved by the tame phase $\Phi \in \mathcal{C}^\infty(]-T,T[, \Gamma^2(\mathbb{R}^{2d}))$. Let χ_t be the related tame canonical transformation in (6.119). Let I be a continuous linear operator $\mathcal{S}(\mathbb{R}^d) \to \mathcal{S}'(\mathbb{R}^d)$. Then the following statements are equivalent:*
(i) *$I = I(\sigma_t, \Phi_{\chi_t})$ is a FIO of type I for some $\sigma_t \in M^\infty_{1\otimes v_s}(\mathbb{R}^{2d})$ such that*

$$\|\sigma_t\|_{M^\infty_{1\otimes v_s}} \leq H(t) \in \mathcal{C}(]-T,T[). \tag{6.127}$$

(ii) *$I \in \mathrm{FIO}(\chi_t, s)$ and the constant $C = C(t)$ in (6.77) is in $\mathcal{C}(]-T,T[)$.*

Proof. First we prove (ii) $\Longrightarrow$ (i). Assume $I \in \mathrm{FIO}(\chi_t, s)$ and

$$|\langle I\pi(w)g, \pi(z))g\rangle| \leq C(t)\langle z - \chi_t(w)\rangle, \tag{6.128}$$

with $C(t)$ positive continuous function on $]-T, -T[$. Setting $w = (x, \eta)$ and $z = (x', \eta')$, using the fact that each component of the mapping $\chi_t(y, \eta)$ and its inverse is in $\mathcal{C}^\infty(]-T, T[, \Gamma^1(\mathbb{R}^{2d}))$, we can control the Lipschitz constants of χ_t and χ_t^{-1} by continuous constants of t so that the equivalence of Lemma 6.4.2 becomes

$$|\nabla_x\Phi(t, x', \eta) - \eta'| + |\nabla_\eta\Phi(x', \eta) - x| \asymp_t |\chi_1(t, x, \eta) - x'| + |\chi_2(t, x, \eta) - \eta'|$$

for every $x, x', \eta, \eta' \in \mathbb{R}^d$ and the implicit constants in the equivalence $\asymp_t$ are continuous with respect to $t \in]-T, T[$. This reduces the study to showing that if the operator $I(\sigma_t, \Phi(t,\cdot))$, with $\Phi(t,\cdot)$ being the phase related to χ_t in (6.119) and satisfying (6.117), fulfills the estimate

$$|\langle I(\sigma_t, \Phi(t,\cdot))\pi(x,\eta)g, \pi(x',\eta')g\rangle| \leq C(t)\langle \nabla_x\Phi(t,x',\eta) - \eta', \nabla_\eta\Phi(t,x',\eta) - x\rangle^{-s} \tag{6.129}$$

with $x, x', \eta, \eta' \in \mathbb{R}^d$, $t \in]-T, T[$, then

$$\|\sigma_t\|_{M^\infty_{1\otimes v_s}} \leq C(t), \quad t \in]-T, T[. \tag{6.130}$$

For $z, w \in \mathbb{R}^{2d}$, let $\Phi_{2,z}(t,\cdot)$ be the remainder in the second order Taylor expansion of the phase $\Phi(t,\cdot)$, i. e.,

$$\Phi_{2,z}(t, w) = 2\sum_{|\alpha|=2} \int_0^1 (1-\tau)\partial^\alpha\Phi(t, z + \tau w)\, d\tau \frac{w^\alpha}{\alpha!}.$$

For a given window $g \in \mathcal{S}(\mathbb{R}^d)$, we set

$$\Psi_z(t, w) = e^{2\pi i\Phi_{2,z}(t,w)}(\overline{g} \otimes \widehat{g})(w).$$

Then the fundamental relation between the Gabor matrix of an FIO and the STFT of its symbol in (6.35) can be rephrased in this framework as

$$|\langle I\pi(x,\eta)g, \pi(x',\eta')g\rangle| = |V_{\Psi_{(x',\eta)}}\sigma_t((x',\eta), (\eta' - \nabla_x\Phi(t,x',\eta), x - \nabla_\eta\Phi(t,x',\eta)))|.$$

Writing $u = (x', \eta)$, $v = (\eta', x)$, (6.129) translates into

$$|V_{\Psi_u(t,\cdot)}\sigma_t(u, v - \nabla\Phi(t,u))| \le C(t)\langle v - \nabla\Phi(t,u)\rangle^{-s},$$

and then into the estimate

$$\sup_{(u,w)\in\mathbb{R}^{2d}\times\mathbb{R}^{2d}} \langle w\rangle^s |V_{\Psi_u(t,\cdot)}\sigma_t(u,w)| \le C(t). \tag{6.131}$$

We have already shown for the time independent case $\Psi_u(t,\cdot) = \Psi_u(\cdot)$ that the set of windows Ψ_u possesses a joint time–frequency envelope, cf. Proposition 6.4.8. This property allows writing $\sigma \in M^\infty_{1\otimes v_s}(\mathbb{R}^{2d})$ if and only if $\sup_{u\in\mathbb{R}^{2d}} |V_{\Psi_u}\sigma| \in L^\infty_{1\otimes v_s}(\mathbb{R}^{4d})$ with

$$\|\sigma\|_{M^\infty_{1\otimes v_s}} \asymp \Big\| \sup_{u\in\mathbb{R}^{2d}} |V_{\Psi_u}\sigma| \Big\|_{L^\infty_{1\otimes v_s}} \tag{6.132}$$

(cf. Proposition 6.4.8 (ii)). The proof of the previous equivalence passes through several lemmas. We point out that the crucial element of the equivalence is a control of $|V_\Psi e^{2\pi i\Phi_{2,z}}(u,w)|$, with $\Psi \in \mathcal{S}(\mathbb{R}^{2d}) \setminus \{0\}$ fixed, by a polynomial $p_\alpha(\partial\Phi_{2,z}(\zeta))$ of derivatives of $\Phi_{2,z}$ of degree at most $|\alpha|$ times a factor that does not depend on t. Since $\Phi(t,\cdot) \in C^\infty(]-T,T[,\Gamma^2(\mathbb{R}^{2d}))$, we can control the polynomial by a continuous function of t and in the end obtain that the equivalence (6.132) depends continuously on t, which together with (6.131) gives (6.130).

(i) $\Longrightarrow$ (ii) If $I = I(\sigma_t, \Phi_\chi)$ is an FIO of type I for $\Phi(t,\cdot)$ and χ_t in (6.119) and some $\sigma_t \in M^\infty_{1\otimes v_s}(\mathbb{R}^{2d})$ which satisfies (6.127), then essentially reading backwards the arguments above gives $I(\sigma_t, \Phi_\chi) \in FIO(\chi_t, s)$ with $C(t)$ being a continuous function of t. □

6.6.2 Unperturbed Schrödinger equations

The previous theory applies in the study of the Cauchy problem for linear Schrödinger equations. First, consider the unperturbed case

$$\begin{cases} i\partial_t u + Au = 0, \\ u(0,x) = u_0(x), \end{cases} \tag{6.133}$$

with $x \in \mathbb{R}^d$, $u_0 \in \mathcal{S}(\mathbb{R}^d)$. The operator $A = a(x,D) \in G^{2,cl}$ is a formally self-adjoint pseudodifferential operator in the Kohn–Nirenberg form. This means that the symbol $a \in \Gamma^{2,cl}(\mathbb{R}^{2d})$ has the expansion

$$a(x,\xi) \sim \sum_{j=0}^{\infty} a_{2-j}(x,\xi),$$

where the principal symbol $a_2(x,\xi)$ is real-valued, since A is self-adjoint. The problem (6.133) is forward and backward well-posed in $\mathcal{S}(\mathbb{R}^d)$ and the corresponding evolution operator e^{itA}, acting from $\mathcal{S}(\mathbb{R}^d)$ into $\mathcal{S}(\mathbb{R}^d)$, extends to L^2-isometries [187].

The classical evolution (6.113) has the solution $(x(t),\xi(t)) = \chi_t(y,\eta)$ in (6.116) and, for a suitable $T > 0$ and $t \in]-T,T[$, the evolution operator e^{itA} can be well approximated by an FIO of type I, as expressed in [187, Proposition 3.1.1] for the special case of elliptic operators (that is, operators whose corresponding principal symbols satisfy (6.112)), but still valid without the assumption (6.112), as observed in [81, Section 5.3]. In our framework the result [187, Proposition 3.1.1] can be rephrased as follows.

Proposition 6.6.14. *Given the Cauchy problem* (6.133) *with* $a(x,D)$ *as above, then there exists a* $T > 0$, *a symbol* $\sigma(t,x,\eta) \in \mathcal{C}^\infty(]-T,T[,\Gamma^0(\mathbb{R}^{2d}))$, *and a real-valued phase function* $\Phi \in \mathcal{C}^\infty(]-T,T[,\Gamma^2(\mathbb{R}^{2d}))$ *satisfying* (6.117) *and* (6.118) *such that the evolution operator can be written as*

$$(e^{itA}u_0)(t,x) = (F_t u_0)(t,x) + (R_t u_0)(t,x), \tag{6.134}$$

where F_t *is the FIO of type I*

$$(F_t u_0)(t,x) = \int_{\mathbb{R}^d} e^{2\pi i \Phi(t,x,\eta)} \sigma(t,x,\eta)\widehat{u_0}(\eta)\, d\eta \tag{6.135}$$

and the operator R_t *has kernel in* $\mathcal{C}^\infty(]-T,T[,\mathcal{S}(\mathbb{R}^{2d}))$ *(thus* R_t *is regularizing, i. e.,* $R_t : \mathcal{S}'(\mathbb{R}^d) \to \mathcal{S}(\mathbb{R}^d)$*).*

This result says that in an interval $]-T,T[$ the propagator e^{itA} can be represented by a type I FIO F_t up to an error, which, however, is a regularizing operator.

Remark 6.6.15. We observe that the function $\Phi(t,\cdot)$ of Proposition (6.6.14) and the related canonical transformation χ_t in (6.116) are tame, with Lipschitz constants of χ_t and its inverse that can be controlled by a continuous function of t on the interval $]-T,T[$ and so can be chosen uniform with respect to t on $]-T,T[$.

We will show that if we replace the type I FIO F_t by a more general operator in the classes $\mathrm{FIO}(\chi_t,s)$, we will be able to remove the error R_t in (6.134). Precisely, we can state the following result.

Proposition 6.6.16. *Under the assumptions of Proposition* 6.6.14, *we have*

$$e^{itA} \in \cap_{s\geq 0}\mathrm{FIO}(\chi_t,s), \quad t \in]-T,T[,$$

where χ_t *is defined in* (6.116). *Moreover, for every* $s \geq 0$, *there exists* $C(t) = C_s(t) \in \mathcal{C}(]-T,T[)$ *such that, for every* $g \in \mathcal{S}(\mathbb{R}^d)$, *the Gabor matrix satisfies*

$$|\langle e^{itA}\pi(w)g,\pi(z)g\rangle| \leq C(t)\langle z-\chi_t(w)\rangle^{-s}, \quad w,z \in \mathbb{R}^{2d}. \tag{6.136}$$

Proof. By Proposition 6.6.14, there exists a $T > 0$ such that the evolution e^{itA} can be written as (6.134), where F_t is a type I FIO with symbol in $\sigma(t,x,\eta) \in \mathcal{C}^\infty(]-T,T[,\Gamma^0)$ and phase $\Phi(t,\cdot)$ in (6.117). Since

$$\mathcal{C}^\infty(]-T,T[,\Gamma^0) \subset \mathcal{C}^\infty(]-T,T[,S^0_{0,0}) = \mathcal{C}^\infty\Big(]-T,T[,\bigcap_{s\geq 0} M^\infty_{1\otimes v_s}\Big),$$

we can find $C > 0$ and $k_1 = k_1(s) \in \mathbb{N}$ such that

$$\|\sigma(t,\cdot)\|_{M^{\infty}_{1\otimes v_s}} \le C \sum_{|\alpha|\le k_1} \|\partial^{\alpha}\sigma(t,\cdot)v_{|\alpha|}\|_{\infty}$$

where $\sum_{|\alpha|\le k_1} \|\partial^{\alpha}\sigma(t,\cdot)v_{-|\alpha|}\|_{\infty} \in \mathcal{C}(]-T,T[)$ by assumption. Hence the characterization of Theorem 6.6.13 gives $F_t \in \cap_{s\ge 0}\mathrm{FIO}(\chi_t, s)$, $t \in\,]-T,T[$, where the canonical transformation χ_t is defined in (6.116) and related with $\Phi(t,\cdot)$ by (6.119) and

$$|\langle F_t\pi(w)g, \pi(z)g\rangle| \le C(t)\langle z-\chi_t(w)\rangle^{-s}$$

with $C(t) \in \mathcal{C}(]-T,T[)$.

Fix now $g \in \mathcal{S}(\mathbb{R}^d)$ with $\|g\|_2 = 1$ so that the inversion formula (1.56) becomes $\mathrm{Id} = V_g^* V_g$ and we can write $R_t = V_g^* V_g R_t V_g^* V_g$. Since R_t is a regularizing operator, for $T_t := V_g R_t V_g^*$, the following diagram is commutative:

$$\begin{array}{ccc} \mathcal{S}'(\mathbb{R}^{2d}) & \xrightarrow{\;T_t\;} & \mathcal{S}(\mathbb{R}^{2d}) \\ \big\downarrow V_g^* & & \big\uparrow V_g \\ \mathcal{S}'(\mathbb{R}^{d}) & \xrightarrow{\;R_t\;} & \mathcal{S}(\mathbb{R}^{d}) \end{array}$$

This means that the linear operator $T_t : \mathcal{S}'(\mathbb{R}^{2d}) \to \mathcal{S}(\mathbb{R}^{2d})$ is regularizing as well, and so its kernel $k_t(w,z) = \langle R_t\pi(w)g, \pi(z)g\rangle \in \mathcal{S}(\mathbb{R}^{4d})$ satisfies

$$|k_t(w,z)| = |\langle R_t\pi(w)g, \pi(z)g\rangle| \le \langle z\rangle^{-N}\langle w\rangle^{-N}, \quad \forall z, w \in \mathbb{R}^{2d},\ \forall N \in \mathbb{N}. \tag{6.137}$$

The previous estimates yield $R_t \in \mathrm{FIO}(\chi, s)$, for every bi-Lipschitz mapping χ and every $s \ge 0$. Indeed,

$$\langle z-\chi(w)\rangle \le \langle z\rangle\langle\chi(w)\rangle \asymp \langle z\rangle\langle w\rangle,$$

and, choosing $N \ge s$ in (6.137), we obtain

$$|k_t(w,z)| \lesssim \langle z\rangle^{-N}\langle w\rangle^{-N} \lesssim \langle z-\chi(w)\rangle^{-s}.$$

Finally, if $\sigma(R_t)(z)$ is the Kohn–Nirenberg symbol of R_t, using the fact that $\mathcal{S}(\mathbb{R}^{2d}) \subset S^0_{0,0}$ with continuous embedding for every $s \ge 0$, we find $C > 0$ and $k_2 \in \mathbb{N}$ such that

$$\|\sigma(R_t)\|_{M^{\infty}_{1\otimes v_s}} \le C \sum_{|\alpha+\beta|\le k_2} \|z^{\alpha}\partial_z^{\beta}\sigma(R_t)(z)\|_{\infty} \in \mathcal{C}(]-T,T[).$$

Using Theorem 6.6.13 with χ_t in (6.116), which is tame for $t \in\,]-T,T[$, we find $C(t) \in \mathcal{C}(]-T,T[)$ such that

$$|\langle R_t\pi(w)g, \pi(z)g\rangle| \le C(t)\langle z-\chi_t(w)\rangle^{-s}.$$

Finally, the claim follows since $\mathrm{FIO}(\chi, s)$ are linear spaces

$$\begin{aligned}|\langle e^{itA}\pi(w)g, \pi(z)g\rangle| &\le |\langle F_t\pi(w)g, \pi(z)g\rangle| + |\langle R_t\pi(w)g, \pi(z)g\rangle| \\ &\le C(t)\langle z - \chi_t(w)\rangle^{-s},\end{aligned}$$

which gives (6.136). □

The previous proposition gives an approximation of e^{itA} for $|t| < T$. Using the group property of the propagator e^{itA} Helffer in [187, page 139] describes how to obtain an approximation of e^{itA} for every $t \in \mathbb{R}$. Indeed, a classical trick, jointly with the group property of e^{itA}, applies. We consider $T_0 < T/2$ and define

$$I_h =]hT_0, (h+2)T_0[, \quad h \in \mathbb{Z}.$$

For $t \in I_h$, by the group property of e^{itA},

$$e^{itA} = e^{i(t-hT_0)A}\big(e^{i(hT_0)A/|h|}\big)^{|h|}, \tag{6.138}$$

and, using Proposition 6.6.14, one can write

$$e^{itA} - F_{t-hT_0}(F_{\frac{h}{|h|}T_0})^{|h|} \in \mathcal{C}^{\infty}(I_h, \mathcal{L}(\mathcal{S}', \mathcal{S})). \tag{6.139}$$

In general, e^{itA} or even the composition $F_{t-hT_0}(F_{\frac{h}{|h|}T_0})^{|h|}$ cannot be represented as a type I FIO in the form (6.135). We shall prove below that the evolution e^{itA} is in the class $\bigcap_{s\ge 0}\mathrm{FIO}(\chi_t, s)$ for every $t \in \mathbb{R}$, with χ defined in (6.116), so that this class is proven to be the right framework for describing the evolution e^{itA}.

Theorem 6.6.17. *Given the Cauchy problem* (6.133) *with* $A = a(x, D)$ *as above, consider the mapping* χ_t *defined in* (6.116). *Then*

$$e^{itA} \in \bigcap_{s\ge 0}\mathrm{FIO}(\chi_t, s), \quad t \in \mathbb{R},$$

and for every $s > 2d$ *there exists* $C(t) \in \mathcal{C}(\mathbb{R})$ *such that*

$$|\langle e^{itA}\pi(w)g, \pi(z)g\rangle| \le C(t)\langle z - \chi_t(w)\rangle^{-s}, \quad w, z \in \mathbb{R}^{2d},\ t \in \mathbb{R}. \tag{6.140}$$

Proof. We fix $T_0 < T/2$ as above. For $t \in \mathbb{R}$, there exists an $h \in \mathbb{Z}$ such that $t \in I_h$. Using Proposition 6.6.16, for $t_1 = t - hT_0 \in]-T, T[$, we have that $e^{it_1A} \in \mathrm{FIO}(\chi_{t_1}, s)$ and, for $t_2 = \frac{h}{|h|}T_0 \in]-T, T[$, $e^{it_2A} \in \mathrm{FIO}(\chi_{t_2}, s)$, for every $s \ge 0$, and there exists a continuous function $C(t)$ on $]-T, T[$ such that (6.136) is satisfied for $t = t_1$ and $t = t_2$. Using the algebra property (6.123), for every $s > 2d$,

$$e^{it_1A}(e^{it_2A})^{|h|} \in \mathrm{FIO}(\chi_{t_1} \circ (\chi_{t_2})^{|h|}, s),$$

and the group law (6.115) for $\chi_t(y,\eta) = S_0(t)(y,\eta)$ gives

$$\chi_{t_1} \circ (\chi_{t_2})^{|h|} = \chi_{t_1+|h|t_2} = \chi_t,$$

as expected, hence, using (6.125), we obtain that the Gabor matrix of the product $e^{it_1A}(e^{it_2A})^{|h|}$ is controlled by a continuous function $C_h(t)$ on I_h. Finally, from the estimate

$$|\langle e^{itA}\pi(w)g, \pi(z)g\rangle| \le C_h(t)\langle z - \chi_t(w)\rangle^{-s}, \quad t \in I_h,$$

with $C_h \in \mathcal{C}(I_h)$, it is easy to construct a new continuous controlling function $C(t)$ on $\mathbb{R}$ such that (6.140) is satisfied. □

6.6.3 Schrödinger equations with bounded perturbations

We now study the Cauchy problem for linear Schrödinger equations of the type

$$\begin{cases} i\frac{\partial u}{\partial t} + Hu = 0, \\ u(0,x) = u_0(x), \end{cases} \tag{6.141}$$

with $t \in \mathbb{R}$ and the initial condition $u_0 \in \mathcal{S}(\mathbb{R}^d)$. We consider a Hamiltonian of the form

$$H = a(x,D) + \sigma(x,D), \tag{6.142}$$

where $A = a(x,D)$ is the pseudodifferential operator satisfying (6.133), whose corresponding propagator $e^{itA} \in \bigcap_{s\ge0} \mathrm{FIO}(\chi_t, s)$, for $t \in \mathbb{R}$, as shown in the preceding subsection.

The perturbation $B = \sigma(x,D)$ is a pseudodifferential operator with a symbol $\sigma \in M^{\infty}_{1\otimes v_s}(\mathbb{R}^{2d})$, $s > 2d$. This last requirement implies the boundedness of B on $M_\mu(\mathbb{R}^d)$ for a weight μ as in the assumptions of Theorem 6.6.12 (with $\chi = \mathrm{Id}$) (see also [162] using $M^{\infty}_{1\otimes v_s}(\mathbb{R}^{2d}) \subset M^{\infty,1}(\mathbb{R}^{2d})$, $s > 2d$), and in particular on $L^2(\mathbb{R}^d)$. Hence, $H = A + B$ is a bounded perturbation of the generator A of a unitary group by [249], and H is the generator of a well-defined (semi)group. We shall heavily use the theory of operator semigroups, referring to the textbooks [249] and [116] for an introduction on the topic. Our result, containing Theorem 6.6.1, is as follows.

Theorem 6.6.18. *Let $s > 2d$. Consider the Cauchy problem* (6.141) *with $A = a(x,D)$ and $B = \sigma(x,D)$ as above. Let χ_t be the mapping defined in* (6.116). *Then the solution can be written as*

$$e^{itH} = e^{itA}Q(t) = \tilde{Q}(t)e^{itA} \in \mathrm{FIO}(\chi_t, s), \quad t \in \mathbb{R},$$

where $Q(t)$ and $\tilde{Q}(t)$ are pseudodifferential operators with symbols in $M^{\infty}_{1\otimes v_s}(\mathbb{R}^{2d})$ and the continuous Gabor matrix satisfies

$$|\langle e^{itH}\pi(w)g, \pi(z)g\rangle| \leq C(t)\langle z - \chi_t(w)\rangle^{-s}, \quad w, z \in \mathbb{R}^{2d},$$

for a suitable positive continuous function $C(t)$ on $\mathbb{R}$.

Proof. The pattern is similar to [69, Theorem 4.1]. We show the result on the interval $[0, +\infty[$, for the interval $]-\infty, 0]$ the result is obtained by the previous case by replacing t with $-t$.

The operator A is the generator of a strongly continuous one-parameter group on $L^2(\mathbb{R}^d)$ and $T(t) = e^{itA}$ is the corresponding (semi)group that solves the evolution equation $i\frac{dT(t)}{dt} = AT(t)$. Then e^{itA} is a strongly continuous one-parameter group on $L^2(\mathbb{R}^d)$. As already observed, by the assumptions on the symbol of B, it follows that B is a bounded operator on $L^2(\mathbb{R}^d)$, hence $H = A + B$ is the generator of a strongly continuous one-parameter group $S(t)$ [116]. The perturbed semigroup $S(t) = e^{itH}$ satisfies an abstract Volterra equation

$$S(t)f = T(t)f + \int_0^t T(t-s)BS(s)f\,ds = T(t)\left(\mathrm{Id} + \int_0^t T(-s)BT(s)T(-s)S(s)\,ds\right)f \tag{6.143}$$

for every $f \in L^2(\mathbb{R}^d)$ and $t \geq 0$. If we define $Q(t) = T(-t)S(t)$, then, by (6.143), $Q(t)$ satisfies the Volterra equation

$$Q(t) = \mathrm{Id} + \int_0^t T(-s)BT(s)Q(s)\,ds, \tag{6.144}$$

where the integral is to be understood in the strong sense. Now write $B(s) = T(-s)BT(s)$, then the solution of (6.144) can be written as a so-called *Dyson–Phillips expansion* ([249, X.69] or [116, Chapter 3, Theorem 1.10])

$$Q(t) = \mathrm{Id} + \sum_{n=1}^{\infty}(-i)^n \int_0^t\int_0^{t_1}\cdots\int_0^{t_{n-1}} B(t_1)B(t_2)\cdots B(t_n)\,dt_1\cdots dt_n := \sum_{n=0}^{\infty} Q_n(t). \tag{6.145}$$

We shall show that $Q(t)$ is a pseudodifferential operator with symbol in $M^{\infty}_{1\otimes v_s}(\mathbb{R}^{2d})$. For $\tau \in [0, t]$, the algebra property (6.123) gives

$$B(\tau) = e^{i(-\tau)A}Be^{i\tau A} \in \mathrm{FIO}(\chi_{-\tau} \circ \mathrm{Id} \circ \chi_\tau, s) = \mathrm{FIO}(\mathrm{Id}, s)$$

since $\chi_{-\tau} \circ \mathrm{Id} \circ \chi_\tau = \chi_{-\tau} \circ \chi_\tau = S_0(0) = \mathrm{Id}$ by (6.115). Moreover, $e^{i(\pm\tau)A}$ satisfies (6.140), so that, using (6.125) with $n = 3$, $T^{(1)} = e^{i(-\tau)A}$, $T^{(2)} = B$, $T^{(3)} = e^{i\tau A}$, and $\chi = \mathrm{Id}$, we can write

$$|\langle B(\tau)\pi(w)g, \pi(z)g\rangle| \leq C(\tau)\langle z - w\rangle^{-s}, \tag{6.146}$$

for a new continuous function $C(\tau)$ on $\mathbb{R}$. Using (6.125) again for the composition of pseudodifferential operators $\prod_{j=1}^{n} B(t_j)$, we obtain

$$\left|\left\langle \prod_{j=1}^{n} B(t_j)\pi(w)g, \pi(z)g \right\rangle\right| \leq C_0 C(t_1)\cdots C(t_n)\langle z-w\rangle^{-s},$$

with $C(t) \in \mathcal{C}(\mathbb{R})$ in (6.146).

We now show that $Q_n(t)$ is a pseudodifferential operator with symbol in $M^{\infty}_{1\otimes v_s}(\mathbb{R}^{2d})$. We control the Gabor matrix of $Q_n(t)$ as follows:

$$|\langle Q_n(t)\pi(w)g, \pi(z)g\rangle| \leq C_0 \int_0^t \int_0^{t_1} \cdots \int_0^{t_{n-1}} C(t_1)\cdots C(t_n) dt_1 \cdots dt_n \langle z-w\rangle^{-s}.$$

Defining

$$H(t) = \max_{\tau\in[0,t]} C(\tau) \in \mathcal{C}(\mathbb{R}),$$

we obtain

$$|\langle Q_n(t)\pi(w)g, \pi(z)g\rangle| \leq C_0 H(t)^n \frac{t^n}{n!}\langle z-w\rangle^{-s}.$$

Finally, setting $\tilde{H}(t) = tH(t) \in \mathcal{C}(\mathbb{R})$,

$$\begin{aligned} |\langle Q(t)\pi(w)g, \pi(z)g\rangle| &\leq \sum_{n=0}^{\infty} |\langle Q_n(t)\pi(w)g, \pi(z)g\rangle| \langle z-w\rangle^{-s} \\ &\leq C_0 \sum_{n=0}^{\infty} \frac{\tilde{H}(t)^n}{n!}\langle z-w\rangle^{-s} \\ &= C(t)\langle z-w\rangle^{-s}, \end{aligned}$$

for a new function $C(t) \in \mathcal{C}(\mathbb{R})$. This gives, by Theorem 6.6.13, that $Q(t) \in FIO(\mathrm{Id}, s)$, that is, Q is a pseudodifferential operator with symbol in $M^{\infty}_{1\otimes v_s}(\mathbb{R}^{2d})$. Finally, the algebra property again gives

$$e^{itH} = T(t)C(t) \in FIO(\chi_t, s),$$

and the estimate (6.123) gives that the Gabor matrix of e^{itH} is controlled by a continuous function $C(t)$ on $\mathbb{R}$. □

Consequently, the Schrödinger equation preserves the phase-space concentration, as expressed by the following result.

Corollary 6.6.19. *Let $0 \leq r < s - 2d$, and $\mu \in \mathcal{M}_{v_r}$. If the initial condition $u_0 \in M^p_{\mu\circ\chi}$, $1 \leq p \leq \infty$, then $u(t,\cdot) = e^{itH}u_0 \in M^p_\mu$, for all $t \in \mathbb{R}$.*

Proof. It follows immediately from Theorems 6.6.18 and 6.6.12. □

Using $v_r \circ \chi \asymp v_r$, we observe that the Schrödinger evolution preserves the phase-space concentration $M^p_{v_r}$ of the initial condition u_0. In other words, the time evolution leaves $M^p_{v_r}$ invariant.

Corollary 6.6.20. *Let $|r| < s - 2d$. If the initial condition $u_0 \in M^p_{v_r}$, $1 \le p \le \infty$, then $u(t, \cdot) = e^{itH}u_0 \in M^p_{v_r}$, for all $t \in \mathbb{R}$.*

Proof. The result is a special case of Corollary 6.6.19 once we prove, for $r > 0$, that v_q is v_r-moderate if and only if $|q| \le r$. But this is an easy consequence of Peetre's inequality

$$\langle z + \zeta\rangle^q \le \langle z\rangle^{|q|}\langle \zeta\rangle^q. \qquad \square$$

From Corollaries 6.6.19 and 6.6.20, we recapture (6.104) in Theorem 6.6.3.

6.6.4 Propagation of singularities

We have reached the last topic of this book concerning the Gabor wave front set and its generalization in Definition 6.6.2.

Proposition 6.6.21. *Let $f \in \mathcal{S}'(\mathbb{R}^d)$, $r > 0$. Then*
(i) *The definitions of $\mathrm{WF}^{p,r}_G(f)$ and $\mathrm{WF}_G(f)$ do not depend on the choice of the window g.*
(ii) *$f \in M^p_{v_r}(\mathbb{R}^d)$ if and only if $\mathrm{WF}^{p,r}_G(f) = \emptyset$. Similarly, $f \in \mathcal{S}(\mathbb{R}^d)$ if and only if $\mathrm{WF}_G(f) = \emptyset$.*

The proof of (i) will be given later, as a consequence of more general arguments. The proof of (ii) follows easily from the compactness of the sphere $\mathcal{S}^{2d-1}$ and the characterization of the Schwartz class using modulation spaces in (2.28).

The following statement gives the second part of Theorem 6.6.3.

Theorem 6.6.22. *Under the assumptions of Theorem 6.6.18, for $u_0 \in M^p_{v_{-r}}(\mathbb{R}^d)$, $1 \le p \le \infty$, $0 < 2r < s - 2d$, we have*

$$\mathrm{WF}^{p,r}_G(e^{itH}u_0) = \chi_t(\mathrm{WF}^{p,r}_G(u_0)). \tag{6.147}$$

Proof. We shall prove that $\mathrm{WF}^{p,r}_G(e^{itH}u_0) \subset \chi_t(\mathrm{WF}^{p,r}_G(u_0))$ for any $t \in \mathbb{R}$. Then, by applying the inclusion to $v_0 = e^{-itH}u_0$, the opposite inclusion will follow, and (6.147) will be proved.

Fixing $t \in \mathbb{R}$, we assume $z_0 \notin \chi_t(\mathrm{WF}^{p,r}_G(u_0))$. Since χ_t is a homogeneous diffeomorphism for large $|z|$, this is equivalent to saying that $w_0 = \chi_t^{-1}(z_0)$ does not belong to $\mathrm{WF}^{p,r}_G(u_0)$. Therefore, for a sufficiently small open conic neighborhood $\Gamma_{w_0} \subset \mathbb{R}^{2d} \setminus 0$ of w_0, we have

$$\int_{\Gamma_{w_0}} |V_g u_0(w)|^p \langle w\rangle^{pr} < \infty. \tag{6.148}$$

Note also that, in view of the assumption $u_0 \in M^p_{v_{-r}}(\mathbb{R}^d)$, we have

$$\int_{\mathbb{R}^d} |V_g u_0(w)|^p \langle w\rangle^{-pr} < \infty. \tag{6.149}$$

Now using Theorem 6.6.18, we can write

$$V_g(e^{itH}u_0)(z) = \int_{\mathbb{R}^{2d}} k(t,w,z)V_g u_0(w)\,dw \tag{6.150}$$

with

$$|k(t,w,z)| \le C(t)\langle z - \chi_t(w)\rangle^{-s}, \quad w,z \in \mathbb{R}^{2d}. \tag{6.151}$$

We have to show that $z_0 \notin \mathrm{WF}^{p,r}_G(e^{itH}u_0)$. To this end, take an open conic neighborhood Γ'_{z_0} of z_0, such that $\overline{\Gamma'_{z_0}} \subset \chi_t(\Gamma_{w_0})$. This implies that, for $z \in \Gamma'_{z_0}$ and $w \notin \Gamma_{w_0}$, we have

$$\langle z - \chi_t(w)\rangle \gtrsim \max\{\langle z\rangle, \langle w\rangle\}, \tag{6.152}$$

since χ_t is a Lipschitz diffeomorphism. Using (6.150) and (6.151), we estimate

$$|\langle z\rangle^r V_g(e^{itH}u_0)(z)| \lesssim \int_{\mathbb{R}^{2d}} I(z,w)\,dw, \tag{6.153}$$

with

$$I(z,w) = \langle z\rangle^r \langle z - \chi_t(w)\rangle^{-s} |V_g u_0(w)|. \tag{6.154}$$

To show $z_0 \notin \mathrm{WF}^r_G(e^{itH}u_0)$, it will be sufficient to establish that

$$\Big\| \int_{\mathbb{R}^{2d}} I(\cdot,w)\,dw \Big\|_{L^p(\Gamma'_{z_0})} < \infty.$$

First, we estimate $\int_{\mathbb{R}^{2d}} I(z,w)\,dw$ for $z \in \Gamma'_{z_0}$. We split the domain of integration into two domains Γ_{w_0} and $\mathbb{R}^{2d} \setminus \Gamma_{w_0}$. In $\mathbb{R}^{2d} \setminus \Gamma_{w_0}$, we use (6.152) to obtain

$$\begin{aligned}
\int_{\mathbb{R}^{2d}\setminus\Gamma_{w_0}} I(z,w)\,dw &\le \int_{\mathbb{R}^{2d}\setminus\Gamma_{w_0}} \langle z\rangle^r \langle w\rangle^r \langle z - \chi_t(w)\rangle^{-s} \frac{|V_g u_0(w)|}{\langle w\rangle^r}\,dw \\
&\lesssim \int_{\mathbb{R}^{2d}\setminus\Gamma_{w_0}} \langle z - \chi_t(w)\rangle^{2r-s} \frac{|V_g u_0(w)|}{\langle w\rangle^r}\,dw \\
&\lesssim \Big(\langle\cdot\rangle^{2r-s} * \frac{|V_g u_0(\cdot)|}{\langle\cdot\rangle^r} \Big)(z).
\end{aligned}$$

So by (6.149) and using $2r - s < -2d$,

$$\Big\| \int_{\mathbb{R}^{2d}\setminus\Gamma_{w_0}} I(\cdot, w)\, dw \Big\|_{L^p(\Gamma'_{z_0})} \lesssim \|\langle\cdot\rangle^{2r-s}\|_{L^1(\mathbb{R}^{2d})} \| |V_g u_0| \langle\cdot\rangle^{-r} \|_{L^p(\mathbb{R}^{2d})} < \infty.$$

In the domain Γ_{w_0}, we have

$$\begin{aligned}\int_{\Gamma_{w_0}} I(z,w)\, dw &\leq \int_{\Gamma_{w_0}} \langle z\rangle^r \langle w\rangle^{-r} \langle z - \chi_t(w)\rangle^{-r} \langle z - \chi_t(w)\rangle^{r-s} |V_g u_0(w)| \langle w\rangle^r \, dw \\ &\lesssim \int_{\Gamma_{w_0}} \langle z - \chi_t(w)\rangle^{r-s} |V_g u_0(w)| \langle w\rangle^r \, dw \\ &\lesssim \langle \chi_t^{-1}(\cdot)\rangle^{r-s} * (\mathrm{Char}_{\Gamma_{w_0}} \cdot |V_g u_0| \langle\cdot\rangle^r)(z)\end{aligned}$$

where $\mathrm{Char}_{\Gamma_{w_0}}$ is the characteristic function of the set Γ_{w_0}. The assumption (6.148) yields the estimate

$$\begin{aligned}\Big\| \int_{\Gamma_{w_0}} I(\cdot, w)\, dw \Big\|_{L^p(\Gamma'_{z_0})} &\lesssim \|\langle \chi_t^{-1}(\cdot)\rangle^{r-s}\|_{L^1(\mathbb{R}^{2d})} \| |V_g u_0| \langle\cdot\rangle^r \|_{L^p(\Gamma_{w_0})} \\ &\asymp \|\langle\cdot\rangle^{r-s}\|_{L^1(\mathbb{R}^{2d})} \| V_g u_0 \langle\cdot\rangle^r \|_{L^p(\Gamma_{w_0})} < \infty,\end{aligned}$$

for χ_t is a bi-Lipschitz diffeomorphism and $r - s < 2r - s < -2d$. This concludes the proof. □

The preceding arguments apply with small changes in the proof of (6.108). Let us detail the proof for sake of clarity.

Proof of (6.108). As in the previous proof, it is enough to show $\mathrm{WF}_G(e^{itH} u_0) \subset \chi_t(\mathrm{WF}_G(u_0))$ for any $t \in \mathbb{R}$. We have to prove that for every $u_0 \in \mathcal{S}'(\mathbb{R}^d)$ and $z_0 \in \mathbb{R}^d$, $z_0 \neq 0$, the assumption $z_0 \notin \chi_t(\mathrm{WF}_G u_0)$ implies $z_0 \notin \mathrm{WF}_G(e^{itH} u_0)$. Arguing as before, we have that the estimates (6.148) are satisfied for every $r > 0$ in a cone Γ_{w_0} independent of r. Now recall from (2.28) that $\mathcal{S}'(\mathbb{R}^d) = \bigcup_{s\geq 0} M^\infty_{v_{-s}}(\mathbb{R}^d)$. Therefore, $u_0 \in M^\infty_{v_{-r_0}}(\mathbb{R}^d)$ for some $r_0 \geq 0$. Since $\sigma \in S^0_{0,0} = \bigcap_{s\geq 0} M^\infty_{1\otimes v_s}(\mathbb{R}^{2d})$ by (6.72), we have $\sigma \in M^\infty_{1\otimes v_s}(\mathbb{R}^{2d})$ for every $s \geq 0$. We may then apply the arguments in the preceding proof with $s > 2r + 2d > 2r_0 + 2d$ and obtain the expected estimates (6.107) for any $r > 0$. By observing that the choice of the cone Γ'_{z_0} does not depend on r, the proof is concluded. □

Proof of Proposition 6.6.21 (i). We prove the independence of the definition of $\mathrm{WF}^{p,r}_G(f)$ on the choice of the window g. The independence of $\mathrm{WF}_G(f)$ is attained similarly.

We assume the estimate for $V_g f$ in (6.103) is satisfied for some fixed $g \in \mathcal{S}(\mathbb{R}^d)\setminus\{0\}$ and some conic neighborhood Γ_{z_0}, and we want to prove that the estimate holds for

$V_h f$, where $h \in \mathcal{S}(\mathbb{R}^d) \setminus \{0\}$ is fixed arbitrary, after possibly shrinking Γ_{z_0}. To this end, we use Lemma 1.2.29 which gives

$$|V_h f(z)| \lesssim (|V_g f| * |V_h g|)(z).$$

Since $V_h g \in \mathcal{S}(\mathbb{R}^{2d})$ for $g, h \in \mathcal{S}(\mathbb{R}^d)$, we have that for every $s \geq 0$,

$$|V_h f(z)| \lesssim \int_{\mathbb{R}^{2d}} \langle z - w \rangle^{-s} |V_g f|(w)\, dw.$$

We know that $f \in M^p_{v_{-r_0}}(\mathbb{R}^d)$, for some $r_0 \geq 0$. Taking then $s > \max\{r, r_0 + 2d\}$, the arguments in the proof of Theorem 6.6.22 apply with $\chi_t = \mathrm{Id}$, $w_0 = z_0$. □

Proposition 6.6.23. *Let $\sigma \in M^\infty_{1\otimes v_s}(\mathbb{R}^{2d})$, $s > 2d$ and $0 < 2r < s - 2d$. Then for every $f \in M^p_{v_{-r}}(\mathbb{R}^d)$, we have*

$$\mathrm{WF}^{p,r}_G(\sigma(x, D) f) \subset \mathrm{WF}^{p,r}_G(f). \tag{6.155}$$

If $\sigma \in S^0_{0,0}$, then for every $f \in \mathcal{S}'(\mathbb{R}^d)$,

$$\mathrm{WF}_G(\sigma(x, D) f) \subset \mathrm{WF}_G(f). \tag{6.156}$$

Proof. If $\sigma \in M^\infty_{1\otimes v_s}(\mathbb{R}^{2d})$, then, from Theorems 6.3.3 and 6.3.5, we have that the Gabor matrix $k(w, z)$ of $\sigma(x, D)$ satisfies

$$|k(w, z)| \lesssim \langle z - w \rangle^{-s}, \quad w, z \in \mathbb{R}^{2d},$$

so that $\sigma(x, D) \in \mathrm{FIO}(\chi, s)$ with $\chi = \mathrm{Id}$. The arguments of the proof of the Theorem 6.6.22 then apply with $w_0 = z_0$. The proof of (6.156) is similar. □

We end this subject with some examples of Schrödinger equations.

Addressing first nonexpert readers, we present some properties of $\mathrm{WF}_G(f)$ and treat in this framework the free particle and the harmonic oscillator with smooth potentials, cf. Examples 6.6.25, 6.6.26, and 6.6.27. The final Example 6.6.28 concerns nonsmooth potentials.

Proposition 6.6.24. *Let $f \in \mathcal{S}'(\mathbb{R}^d)$. Then*

(i) $\mathrm{WF}_G(\pi(z_0) f) = \mathrm{WF}_G(f)$ *for every* $z_0 = (x_0, \xi_0) \in \mathbb{R}^{2d}$.

(ii) *Let δ_{x_0} be the Dirac distribution at the point $x_0 \in \mathbb{R}^d$. Then*

$$\mathrm{WF}_G(\delta_{x_0}) = \{z = (x, \xi), x = 0, \xi \neq 0\}$$

independently of x_0.

(iii) *Let ξ_0 be fixed in $\mathbb{R}^d$. Then*

$$\mathrm{WF}_G(e^{2\pi i x \cdot \xi_0}) = \{z = (x, \xi), x \neq 0, \xi = 0\}$$

independently of ξ_0.

(iv) *Let $c \in \mathbb{R}$, $c \neq 0$, be fixed. Then*

$$\mathrm{WF}_G(e^{\pi i c x^2}) = \{z = (x,\xi),\ x \neq 0,\ \xi = cx\}.$$

Proof. The proof of (i) is a consequence of Proposition 6.6.23, since

$$\pi(z_0) = M_{\xi_0} T_{x_0} = \sigma(x,D),$$

with $\sigma(x,D)$ being a pseudodifferential operator with symbol

$$\sigma(x,\xi) = e^{2\pi i(x\cdot\xi_0 - x_0\cdot\xi)} \in S^0_{0,0}.$$

Concerning (ii), we are left to compute $\mathrm{WF}_G(\delta)$ since

$$\mathrm{WF}_G(\delta_{x_0}) = \mathrm{WF}_G(T_{x_0}\delta) = \mathrm{WF}_G(\delta)$$

by claim (i). On the other hand, $V_g(\delta)(x,\xi) = \overline{g(-x)}$. Hence in a small conic neighborhood $\Gamma \subset \mathbb{R}^{2d}$ of the ray $x = t\xi$, $t \in \mathbb{R}$, $\xi \neq 0$, we have rapid decay of $g(-t\xi)$ except for $t = 0$, giving the claim.

To prove (iii), we proceed similarly as before. From claim (i), we obtain that

$$\mathrm{WF}_G(e^{2\pi i x\cdot\xi_0}) = \mathrm{WF}_G(M_{\xi_0}1) = \mathrm{WF}_G(1).$$

On the other hand, $|V_g 1(x,\xi)| = |M_{-x}\hat{g}(-\xi)|$ so that $|V_g 1(x,\xi)| = |\hat{g}(-\xi)|$ and the arguments when proving claim (ii) give the desired result.

We now prove (iv). We use the Gaussian $g(x) = e^{-\pi x^2}$ as a window for the STFT $V_g f$ with $f(x) := e^{\pi i c x^2}$. Then standard computations (see also [14, Theorem 14]) give

$$|V_g f(x,\xi)| = (1+c^2)^{-d/4} e^{-\pi\frac{(\xi - cx)^2}{1+c^2}}.$$

The right-hand side is rapidly decaying in any open cone of $\mathbb{R}^{2d}$ excluding the line $\xi - cx = 0$. This concludes the proof of the proposition. □

Example 6.6.25 (The free particle). Consider the Cauchy problem for the Schrödinger equation in (5.11) (cf. Section 6.2.3.1). Namely,

$$\begin{cases} i\partial_t u + \Delta u = 0, \\ u(0,x) = u_0(x), \end{cases}$$

with $x \in \mathbb{R}^d$, $d \geq 1$. The explicit formula for the solution in terms of the kernel is given by (5.21), with K_t in (5.22). In terms of classical FIO,

$$u(t,x) = \int_{\mathbb{R}^d} e^{2\pi i(x\cdot\eta - 2\pi t\eta^2)}\widehat{u_0}(\eta)\, d\eta,$$

where the Gabor matrix with window function $g(x) = e^{-\pi x^2}$ is controlled by

$$|k(w,z)| \leq Ce^{-\epsilon(z-\chi_t(w))^2},$$

for suitable constants $C > 0$ and $\epsilon > 0$ and where, for $w = (y,\eta)$,

$$(x,\xi) = \chi_t(y,\eta) = (y + 4\pi t\eta, \eta). \tag{6.157}$$

Beside the effectiveness in numerical analysis already detailed in Section 6.2.3.1, this expression emphasizes the microlocal properties of the propagator. Let us test the propagator of the Gabor wave front set on some particular initial data. If $u_0 = \delta$ then $u(t,x) = K_t(x)$ by (5.21). This is coherent with (6.108) and (6.157), since from Proposition 6.6.24, (iv) and (ii), we have

$$\begin{aligned}\mathrm{WF}_G(u(t,x)) &= \mathrm{WF}_G(K_t) = \{(x,\xi),\ x = 4\pi t\xi,\ \xi \neq 0\} \\ &= \chi_t(\mathrm{WF}_G(\delta)) = \chi_t(\{(y,\eta),\ y = 0, \eta \neq 0\}).\end{aligned}$$

We remark a similar propagation for the initial datum

$$u_0 = K_{-1}(t) = (-4\pi i)^{-d/2} e^{-ix^2/4},$$

for which we have $u_{t=1} = \delta$. Instead, for $u_0 = e^{2\pi i x\cdot\xi_0}$, with $\xi_0 \in \mathbb{R}^d$, we have

$$u(t,x) = e^{-4\pi^2 it\xi_0^2} e^{2\pi i x\cdot\xi_0},$$

and in this case the Gabor wave front set is stuck

$$\mathrm{WF}_G(u(t,x)) = \mathrm{WF}_G(u_0) = \{(x,0),\ x \neq 0\},$$

by Proposition 6.6.24 (iii) and (6.157).

Example 6.6.26 (The harmonic oscillator). Consider the Cauchy problem for the harmonic oscillator in (6.68), with solution in terms of an FIO type I detailed in (6.69).

The Gabor matrix with Gaussian window $g(x) = e^{-\pi x^2}$ can be explicitly computed as

$$|k(w,z)| = 2^{-\frac{d}{2}} e^{-\frac{\pi}{2}(z-\chi_t(w))^2}, \tag{6.158}$$

where the canonical transformation is defined in (6.70). Observe that the expression (6.158) is meaningful for every $t \in \mathbb{R}$.

We may test (6.108) on the initial datum $u_0(x) = 1$, giving for $t < \pi/2$,

$$u(t,x) = (\cos t)^{-d/2} e^{\pi i \tan t x^2}.$$

From Proposition 6.6.24, (iii) and (iv), we have coherently with (6.70) that

$$\begin{aligned}\mathrm{WF}_G(u(t,x)) &= \{(x,\xi),\ x = (\cos t)y,\ \xi = (\sin t)y, y \neq 0\} \\ &= \chi_t(\mathrm{WF}_G(1)) = \chi_t(\{(y,\eta),\ y \neq 0, \eta = 0\}).\end{aligned}$$

Example 6.6.27 (Smooth potentials). We now consider the presence in Example 6.6.25 of a potential with symbol in the class $S^0_{0,0}$. Consider the case

$$e^{-2\pi i x_0\cdot\xi}, \quad x_0 \in \mathbb{R}^d \text{ fixed.}$$

The related pseudodifferential operator $\sigma(D)$ is the translation operator

$$\sigma(D)f(x) = T_{x_0}f(x) = f(x - x_0),$$

which does not preserve the singular support. Consider first the equation

$$\begin{cases} i\partial_t u + \sigma(D)u = 0, \\ u(0,x) = u_0(x). \end{cases}$$

The solution is given by

$$u(t,x) = e^{itT_{x_0}}u_0(x) = \int_{\mathbb{R}^d} e^{2\pi i x\cdot\xi}\exp(ite^{-2\pi i x_0\cdot\xi})\widehat{u_0}(\xi)\,d\xi.$$

Despite the nasty oscillations, the symbol of the solution operator belongs to $S^0_{0,0}$ and, from Proposition 6.6.23, we have for every fixed $t \in \mathbb{R}$,

$$\mathrm{WF}_G(e^{itT_{x_0}}u_0) = \mathrm{WF}_G(u_0),$$

the identity being granted by the fact that $T_{x_0}^{-1} = T_{-x_0}$. Note that the singular support can be expanded. In fact, taking $u_0 = \delta$, we have

$$e^{itT_{x_0}}\delta = \sum_{n=0}^{\infty}\frac{(it)^n}{n!}\delta_{nx_0} \in \mathcal{S}'(\mathbb{R}^d)$$

so that $\operatorname{sing\,supp} e^{itT_{x_0}}\delta = \{nx_0\}_{n\in\mathbb{Z}_+}$ as soon as $t \neq 0$, whereas

$$\mathrm{WF}_G(e^{itT_{x_0}}\delta) = \mathrm{WF}_G(\delta) = \{(0,\xi),\ \xi \neq 0\}.$$

Adding now the potential $\sigma(D)$ to the free particle in Example 6.6.25, we have the Schrödinger equation with space-delay

$$\begin{cases} i\partial_t u + \Delta u + T_{x_0}u = 0, \\ u(0,x) = u_0(x). \end{cases}$$

Since the operators $e^{it\Delta}$ and T_{x_0} commute, the arguments of Section 6.6.3 provide as propagator $e^{itT_{x_0}}e^{it\Delta}$, that is, the convolution with

$$\sum_{n=0}^{\infty}\frac{(it)^n}{n!}K_t(x - nx_0) \in \mathcal{S}'(\mathbb{R}^d),$$

where K_t is defined in (5.22). The Gabor propagation is the same as in Example 6.6.25.

From a physical point of view, it is perhaps more natural to consider the case when the potential depends on x alone, for example,

$$\begin{cases} i\partial_t u + \Delta u + M_{\xi_0} u = 0, \\ u(0,x) = u_0(x), \end{cases}$$

with $M_{\xi_0} u_0 = e^{2\pi i x\cdot\xi_0} u_0$, ξ_0 fixed in $\mathbb{R}^d$. Notice that now the operators $e^{it\Delta}$ and M_{ξ_0} do not commute and, proceeding as in Section 6.6.3 with the perturbation $Bu = M_{\xi_0} u$, we have first to consider

$$B(t) = e^{-it\Delta} e^{2\pi i x\cdot\xi_0} e^{it\Delta}.$$

Omitting further explicit computations, we obtain

$$B(t) = e^{4\pi^2 i \xi_0^2 t} M_{\xi_0} T_{-4\pi t\xi_0}. \tag{6.159}$$

In principle, one could then continue the computation of the pseudodifferential operator $Q(t)$ in (6.145) explicitly, and the solution operator will be $e^{it\Delta} Q(t)$.

Observe in (6.159) the presence of the translation factor $T_{4\pi t\xi_0}$, providing the same phenomena as before.

Example 6.6.28 (Nonsmooth potentials). As examples of admissible nonsmooth potentials, consider first a nonpolynomial homogeneous function $h(z)$, $z = (x,\xi)$, $h(\lambda z) = \lambda^r h(z)$ for $z \neq 0$, $\lambda > 0$, $r > 0$, with $h \in C^\infty(\mathbb{R}^{2d} \setminus \{0\})$, and take then as potential any function $\sigma(z) = h(z)$, for $|z| \leq 1$, and $h(z) \in S^0_{0,0}$ for $|z| \geq 1$. This potential satisfies $\sigma \in M^\infty_{1\otimes v_{r+2d}}(\mathbb{R}^{2d})$. In fact, we may limit the analysis to the singularity at the origin. From Proposition 6.6.6 we have, for $\psi \in \mathcal{S}(\mathbb{R}^{2d})$,

$$|V_\psi \sigma(z,\zeta)| \leq C\langle\zeta\rangle^{-r-2d}, \quad z,\zeta \in \mathbb{R}^{2d}. \tag{6.160}$$

We may now return to the discussion about the smoothness at the origin of the Hamiltonian $a(z)$ in the Introduction. Consider $h(z)$ real-valued nonpolynomial homogeneous of degree 2, $h \in C^\infty(\mathbb{R}^{2d} \setminus \{0\})$, just to give an example

$$h(x,\xi) = (|x|^4 + |\xi|^4)^{1/2}.$$

We can include in our analysis the equation

$$\begin{cases} i\partial_t u + h(x,D)u = 0, \\ u(0,x) = u_0(x), \end{cases} \tag{6.161}$$

by absorbing the singularity at the origin into the potential. Namely, take $\varphi \in C_0^\infty(\mathbb{R}^d)$, $0 \leq \varphi(z) \leq 1$, $\varphi(z) = 1$ for $|z| \leq 1$, $\varphi(z) = 0$ for $|z| \leq 2$, and split

$$h(z) = a(z) + \sigma(z), \quad a(z) = (1 - \varphi(z))h(z), \quad \sigma(z) = \varphi(z)h(z).$$

At this moment $a(z)$ satisfies the assumptions in the Introduction and the potential σ belongs to $M^{\infty}_{1\otimes v_{2+2d}}(\mathbb{R}^{2d})$, in view of (6.160). We may then apply Theorem 6.6.3 to the Cauchy problem (6.161). Note that the result of propagation should be limited to $u_0 \in M^p_{v_{-r}}(\mathbb{R}^d)$ and $\mathrm{WF}^{p,r}_G(e^{itH}u_0)$ with $0 < r < 1$.

Finally, we present an example of a nonsmooth potential depending on x alone, namely in dimension $d = 1$,

$$\sigma(x,\xi) = |\sin x|^{\mu}, \quad \mu > 1,\ x,\xi \in \mathbb{R}.$$

By Corollary 6.6.5, $\sigma \in M^{\infty}_{1\otimes v_{\mu+1}}(\mathbb{R}^2)$. So consider, for instance, the perturbed harmonic oscillator in (6.106). From Theorem 6.6.3 we have that the Cauchy problem is well-posed for $u_0 \in M^p_{v_r}(\mathbb{R})$, $|r| < \mu - 2$, and the propagation of $\mathrm{WF}^{p,r}_G(u(t,\cdot))$ for $t \in \mathbb{R}$ takes place as in Example 6.6.26 for $0 < r < \mu/2 - 1$.

For further applications to the Gabor wave front set to regularity problems and propagation of singularities, we refer to [237, 238].

Finally, we underline that time–frequency analysis has been successfully applied to the theory of path integrals. This subject is out of the scope of this book, but it has many challenging open questions. We refer the interested reader to the works [235, 239].

A Appendix

In this section we list some basic results from functional analysis, we have used throughout this book. Most of the topics are from the text [183]. Additional references are the textbooks [37, 57, 146, 160, 254, 268].

Theorem A.0.1 (Minkowski's integral inequality). *Suppose that $(S_1, \mathcal{M}_1, \mu_1)$ and $(S_2, \mathcal{M}_2, \mu_2)$ are two σ-finite measure spaces and $F : S_1 \times S_2 \to \mathbb{R}$ is a $(\mathcal{M}_1 \otimes \mathcal{M}_2)$-measurable function.*
(i) *If $F \geq 0$, $1 \leq p < \infty$, then*

$$\left(\int_{S_1}\left(\int_{S_2} F(x,y)\,d\mu_2(y)\right)^p d\mu_1(x)\right)^{\frac{1}{p}} \leq \int_{S_2}\left(\int_{S_1} F(x,y)^p\,d\mu_1(x)\right)^{\frac{1}{p}} d\mu_2(y).$$

(ii) *If $1 \leq p \leq \infty$, $F(\cdot,y) \in L^p(S_1)$ for a. e. y, and the function $y \to \|F(\cdot,y)\|_p$ is in $L^1(S_2)$, then $F(x,\cdot) \in L^1(S_2)$ for a. e. x, the function $x \mapsto \int_{S_2} F(x,y)\,d\mu_2(y)$ is in $L^p(S_1)$, and*

$$\left\|\int_{S_2} F(\cdot,y)\,d\mu_2(y)\right\|_p \leq \int_{S_2} \|F(\cdot,y)\|_p\,d\mu_2(y).$$

The closed graph theorem provides a convenient means to test whether a linear operator acting between Banach spaces is continuous.

Theorem A.0.2 (Closed graph theorem). *Let X and Y be Banach spaces. If $A : X \to Y$ is linear, then the following are equivalent:*
(i) *A is continuous.*
(ii) *The graph $A = \{(f, Af) : f \in X\}$ is a closed subset of $X \times Y$.*
(iii) *If $f_n \to f$ in X and $Af_n \to g$ in Y, then $g = Af$.*

Theorem A.0.3 (Uniform boundedness principle). *Let X be a Banach space and Y a normed space. If $\{A_i\}$ is a collection of linear bounded operators from X to Y such that*

$$\forall f \in X \quad \sup_i \|A_i f\| < \infty,$$

then

$$\sup_i \|A_i\| < \infty.$$

Theorem A.0.4. *Let H, K be Hilbert spaces and $A \in \mathcal{B}(H,K)$. We denote by $R(A)$ the range of the operator A. Then the following are equivalent:*
(i) *There exists $C > 0$ such that*

$$\|x\|_H \leq C\|Ax\|_K, \quad \forall x \in H. \tag{A.1}$$

https://doi.org/10.1515/9783110532456-008

(ii) $\ker(A) = \{0\}$ *and* $R(A)$ *is closed.*
(iii) $A^* : K \to H$ *is surjective, i. e.,* $R(A^*) = H$.

Proof. (i) $\Longrightarrow$ (ii) If $Ax = 0$ then by (A.1), $x = 0$, that is, $\ker(A) = \{0\}$. Let us prove that $R(A)$ is closed. Consider $\{y_n\} \subset R(A)$ such that $y_n \to y$, then there exists a sequence $\{x_n\} \subset H$ such that $Ax_n = y_n$. By (A.1),

$$\|x_m - x_n\| \le C\|Ax_m - Ax_n\| = \|y_m - y_n\|,$$

so that, since $\{y_n\}$ is a Cauchy sequence, so is $\{x_n\}$. By the completeness of H, $x_n \to x$ and $Ax_n \to Ax$ for A is bounded. So $Ax = y$, as desired.

(ii) $\Longrightarrow$ (iii) Observe that $R(A)$ is closed if and only if $\ker(A)^\perp = R(A^*)$ (cf. the next Definition A.1.8 and [37, Theorem 2.19]). Using the assumption $\ker(A) = \{0\}$, we infer $R(A^*) = H$, that is, A^* is surjective.

(iii) $\Longrightarrow$ (i) Since $\ker(A) = R(A^*)^\perp$ (see [37, Corollary 2.18]), by the surjectivity of A^* we obtain that A is injective. Since $R(A^*)$ is closed also $R(A)$ is so (see [37, Theorem 2.19]). Hence $A : H \to R(A)$ is a topological isomorphism; this means that if we call $A^{-1} : R(A) \to H$ its pseudoinverse, then A^{-1} is bounded, that is, (A.1) holds. □

Recall that a linear bounded self-adjoint operator on a Hilbert space H is called *positive* if

$$\langle Tx, x\rangle \ge 0, \quad \forall x \in H.$$

Lemma A.0.5 (Generalized Cauchy–Schwarz inequality). *If* $T : H \to H$ *is a linear bounded self-adjoint positive operator on a Hilbert space* H*, then*

$$\left|\langle Tx, y\rangle\right|^2 \le \langle Tx, x\rangle\langle Ty, y\rangle, \quad \forall x, y \in H. \tag{A.2}$$

The following result defines the square root of a positive operator (for a proof, see, e. g., [183, Theorem 2.18]).

Theorem A.0.6. *Let* T *be a linear bounded self-adjoint operator on a Hilbert space* H*. If* T *is positive, then there exists a unique positive operator, denoted by* $T^{1/2}$*, such that* $(T^{1/2})^2 = T$*. Moreover,* $T^{1/2}$ *commutes with any operator which commutes with* T*.*

Remark A.0.7. If a positive operator T is a topological isomorphism, then it is easy to check that T^{-1} is a positive operator as well. Hence, by Theorem A.0.6, the operator T^{-1} has a positive square root $T^{-1/2}$. Let us check that $(T^{1/2})^{-1} = T^{-1/2}$. First, observe that, since T and T^{-1} commute, by Theorem A.0.6, also $T^{1/2}$ and $T^{-1/2}$ commute. Denoting by I the identity operator,

$$\begin{aligned} I = TT^{-1} &= (T^{1/2})^2(T^{-1/2})^2 = T^{1/2}T^{-1/2}T^{1/2}T^{-1/2} \\ &= (T^{1/2}T^{-1/2})^2. \end{aligned}$$

Since $I^2 = I$ and by the uniqueness of the square-root (cf. Theorem A.0.6), we deduce $T^{1/2}T^{-1/2} = I$. Analogously, it can be shown that $T^{-1/2}T^{1/2} = I$.

Theorem A.0.8. *If $T \in B(H)$ (that is, T is a bounded linear operator on the Hilbert space H) is a self-adjoint operator, then*

$$\|T\| = \sup_{\|x\|=1} |\langle Tx, x\rangle|.$$

If T is positive then

$$\|T\| = \sup_{\|x\|=1} \langle Tx, x\rangle.$$

Theorem A.0.9 (The density principle). *Let B_1, B_2 be Banach spaces. Let X be a dense subspace of B_1. Assume that $A : X \to B_2$ is a (conjugate-)linear operator that satisfies the inequality*

$$\|Af\|_{B_2} \le C\|f\|_{B_1}, \quad \forall f \in X, \tag{A.3}$$

for some $C > 0$. Then A extends to a bounded operator from B_1 to B_2 and (A.3) *holds for all f in B_1.*

Likewise, $A : X \to B_2$ extends to a bounded operator on B_1 if it satisfies

$$|\langle Af, h\rangle| \le C\|f\|_{B_1}\|h\|_{B_2^*}, \quad \forall f \in X, \quad \forall h \in Y \tag{A.4}$$

where Y is a dense subset of B_2^*.

Proof. Assume (A.3). For every $f \in B_1$ there exists a sequence $\{f_n\} \subset X$ such that $\lim_{n\to\infty} \|f - f_n\|_{B_1} = 0$. So $\{f_n\}$ is a Cauchy sequence and

$$\|Af_m - Af_n\|_{B_2} = \|A(f_m - f_n)\|_{B_2} \le C\|f_m - f_n\|_{B_1},$$

by (A.3). Thus $\{Af_n\} \subset B_2$ is a Cauchy sequence and, by the completeness of B_2, converges, therefore we may define $Af := \lim_{n\to\infty} Af_n$. As a consequence, relation (A.3) is satisfied for every $f \in B_1$. The definition does not depend on the choice of the approximating sequence. Indeed, choose $\{g_n\} \subset X$ such that $g_n \to f$ and set $\tilde{A}f := \lim_{n\to\infty} Ag_n$, then

$$\begin{aligned}
\|\tilde{A}f - Af\|_{B_2} &= \|\tilde{A}f - Ag_n + Ag_n - Af_n + Af_n - Af\|_{B_2} \\
&\le \|\tilde{A}f - Ag_n\|_{B_2} + \|A(g_n - f_n)\|_{B_2} + \|Af_n - Af\|_{B_2} \\
&\le \|\tilde{A}f - Ag_n\|_{B_2} + C\|g_n - f_n\|_{B_1} + \|Af_n - Af\|_{B_2} \to 0
\end{aligned}$$

for $n \to +\infty$, hence $\tilde{A}f = Af$. The extended mapping A is linear: for every $\lambda, \mu \in \mathbb{C}$, $f, g \in B_1$ there exist sequences $\{f_n\}, \{g_n\} \subset X$ such that $f_n \to f, g_n \to g$, for $n \to +\infty$ and so

$$A(\lambda f_n + \mu g_n) = \lambda A(f_n) + \mu A(g_n),$$

and, taking the limit as $n \to +\infty$, we have $A(\lambda f + \mu g) = \lambda Af + \mu Ag$.

Now, fix $f \in X$ and consider the mapping $h \mapsto \langle Af, h\rangle$. This mapping is (conjugate-)linear and by assumption (A.4) is bounded on Y, a dense subset of B_2^*. Hence, by the previous part of the proof, it extends to a (conjugate-)linear mapping on B_2^* and fulfills (A.4) for every $h \in B_2^*$. Finally, using $\|Af\|_{B_2} = \sup_{\|h\|_{B_2^*}=1} |\langle Af, h\rangle|$ and (A.4), we obtain $\|Af\|_{B_2} \le C\|f\|_{B_1}$ for every $f \in X$; the first part of the proof allows extending the previous estimate to every $f \in B_1$, yielding the claim. □

Theorem A.0.10. *Let $Q : H \times H \to \mathbb{C}$ be a bounded sesquilinear form on a Hilbert space H, in the sense that $Q(f,g)$ is linear in the first component, conjugate linear in the second, and verifies the estimate*

$$|Q(f,g)| \le C\|f\|_H\|g\|_H, \quad \forall f,g \in H.$$

Then there exists a unique bounded operator A with the operator norm $\|A\| \le C$ such that

$$Q(f,g) = \langle Af, g\rangle$$

for all $f,g \in H$.

A.1 Basis theory and series expansions

Here we shall recall definitions and basic properties of this subject, borrowed from the textbook [183].

A.1.1 Bases and orthonormal sequences

Definition A.1.1. A sequence $\{x_n\}$ of vectors in a Banach space X is a basis for X if every vector $x \in X$ can be written as

$$x = \sum_{n=1}^{\infty} c_n(x)x_n, \tag{A.5}$$

for a *unique* choice of scalars $(c_n(x))$.

Observe that this definition implies that the coefficients $c_n(x)$ are linear functionals, also called coefficient functionals.

Definition A.1.2. A sequence $\{x_n\}$ of vectors in a Banach space X is complete if

$$\operatorname{span}\{x_n\} = \left\{\sum_{n=1}^{N} c_n x_n,\ c_n \in \mathbb{C}\right\},$$

that is, the set of all finite linear combinations of elements of $\{x_n\}$ is dense in X.

If $\{x_n\}$ is a basis, for every x, we can find a sequence of scalars (c_n) such that $S_N = \sum_{n=1}^{N} c_n x_n \to x$, for $N \to +\infty$, and this means that the span of $\{x_n\}$ is dense in X. Therefore every basis is a complete sequence. The converse is not true in general, as shown in the following example.

Example A.1.3. For $a, b \in \mathbb{R}$, $a < b$, consider the space $\mathcal{C}([a,b])$ with uniform norm. Weierstrass approximation theorem says that for every $f \in \mathcal{C}([a,b])$ and $\epsilon > 0$ there exists a polynomial $p(x) = \sum_{n=0}^{k} c_n x^n$ such that $\|f - p\|_\infty < \epsilon$. So the set of monomials $\{x^n\}$ is a complete sequence in $\mathcal{C}([a,b])$, but not a basis. Indeed, not every function in $\mathcal{C}([a,b])$ can be written as $f(x) = \sum_{n=0}^{\infty} c_n x^n$ with uniform convergence, cf. [183].

Example A.1.4. A standard basis for the space $c_0(\mathbb{N}) = \{(a_n) : \lim_{n\to+\infty} |a_n| = 0\}$, the space of sequences that decay to zero at infinity, with the uniform norm, is given by $\{\delta_n\}_{n\in\mathbb{N}}$, where $\delta_n = (0, 0, \dots, 0, 1, 0, \dots, 0, \dots)$, with the 1 in the nth component. The same sequence $\{\delta_n\}$ is also a basis for the space of sequences $\ell^p(\mathbb{N})$, $1 \le p \le \infty$, where $\|(a_n)\|_{\ell^p} = (\sum_{n=0}^{\infty} |a_n|^p)^{1/p}$.

Let H be a Hilbert space with inner product $\langle\cdot,\cdot\rangle$. Then we recall the following definitions.

Definition A.1.5. Let $\{x_n\}$ be a sequence in a Hilbert space H.
(i) $\{x_n\}$ is an *orthogonal* sequence if $\langle x_m, x_n\rangle = 0$ whenever $m \neq n$.
(ii) $\{x_n\}$ is an *orthonormal* (in short, o. n.) sequence if $\langle x_m, x_n\rangle = \delta_{m,n}$, that is, $\{x_n\}$ is an *orthogonal* sequence and $\|x_n\| = 1$ for every n.
(iii) An orthonormal sequence $\{x_n\}$ is an *orthonormal basis* if it is both orthonormal and a basis.

When $\langle x, y\rangle = 0$, we write $x \perp y$.

Proposition A.1.6. *Let H be a Hilbert space and let $x, y \in H$. Then*
(i) (Polar identity) $\|x+y\|^2 = \|x\|^2 + 2\Re(\langle x,y\rangle) + \|y\|^2$.
(ii) (Pythagorean theorem) *If* $\langle x,y\rangle = 0$, *then* $\|x+y\|^2 = \|x\|^2 + \|y\|^2$.
(iii) (Parallelogram law) $\|x+y\|^2 + \|x-y\|^2 = 2(\|x\|^2 + \|y\|^2)$.
(iv) (Cauchy–Schwarz inequality) $|\langle x,y\rangle| \le \|x\|\|y\|$, *for all* $x, y \in H$.
(v) $\|x\| = \langle x,x\rangle^{1/2}$ *is a norm on* H.
(vi) $\|x\| = \sup_{\|y\|=1} |\langle x,y\rangle|$.

Corollary A.1.7. *Let H be a Hilbert space. Then we have*
(i) (Continuity of the inner product) *If* $x_n \to x \in H$ *and* $y_n \to y \in H$, *then* $\langle x_n, y_n\rangle \to \langle x,y\rangle$.
(ii) *If the series* $\sum_n x_n$ *converges to* $x \in H$, *then for any* $y \in H$ *we have*

$$\langle x,y\rangle = \left\langle \sum_n x_n, y\right\rangle = \sum_n \langle x_n, y\rangle.$$

Definition A.1.8. Let A be a subset (not necessarily closed) of a Hilbert space H. The orthogonal complement of A is

$$A^\perp = \{x \in H : x \perp A\} = \{x \in H : \langle x, y\rangle = 0 \ \forall y \in A\}.$$

Remark A.1.9. Recall that $A^\perp$ is always a closed subset of H, even if A is not.

The following theorem summarizes some basic results related to convergence of series of orthonormal vectors (for the proof we refer, for instance, to [183, Theorem 1.49]).

Theorem A.1.10. *If $\{x_n\}$ is an o. n. sequence in a Hilbert space H, then we have*
(i) (Bessel inequality) $\sum_n |\langle x, x_n\rangle|^2 \le \|x\|^2$ *for every* $x \in H$.
(ii) *If* $H \ni x = \sum_n c_n x_n$, *then* $c_n = \langle x, x_n\rangle$, *for every* n.
(iii) $\sum_n c_n x_n$ *converges* $\iff \sum_n |c_n|^2 < \infty$.
(iv) $x \in \overline{\operatorname{span}}\{x_n\} \iff x = \sum_n \langle x, x_n\rangle x_n$.
(v) *If* $x \in H$ *then* $p = \sum_n \langle x, x_n\rangle x_n$ *is the orthogonal projection of* x *onto* $\overline{\operatorname{span}}\{x_n\}$.

Hence, if an o. n. sequence $\{x_n\}$ in H is complete, that is, $\overline{\operatorname{span}}\{x_n\} = H$, then every $x \in H$ can be uniquely written as $x = \sum_n \langle x, x_n\rangle x_n$, by Theorem A.1.10(ii)–(iv), that is, $\{x_n\}$ is a basis. In what follows we present several equivalent conditions for a sequence $\{x_n\}$ to be an o. n. basis (see [183, Theorem 1.50]).

Theorem A.1.11. *If $\{x_n\}$ is an o. n. sequence in a Hilbert space H, then the following conditions are equivalent:*
(i) $\{x_n\}$ *is complete in* H.
(ii) $\{x_n\}$ *is a basis for* H, *that is, for every* $x \in H$ *there exists a unique sequence of scalars* (c_n) *such that* $x = \sum_n c_n x_n$.
(iii) $x = \sum_n \langle x, x_n\rangle x_n$, *for every* $x \in H$.
(iv) (Plancherel's equality) $\|x\|^2 = \sum_n |\langle x, x_n\rangle|^2$.
(v) (Parseval's equality) $\langle x, y\rangle = \sum_n \langle x, x_n\rangle \langle x_n, y\rangle$.

A.1.2 Convergence of series

We recall the definition of convergent series in normed vector spaces. In what follows we shall use the notation $\{x_n\}$ for a sequence in a normed vector space, whereas (c_n) for a sequence of scalars.

Definition A.1.12. Let $\{x_n\}$ be a sequence in a normed linear space X.
(i) The series $\sum_{n=1}^\infty x_n$ is *convergent* to $x \in X$ if the partial sums $X_N = \sum_{n=1}^N x_n$ converge to x in the X-norm:

$$\forall \epsilon > 0,\ \exists N_0 > 0, \forall N \ge N_0, \quad \left\| \sum_{n=1}^N x_n - x \right\| < \epsilon.$$

(ii) The series $\sum_{n=1}^{\infty} x_n$ is *unconditionally convergent* if $\sum_{n=1}^{\infty} x_{\sigma(n)}$ is convergent for every permutation σ of $\mathbb{N}$.
(iii) The series $\sum_{n=1}^{\infty} x_n$ is *absolutely convergent* if $\sum_{n=1}^{\infty} \|x_n\| < \infty$.
(iv) If a series $\sum_{n=1}^{\infty} x_n$ converges but not unconditionally, we say that $\sum_{n=1}^{\infty} x_n$ is *conditionally convergent*.

The following result is a simplified version of the celebrated Dirichlet and Riemann–Dini theorems.

Lemma A.1.13. *If (c_n) is a sequence of real or complex scalars, then*

$$\sum_{n=1}^{\infty} c_n \text{ is absolutely convergent} \iff \sum_{n=1}^{\infty} c_n \text{ is unconditionally convergent.}$$

Proof. "$\Longrightarrow$" Suppose $\sum_{n=1}^{\infty} |c_n| < \infty$. Choose any $\epsilon > 0$, then there exists $N_0 > 0$ such that for every $N > M \geq N_0$, $\sum_{n=M+1}^{N} |c_n| < \epsilon$ (the sequence of partial sums is a Cauchy sequence). Note that, since $|\sum_{n=M+1}^{N} c_n| \leq \sum_{n=M+1}^{N} |c_n|$, this implies that $\sum_{n=1}^{\infty} c_n$ is convergent.

For any permutation σ of $\mathbb{N}$, set

$$N_1 = \max\{\sigma^{-1}(1), \ldots, \sigma^{-1}(N_0)\}.$$

Then for every $N > M \geq N_1$, take $n = M+1, \ldots, N$ and observe that $n > N_1$ implies $n \neq \sigma^{-1}(1), \ldots, \sigma^{-1}(N_0)$ so $\sigma(n) > N_0$. In particular,

$$K = \min\{\sigma(M+1), \ldots, \sigma(N)\} > N_0, \quad L = \max\{\sigma(M+1), \ldots, \sigma(N)\} \geq K > N_0.$$

Hence

$$\left|\sum_{n=M+1}^{N} c_{\sigma(n)}\right| \leq \sum_{n=M+1}^{N} |c_{\sigma(n)}| \leq \sum_{n=K}^{L} |c_n| < \epsilon,$$

that is, $\sum_{n=1}^{N} c_{\sigma(n)}$ is a Cauchy sequence and, by the completeness of $\mathbb{C}$ (or $\mathbb{R}$), $\sum c_{\sigma(n)}$ converges.

"$\Longleftarrow$" Assume first that $\sum_{n=1}^{\infty} c_n$ is a sequence of *real* scalars that does not converge absolutely. Define by (p_n) and (q_n) the sequences of nonnegative and negative terms in the given order, respectively (note that (p_n) or (q_n) may also be a finite sequence). If both $\sum_n p_n$ and $\sum_n q_n$ converge, then $\sum_n |c_n| = \sum_n p_n - \sum_n q_n < \infty$, yielding a contradiction. So either $\sum_n p_n$ or $\sum_n q_n$ must diverge.

Suppose that $\sum_n p_n$ diverges. Since $p_n \geq 0$ for every n, there exists $m_1 \in \mathbb{N}$ such that

$$p_1 + \cdots + p_{m_1} > 1.$$

There exists $m_2 > m_1$ such that

$$p_1 + \cdots + p_{m_1} + q_1 + p_{m_1+1} + \cdots + p_{m_2} > 2;$$

there exists $m_3 > m_2$ such that

$$p_1 + \cdots + p_{m_1} + q_1 + p_{m_1+1} + \cdots + p_{m_2} + q_2 + p_{m_2+1} + \cdots + p_{m_3} > 3,$$

and so on. We have constructed a rearrangement of $\sum_n c_n$ that does not converge. Since $\sum_n c_n$ converges unconditionally, this is a contradiction. A similar proof applies if $\sum_n q_n$ is divergent.

Thus we have shown that if $\sum_n c_n$ is a series of real scalars that converges unconditionally, then it must converge absolutely.

Finally, let $\sum_n c_n$ be a series of complex scalars $c_n = a_n + ib_n$ that converges unconditionally. Since

$$\sum_n c_{\sigma(n)} = \sum_n a_{\sigma(n)} + i \sum_n b_{\sigma(n)}$$

is convergent, both $\sum_n a_{\sigma(n)}$ and $\sum_n b_{\sigma(n)}$ converge. So $\sum_n a_n$ and $\sum_n b_n$ converge unconditionally and, by the previous step, absolutely. Finally,

$$\sum_n |c_n| = \sum_n |a_n + ib_n| \leq \sum_n |a_n| + \sum_n |b_n| < \infty.$$

So $\sum_n c_n$ is absolutely convergent, and we are done. □

Example A.1.14. The alternating harmonic series $\sum_{n=1}^{\infty}(-1)^n/n$ converges to $\log(1/2)$. However, it does not converge absolutely, so it cannot converge unconditionally.

Lemma A.1.15. *Let $\{x_n\}$ be a sequence in a Banach space X.*

$$\text{If } \sum_n x_n \text{ converges absolutely} \Longrightarrow \sum_n x_n \text{ converges unconditionally.}$$

Proof. Assume $\sum_n \|x_n\| < \infty$. Then for every $\epsilon > 0$ there exists an N_0 such that for every $N > M \geq N_0$,

$$\left\| \sum_{n=M+1}^{N} x_n \right\| \leq \sum_{n=M+1}^{N} \|x_n\| < \epsilon,$$

hence the sequence of partial sums $\{S_N\}$, with $S_N = \sum_{n=1}^{N} x_n$, is a Cauchy sequence and converges to $x \in X$ by the completeness of X. Take now any permutation σ of $\mathbb{N}$. We can repeat the previous argument for $\sum_n \|x_{\sigma(n)}\| < \infty$, by Lemma A.1.13. Therefore $\sum_n x_n$ is unconditionally convergent. □

However, unconditionally convergence does not imply absolutely convergence, as shown in the following example.

Example A.1.16. Consider the Hilbert space of 1-periodic, square integrable functions $L^2([0,1])$. An orthonormal basis is the Fourier sequence $\{e^{2\pi int}\}$, $n \in \mathbb{Z}$. We consider the series $\sum_{n=1}^{\infty} c_n e^{2\pi int}$, with $c_n = 1/n$. Observe that

$$\left\| \sum_{n=1}^{\infty} c_n e^{2\pi int} \right\|_2^2 = \sum_{n=1}^{\infty} \frac{1}{n^2} < \infty.$$

Since the series of scalars $\sum_{n=1}^{\infty} \frac{1}{n^2}$ is absolutely convergent, it is also unconditionally convergent by Lemma A.1.13. This implies that $\sum_{n=1}^{\infty} c_n e^{2\pi int}$ is unconditionally convergent. However, the previous series is not absolutely convergent; indeed,

$$\sum_{n=1}^{\infty} \|c_n e^{2\pi int}\|_2 = \sum_{n=1}^{\infty} |c_n| = \sum_{n=1}^{\infty} \frac{1}{n} = +\infty,$$

since $\|e^{2\pi int}\|_2 = 1$.

Definition A.1.17. A *directed set* is a nonempty set I with a relation $\leq$ on I such that
(i) $\leq$ is reflexive: $i \leq i$, for $i \in I$;
(ii) $\leq$ is transitive: $i \leq j$ and $j \leq k \Rightarrow i \leq k$; and
(iii) for any $i, j \in I$ there exists some $k \in I$ such that $i \leq k$ and $j \leq k$ (every pair of elements has an upper bound).

Definition A.1.18. A **net** in X is a sequence $\{x_i\}_{i\in I}$ of elements x_i in X, indexed by a directed set I.

The set of natural numbers $\mathbb{N}$ is a directed set and hence every sequence $\{x_n\}_{n\in\mathbb{N}}$ is an example of net. A more interesting example is as follows. Consider the set of *all finite subset of* $\mathbb{N}$:

$$I = \{F \subset \mathbb{N};\ F \text{ is finite}\},$$

ordered by inclusion. Then I is a directed set.

Definition A.1.19 (Convergence of a net). Let X be a topological space and $\{x_i\}_{i\in I}$ a net in X. We say that the sequence $\{x_i\}_{i\in I}$ converges to $x \in X$ if for every open set U containing x there exists $i_0 \in I$ such that for every $i \in I$, $i_0 \leq i$, we have $x_i \in U$. In this case we write $x = \lim_i x_i$.

Observe that we shall work with normed vector spaces X, so that it is enough to check the previous convergence on the open balls $B_\epsilon(x) = \{y \in X : \|y - x\| < \epsilon\}$.

Given a formal series $\sum_n x_n$ (there is no requirement that the series converges in any sense), the associated net of all possible partial sums is given by

$$\left\{ \sum_{n\in F} x_n \right\}_{F\in I} = \left\{ \sum_{n\in F} x_n,\ F \subset \mathbb{N},\ F \text{ finite} \right\}, \tag{A.6}$$

with I being finite subsets of a index set. Note that the index set is not always $\mathbb{N}$ but could be a lattice (for example, in Gabor analysis).

Definition A.1.19 in this case can be rephrased as follows.

Definition A.1.20. Let X be a Banach space and $x_n, x \in X$. We say that the series $\sum_n x_n$ converges to x with respect to the directed set of finite subsets of $\mathbb{N}$ if

$$\forall \epsilon > 0 \, \exists \text{ finite } F_0 \subset \mathbb{N}, \, \forall \text{ finite } F \supseteq F_0, \quad \left\| x - \sum_{n \in F} x_n \right\| < \epsilon.$$

In this case *we write* $x = \lim_F \sum_{n \in F} x_n$.

We now present several equivalent formulations for unconditional convergence in Banach spaces.

Theorem A.1.21. *Given a sequence $\{x_n\}$ in a Banach space X, the following are equivalent:*

(i) $\sum_n x_n$ *converges unconditionally.*

(ii) $\lim_F \sum_{n \in F} x_n$ *exists.*

(iii) *For every $\epsilon > 0$, there exists $N \in \mathbb{N}_+$ such that for every finite $F \subset \mathbb{N}$ with* $\min(F) > N$, $\| \sum_{n \in F} x_n \| < \epsilon$.

Proof. (i) $\Longrightarrow$ (ii) Suppose $x = \sum_n x_n$ and let us prove that $x = \lim_F \sum_{n \in F} x_n$. We argue by contradiction. Assume that there exists $\epsilon > 0$ such that

$$\forall \text{ finite } F_0 \subset \mathbb{N}, \, \exists F_1 \supseteq F_0 : \quad \left\| x - \sum_{n \in F_1} x_n \right\| \geq \epsilon. \tag{A.7}$$

Since $\sum_n x_n$ converges, there exists $M_1 \in \mathbb{N}$ such that for every $N \geq M_1$,

$$\left\| x - \sum_{n=1}^{N} x_n \right\| < \epsilon/2.$$

Define $F_1 = \{1, \dots, M_1\}$, then by (A.7) there exists $G_1 \supseteq F_1$ such that $\|x - \sum_{n \in G_1} x_n\| \geq \epsilon$. Consider $M_2 = \max(G_1)$ and define $F_2 = \{1, \dots, M_2\}$, then by (A.7) there exists $G_2 \supseteq F_2$ such that $\|x - \sum_{n \in G_2} x_n\| \geq \epsilon$. Proceeding this way, we construct a sequence of finite sets such that $F_1 \subseteq G_1 \subseteq F_2 \subseteq G_2 \subseteq \cdots \subseteq F_N \subseteq G_N \subset \cdots$ and

$$\left\| x - \sum_{n \in F_N} x_n \right\| < \epsilon/2, \quad \left\| x - \sum_{n \in G_N} x_n \right\| \geq \epsilon.$$

Hence

$$\begin{aligned} \left\| \sum_{n \in G_N \setminus F_N} x_n \right\| &= \left\| \sum_{n \in G_N} x_n - \sum_{n \in F_N} x_n \right\| \\ &\geq \left\| \sum_{n \in G_N} x_n - x \right\| - \left\| x - \sum_{n \in F_N} x_n \right\| \\ &> \epsilon - \epsilon/2 = \epsilon/2. \end{aligned}$$

In particular, F_N must be a proper subset of G_N. Let σ be a permutation of $\mathbb{N}$ obtained by enumerating in turn the elements of F_1, then $G_1 \setminus F_1$, then $F_2 \setminus G_1$, then $G_2 \setminus F_2$, etc. Then, for each N, we have

$$\left\| \sum_{n=|F_N|+1}^{|G_N|} x_{\sigma(n)} \right\| = \left\| \sum_{n \in G_N \setminus F_N} x_n \right\| > \epsilon/2.$$

Since $|F_N|, |G_N| \to +\infty$ as N increases, we see that $\sum_{n=1}^{N} x_{\sigma(n)}$ is not a Cauchy sequence and cannot converge, and this is a contradiction.

(ii) $\Longrightarrow$ (iii) Assume that $\lim_F \sum_{n\in F} x_n = x \in X$. Choose any $\epsilon > 0$. Then there exists a finite set F_0 such that for every finite $F \subseteq F_0$, $\|x - \sum_{n\in F} x_n\| < \epsilon/2$. Set $N = \max(F_0)$. For any finite set G with $\min(G) > N$, we have $F_0 \cap G = \emptyset$ and

$$\begin{aligned} \left\| \sum_{n\in G} x_n \right\| &\le \left\| x - \sum_{n \in F_0} x_n \right\| + \left\| \sum_{n\in F_0 \cup G} x_n - x \right\| \\ &< \epsilon/2 + \epsilon/2 = \epsilon. \end{aligned}$$

(iii) $\Longrightarrow$ (i) Choose any permutation σ of $\mathbb{N}$. We have to prove that $\sum_n x_{\sigma(n)}$ converges in X. So we shall show that the sequence of partial sums $\{\sum_{n=1}^{N} x_\sigma(n)\}$ is a Cauchy sequence. Consider an $\epsilon > 0$, and let N be the number in claim (iii). Define

$$N_0 = \max\{\sigma^{-1}(1), \dots, \sigma^{-1}(N)\}.$$

Choose $L > K \ge N_0$ and define $F = \{\sigma(K+1), \dots, \sigma(L)\}$. Then for every $k \ge K+1$, we have $k > N_0$ and hence $\sigma(k) \ne 1, \dots, N$, so $\sigma(k) > N$ and, finally, $\min(F) > N$. Thus

$$\left\| \sum_{n=K+1}^{L} x_{\sigma(n)} \right\| = \left\| \sum_{n\in F} x_n \right\| < \epsilon. \qquad \square$$

We can now show that if $\sum_n x_n$ is unconditionally convergent, then its limit does not depend on the permutation σ.

Corollary A.1.22. *Let $\{x_n\}$ be a sequence in a Banach space. If $\sum_n x_n$ is unconditionally convergent, then*

$$\sum_n x_{\sigma(n)} = \sum_n x_n,$$

for every permutation σ of $\mathbb{N}$.

Proof. Suppose $\sum_n x_n$ unconditionally convergent. Then by Theorem A.1.21 (ii), $\lim_F \sum_{n\in F} x_n = x \in X$. Consider any permutation σ and any $\epsilon > 0$. Then, by Definition A.1.20, there exists a finite set $F_0 \subset \mathbb{N}$ such that

$$\forall F \text{ finite}, \quad F \supseteq F_0 \Longrightarrow \left\| x - \sum_{n\in F} x_n \right\| < \epsilon. \tag{A.8}$$

Choose $N_0 \in \mathbb{N}$ large enough such that $F_0 \subset \{\sigma(1), \dots, \sigma(N_0)\}$. For $N \geq N_0$, define $F = \{\sigma(1), \dots, \sigma(N)\}$. Then $F \supseteq F_0$ and, by equation (A.8),

$$\left\| x - \sum_{n=1}^{N} x_{\sigma(n)} \right\| = \left\| x - \sum_{n \in F} x_n \right\| < \epsilon,$$

as desired. □

We shall present a necessary condition for unconditional convergence of series in Hilbert spaces. This is expressed by Orlicz's theorem. For the proof we refer, e. g., to [183, Theorem 3.16].

Theorem A.1.23 (Orlicz's theorem). *If* $\{x_n\}$ *is a sequence in a Hilbert space, then*

$$\sum_n x_n \text{ converges unconditionally} \Longrightarrow \sum_n \|x_n\|^2 < \infty. \tag{A.9}$$

Orlicz's theorem does not extend to Banach spaces, in general. Further, its converse is not true, in general. An example is provided below.

Example A.1.24. Let H be a Hilbert space and fix $x \in H$ with $\|x\| = 1$. Then

$$\left\| \sum_{n=M+1}^{N} c_n x \right\| = \left| \sum_{n=M+1}^{N} c_n \right| \|x\| = \left| \sum_{n=M+1}^{N} c_n \right|.$$

So $\sum_n c_n x$ converges (unconditionally) in H if and only if $\sum_n c_n$ converges (unconditionally) as series of scalars. Therefore, if $(c_n) \in \ell^2$ is such that $\sum_n c_n$ converges conditionally, then also $\sum_n c_n x$ converges conditionally. However, we have

$$\sum_n \|c_n x\|^2 = \sum_n |c_n|^2 = \|(c_n)\|_{\ell^2} < \infty.$$

For example, take $c_n = (-1)^n/n$.

We now come back to the concept of basis in a Banach space. Some types of bases with extra properties are collected in the following definition.

Definition A.1.25. Let $\{x_n\}$ be a basis for a Banach space X.
(i) $\{x_n\}$ is an *unconditional basis* if the series in (A.5) is unconditionally convergent, for each $x \in X$. If this is not the case, we call $\{x_n\}$ a *conditional basis*.
(ii) $\{x_n\}$ is an *absolutely convergent basis* if the series in (A.5) converges absolutely, for each $x \in X$.
(iii) $\{x_n\}$ is a *bounded basis* if $\{x_n\}$ is norm-bounded, both above and below, that is,

$$0 < \inf \|x_n\| \leq \sup \|x_n\| < \infty.$$

(iv) $\{x_n\}$ is a *normalized basis* if $\|x_n\| = 1$, for every n.

An example of unconditional basis is as follows. Consider $L^2([0,1])$, with the o. n. basis $\{e^{2\pi int}\}_{n\in\mathbb{Z}}$. Then, for every $f \in L^2([0,1])$,

$$\|f\|^2_{L^2([0,1])} = \left\|\sum_{n\in\mathbb{Z}} c_n e^{2\pi int}\right\|^2_{L^2([0,1])} = \sum_{n\in\mathbb{Z}} |c_n|^2,$$

where $c_n = \langle f, e^{2\pi int}\rangle$ is the nth Fourier coefficient. Since $\sum_{n\in\mathbb{Z}} |c_n|^2$ is absolutely convergent, and hence unconditionally convergent, it follows that $\{e^{2\pi int}\}_{n\in\mathbb{Z}}$ is an unconditional basis. More generally, every o. n. basis $\{x_n\}_n \subset H$, with H a Hilbert space, is an unconditional basis.

We now provide an example of absolutely convergent basis. Consider the Banach space $\ell^1(\mathbb{Z})$ with the basis $\{\delta_n\}_{n\in\mathbb{Z}}$, where we recall that δ_n is the sequence having elements all equal zero but the nth element that is one. Take any sequence $c = (c_n)_n \in \ell^1$, hence $c = \sum_n c_n\delta_n$ and $\|c\|_{\ell^1} = \sum_n |c_n|$. Now,

$$\sum_n \|c_n\delta_n\|_{\ell^1} = \sum_n |c_n|\|\delta_n\|_{\ell^1} = \sum_n |c_n|.$$

That is, $\{\delta_n\}_{n\in\mathbb{Z}}$ is an absolutely convergent basis.

An example of normalized basis (hence, in particular, a bounded basis) is provided by every o. n. basis in a Hilbert space.

In particular, if we restrict our attention to Hilbert spaces, than we recall the following definition.

Definition A.1.26 (Riesz basis). A sequence $\{x_n\}$ for a Hilbert space H is a *Riesz basis* if it is equivalent to some (and hence every) orthonormal basis for H. That is, there exists a topological isomorphism $T : H \to H$ and there exists an o. n. basis $\{e_n\}$ for H such that

$$Te_n = x_n \quad \forall n.$$

A.2 Some results concerning real interpolation theory

Here we collected some results of real interpolation theory which are used in the proof of the Strichartz estimates (see Chapter 5). We use the notation and terminology of [296]. Let $(X, \mathcal{B}, \mu)$ be a measure space, where X is a set, $\mathcal{B}$ a σ-algebra, and μ a positive σ-finite measure. If A is a Banach space, $1 \le p \le \infty$, then we shall write $L^p(A)$ for the usual vector-valued L^p spaces in the sense of the Bochner integral.

Proposition A.2.1. *Let $\{A_0, A_1\}$ be an interpolation couple. For every $1 \le p_0, p_1 < \infty$, $0 < \theta < 1$, $1/p = (1-\theta)/p_0 + \theta/p_1$. and $p \le q$, we have*

$$\ell^p((A_0, A_1)_{\theta,q}) \hookrightarrow (\ell^{p_0}(A_0), \ell^{p_1}(A_1))_{\theta,q}. \tag{A.10}$$

Proof. Set $\eta = p\theta/p_1$ and $q = pr$, $r \geq 1$. It follows from [296, Theorem 1.4.2, page 29], that, given $c = \{c_j\} \in \ell^p(A_0) + \ell^p(A_1)$, we have

$$\|c\|^p_{(\ell^{p_0}(A_0),\ell^{p_1}(A_1))_{\theta,q}} \asymp \left\| t^{-\eta} \inf_{\substack{a+b=c \\ a\in\ell^{p_0}(A_0),\ b\in\ell^{p_1}(A_1)}} \|a\|^{p_0}_{\ell^{p_0}(A_0)} + t\|b\|^{p_1}_{\ell^{p_1}(A_1)} \right\|_{L^r(\mathbb{R}_+,\frac{dt}{t})}.$$

Hence,

$$\begin{aligned}\|c\|^p_{(\ell^{p_0}(A_0),\ell^{p_1}(A_1))_{\theta,q}} &\asymp \left\| t^{-\eta} \inf_{\substack{a+b=c \\ a\in\ell^{p_0}(A_0),\ b\in\ell^{p_1}(A_1)}} \sum_{j\geq 0} \left(\|a_j\|^{p_0}_{A_0} + t\|b_j\|^{p_1}_{A_1}\right) \right\|_{L^r(\mathbb{R}_+,\frac{dt}{t})} \\ &= \left\| t^{-\eta} \sum_{j\geq 0} \inf_{\substack{a_j+b_j=c_j \\ a_j\in A_0,\ b_j\in A_1}} \left(\|a_j\|^{p_0}_{A_0} + t\|b_j\|^{p_1}_{A_1}\right) \right\|_{L^r(\mathbb{R}_+,\frac{dt}{t})}.\end{aligned} \tag{A.11}$$

By Minkowski's inequality, we deduce

$$\begin{aligned}\|c\|^p_{(\ell^{p_0}(A_0),\ell^{p_1}(A_1))_{\theta,q}} &\lesssim \sum_{j\geq 0} \left\| t^{-\eta} \inf_{\substack{a_j+b_j=c_j \\ a_j\in A_0,\ b_j\in A_1}} \left(\|a_j\|^{p_0}_{A_0} + t\|b_j\|^{p_1}_{A_1}\right) \right\|_{L^r(\mathbb{R}_+,\frac{dt}{t})} \\ &\asymp \|c\|^p_{\ell^p((A_0,A_1)_{\theta,q})}.\end{aligned} \tag{A.12}$$

□

Consider now a partition of unity[1] given by functions $\phi_\alpha \in \mathcal{D}(\mathbb{R}^d)$, $\alpha \in \mathbb{Z}^d$, with $\phi_\alpha = T_\alpha\phi$, $\operatorname{supp}\phi \subset [-2,2]^d$. Thus $\operatorname{supp}\phi_\alpha \subset \alpha + [-2,2]^d$. Let then $\psi \in \mathcal{D}(\mathbb{R}^d)$, $\psi = 1$ on $[-2,2]^d$ and $\psi = 0$ away from $[-4,4]^d$, and set $\psi_\alpha = T_\alpha\psi$. Observe that there is a constant C_d such that

$$\forall\alpha \in \mathbb{Z}^d,\ \#\{\beta \in \mathbb{Z}^d : \operatorname{supp}\psi_\beta \cap \operatorname{supp}\phi_\alpha \neq \emptyset\} \leq C_d \tag{A.13}$$

and

$$\forall\alpha \in \mathbb{Z}^d,\ \#\{\beta \in \mathbb{Z}^d : \operatorname{supp}\psi_\alpha \cap \operatorname{supp}\phi_\beta \neq \emptyset\} \leq C_d. \tag{A.14}$$

The linear operators

$$S : L^1_{\mathrm{loc}} \to (L^1_{\mathrm{loc}})^{\mathbb{Z}^d}, \quad R : (L^1_{\mathrm{loc}})^{\mathbb{Z}^d} \to L^1_{\mathrm{loc}},$$

defined by

$$Sf = \{f\phi_\alpha\}_\alpha, \quad R(\{u_\alpha\}_\alpha) = \sum_\alpha u_\alpha\psi_\alpha, \tag{A.15}$$

enjoy the following properties (see [122, Remark 2.2]).

---

1 Such a partition of unity can be constructed as follows. Take $\chi \in \mathcal{D}(\mathbb{R}^d)$, $0 \leq \chi \leq 1$, $\chi = 1$ on $[-1,1]^d$, $\chi = 0$ away from $[-2,2]^d$. Set $\Phi(x) = \sum_{\alpha\in\mathbb{Z}^d}\chi(x-\alpha)$. Since the sum is locally finite, Φ is well defined and smooth. Moreover, $\Phi(x+\beta) = \Phi(x)$ for every $\beta \in \mathbb{Z}^d$, and also $\Phi \geq 1$. Hence it suffices to take $\phi_\alpha = T_\alpha\chi/\Phi = T_\alpha(\chi/\Phi)$.

Proposition A.2.2. *We have* $RS = \mathrm{Id}$ *on* L^1_{loc} *and, for every local component* B*, as at the beginning of this section, and every* $p \geq 1$*, we have*

$$S : W(B, L^p) \to \ell^p(B) \tag{A.16}$$

and

$$R : \ell^p(B) \to W(B, L^p) \tag{A.17}$$

continuously.

Proof. The equality $RS = \mathrm{Id}$ on L^1_{loc} is clear, whereas (A.16) follows at once from Remark 4 of [122]. To prove (A.17), we observe that

$$\begin{aligned}
\|R(\{u_\alpha\}_\alpha)\|^p_{W(B,L^p)} &\asymp \sum_\beta \Big\| \sum_\alpha u_\alpha \psi_\alpha \phi_\beta \Big\|^p_B \quad \text{(by Remark 4 of [122])} && \text{(A.18)}\\
&\lesssim \sum_\beta \sum_\alpha \|u_\alpha \psi_\alpha \phi_\beta\|^p_B \quad \text{(by (A.13))} && \text{(A.19)}\\
&\lesssim \sum_\alpha \sum_\beta \|\psi_\alpha \phi_\beta\|^p_{\mathcal{F}L^1} \|u_\alpha\|^p_B && \text{(A.20)}\\
&\lesssim \|\{u_\alpha\}_\alpha\|^p_{\ell^p(B)}. && \text{(A.21)}
\end{aligned}$$

In the last inequality we used the fact that, by (A.14),

$$\sum_\beta \|\psi_\alpha \phi_\beta\|^p_{\mathcal{F}L^1} \lesssim \sum_{\beta : \operatorname{supp} \psi_\alpha \cap \operatorname{supp} \phi_\beta \neq \emptyset} \|\psi_\alpha\|^p_{\mathcal{F}L^1} \|\phi_\beta\|^p_{\mathcal{F}L^1} \leq C_d \|\psi_0\|^p_{\mathcal{F}L^1} \|\phi_0\|^p_{\mathcal{F}L^1}. \qquad \square$$

Proposition A.2.3. *Given two local components* B_0, B_1 *as at the beginning of this section, for every* $1 \leq p_0, p_1 < \infty$, $0 < \theta < 1$, $1/p = (1-\theta)/p_0 + \theta/p_1$*, and* $p \leq q$*, we have*

$$W((B_0, B_1)_{\theta,q}, L^p) \hookrightarrow (W(B_0, L^{p_0}), W(B_1, L^{p_1}))_{\theta,q}.$$

Proof. Let R and S be the operators defined above. Then, by Proposition A.2.2,

$$\begin{aligned}
\|f\|_{(W(B_0,L^{p_0}),W(B_1,L^{p_1}))_{\theta,q}} &= \|RSf\|_{(W(B_0,L^{p_0}),W(B_1,L^{p_1}))_{\theta,q}}\\
&\lesssim \|Sf\|_{(\ell^{p_0}(B_0),\ell^{p_1}(B_1))_{\theta,q}}\\
&\lesssim \|Sf\|_{\ell^p((B_0,B_1)_{\theta,q})}, && \text{(A.22)}\\
&\lesssim \|f\|_{W((B_0,B_1)_{\theta,q},L^p)},
\end{aligned}$$

where for (A.22) we used (A.10). This concludes the proof. $\square$

Proposition A.2.4. *Let* $\{A_0, A_1\}$ *be an interpolation couple. For every* $1 \leq p_0, p_1 < \infty$, $0 < \theta < 1$, $1/p = (1-\theta)/p_0 + \theta/p_1$*, and* $p \leq q$*, we have*

$$L^p((A_0, A_1)_{\theta,q}) \hookrightarrow (L^{p_0}(A_0), L^{p_1}(A_1))_{\theta,q}. \tag{A.23}$$

Proof. We use the fact that the bounded functions with values in $A_0 \cap A_1$, vanishing outside a set with finite measure, are dense in $L^{p_0}(A_0) \cap L^{p_1}(A_1)$ and may be approximated in $L^{p_0}(A_0) \cap L^{p_1}(A_1)$ by simple functions:

$$c(x) = \sum_{j=1}^{N} c^j \chi_{F_j}(x), \quad \mu(F_j) < \infty, \quad c^j \in A_0 \cap A_1, \quad F_j \cap F_k = \emptyset \text{ if } j \neq k.$$

Set $\eta = p\theta/p_1$ and $q = pr$, $r \geq 1$. If $c(x)$ is such a function, it follows from [296, Theorem 1.4.2, page 29] that

$$\begin{aligned}
\|c\|^p_{(L^{p_0}(A_0),L^{p_1}(A_1))_{\theta,q}} &\asymp \Big\| t^{-\eta} \inf_{\substack{a+b=c \\ a\in L^{p_0}(A_0),\ b\in L^{p_1}(A_1)}} \|a\|^{p_0}_{L^{p_0}(A_0)} + t\|b\|^{p_1}_{L^{p_1}(A_1)} \Big\|_{L^r(\mathbb{R}_+,\frac{dt}{t})} \\
&= \Big\| t^{-\eta} \inf_{\substack{a+b=c \\ a\in L^{p_0}(A_0),\ b\in L^{p_1}(A_1)}} \int_X \big(\|a(x)\|^{p_0}_{A_0} + t\|b(x)\|^{p_1}_{A_1}\big) \Big\|_{L^r(\mathbb{R}_+,\frac{dt}{t})} \\
&= \Big\| t^{-\eta} \int_X \inf_{\substack{a(x)+b(x)=c(x) \\ a(x)\in A_0,\ b(x)\in A_1}} \big(\|a(x)\|^{p_0}_{A_0} + t\|b(x)\|^{p_1}_{A_1}\big) \Big\|_{L^r(\mathbb{R}_+,\frac{dt}{t})}. \qquad (A.24)
\end{aligned}$$

In order to justify the last equality, we observe that the inequality "$\geq$" is obvious. As far as the converse inequality "$\leq$" is concerned, we take advantage of the fact that $c(x)$ is a simple function, so that one can find minimizing sequences $a_n(x)$, $b_n(x)$ given by simple functions which are constants where $c(x)$ is, and zero where $c(x) = 0$.

Finally, by Minkowski's integral inequality,

$$\begin{aligned}
\|c\|^p_{(L^{p_0}(A_0),L^{p_1}(A_1))_{\theta,q}} &\lesssim \int_X \Big\| t^{-\eta} \inf_{\substack{a(x)+b(x)=c(x) \\ a(x)\in A_0,\ b(x)\in A_1}} \big(\|a(x)\|^{p_0}_{A_0} + t\|b(x)\|^{p_1}_{A_1}\big) \Big\|_{L^r(\mathbb{R}_+,\frac{dt}{t})} \\
&\asymp \|c\|^p_{L^p((A_0,A_1)_{\theta,q})}. \qquad (A.25)
\end{aligned}$$

□

We finally recall [17, Section 3.13.5(b)].

Lemma A.2.5. *If A_i, B_i, C_i, $i = 0, 1$, are Banach spaces and T is a bilinear operator bounded from*

$$\begin{aligned}
T &: A_0 \times B_0 \to C_0, \\
T &: A_0 \times B_1 \to C_1, \\
T &: A_1 \times B_0 \to C_1,
\end{aligned}$$

then, if $0 < \theta_i < \theta < 1$, $i = 0, 1$, $\theta = \theta_0 + \theta_1$, one has

$$T : (A_0, A_1)_{\theta_0,2} \times (B_0, B_1)_{\theta_1,2} \to (C_0, C_1)_{\theta,1}.$$

Bibliography

[1] G. S. Alberti, S. Dahlke, F. De Mari, E. De Vito, and H. Führ. Recent progress in shearlet theory: systematic construction of shearlet dilation groups, characterization of wavefront sets, and new embeddings. In *Frames and Other Bases in Abstract and Function Spaces*, *Appl. Numer. Harmon. Anal.*, pages 127–160. Birkhäuser/Springer, Cham, 2017.

[2] G. S. Alberti, S. Dahlke, F. De Mari, E. De Vito, and S. Vigogna. Continuous and discrete frames generated by the evolution flow of the Schrödinger equation. *Anal. Appl. (Singap.)*, 15(6):915–937, 2017.

[3] K. Asada and D. Fujiwara. On some oscillatory integral transformations in $L^2(\mathbf{R}^n)$. *Jpn. J. Math. New Ser.*, 4(2):299–361, 1978.

[4] R. Balian. Un principe d'incertitude fort en théorie du signal ou en mécanique quantique. *C. R. Acad. Sci. Paris Sér. II Méc. Phys. Chim. Sci. Univers Sci. Terre*, 292(20):1357–1362, 1981.

[5] F. Bartolucci, F. De Mari, E. De Vito, and F. Odone. Shearlets as multi-scale Radon transforms. *Sampl. Theory Signal Image Process.*, 17(1):1–15, 2018.

[6] F. Bastianoni, E. Cordero, and F. Nicola. Decay and smoothness for eigenfunctions of localization operators. *Preprint*, arXiv:1902.03413, 2019.

[7] G. Battle. Heisenberg proof of the Balian-Low theorem. *Lett. Math. Phys.*, 15(2):175–177, 1988.

[8] D. Bayer, E. Cordero, K. Gröchenig, and S. I. Trapasso. Linear perturbations of the Wigner transform and the Weyl quantization. In *Advances in Microlocal and Time-Frequency Analysis*, *Applied and Numerical Harmonic Analysis*, pages 79–120. Birkhäuser/Springer, Cham, 2020.

[9] R. Beals. Characterization of pseudodifferential operators and applications. *Duke Math. J.*, 44(1):45–57, 1977.

[10] A. Benedek and R. Panzone. The space L^p, with mixed norm. *Duke Math. J.*, 28:301–324, 1961.

[11] J. J. Benedetto, C. Heil, and D. F. Walnut. Differentiation and the Balian-Low theorem. *J. Fourier Anal. Appl.*, 1(4):355–402, 1995.

[12] A. Bényi, L. Grafakos, K. Gröchenig, and K. Okoudjou. A class of Fourier multipliers for modulation spaces. *Appl. Comput. Harmon. Anal.*, 19(1):131–139, 2005.

[13] A. Bényi, K. Gröchenig, C. Heil, and K. Okoudjou. Modulation spaces and a class of bounded multilinear pseudodifferential operators. *J. Oper. Theory*, 54(2):387–399, 2005.

[14] A. Bényi, K. Gröchenig, K. A. Okoudjou, and L. G. Rogers. Unimodular Fourier multipliers for modulation spaces. *J. Funct. Anal.*, 246(2):366–384, 2007.

[15] A. Bényi and K. A. Okoudjou. Time-frequency estimates for pseudodifferential operators. In *Harmonic Analysis, Partial Differential Equations, and Related Topics*, volume 428 of *Contemp. Math.*, pages 13–22. Amer. Math. Soc., Providence, RI, 2007.

[16] F. A. Berezin. Wick and anti-Wick symbols of operators. *Mat. Sb. (N.S.)*, 86(128):578–610, 1971.

[17] J. Bergh and J. Löfström. *Interpolation Spaces. An Introduction*, volume 223 of *Grundlehren der Mathematischen Wissenschaften*. Springer-Verlag, Berlin-New York, 1976.

[18] M. Berra, I. M. Bulai, E. Cordero, and F. Nicola. Gabor frames of Gaussian beams for the Schrödinger equation. *Appl. Comput. Harmon. Anal.*, 43(1):94–121, 2017.

[19] S. Bishop. Mixed modulation spaces and their application to pseudodifferential operators. *J. Math. Anal. Appl.*, 363(1):255–264, 2010.

[20] W. R. Bloom. Strict local inclusion results between spaces of Fourier transforms. *Pac. J. Math.*, 99(2):265–270, 1982.

[21] P. Boggiatto, E. Carypis, and A. Oliaro. Two aspects of the Donoho-Stark uncertainty principle. *J. Math. Anal. Appl.*, 434(2):1489–1503, 2016.

https://doi.org/10.1515/9783110532456-009

[22] P. Boggiatto, E. Carypis, and A. Oliaro. Cohen class of time-frequency representations and operators: boundedness and uncertainty principles. *J. Math. Anal. Appl.*, 461(1):304–318, 2018.

[23] P. Boggiatto, E. Cordero, and K. Gröchenig. Generalized anti-Wick operators with symbols in distributional Sobolev spaces. *Integral Equ. Oper. Theory*, 48(4):427–442, 2004.

[24] P. Boggiatto, G. De Donno, and A. Oliaro. Time-frequency representations of Wigner type and pseudo-differential operators. *Trans. Am. Math. Soc.*, 362(9):4955–4981, 2010.

[25] P. Boggiatto and G. Garello. Pseudo-differential operators and existence of Gabor frames. *J. Pseudo-Differ. Oper. Appl.*, 11(1):93–117, 2020.

[26] P. Boggiatto and J. Toft. Embeddings and compactness for generalized Sobolev-Shubin spaces and modulation spaces. *Appl. Anal.*, 84(3):269–282, 2005.

[27] C. Boiti, D. Jornet, and A. Oliaro. Regularity of partial differential operators in ultradifferentiable spaces and Wigner type transforms. *J. Math. Anal. Appl.*, 446(1):920–944, 2017.

[28] C. Boiti, D. Jornet, and A. Oliaro. The Gabor wave front set in spaces of ultradifferentiable functions. *Monatshefte Math.*, 188(2):199–246, 2019.

[29] C. Boiti, D. Jornet, and A. Oliaro. Real Paley-Wiener theorems in spaces of ultradifferentiable functions. *J. Funct. Anal.*, 278(4):108348, 45, 2020.

[30] J.-M. Bony. Opérateurs intégraux de Fourier et calcul de Weyl-Hörmander (cas d'une métrique symplectique). In *Journées "Équations aux Dérivées Partielles" (Saint-Jean-de-Monts, 1994). Exp. No. IX*, 14. École Polytech., Palaiseau, 1994.

[31] J.-M. Bony. Evolution equations and generalized Fourier integral operators. In *Advances in Phase Space Analysis of Partial Differential Equations*, volume 78 of *Progr. Nonlinear Differential Equations Appl.*, pages 59–72. Birkhäuser Boston, Inc., Boston, MA, 2009.

[32] J.-M. Bony and J.-Y. Chemin. Espaces fonctionnels associés au calcul de Weyl-Hörmander. *Bull. Soc. Math. Fr.*, 122(1):77–118, 1994.

[33] M. Born and P. Jordan. Zur quantenmechanik. *Z. Phys.*, 34:858–888, 1925.

[34] A. Boulkhemair. L^2 estimates for pseudodifferential operators. *Ann. Sc. Norm. Super. Pisa, Cl. Sci. (4)*, 22(1):155–183, 1995.

[35] A. Boulkhemair. Remarks on a Wiener type pseudodifferential algebra and Fourier integral operators. *Math. Res. Lett.*, 4(1):53–67, 1997.

[36] M. Bownik, M. S. Jakobsen, J. Lemvig, and K. A. Okoudjou. On Wilson bases in $L^2(\mathbb{R}^d)$. *SIAM J. Math. Anal.*, 49(5):3999–4023, 2017.

[37] H. Brezis. *Functional Analysis, Sobolev Spaces and Partial Differential Equations, Universitext.* Springer, New York, 2011.

[38] R. C. Busby and H. A. Smith. Product-convolution operators and mixed-norm spaces. *Trans. Am. Math. Soc.*, 263(2):309–341, 1981.

[39] E. Buzano and A. Oliaro. Global regularity of second order twisted differential operators. *J. Differ. Equ.*, 268(12):7364–7416, 2020.

[40] A.-P. Calderón and R. Vaillancourt. On the boundedness of pseudo-differential operators. *J. Math. Soc. Jpn.*, 23:374–378, 1971.

[41] E. J. Candès and L. Demanet. The curvelet representation of wave propagators is optimally sparse. *Commun. Pure Appl. Math.*, 58(11):1472–1528, 2005.

[42] E. J. Candès, L. Demanet, and L. Ying. Fast computation of Fourier integral operators. *SIAM J. Sci. Comput.*, 29(6):2464–2493, 2007.

[43] E. J. Candès and D. L. Donoho. Continuous curvelet transform. I. Resolution of the wavefront set. *Appl. Comput. Harmon. Anal.*, 19(2):162–197, 2005.

[44] M. Cappiello, R. Schulz, and P. Wahlberg. Conormal distributions in the Shubin calculus of pseudodifferential operators. *J. Math. Phys.*, 59(2):021502, 18, 2018.

[45] M. Cappiello and J. Toft. Pseudo-differential operators in a Gelfand-Shilov setting. *Math. Nachr.*, 290(5-6):738–755, 2017.

[46] D. Cardona and M. Ruzhansky. Multipliers for Besov spaces on graded Lie groups. *C. R. Math. Acad. Sci. Paris*, 355(4):400–405, 2017.

[47] E. Carypis and P. Wahlberg. Propagation of exponential phase space singularities for Schrödinger equations with quadratic Hamiltonians. *J. Fourier Anal. Appl.*, 23(3):530–571, 2017.

[48] L. Cattabriga. Alcuni problemi per equazioni differenziali lineri con coefficienti costanti. In *Quaderni dell'Unione Matematica Italiana, Pitagora, Bologna*, volume 24, arXiv:1902.08940, 1983.

[49] O. Christensen. Frame Decomposition in Hilbert Spaces. PhD thesis. Arhus Univ. and Univ. of Vienna, DK/Austria, 1993.

[50] O. Christensen. *An Introduction to Frames and Riesz Bases, Applied and Numerical Harmonic Analysis*. Birkhäuser/Springer, Cham, second edition, 2016.

[51] A. Cohen, W. Dahmen, and R. Devore. Adaptive wavelet schemes for nonlinear variational problems. *SIAM J. Numer. Anal.*, 41(5):1785–1823, 2003.

[52] L. Cohen. Generalized phase-space distribution functions. *J. Math. Phys.*, 7:781–786, 1966.

[53] L. Cohen. *Time-Frequency Analysis, Electrical Engineering Signal Processing*. Prentice Hall PTR, 1995.

[54] R. R. Coifman, G. Matviyenko, and Y. Meyer. Modulated Malvar-Wilson bases. *Appl. Comput. Harmon. Anal.*, 4(1):58–61, 1997.

[55] R. R. Coifman and Y. Meyer. *Au delà des opérateurs pseudo-différentiels*, volume 57 of *Astérisque*. Société Mathématique de France, Paris, 1978. With an English summary.

[56] F. Concetti and J. Toft. Schatten-von Neumann properties for Fourier integral operators with non-smooth symbols. I. *Ark. Mat.*, 47(2):295–312, 2009.

[57] J. B. Conway. *Functions of One Complex Variable*, volume 11 of *Graduate Texts in Mathematics*. Springer-Verlag, New York-Berlin, second edition, 1978.

[58] E. Cordero. Gelfand-Shilov window classes for weighted modulation spaces. *Integral Transforms Spec. Funct.*, 18(11-12):829–837, 2007.

[59] E. Cordero, M. de Gosson, M. Dörfler, and F. Nicola. Generalized Born-Jordan distributions and applications. *Adv. Comput. Math.*, 46:51, 2020.

[60] E. Cordero, M. de Gosson, and F. Nicola. On the invertibility of Born-Jordan quantization. *J. Math. Pures Appl. (9)*, 105(4):537–557, 2016.

[61] E. Cordero, M. de Gosson, and F. Nicola. Semi-classical time-frequency analysis and applications. *Math. Phys. Anal. Geom.*, 20(4):26, 23, 2017.

[62] E. Cordero, M. de Gosson, and F. Nicola. Time-frequency analysis of Born-Jordan pseudodifferential operators. *J. Funct. Anal.*, 272(2):577–598, 2017.

[63] E. Cordero, M. de Gosson, and F. Nicola. On the reduction of the interferences in the Born-Jordan distribution. *Appl. Comput. Harmon. Anal.*, 44(2):230–245, 2018.

[64] E. Cordero, M. de Gosson, and F. Nicola. A characterization of modulation spaces by symplectic rotations. *J. Funct. Anal.*, 278(11):108474, 2020.

[65] E. Cordero, L. D'Elia, and S. I. Trapasso. Norm estimates for τ-pseudodifferential operators in Wiener amalgam and modulation spaces. *J. Math. Anal. Appl.*, 471(1-2):541–563, 2019.

[66] E. Cordero and K. Gröchenig. Time-frequency analysis of localization operators. *J. Funct. Anal.*, 205(1):107–131, 2003.

[67] E. Cordero, K. Gröchenig, and F. Nicola. Approximation of Fourier integral operators by Gabor multipliers. *J. Fourier Anal. Appl.*, 18(4):661–684, 2012.

[68] E. Cordero, K. Gröchenig, F. Nicola, and L. Rodino. Wiener algebras of Fourier integral operators. *J. Math. Pures Appl. (9)*, 99(2):219–233, 2013.

[69] E. Cordero, K. Gröchenig, F. Nicola, and L. Rodino. Generalized metaplectic operators and the Schrödinger equation with a potential in the Sjöstrand class. *J. Math. Phys.*, 55(8):081506, 17, 2014.

[70] E. Cordero and F. Nicola. Metaplectic representation on Wiener amalgam spaces and applications to the Schrödinger equation. *J. Funct. Anal.*, 254(2):506–534, 2008.

[71] E. Cordero and F. Nicola. Some new Strichartz estimates for the Schrödinger equation. *J. Differ. Equ.*, 245(7):1945–1974, 2008.

[72] E. Cordero and F. Nicola. Strichartz estimates in Wiener amalgam spaces for the Schrödinger equation. *Math. Nachr.*, 281(1):25–41, 2008.

[73] E. Cordero and F. Nicola. Boundedness of Schrödinger type propagators on modulation spaces. *J. Fourier Anal. Appl.*, 16(3):311–339, 2010.

[74] E. Cordero and F. Nicola. Pseudodifferential operators on L^p, Wiener amalgam and modulation spaces. *Int. Math. Res. Not.*, (10):1860–1893, 2010.

[75] E. Cordero and F. Nicola. Sharp integral bounds for Wigner distributions. *Int. Math. Res. Not.*, (6):1779–1807, 2018.

[76] E. Cordero and F. Nicola. Kernel theorems for modulation spaces. *J. Fourier Anal. Appl.*, 25(1):131–144, 2019.

[77] E. Cordero, F. Nicola, and L. Rodino. Sparsity of Gabor representation of Schrödinger propagators. *Appl. Comput. Harmon. Anal.*, 26(3):357–370, 2009.

[78] E. Cordero, F. Nicola, and L. Rodino. On the global boundedness of Fourier integral operators. *Ann. Glob. Anal. Geom.*, 38(4):373–398, 2010.

[79] E. Cordero, F. Nicola, and L. Rodino. Time-frequency analysis of Fourier integral operators. *Commun. Pure Appl. Anal.*, 9(1):1–21, 2010.

[80] E. Cordero, F. Nicola, and L. Rodino. Time-frequency analysis of Schrödinger propagators. In *Evolution Equations of Hyperbolic and Schrödinger Type*, volume 301 of *Progr. Math.*, pages 63–85. Birkhäuser/Springer Basel AG, Basel, 2012.

[81] E. Cordero, F. Nicola, and L. Rodino. Schrödinger equations in modulation spaces. In *Studies in Phase Space Analysis with Applications to PDEs*, volume 84 of *Progr. Nonlinear Differential Equations Appl.*, pages 81–99. Birkhäuser/Springer, New York, 2013.

[82] E. Cordero, F. Nicola, and L. Rodino. Exponentially sparse representations of Fourier integral operators. *Rev. Mat. Iberoam.*, 31(2):461–476, 2015.

[83] E. Cordero, F. Nicola, and L. Rodino. Gabor representations of evolution operators. *Trans. Am. Math. Soc.*, 367(11):7639–7663, 2015.

[84] E. Cordero, F. Nicola, and L. Rodino. Propagation of the Gabor wave front set for Schrödinger equations with non-smooth potentials. *Rev. Math. Phys.*, 27(1):1550001, 33, 2015.

[85] E. Cordero, F. Nicola, and L. Rodino. Wave packet analysis of Schrödinger equations in analytic function spaces. *Adv. Math.*, 278:182–209, 2015.

[86] E. Cordero, F. Nicola, and S. I. Trapasso. Almost diagonalization of τ-pseudodifferential operators with symbols in Wiener amalgam and modulation spaces. *J. Fourier Anal. Appl.*, 25(4):1927–1957, 2019.

[87] E. Cordero, S. Pilipović, L. Rodino, and N. Teofanov. Localization operators and exponential weights for modulation spaces. *Mediterr. J. Math.*, 2(4):381–394, 2005.

[88] E. Cordero and L. Rodino. Time-frequency analysis: function spaces and applications. *Note Mat.*, 31(1):173–189, 2011.

[89] E. Cordero, A. Tabacco, and P. Wahlberg. Schrödinger-type propagators, pseudodifferential operators and modulation spaces. *J. Lond. Math. Soc. (2)*, 88(2):375–395, 2013.

[90] E. Cordero and S. I. Trapasso. Linear perturbations of the Wigner distribution and the Cohen class. *Anal. Appl. (Singap.)*, 18(3):385–422, 2020.

[91] E. Cordero and D. Zucco. Strichartz estimates for the Schrödinger equation. *CUBO*, 12(3):213–239, 2010.
[92] E. Cordero and D. Zucco. Strichartz estimates for the vibrating plate equation. *J. Evol. Equ.*, 11(4):827–845, 2011.
[93] H. O. Cordes. On compactness of commutators of multiplications and convolutions, and boundedness of pseudodifferential operators. *J. Funct. Anal.*, 18:115–131, 1975.
[94] A. Córdoba and C. Fefferman. Wave packets and Fourier integral operators. *Commun. Partial Differ. Equ.*, 3(11):979–1005, 1978.
[95] S. Coriasco. Fourier integral operators in SG classes. II. Application to SG hyperbolic Cauchy problems. *Ann. Univ. Ferrara, Sez. 7: Sci. Mat.*, 44:81–122 (1999), 1998.
[96] W. Craig, T. Kappeler, and W. Strauss. Microlocal dispersive smoothing for the Schrödinger equation. *Commun. Pure Appl. Math.*, 48(8):769–860, 1995.
[97] W. Czaja and Z. Rzeszotnik. Pseudodifferential operators and Gabor frames: spectral asymptotics. *Math. Nachr.*, 233/234:77–88, 2002.
[98] A. Dasgupta and M. Ruzhansky. The Gohberg lemma, compactness, and essential spectrum of operators on compact Lie groups. *J. Anal. Math.*, 128:179–190, 2016.
[99] I. Daubechies. Time-frequency localization operators: a geometric phase space approach. *IEEE Trans. Inf. Theory*, 34(4):605–612, 1988.
[100] I. Daubechies. The wavelet transform, time-frequency localization and signal analysis. *IEEE Trans. Inf. Theory*, 36(5):961–1005, 1990.
[101] I. Daubechies. *Ten Lectures on Wavelets*, volume 61 of *CBMS-NSF Regional Conference Series in Applied Mathematics*. Society for Industrial and Applied Mathematics (SIAM), Philadelphia, PA, 1992.
[102] I. Daubechies, A. Grossmann, and Y. Meyer. Painless nonorthogonal expansions. *J. Math. Phys.*, 27(5):1271–1283, 1986.
[103] I. Daubechies, S. Jaffard, and J.-L. Journé. A simple Wilson orthonormal basis with exponential decay. *SIAM J. Math. Anal.*, 22(2):554–573, 1991.
[104] I. Daubechies and A. J. E. M. Janssen. Two theorems on lattice expansions. *IEEE Trans. Inf. Theory*, 39(1):3–6, 1993.
[105] M. A. de Gosson. *Symplectic Methods in Harmonic Analysis and in Mathematical Physics*, volume 7 of *Pseudo-Differential Operators. Theory and Applications*. Birkhäuser/Springer Basel AG, Basel, 2011.
[106] M. A. de Gosson. Born-Jordan quantization and the equivalence of the Schrödinger and Heisenberg pictures. *Found. Phys.*, 44(10):1096–1106, 2014.
[107] M. A. de Gosson. The canonical group of transformations of a Weyl-Heisenberg frame; applications to Gaussian and Hermitian frames. *J. Geom. Phys.*, 114:375–383, 2017.
[108] M. A. de Gosson and F. Luef. Preferred quantization rules: Born-Jordan versus Weyl. The pseudo-differential point of view. *J. Pseudo-Differ. Oper. Appl.*, 2(1):115–139, 2011.
[109] J. Delgado and M. Ruzhansky. L^p-nuclearity, traces, and Grothendieck-Lidskii formula on compact Lie groups. *J. Math. Pures Appl. (9)*, 102(1):153–172, 2014.
[110] L. D'Elia and S. I. Trapasso. Boundedness of pseudodifferential operators with symbols in Wiener amalgam spaces on modulation spaces. *J. Pseudo-Differ. Oper. Appl.*, 9(4):881–890, 2018.
[111] R. A. DeVore and V. N. Temlyakov. Some remarks on greedy algorithms. *Adv. Comput. Math.*, 5(2-3):173–187, 1996.
[112] T. Dobler. Wiener amalgam spaces over locally compact groups. Master Thesis. University of Vienna, Austria, 1989.
[113] R. J. Duffin and A. C. Schaeffer. A class of nonharmonic Fourier series. *Trans. Am. Math. Soc.*, 72:341–366, 1952.

[114] J. J. Duistermaat and V. W. Guillemin. The spectrum of positive elliptic operators and periodic bicharacteristics. *Invent. Math.*, 29(1):39–79, 1975.
[115] J. J. Duistermaat and L. Hörmander. Fourier integral operators. II. *Acta Math.*, 128(3-4):183–269, 1972.
[116] K.-J. Engel and R. Nagel. *A Short Course on Operator Semigroups*, *Universitext*. Springer, New York, 2006.
[117] H. G. Feichtinger. Modulation spaces on locally compact Abelian groups. Technical report, University of Vienna, 1983.
[118] H. G. Feichtinger. Segal algebras that are not character invariant. *Chin. J. Math.*, 7(1):55–59, 1979.
[119] H. G. Feichtinger. Un espace de Banach de distributions tempérées sur les groupes localement compacts abéliens. *C. R. Acad. Sci. Paris Sér. A-B*, 290(17):A791–A794, 1980.
[120] H. G. Feichtinger. Banach spaces of distributions of Wiener's type and interpolation. In *Functional Analysis and Approximation (Oberwolfach, 1980)*, volume 60 of *Internat. Ser. Numer. Math.*, pages 153–165. Birkhäuser, Basel-Boston, Mass., 1981.
[121] H. G. Feichtinger. A characterization of minimal homogeneous Banach spaces. *Proc. Am. Math. Soc.*, 81(1):55–61, 1981.
[122] H. G. Feichtinger. Banach convolution algebras of Wiener type. In *Functions, Series, Operators, Vol. I, II (Budapest, 1980)*, pages 509–524. North-Holland, Amsterdam, 1983.
[123] H. G. Feichtinger. Atomic characterizations of modulation spaces through Gabor-type representations. *Rocky Mt. J. Math.*, 19(1):113–125, 1989. Constructive Function Theory–86 Conference (Edmonton, AB, 1986).
[124] H. G. Feichtinger. Generalized amalgams, with applications to Fourier transform. *Can. J. Math.*, 42(3):395–409, 1990.
[125] H. G. Feichtinger. Wiener amalgams over Euclidean spaces and some of their applications. In *Function Spaces (Edwardsville, IL, 1990)*, volume 136 of *Lecture Notes in Pure and Appl. Math.*, pages 123–137. Dekker, New York, 1992.
[126] H. G. Feichtinger. Modulation spaces on locally compact Abelian groups. In Krishna, M., Radha, R., & Thangavelu, S. (eds.), *Wavelets and Their Applications*, pages 99–140, Chennai, Jan. 2002, Allied Publishers (New Delhi), 2003.
[127] H. G. Feichtinger. Modulation spaces: looking back and ahead. *Sampl. Theory Signal Image Process.*, 5(2):109–140, 2006.
[128] H. G. Feichtinger. Elements of postmodern harmonic analysis. In *Operator-Related Function Theory and Time-Frequency Analysis*, volume 9 of *Abel Symp.*, pages 77–105. Springer, Cham, 2015.
[129] H. G. Feichtinger, C. C. Graham, and E. H. Lakien. Nonfactorization in commutative, weakly selfadjoint Banach algebras. *Pac. J. Math.*, 80(1):117–125, 1979.
[130] H. G. Feichtinger and P. Gröbner. Banach spaces of distributions defined by decomposition methods. I. *Math. Nachr.*, 123:97–120, 1985.
[131] H. G. Feichtinger and K. Gröchenig. Banach spaces related to integrable group representations and their atomic decompositions. I. *J. Funct. Anal.*, 86(2):307–340, 1989.
[132] H. G. Feichtinger and K. Gröchenig. Banach spaces related to integrable group representations and their atomic decompositions. II. *Monatshefte Math.*, 108(2-3):129–148, 1989.
[133] H. G. Feichtinger and K. Gröchenig. Gabor wavelets and the Heisenberg group: Gabor expansions and short time Fourier transform from the group theoretical point of view. In *Wavelets*, volume 2 of *Wavelet Anal. Appl.*, pages 359–397. Academic Press, Boston, MA, 1992.
[134] H. G. Feichtinger and K. Gröchenig. Gabor frames and time-frequency analysis of distributions. *J. Funct. Anal.*, 146(2):464–495, 1997.

[135] H. G. Feichtinger, K. Gröchenig, and D. Walnut. Wilson bases and modulation spaces. *Math. Nachr.*, 155:7–17, 1992.

[136] H. G. Feichtinger and A. J. E. M. Janssen. Validity of WH-frame bound conditions depends on lattice parameters. *Appl. Comput. Harmon. Anal.*, 8(1):104–112, 2000.

[137] H. G. Feichtinger and N. Kaiblinger. Varying the time-frequency lattice of Gabor frames. *Trans. Am. Math. Soc.*, 356(5):2001–2023, 2004.

[138] H. G. Feichtinger and W. Kozek. Quantization of TF lattice-invariant operators on elementary LCA groups. In *Gabor Analysis and Algorithms, Appl. Numer. Harmon. Anal.*, pages 233–266. Birkhäuser Boston, Boston, MA, 1998.

[139] H. G. Feichtinger and K. Nowak. A first survey of Gabor multipliers. In *Advances in Gabor Analysis, Appl. Numer. Harmon. Anal.*, pages 99–128. Birkhäuser Boston, Boston, MA, 2003.

[140] H. G. Feichtinger and G. Zimmermann. A Banach space of test functions for Gabor analysis. In *Gabor Analysis and Algorithms, Appl. Numer. Harmon. Anal.*, pages 123–170. Birkhäuser Boston, Boston, MA, 1998.

[141] C. Fernández and A. Galbis. Compactness of time-frequency localization operators on $L^2(\mathbb{R}^d)$. *J. Funct. Anal.*, 233(2):335–350, 2006.

[142] C. Fernández, A. Galbis, and J. Martínez. Multilinear Fourier multipliers related to time-frequency localization. *J. Math. Anal. Appl.*, 398(1):113–122, 2013.

[143] C. Fernández, A. Galbis, and E. Primo. Compactness of Fourier integral operators on weighted modulation spaces. *Trans. Am. Math. Soc.*, 372(1):733–753, 2019.

[144] V. Fischer and M. Ruzhansky. *Quantization on Nilpotent Lie Groups*, volume 314 of *Progress in Mathematics*. Birkhäuser/Springer, Cham, 2016.

[145] G. B. Folland. *Harmonic Analysis in Phase Space*, volume 122 of *Annals of Mathematics Studies*. Princeton University Press, Princeton, NJ, 1989.

[146] G. B. Folland. *Real Analysis, Pure and Applied Mathematics (New York)*. John Wiley & Sons, Inc., New York, second edition, 1999. Modern techniques and their applications, A Wiley-Interscience Publication.

[147] J. J. F. Fournier. Local complements to the Hausdorff-Young theorem. *Mich. Math. J.*, 20:263–276, 1973.

[148] J. J. F. Fournier and J. Stewart. Amalgams of L^p and l^q. *Bull., New Ser., Am. Math. Soc.*, 13(1):1–21, 1985.

[149] M. Frank, V. I. Paulsen, and T. R. Tiballi. Symmetric approximation of frames and bases in Hilbert spaces. *Trans. Am. Math. Soc.*, 354(2):777–793, 2002.

[150] M. Frazier and B. Jawerth. Decomposition of Besov spaces. *Indiana Univ. Math. J.*, 34(4):777–799, 1985.

[151] M. Frazier and B. Jawerth. A discrete transform and decompositions of distribution spaces. *J. Funct. Anal.*, 93(1):34–170, 1990.

[152] D. Gabor. Theory of communication. *J. IEE*, 93(III):429–457, 1946.

[153] Y. V. Galperin and S. Samarah. Time-frequency analysis on modulation spaces $M_m^{p,q}$, $0 < p, q \leq \infty$. *Appl. Comput. Harmon. Anal.*, 16(1):1–18, 2004.

[154] C. Garetto and M. Ruzhansky. Wave equation for sums of squares on compact Lie groups. *J. Differ. Equ.*, 258(12):4324–4347, 2015.

[155] I. M. Gelfand and G. E. Shilov. *Generalized Functions. Vol. 3: Theory of Differential Equations*. Translated from the Russian by Meinhard E. Mayer. Academic Press, New York-London, 1967.

[156] I. M. Gelfand and G. E. Shilov. *Generalized Functions*, volume 2. Academic Press [Harcourt Brace Jovanovich, Publishers], New York-London, 1968 [1977]. Spaces of fundamental and generalized functions, Translated from the Russian by Morris D. Friedman, Amiel Feinstein and Christian P. Peltzer.

[157] J. Ginibre and G. Velo. Smoothing properties and retarded estimates for some dispersive evolution equations. *Commun. Math. Phys.*, 144(1):163–188, 1992.
[158] S. Graffi and L. Zanelli. Geometric approach to the Hamilton-Jacobi equation and global parametrices for the Schrödinger propagator. *Rev. Math. Phys.*, 23(9):969–1008, 2011.
[159] K. Gröchenig. Describing functions: atomic decompositions versus frames. *Monatshefte Math.*, 112(1):1–42, 1991.
[160] K. Gröchenig. *Foundations of Time-Frequency Analysis, Applied and Numerical Harmonic Analysis*. Birkhäuser Boston, Inc., Boston, MA, 2001.
[161] K. Gröchenig. A pedestrian's approach to pseudodifferential operators. In *Harmonic Analysis and Applications, Appl. Numer. Harmon. Anal.*, pages 139–169. Birkhäuser Boston, Boston, MA, 2006.
[162] K. Gröchenig. Time-frequency analysis of Sjöstrand's class. *Rev. Mat. Iberoam.*, 22(2):703–724, 2006.
[163] K. Gröchenig and C. Heil. Modulation spaces and pseudodifferential operators. *Integral Equ. Oper. Theory*, 34(4):439–457, 1999.
[164] K. Gröchenig and Y. Lyubarskii. Gabor (super)frames with Hermite functions. *Math. Ann.*, 345(2):267–286, 2009.
[165] K. Gröchenig and Z. Rzeszotnik. Banach algebras of pseudodifferential operators and their almost diagonalization. *Ann. Inst. Fourier (Grenoble)*, 58(7):2279–2314, 2008.
[166] K. Gröchenig, Z. Rzeszotnik, and T. Strohmer. Convergence analysis of the finite section method and Banach algebras of matrices. *Integral Equ. Oper. Theory*, 67(2):183–202, 2010.
[167] K. Gröchenig and S. Samarah. Nonlinear approximation with local Fourier bases. *Constr. Approx.*, 16(3):317–331, 2000.
[168] K. Gröchenig and T. Strohmer. Pseudodifferential operators on locally compact abelian groups and Sjöstrand's symbol class. *J. Reine Angew. Math.*, 613:121–146, 2007.
[169] K. Gröchenig and J. Toft. Isomorphism properties of Toeplitz operators and pseudo-differential operators between modulation spaces. *J. Anal. Math.*, 114:255–283, 2011.
[170] K. Gröchenig and J. Toft. The range of localization operators and lifting theorems for modulation and Bargmann-Fock spaces. *Trans. Am. Math. Soc.*, 365(8):4475–4496, 2013.
[171] K. Gröchenig and G. Zimmermann. Spaces of test functions via the STFT. *J. Funct. Spaces Appl.*, 2(1):25–53, 2004.
[172] A. Grossmann, G. Loupias, and E. M. Stein. An algebra of pseudodifferential operators and quantum mechanics in phase space. *Ann. Inst. Fourier (Grenoble)*, 18(fasc. 2):343–368 viii (1969), 1968.
[173] K. Guo and D. Labate. Optimally sparse multidimensional representation using shearlets. *SIAM J. Math. Anal.*, 39(1):298–318, 2007.
[174] K. Guo and D. Labate. Sparse shearlet representation of Fourier integral operators. *Electron. Res. Announc. Math. Sci.*, 14:7–19, 2007.
[175] K. Guo and D. Labate. Representation of Fourier integral operators using shearlets. *J. Fourier Anal. Appl.*, 14(3):327–371, 2008.
[176] W. Guo, J. Chen, D. Fan, and G. Zhao. Characterizations of some properties on weighted modulation and Wiener amalgam spaces. *Mich. Math. J.*, 68(3):451–482, 2019.
[177] W. Guo, H. Wu, and G. Zhao. Inclusion relations between modulation and Triebel-Lizorkin spaces. *Proc. Am. Math. Soc.*, 145(11):4807–4820, 2017.
[178] A. Hassell and J. Wunsch. The Schrödinger propagator for scattering metrics. *Ann. Math. (2)*, 162(1):487–523, 2005.
[179] C. Heil. *Wiener amalgam spaces in generalized harmonic analysis and wavelet theory*. ProQuest LLC, Ann Arbor, MI, 1990. Thesis (Ph.D.)–University of Maryland, College Park.

[180] C. Heil. Integral operators, pseudodifferential operators, and Gabor frames. In *Advances in Gabor Analysis, Appl. Numer. Harmon. Anal.*, pages 153–169. Birkhäuser Boston, Boston, MA, 2003.

[181] C. Heil. An introduction to weighted Wiener amalgams. In Krishna, M., Radha, R., & Thangavelu, S. (eds.), *Wavelets and Their Applications*, pages 183–216, Chennai, Jan. 2002, Allied Publishers (New Delhi), 2003.

[182] C. Heil. History and evolution of the density theorem for Gabor frames. *J. Fourier Anal. Appl.*, 13(2):113–166, 2007.

[183] C. Heil. *A Basis Theory Primer, Applied and Numerical Harmonic Analysis*. Birkhäuser/Springer, New York, expanded edition, 2011.

[184] C. Heil, J. Ramanathan, and P. Topiwala. Linear independence of time-frequency translates. *Proc. Am. Math. Soc.*, 124(9):2787–2795, 1996.

[185] C. Heil, J. Ramanathan, and P. Topiwala. Singular values of compact pseudodifferential operators. *J. Funct. Anal.*, 150(2):426–452, 1997.

[186] C. Heil and D. F. Walnut. Continuous and discrete wavelet transforms. *SIAM Rev.*, 31(4):628–666, 1989.

[187] B. Helffer. *Théorie spectrale pour des opérateurs globalement elliptiques*, volume 112 of *Astérisque*. Société Mathématique de France, Paris, 1984. With an English summary.

[188] B. Helffer and D. Robert. Comportement asymptotique précisé du spectre d'opérateurs globalement elliptiques dans $\mathbf{R}^n$. *C. R. Acad. Sci. Paris Sér. I Math.*, 292(6):363–366, 1981.

[189] F. Holland. Harmonic analysis on amalgams of L^p and 1^q. *J. Lond. Math. Soc. (2)*, 10:295–305, 1975.

[190] L. Hörmander. On the division of distributions by polynomials. *Ark. Mat.*, 3:555–568, 1958.

[191] L. Hörmander. Fourier integral operators. I. *Acta Math.*, 127(1-2):79–183, 1971.

[192] L. Hörmander. The Weyl calculus of pseudodifferential operators. *Commun. Pure Appl. Math.*, 32(3):360–444, 1979.

[193] L. Hörmander. *The Analysis of Linear Partial Differential Operators. III*, volume 274 of *Grundlehren der Mathematischen Wissenschaften [Fundamental Principles of Mathematical Sciences]*. Springer-Verlag, Berlin, 1985. Pseudodifferential operators.

[194] L. Hörmander. *The Analysis of Linear Partial Differential Operators. I, Springer Study Edition*. Springer-Verlag, Berlin, second edition, 1990. Distribution theory and Fourier analysis.

[195] L. Hörmander. Quadratic hyperbolic operators. In *Microlocal Analysis and Applications (Montecatini Terme, 1989)*, volume 1495 of *Lecture Notes in Math.*, pages 118–160. Springer, Berlin, 1991.

[196] L. Hörmander. *The Analysis of Linear Partial Differential Operators. II, Classics in Mathematics*. Springer-Verlag, Berlin, 2005. Differential Operators with Constant Coefficients.

[197] R. Howe. Quantum mechanics and partial differential equations. *J. Funct. Anal.*, 38(2):188–254, 1980.

[198] R. L. Hudson. When is the Wigner quasi-probability density non-negative? *Rep. Math. Phys.*, 6(2):249–252, 1974.

[199] K. Husimi. Some formal properties of the density matrix. *Proc. Phys. Math. Soc. Jpn.*, 22(2):264–314, 1940.

[200] K. Ito. Propagation of singularities for Schrödinger equations on the Euclidean space with a scattering metric. *Commun. Partial Differ. Equ.*, 31(10-12):1735–1777, 2006.

[201] K. Ito and S. Nakamura. Singularities of solutions to the Schrödinger equation on scattering manifold. *Am. J. Math.*, 131(6):1835–1865, 2009.

[202] M. S. Jakobsen. On a (no longer) new Segal algebra: a review of the Feichtinger algebra. *J. Fourier Anal. Appl.*, 24(6):1579–1660, 2018.

[203] A. J. E. M. Janssen. Bargmann transform, Zak transform, and coherent states. *J. Math. Phys.*, 23(5):720–731, 1982.
[204] A. J. E. M. Janssen. The Zak transform: a signal transform for sampled time-continuous signals. *Philips J. Res.*, 43(1):23–69, 1988.
[205] K. Johansson, S. Pilipović, N. Teofanov, and J. Toft. Gabor pairs, and a discrete approach to wave-front sets. *Monatshefte Math.*, 166(2):181–199, 2012.
[206] T. Kato. Linear evolution equations of "hyperbolic" type. *J. Fac. Sci. Univ. Tokyo Sect. I*, 17:241–258, 1970.
[207] T. Kato. Boundedness of some pseudo-differential operators. *Osaka J. Math.*, 13(1):1–9, 1976.
[208] M. Keel and T. Tao. Endpoint Strichartz estimates. *Am. J. Math.*, 120(5):955–980, 1998.
[209] S. Kim and Y. Koh. Strichartz estimates for the magnetic Schrödinger equation with potentials V of critical decay. *Commun. Partial Differ. Equ.*, 42(9):1467–1480, 2017.
[210] S. Kim, Y. Koh, and I. Seo. Strichartz estimates for the Schrödinger propagator in Wiener amalgam spaces. *J. Math. Anal. Appl.*, 478(1):236–248, 2019.
[211] M. Kobayashi. Modulation spaces $M^{p,q}$ for $0 < p, q \leq \infty$. *J. Funct. Spaces Appl.*, 4(3):329–341, 2006.
[212] M. Kobayashi. Dual of modulation spaces. *J. Funct. Spaces Appl.*, 5(1):1–8, 2007.
[213] J. J. Kohn and L. Nirenberg. An algebra of pseudo-differential operators. *Commun. Pure Appl. Math.*, 18:269–305, 1965.
[214] S. G. Krantz and H. R. Parks. *The Implicit Function Theorem*. Modern Birkhäuser Classics. Birkhäuser/Springer, New York, 2013. History, theory, and applications, Reprint of the 2003 edition.
[215] G. Kutyniok and D. Labate. Resolution of the wavefront set using continuous shearlets. *Trans. Am. Math. Soc.*, 361(5):2719–2754, 2009.
[216] D. Labate. Pseudodifferential operators on modulation spaces. *J. Math. Anal. Appl.*, 262(1):242–255, 2001.
[217] P. D. Lax. Asymptotic solutions of oscillatory initial value problems. *Duke Math. J.*, 24:627–646, 1957.
[218] E. H. Lieb. Integral bounds for radar ambiguity functions and Wigner distributions. *J. Math. Phys.*, 31(3):594–599, 1990.
[219] S. Lojasiewicz. Sur le problême de la division. *Stud. Math.*, 18:87–136, 1959.
[220] F. Low. Complete sets of wave packets. In *A Passion for Physics-Essays in Honor of Geoffrey Chew*, pages 17–22. World Scientific, Singapore, 1985.
[221] Y. I. Lyubarskii. Frames in the Bargmann space of entire functions. In *Entire and Subharmonic Functions*, volume 11 of *Adv. Soviet Math.*, pages 167–180. Amer. Math. Soc., Providence, RI, 1992.
[222] A. Martinez, S. Nakamura, and V. Sordoni. Analytic smoothing effect for the Schrödinger equation with long-range perturbation. *Commun. Pure Appl. Math.*, 59(9):1330–1351, 2006.
[223] A. Martinez, S. Nakamura, and V. Sordoni. Analytic wave front set for solutions to Schrödinger equations. *Adv. Math.*, 222(4):1277–1307, 2009.
[224] N. H. McCoy. On the function in quantum mechanics which corresponds to a given function in classical mechanics. *Proc. Natl. Acad. Sci.*, 18(11):674–676, 1932.
[225] R. B. Melrose. Spectral and scattering theory for the Laplacian on asymptotically Euclidian spaces. In *Spectral and Scattering Theory (Sanda, 1992)*, volume 161 of *Lecture Notes in Pure and Appl. Math.*, pages 85–130. Dekker, New York, 1994.
[226] B. S. Mitjagin. Nuclearity and other properties of spaces of type *S*. *Trudy Moskov. Mat. Obšč.*, 9:317–328, 1960.
[227] R. Mizuhara. Microlocal smoothing effect for the Schrödinger evolution equation in a Gevrey class. *J. Math. Pures Appl. (9)*, 91(2):115–136, 2009.

[228] S. Molahajloo, K. A. Okoudjou, and G. E. Pfander. Boundedness of multilinear pseudo-differential operators on modulation spaces. *J. Fourier Anal. Appl.*, 22(6):1381–1415, 2016.
[229] S. Molahajloo and G. E. Pfander. Boundedness of pseudo-differential operators on L^p, Sobolev and modulation spaces. *Math. Model. Nat. Phenom.*, 8(1):175–192, 2013.
[230] M. Măntoiu and M. Ruzhansky. Pseudo-differential operators, Wigner transform and Weyl systems on type I locally compact groups. *Doc. Math.*, 22:1539–1592, 2017.
[231] M. Nagase. The L^p-boundedness of pseudo-differential operators with non-regular symbols. *Commun. Partial Differ. Equ.*, 2(10):1045–1061, 1977.
[232] S. Nakamura. Propagation of the homogeneous wave front set for Schrödinger equations. *Duke Math. J.*, 126(2):349–367, 2005.
[233] S. Nakamura. Semiclassical singularities propagation property for Schrödinger equations. *J. Math. Soc. Jpn.*, 61(1):177–211, 2009.
[234] F. Nicola. Phase space analysis of semilinear parabolic equations. *J. Funct. Anal.*, 267(3):727–743, 2014.
[235] F. Nicola. Convergence in L^p for Feynman path integrals. *Adv. Math.*, 294:384–409, 2016.
[236] F. Nicola and L. Rodino. *Global Pseudo-Differential Calculus on Euclidean Spaces*, volume 4 of *Pseudo-Differential Operators. Theory and Applications*. Birkhäuser Verlag, Basel, 2010.
[237] F. Nicola and L. Rodino. Global regularity for ordinary differential operators with polynomial coefficients. *J. Differ. Equ.*, 255(9):2871–2890, 2013.
[238] F. Nicola and L. Rodino. Propagation of Gabor singularities for semilinear Schrödinger equations. *Nonlinear Differ. Equ. Appl.*, 22(6):1715–1732, 2015.
[239] F. Nicola and S. I. Trapasso. On the pointwise convergence of the integral kernels in the Feynman-Trotter Formula. *Commun. Math. Phys.*, 376:2277–2299, 2020.
[240] J. Peetre. *New Thoughts on Besov Spaces*, volume 1 of *Duke University Mathematics Series*. Mathematics Department, Duke University, Durham, N.C., 1976.
[241] G. E. Pfander. Sampling of operators. *J. Fourier Anal. Appl.*, 19(3):612–650, 2013.
[242] S. Pilipović. Tempered ultradistributions. *Boll. Unione Mat. Ital., B (7)*, 2(2):235–251, 1988.
[243] S. Pilipović and N. Teofanov. Pseudodifferential operators on ultra-modulation spaces. *J. Funct. Anal.*, 208(1):194–228, 2004.
[244] S. Pilipović, N. Teofanov, and J. Toft. Micro-local analysis with Fourier Lebesgue spaces. Part I. *J. Fourier Anal. Appl.*, 17(3):374–407, 2011.
[245] K. Pravda-Starov, L. Rodino, and P. Wahlberg. Propagation of Gabor singularities for Schrödinger equations with quadratic Hamiltonians. *Math. Nachr.*, 291(1):128–159, 2018.
[246] J. Ramanathan and P. Topiwala. Time-frequency localization via the Weyl correspondence. *SIAM J. Math. Anal.*, 24(5):1378–1393, 1993.
[247] J. Rauch. *Partial Differential Equations*, volume 128 of *Graduate Texts in Mathematics*. Springer-Verlag, New York, 1991.
[248] H. Rauhut. Coorbit space theory for quasi-Banach spaces. *Stud. Math.*, 180(3):237–253, 2007.
[249] M. Reed and B. Simon. *Methods of Modern Mathematical Physics. II. Fourier Analysis, Self-Adjointness*. Academic Press [Harcourt Brace Jovanovich, Publishers], New York-London, 1975.
[250] L. Robbiano and C. Zuily. Microlocal analytic smoothing effect for the Schrödinger equation. *Duke Math. J.*, 100(1):93–129, 1999.
[251] L. Robbiano and C. Zuily. Analytic theory for the quadratic scattering, wave front set and application to the Schrödinger equation. In *Hyperbolic Problems and Related Topics, Grad. Ser. Anal.*, pages 319–339. Int. Press, Somerville, MA, 2003.
[252] R. Rochberg and K. Tachizawa. Pseudodifferential operators, Gabor frames, and local trigonometric bases. In *Gabor Analysis and Algorithms, Appl. Numer. Harmon. Anal.*, pages 171–192. Birkhäuser Boston, Boston, MA, 1998.

[253] L. Rodino and P. Wahlberg. The Gabor wave front set. *Monatshefte Math.*, 173(4):625–655, 2014.

[254] W. Rudin. *Real and Complex Analysis*. McGraw-Hill Book Co., New York, third edition, 1987.

[255] M. Ruzhansky, M. Sugimoto, J. Toft, and N. Tomita. Changes of variables in modulation and Wiener amalgam spaces. *Math. Nachr.*, 284(16):2078–2092, 2011.

[256] M. Ruzhansky and D. Suragan. Uncertainty relations on nilpotent Lie groups. *Proc. A.*, 473(2201):20170082, 12, 2017.

[257] M. Ruzhansky and V. Turunen. *Pseudo-Differential Operators and Symmetries*, volume 2 of *Pseudo-Differential Operators. Theory and Applications*. Birkhäuser Verlag, Basel, 2010. Background analysis and advanced topics.

[258] M. Ruzhansky and J. Wirth. L^p Fourier multipliers on compact Lie groups. *Math. Z.*, 280(3-4):621–642, 2015.

[259] R. Schulz and P. Wahlberg. Microlocal properties of Shubin pseudodifferential and localization operators. *J. Pseudo-Differ. Oper. Appl.*, 7(1):91–111, 2016.

[260] R. Schulz and P. Wahlberg. Equality of the homogeneous and the Gabor wave front set. *Commun. Partial Differ. Equ.*, 42(5):703–730, 2017.

[261] L. Schwartz. *Théorie des Distributions, Publications de l'Institut de Mathématique de l'Université de Strasbourg, No. IX-X. Nouvelle édition, entiérement corrigée, refondue et augmentée*. Hermann, Paris, 1966.

[262] K. Seip. Density theorems for sampling and interpolation in the Bargmann-Fock space. *Bull., New Ser., Am. Math. Soc.*, 26(2):322–328, 1992.

[263] K. Seip and R. Wallstén. Density theorems for sampling and interpolation in the Bargmann-Fock space. II. *J. Reine Angew. Math.*, 429:107–113, 1992.

[264] K. Seongyeon, K. Youngwoo, and S. Ihyeok. Strichartz estimates for the Schrödinger Propagator in Wiener Amalgam Spaces. *Preprint*, arXiv:1902.08940, 2019.

[265] M. A. Shubin. *Pseudodifferential Operators and Spectral Theory, Springer Series in Soviet Mathematics*. Springer-Verlag, Berlin, 1987. Translated from the Russian by Stig I. Andersson.

[266] J. Sjöstrand. An algebra of pseudodifferential operators. *Math. Res. Lett.*, 1(2):185–192, 1994.

[267] J. Sjöstrand. Wiener type algebras of pseudodifferential operators. In *Séminaire sur les Équations aux Dérivées Partielles, 1994–1995*, pages Exp. No. IV, 21. École Polytech., Palaiseau, 1995.

[268] E. M. Stein. *Singular Integrals and Differentiability Properties of Functions*, volume 30 of *Princeton Mathematical Series*. Princeton University Press, Princeton, N.J., 1970.

[269] E. M. Stein. *Harmonic Analysis: Real-Variable Methods, Orthogonality, and Oscillatory Integrals*, volume 43 of *Princeton Mathematical Series*. Princeton University Press, Princeton, NJ, 1993. With the assistance of Timothy S. Murphy, Monographs in Harmonic Analysis, III.

[270] E. M. Stein and G. Weiss. *Introduction to Fourier Analysis on Euclidean Spaces*, volume 32 of *Princeton Mathematical Series*. Princeton University Press, Princeton, N.J., 1971.

[271] T. Strohmer. Pseudodifferential operators and Banach algebras in mobile communications. *Appl. Comput. Harmon. Anal.*, 20(2):237–249, 2006.

[272] M. Sugimoto. L^p-Boundedness of pseudodifferential operators satisfying Besov estimates. I. *J. Math. Soc. Jpn.*, 40(1):105–122, 1988.

[273] M. Sugimoto. L^p-Boundedness of pseudodifferential operators satisfying Besov estimates. II. *J. Fac. Sci. Univ. Tokyo Sect. IA Math.*, 35(1):149–162, 1988.

[274] M. Sugimoto and N. Tomita. The dilation property of modulation spaces and their inclusion relation with Besov spaces. *J. Funct. Anal.*, 248(1):79–106, 2007.

[275] K. Tachizawa. The pseudodifferential operators and Wilson bases. *J. Math. Pures Appl. (9)*, 75(6):509–529, 1996.

[276] T. Tao. Low regularity semi-linear wave equations. *Commun. Partial Differ. Equ.*, 24(3-4):599–629, 1999.
[277] T. Tao. *Nonlinear Dispersive Equations: Local and Global Analysis, CBMS Regional Conference Series in Mathematics*. Amer. Math. Soc., 2006.
[278] T. Tao. Lecture Notes 2 for 247 A: Fourier analysis. www.math.ucla.edu/˜tao, 2007.
[279] D. Tataru. Phase space transforms and microlocal analysis. In *Phase Space Analysis of Partial Differential Equations. Vol. II, Pubbl. Cent. Ric. Mat. Ennio De Giorgi*, pages 505–524. Scuola Norm. Sup., Pisa, 2004.
[280] M. E. Taylor. *Pseudodifferential Operators*. Princeton Univ. Press, Princeton, NJ, 1981.
[281] M. E. Taylor. *Tools for PDE*, volume 81 of *Mathematical Surveys and Monographs*. American Mathematical Society, Providence, RI, 2000. Pseudodifferential operators, paradifferential operators, and layer potentials.
[282] N. Teofanov. Ultradistributions and time-frequency analysis. In *Pseudo-Differential Operators and Related Topics*, volume 164 of *Oper. Theory Adv. Appl.*, pages 173–192. Birkhäuser, Basel, 2006.
[283] N. Teofanov. Continuity and Schatten–von Neumann properties for localization operators on modulation spaces. *Mediterr. J. Math.*, 13(2):745–758, 2016.
[284] H. ter Morsche and P. J. Oonincx. On the integral representations for metaplectic operators. *J. Fourier Anal. Appl.*, 8(3):245–257, 2002.
[285] E. C. Titchmarsh. *Introduction to the Theory of Fourier Integrals*. Chelsea Publishing Co., New York, third edition, 1986.
[286] J. Toft. Continuity properties for modulation spaces, with applications to pseudo-differential calculus. I. *J. Funct. Anal.*, 207(2):399–429, 2004.
[287] J. Toft. Continuity properties for modulation spaces, with applications to pseudo-differential calculus. II. *Ann. Glob. Anal. Geom.*, 26(1):73–106, 2004.
[288] J. Toft. Continuity and Schatten properties for pseudo-differential operators on modulation spaces. In *Modern Trends in Pseudo-Differential Operators*, volume 172 of *Oper. Theory Adv. Appl.*, pages 173–206. Birkhäuser, Basel, 2007.
[289] J. Toft. The Bargmann transform on modulation and Gelfand-Shilov spaces, with applications to Toeplitz and pseudo-differential operators. *J. Pseudo-Differ. Oper. Appl.*, 3(2):145–227, 2012.
[290] J. Toft. Matrix parameterized pseudo-differential calculi on modulation spaces. In *Generalized Functions and Fourier Analysis*, volume 260 of *Oper. Theory Adv. Appl.*, pages 215–235. Birkhäuser/Springer, Cham, 2017.
[291] J. Toft, F. Concetti, and G. Garello. Schatten-von Neumann properties for Fourier integral operators with non-smooth symbols II. *Osaka J. Math.*, 47(3):739–786, 2010.
[292] J. Toft, K. Johansson, S. Pilipović, and N. Teofanov. Sharp convolution and multiplication estimates in weighted spaces. *Anal. Appl. (Singap.)*, 13(5):457–480, 2015.
[293] J. Toft, R. Üster, E. Nabizadeh, and S. Öztop. Continuity and Bargmann mapping properties of quasi-Banach Orlicz modulation spaces. *Preprint*, arXiv:1907.02331, 2020.
[294] S. I. Trapasso. Time-frequency analysis of the Dirac equation. *J. Differ. Equ.*, 269(3):2477–2502, 2020.
[295] F. Trèves. *Topological Vector Spaces, Distributions and Kernels*. Academic Press, New York-London, 1967.
[296] H. Triebel. *Interpolation Theory, Function Spaces, Differential Operators*, volume 18 of *North-Holland Mathematical Library*. North-Holland Publishing Co., Amsterdam-New York, 1978.
[297] H. Triebel. *Spaces of Besov-Hardy-Sobolev Type*. BSB B. G. Teubner Verlagsgesellschaft, Leipzig, 1978. Teubner-Texte zur Mathematik, with German, French and Russian summaries.

[298] H. Triebel. Modulation spaces on the Euclidean n-space. *Z. Anal. Anwend.*, 2(5):443–457, 1983.
[299] J. Ueberberg. Zur Spektralinvarianz von Algebren von Pseudodifferentialoperatoren in der L^p-Theorie. *Manuscr. Math.*, 61(4):459–475, 1988.
[300] J. Ville. Theorie et applications de la notion de signal analytique. *Câbles et Transmissions*, 2(2):61–74, 1948.
[301] P. Wahlberg. Propagation of polynomial phase space singularities for Schrödinger equations with quadratic Hamiltonians. *Math. Scand.*, 122(1):107–140, 2018.
[302] D. F. Walnut. Continuity properties of the Gabor frame operator. *J. Math. Anal. Appl.*, 165(2):479–504, 1992.
[303] B. Wang and C. Huang. Frequency-uniform decomposition method for the generalized BO, KdV and NLS equations. *J. Differ. Equ.*, 239(1):213–250, 2007.
[304] B. Wang and H. Hudzik. The global Cauchy problem for the NLS and NLKG with small rough data. *J. Differ. Equ.*, 232(1):36–73, 2007.
[305] B. Wang, Z. Huo, C. Hao, and Z. Guo. *Harmonic Analysis Method for Nonlinear Evolution Equations. i.* World Scientific Publishing Co. Pte. Ltd., Hackensack, NJ, 2011.
[306] B. Wang, Z. Lifeng, and G. Boling. Isometric decomposition operators, function spaces $E^{\lambda}_{p,q}$ and applications to nonlinear evolution equations. *J. Funct. Anal.*, 233(1):1–39, 2006.
[307] A. Weinstein. A symbol class for some Schrödinger equations on $\mathbf{R}^n$. *Am. J. Math.*, 107(1):1–21, 1985.
[308] F. Weisz. Inversion of the short-time Fourier transform using Riemannian sums. *J. Fourier Anal. Appl.*, 13(3):357–368, 2007.
[309] H. Weyl. *Gruppentheorie und Quantenmechanik.* Wissenschaftliche Buchgesellschaft, Darmstadt, second edition, 1977.
[310] N. Wiener. On the representation of functions by trigonometrical integrals. *Math. Z.*, 24(1):575–616, 1926.
[311] N. Wiener. Tauberian theorems. *Ann. Math. (2)*, 33(1):1–100, 1932.
[312] N. Wiener. *Generalized Harmonic Analysis and Tauberian Theorems.* The M.I.T. Press, Cambridge, Mass.-London, 1966.
[313] N. Wiener. *The Fourier Integral and Certain of Its Applications, Cambridge Mathematical Library.* Cambridge University Press, Cambridge, 1988. Reprint of the 1933 edition, With a foreword by Jean-Pierre Kahane.
[314] E. Wigner. On the quantum correction for thermodynamic equilibrium. *Phys. Rev.*, 40:749–759, Jun 1932.
[315] P. Wojdyllo. Characterization of Wilson systems for general lattices. *Int. J. Wavelets Multiresolut. Inf. Process.*, 6(2):305–314, 2008.
[316] T. H. Wolff. *Lectures on Harmonic Analysis*, volume 29 of *University Lecture Series*. American Mathematical Society, Providence, RI, 2003.
[317] M. W. Wong. *Weyl Transforms*. Springer-Verlag, New York, 1998.
[318] M. W. Wong. *Localization Operators.* Seoul National University Research Institute of Mathematics Global Analysis Research Center, Seoul, 1999.
[319] M. W. Wong. *Wavelet Transforms and Localization Operators*, volume 136 of *Operator Theory: Advances and Applications*. Birkhäuser Verlag, Basel, 2002.
[320] P. M. Woodward. *Probability and Information Theory, with Applications to Radar*. McGraw-Hill Book Co., Inc./Pergamon Press Ltd., New York/London, 1953.
[321] J. Wunsch. Propagation of singularities and growth for Schrödinger operators. *Duke Math. J.*, 98(1):137–186, 1999.
[322] K. Yajima. Existence of solutions for Schrödinger evolution equations. *Commun. Math. Phys.*, 110(3):415–426, 1987.

Index

Index of Notation

De Gruyter Studies in Mathematics

Volume 25
Karl H. Hofmann, Sidney A. Morris
The Structure of Compact Groups, 4th Edition, 2020
ISBN 978-3-11-069595-3, e-ISBN 978-3-11-069599-1,
e-ISBN (ePUB) 978-3-11-069601-1

Volume 74
Alessandro Carbotti, Serena Dipierro, Enrico Valdinoci
Local Density of Solutions to Fractional Equations, 2019
ISBN 978-3-11-066069-2, e-ISBN 978-3-11-066435-5,
e-ISBN (ePUB) 978-3-11-066131-6

Volume 41
Ulrich Knauer, Kolja Knauer
Algebraic Graph Theory. Morphisms, Monoids and Matrices, 2nd Edition, 2019
ISBN 978-3-11-061612-5, e-ISBN 978-3-11-061736-8,
e-ISBN (ePUB) 978-3-11-061628-6

Volume 43
Mark M. Meerschaert, Alla Sikorskii
Stochastic Models for Fractional Calculus, 2nd Edition, 2019
ISBN 978-3-11-055907-1, e-ISBN 978-3-11-056024-4,
e-ISBN (ePUB) 978-3-11-055914-9

Volume 52
Miroslav Pavlović
Function Classes on the Unit Disc, 2nd Edition, 2019
ISBN 978-3-11-062844-9, e-ISBN 978-3-11-063085-5,
e-ISBN (ePUB) 978-3-11-062865-4

Volume 73
Carol Jacoby, Peter Loth
Abelian Groups: Structures and Classifications, 2019
ISBN 978-3-11-043211-4, e-ISBN 978-3-11-042768-4,
e-ISBN (ePUB) 978-3-11-042786-8

Bei Fragen zur Produktsicherheit wenden Sie sich bitte an:
If you have any questions regarding product safety,
please contact:

Walter de Gruyter GmbH
Genthiner Straße 13
10785 Berlin
productsafety@degruyterbrill.com